The Materials Research Society Series

The Materials Research Society Series covers the multidisciplinary field of materials research and technology, publishing across chemistry, physics, biology, and engineering. The Series focuses on premium textbooks, professional books, monographs, references, and other works that serve the broad materials science and engineering community worldwide. Connecting the principles of structure, properties, processing, and performance and employing tools of characterization, computation, and fabrication the Series addresses established, novel, and emerging topics.

Isik C. Kizilyalli • Jung Han • James S. Speck •
Eric P. Carlson
Editors

Gallium Nitride and Related Materials

Device Processing and Materials Characterization for Power Electronics Applications

Springer

Editors
Isik C. Kizilyalli
Stanford University
Palo Alto, CA, USA

Jung Han
Yale University
New Haven, CT, USA

James S. Speck
University of California, Santa Barbara
Santa Barbara, CA, USA

Eric P. Carlson
Booz Allen Hamilton
Washington, DC, USA

ISSN 2730-7360 ISSN 2730-7379 (electronic)
The Materials Research Society Series
ISBN 978-3-031-83055-6 ISBN 978-3-031-83056-3 (eBook)
https://doi.org/10.1007/978-3-031-83056-3

This Springer imprint is published by the registered company Springer Nature Switzerland AG
The registered company address is: Gewerbestrasse 11, 6330 Cham, Switzerland

Preface

Isik C. Kizilyalli, Jung Han, James S. Speck, and Eric P. Carlson

Advances in power electronics have been one of the primary enablers of the unprecedented growth of renewable energy sources on the electrical power grid over the past several decades along with the current evolution toward ever-higher levels of electrification, the process of replacing fossil fuels with electricity as the source of energy. Broadly, no matter which industry is electrified, all will require efficient, power converters and controllers. As a result, power electronic circuits will play an important role in all decarbonization efforts for mitigating climate change and future sustainability.

Power electronics regulates the flow of electric power and is the requisite interface between an electrical source and a load. The source and load can differ in frequency, amplitude, number of phases, and where voltages and currents are converted from one form to another. Some examples of power electronic systems are data center power distribution elements converting and managing grid level voltages to provide power to the server rack, a solar inverter converting 600–1500 V DC power to AC power for connection to the grid, an electric vehicle (EV) motor drive using 400–800V DC battery power to drive a 320–650 V AC motor, a solid-state transformer replacing the 60 Hz "service transformer," that converts available medium or high voltage levels to desired low-voltage AC or DC, direct grid-connected converters, which must withstand voltages ranging from 13 kV to 500 kV, and even a laptop charger converting 90–260V AC power to 19V–3.3A DC power. Estimates suggest that the fraction of electricity processed through some form of power electronics could be as high as 80% by 2030 (including generation and consumption), approximately a twofold increase over the current percentage.

A key component of any (most) power electronic system is the semiconductor switching device which determines the frequencies and power levels at which the electronic system may operate. Along with the gate drivers and controller circuits, the switching device comprises a power electronics system. Historically, immediately following the invention of the germanium-based transistor in 1947, important new classes of silicon-based power semiconductor devices were developed and matured into important enabling components during the following decades. Development of power bipolar junction transistor (BJTs), silicon-controlled thyristor (SCR) and its variants, insulated gate bipolar transistors (IGBTs), and power MOSFETs (Metal Oxide Semiconductor Field Effect Transistors) along with their control schemes and controller technologies over the 50-year period from 1975 to 2025 have resulted in a three-order increase in the volumetric power density of for example industrial motor drives. Silicon (Si) has been the semiconductor material of choice for power devices for quite some time due to cost, ease of processing, and the vast amount of information available about its material properties. Si devices are, however, reaching their operational limits in blocking voltage capability, temperature of operation, and switching frequency due to the intrinsic material properties of Si. Wide-bandgap (WBG) semiconductors (SiC and GaN), with their superior electrical and thermal characteristics, are an attractive emerging alternative to Si in many applications and can enable power converters with higher efficiency and higher power conversion densities.

GaN power semiconductor devices have been the recent focus of intense development efforts due to the promising material properties of the GaN material system. Among the various possible approaches to GaN power electronics, vertical devices on native GaN substrates offer the advantage of better area efficiency due to the ability to grow thick drift regions and improved reliability due to the reduced defect density and the possibility of field control via doping profile optimization and effective device edge-termination. Specifically, there are multiple reasons in developing GaN-based power devices and those that utilize a vertical architecture. (Note that in vertical device architectures the current flow is in the direction perpendicular to the wafer surface and the back-side of the wafer serves as a drain contact and that most Si- or SiC-based power semiconductor devices are vertical architectures.) Firstly, the fundamental material-based power device figure of merit parameters for GaN are better than SiC and significantly better than Si. Secondly, the emergence of the GaN-based solid-state lighting and radio frequency (RF) electronics industries has paved the way in fundamental understanding of substrate technology, epitaxial growth using MOCVD (metalorganic chemical vapor deposition) processes, fabrication methods (etching, metallization, passivation), packaging (die attach, substrate thinning), device test, and reliability methodology. Thirdly, it is generally recognized that lateral device architectures (high electron mobility transistors in the case of GaN) are less well suited to higher power applications, while vertical structures are optimal for achieving higher breakdown voltage (>600 V) coming about from arbitrarily thick electron drift layers, and higher current (>50 A) devices without requiring paralleling (monolithic). Furthermore, vertical structures suffer less from thermal management issues associated with thin film surfaces as power is dissipated in the bulk (volume). Moreover, vertical structures fabricated on low defect density bulk GaN substrates are expected to be more reliable due to the lack of mismatched interfaces, insensitivity to surfaces, and lower defect density in the device volume. Lastly, if selectively placed high-quality p-n junctions can be integrated in a vertical architecture, this in combination with the low defect density n-type drift regions, would enable these devices to have avalanche capability. That is, a vertical GaN

Table 1 Key technical metrics for the selective area p-n junction for the PNDIODES program

Category/metric	Target
Breakdown voltage	$\geq$1200 V
Leakage current	$\leq 1 \times 10^{22129}$ A (@600 V)
Turn-on voltage	2.6–4.4 V
Specific R_{DSON}	<3 mΩ-cm^2
I_{on}/I_{off} ratio	$>10^{10}$
Avalanche capability	No parametric shift after repetitive avalanche testing
Surge capability	>20 A surge capability for 10 μs pulse at 25 °C

device incorporating high-quality p-n junctions (electrically identical to an as-grown planar p-n junction) can go into avalanche breakdown safely without catastrophically failing and without suffering parametric degradation. For power switching and rectifying applications, the ruggedness in breakdown is a critical requirement as supplementary snubber circuits impose a performance and efficiency penalty on the system.

Analysis of market drivers showed that accelerating the commercial uptake of more efficient power converters enabled by WBG devices requires driving down the costs of the components. This led to the launch of the ARPA-E SWITCHES program in 2014 to catalyze the development of vertical GaN devices using innovations in materials and/or device architectures that drive the costs of the devices. However, many SWITCHES project teams experienced a major obstacle in fabricating vertical GaN power electronic devices specifically the lack of a viable GaN selective area doping or selective area epitaxial regrowth process that yields material of sufficiently high quality to enable defect-free p-n junctions on patterned GaN surfaces. In GaN selective area p-type doping proved elusive, because the most obvious approaches, laterally patterned ion implantation with activation or selective area diffusion of p-type dopants (e.g., Mg, Be, Zn), did not produced p-type regions of satisfactory quality (i.e., equivalent to as grown planar p-n junctions). Furthermore, selective area etch and regrowth approaches which can demonstrate p-type regions have not resulted in sufficient electrical performance to be useful in power electronic applications. A breakthrough was needed in this area to enable high-performance vertical GaN transistors. This specific remaining challenge to the SWITCHES program prompted the announcement in late 2016 of the PNDIODES program. The PNDIODES program aimed to develop transformational advances and mechanistic understanding in the process of selective area doping in the III-Nitride wide-bandgap semiconductor material system. The expectation was that this would lead to the demonstration of arbitrarily placed, reliable, contactable, and generally useable p-n junction regions that enable high-performance and reliable vertical power semiconductor devices (Table 1).

The book presents the progress in device processing and materials characterization of GaN and related materials for power electronics applications, mostly based on the output of ARPA-E's focused area programs: SWITCHES, PNDIODES, as well as various OPEN solicitations. The material is organized in eight parts with a total of 28 chapters contributed from invited experts that were part of the said ARPA-E programs. We also requested chapters from a few select experts from around the world who are actively engaged in GaN and related WBG semiconductor research and development. The book includes an overview of GaN power electronic devices and systems, detailed descriptions of GaN bulk substrate technology and epitaxial growth processes, a comprehensive review of the key vertical device processing challenges, in-depth discussions on innovative material characterization techniques developed to understand the device processing challenges, and an overview of approaches to GaN selective area p-type doping with an eye toward mechanistic understanding of the challenges.

We intend this book to be an essential reference for anyone working in either basic research or advanced development of vertical architecture GaN power electronics and technologies, more than likely fabricated on native bulk GaN substrates. In a broader sense, we anticipate this book to become a go-to reference to any scientist and engineer working in any nitride semiconductor material seeking an updated coverage of the state-of-the-art processing and characterization techniques that will push GaN know-how to new materials and device frontiers.

In *Part I—Introduction*, the readers are provided with an overview of vertical architecture GaN power electronic devices based on native bulk GaN substrates. Potential power system applications for vertical GaN power diodes and transistors are included. In the two chapters comprising Part I, motivation for vertical device structures for power electronics applications is given. These are, namely, current and voltage scaling beyond lateral device implementation and broadening the application areas for GaN power devices. The rectifier and switch device development work discussed in these chapters unearthed fundamental gaps in the mechanistic understanding of the basic building blocks for the realization of highly reliable and high-performance vertical devices in GaN.

Chapter 1 "Introduction to Vertical GaN Power Electronics" describes the motivation and potential power systems application areas for vertical GaN diodes and transistors. The chapter describes vertical p–n diodes and vertical junction field effect transistors and other transistor structures fabricated on pseudo bulk low defect density (10^4 to 10^6 cm^{-2}) bulk GaN substrates. The basic building block of most all power semiconductor devices, the p-n junction, is discussed in detail. Topics included in the chapter are power semiconductor device design, edge termination strategies, epitaxial growth and incoming substrate specifications, back side processing and metallization schemes, packaging, temperature studies of devices down to cryogenic regime, impact ionization-based avalanche breakdown, and early reliability studies. ARPA-E focused area Programs SWITCHES and PNDIODES are discussed considering the challenges associated with selective area p-type doping of GaN. This chapter provides the inspiration for the rest of the book, identifying the need for more scientific and fundamental scholarly studies to overcome the roadblocks to enable vertical architecture GaN power semiconductors.

Chapter 2 "Electronic Transport and Advances in Vertical Bulk GaN-Based Power Semiconductor Devices" discusses electronic transport relevant to device performance in the context of GaN vertical diodes and transistors. Fabrication and device physics of Merged PIN-Schottky (MPS) and planar p-n diodes are covered. The impact of edge, mixed-type, and screw dislocations to device reverse leakage current and breakdown voltage are debated. Elimination of nano-pipes during epitaxial growth is elucidated. Transport parameters such as impact ionization rates for electrons and holes in GaN are extracted experimentally from fabricated diode measurements. The chapter reviews in detail vertical GaN transistor (switch) structures that have been implemented. These include the planar and trench Current Aperture Vertical Electron Transistor (CAVET), bi-directional CAVET, trench MOSFET, and Static Induction Transistor (SIT). Basic device building blocks such as drift regions, device edge termination, p-type current blocking layers, device layout options, and other fabrication process challenges are described. Trade-offs between normally-off and normally-on (cascode implementation) devices are also explored.

In *Part II—Bulk GaN Substrates*, four chapters are included to introduce readers into the main techniques for producing bulk GaN substrates. Bulk GaN growth by acidic ammonothermal and basic ammonothermal growth is covered. An overview of the different hydride vapor phase epitaxy (HVPE) approaches for producing bulk GaN substrates is included. The quality of bulk GaN substrates grown by various methods are assessed by advanced x-ray diffraction techniques.

Chapter 3 "Bulk GaN" provides an overview of different bulk GaN growth techniques including vapor phase growth, liquid phase growth, and solvothermal growth. The chapter focuses on a new low pressure acidic ammonothermal growth process for GaN that enables scaling to large autoclaves. Using this technique, the authors demonstrate four-inch GaN crystals with dislocation density in the 10^3–10^4 cm^{-2} range. The chapter presents reactor design and process issues in scaling ammonothermal growth for mass production.

Chapter 4 "Near Equilibrium Ammonothermal (NEAT) Growth of Low-Dislocation GaN for Power Devices" provides an overview of the ammonothermal method and then introduces growth at low thermodynamic driving force and the subsequent advantages for bulk crystal quality. The authors focus on basic growth conditions. They demonstrate the importance of high-quality seeds to avoid cracking in the bulk boules. Oxygen is a major impurity in ammonothermal growth and by process improvement crystals with oxygen concentration in the low 10^{18} cm^{-3} range are demonstrated with low optical absorption. The authors demonstrate high-quality two-inch and four-inch crystals using the NEAT process with threading dislocation densities in the low to mid 10^5 cm^{-2} range.

Chapter 5 "Use of HVPE for Growth of Bulk GaN Wafers and GaN Epilayers for Power Electronics Applications" presents a broad overview of the HVPE growth technique and its variants. This includes a brief history of the HVPE growth technique and its application to GaN. The chapter discusses the commercial production of bulk GaN substrates by HVPE, where the technique dominates, including considerations in the reactor design to accommodate long growths for wafer production. Additionally, the use of HVPE as an alternative device epitaxial growth technique is discussed, particularly for thick drift layers needed in high voltage power electronics devices. The chapter concludes with a brief discussion of emerging HVPE-adjacent techniques, which have been proposed for bulk GaN and GaN-based epitaxy.

Chapter 6 "X-ray Topography and High-Resolution X-ray Diffraction Characterization of GaN Materials for Power Electronics Applications" reports on the evaluation of GaN substrates and epilayers using X-ray topography and X-ray diffraction techniques. The evaluation showed ammonothermal grown GaN substrates have the best quality with the lowest dislocation densities and remain the most promising for vertical device fabrication. HVPE wafers suffer from relatively high but uniformly distributed dislocation densities while patterned HVPE wafers have a nonuniform dislocation distribution with large areas containing low threading dislocation densities. Ion implantation into the GaN substrates induced strain with a depth profile correlated to the implanted doping profile but annealing can effectively recover the damage with some optimization required for higher doping concentrations. Inductively coupled plasma etching introduces strain in the GaN material that can be partially recovered by follow-up non-plasma etching.

In *Part III—Epitaxial Growth of Gallium Nitride and Processing*, five chapters are included that discuss MOCVD and MBE growth of thick drift layers and PN diodes. The chapters also include a survey of wafer scale characterization, high-voltage diodes exhibiting avalanche breakdown, and novel layer transfer technologies.

Chapter 7 "Laser-Assisted MOCVD GaN and the Development of Vertical GaN-on-GaN PN Diodes" presents a novel laser-assisted MOCVD (LA-MOCVD) technique in which ~9.2 μm laser light directly couples to NH_3 vibrational modes to improve the NH_3 cracking efficiency. The LA-MOCVD technique results in reduced carbon incorporation. Combining the new MOCVD method with large miscut bulk GaN substrates show record mobility as high as 1275 cm^2/Vs. PN diodes with designs for edge termination including beveled mesa etching, implanted guard rings, and high κ dielectrics are shown. The authors demonstrate 7.8 kV PN diodes that incorporate thick LA-MOCVD grown drift layers and advanced field management.

Chapter 8 "MOCVD Growth of GaN Drift Layers on Bulk GaN Substrates for Power Electronic Devices" presents characterization of substrates from different vendors with either uniform low dislocation density or nonuniform dislocation density. 40–60 μm thick MOCVD epitaxial layers are demonstrated with net donor doping (N_D–N_A) in the low to mid 10^{15} cm^{-3} range. 6.0 kV PN diodes with 50 μm drift layers are realized.

Chapter 9 "MBE Grown GaN P-N Diodes" reviews the use of MBE technology for III-nitrides electronic devices. MBE is known to provide exceptionally clean growth background with atomic-level control of growth kinetics. The authors discuss the operating principles, growth mechanisms, and trade-offs between two common techniques for nitride MBE: plasma-assisted and ammonia-based. Toward GaN vertical power diodes, the authors demonstrate MBE-grown GaN with very low concentrations of background impurities and high mobilities. To achieve a thick drift layer for high-breakdown diode operation, the authors employ an indium surfactant to stabilize the surface morphology while maintaining a high growth rate of ~μm/hr.

Chapter 10 "Gallium Nitride Wafer Metrology and Device Processing" reviews metrology and device processing techniques to enable vertical GaN PiN diodes. This includes a discussion on wafer metrology techniques, such as mercury probe mapping, optical profilometry, and X-ray, as a means to evaluate both as-received wafers and epitaxial layers grown on wafers. Vertical GaN PiN diode processing is discussed including epitaxial layer design, device fabrication steps, edge termination processes, and process validation through device testing. A brief discussion of packaging and electrical stress protocols to identify fundamental failure mechanisms is included.

Chapter 11 "Layer Transfer of Gallium Nitride by Controlled Spalling" describes a novel fracture-based layer transfer method of GaN called controlled spalling using a very thick Ni stressor layer. The details of the equipment and the controlled spalling procedures are discussed as well as the unique challenges associated with the transfer method on the relatively hard GaN material system. The authors demonstrate that wafer-scale layer transfer from bulk GaN wafers can be carried out with the material quality preserved.

In *Part IV—Selective Area Doping, Processes, and Characterization*, seven chapters are included to provide a snapshot of different approaches and challenges encountered in the selective area doping of GaN.

Chapter 12 "In Situ TBCl Etching and Selective-Area Growth and Doping of GaN" introduces a novel precursor tertiarybutylchloride (TBCl) for in situ etching of GaN in MOCVD growth. Through systematic experiments, the authors show the in situ removal of dry etch damage in PN diodes. The authors also demonstrate the remaining challenges of unintentional silicon removal. Nonplanar regrowth of p-GaN in etched trenches is studied. The authors explain the origin of nonuniform Mg incorporation based on concepts of kinetic Wulff plots.

Chapter 13 "Selective Area Regrowth and Doping for Vertical GaN Power Devices" reports the combination of ex situ etching and MOCVD regrowth to achieve selective area doping. The authors employ a broad range of characterization techniques to correlate the observed tunnel junction-like leakage with the presence of contaminants and etching-related defects. The authors introduce two mitigating solutions to ex situ selective area etching, including a multi-step dry-etching to reduce the ion damage, and the use of atomic layer etching by exposing GaN surface successively to oxygen plasma, hydrogen fluoride gas, and organometallic precursors. Promising results from regrowth diodes and junction FETs are also included to demonstrate the feasibility.

Chapter 14 "Acceptor Dopant Ion Implant and Gyrotron Rapid Thermal Annealing of GaN" investigates a novel nonequilibrium or metastable heating process to support implant-based selective area doping. The authors discuss the criteria in selecting microwave energy sources and articulate the merits of direct energy transfer to GaN samples at gyrotron frequency (60 GHz). Photoluminescence is used to determine the removal of implant damages and activation of Mg acceptors. While the metastable annealing technique still causes noticeable Mg diffusion, the authors demonstrate a process to suppress diffusion (by at least 10×) through a multi-step, multi-temperature annealing methodology.

Chapter 15 “Ion Implantation for p-Type Conductivity in GaN” presents the authors’ effort in achieving ultrahigh pressure annealing of Mg-implanted samples for selective area doping. This chapter contains a brief overview of various alternative approaches to selective doping of GaN, and the historical challenges in overcoming the decomposition of GaN during high-temperature post-implant annealing. By providing an annealing environment with ultrahigh N_2 pressure (~GPa) and immersing the implanted samples in GaN powders, the authors demonstrate the removal of implant damages by annealing while preserving the surface morphology. The authors present very promising results in transport measurement of annealed p-GaN samples and nearly ideal junction barrier Schottky (JBS) diodes.

Chapter 16 “High Energy Ion Implantation Using Electrostatic Accelerators” covers high energy ion implantation developed at the Brookhaven National Laboratory (BNL). The High Energy Implanter has demonstrated uniform implantation in SiC from the surface to a depth of >12 μm by combining a high-voltage Tandem Van de Graaff with a stationary aluminum energy absorber. To implant ions near the surface of the SiC wafer requires the use of an energy absorber to degrade the ion energy from 10’s of MeV to a few keV. An upgrade to the present system allows ion implantation while the wafers are being heated to reduce ion implant damage via in situ annealing. This technology is used for fabricating SiC vertical super-junction transistors for medium voltage applications.

Chapter 17 “Solid-State Diffusion in GaN” introduces a novel approach in tackling the challenge of selective area doping of GaN using solid-state Mg diffusion. The authors analyze bilayer designs to simultaneously extract Ga atoms to form Ga-vacancies, and to diffuse Mg to substitutional sites through exchange. First principle calculation is given to compare formation and binding energies of possible species and diffusion pathways. The authors test different metals for the gallidation process and demonstrate diffusion and activation of Mg, at least to a level to compensate the high n-type background. Photon-assisted activation of Mg acceptors is also conducted to provide evidence of p-type conductivity. To explore the effect of surface Fermi level on the diffusion barrier, the authors also report in situ field-assisted diffusion and show a noticeable improvement in Mg diffusion.

Chapter 18 “Transmutation Doping of GaN” introduces n-type doping of GaN through transmutation of the Ga into Ge via neutron irradiation. Uniform doping concentrations of 10^{16}–10^{18} Ge atoms/cm^3 were achieved throughout the GaN wafer. The authors also observed production of ^{14}C from transmutation of the N atoms. The generated germanium acts as an n-type dopant and carbon behaves as a p-type dopant. The ^{14}C production rate could be lowered by adjusting the neutron energy level due to the lowered neutron capture cross section of N at higher energies. Neutrons can cause damage to GaN crystals and introduce new defects since neutrons are mass-and energy-containing particles. A nitrogen-rich thermal annealing process mitigated the radiation-induced damage. The authors also investigated other radiation sources including proton and gamma beams for transmutation of the Ga into Zn as a p-type dopant.

In *Part V—Materials Characterization Techniques*, we include four chapters devoted to advanced (spatially, temporally, and/or spectroscopically resolved) characterization techniques that have been brought to bear in unraveling the defect properties in GaN. Evaluation of any processing development and device performance requires detailed knowledge of defects and impurities.

Chapter 19 "Deep Level Defect Spectroscopy and Electron Beam-Induced Current Characterization of GaN Junctions" describes the characterization of defects in GaN due to plasma etching, their deleterious impact on the electrical performance of vertical p-n diodes and methods to remove these defects in p-n diodes formed by etching and subsequent regrowth. Deep level defect energy, concentration, and depth profile is measured as a function of etch method and post-etch treatment using photo-capacitance, and these data correlate well with reverse bias diode performance. Photo-capacitance studies of both c-plane and m-plane GaN provide insight to etch-induced defectivity for both the basal plane and sidewalls relevant to patterned selective area regrowth. Electron beam-induced current (EBIC) can also assess the influence of defects by measuring the minority carrier diffusion length for both the p-and n-type regions of etched-and-regrown p-n diodes. Methods to improve the accuracy of carrier diffusion length measurements using EBIC and to improve the spatial resolution of EBIC in vertical p-n diodes are discussed.

Chapter 20 "Ion Beam Analysis of GaN Surfaces and Interfaces" reports the use of ion-beam and supplemental electron-beam probes to characterize GaN surfaces and regrowth interfaces that have been subject to different etching and cleaning processes. The authors discuss the principles of ion channeling and elastic recoil, and present new understanding of relationships between structure, chemistry, and electronic states at GaN surfaces and interfaces intended for use in vertical GaN p-i-n devices. The authors compare and contrast GaN surfaces and interfaces prepared using epitaxy/etching/regrowth, ion implantation plus gyrotron annealing, and electric-field assisted diffusion, suggesting promising avenues for future investigations.

Chapter 21 "Structural and Chemical Defect Characterization for Selectively Doped GaN" presents advanced x-ray characterization of the structure of GaN substrates produced by HVPE with bundled dislocations in core regions and relatively perfect GaN away from the core regions. The high-quality GaN in structural and electrical measurements is confirmed away from the core regions. Mg implantation studies are presented which include high temperature annealing to recover the structural quality of the pristine unimplanted GaN.

Chapter 22 "Secondary Ion Mass Spectroscopy (SIMS) Analysis of GaN Epitaxial Films" provides an overview of SIMS and related techniques to quantify intentional dopant and unintentional dopant concentrations in epitaxial layers. The technique provides insight into microstructural features such as overgrowth of nonplanar surfaces. The technique of Point-by-Point CORrected (PCOR)-SIMS is introduced to overcome artifacts in accurate impurity concentration analysis.

In *Part VI—Gallium Nitride Materials Properties for Power Electronics Devices*, three chapters are included that discuss the fundamental materials properties of GaN in relation to its use in power electronics including thermal conductivity, breakdown, and high field electronic transport.

Chapter 23 "Intrinsic and Extrinsic Thermal Conductivity of Gallium Nitride and Silicon Carbide" provides a comprehensive review concerning the thermal transport properties of both GaN and SiC semiconductors. The authors discuss phonon transport models and applied to narrow-, wide-, and ultrawide-bandgap semiconductors. Both intrinsic (phonon-phonon) and extrinsic (phonon-defect) scattering factors are treated. The authors review a non-contact and efficient thermal metrology based on thermoreflectance measurement and compare the tradeoffs among the three modalities (Time-Domain Thermoreflectance, Frequency Domain Thermoreflectance, and Steady State Thermoreflectance). Experimental data of thermoreflectance measurement of GaN with both intrinsic and extrinsic variables are presented, which serve as an important foundation in the design of GaN vertical power electronics.

Chapter 24 "Breakdown Phenomena in GaN" provides an overview of breakdown phenomena under high electric field in GaN. Two different mechanisms, Zener breakdown and avalanche breakdown, are discussed in detail. Zener breakdown is due to band-to-band tunneling and avalanche breakdown in GaN is caused by impact ionization. After reviewing explanation of basic theories, recent experimental reports on these phenomena by using high-quality GaN p-n junctions grown on GaN (0001) bulk substrates are reviewed. Material parameters to describe Zener and avalanche breakdown phenomena are extracted. These material parameters can be used to predict junction reverse-bias leakage current and breakdown voltage and are critical for design and optimization of vertical GaN device structures.

Chapter 25 "Theoretical Modeling of Avalanche Currents in GaN" discusses the theoretical modeling of the local theory of impact ionization and avalanche generation in GaN materials and devices from a charge transport perspective. New analytical expressions are developed for particle densities and current densities in the local approximation and its implication for concepts such as the relationship between breakdown field and layer thickness (demonstrating that critical field is not an intrinsic material property). A discussion of nonlocal transport in the context of entirely new analytic solutions of the Boltzmann transport equation is included for particles injected from a thermal distribution into a region of uniform or linear field profile, demonstrating the origins of a characteristic relaxation length, for which the "dead space" is a specific case, and a critique of semi-analytic models attempting to account for dead space in avalanche current modeling

In *Part VII— Related Wide-Band gap Materials and Devices*, two chapters explore nitride materials that are in an earlier stage of development but show promise for use as power semiconductor devices along with other applications such as detection and integrated photonics.

Chapter 26 "Cubic GaN: Growth, Characterization, and Applications" introduces this crystal form of GaN which is considered as a promising candidate for next-generation power electronics including the realization of normally off integrated high-electron-mobility transistors or lateral super-junction transistors. However, the metastability of cubic GaN (and its alloys) prevents the material advantage from being translated into functional devices. Innovations in crystal growth on U-grooved Si (100) lead to phase-pure cubic GaN that is Complementary Metal–Oxide–

Semiconductor (CMOS) compatible and suitable for power electronics as well as integrated photonics applications. Future development in selective area p-type doping in cubic GaN has the potential to catalyze the power semiconductor industry.

Chapter 27 "Development of Wafer-Scale h-BN Quasi-Bulk Crystals" introduces the readers to crystalline and optical qualities of h-BN quasi-bulk crystals produced by HVPE. It is demonstrated that h-BN quasi-bulk crystals with a thickness around 100 μm have a good stacking sequence and long-range order in the c-axis. A thermal neutron detector of 1 cm^2 in detection area fabricated from a 100 μm thick ^{10}B-enriched h-BN wafer was shown to deliver a record high detection efficiency of 60% at a bias voltage of 450 V. It is posited that h-BN is emerging as an important ultra-wide bandgap semiconductor material for a broad range of applications in deep UV photonics, high power electronics, high efficiency neutron detectors, substrates for 2D devices, solid-state host for single photon emitters, and quantum defect qubits.

In *Part VIII—Future Outlook*, we include Chap. 28 "Outlook of Wide Bandgap Semiconductors and Power Electronics in the Energy Transition Towards a More Electric Future" to briefly discuss advances in power electronics as one of the primary enablers of the unprecedented growth of renewable energy sources on the electrical power grid over the past several decades along with the current evolution toward ever-higher levels of electrification. The chapter includes future research and development directions in materials science, characterization, and physical electronic device architectures to push the limits of power electronics to utilize wide bandgap semiconductors that go beyond Si and SiC. The areas that still need development are discussed along with potential high impact application areas.

To broaden and unlock the true potential of GaN for power electronics, vertical device architectures are needed. To achieve vertical GaN power devices, progress is required in bulk substrate technology, thick epitaxial growth with controlled low doping levels with negligible compensation, low damage etch and regrowth technologies, ion implantation and anneal processes, and most critically in the general area of selective area p-type doping capability that yields high-quality p-n junctions. Advances in all these areas are necessary at the fundamental material level to achieve the desirable device and system level performance with near perfect reliability. Primarily, previous publications have focused on optical and lateral GaN devices looking at device operation, characterization, and reliability.

In summary, this book aims to fill the gap by addressing in-depth the materials processing challenges for vertical devices based on bulk GaN substrates as well as describing innovative materials characterization techniques that elucidate the processing challenges. The material presented here is unique in its technological focus on vertical GaN devices and distinct in providing a forward-looking discussion of a broad range of toolkits necessary to realize vertical and next-generation devices. We intend this book to be an essential reference for anyone working in either basic research or advanced development of vertical architecture GaN power electronics and technologies, more than likely fabricated on native bulk GaN substrates. In a broader sense, we anticipate this book to become a go-to reference to any scientist and engineer working in any nitride semiconductor material seeking an updated coverage of the state-of-the-art processing and characterization techniques that will push GaN know-how to new materials and device frontiers.

Finally, we would like to express our gratitude to all the chapter contributors. This book would not be possible without their devoted and exemplary efforts over the years supported through ARPA-E's SWITCHES and PNDIODES Programs and the Agency's various OPEN solicitations. Furthermore, we recognize the close collaborations of our teams and the editors with colleagues in various institutions in Japan and Poland. We would also like to thank the staff at Springer, in particular Dhivya Savariraj, Charles Glaser, Brian Halm, Kaitlyn Bonomo, and Mahalakshmi Saravanan for their help and support. Encouragement from Dr. Ram Devanathan of Pacific Northwest National Labs (PNNL) is sincerely appreciated. Lastly, one of the authors (ICK) gratefully acknowledges the support of his ARPA-E colleagues during the creation and development of the work that led to this book, namely Dr. T. Heidel, S. Kosinski, Dr. P. McGrath, Dr. D. Cunningham, Dr. E. Rohlfing, and Agency Directors honorable Dr. E. Williams, L. Genatowski, and Dr. E. Wang.

Contents

About the Editors

Isik C. Kizilyalli recently served as the Senior Director of Technology (R&D) Sustainability Accelerator within the Stanford Doerr School of Sustainability. Prior to joining the Accelerator, he was the Associate Director for Technology at the Advanced Research Projects Agency – Energy (ARPA-E). In this role, Dr. Kizilyalli oversaw all technology issues relating to ARPA-E's programs, program development, Program Director and Fellow recruitment, and coordinating project management across the Agency. As an ARPA-E Program Director his focus at ARPA-E included power electronics, semiconductor devices, electrification of transportation (aviation, ships, automotive, space) and associated infrastructure, enhanced geothermal systems, subsurface instrumentation and drilling, electric distribution and transmission grids, and grid technologies and resiliency against aging, EMP, space weather, natural disaster, and cyber threats. Before joining ARPA-E, Dr. Kizilyalli served as founder and CEO/CTO of Avogy Inc. and Zolt Inc., venture backed startups focused on power electronics. At the 2015 International Consumer Electronics Show (CES) Zolt Inc. was a CES Best of Innovation Awards Honoree, Best Startup CES finalist (by Engadget), a Top Tech of CES nominee (by Digital Trends), and a Top Pick CES (by Laptop Magazine). Previously, he was with AT&T Bell Laboratories and its spinouts for nearly 17 years, followed by Nitronex Corporation, and solar PV startup Alta Devices where his team holds the world record for single junction solar cell conversion efficiency. Dr. Kizilyalli was elected a Fellow of the Institute of Electrical and Electronics Engineers (IEEE) in 2007 for his contributions to "Integrated Circuit Technology." He also received the Bell Laboratories' Distinguished Member of Technical Staff award and the Best Paper Award at the International Symposium on Power Semiconductors and Integrated Circuits (ISPSD) in 2013. Dr. Kizilyalli holds his BS in Electrical Engineering, MS in Metallurgy, and PhD in Electrical Engineering all from the University of Illinois (Urbana). He has published a Springer-Nature book on Direct Current Fault Protection, more than 100 papers, and holds 127 issued US patents.

Jung Han is the William Norton Professor in Technological Innovation and a Professor of Electrical Engineering at Yale University. Prof. Han focuses his research on wide-bandgap semiconductor materials, optoelectronic and microelectronic devices, nanoscale materials and devices, semiconductor epitaxy, and hybrid organic-inorganic semiconductors. Prior to joining Yale in 2001, Prof. Han was a senior technical staff member at Sandia National Laboratories where he pioneered the growth of AlGaN for UV emitters and the use of in situ diagnostics during crystal growth, the latter has since been adopted worldwide by the LED industry. Prof. Han has served as a reviewer, panelist, and advisory board member for agencies including the National Science Foundation, Department of Energy, and Department of Defense. He has published more than 300 papers in peer-reviewed journals, and has served as editor of four books and special journal issues. He holds more than 20 US patents and is the co-founder of two startup companies, Saphlux and Ganvix. Prof. Han has received numerous awards including a Department of Commerce R&D 100 Award, Materials Research Society Ribbon Award, and The IEEE Electromagnetic Compatibility Society Best Paper Award. He is a member of the Connecticut Academy of Science and Engineering, and a fellow of the Institute of Physics and the Institute of Electrical and Electronic Engineers (IEEE). Prof. Han graduated from National Taiwan University and earned his MS and PhD in electrical engineering from Purdue University. His graduate and postdoctoral research focused on the crystal growth of II–VI ZnSe compound by molecular beam epitaxy which led to the one of the first demonstrations of solid-state blue-green diode laser.

James S. Speck is a Professor in the Materials Department at the University of California Santa Barbara (UCSB). At UCSB, his early work focused on epitaxial oxide films on semiconductors, ferroelectric thin films, and strain relaxation in highly misfitting epitaxial systems. He has worked extensively on the materials science of GaN and related alloys. Major aspects of his work on nitrides include elucidating basic growth modes and defect generation, the development of MBE growth of GaN, and the development of nonpolar and semipolar GaN. Prof. Speck received the Quantum Device Award with Umesh Mishra from the International Symposium on Compound Semiconductors in 2007, he was named an inaugural Materials Research Society Fellow in 2008, and received the Japanese Journal of Applied Physics Best Paper Award in 2008. In 2009 he became an American Physical Society Fellow. In 2010 he received the IEEE Photonics Society Aron Kressel Award with Steve DenBaars for their work on nonpolar and semipolar GaN-based materials and devices. In 2007, Prof. Speck and his longtime collaborators Steve DenBaars and Shuji Nakamura founded Santa Barbara-based start-up companies Kaai and Soraa to commercialize their work on nonpolar and semipolar nitrides. In 2016 he was elected as a Fellow of the National Academy of Inventors. Prof. Speck has over 950 publications in the referred archival literature. He received his BSME degree in metallurgical engineering from the University of Michigan in 1983 and his SM and ScD in materials science from the Massachusetts Institute of Technology in 1985 and 1989, respectively.

Eric P. Carlson is currently a Senior Lead Scientist at Booz Allen Hamilton where he supports the U.S. Department of Energy's Advanced Research Projects Agency-Energy (ARPA-E) as a technical support contractor. Prior to joining Booz Allen Hamilton, he served 12 years as a Research Specialist in the Compound Semiconductor Solutions Division at Dow Corning Corporation where he developed crystal growth, epitaxy, and characterization of wide bandgap semiconductor materials. Before joining Dow Corning Corporation, Dr. Carlson served as the Director of GaN R&D at Kyma Technologies where he was developing growth processes of bulk GaN wafers. Dr. Carlson is the author and co-author of more than 50 scientific publications and holds 6 patents. He received his BS and PhD in materials science and engineering from North Carolina State University in 1993 and 2001, respectively.

Contributors

Ozgur Aktas Transphorm Inc., Goleta, CA, USA

Noah Allen Lawrence Livermore National Laboratory, Livermore, CA, USA

Andrew A. Allerman Sandia National Laboratories, Albuquerque, NM, USA

Travis J. Anderson University of Florida, Gainesville, FL, USA

Andrew Aragon Intel, Rio Rancho, NM, USA

Andrew M. Armstrong Sandia National Laboratories, Albuquerque, NM, USA

Can Bayram The Grainger College of Engineering, University of Illinois Urbana-Champaign, Urbana, IL, USA

Stephen W. Bedell IBM T. J. Watson Research Center, Yorktown Heights, NY, USA

Zachary J. Biegler University of California, Santa Barbara, CA, USA

Andrew Binder Sandia National Laboratories, Albuquerque, NM, USA

Michal Bockowski Institute of High-Pressure Physics, Warsaw, Poland

Temel Buyuklimanli Eurofins | EAG laboratories, Sunnyvale, CA, USA

David G. Cahill Department of Materials Science and Engineering, Materials Research Laboratory, University of Illinois, Urbana, IL, USA

Charles Carlson Brookhaven National Laboratory, Upton, NY, USA

Eric P. Carlson Booz Allen Hamilton, Washington, DC, USA

Alexander S. Chang Department of Material Science and Engineering, Northwestern University, Evanston, IL, USA

Cheng Chang Rice University, Houston, TX, USA

Guanjie Cheng University of Michigan, Ann Arbor, MI, USA

Srabanti Chowdhury Stanford University, Stanford, CA, USA

Rongming Chu The Pennsylvania State University, University Park, PA, USA

Ramón Collazo Department of Materials Science and Engineering, North Carolina State University, Raleigh, NC, USA

Joshua J. P. Cooper University of Michigan, Ann Arbor, MI, USA

Michael Dudley Stony Brook University, Stony Brook, NY, USA

Esmat Farzana University of California, Santa Barbara, CA, USA
Iowa State University, Ames, IA, USA

Daniel Feezell University of New Mexico, Albuquerque, NM, USA

Sam Frisone University of Michigan, Ann Arbor, MI, USA

Clint Frye Lawrence Livermore National Laboratory, Livermore, CA, USA

Houqiang Fu Arizona State University, Tempe, AZ, USA

Kai Fu The University of Utah, Salt Lake City, UT, USA

Kenji Fujito Mitsubishi Chemical Co., Tokyo, Japan

James C. Gallagher U.S. Naval Research Laboratory, Washington, DC, USA

Reza Ghandi GE Aerospace Research, Niskayuna, NY, USA

Rachel S. Goldman University of Michigan, Ann Arbor, MI, USA

Mark Goorsky Materials Science and Engineering, University of California Los Angeles, Los Angeles, CA, USA

Jung Han Department of Electrical Engineering, Yale University, New Haven, CT, USA
Yale University, New Haven, CT, USA

Tadao Hashimoto SixPoint Materials, Inc., Buellton, CA, USA

Jiaheng He University of Michigan, Ann Arbor, MI, USA

Karl D. Hobart U.S. Naval Research Laboratory, Washington, DC, USA

Kenny Huynh Information Sciences Institute, University of Southern California, Marina del Rey, CA, USA

Satoru Izumisawa Mitsubishi Chemical Co., Tokyo, Japan

Alan G. Jacobs U.S. Naval Research Laboratory, Washington, DC, USA

Hongxing Jiang Texas Tech University, Lubbock, TX, USA

Robert Kaplar Sandia National Laboratories, Albuquerque, NM, USA

Daryl Key SixPoint Materials, Inc., Buellton, CA, USA

Isik C. Kizilyalli Stanford University, Palo Alto, CA, USA

Robert Koch University of Applied Sciences Jena, Jena, Germany

Florian Krippendorf mi2-factory GmbH, Jena, Germany

Thomas Kubley Brookhaven National Laboratory, Upton, NY, USA

Kyoung Kweon Lawrence Livermore National Laboratory, Livermore, CA, USA

Jae W. Kwon Infinity Power LLC, San Diego, CA, USA

Lincoln J. Lauhon Department of Material Science and Engineering, Northwestern University, Evanston, IL, USA

Jacob H. Leach Kyma Technologies, Raleigh, NC, USA

Jaekwon Lee The Grainger College of Engineering, University of Illinois Urbana-Champaign, Urbana, IL, USA

François Léonard Sandia National Laboratories, Livermore, CA, USA

Edward Letts SixPoint Materials, Inc., Buellton, CA, USA

Bingjun Li Department of Electrical Engineering, Yale University, New Haven, CT, USA

Michael Liao Apex Microdevices, West Chester, OH, USA

Jingyu Lin Texas Tech University, Lubbock, TX, USA

Wu Lu Department of Electrical and Computer Engineering, The Ohio State University, Columbus, OH, USA

Yang Lu Department of Mechanical and Aerospace Engineering, The Hong Kong University of Science and Technology, Hong Kong, China

James Spencer Lundh U.S. Naval Research Laboratory, Washington, DC, USA

Nadeemullah A. Mahadik U.S. Naval Research Laboratory, Washington, DC, USA

Benjamin McEwen State University of New York, University at Albany, Albany, NY, USA

Vincent Meyers Sandia National Laboratory, Albuquerque, NM, USA

Yutaka Mikawa Mitsubishi Chemical Co., Tokyo, Japan

Dinusha Herath Mudiyanselage Arizona State University, Tempe, AZ, USA

Balaji Raghothamachar Stony Brook University, Stony Brook, NY, USA

Imteaz Rahaman The University of Utah, Salt Lake City, UT, USA

Pramod Reddy Adroit Materials Inc., Cary, NC, USA

Anthony Rice Sandia National Laboratories, Albuquerque, NM, USA

Emma Rocco U.S. Naval Research Laboratory, Washington, DC, USA

Michael Rub mi2-factory GmbH, Jena, Germany
University of Applied Sciences Jena, Jena, Germany

F. Shadi Shahedipour-Sandvik State University of New York, University at Albany, Albany, NY, USA

Zlatko Sitar Department of Materials Science and Engineering, North Carolina State University, Raleigh, NC, USA

James S. Speck University of California, Santa Barbara, CA, USA

Dannie Steski Brookhaven National Laboratory, Upton, NY, USA

Jun Suda Department of Electronics, Nagoya University, Nagoya, Aichi, Japan

A. Alec Talin Sandia National Laboratories, Livermore, CA, USA

Vijay Gopal Thirupakuzi Vangipuram Department of Electrical and Computer Engineering, The Ohio State University, Columbus, OH, USA

Joel Varley Lawrence Livermore National Laboratory, Livermore, CA, USA

Lars Voss Lawrence Livermore National Laboratory, Livermore, CA, USA

Yekan Wang Intel Corp, Chandler, AZ, USA

Yibo Xu Department of Electrical and Computer Engineering, The Ohio State University, Columbus, OH, USA

Luke Yates Sandia National Laboratories, Albuquerque, NM, USA

P. D. Yoder Georgia Institute of Technology, Atlanta, GA, USA

Zhirong Zhang University of California Berkeley, Berkeley, CA, USA

Hongping Zhao Department of Electrical and Computer Engineering, The Ohio State University, Columbus, OH, USA
Department of Materials Science and Engineering, The Ohio State University, Columbus, OH, USA

Yuji Zhao Rice University, Houston, TX, USA

Qiye Zheng Department of Mechanical and Aerospace Engineering, The Hong Kong University of Science and Technology, Hong Kong, China

Part I
Introduction

Chapter 1
Introduction to Vertical GaN Power Electronics

Isik C. Kizilyalli and Eric P. Carlson

1 Introduction to Vertical GaN Power Devices

The building blocks comprising a power electronics system are power semiconductor devices, gate drivers, and controller circuits [1]. The needs of the power semiconductor components of power electronic systems have been well addressed by silicon-based diodes (Schottky diodes, p-i-n diodes) and transistors (metal-oxide-semiconductor field-effect transistor (MOSFET), insulated gate bipolar transistor (IGBT), and silicon-controlled rectifier (SCR) and its variants). Impressive reduction in size and improvements in efficiency, weight, and power density of systems have resulted from the advancing performance of silicon-based power semiconductor devices. However, the performance of devices based on silicon is at the fundamental material limits of silicon. This limitation has resulted in a rapid expansion of efforts to develop wide-bandgap power semiconductor alternatives because further reduction in size and weight, and increase in efficiency in power electronic systems can be achieved by utilizing SiC [2] and GaN [3–6].

There are multiple reasons for developing GaN-based power devices. First, the fundamental (material-based) figure of merit parameters for GaN are better than SiC and significantly better than Si. Specifically, some of the desirable properties associated with GaN and its related alloys and heterostructures include high bandgap energy ($E_G \approx 3.4$ eV) and hence low intrinsic carrier concentration (useful for high-temperature operation), favorable transport properties (large electron mobility and saturation velocity), a high critical breakdown electric field, and high thermal

I. C. Kizilyalli (✉)
Stanford University, Palo Alto, CA, USA
e-mail: kizilyalli@ieee.org

E. P. Carlson
Booz Allen Hamilton, Washington, DC, USA
e-mail: carlson_eric2@bah.com

I. C. Kizilyalli et al. (eds.), *Gallium Nitride and Related Materials*, The Materials Research Society Series, https://doi.org/10.1007/978-3-031-83056-3_1

conductivity [7–11]. Second, the emergence of the GaN-based solid-state lighting [12–14] and radio frequency (RF) electronics [15] industries has paved the way for a fundamental understanding of substrate technology, epitaxial growth using MOCVD (metalorganic chemical vapor deposition) processes, fabrication methods (etching, metallization), packaging (die attach, substrate thinning), test, and reliability [16–18].

For power electronics, the power semiconductor device figure of merit captures the trade-off between two modes of operation for a power semiconductor device. In the conduction mode, it is desired that the device exhibits a low resistance as described by its specific resistance (R_{sp}), while in the blocking mode, the device supports a large breakdown voltage (BV). These conflicting requirements (for a unipolar power device) have been captured and formulated [19] as $BV^2/R_{sp} \sim \mu_n E_C^3$, where μ_n is the mobility for electrons and E_C is the critical electric field at which breakdown occurs. The advantage of wide band-gap semiconductor devices over silicon devices arises from the cubed dependence of the figure of merit on the critical electric field. While the critical electric field for silicon is 0.3 MV/cm, we estimate a value of at least 3.5 MV/cm for bulk GaN and possibly as high as 3.75 MV/cm. A detailed discussion is given in [19–22]. This value for the critical electric field combined with an electron mobility of 900–1800 cm^2/V-s form the basis of the "theoretical limits" for GaN. It is acknowledged that the concept of a critical electric field in the context of nonuniform electric fields and considering the complicated nonisotropic band-structure of GaN ([23–25], see also Chap. 25) is overly simplistic and misleading. We choose to use it here since the power semiconductor literature has adopted its use for benchmarking purposes. Table 1.1 summarizes the material parameters for semiconductor materials practical and currently suitable for power applications, namely Si, SiC, and GaN.

Table 1.1 Fundamental material parameters for Si, SiC, GaN [8–10, 20–22], and diamond

		Si	4H-SiC	GaN (2 DEG)	GaN (Bulk)	Diamond
Band gap	E_g (eV)	1.1	3.26	3.39	3.39	5.45
Intrinsic concentration	n_i (cm^{-3})	1.5×10^{10}	8.2×10^{-9}	1.9×10^{-10}	1.9×10^{-10}	1.6×10^{-27}
Electron mobility (low)	μ_n (cm^2/V-s)	1350	700	1000	500	1900
Electron mobility (high)	μ_n (cm^2/V-s)	1450	950	2000	1200	4000
Electron saturation velocity	v_{sat} (10^7 cm/s)	1	2	2.5	2.5	2.7
Breakdown electric field	E_{br} (MV/cm)	0.3	2	3.3	3.5–3.75	5.6
Thermal conductivity	K (W/cm-°K)	1.5	4.9	1.3	2.3	20

Details of thermal conductivity measurements can be further found in Chap. 23

SiC-based Schottky diodes and transistors have already been commercialized and are commonly utilized in applications that demand higher efficiency [26]. However, the performance and reliability of lateral GaN-based devices have fallen short of their potential. This is because GaN layers grown on mismatched foreign substrates (e.g., Si or SiC) create difficulties in realizing vertical device structures and have higher defect densities ($>10^8$ to 10^9 cm^{-2}). Low defect density is important in power devices because it can affect the performance characteristics (e.g., breakdown voltage and off-state leakage current), yield, and reliability (high-temperature reverse bias and operating life, avalanche ruggedness). For GaN/InGaN lasers, significant improvements in performance and reliability have been demonstrated using low defect density bulk GaN substrates [12, 14]. It is generally recognized that lateral device architectures (high electron mobility transistors) are less well suited to power applications, while vertical structures are optimal for achieving higher breakdown voltage and current devices without paralleling (monolithic), suffer less from thermal management issues associated with thin film surfaces, and yield more die on a wafer. It should be noted that all high voltage (BV $>$ 600 V) Si or SiC-based devices are vertical architectures. Furthermore, vertical structures fabricated on bulk GaN substrates are expected to be more reliable due to the lack of mismatched interfaces and insensitivity to surfaces.

The reason that there have been so few reports of vertical GaN power devices has been the lack of availability of low defect density ($<10^6$ cm^{-3}) bulk GaN substrates. A related reason is the assumption that bulk GaN substrates are too expensive to be commercially viable. However, currently, there are already multiple suppliers of bulk GaN substrates. The material quality has been improving rapidly due to research and development [27–29] as also described in Chaps. 3, 4, and 5. Also, the demand and scale of the optoelectronics industry are driving the prices down. Table 1.2 summarizes first-order materials attributes such as defect density, lattice

Table 1.2 Comparison of materials growth and device fabrication

Attributes	GaN on Si	GaN on SiC	GaN on Bulk-GaN
Defect density (cm^{-2})	10^9	10^8 to 10^9	10^3 to 10^6
Lattice mismatch (%)	17	3.5	0
CTE mismatch (%)	54	25	0
Layer thickness (μm)	< 10	< 10	> 40
Avalanche capability	No	No	Yes
Device types	Lateral	Lateral	Vertical and Lateral
Microscopy and growth	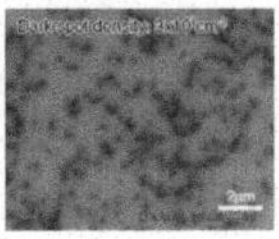	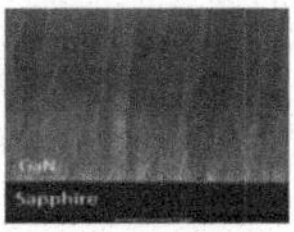	 

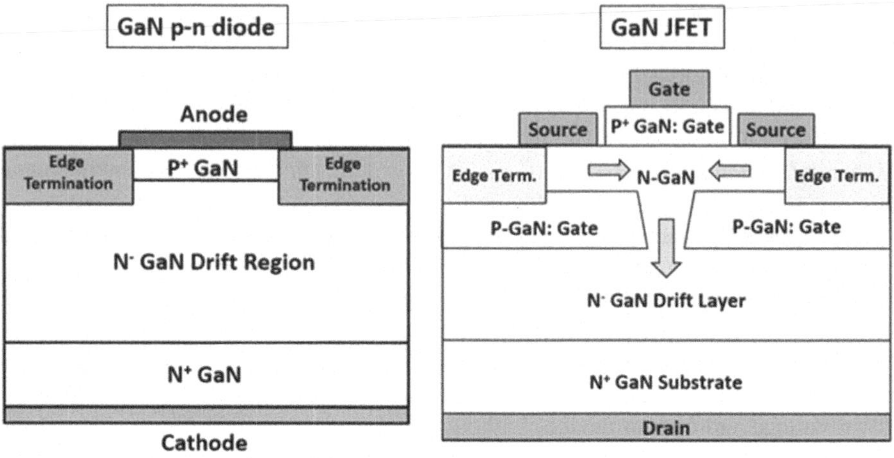

Fig. 1.1 Illustration of a GaN vertical p-n diode and junction field-effect transistor on bulk GaN

mismatch, coefficient of thermal expansion (CTE) mismatch, safe breakdown limited operation, avalanche capability, and includes microscopic images for GaN grown on Si, SiC, and bulk GaN. Plan-view cathode-luminescence (CL) imaging reveals that the threading dislocation density in the films grown over bulk GaN substrates is limited by the defect density in the substrate and many orders of magnitude smaller compared with material grown on silicon, sapphire, or SiC substrates. Also, the ability to grow arbitrarily thick drift layers (up to 50 μm) on native substrates allows the realization of high voltage (>10 kV) vertical diodes and transistors.

The salient features of vertical power semiconductor devices can be explained by the illustrations in Fig. 1.1. The device on the left is a p-i-n diode, while the one shown on the right is a vertical junction field effect transistor.

In both devices, there are three regions of interest. First, the core of the device; in the case of the diode this is the p-n junction, for this specific transistor shown here, it is the device channel, gate regions, and the aperture. In structures that utilize submicron fin-shaped channels such as the fin power field-effect-transistor structure, the reader can imagine replacing the cores described above with the vertical fins [30] as described later and in detail in Chap. 2. Second, the lightly doped drift region, and third, the edge termination/isolation region that is common to most all device types. The maximum breakdown voltage of the device is determined by the design of the n-type drift-layer layer doping and the thickness, along with the edge termination scheme.

High electron mobility at low doping levels in the drift region allows the same region supporting high voltage in the off-state to conduct current in the on-state with a superior tradeoff compared to other materials. Power device figure of merit can be used to estimate the drift layer doping and thickness assuming an infinite planar junction, ideal edge termination, one-dimensional electron transport with $E_C = 3.3$ MV/cm and $\mu_n = 1150$ cm^2/V-s. For example, to achieve a reverse-bias blocking voltage of 5–10 kV with the optimum on-state conductance, the n-type drift

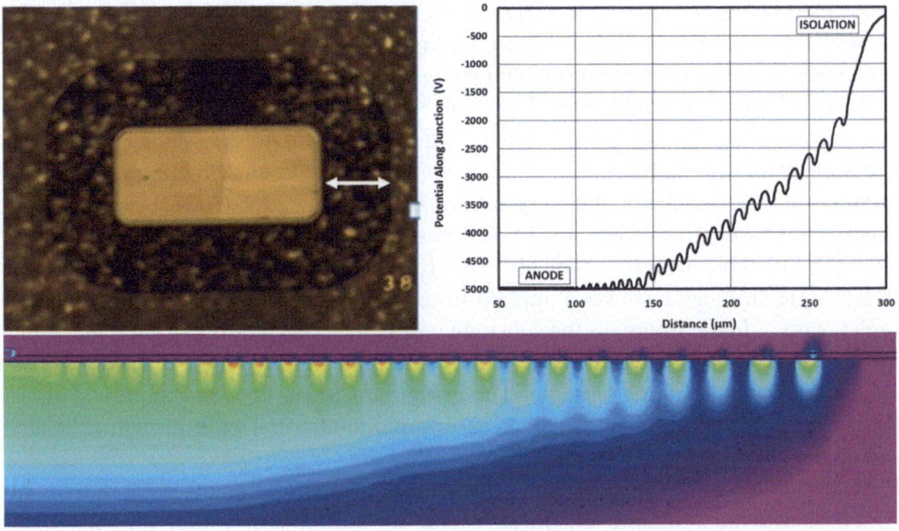

Fig. 1.2 Optimization of edge termination structure for a 5 kV fabricated p-n diode. The 2-D simulated electric field (bottom) and the electrostatic potential are plotted along the white arrow [31]

region thickness needs to be about 40 μm with a net doping level (N_D-N_A) in the range of low 10^{15} cm^{-3}. While to design a device with a breakdown voltage of 1200–1700 V, ideally a drift layer of 10 μm and a net doping of 1.5×10^{16} cm^{-3} are sufficient. To take full advantage of the high critical electric field for the onset of nondestructive avalanche breakdown in GaN, it is essential to manage the electric field at the edges (corner) of the device. An edge termination structure spreads the potential applied to the top metals over a distance (thicker than the drift region). The technique of smoothing the equipotential contours at the edges of the reverse-biased p-n junction device results in manageable surface electric fields (gradient of the electrostatic voltage) to prevent the device from premature breakdown [31–34] as shown in Fig. 1.2.

In the work presented here, we use a version of junction termination extension developed for vertical Si and SiC power devices but adapted for GaN using GaN-specific processes [21, 22, 31–34]. In any case, perfect geometric planes are impossible to achieve, and a near-ideal edge termination structure for GaN vertical devices requires further development with the limitations to be discussed later in the chapter.

2 Bulk GaN Substrates and Epitaxial Growth

Demand for bulk GaN substrates was initially driven by laser diodes. Studies at Nichia and Sony established that the laser reliability was critically dependent on the threading dislocation density (TDD); and that a TDD $<10^6$ cm^{-2} was required for long-lived lasers [12, 14]. This TDD is 3 orders of magnitude lower than that of the GaN-on-sapphire films used for light-emitting diodes and could only be achieved by localized dislocation filtering techniques such as epitaxial lateral overgrowth (ELOG) or with bulk GaN. However, it is exceedingly difficult to grow bulk-GaN crystals in the traditional manner from a liquid melt, due to the very high nitrogen pressure needed to stabilize the material. Instead, most commercial bulk-GaN substrates are derived from thick crystalline films, grown by hydride vapor phase epitaxy on nonnative substrates, and subsequently removed from the substrate. The TDD of these wafers is typically 10^6–10^7 cm^{-2}; and sophisticated dislocation-gettering techniques have been applied to reduce TDD to 10^4 cm^{-2} over a limited area (~mm^2). In the future, as the substrate cost comes down, it is expected that bulk-GaN substrates will become more widely adopted for large-area vertical devices, both electronic and optoelectronic. The benefits of lower cost and improved structural quality will accrue from advances in ammonothermal bulk growth [28] also covered in Chaps. 3 and 4 of this book.

The p-n junction device structures discussed in this chapter were grown by low-pressure metalorganic chemical vapor deposition (MOCVD) on bulk GaN substrates. The 0.5 μm p-type layers are doped with Mg, at a concentration of 2×10^{19} cm^{-3}; and a higher-doped layer at the surface for reduced contact resistance. P-type material has been characterized by Hall measurements, which indicate a room-temperature hole concentration of 4×10^{17} cm^{-3} and a mobility of 12 cm^2/V-sec. The net doping (N_D-N_A) determines the conductivity in forward bias and the depletion depth in reverse bias. Consequently, for kV-level operation, net doping control in the n-type drift layer in the range of 10^{15}–10^{16} cm^{-3} is crucial. During MOCVD growth of GaN, low-n-type-doping is made difficult by contamination from carbon. Carbon is inherent in the MOCVD growth chemistry, supplied by the organic gallium precursor, trimethylgallium Ga-$(CH_3)_3$. Since carbon assumes a deep-acceptor state in GaN, it compensates intentional donors, typically Si (or Ge). During the MOCVD growth process, the carbon background is minimized using a hydride nitrogen precursor, ammonia (NH_3), which yields both chemically active nitrogen and hydrogen upon pyrolysis. The hydrogen liberated by ammonia pyrolysis eliminates the methyl radicals from the trimethylgallium in the form of methane, which is stable and chemically inert in the process. Thereby, to the extent that this reaction is complete, the carbon contamination is eliminated in MOCVD-grown GaN films. However, in practice, some residual carbon remains in a small concentration. This carbon background is minimized by growing under conditions that represent a large partial pressure of active ammonia. These conditions may include a combination of high growth pressure, high growth temperature, and a high V:III ratio. Due to the chemical stability of ammonia, the V:III ratios used for GaN MOCVD growth are particularly high, typically greater than 1000. By growing under optimal MOCVD conditions, the compensating carbon acceptor background may be reduced to the 10^{15}–10^{16} cm^{-3} range. Thus, the drift layer films are lightly n-type-doped and closely compensated with unintentional carbon, with net n-type doping of 2–5 $\times$ 10^{15} to 2×10^{16} cm^{-3}. Note that silane (diluted 100 ppm in nitrogen) for n-type doping was used. At these low net-doping values, it is difficult

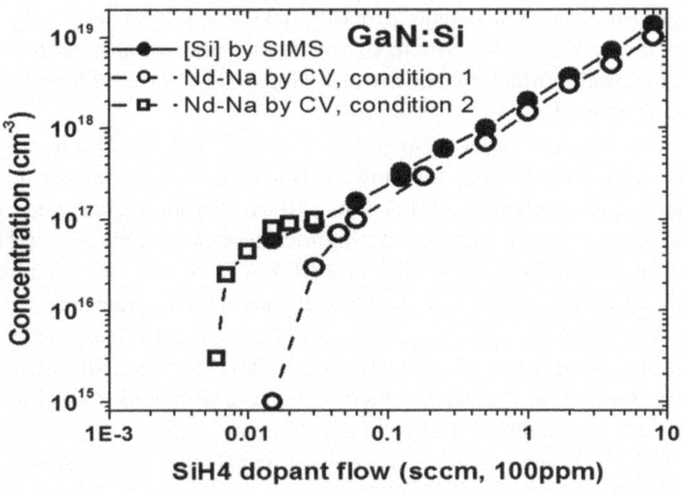

Fig. 1.3 Net doping and silicon concentration as a function of the silane flow [22]

to assess the impurity concentrations in the traditional manner, by secondary ion mass spectroscopy (SIMS). The readers are asked to refer to Chap. 22 for recent advances in SIMS for GaN. Rather, it is most sensible to estimate the carbon concentration from its compensation of the intentional and well-controlled silicon donors. This is because the background concentrations of silicon, oxygen, and hydrogen in SIMS analysis are often in the 10^{16} cm^{-3} range. The compensating effect is shown in Fig. 1.3.

This shows as a function of the silane doping flow, the silicon concentration measured by SIMS (solid circles), compared to the net doping (N_D-N_A) for two different growth conditions (open circles and squares). At high silicon concentrations, the silicon and net-doping concentrations are well matched. However, at low concentrations, the two characteristics deviate. The compensation from unintentional carbon acceptors is responsible for this deviation, which reduces the net doping (N_D-N_A) from the Si (donor N_D) concentration. From the characteristics in Fig. 1.3, we may infer that for growth condition 1 (open circles), the concentration of unintended carbon acceptors is approximately 5×10^{16} cm^{-3}, while for improved growth condition 2 (open squares), the concentration of unintended carbon acceptors is approximately 1–2 $\times$ 10^{16} cm^{-3} (about 1 part per million).

Substrate orientation plays a significant role in the device's performance and reliability. A nominal c-orientation (0001), with a slight inclination (0.2° to 0.6°) toward the m plane is preferred, similar (but not identical) to the case for near-UV LEDS and laser diodes grown on bulk c-oriented-GaN substrates. For the case of on-axis growth (i.e., no misorientation), large hexagonal hillocks are naturally formed during growth documented in [35, 36]. Hillocks formed by MOCVD epitaxial growth of GaN on an exact-(0001)-oriented bulk GaN substrate are easily visible. These hillocks have been attributed to spiral growth around a screw dislocation. By introducing a slight miscut of several tenths of a degree, the surface is made vicinal, with step edge density proportional to the miscut angle. This increases the velocity of the step-flow growth, overwhelming the spiral growth and thereby producing a smooth surface as shown in a Nomarski micrograph image covering 1 mm^2 of GaN (Fig. 1.4).

Vertical p-n junction diode testing has revealed that the elimination of these hillocks and surface morphology (as shown in Substrate B) is mandatory for achieving low reverse leakage currents (<1 μA for a 10 A forward current rated device) at voltages larger than 500 V and passing high-temperature reverse bias (HTRB) reliability stress tests. Surface preparation and MOCVD epi growth initiation are also found to be critical. Reliability and yield models based on accurate spatial mapping of the surface morphology have been developed and serve as an effective screen pre- and post-processing of devices [37, 38]. The GaN device comprising all the vertical devices discussed in this chapter, if otherwise stated, were epitaxially grown on full 2-inch bulk GaN substrates with a defect density in the range of 10^4 to 3×10^6 cm^{-2}. Plain-view cathode-luminescence (CL) imaging reveals that the threading dislocation density in the films grown over bulk GaN substrates is limited by the defect density in the substrate and many orders of magnitude smaller compared with material grown on silicon, sapphire, or SiC substrates (Fig. 1.5).

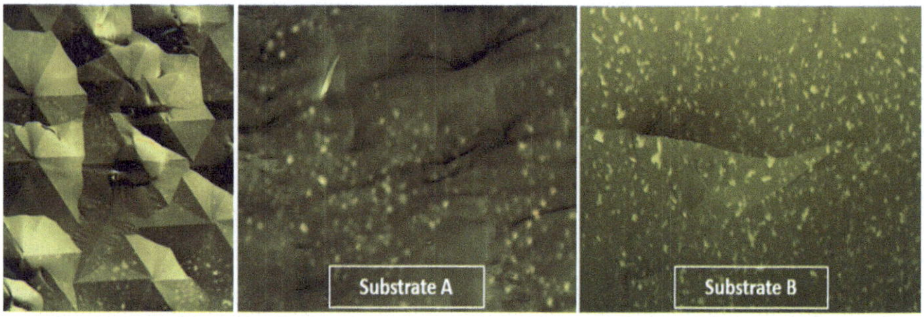

Fig. 1.4 Nomarski image of surface morphology observed on devices grown on GaN substrates. The image on the left demonstrates hillocks formed on the GaN surface in growth on low miscut angle substrates. The white speckles are from the back side reflection [22]

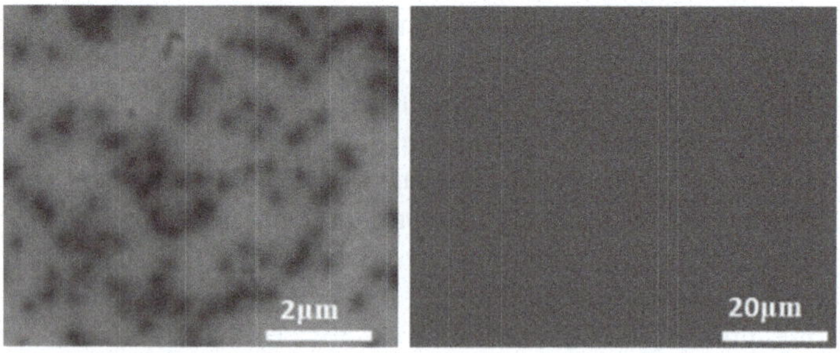

Fig. 1.5 CL images of GaN grown on sapphire (left) and bulk GaN (right) starting material defect density of 10^4 cm^{-2}. Note the scale change for the image on the right from the left [22]

3 Device Fabrication Process

In the following section the device fabrication process is presented for the p-n diodes as well as the vertical transistor structures on bulk GaN substrates that were graphically depicted in Fig. 1.1. After p-n structures are epitaxially grown by MOCVD, the p-n diode wafer process utilizes e-beam deposition and patterning of Pd/Pt-based metallization scheme to contact the p-type GaN to form the anode. In the processing of both the diodes and the transistors, SiN_x is used as the passivation layer.

For the transistor process, narrow trenches are etched using a Cl-based chemistry in an inductively coupled plasma (ICP) process. An n-type GaN layer is regrown by MOCVD to fill the trenches to form a channel above the patterned p-type GaN layer (buried gate) and the aperture for the vertical current flow. The p-type GaN layer deposited during the same MOCVD growth run as the channel n-type layer is later patterned to form the top gate. The thickness and doping of the n-GaN layer in the channel region were designed to target the threshold voltage of about $V_{th} = -20$ V targeting a cascode solution. The length of the channel region was chosen to be sufficient to preclude the effect of the drain voltage on the potential barrier in the channel in the off-state. The SEM cross-section of the transistor is shown in Fig. 1.6 (top). The buried p-body layer can be contacted using an appropriate metal for ohmic contact to p-type GaN, such as Pd, Pt, or Sc. In these devices, a Pd-based stack was

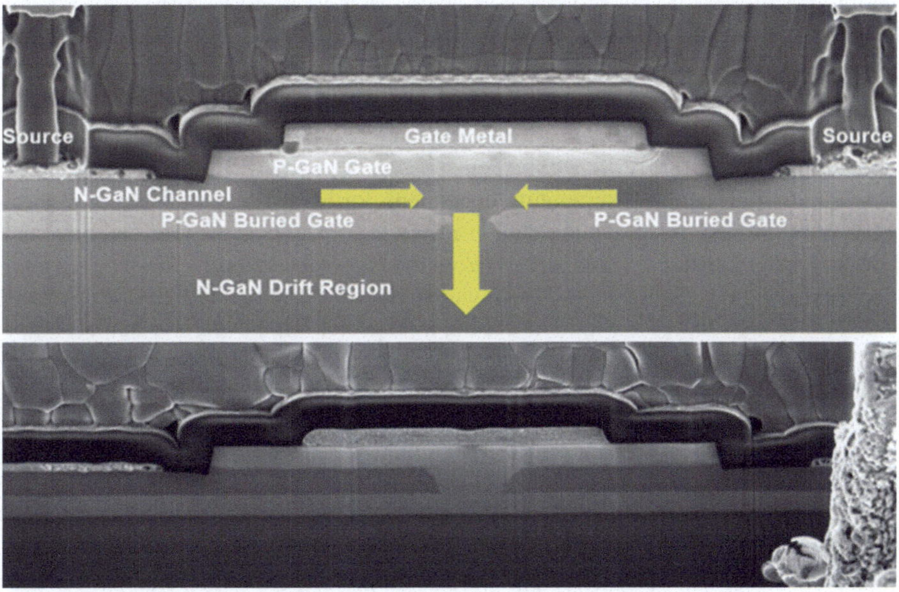

Fig. 1.6 Cross-section SEM of the junction field-effect transistor on bulk GaN. The top image shows the channel, gate p-GaN regions (light color), and the aperture. The bottom figure shows the p-GaN plug. Arrows show the transistor on-state current flow. (Avogy Inc., private communication)

utilized. Since it is very difficult to achieve good Ohmic contacts to ICP etched p-GaN surfaces, a p-GaN plug process is used as shown in Fig. 1.6 (bottom). The source electrode is Ti/Al based. A SiN_x-based film serves as an interlayer dielectric (ILD) that is deposited by plasma-enhanced chemical vapor deposition (PECVD). Via contact, holes were patterned for both the source and the gate.

Finally, 4 μm thick source and gate pad metals (Au-based) are deposited for wire bonding and current spreading. A similar pad metal layer structure is used for the p-n diode anode contact as well. A backside contact (drain or cathode) was formed by evaporating proprietary metallization onto the bottom of an n^+-type GaN substrate after the substrate was lapped, polished, and ICP etched down to 6 mils (150–180 μm).

The thinning process is crucial in reducing the electrical and thermal resistance of the device. After on-wafer electrical screening, the devices are diced, and placed in standard power packages (for example, TO220, TO247, or power modules) for electrical characterization, cryogenic measurement, avalanche capability assessment, or other battery of reliability testing.

4 Vertical Power PN Diodes Based on Bulk GaN

The GaN p-n junction serves as a basic building block for most (if not all) high-voltage (BV > 600 V) GaN vertical transistors as described above. Furthermore, it is the ideal structure to study and extract preliminary information on substrate material quality, doping, defects and generation-recombination rates, avalanche and impact ionization, critical electric field, Ohmic contacts, hole and electron mobility, minority carrier lifetimes, temperature effects, and reliability on an emerging material system. To this end, we utilize various vertical GaN p-n diodes optimized for device operation covering 600 V to 4 kV.

First, the measured temperature-dependent forward I-V curves of a typical 1700 V optimized p-n diode are shown in Fig. 1.7. The drift region thickness and doping are 18 μm and 1×10^{16} cm^{-3} (from the Hg-probe and device C-V measurement extraction), respectively. The device area is 0.72 mm^2. The turn-on voltage (~3 V) is seen to decrease with temperature as expected. However, it is also observed that the forward current improves markedly with increasing temperature.

To understand this phenomenon, the Ohmic contact and sheet resistance of the Mg-doped (deep acceptor ionization) p-GaN layer were studied. It was found that while the resistivity of the n-type drift region increases due to reduced electron mobility, the resistivity of the p-GaN and the ohmic contact to p-GaN decrease significantly with increasing temperature. The root cause for this behavior is that the deep Mg acceptor level provides more holes at elevated temperatures. Magnesium freeze-out effect will be described in more detail in Sect. 1.6 of this chapter. For this 0.7 mm^2 area device, the Pd/Pt p-Ohmic contact resistance contribution to the overall ON-state resistance is ~100 mΩ at 30 °C (comparable to the total device resistance of 200 mΩ at $V_F = 5$ V) dropping to less than 40 mΩ at 150 °C.

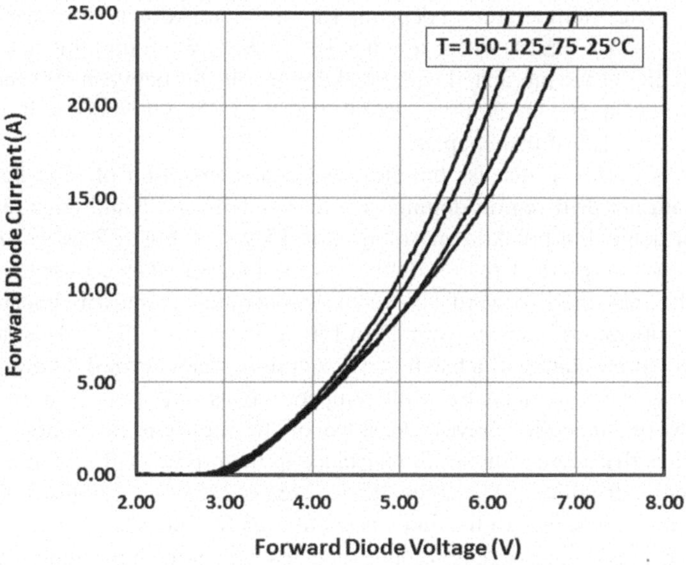

Fig. 1.7 Measured forward I-V characteristics of a 1700 V optimized p-n diode [22]

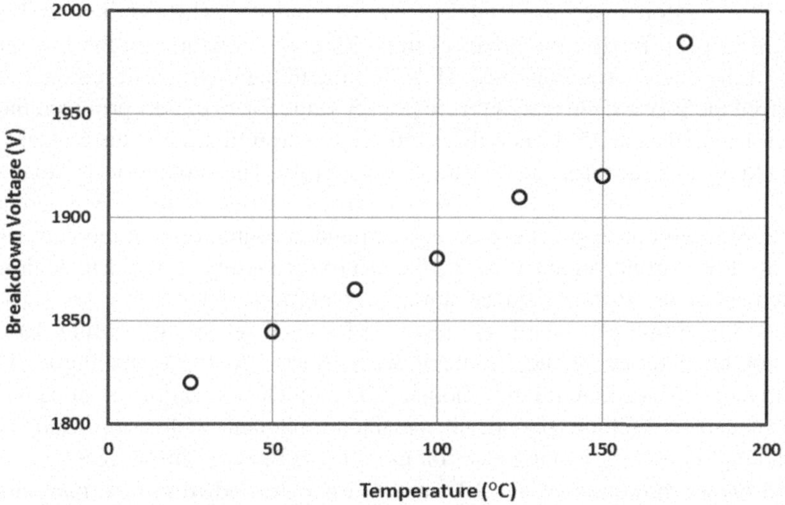

Fig. 1.8 Temperature dependence of breakdown voltage in the GaN p-n diode [22]

Further, the 0.72 mm^2 area GaN diodes were driven into nondestructive avalanche breakdown at 2600 V. The measured positive temperature coefficient of breakdown voltage proved that the breakdown mechanism in reverse-biased diodes is due to impact ionization initiated nondestructive avalanche. Figure 1.8 confirms that increasing the temperature delays the onset of impact ionization due to phonon scattering. The temperature dependence of breakdown voltage is BV $(T) \approx BV_{25\,^\circ C}(1 + \alpha \Delta T)$, where $\alpha \approx 6 \times 10^{-4}\,^\circ K^{-1}$. We found this to be roughly valid for all diodes we fabricated with breakdown voltages between 600 and 2500 V. The topic of avalanche breakdown capability of the GaN p-n diodes will be covered in detail in Sect. 1.5 of this chapter.

Next, it is useful to discuss the electrical characterization of long base diodes (40 μm) with net drift region doping of 5 to 6 × 10^{15} cm^{-3} and edge termination suitable for achieving breakdown voltages of 4 kV (see Fig. 1.2 as well). In these structures, the n-type drift region dominates the total resistance over the p-Ohmic contact. The measured forward I-V versus temperature is shown (linear scale) in Fig. 1.8. Temperature is swept from 25 to 150 °C in steps of 25 °C. This device was chosen as a representative of a batch of a dozen fully characterized diodes. The turn-on voltage is seen to decrease with temperature as expected, and in this case (contrary to the scenario above) the slope of the high current region decreases. This indicates that increasing series resistance in the n-type drift region is dominating all other contributions. The active device area is 235 μm × 470 μm = 0.11 mm^2. The differential resistance of the diode is about 2.25 Ω. The p-Ohmic contribution is only 0.35 Ω. Fabrication of these diodes represents a unique opportunity to estimate the electron mobility in the low ionized impurity scattering limit and self-consistently extract hole minority carrier lifetimes using TCAD models. This is because the forward I-V curves are more dominated by the p-n junction and the n-type drift region resistance and less by the contacts. The detailed analysis is provided in [22]. Firstly, the effect of minority carrier lifetime on the low current region of the device was examined. Hole lifetime in the n-type drift region strongly influenced the forward current in this region. A value for τ_h of 2 ns provided the best fit to measured data at 25 °C as well as 150 °C. Electron lifetime in the p^+-GaN layer was set to 0.7 ns in accordance with the literature [39]. The simulation and extraction results were insensitive to the electron minority carrier lifetime. Mg-activation/ionization/freeze-out in p-GaN was included and accounted for in the simulations. Once the low current region was set, the electron mobility in the drift region was adjusted to fit the drift resistance dominated portion of the curve, as shown in Fig. 1.9. The fitting procedure was repeated for each set of temperature data. The value for temperature in the simulator was adjusted to the temperature of each measurement to account for the modeled incomplete ionization of dopants. The minority carrier lifetime parameter remained invariant with temperature while maintaining a good fit, leaving electron mobility as the only fitting factor.

Drift layer net doping of 6 × 10^{15} cm^{-3} was extracted from C-V measurement first, using Hg probe post epitaxial regrowth (prior to p-GaN activation) and second direct measurement of the diode. The drift layer thickness (40 μm) was estimated by known growth rates in the MOCVD reactor and verified with a caliper on starting substrates and after growth.

The temperature-dependent mobility used to fit all the data is shown in Fig. 1.10. A functional dependence of ~$T^{-3/2}$ is extracted indicative of phonon scattering being the mechanism limiting mobility in the low-ionized impurity scattering limit. The room temperature mobility for electrons in low defect and lightly doped GaN could

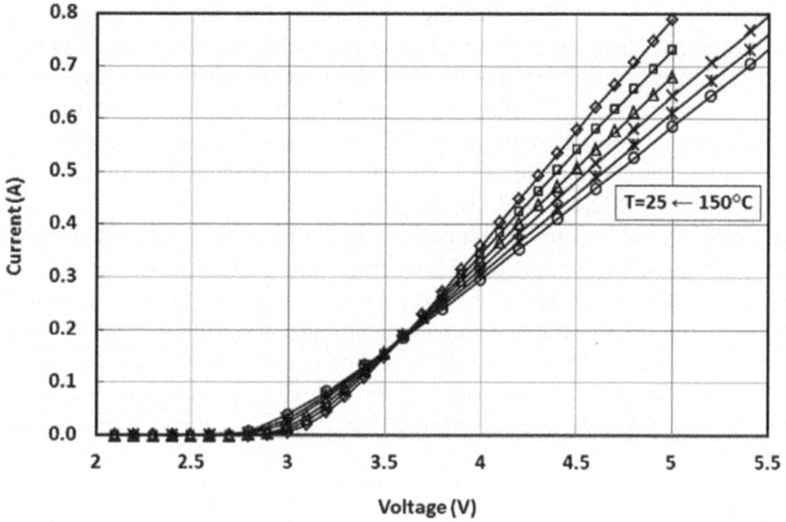

Fig. 1.9 Measured and TCAD I-V data fits for the 40 μm drift diode for 25–150 °C, $\tau_h = 2$ ns [22]

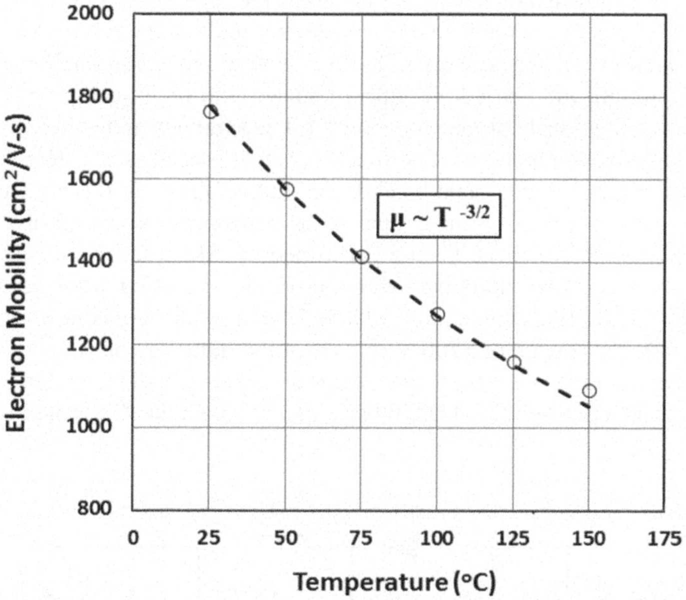

Fig. 1.10 Extracted electron mobility as a function of temperature. A functional dependence of ~$T^{-3/2}$ is extracted indicative of phonon scattering being the mechanism limiting mobility in the low-ionized impurity scattering limit [22]

be as high as 1750 cm^2/V-s with the assumption that the hole minority carrier lifetime is temperature and bias invariant at $\tau_h \approx 2$ ns (conductivity modulation in the drift region is negligible). The readers are encouraged to see Chap. 19 of this book for the measurement of minority carrier diffusion lengths for p-n diodes using the electron beam induced current (EBIC) method. The measured devices show breakdown voltages larger than 4 kV with an area differential specific on-resistance (R_{sp}) of less than 3 mΩ-cm^2 [22, 31, 40, 41].

Last, for GaN devices to be competitive with their silicon and SiC counterparts for power electronics applications, device forward current levels need to exceed 10 amps and reach up to 100s of amps. Fabricated large-area (16 mm^2) vertical GaN p-n diodes have demonstrated breakdown voltages exceeding 700 V and pulsed (100 μs) currents of 400 A (2.5 kA/cm^2). For details, the readers are referred to [22, 42]. Future studies of large area (>1 cm^2) devices pushing the device forward currents to 100 A-kA regime are needed for a broader adoption of vertical GaN devices for power.

5 Avalanche Capability of Vertical GaN p-n Junctions

For power switching and rectifying applications, the ruggedness in the breakdown is a critical requirement as supplementary snubber circuits impose a performance and efficiency penalty on the system [43–45]. Contrary to common belief, there is avalanche multiplication based on impact ionization in high-quality GaN p-n junctions. This was previously discussed in Sect. 1.4, also demonstrated in [21, 22]. Even though a positive temperature coefficient of the breakdown voltage has been observed from quasi-steady-state breakdown studies, the peak reverse current densities have been necessarily small due to the long pulse times involved. In this section, we present avalanche testing performed by Aktas and Kizilyalli [46] on a custom-built unclamped inductive switching (UIS) test setup whose schematic is shown in Fig. 1.11. The setup employed a SiC metal-oxide-semiconductor transistor with a breakdown voltage of 1700 V to control the inductor current.

The device under test is a 0.36 mm^2 active area GaN p-n diode with a breakdown voltage of 1200 V housed in a TO220 package with a median ON state resistance of

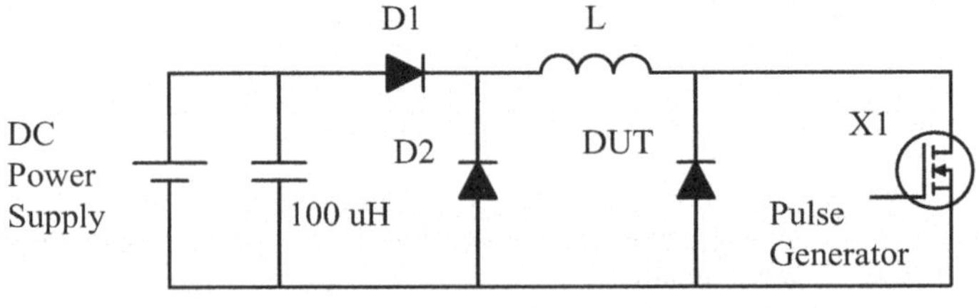

Fig. 1.11 Avalanche test circuit. D1/D2 are Schottky diodes, X1 is the switch controlled by an external pulse generator, L is the load inductor, and DUT is the device under test

0.5 Ohms. The junction termination extension of the GaN diode enabled a breakdown voltage close to the design limits. The devices exhibited a median ON voltage of 3.0 V and 3.6 V at forward currents of 10 mA and 1 A, respectively. Parametric testing of the devices was performed by using the FocusTest FTI-1000 tester (ATE), and devices passing breakdown voltage and ON resistance criteria were accepted for avalanche testing. Temperature-dependent measurements of breakdown voltage under constant reverse current were made using a Keithley 2657A source-measure unit (SMU). The temperature-dependent breakdown voltage and avalanche voltage measurements were made by attaching the device to a temperature-controlled (forced air-cooled) heat sink, and the device temperature was measured continuously throughout each temperature-dependent test for each device by means of a thermocouple attached to the TO-220 package tab.

Repetitive avalanche testing was performed by applying periodic avalanche current pulses to the device at a repetition rate of 5 kHz 10^5 times. To test the limits of repetitive avalanche ruggedness, the repetitive avalanche current was increased in a step-stress manner after each set of 10^5 pulses. Parametric testing of the devices was performed on incoming devices and after each step of repetitive avalanche testing. Figure 1.12 shows the resulting avalanche current waveform on the DUT in red and the avalanche voltage waveform in yellow. For this device and pulse condition, the peak reverse diode current was 8 A.

From Fig. 1.12, it can be observed that the reverse voltage exhibits an increase while the reverse current is decreasing. The physical mechanism that is causing the reserve voltage to increase while the reverse (avalanche) current is decreasing is attributed to the heating of the device because of energy deposited within the active region. The temperature dependence of the breakdown voltage (V_R) as measured by a K2657A SMU, and the avalanche voltage (V_A) as measured on the avalanche test

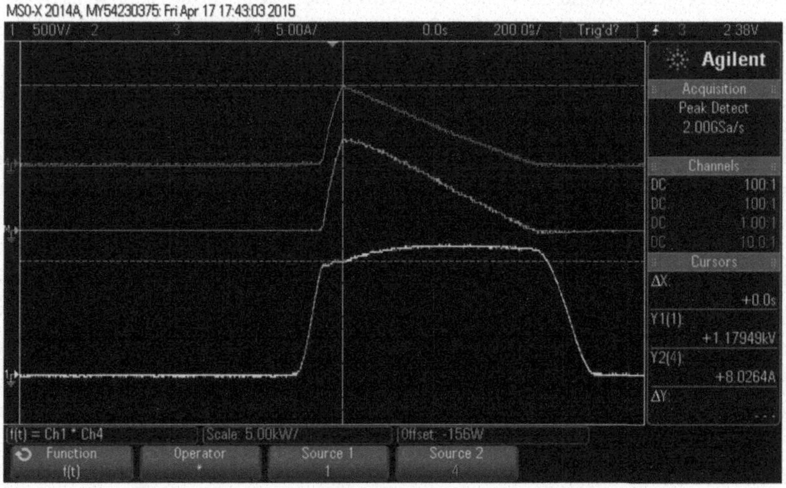

Fig. 1.12 Scope waveforms displaying DUT current (red), voltage (yellow), and purple (power) [46]

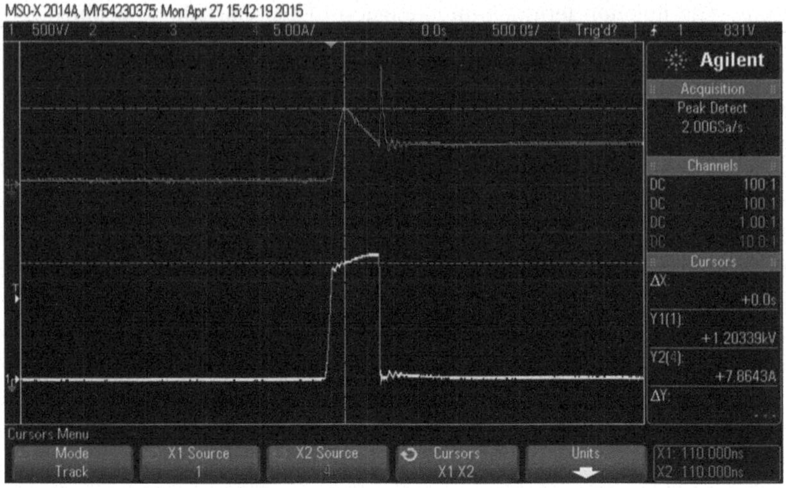

Fig. 1.13 Destructive avalanche waveform [46]

system exhibits a positive temperature coefficient for all reverse currents [46]. In Fig. 1.12 the purple waveform is the instantaneous power on the device calculated by the multiplication of the voltage and current. As can be seen, the peak power is about 10 kW, corresponding to an area-normalized power dissipation of 2.8 MW/cm^2. The energy dissipated on the device is extracted from integrating the power waveform.

In contrast, Fig. 1.13 is a destructive avalanche waveform for a similar device. The destructive avalanche waveform shown in Fig. 1.13 indicates that the damage does not occur at the beginning of the current pulse, but after dissipation of some amount of energy as would be consistent with a thermally induced destruction of the device. In Fig. 1.13 we define the time from the onset of the avalanche to the destruction of the device as breakdown time (typically 1–10 μs).

Figure 1.14 shows the number of parts that survive each test step. There are several devices failing at small repetitive avalanche currents even though the median current that can be handled (with 100 μH inductor) was about 10 A. The early failures during repetitive testing are attributed to the thermal load induced on the device by the average power dissipation and the avalanche capability is strongly impacted by the transient nature of the heat transfer.

Statistics of the reverse current, on-resistance, and forward voltage of the devices surviving each stress step were studied. From Fig. 1.15 it is seen that fresh devices with higher reverse leakage currents are getting eliminated (failing) at smaller stress currents during the step-stress test. A careful study of Fig. 1.15 shows that the devices with higher leakage have a higher probability of elimination. However, the study found no correlation to the device's ON-resistance or forward-voltage. Furthermore, the devices that survived and continued with the testing did not exhibit a significant shift in the forward voltage, ON resistance, or breakdown voltage.

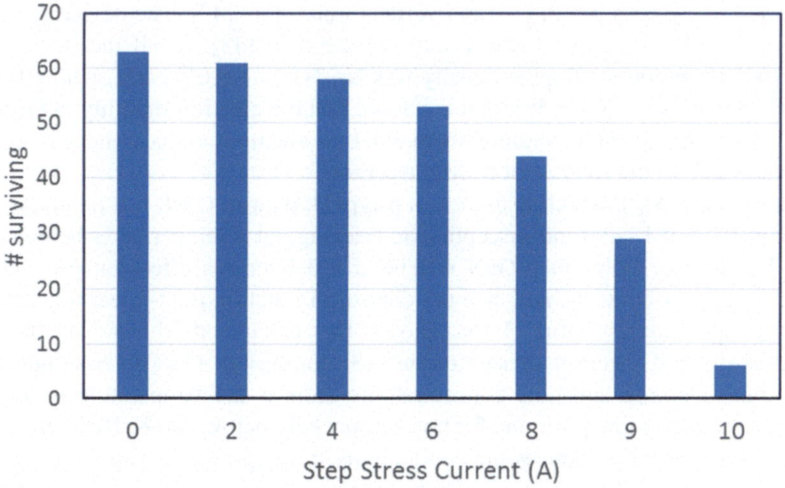

Fig. 1.14 Number of devices surviving after each step-stress current level from 0 to 10A. "0" indicates the incoming device population for the study [46]

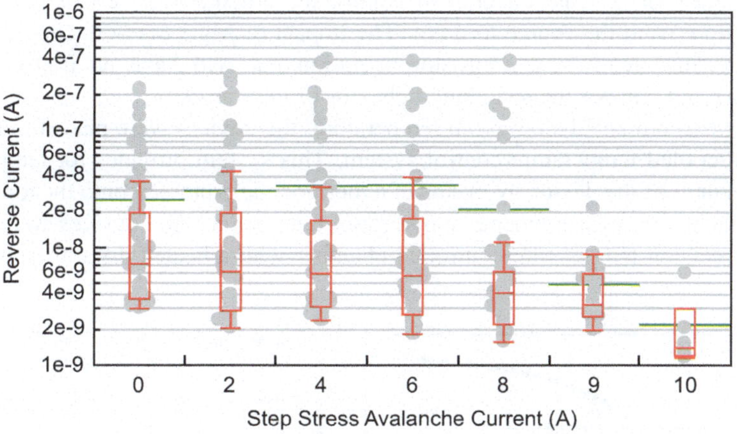

Fig. 1.15 Statistics of reverse current at a reverse bias of 600 V for devices surviving after each repetitive avalanche stress step. "0 A" indicates the incoming population [46]

6 GaN Diodes and Transistors at Cryogenic Temperatures

It is expected that power electronic systems are operational in a large range of environmental conditions. For example, the temperature range of power electronics for automotive applications can range from −60 °C to 150 °C (213 K to 423 K). To assess whether vertical GaN devices relying on p-n junctions are suitable for low-temperature applications (down to 77 K), the study described below was conducted using both p-n diodes and vertical transistors [47]. The device structures were designed for a breakdown voltage of 1200 V [48]. In all the devices, the nominal drift region net doping density was N_D-$N_A = 1\times10^{16}$ cm^{-3}, while the drift layer thickness was 15 μm. While it is known that the electron mobility in GaN will improve with reduced temperature, there will be some limitations to the performance of devices due to the freeze-out of Mg acceptors.

Magnesium (Mg), which is used as a p-type dopant in GaN, is a relatively deep acceptor ($E_A \approx 0.18$ eV) and susceptible to freezing out at temperatures below 200 K [49]. The loss of holes in p-GaN can have a deleterious effect on p-n junction behavior, p-GaN contacts, and channel control in junction field-effect transistors at lower temperatures. Figure 1.16a shows the calculated change in the hole (Mg-doping) and electron concentrations (Si-doping) in GaN over temperature. While both electron and hole concentrations drop as the temperature is reduced, the effect is much more profound for holes, especially below 200 K. Reduction in the concentration of holes reduces the conductivity in the p-regions and interferes with our ability to form low-resistance ohmic contacts as shown in Fig. 1.16b. On the other hand, at higher temperatures (e.g. 423 K) the contact resistance to p-GaN improves significantly the contribution to the improved performance of GaN p-n diodes at elevated temperatures. For the junction field-effect transistor, the p-GaN regions are used as gate material to control the potential in the channel and the current flow from the source to drain. The impact of Mg freeze-out should be less severe during steady-state transistor operation but will have an impact during transients and hamper the switching behavior of the transistor.

For measurements below room temperature, the devices were placed in a liquid nitrogen-cooled temperature control system. This system provides for temperature measurement of the device by means of a thermocouple pressed onto the top-side of the package. For measurements above room temperature, the devices were placed into a precision temperature-controlled chamber with a heated dry nitrogen stream

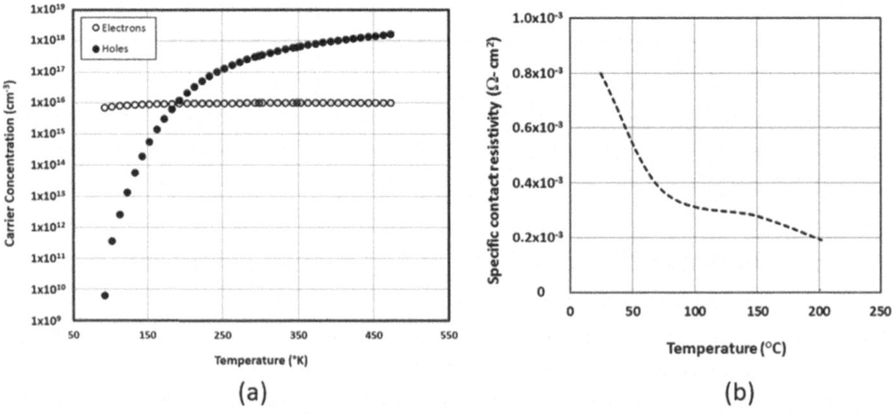

Fig. 1.16 (**a**) Mg freeze-out effect and hole concentration and (**b**) measured contact resistance versus temperature for p-type GaN [22, 47]

maintaining the temperature of the device at the set-point. In all measurements, the temperature control accuracy was limited to ±15 °C due to measurement-induced errors and temperature variations caused by self-heating.

6.1 Electrical Results for GaN p-n Diodes

I-V and C-V (not shown here) measurements from 79 K (−194 °C) to 423 K (150 °C) were performed on GaN p-n diodes housed in TO220 power packages. The active die area (anode metal) is 0.72 mm^2. Figure 1.17 shows the diode forward current in a linear (a) and a logarithmic (b) scale. As expected, the p-n diode turns on at 3 V at room temperature, and with decreasing temperatures the turn-on voltage occurs at a higher voltage. The GaN p-n diode is functional at 80 K but suffers from the lack of holes available for recombination or injection as the forward bias is increased. As the diode temperature is increased from room temperature to 423 K (150 °C), two competing factors play out. The contribution to resistance from the p-GaN and p-GaN ohmic contact decreases while the drift region resistivity increases. Hence, the overall p-n diode performance is not negatively impacted by the increasing temperature. This behavior has been discussed in Ref. [22] in detail. These diodes can deliver 10–20 A of current at room temperature as the forward voltage is increased [21, 22].

In Fig. 1.18a the I-V characteristics of a reverse-biased GaN p-n diode are swept between 80 and 415 K. By lowering the ambient temperature the reverse leakage current of the diode can be improved (reduced). This behavior is typical of leakage current generated from recombination centers in the depletion region. At higher voltages, the diode undergoes avalanche breakdown, which occurs due to impact ionization. The current was limited to 5 μA during this test. The breakdown voltage versus temperature is plotted in Fig. 1.18b. It should be noted that while the leakage current due to recombination increases with temperature, the device breakdown

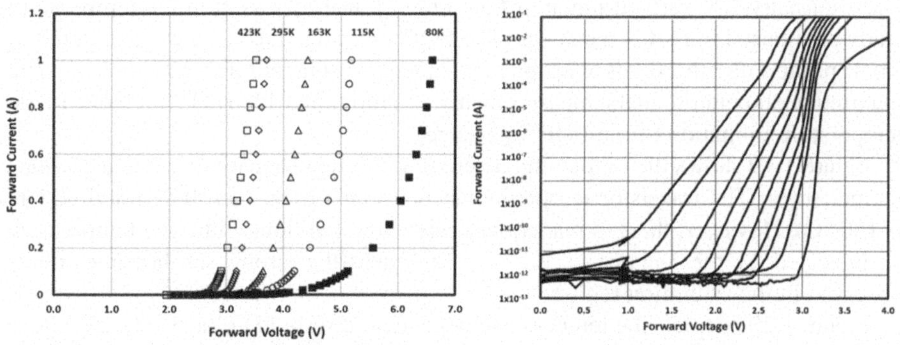

Fig. 1.17 Forward I-V characteristics of p-n diodes on a linear (left) and logarithmic scale (right) at 423, 362, 295, 250, 228, 196, 169, 151, 129, and 80 K [47]

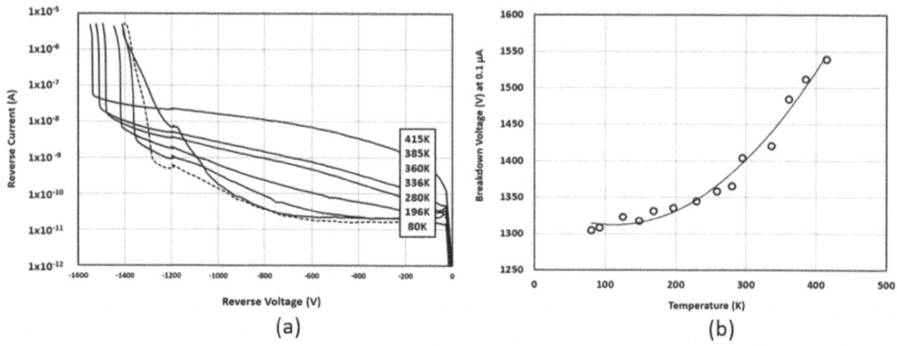

Fig. 1.18 (**a**) Reverse I-V curves and (**b**) breakdown voltage versus the ambient temperature [47]

voltage increases with temperature. This result is achieved because the p-n junctions are fabricated on high-quality (lower defect density) GaN substrates. The device breakdown voltage is about 1400 V at room temperature. At temperatures below about 200 K, a soft breakdown is observed. We attribute this to the Mg freeze-out effect.

For power switching and rectifying applications, the ruggedness in breakdown is a critical requirement as supplementary snubber circuits impose a performance and efficiency penalty on the system. Inductive avalanche capability and ruggedness characterization of vertical GaN p-n junction diodes grown on native GaN substrates discussed earlier in the chapter demonstrate that the diodes based on the technology described here can sustain reverse currents as high as 10 A (2.8 kA/cm^2 for A = 0.36 mm^2) and in breakdown. Inductive avalanche testing of GaN p-n diodes down to cryogenic temperatures should be performed for completeness in the future.

6.2 *Electrical Results for Junction Field-Effect Transistors*

GaN normally-ON vertical junction field effect transistors used in the temperature study are housed in PQFN packages. The PQFN packaged devices were soldered to small printed circuit boards with pins designed for standard TO-220 sockets. The maximum temperature in the measurements was limited to 423 K (150 °C) due to the temperature limitation of the PQFN package.

Figure 1.19 shows the output characteristics of a representative device at various temperatures. The transistor is operational between 123 K (−150 °C) and 423 K (+150 °C). However, there is inadequate gate control at liquid nitrogen temperature as indicated by the leakage at $V_G = -30$ V and the strong self-heating effects observed in the saturation region.

Figure 1.20a depicts the temperature-dependent threshold voltage extracted from (I_D–V_G) measurements at a drain bias of 5 V. The inadequate gate control seen in Fig. 1.19d is then attributed to the strong temperature dependence of the threshold

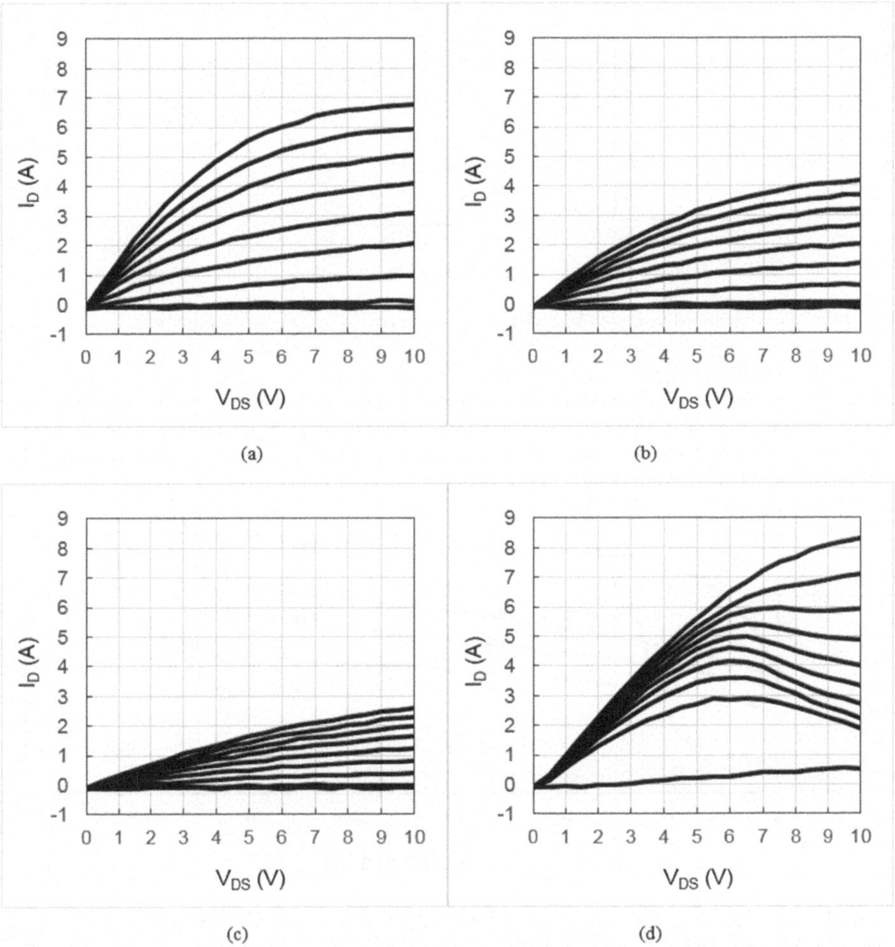

Fig. 1.19 Output characteristics of the packaged devices at (a) 123 K, (b) 300 K, (c) 423 K, and (d) 77 K. The measurements were performed using 300 μs voltage pulses. The gate bias was stepped at the following voltage values: −30, −20, −17.5, −15, −12.5, −10, −7.5, −5, −2.5, and 0 V [47]

voltage below −100 °C and the shifting of the threshold voltage below −30 V at 77 K.

From the output characteristics shown in Fig. 1.19, the on-resistance (R_{DSON}) (resistance at $V_D = 1$ V/$V_G = 0$ V) and the maximum drain current (I_{DP}) (drain current at $V_D = 10$ V, $V_G = 0$ V) can be plotted versus temperature. Figure 1.20a–c shows the temperature characteristics of the threshold voltage (V_{th}), $R_{DS,ON}$, and I_{DP} plotted as above. The ON resistance is exhibiting a minima at 123 K (−150 °C).

The increased resistance at 77 K as compared to 123 K is attributed to the nonlinear onset of the current-voltage curve observable in Fig. 1.20d indicating the presence of a barrier in the conduction path. Beyond 123 K, the R_{DSON} is exhibiting

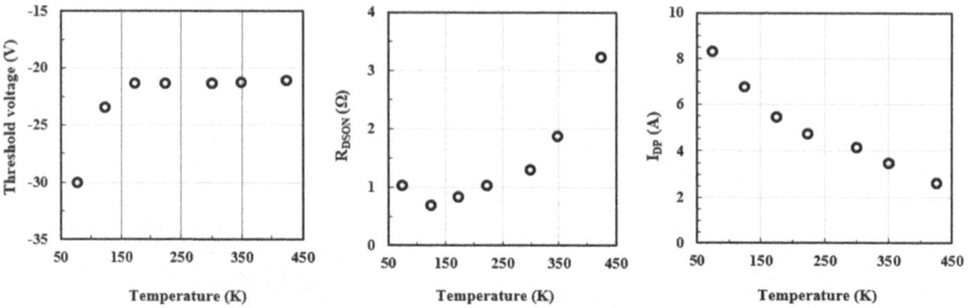

Fig. 1.20 (**a**) Threshold voltage (V_{th}), (**b**) R_{DSON}, and (**c**) $I_{D,\ max}$ versus temperature [47]

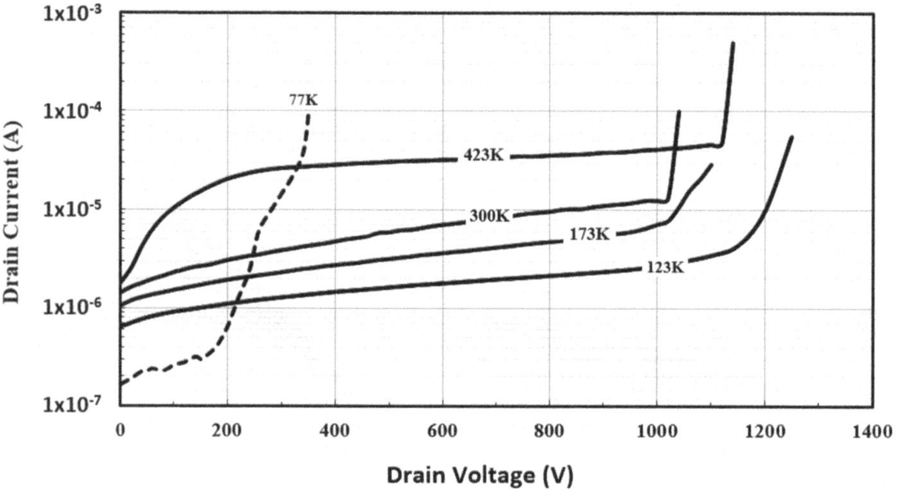

Fig. 1.21 Leakage current characteristics of a representative device at various temperatures. The data with the solid lines were measured using $V_g = -30$ V. The dotted line used $V_g = -40$ V [47]

an increase in a manner faster than dictated by the $T^{-1.5}$ dependency of phonon scattering-dominated mobility. Similarly, the reduction in I_{DP} is much faster than the reduction dictated by saturation velocity with increasing temperature. The device's active area is about 0.4 mm^2 which translates to a current density of 1.0 kA/cm^2 and a specific resistivity of 5 mΩ-cm^2 at room temperature.

The temperature-dependent leakage data shown in Fig. 1.21 show that the devices can be functional between 123 K (−150 °C) and 423 K (+150 °C). However, at 77 K (−196 °C), the devices cannot sustain high drain biases even when the gate bias is reduced to $V_G = -40$ V. The $V_G = -40$ V leakage data presented in Fig. 1.21 indicate that the issue with the transistor operation below 123 K (−150 °C) is related to the strong reduction in the breakdown voltage of the p-n junction diode observed at similarly low temperatures. Figure 1.8 also demonstrates that at temperatures below the room temperature, the breakdown voltage exhibits a negative temperature coefficient and the reverse leakage characteristic has a soft breakdown. Based on these negative temperature coefficients and the soft breakdown, it is considered that at reduced temperatures, the voltage-blocking capability is limited by defect-induced leakage. At room temperature and above, the breakdown voltage has a positive temperature coefficient and the breakdown is sharp suggesting that the breakdown may be brought on by impact ionization-based avalanche multiplication. Although much work must be performed to confirm this. The strong increase in leakage at high temperatures seen in Fig. 1.21 is another factor that may limit the operational range of the device. From the analysis of the 900 V leakage data, it can be inferred that the leakage current is dominated by defects with an activation energy of about 1 eV.

In summary, there is good evidence that vertical GaN devices (diodes and transistors) utilizing p-n junctions are suitable for most practical applications including automotive (210 K $<$ T $<$ 423 K). While devices are functional at cryogenic temperatures (77 K), there may be some limitations to their performance due to the freeze-out of Mg acceptors. Significant improvement in device $R_{DS,ON}$ and I_{DP} is measured (as expected) in temperature regimes where *Mg* freeze-out effects do not dominate.

7 Reliability Studies

In the following section, the results of reverse leakage tests, high-temperature reverse bias (HTRB), high-temperature operating life (HTOL), temperature humidity bias (THB), temperature cycling (TC), and HTRB for vertical JFET devices packaged in standard power packages (TO 220) are presented [37, 50].

In the case of p-n diode reliability studies, each wafer lot typically contains 3 full 2-inch wafers with about 800 diodes/wafer rated at 10 A and 1200 V. The devices were thinned to about 6 mils and housed in TO220 power packages. The p-n diode areas in this study are 0.72 mm^2. To reach a reasonable level of process maturity, nearly 60 lots and about 200 2-inch wafers were processed, stressed, and failure mechanisms were analyzed. Such early reliability studies fed into the refinements in the substrate specifications, growth condition optimization, fabrication process details including the backside metallization, die attach and packaging, and testing strategies. During this period many problems were identified, solved, and documented in the literature [22, 37, 50]. For example, in Sect. 1.2 of this chapter we have touched on growth and first-order reliability-related issues connected with the substrate quality and the substrate miscut angle specifications.

High-temperature reverse bias (HTRB) tests on the p-n diodes were conducted at 150 and 175 °C and 960 V reverse bias. Device forward resistance, reverse leakage current, breakdown voltage, and forward voltage at various diode currents were recorded at set time points. Failure was defined as a device drifting out of product specification or a catastrophic event (short or open) took place. Once mechanical failures such as dielectric delamination and metal lifting were solved by process integration methods, the remaining failures were related to substrate and epitaxial layer quality. HTRB stress test failures were mostly eliminated by instituting strict substrate specifications. The remaining few problematic dies could be screened out by visual inspections and using electrical measurements at wafer level looking for the anomalous increase in reverse current at higher reverse bias voltages (described earlier). Multiple substrate vendors that were evaluated eventually provided material that could be used to make p-n diodes that routinely passed HTRB stress tests at 150 and 175 °C and for 1000–2500 hours [37].

Temperature humidity bias (HTB) tests challenge the hermetic properties of the die and the package. Initial experiments showed a few failures at 850 hours due to delamination caused by water ingress from the edge of the die. The tests were repeated after some development work on the encapsulation dielectric layer. Subsequent tests showed no failures.

Temperature cycling tests evaluate the quality of the die attach. Considerable work was made to optimize the back Ohmic metal stack to the n^+ GaN substrate and the die attach process. After going through multiple learning cycles, a solution was identified and the temperature cycling test was passed. Using this developed back metal stack and die attach process, the p-n diodes also held up to current densities exceeding 2.5 kA/cm^2 pulsed and 600 A/cm^2 DC.

In the high-temperature operating life (HTOL) test, the p-n junction is forward biased and the device self-heats to the desired junction temperature. A preliminary on-wafer test showed that the devices were robust to junction temperatures up to 275 °C while being stressed at 20 kA/cm^2. It was noted that the device's forward stress voltage does not shift over time with stress. This result (no shift in forward characteristics) was later verified on packaged product diodes after a significant development effort in the back metal and die attach process was made. For example, early on it was found that using conductive epoxies as a die attach material resulted in the degradation of the epoxy due to exposure to UV light generated by the forward-biased p-n junction and the gradual increase in the packaged device resistance. Also, certain back metal stacks degraded over time during HTOL tests, resulting in large shifts in forward resistance. Certain HTOL failures manifested themselves in shifts in the p-n diode reverse current characteristics. These failures were not accelerated by temperature but were more correlated with the accumulated stress the junction experienced. The problem was narrowed down to the substrate vendors but a mechanistic understanding was never reached.

Lastly, the aging characteristics of normally on vertical JFETs are discussed under HTRB conditions. The device structure is the one discussed in Sects. 1.3 and 1.6.2. The cumulative plots shown in Fig. 1.22 compare the distributions for threshold voltage, drain-source on-resistance, and forward drain current before and after 788 hours of stress. Early device failures have been excluded from the distribution. As expected from vertical structures fabricated on bulk GaN substrates, no shift is observed in these parameters due to the lack of mismatched interfaces and sensitivity to surfaces (i.e., GaN HEMTs).

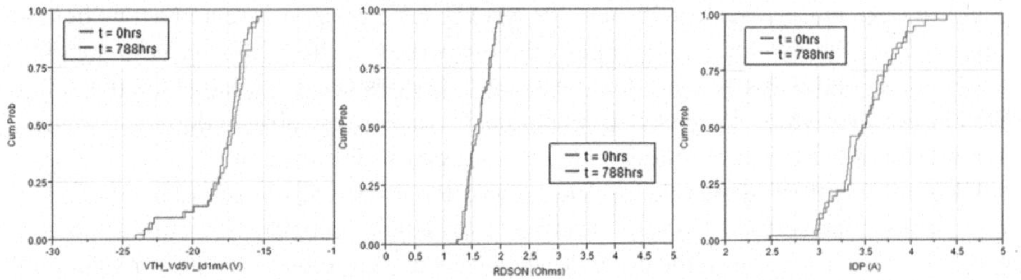

Fig. 1.22 Cumulative plots for critical transistor parameters before and after HTRB stress [37]

For technology commercialization and adoption, the authors believe that significant work is needed to elucidate all the degradation mechanisms that are unique to this material system and to the singular fabrication processes needed here.

8 ARPA-E SWITCHES Program

The U.S. Department of Energy's Advanced Research Project Agency—Energy (ARPA-E) has invested in WBG semiconductors to enable a new generation of power semiconductor devices that far exceed the performance of silicon-based devices. From materials and devices to modules and circuits to application-ready systems integration, ARPA-E projects have demonstrated the potential of WBG semiconductors to lower the cost of high-efficiency power electronics to enable broad adoption in energy applications [51]. Analysis of market drivers showed that widespread adoption would require driving down the costs of the components, which led to the creation of the SWITCHES (Strategies for Wide Bandgap, Inexpensive Transistors for Controlling High-Efficiency Systems) program [52] in 2014. The SWITCHES program was aimed at the key materials and device fabrication and architecture issues that drive costs for WBG devices. The goal was to enable the development of high-voltage (>1200 V), high-current (100 A) single-die WBG power semiconductor devices that would have the potential to reach functional cost parity with Si power transistors while also offering breakthrough relative circuit performance. The SWITCHES technologies would reduce the barriers to ubiquitous deployment of low-loss WBG power semiconductor devices in many cost-sensitive applications. Before the envisioning of the SWITCHES program, the majority of GaN power device development had been directed toward lateral architectures. There were simply no vertical GaN devices commercially or even as prototypes available. The lateral devices suffered from well-known issues such as current-collapse, dynamic on-resistance, inability to support avalanche breakdown [53, 54], and commercially accepted usable breakdown voltages no greater than 650 V. Vertical devices on the other hand had the possibility to realize the material-limited potential of GaN including true avalanche-limited breakdown. Tackling the device design and fabrication barrier, the SWITCHES program created a new field of vertical GaN device designs to exploit the properties of the WBG semiconductor.

Potential pathways toward low-cost GaN devices were investigated under the SWITCHES program. Projects (for example, Cornell) were able to demonstrate near theoretical, high-power vertical GaN p-n diodes exhibiting breakdown voltages >4 kV and figures-of-merit (V_{BR}^2/R_{ON}) greater than 3 GW/cm^2 [31, 41, 55]. It was shown that vertical GaN devices were avalanche capable, indicating the ruggedness of such devices in the breakdown, a critical requirement for power switching and rectifying applications. The projects demonstrated 80% process yield for the large area p-n junctions indicating the pathway toward ¢10/A for GaN devices is promising. With the current cost of 100 mm GaN wafers and a die size of 12–16 mm^2 for a 100 A, 1200 V device, vertical GaN devices should be capable of reaching the cost range of <¢10/A. Under the SWITCHES program, the University of California, Santa Barbara, demonstrated a modified vertical trench MOSFET, named OG-FET, which takes advantage of a regrown unintentionally doped GaN interlayer followed by an in situ dielectric deposited in the trench for enhanced electron mobility. For a single unit cell OG-FET, V_{BR} as high as 700 V corresponding to a breakdown electric field of 1.4 MV/cm was reported with an $R_{on,sp}$ of 0.98 mΩ-cm^2 [56]. Columbia University and MIT demonstrated a vertical fin power field-effect-transistor structure (VFET) on bulk GaN substrates. The VFET consists of fin-shaped channels etched into an 8-μm-thick n-doped GaN drift layer surrounded by metal gate pads that pinch off the channel. Fabricated VFETs demonstrated a threshold voltage of 1 V, 10^{11} on/off current ratio, and a blocking voltage of 800 and 1200 V [57, 58]. Vertical trench MOSFETs were demonstrated by HRL using an AlN/SiN dielectric stack employed as the gate "oxide" yielding a device with a threshold voltage of 4.8 V, blocking voltage of 600 V at a gate bias of 0 V, and on-resistance of 1.7 Ω at a gate bias of 10 V [59]. Avogy demonstrated 2.5 A vertical transistors using buried p-layers and a hexagonal layout with breakdown voltages exceeding 1.5-kV and a specific resistance of 2.2 mΩ-cm^2 [3].

A barrier in GaN device fabrication experienced by many SWITCHES project teams was the lack of a selective area p-type doping process in GaN. To confront this barrier, the PNDIODES (Power Nitride Doping Innovation Offers Devices Enabling SWITCHES) program was launched in 2017 [60] which forms the basis for this book.

9 Concluding Remarks

This chapter introduced vertical p-n diodes and vertical junction field effect transistors and other transistor structures fabricated on pseudo-bulk low defect density (10^4–10^6 cm^{-2}) bulk GaN substrates. The basic building block of most semiconductor devices, the p-n junction, was discussed in detail. Topics included in the chapter are device design, edge termination strategies, epitaxial growth and incoming substrate specifications, backside processing and metallization schemes, packaging, temperature studies down to cryogenic regime, impact ionization-based avalanche breakdown, and reliability studies.

Acknowledgments The authors acknowledge significant contributions from D. Bour, J. Kim, D. Disney, H. Nie, O. Aktas, T. Prunty, R. Brown, J. Shealy, L. Romano, B. Alvarez, A.P. Edwards, Q. Diduck, B.M. Kayes, P. Bui-Quang, H. Bhatia, H. Shah, S. Nafis, G. Ye, R. Orr, K. Patel, A. Agarwal, T. Heidel, and L.J. Petersen and partial funding by the U.S. Department of Energy Advanced Research Projects Agency (ARPA-E).

References

1. N. Mohan, *Power Electronic: A First Course* (Wiley, 2012)
2. J. Palmour, J. Zhang, M.K. Das, R. Callanan, A. Agarwal, D. Grider, SiC power devices for smart grid systems, in *Proceedings of IEEE International Conference on Power Electronics*, (2010), pp. 1006–1013
3. H. Nie, Q. Diduck, B. Alvarez, A.P. Edwards, B.M. Kayes, M. Zhang, G. Ye, T. Prunty, D. Bour, I.C. Kizilyalli, 1.5 kV and 2.2 mΩ-cm2 vertical GaN transistors on bulk-GaN substrates. IEEE Electron Device Lett. **35**, 939–941 (2014)
4. Y. Uemoto, D. Shibata, M. Yanagihara, H. Isida, H. Matsuo, S. Nagai, N. Batta, M. Li, T. Ueda, T. Tanaka, D. Ueda, 8300V blocking voltage AlGaN/GaN power HFET with thick poly-AlN passivation. IEDM Tech. Dig., 861–864 (2007)
5. Y. Hatakeyama, K. Nomoto, N. Kaneda, T. Kawano, T. Mishima, T. Nakamura, Over 3.0 GW/cm2 figure-of-merit GaN p-n junction diodes on free-standing GaN substrates. IEEE Electron Device Lett. **32**(12), 1674–1676 (2011)
6. B. Lu, T. Palacios, High breakdown (>1500V) AlGaN/GaN HEMTs by substrate transfer technology. IEEE Electron Device Lett. **31**, 951–953 (2010)
7. J.D. Albrecht, R. Wang, P. Ruden, M. Farahmand, K.F. Brennan, Electron transport characteristics of GaN for high temperature device modeling. J. Appl. Phys. **83**, 4777–4781 (1998)
8. S. Kasap, P. Capper (eds.), *Springer Handbook of Electronic and Photonic Materials* (Springer, New York, NY, 2006)
9. T.P. Chow, *GaN Power Devices* (International Symposium on Power Semiconductor Devices and ICs, Short Course, 2008), pp. 159–186
10. R. Quay, *Gallium Nitride Electronics* (Springer, 2008)
11. C. Mion, J. Muth, E. Preble, D. Hanser, Accurate dependence of gallium nitride thermal conductivity on dislocation density. Appl. Phys. Lett. **89**, 092123–1–092123–3 (2006)
12. S. Nakamura et al., High-power, long-lifetime InGaN/GaN/AlGaN-based laser diodes grown on pure GaN substrates. Jpn. J. Appl. Phys. **37**(3B), L309–L312 (1998)
13. C.A. Hurni, A. David, M.J. Cich, R.I. Aldaz, B. Ellis, K. Huang, A. Tyagi, R.A. DeLille, M.D. Craven, F.M. Steranka, M.R. Krames, Bulk GaN flip-chip violet light-emitting diodes with optimized efficiency for high-power operation. Appl. Phys. Lett. **106**, 031101 (2015)
14. S. Tomiya et al., Dislocation related issues in the degradation of GaN-based laser diodes. IEEE J. Sel. Topics Quant. Electron. **10**(6), 1277–1286 (2004)
15. R. Therrien, S. Singhal, J. Johnson, W. Nagy, R. Borges, I.C. Kizilyalli, K. Linthicum, et al., A 36 mm GaN-on-Si HFET producing 368 W at 60 V with 70% efficiency, in *IEEE International Electron Devices Meeting (IEDM)*, (2005), pp. 577–580
16. M. Kanechika, T. Kachi, et al., A vertical insulated gate AlGaN/GaN heterojunction field effect transistor. Jpn. J. Appl. Phys. **46**(21), L503–L505 (2007)
17. T. Oka et al., Vertical GaN-based trench metal oxide semiconductor field-effect transistors on a free-standing GaN substrate with blocking voltage of 1.6kV. Appl. Phys. Exp. **7**, 021002 (2014)

18. S. Singhal, J. Roberts, P. Rajagopal, A. Hanson, R. Therrien, J. Johnson, I.C. Kizilyalli, K. Linthicum, GaN-on-Si failure mechanisms and reliability improvements, in *IEEE International Reliability Physics Symposium (IRPS)*, (2006), pp. 95–98
19. K. Shenai, R. Scott, B. Baliga, Optimum semiconductors for high-power electronics. IEEE Trans. Electron Devices **36**(9), 1811–1823 (1989)
20. A.M. Ozbek, B.J. Baliga, Planar nearly ideal edge-termination technique for GaN devices. IEEE Electron Device Lett. **32**, 300–302 (2011)
21. I.C. Kizilyalli, A. Edwards, H. Nie, D. Disney, D. Bour, High voltage vertical GaN PN diodes with avalanche capability. IEEE Trans. Electron Devices **60**, 3067–3070 (2013)
22. I.C. Kizilyalli, A.P. Edwards, O. Aktas, T. Prunty, D. Bour, Vertical power PN diodes based on bulk GaN. IEEE Trans. Electron Devices **62**(2), 414–422 (2015)
23. S. Sridharan, A. Christensen, A. Venkatachalam, S. Graham, P.D. Yoder, Temperature- and doping dependent anisotropic stationary electron velocity in wurtzite GaN. IEEE Electron Device Lett. **32**, 1522–1524 (2011)
24. I. Oguzman, E. Bellotti, K. Brennan, Theory of hole initiated impact ionization in bulk zincblende and wurtzite GaN. J. Appl. Phys. **81**, 7827–7834 (1997)
25. S. Sridharan, P. Yoder, S. Shen, J. Ryou, R. Dupuis, Geiger mode simulation of GaN homojunction avalanche photodetectors. Phys. Status Solidi **6**, S662–S665 (2009)
26. V. Veliadis, Barriers to SiC power semiconductor device commercialization. PRiME **MA2024-02** (2024. (October 6–11, 2024))
27. K. Motoki, Development of gallium nitride substrates. SEI Tech. Rev. **70**, 28–35 (2010)
28. D. Ehrentraut, E. Meissner, M. Bockowski (eds.), *Technology of Gallium Nitride Crystal Growth* (Springer, 2001)
29. Y. Otoki et al., Impact of crystal-quality improvement of epitaxial wafers on RF and power switching devices by utilizing VAS-method grown GaN substrates with low-density and uniformly distributed dislocations, in *Digest CS MANTECH Conference*, (2013), pp. 109–112
30. H. Fu, K. Fu, S. Chowdhury, T. Palacios, Y. Zhao, Vertical GaN power devices: Device principles and fabrication technologies-Part I. IEEE Trans. Electron Devices **68**(7), 3200–3211 (2021)
31. I.C. Kizilyalli, T. Prunty, O. Aktas, 4-kV and 2.8-mΩ-cm2 vertical GaN p-n diodes with low leakage currents. IEEE Electron Device Lett. **36**(10), 1073–1075 (2015)
32. A. Bolotnikov, P. Muzykov, Q. Zhang, A.K. Agarwal, T.S. Sudarshan, Junction termination extension implementing drive-in diffusion of boron for high-voltage SiC devices. IEEE Trans. Electron Devices **57**, 1930–1935 (2007)
33. D. Disney, H. Nie, I.C. Kizilyalli, et al., Method and system for fabricating floating guard rings in GaN materials. U.S. Patent 8,741,707 (2014)
34. H. Nie, A. Edwards, R. Brown, et al., Method and system for junction termination in GaN materials using conductivity modulation. U.S. Patent 8,716,716 (2014)
35. R.M. Farrell, E.C. Young, F. Wu, S.P. DenBaars, J.S. Speck, Materials and growth issues for high-performance nonpolar and semipolar light-emitting devices. Semicond. Sci. Technol. **27**, 024001 (2012)
36. F. Oehler, T. Zhu, S. Rhode, M.J. Kappers, C.J. Humphreys, R.A. Oliver, Surface morphology of homoepitaxial c-plane GaN: Hillocks and ridges. J. Crystal Growth **383**, 12–18 (2013)
37. I.C. Kizilyalli, P. Bui-Quang, D. Disney, H. Bhatia, O. Aktas, Reliability studies of vertical GaN devices based on bulk GaN substrates. Microelectron. Reliab. **55**, 1654–1661 (2015)
38. O. Aktas, X. Xin, T. Prunty, M. Raj, P. Bui-Quang, M.P. D'Evelyn, D.S. Kamber, W. Jiang, I.C. Kizilyalli, Vertical power semiconductor devices based on bulk GaN substrates, in *Proc. CS MANTECH, USA*, (2015), pp. 301–304
39. K. Kumakura et al., Minority carrier diffusion length in GaN: Dislocation density and doping concentration dependence. Appl. Phys. Lett. **86**, 052105-1–052105-3 (2005)
40. I.C. Kizilyalli, A.P. Edwards, H. Nie, T. Prunty, D. Bour, D. Disney, 3.7kV vertical GaN PN diodes. IEEE Electron Device Lett. **35**(2), 247–249 (2014)
41. I.C. Kizilyalli, T. Prunty, X. Xin, M. Raj, P. Bui-Quang, O. Aktas, Vertical GaN diode and transistors for power electronics, in *GOMACTech Conference*, (St. Louis, MO, 2015)

42. I.C. Kizilyalli, A.P. Edwards, H. Nie, P. Bui-Quang, D. Disney, D. Bour, 400A vertical GaN PN diode with breakdown voltage of 700V. IEEE Electron Device Lett. **35**, 654–656 (2014)
43. X. Huang, G. Wang, L. Jiang, A.Q. Huang, Ruggedness analysis of 600V 4H-SiC JBS diodes under repetitive avalanche conditions, in *Proc. Applied Power Electronics Conference and Exposition (APEC)*, (2012), pp. 1688–1691
44. R.W. Erickson, D. Maksimovic, *Fundamentals of Power Electronics* (Kluwer Academic Publishers, New York, NY, 2004)
45. H. Shah, S. Oknaian, E. Persson, H. Rongjun, From planar to trench -Evaluation of ruggedness across various generations of power MOSFETs and implications on in-circuit performance, in *Proc. Applied Power Electronics Conference and Exposition (APEC)*, (2011), pp. 485–490
46. O. Aktas, I.C. Kizilyalli, Avalanche capability of vertical GaN p-n junctions on bulk GaN substrates. IEEE Electron Device Lett. **36**(9), 890–892 (2015)
47. I.C. Kizilyalli, O. Aktas, Characterization of vertical GaN p-n diodes and junction field effect transistors on bulk GaN down to cryogenic temperatures. Semicond. Sci. Technol. **30**(12), 12401 (2015)
48. D. Disney, H. Nie, A. Edwards, D. Bour, H. Shah, I.C. Kizilyalli, Vertical power diodes in bulk GaN, in *IEEE International Symposium on Power Semiconductor Devices and ICs (ISPSD), Kanazawa, Japan*, (2013), pp. 59–62
49. W. Götz, N.M. Johnson, J. Walker, D.P. Bour, Activation of acceptors in Mg-doped, p-type GaN. MRS Proc. **423**, 595–600 (1996)
50. O. Aktas, I.C. Kizilyalli, et al., Vertical power semiconductor devices based on bulk GaN substrates, in *Proc. CS MANTECH, Scottsdale, AZ, USA*, (2015), pp. 301–304
51. I.C. Kizilyalli, Y.A. Xu, E. Carlson, J. Manser, D.W. Cunningham, 5th Work-shop on Wide Bandgap Power Devices and Applications (WiPDA), Albuquerque, NM (2017)
52. SWITCHES, ARPA-E Program Listing Information. Available at: https://arpa-e.energy.gov/?q=program-listing
53. E. Zanoni, M. Meneghini, A. Chini, D. Marcon, G. Meneghesso, AlGaN/GaN-based HEMTs failure physics and reliability: Mechanisms affecting gate edge and schottky junction. IEEE Trans. Electron Devices **60**, 3119–3131 (2013)
54. P. Moens, A. Banerjee, P. Coppens, F. Declercq, M. Tack, AlGaN/GaN power device technology for high current (100+ A) and high voltage (1.2 kV), in *Proceedings of the 2016 28th International Symposium on Power Semiconductor Devices and ICs (ISPSD)*, (2016) Prague CZ
55. K. Nomoto, B. Song, Z. Hu, M. Zhu, M. Qi, N. Kaneda, T. Mishima, T. Nakamura, D. Jena, H. Xing, 1.7-kV and 0.55-mΩ·cm2 GaN p-n diodes on bulk GaN substrates with avalanche capability. IEEE Electron Device Lett. **37**, 161–164 (2016)
56. D. Ji, C. Gupta, A. Agarwal, S. Chan, C. Lund, W. Li, S. Keller, U. Mishra, S. Chowdhury, Large-area in-situ oxide, GaN interlayer-based vertical trench MOSFET (OG-FET). IEEE Electron Device Lett. **39**, 711 (2018)
57. M. Sun, Y. Zhang, X. Gao, T. Palacios, IEEE Electron Device Lett. **38**, 509 (2017)
58. Y. Zhang, M. Sun, J. Perozek, Z. Liu, A. Zubair, D. Piedra, N. Chowdhury, X. Gao, K. Shepard, T. Palacios, Large area 1.2 kV GaN vertical power FinFETs with a record switching figure-of-merit. IEEE Electron Device Lett. **40**, 75–78 (2019)
59. R. Li, Y. Cao, M. Chen, R. Chu, 600 V/ 1.7 Ω normally-off GaN vertical trench metal–oxide–semiconductor field-effect transistor. IEEE Electron Device Lett. **37**, 1466–1469 (2016)
60. PNDIODES, ARPA-E Program Listing Information. Available at: https://arpa-e.energy.gov/?q=program-listing

Chapter 2
Electronic Transport and Advances in Vertical Bulk GaN-Based Power Semiconductor Devices

Srabanti Chowdhury, Ozgur Aktas, and Rongming Chu

1 Introduction

Wide bandgap (WBG) semiconductors present a pathway to push the limits of power conversion efficiency beyond that available from silicon (Si)-based devices, enabling significant energy savings. Recent progress in gallium nitride (GaN)-based power electronic devices is compelling. Reducing conversion losses is not only critical for minimizing consumption of limited resources but it also enables new compact architectures that form the basis for a new industry offering increased power conversion performance at reduced system costs. This is because GaN devices enable power electronics with (1) higher efficiency at a higher frequency of operation and (2) higher efficiency over a wider range of operating temperatures compared with what is possible with Si, which is approaching its physical material limit in power conversion. High-efficiency operation at higher operating frequency reduces the size, weight, and cost of the overall system by reducing the size of the passive components and the heat sink.

Lateral GaN on Si HEMTs has enabled over 99% efficient converters at 800 V at 100 kHz. Recent results have shown that an average of 5% improvement in overall system efficiency was possible by using GaN (lateral) switched inverters to drive an induction motor. While lateral GaN devices are more mature in technology and have

S. Chowdhury
Stanford University, Stanford, CA, USA
e-mail: srabanti@stanford.edu

O. Aktas
Transphorm Inc., Goleta, CA, USA
e-mail: ozgur.aktas@transphormusa.com

R. Chu (✉)
The Pennsylvania State University, University Park, PA, USA
e-mail: ruc634@psu.edu

I. C. Kizilyalli et al. (eds.), *Gallium Nitride and Related Materials*, The Materials Research Society Series, https://doi.org/10.1007/978-3-031-83056-3_2

entered the medium power conversion market (up to 10 kW), vertical GaN devices are evolving to address high power conversion (10 kW–10 MW). Device researchers are exploring vertical GaN devices and viable manufacturing technologies to be able to define and sustain a roadmap for these devices.

Traditionally, power semiconductor devices are vertical devices, in which the source/gate electrodes and the drain electrodes are located on the top surface and backside of the wafer, respectively. The off-state voltage is dropped across the lightly doped drift layer and is primarily determined by the doping density and the thickness of the drift layer. The on-state current conducts vertically through the drift layer, scaling linearly with the device area.

The vertical device topology features several advantages relative to the lateral devices. First, the vertical current flow eliminates the need to carry current horizontally through metal interconnects, which often results in inefficient utilization of the chip area in lateral devices. Second, the vertical design only requires E-field termination at the outside edge of the device, versus the need for E-field shaping within every unit cell in the case of lateral devices. Third, with proper edge termination, the vertical device structure has a high E-field region buried within the semiconductor bulk, enabling avalanche capability. For these reasons, there have been significant research efforts advancing the GaN vertical device technology.

These research efforts have led to exciting device characteristics, revealed key technical challenges, and resulted in learning about electronic transport in vertical bulk GaN-based power semiconductor devices. This chapter will review these learnings in the context of vertical diodes and transistors.

2 Vertical Diodes

Following the demonstration of vertical GaN diodes with excellent characteristics [1, 2], recent research in GaN vertical rectifiers has developed a range of technological capabilities that would support the development of commercially viable vertical GaN rectifier products that can compete in the marketplace against other technological competitors. We can list the following recent developments among these capabilities:

1. Various methods for edge-termination including bevel-etch, field-plates on mesa, JTE, and H-passivation-based guard-rings were reviewed in a 2021 TED paper [3]. Since 2021, intense research on edge-termination structures has continued and the structures reviewed in this 2021 paper were further optimized:

    - Hybrid guard-ring/JTE field control structures were advanced to achieve improved processing tolerance [4–6].
    - Guard ring structures manufactured by N implant were shown to be avalanche capable [7, 8].

- Breakdown voltage and leakage of mesa-based structures were improved using low-damage wet-etch [9] or implant-based passivation of the etch damage [10].
- Experimental and simulation-based optimization of two-zone step-etched edge termination structures was performed [11, 12].
- Bevel mesa edge termination structures were studied for avalanche uniformity [13] and demonstrated to have UIS stability [14]. Bevel mesa etch structures were further optimized to produce low-angle bevel structures [15]. Bevel implant edge structures were implemented and optimized to enable avalanche breakdown [16].

Methods using H-plasma for passivation of p-GaN were further developed:

- H plasma-based implementation of guard rings [17].
- H implant to reduce field along the mesa and reduce leakage [18].
- H plasma implementation of JTE [19].
- H plasma passivation of p-GaN for implementation of field plates [20].

H-passivation of p-GaN offers an alternative approach for the implementation of edge-termination structures that may have advantages from a manufacturability perspective; however, the long-term stability of the H-passivation-based edge-termination structures has not yet studied.

The maturation of mesa, guard-ring, JTE-based edge-termination structures is important as this enables manufacturable process margins enabling increased yield and reproducibility, which in turn makes it easier to optimize the rest of the device parameters in an environment where the device performance and reliability are influenced by substrate quality, epitaxy quality, and uniformity.

2. Methods for selective area doping were reviewed in a 2021 paper by Kizilyalli and Carlson [21] with the conclusion that while Mg incorporation and activation were achieved by implantation, contact resistance and leakage remained a concern. Further research focusing on improved annealing cap layer design demonstrated that Mg implant damage-related leakage current can be suppressed to the extent that MPS diodes would be feasible [22]. Another line of research focusing on ultra-high-pressure annealing for activation demonstrated functional devices of increasing complexity and performance [23–27]. However, the scalability of ultra-high-pressure annealing to 2″ and larger wafers remains a critical technological challenge limiting the usability of this technique.
3. Methods to fabricate edge-terminated Schottky contacts with improved leakage [28, 29]. Schottky contact based rectifiers are attractive as they provide zero reverse recovery charge, a low turn-on voltage, and a relatively simpler fabrication procedure. The recent incorporation of edge-termination structures that avoid p-GaN and an improved understanding of the handling of GaN before deposition of the Schottky contact metal may enable commercialization of Schottky rectifiers.

4. Fabrication procedures that enable MPS rectifiers [22, 25–27]. MPS structures were fabricated by Mg-implantation which avoids the need to fabricate Schottky contacts on etched surfaces or the need to regrow p-GaN. Mg-implanted MPS rectifiers were demonstrated to have surge and avalanche robustness. With further optimization of the annealing cap and annealing conditions, further improvement of the device characteristics can be expected [30–32]. MPS diodes were also implemented using UHPA but again the scalability of this technique to 2″ and larger wafers limits the usability of this method.
5. Polarization doping enabling p-type GaN without Mg doping and the associated need for Mg activation anneal [33, 34]. Polarization doping provides a path to implement pn-junctions that provide longer minority carrier lifetimes which may help reduce drift-region resistive via improved forward bias conductivity modulation. Polarization doping may also be useful to avoid some of the forward bias instabilities that have been attributed to high-defect density in p-GaN.
6. Methods to fabricate high-performance devices either on GaN, Si, sapphire, or SiC [8, 19, 35, 36]. While free-standing substrates have been improving, their cost remains high and the scalability to large diameter wafers with low-cost and high-quality remains a challenge. The ability to fabricate GaN diodes on non-native substrates may enable the commercialization of GaN-based rectifiers. GaN-on-Si has seen the most attempts to fabricate pseudo-vertical GaN rectifiers with 1200 V demonstrated. GaN-on-Sapphire provides an alternative path that provides a substrate that is less susceptible to bow and breakage concerns. Recent developments include the growth of GaN-on-SiC diodes which incorporate SiC as part of the drift region thus reducing the thickness of GaN that needs to be grown. GaN-on-QST provides a further alternative that may be easier to scale to 8″, especially in combination with HVPE-grown GaN.
7. Cost-effective HVPE to grow thick high-purity GaN with high mobility and high voltage blocking capability [37–39]

Further, significant strides were made in the following topics that enable the design of optimized device structures (Fig. 2.1):

- Experimental estimates for impact ionization coefficients and breakdown characteristics [40–43].

- Understanding of conduction via Monte Carlo and classical methods [44–46].
- Understanding of mobility, lifetime, and trapping in various impurities and native defects [47–52]. Lifetime analysis indicates results correlated to TD density [34, 47, 48].

Since GaN is either grown heteroepitaxially or on native-GaN substrates with high dislocation density, the impact of threading dislocations on the conduction characteristics has been studied in detail over the years with increasing sophistication as the quality of substrates and epitaxy improved [50–60]. The results indicate that most of the threading dislocations in GaN are either edge or mixed type and that

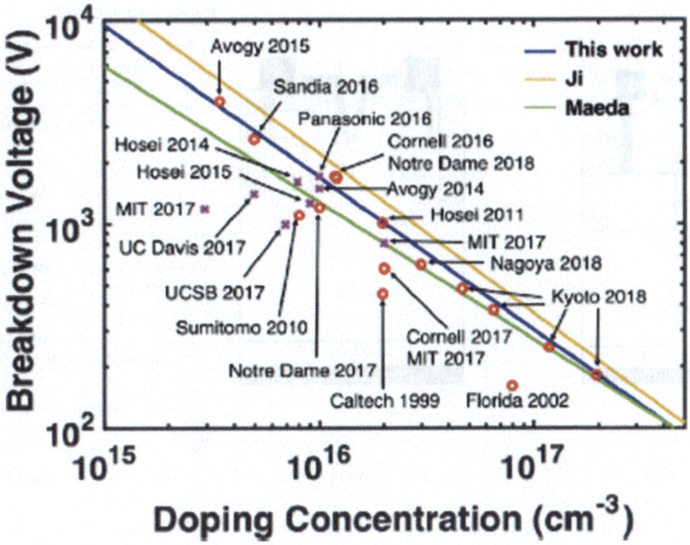

Fig. 2.1 Dependence of breakdown voltage on doping concentration (Ref. [42])

these do not contribute to the reverse leakage or breakdown [51, 53, 56]. Screw-type threading dislocations are associated with leakage, with open-core screws being identified as being responsible for leakage. The Ga-rich and low-pressure growth conditions are identified as drivers of formation for nano-pipes, whereas N-rich and high-pressure growth conditions reduce nano-pipe density and result in closed core structures. An open-core screw dislocation is identified as the source of the EM bright spot, whereas a closed-core screw is also found to have high leakage.

3 Vertical Transistors

3.1 Current Aperture Vertical Electron Transistors (CAVET)

CAVETs (see Fig. 2.2a, b) are vertical GaN devices that take full advantage of the polarization-based 2DEG in their design. Comprised of a 2DEG-induced channel to carry the current, a CAVET needs a thick homoepitaxially grown drift region to hold the blocking voltage. There have been two types of approaches to fabricating CAVETs, which are discussed in the following sections.

3.1.1 Planar CAVET

In 2004 quasi-vertical transistors, in the form of a CAVET, were first demonstrated on a sapphire substrate keeping RF application in mind [61]. Later in 2008, CAVET

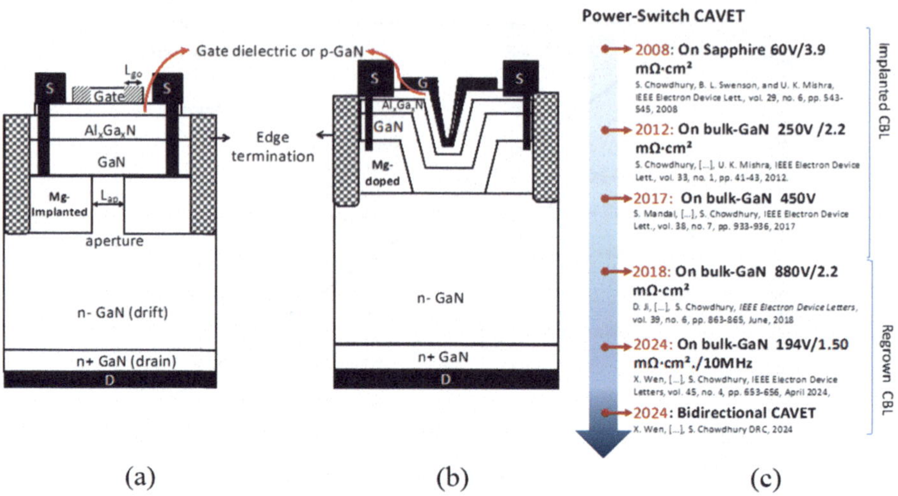

Fig. 2.2 (**a**) Cross-section of a planar CAVET, (**b**) trench-CAVET, and (**c**) various milestones achieved in a CAVET

was re-designed for power switching applications and proof of concept of both D-mode and E-mode was demonstrated on sapphire substrate [62] blocking 75 V with a 0.5 μm of the n-GaN ([Si] = 5×10^{16} cm^{-3}) drift region. The device structure began with n-GaN/n + GaN epilayers, with the top layer implanted with Mg at room temperature to form the current blocking layer (CBL). The CBL plays a crucial role in blocking current, similar to the p-region in a Si DMOSFET. An aperture region was etched, and an AlGaN/GaN HEMT structure was regrown by MOCVD to refill the aperture region, serving as the channel. The design of the GaN planar CAVET was centered around two important functionalities:

1. *Efficacy of Mg-implanted Blocking Layer*: The activation of Mg in GaN requires temperatures over 1400 °C under high nitrogen overpressure to prevent decomposition. In these CAVETs, implanted-Mg activation was not directly performed. However, the regrowth of AlGaN/GaN layers or subsequent source-contact implantation activation involved high-temperature conditions, which led to some activation of the Mg-implanted CBL.
2. *Well-Controlled Si-doped Drift Layer*: This layer is crucial for managing the majority of the off-state voltage drop.

While CAVETs on sapphire provided the proof-of-concept [62], several years later, in 2011, the first power GaN CAVET based on a bulk GaN substrate was reported, blocking over 250 V [63]. The CBL in this device was formed by Mg-ion-implantation. The device process began with MOCVD-grown n-GaN on a conductive bulk GaN substrate. Selective area Mg-ion-implantation was then implemented to form the CBL to block current from the aperture region. Finally, the top AlGaN/GaN HEMT structure was regrown by MBE. The MBE regrown layers were a critical feature adopted to suppress diffusion of Mg into the over-grown channel layers. The effective gate of the device is indicated by the shaded region. Sources are connected to the CBL to offer a discharge path to the traps in the CBL.

During the on-state, electrons from the source go through the 2DEG channel to the drift region and are finally collected by the drain electrode. The overlap between the gate electrode and the CBL is the effective channel length. When a negative gate bias is applied, the transistor is in its off state. The p-n junction is formed by p-type (or Mg-implanted) CBL and the drift region blocks the off-state voltages. Since the first high voltage CAVET blocked a 300 V [64] breakdown voltage and a specific on-resistance ($R_{on,sp}$) of 2.2 m$\Omega\cdot$cm^2, a lot of notable achievements have been reported as illustrated in Fig. 2.2c.

3.1.2 Design of a CAVET

A. *Normally Off Operation*

CAVETs are normally on devices similar to GaN HEMTs. Techniques typically applied to HEMTs can be applied to CAVETs to render them normally off. To date, three techniques have been proposed for the CAVETs: (1) cascode configuration, (2) p-GaN gate layer, and (3) gate recess structure.

1. *Cascode Configuration*: The cascode configuration is designed with a low-voltage normally off Si MOSFET as the input and a high-voltage normally on CAVET as the output. The forward and reverse conduction of the cascoded CAVET is shown. In forward conduction, the total Ron is the sum of Ron (MOSFET) and Ron (CAVET). In reverse conduction, the current flows through the body diode of the Si MOSFET and the CAVET channel, while the body diode of the CAVET is not conductive.
2. *p-GaN Gate Layer*: The CAVET structure with a p-GaN gate layer achieves a threshold voltage of 1 V and 2.9 V under zero-gate bias. Avogy demonstrated the possibility of junction-gated vertical transistor in step toward commercializing as shown in Fig. 2.3 [65].
3. *Gate Recess Structure*: In the gate recess CAVET structure, electrons in the sidewall channel are depleted by the p-GaN base region, resulting in normally off behavior. The threshold voltage depends on the sidewall angle, with a straight sidewall ($>$ 90 degrees) yielding a threshold voltage greater than 1 V.

B. *High Breakdown Voltage*

Ideally, the CAVET uses a p-n junction formed by the p-GaN base region and the n-GaN drift region to block high voltage. In the off-state, a positive voltage is applied to the cathode, resulting in a triangular-shaped electric field distribution along the depletion region. Using Poisson's equation, the maximum electric field (E_{max}) and breakdown voltage (V_{BR}) can be calculated, indicating the breakdown characteristics of the vertical GaN transistors.

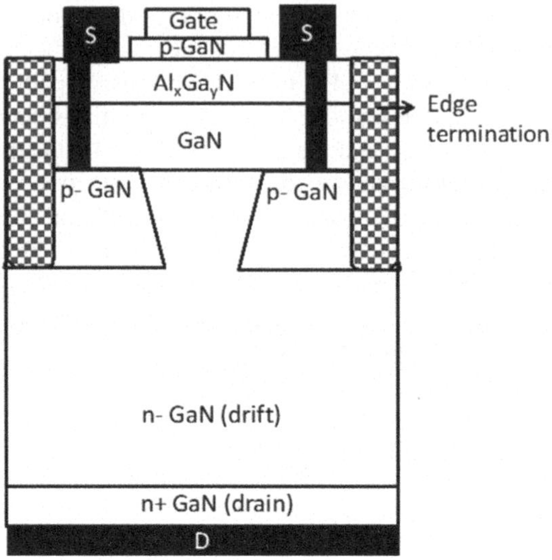

Fig. 2.3 (redrawn) Cross-sectional picture of a vertical JFET (Avogy Inc., Ref. 65)

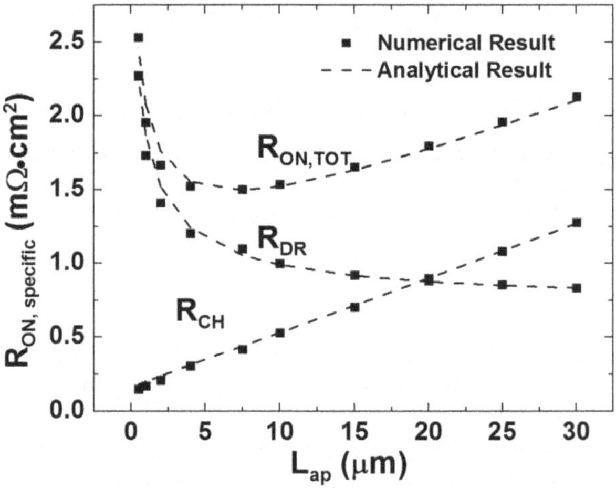

Fig. 2.4 $R_{ON,TOT}$, R_{DR} and R_{CH} as a function of L_{ap}. The solid squares indicate the results obtained numerically, while the dash curves show the analytical result (channel electron mobility: 1500 cm^2/Vs; bulk GaN electron mobility: 900 cm^2/Vs). Total R_{on} as a function of the electron mobility in the drift region is listed. Bulk GaN mobility will play a key role in distinguishing GaN vertical devices from SiC counterpart [66]

C. *Low On-state Resistance*

The ideal R_{onsp} of the GaN CAVET is influenced by the channel electron mobility for low breakdown voltage device designs (< 1000 V) and by the bulk GaN mobility for higher voltage devices (> 1000 V). The total on-state resistance (R_{ON-TOT}) is lowest between an aperture length (L_{ap}) of 4 μm and 5 μm, with a minimum of 1.5 $m\Omega \cdot cm^2$ as shown in Fig. 2.4

D. *Dispersion-Free Behavior*

With a CBL of implanted Mg, CAVETs have always demonstrated dispersion-free behavior. The pulsed I-V characteristics confirmed the absence of DC-RF dispersion (pulsed I-V measurements performed down to 5 μs pulse widths), evidence of the effectiveness of the buried high electric field region within the drift layer. This buried peak electric field alleviates surface-related dispersion effects, enhancing the overall reliability and performance of the device [67].

Implanted Mg-based CBL has been even demonstrated on nitrogen-polar CAVET with similar dispersion-less behavior. It is a key advantage of vertical GaN devices which helps make a case for it over and above lateral HEMT where field plates are mandatory for high voltage blocking [68].

3.2 Trench CAVET

The conventional GaN CAVET has faced fabrication challenges, particularly due to the diffusion of Mg species into the channel and the reliance on low-temperature regrowth processes. To address these issues, a trench gate structure was introduced into the conventional CAVET. Instead of using Mg-implanted p-GaN for the CBL, the trench CAVET adopts MOCVD-grown Mg-doped p-GaN as the CBL material. The trench sidewall angle determines the polarization scale of the channel, with a 90-degree angle indicating a nonpolar plane and a 45-degree angle indicating a semi-polar plane.

The trench CAVET functions similarly to a trench MOSFET but utilizes the regrown AlGaN/GaN on the trench sidewall as the channel. The channel mobility in a trench CAVET can be as high as 1690 cm^2/Vs due to the 2DEG structure [69].

The first MIS gate trench CAVET, reported in 2016, had a breakdown voltage of 225 V [70], later improved to over 880 V in blocking voltage [71]. The p-GaN gate structure is widely used in normally off lateral GaN HEMTs, achieving a positive threshold voltage of 2.5–2.9 V, and was adopted by industry practices [66, 72]. Panasonic demonstrated 1.7 kV blocking with C-based layers deployed on top of Mg-doped CBL [72].

Trench CAVETs have shown significant advancements over conventional designs, offering improved performance, and addressing key fabrication challenges. A 100-nm LT-GaN layer was grown at 750 °C on p-GaN to prevent Mg out-diffusion, which played a remarkable role leading to high current density [73, 74]. The devices showed low gate leakage current and high on/off ratios. Below are the key developments and advancements:

1. *High Voltage and Low On-State Resistance*: The development of an 880 V trench CAVET demonstrated a high breakdown voltage and a low specific on-resistance ($R_{on,sp}$) of 2.7 $m\Omega \cdot cm^2$. This improvement was achieved through the use of a thicker gate dielectric layer and a low-damage trench etch process. The specific structure included a 15 μm-thick lightly doped drift layer on the bulk GaN substrate and a 300 nm p + GaN current blocking layer (CBL) grown by MOCVD [71]. This suggests that there is more room for improvement with a proper edge termination technique.

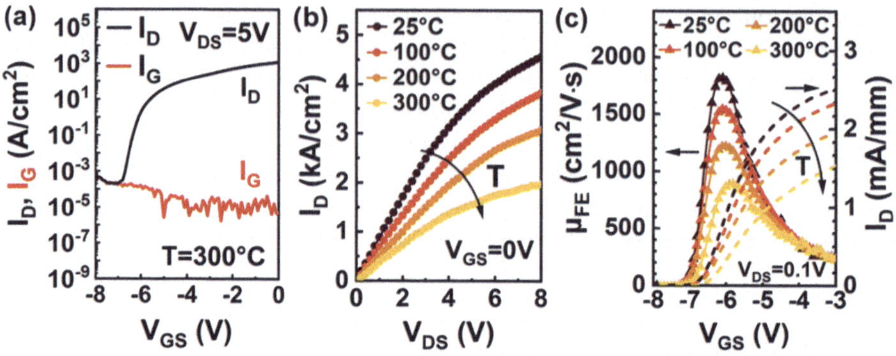

Fig. 2.5 High-temperature DC characteristics of a 10 MHz CAVET (200 V). (**a**) I_D-V_{GS} at 300 °C. (**b**) Temperature-dependent I_D-V_{DS} at $V_{GS} = 0$ V from RT to 300 °C. (**c**) Field effect mobility at varying temperatures, showing a peak value of 1821 cm2/(V·s) at RT and 898 $cm^2/(V{\cdot}s)$ at 300 °C [69]

2. *Improved Channel Mobility and High-Frequency Switching*: Recent innovations included the insertion of a low-temperature (LT) GaN interfacial layer to inhibit Mg out-diffusion from p-GaN, leading to a record-high field effect mobility of 1821 $cm^2/(V{\cdot}s)$ in a regrown AlGaN/GaN channel. This enabled the low-loss, high-frequency switching capability of the fabricated device. The trench CAVETs demonstrated effective scaling up to 3-finger structures with reduced Ron,sp. to 0.98 $m\Omega{\cdot}cm^2$. High-frequency switching up to 10 MHz and high-temperature operation up to 300 °C were successfully demonstrated [69].
3. *Enhanced Fabrication Techniques*: Improvements in the fabrication process included the use of a low-power dry etching technique with Cl_2/BCl_3 gases, UV Ozone + HF chemical treatment to reduce interfacial impurities, and the regrowth of high-quality AlGaN/GaN channel layers. The devices also utilized in situ Si3N4 as the gate dielectric, contributing to low hysteresis and high device reliability [75].
4. *High-Temperature Operation*: DC characteristics (see Fig. 2.5) at elevated temperatures showed that the off-state leakage current increased significantly at 300 ° C, primarily due to higher gate leakage. Despite this, the devices maintained proper functionality up to 300 °C, demonstrating their potential for high-temperature applications. The field-effect mobility in the regrown channel was measured at varying temperatures, with a peak value of 1821 $cm^2/(V{\cdot}s)$ at room temperature, reducing to 898 $cm^2/(V{\cdot}s)$ at 300 °C [69].

These advancements highlight the potential of trench CAVETs for high-power and high-frequency applications, providing significant improvements in performance, reliability, and fabrication techniques. The ability to maintain high performance under elevated temperatures and high switching frequencies makes them suitable for next-generation power electronics.

3.3 *Bidirectional GaN-on-GaN Trench Current Aperture Vertical Electron Transistor (Bi-CAVET)*

Bidirectional switches (BSWs) are pivotal in various power electronics applications, such as solid-state circuit breakers (SSCBs) and AC-AC matrix converters. Unlike traditional unidirectional switches, four-quadrant BSWs enable efficient current flow and voltage blocking in both directions. Gallium nitride (GaN), with its superior electrical properties—including a high critical electric field, high saturation velocity, and high carrier mobility—presents substantial advantages over silicon (Si) for high-power and high-efficiency devices. The innovative Bi-CAVET integrates two adjacent trench CAVETs on the same GaN substrate. Each trench CAVET includes magnesium (Mg)-doped p-GaN layers as CBL to prevent current flow in the off-state [76]. The 2DEG channel, generated by the regrown AlGaN/GaN heterostructure, offers low on-state resistance and high-frequency switching capabilities. This combination makes Bi-CAVET highly suitable for applications requiring high power and efficiency.

Recently, the development and experimental validation of a monolithic-integrated bidirectional GaN-on-GaN trench Bi-CAVET were reported [76]. This advanced device, fabricated on a GaN substrate, integrates a highly conductive two-dimensional electron gas (2DEG) channel that exhibits an impressive peak field effect mobility (μ_{FE}) of 1792 $cm^2/(V \cdot s)$. The Bi-CAVET showcases symmetrical I-V characteristics in bidirectional mode, achieving a specific on-resistance ($R_{on,sp}$) of 2.5 $m\Omega \cdot cm^2$ and a breakdown voltage of 200 V in both the first and third quadrants. Furthermore, in single-directional mode, the Bi-CAVET facilitates current conduction in one direction with a turn-on offset voltage of 1 V. The reverse conduction of the trench CAVET was rigorously analyzed, revealing that the primary path for reverse current flow is through the 2DEG channel.

3.4 *Trench MOSFET*

GaN Trench MOSFETs offer a normally off solution, addressing a significant drawback of GaN HEMT-based designs. There are two types of trench MOSFETs reported so far: (1) Non-regrowth-based MOSFET and (2) regrowth-based MOSFET (OGFET).

Non-regrowth-based MOSFETs

Since the development of dry etch technology, the trench MOSFET has been a dominant device structure for power electronics. The first GaN MOSFET [77], reported in 2007, had a high threshold voltage of 5.1 V. Later, a 1.6 kV device was developed in 2014. The GaN MOSFET structure includes a main p-n junction between the source and drain, with the channel located at the etched sidewall formed by the MOS structure's inversion layer.

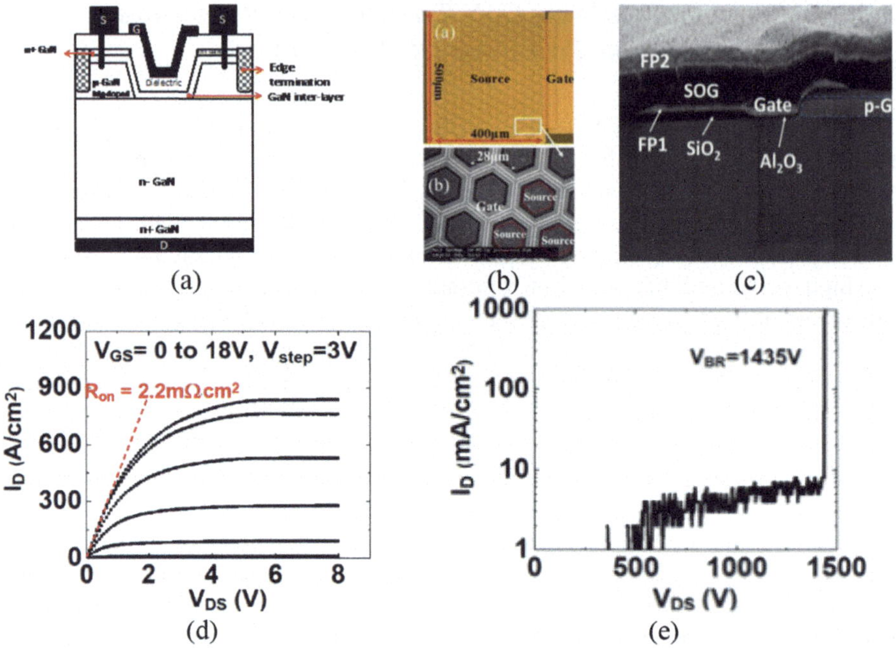

Fig. 2.6 (**a**) A (regrown channel) OGFET; (**b**) cross-sectional SEM view in (**b**) top-view of a large area OGFET. (**c**) SEM image of a 2FP-OGFET. (**d**) I-V characteristics of fabricated single unit cell OGFET with a saturation current density of 850 A/cm^2 and $R_{on,sp}$ of 2.2 mΩ·cm^2. (**e**) >1.4 kV breakdown [80]

Regrowth-Based MOSFET (OGFET)

The GaN OGFET, with its key features and device characteristics, as shown in Fig. 2.6, is a modified trench MOSFET that uses an unintentional doped (UID) GaN interlayer as the channel region, enhancing electron mobility and gate oxide reliability [77–81]. The first OGFET results, reported in 2016, showed a 60% $R_{on,sp}$ reduction while maintaining a threshold voltage >2 V. In 2017, further developments demonstrated OGFETs with high performance, low $R_{on,sp}$, and significant on-state performance improvements, including some modest scaling demonstrations [82].

The switching waveforms of the OGFET during turn-off and turn-on transients in a double-pulse test circuit show a total turn-off time of ~30 ns and turn-on time of 47 ns. The comparison of simulated cascoded CAVET, simulated GaN OGFET, and SiC MOSFETs indicates that GaN vertical devices exhibit fast switching speeds and lower energy loss, highlighting their potential for power electronics applications. The experimental demonstration has shown 1 MHz [83] and 10 MHz [69] switching making it more promising for the future.

3.5 Static Induction Transistor (SIT)

The concept of Static Induction Transistor (SIT) was introduced by Nishizawa in 1975, with experimental SITs demonstrating the predicted space-charge injection mode in their drain current behavior. A typical SIT features a recessed gate structure with bottom and sidewall contacts, forming a class of short-channel field effect transistors. In these devices, the current flows vertically between the source and the drain, regulated by an electrostatic potential barrier created by two opposed gates. This configuration results in triode-like unsaturated output characteristics. Silicon carbide (SiC) SITs are noted for their potential in high-power and high-frequency applications due to their high electron mobility and saturation velocity. GaN SITs, with even higher electron mobility and saturation velocity than SiC, are promising for even higher frequency and output power operations. Recent advancements have led to the successful design and fabrication of GaN SITs using a self-aligned process, showcasing their potential for advanced electronic applications. Typical SIT features a recessed gate structure with contacts on both the bottom and sidewalls, which is a distinguishing characteristic of these transistors. As a class of short-channel FETs, SITs operate by controlling the current flow vertically between the source and the drain. This flow is regulated by an electrostatic potential barrier induced by two opposed gates, resulting in unique triode-like unsaturated output characteristics. This means that, unlike conventional transistors, SITs can handle higher currents without saturating, making them particularly useful in certain high-performance applications.

One of the most notable materials used in SITs is silicon carbide (SiC), which has shown tremendous potential for high-power and high-frequency applications. SiC SITs benefit from high electron mobility and a high saturation velocity, enabling them to perform efficiently under demanding conditions. However, gallium nitride (GaN) SITs take these advantages even further. GaN offers even higher electron mobility and saturation velocity than SiC, which allows GaN SITs to operate at higher frequencies with greater output power. This makes GaN SITs ideal for applications requiring extremely high performance [83, 84]. In recent advancements, GaN SITs have been successfully designed and fabricated using a self-aligned process. This method enhances the precision and performance of the transistors, paving the way for their use in cutting-edge electronic devices. These developments highlight the ongoing innovation in SIT technology and its growing importance in the field of electronics, especially in applications demanding high power and frequency.

A recent study explored the performance of GaN vertical SIT (aka FinFETs) for its resemblance with fin-based structures, but operationally behaves like an SIT on a bulk GaN substrate, focusing on self-heating and DC-RF dispersion [85]. Figure 2.7 illustrates the device cross-section and representative device data. High performance was achieved with a drain current density exceeding 175 kA/cm^2 and a low gate leakage of less than 1×10^{-17} kA/cm^2 through low-temperature post-gate processes, including argon plasma-enhanced Ohmic contact. A 300 nm fin-width single-finger device demonstrated a specific on-resistance of 0.030 mΩ-cm^2 at a drain on-current

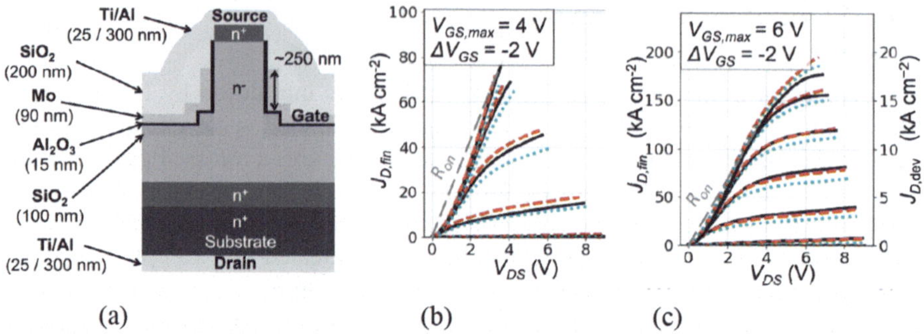

Fig. 2.7 (**a**) The cross-sectional schematic of the transistor fin unit; (**b**) 300 nm 1-finger devices; and (**c**) 300 nm 25-finger devices with 3 μm fin-to-fin pitch, normalized by the total active fin area (left y-axis) and the total device area [85]

of 119 kA/cm^2, while a 25-finger device with the same fin width and a 3 μm fin-to-fin pitch showed an on-resistance of 0.043 mΩ-cm^2, one of the lowest reported values [86]. Self-heating effects increased with decreasing fin width and more fingers, with the maximum deviation in drain current rising from 6% for the 300 nm 1-finger device to 20% for the 25-finger 150 nm device. Low DC-RF dispersion was noted in devices more than 2.5 mm away from the wafer edge, although it did not strongly correlate with fin width or multi-finger scaling. Improved specific on-resistance and breakdown voltages were observed with increased fin width, while multi-finger scaling slightly increased specific on-resistance due to non-linearly scaling resistive components. These findings underscore the potential of GaN vertical FinFETs for high-power and high-frequency applications, emphasizing the need to manage self-heating and DC-RF dispersion in device design.

4 Concluding Remarks

As shown in the previous sections, bulk GaN-based vertical devices hold great promise in becoming a competitive power switching technology. Three metrics are most important: off-state voltage blocking, on-state conduction loss, and switching capabilities. As discussed in Sect. 2, the off-state blocking voltage is primarily determined by the doping density of the drift layer, the thickness of the drift layer, and the edge termination structure. A variety of edge termination strategies have been developed such that the blocking voltage of vertical diodes and transistors is not limited by E-field crowding near the edge of the device.

On-state conduction loss of GaN vertical devices is inherently limited by the drift region. Thanks to the high-electron-mobility and high-critical-field drift layer, GaN vertical PN diodes with very competitive specific on-resistance values have been reported. Being a direct bandgap semiconductor, the minority carrier recombination lifetime is short, in the order of nanoseconds [87]. Therefore, conductivity modulation due to minority carrier injection is often ignored; and the GaN vertical PN diodes are often treated as unipolar devices. There have even been reports of specific on-resistance even lower than the 1D unipolar limit [88]. The lower-than-theoretical-limit on-resistance was attributed to photon recycling effects [89]. Photons from electron-hole pair recombination stay within the drift region, generating photo carriers, thereby further reducing on-resistance. Despite the low on resistance, GaN PN diodes have a large turn-on voltage of ~3 V, arising from the PN junction built-in potential. The turn-on voltage is an important factor in diode conduction loss. This is a major bottleneck for low-voltage-rating devices, where the on-voltage drop across the drift layer is small as compared with the built-in potential. However, for medium- and high-voltage applications, the on-voltage drop across the drift layer increases, and the built-in potential becomes less important. Schottky diodes feature lower turn-on voltage and are preferred for low conduction loss. High-reverse leakage of the GaN Schottky junction [90] remains a major obstacle to the maturation of GaN vertical Schottky didoes. For the GaN vertical transistors, the channel resistance plays an important role in the total on-resistance, as discussed in Sect. 3. An optimum gate structure is still in the search to simultaneously enable positive threshold voltage, large gate over drive capability, high channel electron mobility, low gate leakage, and long-term reliability.

Preliminary switching testing has shown encouraging results. Under a high dv/dt slew rate, there is a risk of false turn-on, due to the potential drop between the body contact and the channel while discharging the output capacitor. The relatively high resistance in the p-GaN base layer and in the body contact signifies the risk. Further studies are needed to quantify the dv/dt capability.

References

1. H. Ohta, N. Asai, F. Horikiri, Y. Narita, T. Yoshida, T. Mishima, 4.9 kV breakdown voltage vertical GaN p-n junction diodes with high avalanche capability. Jpn. J. Appl. Phys. **58**(C), SCCD03 (2019). https://doi.org/10.7567/1347-4065/ab0cfa
2. I.C. Kizilyalli, T. Prunty, O. Aktas, 4-kV and 2.8-m omega-cm(2) vertical GaN p-n diodes with low leakage currents. IEEE Electron Device Lett. **36**(10), 1073–1075 (2015). https://doi.org/10.1109/LED.2015.2474817
3. H. Fu, K. Fu, S. Chowdhury, T. Palacios, Y. Zhao, Vertical GaN power devices: device principles and fabrication technologies-part I. IEEE Trans. Electron Devices **68**(7), 3200–3211 (2021). https://doi.org/10.1109/TED.2021.308323
4. J.S. Lundh et al., Experimental validation of robust hybrid edge termination structures in vertical GaN p-i-n diodes with avalanche capability. IEEE Electron Device Lett. **45**(5), 873–876 (2024). https://doi.org/10.1109/LED.2024.3377532
5. M.A. Ebrish et al., Study of anode doping and avalanche in foundry compatible 1.2 kV vertical GaN PiN diodes. Appl. Phys. Express **16**(11), 116501 (2023). https://doi.org/10.35848/1882-0786/ad0655
6. P. Pandey et al., Hybrid edge termination for high-voltage vertical GaN devices: empirical validation and robust processing tolerance. IEEE Trans. Electron Devices **71**, 3552 (2024). https://doi.org/10.1109/TED.2024.3383419

7. Y. Wang et al., Implanted guard ring edge termination with avalanche capability for vertical GaN devices. IEEE Trans. Electron Devices **71**(3), 1481–1487 (2024). https://doi.org/10.1109/TED.2023.3321010
8. X. Guo et al., 1200-V GaN-on-Si quasi-vertical p-n diodes. IEEE Electron Device Lett. **43**(12), 2057–2060 (2022). https://doi.org/10.1109/LED.2022.3219103
9. N. Asai, H. Ohta, F. Horikiri, Y. Narita, T. Yoshida, T. Mishima, Impact of damage-free wet etching process on fabrication of high breakdown voltage GaN p-n junction diodes. Jpn. J. Appl. Phys. **58**(C), SCCD05 (2019). https://doi.org/10.7567/1347-4065/ab0401
10. D. Ji, S. Li, B. Ercan, C. Ren, S. Chowdhury, Design and fabrication of ion-implanted moat etch termination resulting in 0.7 m Ω·cm2/1500 V GaN diodes. IEEE Electron Device Lett. **41**(2), 264–267 (2020). https://doi.org/10.1109/LED.2019.2960349
11. A.T. Binder et al., Optimization of step-etched junction termination extensions for vertical GaN devices. IEEE Trans. Electron Devices **71**(3), 1541–1545 (2024). https://doi.org/10.1109/TED.2023.3344057
12. A. Taube, M. Kaminski, Optimization of two-zone step-etched junction termination structures for vertical GaN power devices. Phys. Status Solidi **221** (2024). https://doi.org/10.1002/pssa.202400075
13. K. Zeng, S. Chowdhury, B. Gunning, R. Kaplar, T. Anderson, Study on avalanche uniformity in 1.2KV GaN vertical PIN diode with bevel edge-termination, in *2021 IEEE International Reliability Physics Symposium (IRPS), in International Reliability Physics Symposium*, (IEEE, 2021). https://doi.org/10.1109/IRPS46558.2021.9405165
14. B. Shankar et al., Study of avalanche behavior in 3 kV GaN vertical P-N diode under UIS stress for edge-termination optimization, in *2022 IEEE International Reliability Physics Symposium (IRPS), in International Reliability Physics Symposium*, (IEEE, 2022). https://doi.org/10.1109/IRPS48227.2022.9764525
15. J. Tarenko et al., Fabrication of ultralow-bevel angle mesa structures for vertical GaN devices. Phys. Status Solidi **221** (2024). https://doi.org/10.1002/pssa.202400079
16. M. Xiao et al., Robust avalanche in 1.7 kV vertical GaN diodes with a single-implant bevel edge termination. IEEE Electron Device Lett. **44**(10), 1616–1619 (2023). https://doi.org/10.1109/LED.2023.3302312
17. H. Fu et al., High voltage vertical GaN p-n diodes with hydrogen-plasma based guard rings. IEEE Electron Device Lett. **41**(1), 127–130 (2020). https://doi.org/10.1109/LED.2019.2954123
18. K. Fu, Z. He, C. Yang, J. Zhou, H. Fu, Y. Zhao, GaN-on-GaN p-i-n diodes with avalanche capability enabled by eliminating surface leakage with hydrogen plasma treatment. Appl. Phys. Lett. **121**(9) (2022). https://doi.org/10.1063/5.0107677
19. X. Liu et al., GaN-on-Si Quasi-vertical p-n diode with junction termination extension based on hydrogen plasma treatment and diffusion. IEEE Trans. Electron Devices **70**(4), 1636–1640 (2023). https://doi.org/10.1109/TED.2023.3247366
20. Z. Bian, K. Zeng, S. Chowdhury, 2.8 kV avalanche in vertical GaN PN diode utilizing field plate on hydrogen passivated P-layer. IEEE Electron Device Lett. **43**(4), 596–599 (2022). https://doi.org/10.1109/LED.2022.3149748
21. I.C. Kizilyalli, E.P. Carlson, (Invited) Selective Area Doping in Gallium Nitride: A Retrospective of the ARPA-E PNDIODES Program. ECS Meet. Abstr. **MA2021-02**(34), 981–981 (2021). https://doi.org/10.1149/ma2021-0234981mtgabs
22. F. Zhou et al., Demonstration of avalanche and surge current robustness in GaN junction barrier Schottky diode with 600-V/10-A switching capability. IEEE Trans. Power Electron. **36**(11), 12163–12167 (2021). https://doi.org/10.1109/TPEL.2021.3076694
23. M. Matys et al., Design and demonstration of nearly-ideal edge termination for GaN p-n junction using Mg-implanted field limiting rings. Appl. Phys. Express **14**(7), 074002 (2021). https://doi.org/10.35848/1882-0786/ac0b09
24. M. Matys et al., Mg-implanted bevel edge termination structure for GaN power device applications. Appl. Phys. Lett. **118**(9) (2021). https://doi.org/10.1063/5.0039183

25. M. Matys, K. Kitagawa, T. Narita, T. Uesugi, J. Suda, T. Kachi, Mg-implanted vertical GaN junction barrier Schottky rectifiers with low on resistance, low turn-on voltage, and nearly ideal nondestructive breakdown voltage. Appl. Phys. Lett. **121**(20) (2022). https://doi.org/10.1063/5.0106321
26. M. Matys et al., Design and fabrication of vertical GaN junction barrier Schottky rectifiers using Mg ion implantation. Jpn. J. Appl. Phys. **62**(SN), N (2023). https://doi.org/10.35848/1347-4065/acec69
27. S.R. Stein et al., Analysis of vertical GaN JBS and p-n diodes by Mg ion implantation and ultrahigh-pressure annealing. IEEE Trans. Electron Devices **71**(3), 1494–1501 (2024). https://doi.org/10.1109/TED.2023.3339592
28. W. Qi et al., 0.58 mΩ·cm^2/523 V GaN vertical Schottky barrier diode with 15.6 kA/cm^2; Surge current enabled by laser lift-off/ annealing and N-ion implantation. IEEE Electron Device Lett. **45**(6), 964–967 (2024). https://doi.org/10.1109/LED.2024.3390125
29. Y. Irokawa, A. Ohi, T. Nabatame, Y. Koide, Pt/GaN Schottky barrier height lowering by incorporated hydrogen. ECS J. Solid State Sci. Technol. **13**(4) (2024). https://doi.org/10.1149/2162-8777/ad3959
30. M.A. Ebrish et al., Process optimization for selective area doping of GaN by ion implantation. J. Electron. Mater. **50**(8), 4642–4649 (2021). https://doi.org/10.1007/s11664-021-08984-8
31. A.G. Jacobs, J.A. Spencer, J.K. Hite, K.D. Hobart, T.J. Anderson, B.N. Feigelson, Novel codoping moiety to achieve enhanced P-type doping in GaN by ion implantation. Phys. Status Solidi **220**(16), SI (2023). https://doi.org/10.1002/pssa.202200848
32. T. Nakamura, M. Yoshino, T. Toyabe, A. Yasuda, Breakdown characteristics of GaN DMISFETs fabricated via Mg, Si and N triple ion implantation. Micromachines **15**(1) (2024). https://doi.org/10.3390/mi15010147
33. K. Nomoto et al., Distributed polarization-doped GaN p-n diodes with near-unity ideality factor and avalanche breakdown voltage of 1.25 kV. Appl. Phys. Lett. **120**(12) (2022). https://doi.org/10.1063/5.0083302
34. T. Kumabe, S. Kawasaki, H. Watanabe, S. Nitta, Y. Honda, H. Amano, Space-charge profiles and carrier transport properties in dopant-free GaN-based p-n junction formed by distributed polarization doping. Phys. Status Solidi-Rapid Res. Lett. **16**(7) (2022). https://doi.org/10.1002/pssr.202200127
35. Y. Li, S. Yang, K. Liu, K. Cheng, K. Sheng, B. Shen, Fully-vertical GaN-on-SiC Schottky barrier diode: role of conductive buffer structure. IEEE Trans. Electron Devices **70**(2), 619–626 (2023). https://doi.org/10.1109/TED.2022.3227227
36. J. Li, R. Zhu, K.M. Wong, K.M. Lau, Fully vertical GaN-on-SiC p-i-n diodes with BFOM of 2.89 GW/cm2. IEEE J. Electron Devices Soc. **12**, 318–321 (2024). https://doi.org/10.1109/JEDS.2024.3386857
37. K. Ohnishi et al., Vertical GaN p(+)-n junction diode with ideal avalanche capability grown by halide vapor phase epitaxy. Appl. Phys. Lett. **119**(15) (2021). https://doi.org/10.1063/5.0066139
38. V. Talesara, Y. Zhang, Z. Chen, H. Zhao, W. Lu, Design and development of 1.5 kV vertical GaN pn diodes on HVPE substrate. J. Mater. Res. **36**(24), 4919–4926 (2021). https://doi.org/10.1557/s43578-021-00435-8
39. X. Liu et al., Vertical GaN Schottky barrier diode with record high figure of merit (1.1 GW/cm2) fully grown by hydride vapor phase epitaxy. IEEE Trans. Electron Devices **70**(7), 3748–3753 (2023). https://doi.org/10.1109/TED.2023.3279059
40. D. Ji, B. Ercan, S. Chowdhury, Experimental determination of impact ionization coefficients of electrons and holes in gallium nitride using homojunction structures. Appl. Phys. Lett. **115**(7) (2019). https://doi.org/10.1063/1.5099245
41. L. Cao et al., Experimental characterization of impact ionization coefficients for electrons and holes in GaN grown on bulk GaN substrates. Appl. Phys. Lett. **112**(26) (2018). https://doi.org/10.1063/1.5031785

42. L. Cao et al., Temperature dependence of electron and hole impact ionization coefficients in GaN. IEEE Trans. Electron Devices **68**(3), 1228–1234 (2021). https://doi.org/10.1109/TED.2021.3054355
43. T. Maeda et al., Breakdown electric field of GaN p+-n and p-n+ junction diodes with various doping concentrations. IEEE Electron Device Lett. **43**(1), 96–99 (2022). https://doi.org/10.1109/LED.2021.3125328
44. F. Bertazzi, M. Moresco, E. Bellotti, Theory of high field carrier transport and impact ionization in wurtzite GaN. Part I: A full band Monte Carlo model. J. Appl. Phys. **106**(6) (2009). https://doi.org/10.1063/1.3213363
45. W. Miyazaki, H. Tanaka, N. Mori, Full-band Monte Carlo analysis of strain effects on carrier transport in GaN. Jpn. J. Appl. Phys. **63**(2) (2024). https://doi.org/10.35848/1347-4065/ad1005
46. V. Maurya et al., Electrical transport characteristics of vertical GaN Schottky-barrier diode in reverse bias and its numerical simulation. Energies **16**(14) (2023). https://doi.org/10.3390/en16145447
47. T. Kumabe, S. Kawasaki, H. Watanabe, Y. Honda, H. Amano, Electron lifetime and diffusion coefficient in dopant-free p-type distributed polarization doped AlGaN. Appl. Phys. Lett. **123**(25) (2023). https://doi.org/10.1063/5.0180062
48. K. Mochizuki, H. Ohta, F. Horikiri, T. Mishima, Estimation of Shockley-Read-Hall lifetime in homoepitaxial n-GaN on low-dislocation-density GaN substrates prepared by hydride vapor phase epitaxy and maskless 3D. Phys. Status Solidi B **259**(2) (2022). https://doi.org/10.1002/pssb.202100215
49. C. Koller et al., Trap-related breakdown and filamentary conduction in carbon doped GaN. Phys. Status Solidi B **256**(6), SI (2019). https://doi.org/10.1002/pssb.201800527
50. H. Ohta, N. Asai, F. Horikiri, Y. Narita, T. Yoshida, T. Mishima, Impact of threading dislocations in GaN p-n diodes on forwardI-Vcharacteristics. Jpn. J. Appl. Phys. **59**(10), 106503 (2020)
51. H. Ohta, N. Asai, F. Horikiri, Y. Narita, T. Yoshida, T. Mishima, Breakdown phenomenon dependences on the number and positions of threading dislocations in vertical p-n junction GaN diodes. Jpn. J. Appl. Phys. **60**(SB) (2021). https://doi.org/10.35848/1347-4065/abdccc
52. S. Li, B. Ercan, C. Ren, H. Ikeda, S. Chowdhury, A study on the impact of dislocation density on leakage current in vertical GaN-on-GaN p-n diodes. IEEE Trans. Electron Devices **69**(8), 4206–4211 (2022). https://doi.org/10.1109/TED.2022.3186271
53. T. Nakano et al., Electronic structure analysis of core structures of threading dislocations in GaN, in *2019 Compound Semiconductor Week (CSW)*, (2019). https://doi.org/10.1109/iciprm.2019.8819270
54. T. Nakano et al., Screw dislocation that converts p-type GaN to n-type: microscopic study on Mg condensation and leakage current in p-n diodes. Appl. Phys. Lett. **117**(1) (2020). https://doi.org/10.1063/5.0010664
55. S. Usami et al., Correlation between nanopipes formed from screw dislocations during homoepitaxial growth by metal-organic vapor-phase epitaxy and reverse leakage current in vertical p-n diodes on a free-standing GaN substrates. Jpn. J. Appl. Phys. **58**(C), SCCB24 (2019). https://doi.org/10.7567/1347-4065/ab1250
56. T. Sato et al., Conduction mechanism of Schottky contacts fabricated on etch pits originating from single threading dislocation in a highly Si-doped HVPE GaN substrate. Mater. Sci. Semicond. Process. **167**, 107778 (2023). https://doi.org/10.1016/j.mssp.2023.107778
57. T. Hamachi et al., Comprehensive analysis of current leakage at individual screw and mixed threading dislocations in freestanding GaN substrates. Sci. Rep. **13**(1), 2436 (2023). https://doi.org/10.1038/s41598-023-29458-3
58. Y. Ishikawa et al., Characterization of dislocations at the emission site by emission microscopy in GaN p-n diodes. J. Mater. Sci. **58**(22), 9221–9232 (2023). https://doi.org/10.1007/s10853-023-08596-z
59. W. Kwon et al., Reverse leakage mechanism of dislocation-free GaN vertical p-n diodes. IEEE Electron Device Lett. **44**(7), 1172–1175 (2023). https://doi.org/10.1109/LED.2023.3274306

60. Q. Yang, J. Yang, H. Long, Z. Liu, L. Duan, D. Zhao, Reducing leakage current in GaN-Schottky diodes by suppressing hollow core nanopipes. AIP Adv. **14**(3) (2024). https://doi.org/10.1063/5.0200855
61. I. Ben-Yaacov et al., AlGaN/GaN current aperture vertical electron transistors with regrown channels. J. Appl. Phys. **95**(4), 2073–2078 (2004)
62. S. Chowdhury et al., Enhancement and depletion mode AlGaN/GaN CAVET with Mg-ion-implanted GaN as current blocking layer. IEEE Electron Device Lett. **29**(6), 543–545 (2008)
63. S. Chowdhury et al., CAVET on bulk GaN substrates achieved with MBE-regrown AlGaN/GaN layers to suppress dispersion. IEEE Electron Device Lett. **33**(1), 41–43 (2011)
64. S. Chowdhury et al., Current status and scope of gallium nitride-based vertical transistors for high-power electronics application. Semicond. Sci. Technol. **28**(7), 074014 (2013)
65. H. Nie et al., 1.5-kV and 2.2-mΩcm^2 vertical GaN transistors on bulk-GaN substrates. IEEE Electron Device Lett. **35**(9), 939–941 (2014)
66. D. Ji, S. Chowdhury, Design of 1.2 kV power switches with low RON using GaN-based vertical JFET. IEEE Trans. Electron Devices **62**(8), 2571–2578 (2015)
67. S. Mandal et al., Dispersion free 450-V p GaN-gated CAVETs with Mg-ion implanted blocking layer. IEEE Electron Device Lett. **38**(7), 933–936 (2017)
68. S. Rajabi et al., A demonstration of nitrogen polar gallium nitride current aperture vertical electron transistor. IEEE Electron Device Lett. **40**(6), 885–888 (2019)
69. X. Wen et al., 10 MHz-switching on GaN trench CAVET up to 300 °C operation enabled by high channel mobility. IEEE Electron Device Lett. **45**(4), 653–656 (2024). https://doi.org/10.1109/LED.2024.3360917
70. D. Ji et al., Normally off trench CAVET with active Mg-doped GaN as current blocking layer. IEEE Trans. Electron Devices **64**(3), 805–808 (2017)
71. D. Ji et al., 880 V/2.7 mΩ cm^2 MIS gate trench CAVET on bulk GaN substrates. IEEE Electron Device Lett. **39**(6), 863–865 (2018)
72. D. Shibata et al., 1.7 kV/1.0 mΩcm^2 normally-off vertical GaN transistor on GaN substrate with regrown p-GaN/AlGaN/GaN semipolar gate structure, in *Technical Digest - International Electron Devices Meeting, IEDM*, (2017), pp. 10.1.1–10.1.4
73. K. Lee et al., A systematic study on the efficacy of low temperature-GaN regrown on p-GaN to suppress Mg out-diffusion. Front. Mater. **10**, 1229036 (2023). https://doi.org/10.3389/fmats.2023.1229036
74. X. Wen et al., High current density trench CAVET on bulk GaN substrates with low-temperature GaN suppressing Mg diffusion. Crystals **13**(4), 709 (2023). https://doi.org/10.3390/cryst13040709
75. M. Noshin et al., A systematic study of the regrown interface impurities in unintentionally doped Ga-polar c-plane GaN and methods to reduce the same. Semicond. Sci. Technol. **37**, 075018. https://doi.org/10.1088/1361-6641/ac71bf
76. X. Wen et al., First experimental demonstration of monolithic bidirectional switch using GaN current aperture vertical electron transistor (CAVET), in *IEEE Device Research Conference*, (2024)
77. H. Otake et al., GaN-based trench gate metal oxide semiconductor field effect transistors with over 100 cm^2/(Vs) channel mobility. Jpn. J. Appl. Phys., Part 2 Lett. **46**(25–28), L599 (2007)
78. T. Oka et al., Vertical GaN-based trench metal oxide semiconductor field-effect transistors on a free-standing GaN substrate with blocking voltage of 1.6 kV. Appl. Phys. Express **7**(2) (2014)
79. C. Gupta et al., In situ oxide, GaN interlayer-based vertical trench MOSFET (OG-FET) on bulk GaN substrates. IEEE Electron Device Lett. **38**(3), 353–355 (2017)
80. D. Ji et al., Demonstrating >1.4 kV OG-FET performance with a novel double field-plated geometry and the successful scaling of large-area devices, in *Technical Digest - International Electron Devices Meeting, IEDM*, (2018), pp. 9.4.1–9.4.4
81. M. Ruzzarin et al., Gate stability and robustness of in-situ oxide GaN interlayer based vertical trench MOSFETs (OG-FETs), in *IEEE International Reliability Physics Symposium (IRPS)*, (IEEE, 2019), pp. 1–5

82. D. Ji et al., Large-area in-situ oxide, GaN interlayer-based vertical trench MOSFET (OG-FET). IEEE Electron Device Lett. **39**(5), 711–714 (2018)
83. M. Basler et al., Switching of GaN CAVET with quasi-monolithic integrated HEMT gate driver. IEEE Electron Device Lett. **44**(8), 1332–1335 (2023). https://doi.org/10.1109/LED.2023.3290608
84. W. Li, S. Chowdhury, Design and fabrication of a 1.2kV GaN-based MOS vertical transistor for single chip normally off operation. Phys. Status Solidi **213**(10), 2714–2720 (2016)
85. J. Chun et al., Schottky junction vertical channel GaN static induction transistor with a sub-micrometer fin width. Adv. Electron. Mater. **5**(1), 1800689 (2019)
86. S. Jeong et al., Scaling study on high-current density low-dispersion GaN vertical FinFETs. IEEE Electron Device Lett. **44**(5), 841–844 (2023). https://doi.org/10.1109/LED.2023.3259002
87. Z.Z. Bandić et al., Minority carrier diffusion length and lifetime in GaN. Appl. Phys. Lett. **72**(24), 3166–3168 (1998)
88. Y. Hatakeyama et al., High-breakdown-voltage and low-specific-on-resistance GaN p–n junction diodes on free-standing GaN substrates fabricated through low-damage field-plate process. Jpn. J. Appl. Phys. **52**(2R), 028007 (2013)
89. K. Mochizuki et al., Photon-recycling GaN pn diodes demonstrating temperature-independent, extremely low on-resistance, in *2011 International Electron Devices Meeting*, (IEEE, 2011)
90. Y. Cao et al., Improved performance in vertical GaN Schottky diode assisted by AlGaN tunneling barrier. Appl. Phys. Lett. **108**, 11 (2016)

Part II
Bulk GaN Substrates

Chapter 3
Bulk GaN

Kenji Fujito, Yutaka Mikawa, and Satoru Izumisawa

1 Introduction

The progress of bulk GaN substrates has been driven by the laser diode (LD) market. GaN substrates were initially used in blue-violet LDs for Blu-ray applications. Subsequently, improvements in the quality of GaN substrates and the optimization of device structures have led to longer wavelengths, higher power, and higher efficiency LDs. Subsequently, blue and green LDs are used for various applications, including projectors, laser TVs, machining processes such as copper welding, high-brightness special lighting, and car headlights.

GaN substrates for LDs are mainly made by the hydride vapor phase epitaxy (HVPE) method, with a wafer size of 2 inches and a dislocation density of 10^5–10^6 cm^{-2} being typical. HVPE has the advantage of a relatively high growth rate and the ability to grow at atmospheric pressure. However, the utilization efficiency of raw gas is low, which limits its ability to reduce manufacturing costs. Additionally, the gas flow in the reaction zone must be highly controlled, thereby making it difficult to increase the size of the GaN crystals.

Bulk GaN substrates are also expected to be used as substrates for power device applications due to their high breakdown field and mobility. To realize and popularize GaN substrates for power device applications, wafer sizes of ≥6 inches, dislocation densities of ≤10^3–10^4 cm^{-2}, and cost-effective substrates are required. To achieve these goals, the development of crystal growth methods such as the sodium flux method, which uses molten sodium, and the ammonothermal method, which uses supercritical ammonia, have been becoming more active.

K. Fujito (✉) · Y. Mikawa · S. Izumisawa
Mitsubishi Chemical Co., Tokyo, Japan
e-mail: kenji.fujito.mv@mcgc.com; yutaka.mikawa.mc@mcgc.com; satoru.izumisawa.mf@mcgc.com

I. C. Kizilyalli et al. (eds.), *Gallium Nitride and Related Materials*, The Materials Research Society Series, https://doi.org/10.1007/978-3-031-83056-3_3

2 Overview of Bulk GaN Crystal Growth

GaN decomposes into Ga and N_2 at ~760 °C and normal pressure [1], so there is no liquid phase at normal pressure. Therefore, bulk GaN crystals cannot be grown by the Czochralski method as in the case of Si.

Bulk GaN crystal growth methods can be mainly classified into the following three types:

1. The vapor-phase growth method involving the reaction of group V and group III gases
2. The liquid-phase growth method using a solution or melt
3. The solvothermal method using uses a supercritical state, which is neither a gas nor liquid phase.

2.1 Vapor-Phase Growth Method

The HVPE method is the most common vapor-phase growth method for bulk GaN crystals, and most of the GaN substrates on the market are made using this method. Figure 3.1 shows a schematic of the growth equipment used in the HVPE method.

The HVPE method uses hydrogen and nitrogen as carrier gases. In the first stage of the reaction process, Ga metal and hydrochloric gas react upstream of the quartz reactor at 800–850 °C to form GaCl. In the second stage reaction, GaCl and ammonia are mixed and reacted at 1000–1100 °C downstream, and GaN is deposited on a seed substrate placed near the mixing zone. The upstream and downstream reaction equations are shown below [2].

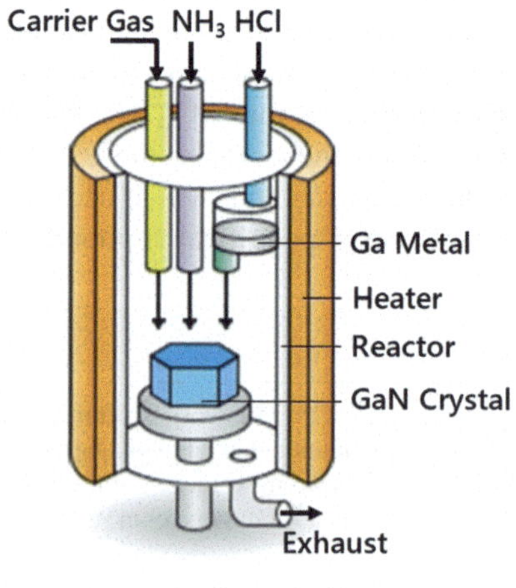

Fig. 3.1 Schematic of crystal growth equipment used in the HVPE method

$$\text{Upstream} \qquad 2Ga(l) + 2HCl(g) \rightarrow 2GaCl(g) + H_2(g) \tag{3.1}$$

$$\text{Downstream} \qquad GaCl(g) + NH_3(g) \rightarrow GaN(s) + HCl(g) + H_2(g) \tag{3.2}$$

$$\text{Side reaction} \qquad HCl(g) + NH_3(g) \rightarrow NH_4Cl(s) \tag{3.3}$$

The HVPE method exhibits a high growth rate of ≥100 μm/h. When obtaining bulk crystals of millimeter-scale thickness, a key challenge is the reaction of hydrochloric gas with ammonia, as indicated in Eq. (3.3). This reaction produces solid NH_4Cl as a by-product, which can cause issues such as clogging the exhaust pipes and necessitates appropriate countermeasures.

Foreign substrates such as sapphire or GaAs are generally used as the seed substrates. When GaN is grown on foreign substrates, many crystal dislocations are introduced due to lattice mismatches. In addition, differences in the thermal expansion coefficient and lattice constants between GaN and the seed substrates cause the wafer to bow.

To solve these problems, the dislocation elimination by the epitaxial growth with the inverse-pyramidal pits (DEEP) method [3], in which high and low dislocation density areas are formed periodically by intentionally concentrating dislocations, and the void-assisted separation (VAS) method [4], in which the crystal quality is improved by forming voids at the interface between the seed substrate and GaN layer, have been developed.

In addition, a method using freestanding GaN substrates as seed substrates has also been applied. In this case, the problems of lattice constant and thermal expansion coefficient differences do not occur. Using high-quality GaN substrates fabricated by the ammonothermal method described below as seed substrates [5], GaN substrates with dislocation densities of 10^3–10^4 cm^{-2} have been realized [6]. This dislocation density is 1–2 orders of magnitude lower than that of GaN substrates obtained from thick growth on foreign substrates.

A tri-halide vapor phase epitaxy (THVPE) method, which is an extension of HVPE and enables faster growth at higher temperatures, has also been proposed [7]. The use of $GaCl_3$ as the source gas enables a growth rate several times higher than that of conventional HVPE at temperatures exceeding 1000 °C, and the crystal diameter can be maintained, or increased, by using the specific adsorption behavior of the source molecules on the crystal surface. THVPE is expected to contribute to the price reduction and diameter expansion of GaN substrates.

An oxide vapor phase epitaxy (OVPE) method, which uses Ga_2O as the raw material gas, is also under development [8, 9]. The reaction of Ga_2O with ammonia is demonstrated in formula (3.4). Since no hydrochloric gas is produced, which produces a solid by-product when reacting with ammonia, the problem of exhaust pipe clogging, which is an issue during long-term growth in the HVPE method, does not occur.

$$Ga_2O\,(g) + 2NH_3(g) \rightarrow 2GaN(s) + H_2O(g) + 2H_2(g) \tag{3.4}$$

Although the oxygen concentration in crystals produced by the OVPE method is extremely high (~10^{20} cm^{-3}), the quality of the crystals is equivalent to that of seed crystals. Furthermore, since oxygen acts as a donor in the GaN crystal, the substrate resistivity is in the 10^{-4} Ωcm range, one order of magnitude lower than that of a typical HVPE substrate [10]. This low-resistivity GaN substrate shows great promise for vertical GaN power device applications.

2.2 *Liquid-Phase Growth Method*

A widely known liquid-phase growth method to obtain bulk GaN crystals is the Na-flux method.

This technique synthesizes GaN crystals in a mixture of molten Ga and Na by dissolving nitrogen in it under pressure [11, 12]. The addition of Na to the Ga melt greatly increases the solubility of nitrogen in the melt, enabling crystal growth at relatively low temperatures and pressures of 800–900 °C and 3–5 MPa, respectively.

The Na-flux method has succeeded in growing high-quality crystals with almost no dislocations from small seed crystals called point seeds [13]. By combining multiple point seeds [14], which are made by bonding multiple point seeds together, with the Flux Film Coated (FFC) method [15], in which a crystal is pulled from the Na-Ga melt in a crucible during crystal growth, GaN crystals larger than 6 inches have also been successfully grown [16].

The Na flux method, however, suffers from the problem of metallic inclusions from the constituents of the melt. As one effort to realize large-diameter bulk GaN crystals without inclusions, development has begun using crystals grown by the Na flux method as seed crystals for the ammonothermal method described below [17].

2.3 *Solvothermal Method*

In the solvothermal method, raw materials are dissolved in a supercritical solvent at high temperature and pressure, and crystals are grown on the seed crystal due to solubility differences resulting from this temperature difference. The hydrothermal method uses supercritical water as the solvent and is used for the crystal growth of quartz and zinc oxide. The ammonothermal method [18] uses supercritical ammonia as the solvent and is used for the growth of nitride crystals such as GaN.

Since a high-pressure atmosphere is required to realize a supercritical state of ammonia, GaN crystal growth is performed in a pressure vessel (autoclave).

The ammonothermal method uses a mineralizer to increase the solubility of the raw material, polycrystalline GaN. Depending on the type of mineralizer, it can be classified into the basic ammonothermal method, which uses alkali amides (e.g., $NaNH_2$ or KNH_2) as mineralizers, and acidic ammonothermal method, which uses ammonium halides (e.g., NH_4Cl or NH_4Br) as mineralizers [18–20].

Although the crystal growth rate is very slow at several tens to several hundreds of μm/day, it is now possible to realize several hundred bulk crystals in one growth run over a period of one to 6 months by scaling up the autoclave. As this method has highly efficient feedstock utilization, this is expected to reduce production costs compared to the HVPE method. Moreover, since the ammonothermal method grows crystals in a state close to thermal equilibrium, extremely high-quality crystals with minimal dislocations and strain have been demonstrated.

Since supercritical ammonia is highly soluble and erosive, and some types of mineralizers are also highly corrosive, appropriate autoclave materials and lining materials must be selected. This makes it possible to avoid autoclave damage and contamination of the crystals with metallic impurities.

The acidic ammonothermal method that the authors are developing is described in detail below.

3 Acidic Ammonothermal Method

The acidic ammonothermal method has advantages compared with the basic ammonothermal method. Namely, acidic growth rates are higher than basic growth rate (basic: 30–50 μm/day, acidic: 100–300 μm/day) and the acidic growth can be realized at lower pressures than basic growth (basic: 300–400 MPa, acidic: 100–250 MPa). These two features are beneficial for mass production and growing larger crystals. Although the acidic method requires lower pressure than the basic method, it requires temperature and pressure conditions (550 ~ 650 °C, ~200 MPa) that are much higher than those of the hydrothermal method used for quartz growth (350 ~ 400 °C, ~150 MPa). Therefore, it is necessary to use Ni-based superalloys with excellent high-temperature strength for the autoclave materials. The acidic ammonothermal method we are using has two types: the high-pressure method with a pressure of about 200 MPa, and the low-pressure method with a pressure of about 100 MPa. The size of the GaN crystals that can be grown is limited by the autoclave size, and the size of the autoclaves that can be manufactured increases with lower working pressure. The maximum size of GaN crystal that can be grown using the high-pressure method is 4 inches. To produce even larger GaN crystals, a large-diameter autoclave is required. The low-pressure method enables a large-diameter autoclave by maintaining the pressure at about 100 MPa, and we are working toward the production of 6-inch GaN crystals.

Figure 3.2 shows a schematic of the crystal growth apparatus used in the high-pressure acidic ammonothermal method.

The crystal growth apparatus comprises an autoclave and a heater. The high-pressure acidic ammonothermal crystal growth procedure is briefly described below. First, the seed (GaN single crystal) is placed at the top of the autoclave, the raw

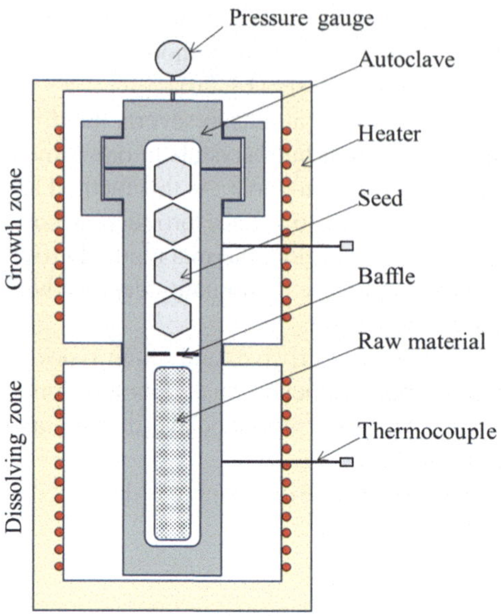

Seed	GaN single crystal
Raw material	GaN poly cryatal
Mineralizer	NH_4X (X: halogen)
Solvent	NH_3
Temperature	550 ~ 650°C
Pressure	100 ~ 250 MPa

Fig. 3.2 Schematic of high-pressure ammonothermal equipment and the growth conditions

material (GaN polycrystals) at the bottom, and a baffle is installed in the center for controlling convection. Next, a mineralizer that has the function of dissolving the GaN raw material into supercritical ammonia is added, and the autoclave is filled and sealed with ammonia as the solvent. Subsequently, by heating using an external heater, the internal pressure increases due to the thermal expansion of ammonia. The pressure is determined by the ammonia fill rate and the temperature. The temperature of the autoclave at the top is maintained lower than that at the bottom.

Figure 3.3 shows the solubility curve for the high-pressure acidic ammonothermal method. The solubility curve is determined by the chemical type and concentration of the mineralizer. In the case of mineralizers used in the high-pressure method, the solubility curve exhibits a positive gradient, demonstrating that GaN solubility increases as the temperature rises. The arrangement of the seed and raw material is reversed depending on the positive or negative gradient of the solubility curve. In the basic and low-pressure acidic ammonothermal methods, the raw material is placed in the upper region, while the seed is placed in the lower region. The temperature difference (ΔT) generates thermal convection, and the raw material dissolved at the bottom is transported to the top via thermal convection. The solubility difference (ΔS) due to the temperature decrease causes the precipitation on the seed surface, resulting in crystal growth. As the opening ratio of the baffle changes the balance between the convection flow rate and temperature difference, it is important to design the baffle appropriately while considering the autoclave size and the number of seeds installed. For example, if the opening ratio is too small, the temperature difference increases; however, the amount of raw material supplied to the seed is insufficient due to reduced transport. Conversely, if the

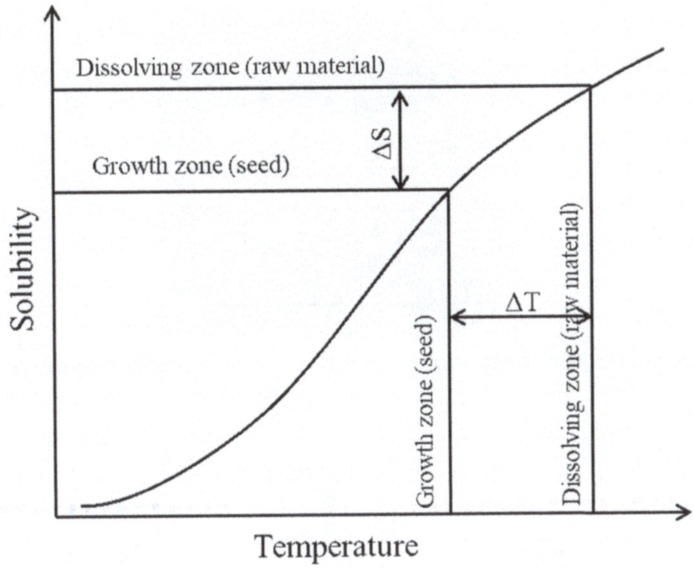

Fig. 3.3 Solubility curve of GaN in high-pressure acidic ammonothermal method

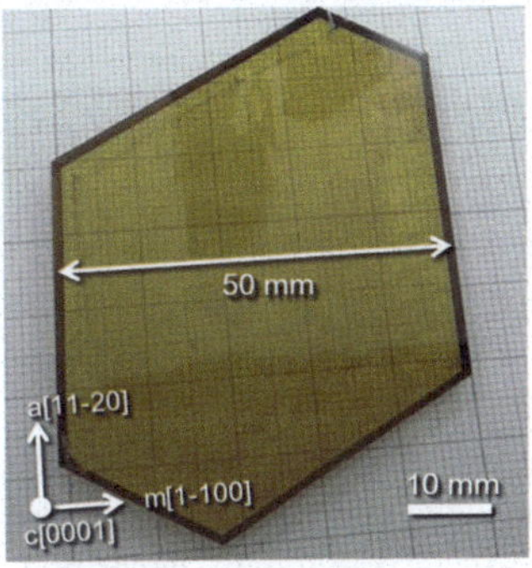

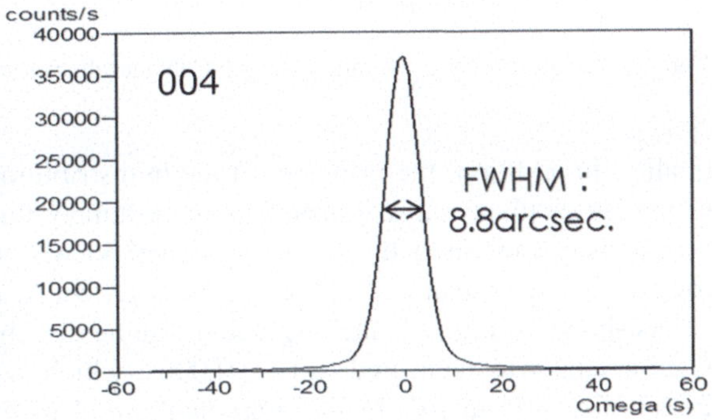

Fig. 3.4 Two-inch c-plane GaN and the X-ray rocking curve on 004 reflection

opening ratio is too large, the amount of transport increases, but it is difficult to create the necessary temperature difference, which can be disadvantageous to crystal growth. The optimization of convection in the autoclave is attempted using computer simulation and is described in the next section.

A typical 2-inch c-plane GaN crystal grown by the high-pressure acidic ammonothermal method and its X-ray rocking curve (XRC) measurements are shown in Fig. 3.4. The external shape of the crystal, reflecting a hexagonal system surrounded by natural planes, is well characterized by the ammonothermal method, which grew under near-equilibrium conditions. The full width at half maximum by XRC measurement is extremely small (8.8 arcsec.), indicating high crystallographic

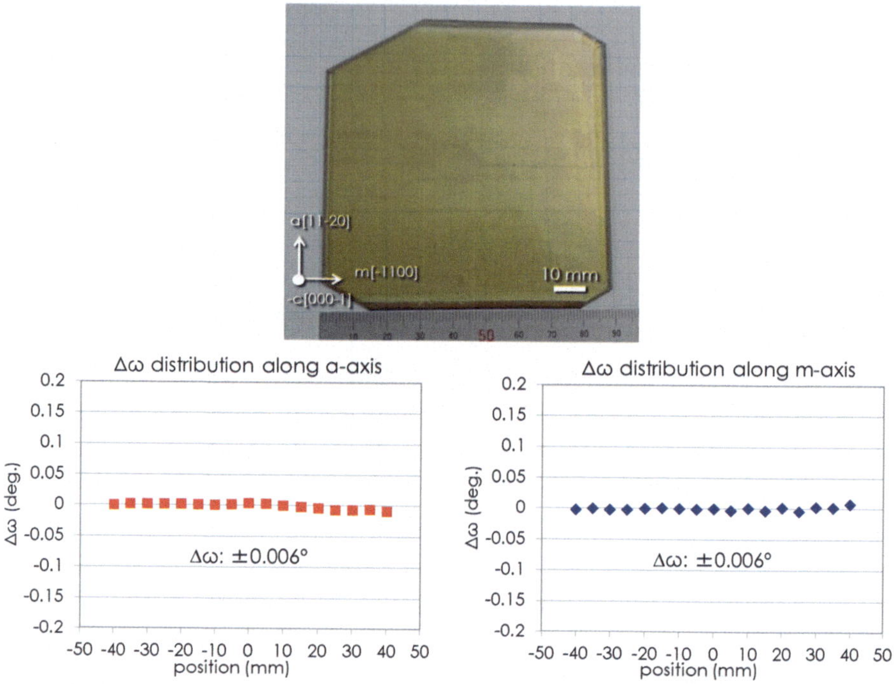

Fig. 3.5 Large size (90 × 90 mm^2) c-plane GaN and the off-angle distributions

quality. In addition, the in-plane off-angle distribution (crystallographic curvature) of the lattice plane was confirmed to be extremely flat around ±0.01°. The dislocation density measured by the etch-pit method was 1.1 × 10^4/cm^2, two orders of magnitude lower than that of GaN crystals grown by the typical HVPE method [6].

As shown in Fig. 3.5, the in-plane off-angle distribution of 90 × 90 mm^2 c-plane GaN crystal was measured to be ±0.006°, which is at least the equivalent of the 2-inch size crystal [6]. In the ammonothermal method, as previously described, growth occurs under near-equilibrium conditions, and unlike the melt method, there is almost no temperature gradient at the growth boundary. Therefore, there is no increase in thermal stress due to the enlargement of the crystal size. This indicates that, if the equipment can be scaled up, it is possible in principle to increase the size of the crystals while maintaining crystallographic quality.

In the ammonothermal method, the size of the autoclave limits the maximum size of crystals, and as mentioned previously, the maximum crystal size that can be grown in the high-pressure method is 4 inches. In collaboration with Tohoku University and The Japan Steel Works, we have found that by using NH_4F as a mineralizer, it is possible to grow crystals having the same crystallinity as that of the high-pressure method even when the pressure is reduced to 100 MPa, half the pressure of the high-pressure method [21–23]. The low pressure has enabled us to increase the size of the autoclave, and we have installed a large autoclave capable of

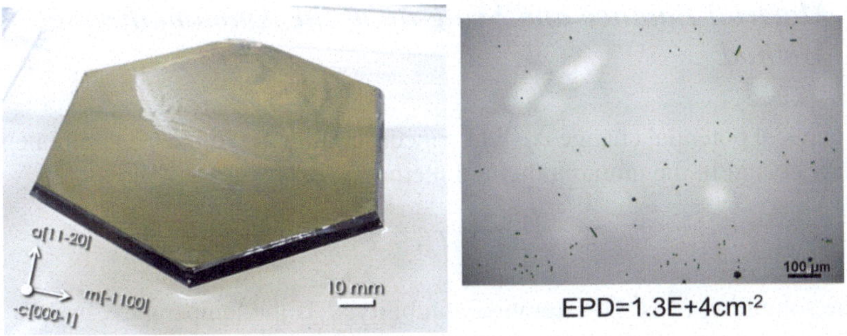

Fig. 3.6 Four-inch GaN grown by low-pressure ammonothermal method and the etch-pit density

growing crystals over 6 inches in size, and are currently carrying out studies toward mass production. Figure 3.6 shows the appearance and etch-pit measurements of a 4-inch crystal grown using the low-pressure acidic ammonothermal method in a production-scale autoclave. The crystal thickness is about 5 mm with flat growth over the entire surface. The dislocation density calculated by etch-pit observation was 1.3×10^4 cm^{-2}, which was equivalent to that of crystals grown by the high-pressure method used as the seed, confirming that the crystal was grown while maintaining a low dislocation density.

4 Mass Production of GaN Substrates

The difficulty in increasing the size and reducing the cost of GaN substrates is primarily due to the inherent principle of the HVPE method. The raw gases GaCl and NH_3, react on the surface of the reactor components, causing the precipitation of polycrystalline GaN. Precise gas flow control and reaction field formation are essential to ensure proper gas mixing only on the seed crystal. Even if the reactor is enlarged to accommodate a large number of seed crystals, it is impossible to form a reaction field that allows for homogeneous growth at the same time, and this is the main reason for the difficulty in realizing the benefits of mass production.

In the ammonothermal method, an equilibrium metastable state can be formed by supersaturating Ga components using NH_3 as the solvent. Therefore, the conditions for GaN crystal growth only onto the seed crystal can be easily achieved. If the concentration and temperature distribution are properly adjusted, uniform crystal growth conditions can be achieved in a large space, and the only restriction on mass production is the size of the equipment. As mentioned above, the low-pressure acidic ammonothermal method alleviates this limitation and greatly improves the mass productivity of GaN crystal growth. To take advantage of this method, it is important to realize a design technology that can appropriately scale up the process and equipment. The basic concept and methodology are described below.

4.1 Material Balance and Scale-up of the Ammonothermal Method

The chemical potential change $\Delta\mu$ while incorporating the solute Ga component into the GaN crystal in the ammonothermal method is expressed as follows:

$$\Delta\mu \approx k_B T(c - c^*)/c^* \quad (3.5)$$

c is the solubility, c^* is the saturation solubility, T is the temperature, and k_B is the Boltzmann constant. If $\Delta\mu$ is too large, GaN nucleates spontaneously, but within an appropriate concentration range, GaN precipitates only on the seed crystal surface. Since in most cases, the crystal growth rate is determined by the rate of solute transport from the solution to the growth surface through the boundary layer, the diffusion flux to the growth surface, or crystal growth rate G, is expressed using the mass transfer coefficient k as follows:

$$G = k(c - c^*) \quad (3.6)$$

The ideal requirement for scale-up is to maintain conditions where the difference between the concentration c near the crystal and the concentration c^* on the crystal surface is uniform in the system and the difference is sufficiently small relative to c^*.

Figure 3.7 shows a conceptual diagram of the material balance in the ammonothermal method. In the steady state of the ammonothermal process, which progresses in a closed system, the dissolution rate S_f and precipitation rate $-S_g$ will be equal, and the concentration difference between the upstream and downstream

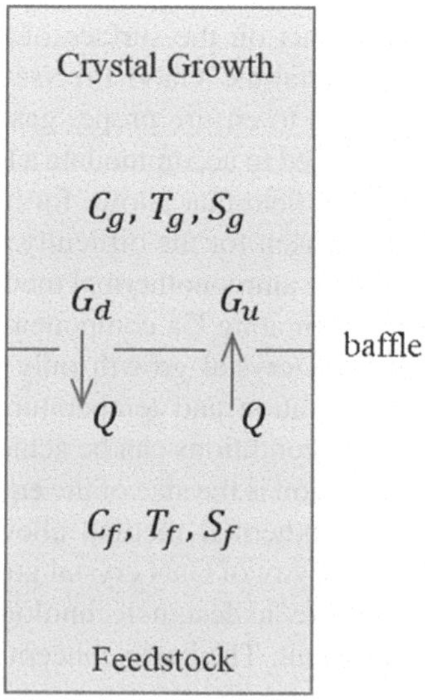

Fig. 3.7 Conceptual diagram of the material balance in the ammonothermal method

Table 3.1 Relationship between scale-up factors and autoclave shape. D is the inner diameter and L is the height

Volume	Inner surface area	Specific surface area	Grashof number
$\propto D^2L$	$\propto DL$	$\propto D^{-1}$	$\propto L^3$

flows at the baffle separating the top and bottom is related to the crystal precipitation rate.

$$S_g + S_f = 0 \tag{3.7}$$

$$Q(C_f - C_g) = -S_g \tag{3.8}$$

Q represents the volumetric flow velocity. As demonstrated in the conceptual diagram, in an ideal system, increasing the deposition rate, i.e., the number of crystals or the growth rate, creates a substantial difference between the growth rates G_d and G_u in the downward and upward flows just above the baffle, and this decreases the growth rate uniformity in the system. There is a trade-off between maintaining uniform crystal growth conditions and maximizing crystal productivity. Achieving both at a high level requires high solubility and a high exchange rate through vertical convection. Therefore, choosing the right type and concentration of mineralizer, as well as the appropriate growth temperature range, is critical to process design.

In a real system, there is a temperature distribution. Therefore, it is necessary to take this into account when scaling up. As shown in Fig. 3.2, heat input and output occur through the surface of the autoclave; therefore, equivalent temperature distributions can be expected at larger scales after scaling up by maintaining a specific surface area for heat transfer. However, since the specific surface area of an autoclave, which is usually cylindrical, is approximately determined by the inner diameter, a constant specific surface area indicates that the autoclave expands only in the longitudinal direction L (Table 3.1). As increasing crystal size is one of the main objectives of scaling up, this guideline is not practical.

4.2 Simulation Techniques for Scale-up

If the specific surface area decreases due to the enlargement of D, the thermal uniformity decreases within the system, and this leads to a reduction in the number of crystals, growth rate, and thus productivity, due to the aforementioned trade-off relationship. The strong nonlinearity of thermal convection with respect to scale and other factors can be practically optimized to minimize the loss due to the scale-up. As these factors are difficult to formulate and design analytically, computational fluid dynamics (CFD) simulations using finite volume methods are a powerful tool for validating the design of equipment and processes [24].

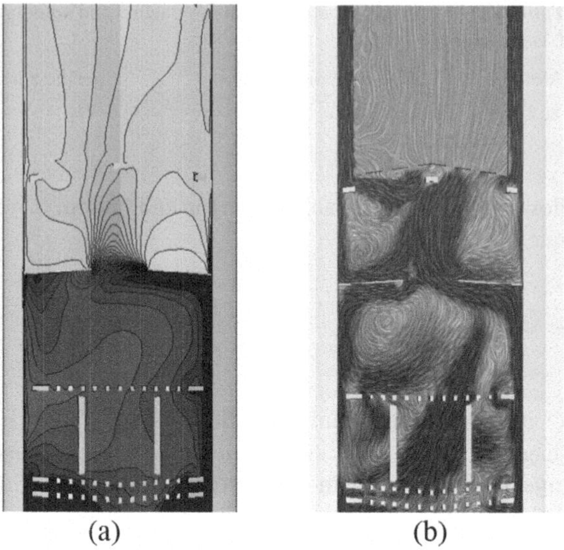

Fig. 3.8 Examples of temperature (**a**) and thermal convection velocity (**b**) distributions inside an autoclave obtained via CFD simulation

In CFD, a model is constructed with temperatures that can be observed or controlled, such as the temperature of an outer wall, as boundary conditions [25]. The thermal convection and temperature distribution of supercritical NH_3 in a closed autoclave can be calculated (Fig. 3.8). By adding the source terms, which are formulated using the experimentally determined temperature dependence of solubility and mass transfer coefficients for precipitation and dissolution, the growth rate distribution in the system can be obtained as a result of the simulation.

The number and arrangement of seed crystals and raw materials, the shape and aperture of the intermediate baffle, and other factors can be optimized using computer simulations. Once a certain autoclave geometry is determined, the conditions for uniformity and maximum growth rate within the system can be studied. For example, the lower limit of the inner diameter of autoclave D is determined by the size of the crystal to be grown. In the case of 6-inch crystal production, the geometry somewhat dictates the range of candidate inner diameters, since an excessively large inner diameter beyond that would sacrifice growth rate and uniformity. For a given candidate value of D, the height L is determined by the maximum ingot size of the alloy that serves as the autoclave material. By performing CFD simulations for various operating conditions, it is possible to calculate the optimized growth conditions for a given autoclave size and determine the maximum expected productivity. These results can be calculated and compared for all candidate autoclave shapes to finalize the equipment and process design. It is evident that semi-empirical optimization of these tasks is impractical, given the time-consuming nature of the process. The manufacturing of autoclaves and the repetition of ammonothermal experiments, which typically span a minimum of 30 days, render this approach infeasible. The design of equipment and processes has become an essential technology.

4.3 Optimization of Crystal Growth Conditions

While recent improvements in parallel computing power mean that the cost of handling these large CFD models is no longer a challenge, there are still benefits to be gained from reducing computational time. In thermal convection, steady flow is difficult to develop, resulting in a long time to take to obtain a converged CFD solution. The inclusion of source terms further increases the computation time. If the optimization of the process described above were to be performed on a brute force basis, it would require tens of thousands of CFD calculations, which, as it is, would take time and incur enormous computational costs that would not differ much from conducting the experiments themselves.

Large-scale multi-objective optimization calculations are being attempted to devise ways to increase the calculation speed in order to extract highly optimal conditions from a wide range. For example, the computational time can be reduced using weakly coupled analyses with separately discretized models for dissolution and precipitation, and by combining machine learning and statistical processing, thus making large-scale multi-objective optimization calculations possible. The Pareto optimal solution obtained from such an analysis is shown in Fig. 3.9.

By analyzing only the area around the Pareto optimal solution in detail, and performing categorization from other aspects, such as crystal quality and stability of control, it is possible to understand the process window and utilize it in the analysis of the experimental results.

Once appropriate conditions have been determined to some extent, detailed optimization can be performed using specific indicators in the system as objective functions. Figure 3.10 compares the growth rate distributions obtained by CFD before and after the optimization process. Improving the uniformity of the growth

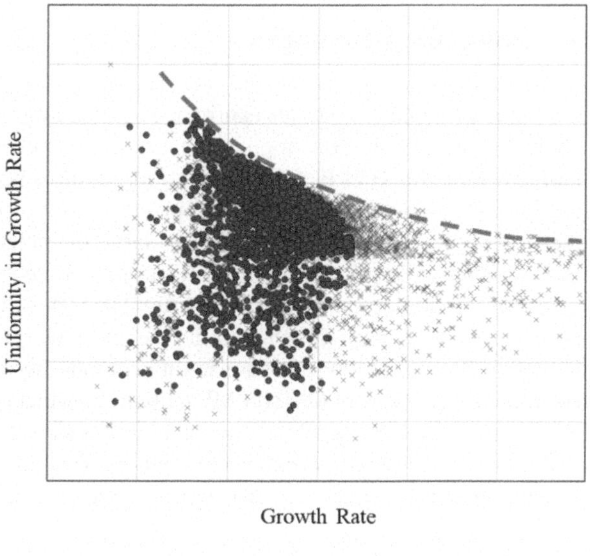

Fig. 3.9 Example of Pareto optimal solution obtained from an optimization calculation of growth rates and their uniformity. Each point corresponds to the results of the simulation under different conditions. The black circles indicate acceptable optimal solutions that can be obtained within the operational range of actual equipment. The "×" mark indicates optimal solutions that cannot be applied to the actual equipment as they are obtained outside the operational range of the actual equipment

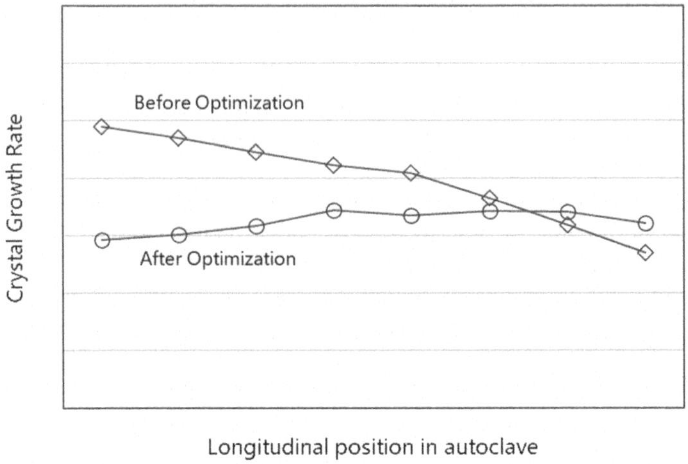

Fig. 3.10 Example of optimizing conditions in a simulation where growth rate distribution is the objective function

rate within the system provides room to increase the average growth rate, further improving productivity. In the development of technology for the mass production of GaN crystals using the ammonothermal method, these simulation and optimization tools are being used in a digital twin manner, not only in the design phase but also in the optimization and analysis of operating conditions. Computational experiments are being conducted in advance to further improve mass production and reduce costs in the future.

5 Concluding Remarks

The market for GaN substrates is primarily expanding in the field of LD applications, where the most common application involves two-inch substrates manufactured mainly using the HVPE method. Various approaches, such as THVPE, OVPE, and Na-flux methods, have been proposed to solve problems that are difficult to overcome with HVPE methods, such as large diameter, low cost, high quality, and mass production required for substrates in power electronics.

The ammonothermal method, in particular, is a technology that can satisfy all of these issues at a high level, and its development for mass production is now underway. The latest low-pressure acidic ammonothermal technology has succeeded in reducing the pressure to ~100 MPa, about half that of conventional methods, and has demonstrated the reproduction of low dislocation density of 10^3–10^4 cm^{-2}, a characteristic of the ammonothermal method. This advancement has made the mass production of large-diameter GaN substrates in large autoclaves a realistic prospect.

In the mass production process of ammonothermal, which is grown for long periods of time in large equipment, optimization methods using CFD and other numerical simulations, as well as machine learning, are being utilized for the design. By substantially reducing experimental trial and error, the development and start-up of mass production of large-diameter GaN substrates have been accelerated.

References

1. S. Porowski, I. Gregory, J. Cryst. Growth **178**, 174 (1997)
2. K. Naniwae, S. Itoh, H. Amano, K. Itoh, K. Hiramatsu, I. Akasaki, J. Cryst. Growth **99**, 381 (1990)
3. K. Motoki, T. Okahisa, S. Nakahata, N. Matsumoto, H. Kimura, K. Kasai, K. Takemoto, M. Uematsu, Y. Ueno, A. Kumagai, H. Koukitu, Seki, J. Cryst. Growth **237–239**, 912 (2002)
4. Y. Oshima et al., Jpn. J. Appl. Phys. **42**, L1 (2003)
5. T. Sochack, M. Amilusik, B. Lucznik, M. Fijalkowski, J.L. Weyher, G. Nowak, B. Sadovyi, G. Kamler, R. Kucharski, M. Iwinska, I. Gregory, M. Bockowski, J. Cryst. Growth **403**, 32 (2014)
6. Y. Mikawa, T. Ishinabe, Y. Kagamitani, T. Mochizuki, H. Ikeda, K. Iso, T. Takahashi, K. Kubota, Y. Enatsu, Y. Tsukada, S. Izumisawa, Proc. SPIE **11280**, 1128002 (2020)
7. T. Yamane, K. Hanaoka, H. Murakami, Y. Kumagai, A. Koukitu, Phys. Status Solidi C **8**, 1471 (2011)
8. A. Miura, S. Shimada, T. Sekiguchi, J. Cryst. Growth **299**, 22 (2007)
9. P. Konkapaka, B. Raghothamachar, M. Dudley, Y. Makarov, M.G. Spencer, J. Cryst. Growth **289**, 140 (2006)
10. J. Takino et al., Jpn. J. Appl. Phys. **58**, SC1043 (2019)
11. H. Yamane, M. Shimada, S.J. Clarke, F.J. DiSalvo, Chem. Mater. **9**, 413 (1997)
12. M. Aoki, H. Yamane, M. Shimada, S. Sarayama, F.J. DiSalvo, Cryst. Growth Des. **1**, 119 (2001)
13. M. Imade, K. Murakami, D. Matsuo, H. Imabayashi, H. Takazawa, Y. Todoroki, A. Kitamoto, M. Maruyama, M. Yoshimura, Y. Mori, Cryst. Growth Des. **12**, 3799 (2012)
14. M. Imanishi, K. Murakami, H. Imabayashi, H. Takazawa, Y. Todoroki, D. Matsuo, M. Maruyama, M. Imade, M. Yoshimura, Y. Mori, Phys. Status Solidi C **10**, 400 (2013)
15. M. Imanishi, K. Murakami, T. Yamada, K. Kakinouchi, K. Nakamura, T. Kitamura, K. Okumura, M. Yoshimura, Y. Mori, Appl. Phys. Express **12**, 045508 (2019)
16. Toyoda Gosei News, Toyoda Gosei Succeeds in Making Larger GaN Substrates for Next-Generation Power Devices, 15 Mar 2022, https://www.toyoda-gosei.com/news/details.php?id=306
17. K. Yamaguchi, Nikkei Xtech article, 1 Feb 2023, https://xtech.nikkei.com/atcl/nxt/column/18/02277/012700003/?P=2
18. R. Dwiliński, R. Doradzinski, J. Garczynski, L. Sierzputowski, A. Puchalski, Y. Kanbara, K. Yagi, H. Minakuchi, H. Hayashi, J. Cryst. Growth **310**, 3911 (2008)
19. K. Yoshida, K. Aoki, T. Fukuda, J. Cryst. Growth **393**, 93 (2014)
20. D. Gogova, P.P. Petrov, M. Buegler, M.R. Wagner, C. Nenstiel, G. Callsen, M. Schmidbauer, R. Kucharski, M. Zajac, R. Dwilinski, M.R. Phillips, A. Hoffmann, R. Fornari, J. Appl. Phys. **113**, 203515 (2013)
21. Q. Bao, M. Saito, K. Hazu, K. Furusawa, Y. Kagamitani, R. Kayano, D. Tomida, K. Qiao, T. Ishiguro, C. Yokoyama, S.F. Chichibu, Cryst. Growth Des. **13**, 4158 (2013)
22. D. Tomida, Q. Bao, M. Saito, K. Kurimoto, F. Sato, T. Ishiguro, S.F. Chichibu, Appl. Phys. Express **11**, 091002 (2018)

23. K. Kurimoto, Q. Bao, Y. Mikawa, K. Shima, T. Ishiguro, S.F. Chichibu, Appl. Phys. Express **15**, 055504 (2022)
24. S. Schimmel, D. Tomida, T. Ishiguro, Y. Honda, S.F. Chichibu, H. Amano, Crystals **11**(4), 356 (2021)
25. S. Schimmel, D. Tomida, M. Saito, Q. Bao, T. Ishiguro, Y. Honda, S.F. Chichibu, H. Amano, Crystals **11**(3), 254 (2021)

Chapter 4
Near-Equilibrium Ammonothermal (NEAT) Growth of Low-Dislocation GaN for Power Devices

Tadao Hashimoto, Edward Letts, and Daryl Key

1 Introduction

In the early time of GaN research and development, GaN-based devices were developed using heteroepitaxy on sapphire substrates because there was no native substrate of GaN. Despite the fact that heteroepitaxial GaN grown by metal organic chemical vapor deposition (MOCVD) contains about 10^9 cm^{-2} dislocations, highly bright blue/green/white light-emitting diodes (LEDs) were successfully developed in the early 1990s [1]. However, many people believe that the reduction of dislocation is indispensable for further advancing GaN device technology. This was the motivation for developing native GaN substrates for GaN-based devices. From the late 1980s to the late 1990s several groups developed hydride vapor phase epitaxy (HVPE), which can grow GaN at a rate fast enough to obtain a free-standing layer [2]. The HVPE technology-enabled "quasi-bulk," free-standing GaN substrates, manufactured by a thick growth of heteroepitaxial GaN followed by the removal of the original substrates [3]. Concurrently, the first room-temperature, continuous wave lasing from InGaN blue laser diodes was demonstrated using a sapphire substrate with a limited lifetime [4]. Soon, it was shown that the use of free-standing GaN substrates having a dislocation density of 10^6 cm^{-2} extended the lifetime through the mitigation of degradation in the active layer [5].

Recently, research and development of GaN-based power transistors have shifted from a horizontal configuration using heteroepitaxial GaN to a vertical configuration using a GaN-on-GaN structure. GaN transistors fabricated on a GaN substrate are reported to have lower leakage current than those on Si or sapphire [6]. The development of low-dislocation GaN substrates using bulk GaN crystals is indispensable for further advancing GaN power electronics.

T. Hashimoto (✉) · E. Letts · D. Key
SixPoint Materials, Inc., Buellton, CA, USA
e-mail: tadao@spmaterials.com

I. C. Kizilyalli et al. (eds.), *Gallium Nitride and Related Materials*, The Materials Research Society Series, https://doi.org/10.1007/978-3-031-83056-3_4

Nevertheless, due to the lack of a liquid phase under atmospheric pressure, the growth of bulk GaN crystals has been extremely difficult. To achieve the growth of bulk GaN crystals, several nonvapor phase methods including high-pressure solution growth [7], flux growth [8, 9], and ammonothermal growth [10, 11] have been researched. The authors have been developing the ammonothermal GaN growth to replicate the successful application of hydrothermal growth to α-quartz production. The ammonothermal method is promising for the mass production of bulk GaN crystals due to the enormous scalability of the crystal growth system. This chapter will review the current status of bulk GaN crystals by the ammonothermal method at SixPoint Materials, Inc.

2 Ammonothermal Method

A method of growing crystals in a supercritical solvent is generally called solvothermal growth. The high-pressure and high-temperature solvent environment needed for solvothermal growth is realized with a high-pressure reactor called an autoclave. Hydrothermal growth of α-quartz which uses supercritical water is the most successfully commercialized solvothermal growth. Solvothermal growth has a scalability advantage over other crystal growth methods. Currently, hydrothermal quartz autoclave can grow over 1000 crystals in one batch.

Ammonothermal growth is a solvothermal growth using ammonia as a solvent. In general, the growth mechanism of GaN in ammonothermal growth is explained as follows: (1) nutrient such as polycrystalline GaN placed in a nutrient region dissolves in supercritical ammonia under high-pressure (100 ~ 400 MPa) and high-temperature (400 ~ 600 °C); (2) the dissolved GaN precursor is transported by convective flow of supercritical ammonia to a seed region; (3) the solvent is supersaturated in the seed region due to a different temperature condition from the nutrient region; and (4) GaN crystallizes on seed crystals.

Schematic drawings of typical ammonothermal growth setups are shown in Fig. 4.1. The nutrient, seed crystals, and internal components are set in the autoclave and heated by external heaters. The heater is divided into multiple regions to maintain the lower region at a higher temperature and the upper region at a lower temperature. The configuration differs depending on the type of mineralizer because the temperature dependence of GaN solubility varies with the mineralizer used in the supercritical ammonia. In the case of basic mineralizers like sodium and an acidic NH_4F mineralizer, the nutrient is located in the upper region of the autoclave and the seed crystals are located in the lower region of the autoclave as shown in Fig. 4.1a [12, 13]. By contrast, the seed crystals are located in the upper region, and the nutrient is located in the lower region for acidic mineralizers (except for NH_4F) as shown in Fig. 4.1b [14]. The authors' group has been developing ammonothermal GaN growth using a basic mineralizer.

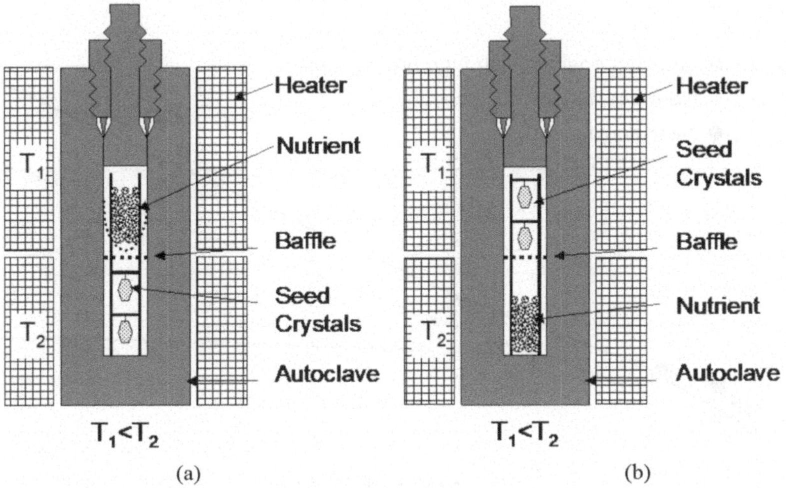

Fig. 4.1 Schematic drawing of ammonothermal GaN growth setup. (**a**) A configuration with retrograde solubility environment using basic mineralizer or NH_4F. (**b**) A configuration with normal solubility environment with acidic mineralizer except for NH_4F

3 Near Equilibrium Ammonothermal (NEAT) Method

Increasing growth rate was the primary focus in the early stages of our development because the typical growth rate of ammonothermal growth was 10 to 100 μm/day. Since the typical growth rate of HVPE is ~100 μm/h, the slow growth rate of the ammonothermal method was regarded as a major disadvantage. However, after a comprehensive growth parameter study using sodium-based mineralizers, we concluded that a faster growth rate is not suited to growing high-quality thick bulk GaN crystals in the ammonothermal method.

To increase the growth rate, it is a common practice to increase the driving force of the growth, i.e., the supersaturation of the solute in the growth region. This results in a higher growth rate on the seed crystals and enhanced spontaneous nucleation on the reactor wall and internal components. As growth proceeds, the polycrystalline GaN deposits on the wall start to consume the majority of the solute delivered to the growth zone. Figure 4.2 shows the total thickness gain of grown GaN crystal as a function of growth duration for three different driving forces. With a high driving force (red dots in Fig. 4.2), the initial growth rate was as high as 300 μm/day; however, the growth practically stopped after 10 days. This was due to the heavy wall deposits under a high driving force. When the driving force was lowered by decreasing the temperature difference between the nutrient zone and the growth zone (green dots and blue dots in Fig. 4.2), wall deposits were greatly reduced, resulting in sustained growth over 90 days (blue dots in Fig. 4.2). To avoid wall deposits, the growth condition was set near the equilibrium condition by decreasing the driving

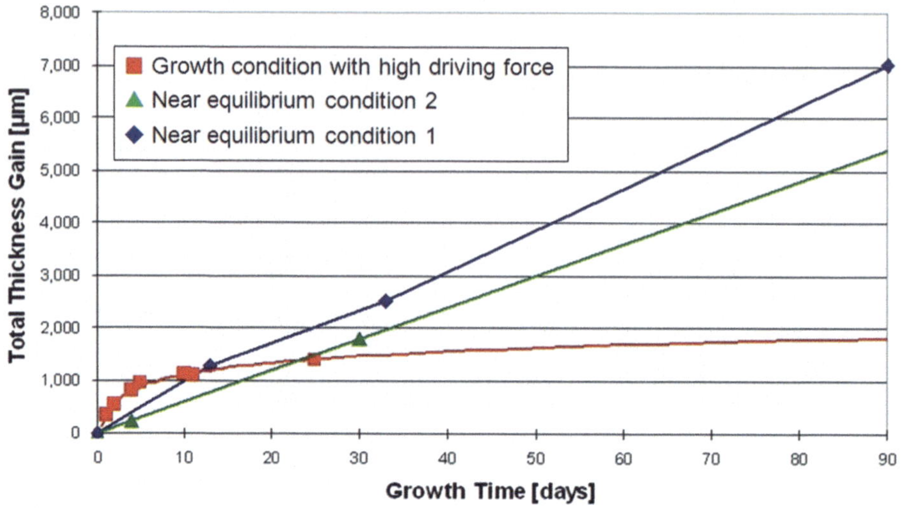

Fig. 4.2 Growth thickness as a function of growth time. A growth condition with high driving force (red dots) results in high growth rate in the early stage of growth, but growth stops in a short duration due to parasitic wall deposits. Growth at near equilibrium conditions (green and blue dots) enabled sustainable growth over a long duration of time

force. The near-equilibrium ammonothermal (NEAT) method is advantageous for growing high-quality thick bulk GaN crystals.

There are a few advantages in the NEAT method compared with conventional ammonothermal growth with a faster growth rate. First, reduced wall deposits enable higher efficiency of nutrient consumption. Since wall deposits are microcrystalline in the shape of hexagonal rods, reusing the wall deposits in successive growth is very challenging. The elimination of wall deposits enables extremely high (>90%) growth efficiency. Second, the microstructure of GaN crystals becomes better than that of seed crystals under the near equilibrium condition. A faster growth rate tends to deteriorate microstructure because the precursor species on the GaN growth surface do not migrate to find the best spot to be incorporated in the crystal. In addition, abnormal growth caused by spontaneous nucleation of GaN on or near the seed crystals is greatly reduced under the near-equilibrium condition. Third, the NEAT method enables consistent extended growth that is preferable in terms of productivity. Frequent opening and closing of the high-pressure reactor are time and labor intensive. Thick growth on a seed crystal enables several substrates to be fabricated from one bulk crystal, reducing the effective seed crystal cost per substrate. The invention of the NEAT method is one of the breakthroughs SixPoint made toward the realization of GaN substrate production by ammonothermal growth.

4 Improvement of Structural Quality

Although the microstructure of GaN was improved by growing in the near-equilibrium condition, the crystal quality often deteriorated due to crack formation. Because of the highly ionic nature of the Ga-N bond, GaN is susceptible to cracking, but highly dislocated GaN did not show high crack susceptibility. However, cracking of the crystal is the most serious challenge in our bulk GaN development. In metallurgy, dislocations in materials stop the propagation of cracks; therefore, it is reasonable that low-dislocation GaN faces more crack problems. Figure 4.3 shows photographs of sliced GaN wafers after bulk growth on various GaN seed crystals. We found that the generation of cracks highly depends on the seed crystals, i.e., the manufacturer of the GaN substrates.

To investigate the correlation between the seed crystal and cracking, X-ray rocking curves were measured at many points along the centerline of seed crystals. Figure 4.4 shows correlations between crack formation and line plots of full-width half maximum (FWHM) of the X-ray rocking curve from 201 diffractions. A high-quality seed (red dots) showed uniform FWHM over the length of the seed, whereas a low-quality seed (green dots) showed scattered FWHM. The resulting bulk GaN crystals grown in the same run showed an obvious difference in crack formation. The bulk GaN grown on a high-quality seed was crack-free, whereas one grown on low-quality seed resulted in a high density of cracks. The results implied that the high density of misaligned in-plane grains is one of the primary culprits of crack formation in bulk GaN growth by the NEAT method.

Fig. 4.3 Photographs of sliced GaN crystals grown on various GaN seed crystals by the NEAT method

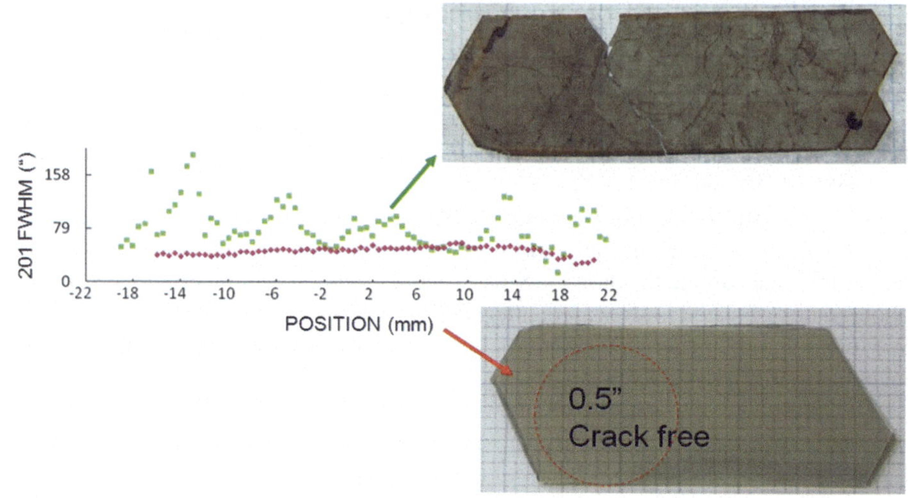

Fig. 4.4 Line plots of FWHM of 201 X-ray rocking curves for a high-quality seed (red dots) and a low-quality seed (green dots) and photographs of resulting bulk GaN crystals grown in the same run on these seeds

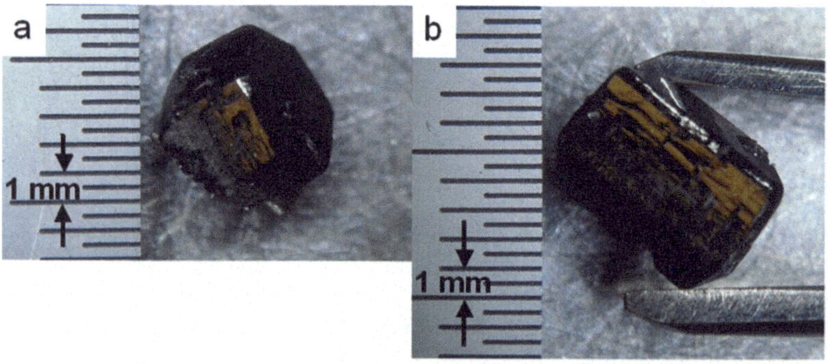

Fig. 4.5 A bulk GaN crystal grown in 2006 by one of the authors at University of California, Santa Barbara

5 Oxygen in NEAT GaN Crystals

In the early time of bulk GaN development by the ammonothermal method, the grown crystals were dark brown. Figure 4.5 shows photographs of the first bulk GaN crystals grown by one of the authors at the University of California Santa Barbara in 2006 [15]. At the time, the development of bulk GaN substrates targeted LEDs as the primary application. Therefore, a reduction of coloration was an urgent task to minimize the absorption of emitted light. Authors have tried many dopants to compensate color centers. It turned out that adding Mg, Ca, or Li reduced colors

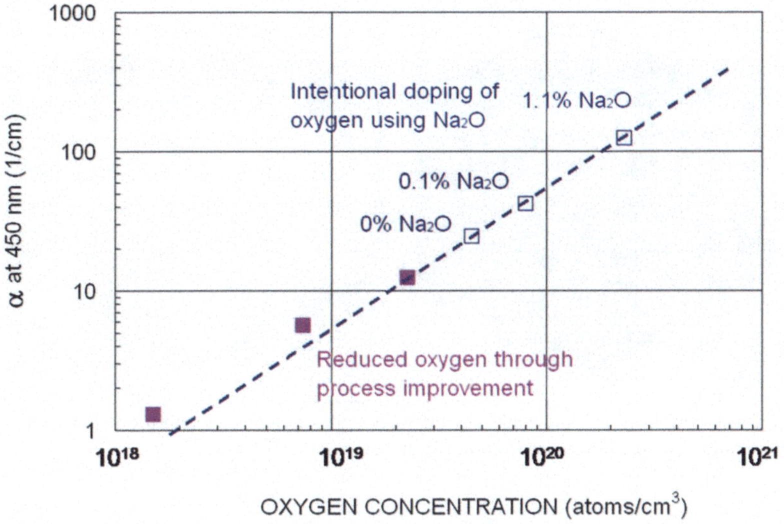

Fig. 4.6 Optical absorption coefficient of GaN at 450 nm as a function of oxygen concentration measured by SIMS

to some extent. After this, intentional doping of oxygen was conducted to confirm that the primary cause of coloration in our material is oxygen.

Figure 4.6 shows the optical absorption coefficient of GaN substrates at 450 nm as a function of oxygen concentration measured by secondary ion mass spectroscopy (SIMS). By adding Na_2O to the growth environment, oxygen was intentionally doped in GaN. The resulting GaN substrates with higher oxygen concentrations showed higher absorption coefficients, confirming that oxygen was the primary cause of the coloration.

By reducing oxygen in the growth environment, GaN crystals with reduced oxygen were also obtained. The optical absorption for the reduced oxygen GaN followed the same trend line as the GaN substrates with intentional oxygen doping as shown in Fig. 4.6. Currently, by taking all possible measures to reduce oxygen and moisture from the growth environment, GaN crystals with an oxygen concentration of low-10^{18} cm^{-3} can be obtained. The minimum optical absorption coefficient at 450 nm is about 1.3 cm^{-1} as shown in Fig. 4.6.

Figure 4.7 shows the comparison of 2″ bulk GaN crystals with an oxygen concentration of mid-10^{19} cm^{-3} (left) and low-10^{18} cm^{-3} (right). This manifests the impact of oxygen concentration on the coloration of the NEAT GaN crystal. Low-10^{18} cm^{-3} oxygen concentration is currently standard at SixPoint.

The free electron concentrations of the NEAT GaN are also affected by the oxygen concentration because oxygen is a donor impurity with an activation energy of 27 meV [16]. The carrier concentrations in NEAT GaN crystals are similar to the oxygen concentration of the crystal. This result agrees with the result of oxygen-doped GaN grown by MOCVD [17]. Since a conductive GaN substrate with carrier concentrations between mid-10^{18} cm^{-3} to high-10^{19} cm^{-3} is preferable for vertical

Fig. 4.7 Photograph of 2″ bulk GaN crystal with high oxygen concentration (left, [O] ~ high-10^{19} cm^{-3}) and low oxygen concentration (right, [O] ~ low-10^{18} cm^{-3})

Fig. 4.8 2″ and 4″ (bottom left) bulk GaN crystals grown in one batch of NEAT growth

power devices, we are not making a strong effort to further reduce oxygen at this moment.

6 Fabrication of 2″ NEAT n-GaN Substrates

Using a 4″ capacity autoclave at SixPoint Materials, Inc., 2″ and 4″ GaN bulk crystals are consistently grown. Figure 4.8 is an example of bulk GaN crystals grown in one batch of NEAT growth. A 2″ wafer prepared from the 2″; bulk GaN crystals is shown in Fig. 4.9.

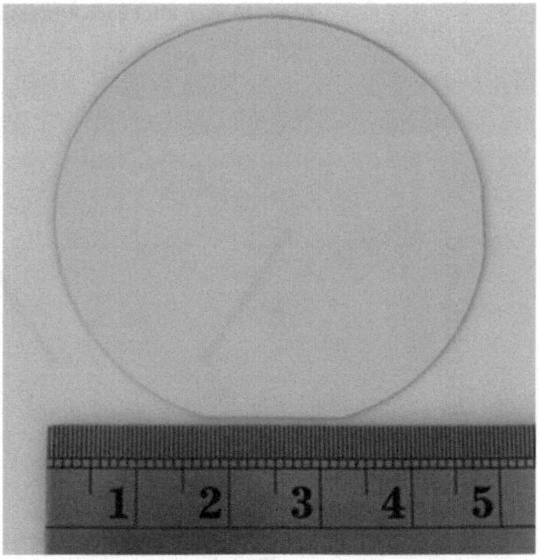

Fig. 4.9 2″ NEAT GaN substrate

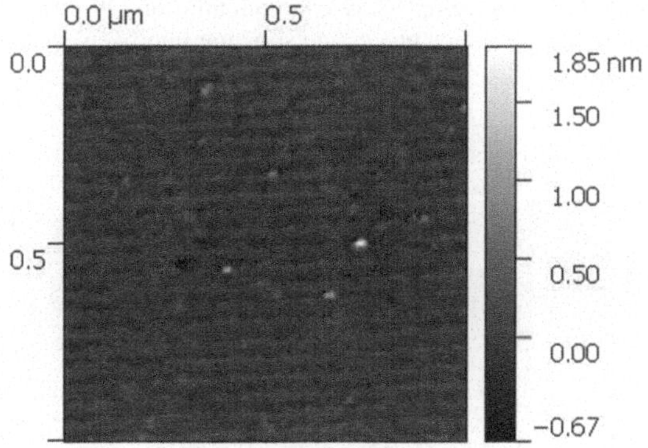

Fig. 4.10 Atomic Force Microscope (AFM) image of the Ga-face of a GaN substrate after CMP

It is well known that mechanical polishing causes subsurface damage in the GaN substrates [18]. Therefore, CMP is the critical finishing step in the fabrication of GaN substrates. After CMP of the Ga surface of the substrate, the surface shows atomic steps as observed with the atomic force microscope (AFM) image in Fig. 4.10. However, the physical smoothness does not guarantee the removal of subsurface damage [19].

To ensure the removal of subsurface damage, we developed an evaluation method for residual subsurface damage using X-ray diffraction. Using a diffraction method is preferable because it is sensitive to atomic distortion in crystals, quick, and nondestructive. To enhance the signal from the surface, a grazing angle X-ray configuration obtained with 114 diffraction peaks for GaN was used. The schematic configuration of 114 diffraction is presented in Fig. 4.11. The incident angle of the X-ray to the c-plane of the GaN surface is about 10.9°. As shown in Fig. 4.11, the shoulder of rocking curves from the GaN surface after each step of surface processing decreases as it proceeds. The full width five thousandth maximum was

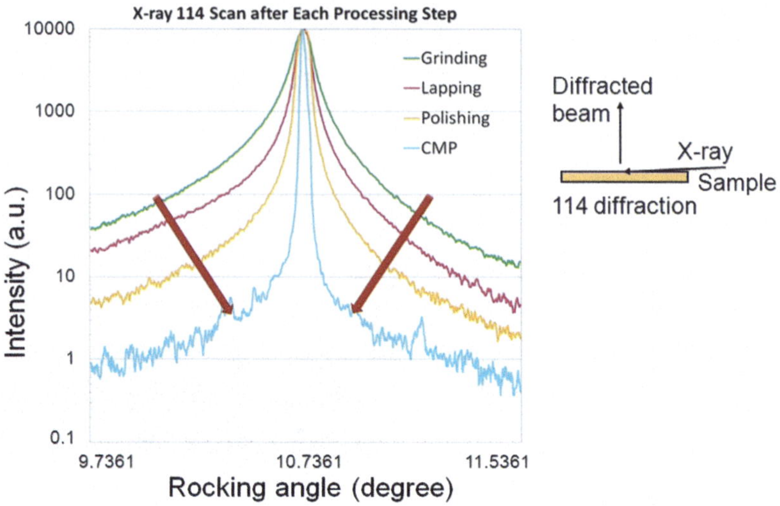

Fig. 4.11 A grazing X-ray configuration for 114 diffraction (right) and 114 rocking curves after each step of surface processing showing improvement in the shoulders of the peak (left)

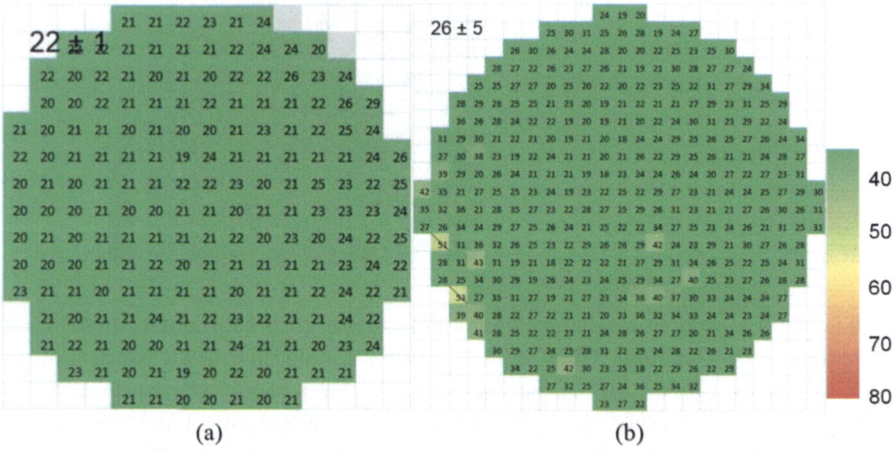

Fig. 4.12 X-ray FWHM map of 2″ NEAT GaN substrate. The number in each square is the FWHM of the rocking curve at that location in arcsec. A diagonal line means there were double peaks in the rocking curve and crossed lines mean there were triple peaks in the rocking curve. (**a**) 002 rocking curve and (**b**) 201 rocking curve

quantified from the rocking curve and compared with the value of as-grown crystals which inherently has no subsurface damage. In this way, the removal of subsurface damage was confirmed without using a time-consuming method like cathodoluminescence (CL). An example of X-ray FWHM maps from 002 and 201 diffractions for a 2″ NEAT n-GaN substrate are shown in Fig. 4.12. The numbers in each cell are the FWHM value at that position. A diagonal line means there were double peaks in the rocking curve and crossed lines indicate there were triple peaks in the rocking curve. The average FWHM from the 002 and 201 rocking curves for this particular substrate were 22 ± 1 and 26 ± 5 arcsec, respectively. The wafer shows uniform properties over the 2″ area. While some dependence on the initial seed exists, x-ray quality is fairly consistent across all samples in a single production growth.

7 Mn Doping for Semi-insulating GaN

Although vertical GaN-on-GaN devices require n-GaN, some applications like high-frequency devices require semi-insulating GaN. In MOCVD or HVPE, Fe or C is typically used for growing semi-insulating GaN [20, 21]. Fe forms an impurity level of about 3.17 eV [22] above the valence band, which is about 0.28 eV from the conductions band. C forms a deep acceptor level of about 0.9 eV [23] above the valence band. On the other hand, Mn forms a mid-gap level of about 1.42 eV above the valence band [24]. A very deep Mn acceptor level is expected to realize a robust semi-insulating nature even at high temperatures.

The initial motivation of developing Mn-doped semi-insulating GaN at SixPoint was to develop a thick semi-insulating GaN crystal for a high-voltage (~100 kV) vertical photoconductive semiconductor switch (PCSS). To ensure semi-insulating nature under high voltage bias, we chose Mn as a dopant. Mn can be doped by using separately synthesized Mn-doped nutrient (GaN) or directly adding metallic Mn in the ammonothermal environment. Both methods can achieve Mn-doped GaN with reasonably uniform Mn concentration along the growth direction.

Figure 4.13 shows a photograph of the Mn-GaN bulk crystal. Mn-doped GaN crystals show a reddish color, and the darkness of the crystal depends on the Mn. The diced and polished cross-section shows uniform coloration indicating uniform Mn concentrations in the crystal (Fig. 4.13b). This particular crystal has a Mn concentration of 1×10^{19} cm^{-3}. Figure 4.14 shows X-ray FWHM maps of the crystal shown in Fig. 4.13. The average FWHM for 002 and 201 rocking curves were 31 ± 5 arcsec

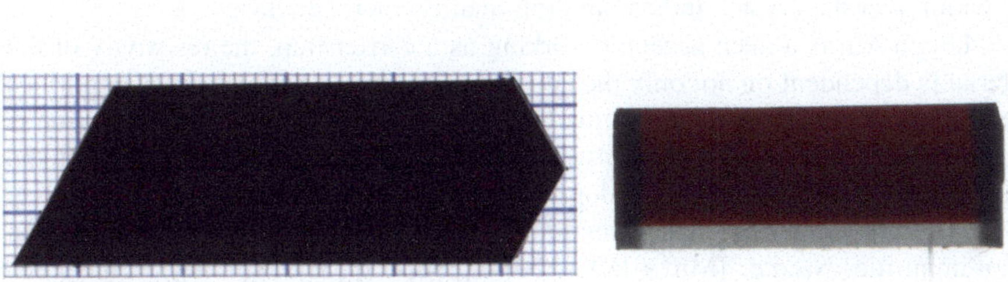

(a) (b)

Fig. 4.13 Photographs of a Mn-doped GaN crystal: (**a**) top view and (**b**) cross-sectional view. The transparent region in (b) is a seed crystal

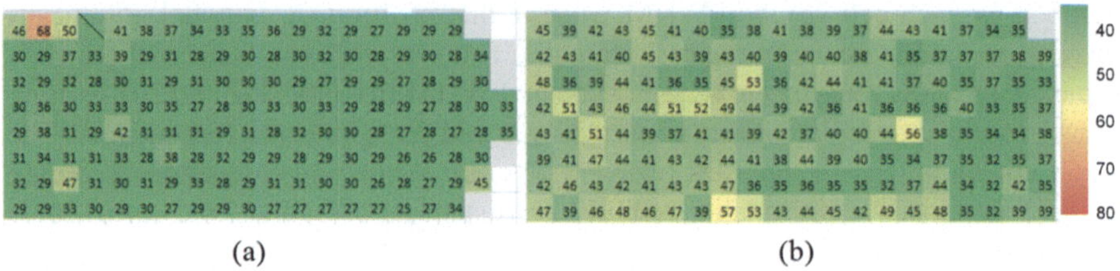

Fig. 4.14 X-ray FWHM map of Mn-doped NEAT GaN crystal. The numbers in each square is the FWHM of the rocking curve at that location in arcsec. A diagonal line means there was double peaks in the rocking curve and crossed lines means there was triple peaks in the rocking curve. (**a**) 002 rocking curve and (**b**) 201 rocking curve

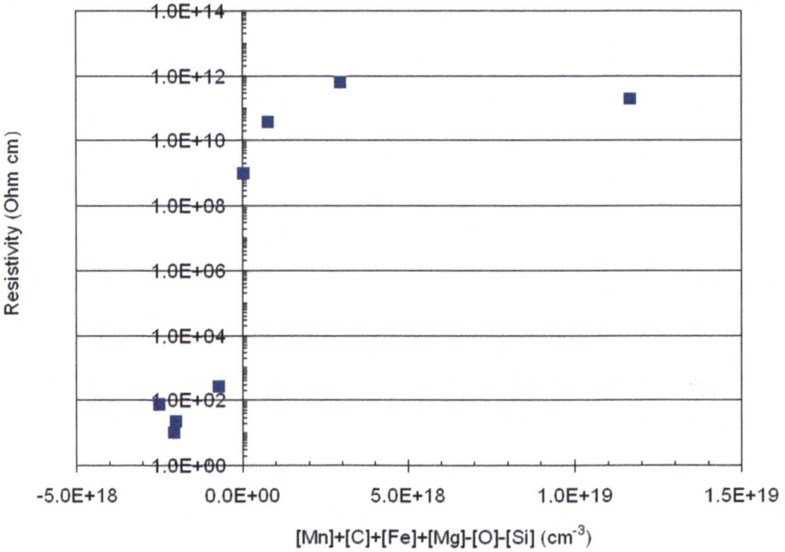

Fig. 4.15 Resistivity of Mn-doped GaN as a function of total effective acceptor impurities

and 41 ± 5 arcsec, respectively. Both the 002 and 201 rocking curves show a narrow line width over the crystal, indicating high-quality microstructure.

Although Mn is a deep acceptor working as a carrier trap, the resistivity of the material is dependent on not only the concentration of Mn but also other impurities like O, Si, and Mg. Particularly, in the NEAT method, the oxygen concentration is fairly high—e.g., typically between low 10^{18} cm^{-3} and high 10^{19} cm^{-3}. If the residual oxygen exceeds the Mn doping level, the material becomes conductive. Figure 4.15 shows resistivity as a function of total acceptor N_A impurities minus total donor impurities N_D, i.e., [Mn] + [C] + [Fe] + [Mg]-[O]-[Si] ($N_A - N_D$). If the value is positive, the material becomes insulating. With [Mn] concentration low-10^{19} cm^{-3} and oxygen concentration low-10^{18} cm^{-3}, the resistivity is typically on the order of $10^{10} \sim 10^{11}$ Ωcm.

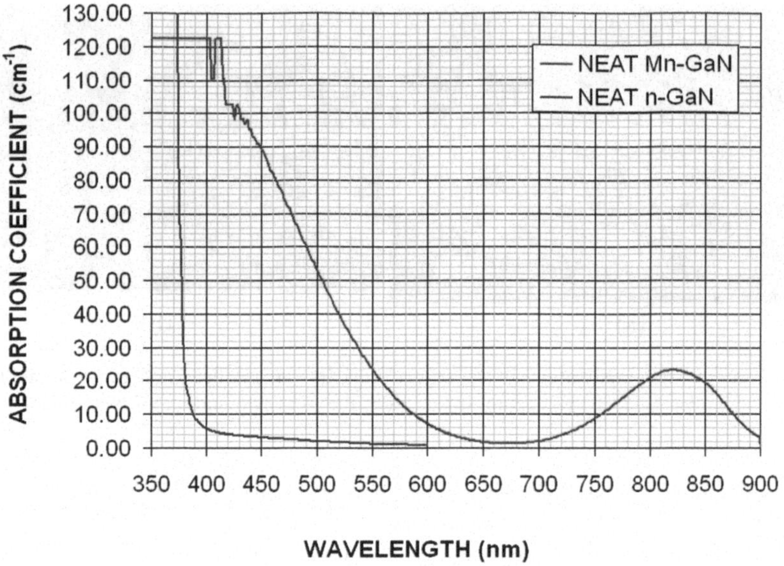

Fig. 4.16 Optical absorption spectra for NEAT Mn-GaN and NEAT n-GaN

Mn in GaN forms a mid-gap level as well as a large band tail, as reported previously [24]. Figure 4.16 shows the optical absorption coefficient of Mn-GaN and n-GaN. The absorption coefficient of Mn-GaN is characterized by a mid-gap peak at about 820 nm and a large band tail. Despite the large band-tail, the Mn-GaN was insulating. Detailed studies of the optical and electrical properties of Mn-GaN is still needed to understand the interaction of Mn-GaN with photons.

8 Toward 4″ GaN Substrate Development

GaN substrates for power devices require a larger wafer size for cost reduction of the devices. NEAT GaN growth using 4″ HVPE seed crystals was conducted to demonstrate 4″ bulk NEAT growth of GaN. Thanks to the excellent scalability of the method, a 4″ GaN crystal was grown together with many 2″ crystals as shown in Fig. 4.8. Figure 4.9 shows an example of a 4″ bulk GaN crystal grown on the 4″ HVPE seed. There is no fundamental difference between 2″ growth and 4″ growth in terms of coloration and microstructures. Due to possible higher strain caused by a slight difference in the lattice constant between the NEAT GaN and HVPE GaN, growth of 4″ bulk GaN may experience higher crack density, although we have not experienced it for a few trials of 4″ GaN growth. The limited availability of 4″ seeds is a bigger limitation than any technical one.

To confirm the reduction of dislocations from the seed crystal, reflection-mode X-ray topography was recorded on the as-grown surface of NEAT-GaN (N-face) and

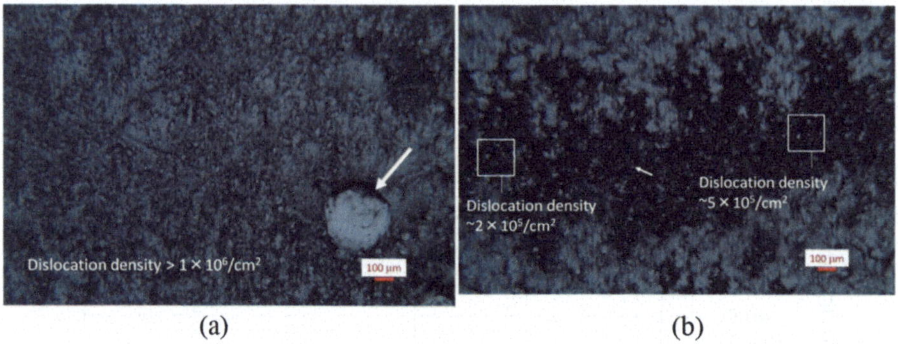

Fig. 4.17 X-ray topography obtained from the 4″ bulk GaN. (**a**) Reflection image from the HVPE seed crystal on Ga-face. (**b**) Reflection image from the as-grown surface of NEAT GaN on the N-face

the original seed surface (Ga-face). There was no growth of GaN on the Ga-face in the NEAT growth because the Ga-face was masked by a metal seed holder. Therefore, the reflection-mode X-ray topograph obtained from the Ga-face provides the defect contrasts of the HVPE GaN seed. Figure 4.17 shows the X-ray topograph from the HVPE GaN surface (Fig. 4.17a) and the NEAT GaN surface (Fig. 4.17b).

The X-ray topograph of HVPE GaN shows a contrast with a high density of dislocations, where multiple white dots are merged together. Considering the size of the contrasts, the dislocation density is greater than 1×10^6 cm^{-2}. In addition, the image in Fig. 4.17a shows a contrast of a physical pit or hole indicated with an arrow. On the other hand, the X-ray topograph obtained from the as-grown NEAT GaN surface showed individual dislocation contrast in the dark band between the white bands caused by lattice bowing. Any pits and holes in the seed disappear as they close in the ammonothermal growth. The dislocation density is in the range of 2–5 $\times 10^5$ cm^{-2}. This demonstrates an order of magnitude reduction of dislocation density in the NEAT GaN crystals. So far, the reduction of dislocation is observed regardless of the crystal diameter. At this moment, the cause of dislocation reduction in the NEAT growth is not well understood. Nevertheless, lower dislocation density is indispensable for applications handling high density of power. NEAT GaN substrates are the best choice to fabricate high-power GaN-on-GaN devices.

9 Concluding Remarks

This chapter covered the development of GaN substrates by the NEAT method. To obtain high-quality GaN crystals in a basic ammonothermal method, GaN crystals are grown in near-equilibrium conditions. The near-equilibrium condition provides sustained growth over a long period of time and reduces dislocation density by about one order of magnitude. Through the reduction of residual oxygen in the growth environment, GaN crystals with an oxygen concentration of about 1.5×10^{18} cm^{-3}

are obtained. The corresponding optical absorption coefficient at 450 nm is about 1.3 cm^{-1}. GaN substrates are fabricated by slicing the bulk GaN crystals with multiple wire saw followed by polishing. After CMP, the removal of subsurface damage is confirmed with grazing angle X-ray diffraction. Semi-insulating GaN can be grown by adding Mn in the NEAT growth environment. Using 4″ HVPE GaN seeds, 4″ NEAT bulk GaN crystals are grown together with numerous 2″ bulk crystals. Regardless of the crystal size, the dislocation density of NEAT GaN crystals is about an order of magnitude lower than that of the HVPE GaN seed crystals. With excellent scalability, the NEAT GaN growth will provide low-cost, large-area, low-dislocation GaN substrates for applications that handle high current density and/or high voltage.

Acknowledgments This work was supported by the Small Business Innovative Research (SBIR) program by the National Science Foundation (IIP-1142356), U.S. Department of Energy (DOE), Advanced Research Program Agency, Energy (ARPA-E), SWITCHES program (DE-AR0000445), ARPA-E OPEN 2018 programs (DE-AR0001008 and DE-AR0001008), U.S. DOE, the Office of Energy Efficiency and Renewable Energy (EERE) under the Advanced Manufacturing Office (AMO), SBIR program (DE-SC0013791), US DOE EERE AMO FY18/FY19 Lab Call (DE-LC-000 L059), the Office of Naval Research (N00014-19-1-2069), and ARPA-E OPEN 2021 program (DE-AR0001562).

The synchrotron radiation experiments were performed at the BL16B2 of Spring-8 with the approval of the Japan Synchrotron Radiation Research Institute (JASRI) (Proposal No.2021B5370). The authors appreciate Prof. Masakazu Kanechika at Nagoya University, and Satoshi Yamaguchi, Kosuke Kitazumi, and Yasuji Kimoto at Toyota Central R&D Labs., Inc. for the measurement of X-ray topography.

References

1. S. Nakamura et al., High-brightness InGaN/AlGaN double-heterostructure blue-green-light-emitting diodes. J. Appl. Phys. **76**, 8189 (1994)
2. K. Naniwae et al., Growth of single crystal GaN substrate using hydride vapor phase epitaxy. J. Cryst. Growth **99**, 381 (1990)
3. K. Motoki et al., Growth and characterization of freestanding GaN substrates. J. Cryst. Growth **237-239**, 912 (2002)
4. S. Nakamura et al., Room-temperature continuous-wave operation of InGaN multi-quantum-well structure laser diodes. Appl. Phys. Lett. **69**, 4056 (1996)
5. S. Nakamura et al., High-power, long-lifetime InGaN/GaN/AlGaN-based laser diodes grown on pure GaN substrate. Jpn. J. Appl. Phys. **37**, L309 (1998)
6. M. Alomari, "GaN HEMTs: The substrate conundrum," Compound Semiconductor Magazine 19th June 2017
7. S. Porowski, J. Cryst. Growth **189/190**, 153 (1998)
8. H. Yamane, M. Shimada, T. Sekiguchi, F.J. DiSalvo, J. Cryst. Growth **186**, 8 (1998)
9. F. Kawamura et al., Jpn. J. Appl. Phys. **42**, L729 (2003)
10. R. Dwilinski, J.M. Baranowski, M. Kaminska, ACTA Phys. Pol. A **90**, 763 (1996)
11. T. Hashimoto, F. Wu, J.S. Speck, S. Nakamura, Jpn. J. Appl. Phys. **46**, L889 (2007)
12. T. Hashimoto et al., Growth of gallium nitride via fluid transport in supercritical ammonia. J. Cryst. Growth **275**, E525 (2005)

13. Q. Bao, Ammonothermal growth of GaN on a self-nucleated GaN seed crystal. J. Cryst. Growth **404**, 168 (2014)
14. Y. Kagamitani et al., Ammonothermal epitaxy of thick GaN film using NH_4Cl mineralizer. Jpn. J. Appl. Phys. **45**, 4018 (2006)
15. T. Hashimoto et al., Growth of bulk GaN crystals by the basic ammonothermal method. Jpn. J. Appl. Phys. **46**, L889 (2007)
16. R.Y. Korotkov et al., Electrical properties of oxygen doped GaN grown by metalorganic vapor phase epitaxy. MRS Internet J. Nitride Semicond. Res. **5**(Suppl1), 301 (2000)
17. M.A. di Forte-Poisson et al., Relationship between physical properties and gas purification in GaN grown by metalorganic vapor phase epitaxy. J. Cryst. Growth **195**, 314 (1998)
18. E. Letts et al., X-ray characterization technique for the assessment of surface damage in GaN wafers. J. Cryst. Growth **501**, 13 (2018)
19. H. Aida et al., Evaluation of subsurface damage in GaN substrate induced by mechanical polishing with diamond abrasives. Appl. Surf. Sci. **292**, 531 (2014)
20. A. Fariza et al., Leakage currents and Fermi-level shifts in GaN layers upon iron and carbon-doping. J. Appl. Phys. **122**, 025704 (2017)
21. J.A. Freitas Jr. et al., Semi-insulating GaN substrates for high-frequency device fabrication. J. Cryst. Growth **310**, 3968 (2008)
22. R. Heitz, Excited states of Fe^{3+} in GaN. Phys. Rev. B **55**, 4382 (1997)
23. J.L. Lyons et al., Effects of carbon on the electrical and optical properties of InN, GaN, and AlN. Phys. Rev. B **89**, 035204 (2014)
24. R.Y. Korotkov, Optical properties of the deep Mn acceptor in GaN:Mn. Appl. Phys. Lett. **80**, 1731 (2002)

Chapter 5
Use of HVPE for Growth of Bulk GaN Wafers and GaN Epilayers for Power Electronics Applications

Jacob H. Leach

1 Introduction

The vapor phase epitaxy of GaN dates back to the late 1960s when Maruska and Tietjen utilized the emerging halide vapor phase epitaxy or "HVPE" technique at RCA to prepare GaN films on sapphire [1]. The HVPE technique had been under development at RCA for several years prior and was successfully utilized to prepare several important compound semiconductors, including gallium/indium arsenide (Ga,In)As, phosphide (Ga,In)P, their alloys, and also gallium antimonide, GaSb [2]. The early GaN crystals grown by HVPE were highly n-type with free electron concentrations above 10^{19} cm^{-3}, with the high values originally being attributed to nitrogen vacancies, but in retrospect, it is more likely that they were due to high unintentional concentrations of oxygen stemming from impurities in the ammonia and HCl used during HVPE growth [3]. GaN was being pursued at that time for its potential as a blue light emitting diode (LED), and several efforts using HVPE were pursued through the mid-1970s to realize pn junctions in order to produce a blue LED using either Zn or Mg as the p-dopant. While blue LEDs were indeed produced during that era at RCA using Zn as the p-dopant [4], the quality of the films and correspondingly the brightness of the diodes were too low to be used for display applications. As such, these efforts along with most GaN-related activities were largely abandoned for the next ~10 years until advancements in GaN crystal quality using the nascent metal-organic chemical vapor deposition (MOCVD) technique were realized. Brighter blue LEDs were demonstrated using MOCVD in the early 1980s [5], renewing interest in GaN and kicking off an explosion in research [6], which culminated in a Nobel prize for the blue LED in 2014. The InGaN-based blue LED was a true revolution in general lighting and revealed the importance of GaN

J. H. Leach (✉)
Kyma Technologies, Raleigh, NC, USA
e-mail: leach@kymatech.com

I. C. Kizilyalli et al. (eds.), *Gallium Nitride and Related Materials*, The Materials Research Society Series, https://doi.org/10.1007/978-3-031-83056-3_5

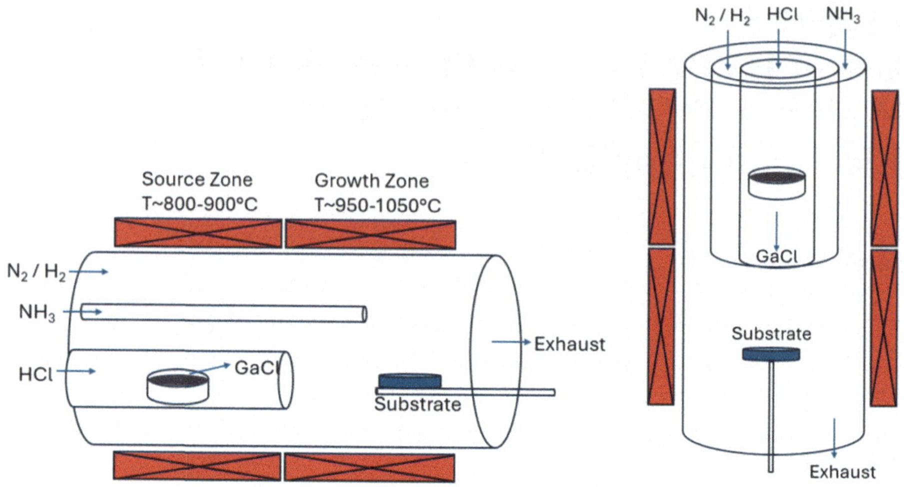

Fig. 5.1 Schematic of HVPE process, vertical and horizontal reactors

and its III-N cousins for a number of optoelectronic and high-power and high-frequency electronic device applications [7].

The HVPE technique for III–V compounds relies on a reaction between a group-III metal halide (typically chlorides are used for III-N growth, e.g., AlCl, $AlCl_3$, GaCl, $GaCl_3$, InCl, or $InCl_3$, although bromides and iodides could in principle be used, particularly for InN and InGaN growth [8]) and a group-V hydride (e.g., NH_3, AsH_3, PH_3, or SbH_3). The group-III compound is typically produced either from a sublimated solid source or more commonly through a pre-reaction within the HVPE system between a pure metal and the halide gas, most commonly HCl for the growth of GaN. Since the group-III precursors are pure metals and not, e.g., metalorganics, the chemical purity of the films can in principle be very high compared to the MOCVD technique when high-purity halides and hydride gases are used. Additionally, the fact that the process takes place near equilibrium and that large fluxes of precursors can be supplied to the wafer surface allows for very high growth rates to be achieved. The general approach for GaN growth is shown schematically in Fig. 5.1 and consists of two primary reactions that take place near thermodynamic equilibrium—the source reaction that produces the GaCl, typically at temperatures between 800 °C and 900 °C:

$$2Ga + 2HCl \leftrightarrow 2GaCl + H_2$$

And the growth reaction that takes place at temperatures between 950 °C and 1050 °C:

$$GaCl + NH_3 \leftrightarrow GaN + HCl + H_2$$

A combination of N_2 and H_2 is typically used as the carrier gases for the reaction precursors. The prevention of parasitic GaN deposition onto the components of the HVPE system and controlling the location of the condensation of the reaction by-product ammonium-chloride (NH_4Cl) are crucially impacted by the reactor design and will impact the number of runs that can be performed, the length of the GaN crystals that can be produced, and the presence of particles and defects in the crystals that are produced using the technique. More details on these practical concerns will be discussed later in the chapter.

Despite not being commonly used as a commercial epitaxial technique for GaN-based devices, recent developments toward using HVPE, along with its variants, as an epitaxial growth technique are challenging the status quo. This notwithstanding, the HVPE technique is quite important to the GaN ecosystem today. Since the HVPE technique remains the dominant technique for producing bulk GaN substrates which are now widely available in 2″ and 4″ diameters, the first portion of this chapter discusses the use of HVPE for bulk GaN wafer manufacturing as well as considerations in the reactor design to accommodate wafer production. Next, efforts to utilize HVPE as an alternative epitaxial technique to the dominant MOCVD method will be discussed. Finally, new advances/modifications to the HVPE technique for the production of both bulk GaN as well as for GaN-based epiwafers will be presented and discussed.

2 HVPE for Bulk GaN Substrates

The lack of a native GaN substrate on which to grow III-N structures has always presented a challenge to researchers seeking to develop GaN-based devices since thermal and lattice mismatches exist when GaN is grown on any of the viable non-native substrates (e.g., silicon, sapphire, SiC, and GaAs). These mismatches give rise to large numbers of extended crystallographic defects in the GaN films, $>10^8$ cm^{-2} depending on substrate and epilayer thickness, and require epigrowers to employ complex buffer layers to attempt to mitigate strain and to reduce the threading dislocation densities (TDD) [9, 10]. It has been a remarkable achievement that GaN-based LEDs and AlGaN/GaN HEMT devices have been able to be commercialized without requiring native GaN substrates. However, more demanding device structures, particularly laser diodes, which require significantly lower defect densities than those that can be realized on non-native substrates [11] and vertical power electronics devices [12] have motivated crystal growers to develop technologies to prepare freestanding GaN wafers.

The primary challenge with preparing bulk GaN crystals is the inability of GaN to be grown by traditional liquid phase crystal growth techniques. Indeed, the temperatures and pressures required to melt the material are expected to be as high as 2200 °

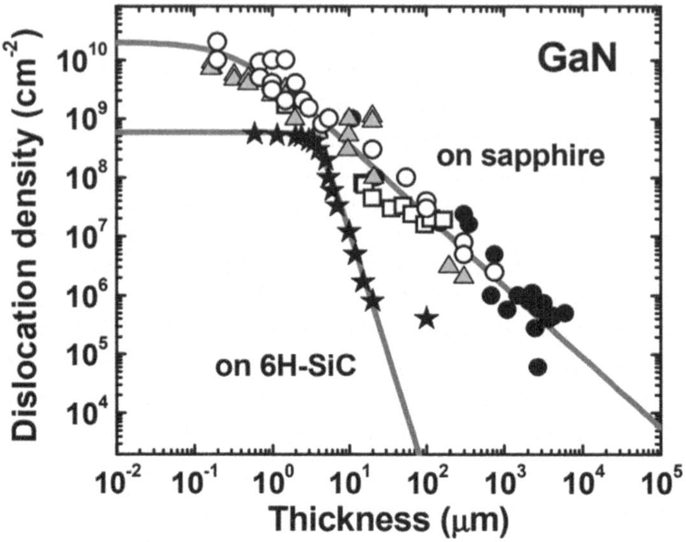

Fig. 5.2 Chart showing the model for dislocation reduction versus GaN thickness and experimental data for HVPE GaN layers. (Figure from Karpov [19] © 2008 Imperial College Press)

C and 6GPa [13], which is obviously an exceedingly difficult technological challenge to overcome. As such, vapor phase (e.g., HVPE) or solution-based techniques (e.g., ammonothermal or Na-flux) have been pursued aggressively over the past several decades and have enjoyed varying degrees of success [14, 15].

HVPE has become the dominant method for the commercial manufacture of freestanding bulk GaN substrates in large part because of its ability to achieve high growth rates (several hundred microns per hour are routine) and high degrees of cleanliness (in terms of unintentional impurities), which allows better control of the doping concentration in the substrates as compared to the solution-based techniques. This is despite the fact that "HVPE GaN substrates" tend to have inferior crystalline quality (in terms of their threading dislocation densities, or TDD) when compared to the so-called "native" or "true-bulk" crystals which are produced from the ammonothermal or Na-flux growth techniques. The defects stem from misfit dislocations generated at the interface of the GaN and the foreign substrate and can be reduced to the ~10^6–10^7 cm^{-2} level with sufficiently thick GaN growth [16]. Such densities of TDs may be acceptable for some devices; however, dislocations with screw components are known sources of leakage in vertical power electronics devices [17], and high densities of dislocations are known to adversely affect the thermal conductivity of the GaN [18].

Figure 5.2 shows the dislocation reduction model prediction [16] as well experimental data of TDD versus layer thickness for HVPE-grown films [19]. It is notable that even when efforts are taken to enhance the dislocation annihilation probability by minimizing the edge types of dislocations as compared to mixed types, the ability to reach the $1 \times 10^3 - 1 \times 10^4$ cm^{-2} level of TDD, which is common for

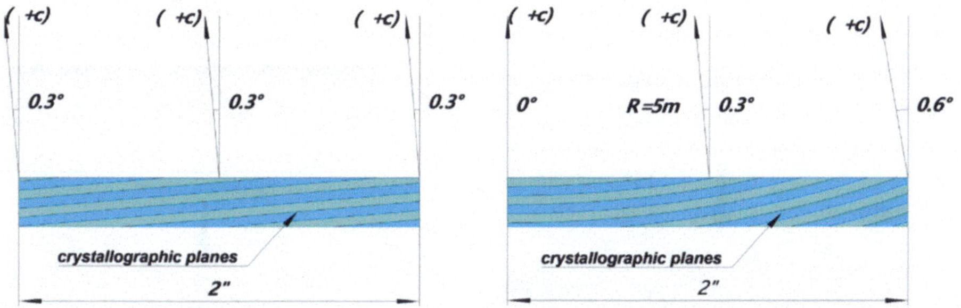

Fig. 5.3 Schematic of GaN crystals with nominal surface miscuts of 0.3° which exhibit (left) high curvature/low lattice lensing or (right) low curvature/high lattice lensing. (Reproduced from Kucharski et al. [15], with the permission of AIP Publishing)

ammonothermal or Na-flux grown crystals, is practically unreachable when growth starts on sapphire substrates. While SiC substrates offer more rapid dislocation reduction with GaN thickness than sapphire, the tensile strain in the GaN causes cracks in the crystals to appear after a few 10 s of microns of growth due to the thermal mismatch between GaN and SiC, also preventing one from reaching the $1 \times 10^3 - 1 \times 10^4$ cm^{-2} level of TDD. Alternatively, the use of lateral epitaxial overgrowth (LEO) or similar approaches can result in crystals with regions of high-quality material (with TDD in the range of 10^5 cm^{-2} or lower) but with isolated regions of lower quality material. In the case of wafers produced using such a pattern, this obviously impacts the size and layout of any devices that are produced thereon.

In addition to challenges with high dislocation densities, the strain-induced grain structure of GaN wafers produced from foreign substrates with different thermal expansion coefficients typically results in "lattice lensing" (or high crystal bending, with radii of curvature on the order of 10 meters or smaller), which manifests as a change in the surface miscut of the substrate across the diameter of the wafer and shown schematically in Fig. 5.3. This lensing effect is worse for more thermally mismatched substrates [20] and results in challenges with subsequent epitaxial growth on these types of substrates (e.g., local variations in dopant incorporation rates [21] or in local variations in the composition of ternary films grown on the substrates [22]). Despite all these challenges, the use of HVPE growth on foreign substrates continues to be the most commonly utilized technique for the production of bulk GaN substrates today.

While some early demonstrations of high structural quality freestanding GaN crystals were demonstrated using the high-pressure high-temperature technique [23], these early crystals were not viable for commercial GaN substrate production due to their small sizes. The first demonstrations of *viable* freestanding GaN wafers (where we have defined "viability" as a process that can produce wafers with diameters of at least 2″) were produced by growing thick films of GaN using HVPE on sapphire substrates, and subsequently removing the GaN from the sapphire after growth using a laser-lift-off (LLO) approach [24]. HVPE-grown GaN wafers produced using LLO

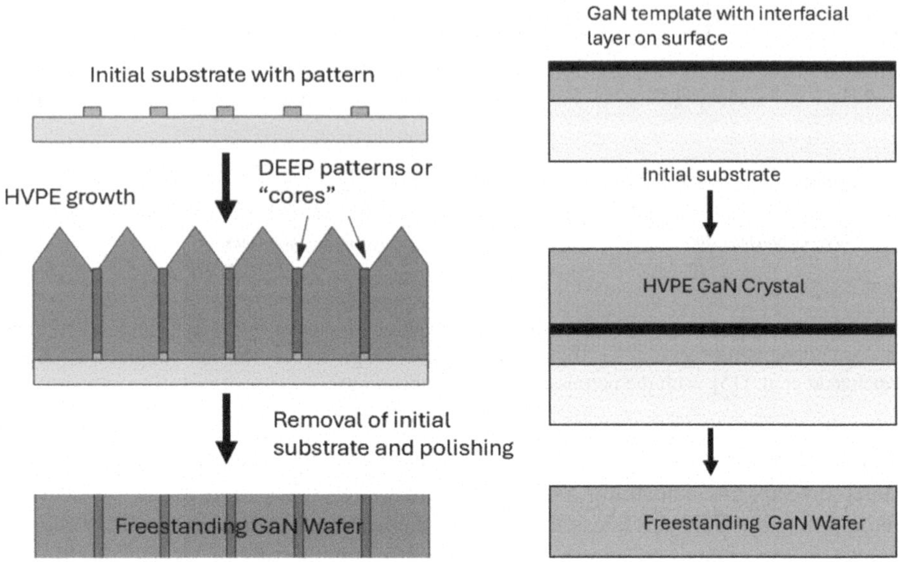

Fig. 5.4 Conceptual schematics for HVPE-grown GaN wafer production processes. The left side shows a patterned approach, specifically the DEEP approach utilized by SEI (modified from Motoki [26]) and the right side shows a general approach using an interfacial layer to promote separation of the GaN crystal from the underlying substrate

or using self-separated crystals allowed early researchers to realize more of the promise of GaN-based lasers [11] and to demonstrate early vertical power devices [12], but the low yields of such approaches, differences in material quality among different producers of wafers, and sustained high costs prevented the mass adoption of bulk GaN wafers for LEDs and HEMTs. Since those early days, developments using HVPE to grow GaN crystals and to produce GaN substrates have continued. The two major approaches for HVPE production of commercially produced GaN wafers today fall into one of two general categories: approaches utilizing patterned substrates on which to grow thick GaN crystals, and nonpatterned approaches that employ one or more various strategies to assist in the process of separating the HVPE-grown GaN crystal from the underlying substrate, but do not consist of patterning per se. This second class of approaches typically results in more uniform wafers than the patterned ones but do not tend to have as low of a TDD as some of the regions of the wafers produced using the patterned approaches. The general concepts are shown schematically in Fig. 5.4.

The most common among the first class of wafers are probably the so-called dislocation elimination by the epitaxial growth with inverse-pyramidal pits or "DEEP Approach" [25] and its later derivatives [26], which were developed by Sumitomo Electric Industries (SEI) and consist of growing on GaAs wafers that have been patterned with regions to prevent growth in order to carry out epitaxial lateral overgrowth (ELO) by HVPE or engineered to have inversion domains (e.g., regions of N-polar GaN in an otherwise Ga-polar material) which tend to accumulate the threading dislocations into small regions. This technique gives rise to a substrate with relatively low dislocation densities over vast regions of the wafer (on the order of 10^4 cm^{-2}) interspersed with small pockets of much lower-quality material (see Fig. 5.4). The GaAs wafer can be chemically removed after growth to yield a freestanding GaN substrate. The second class of approaches utilizes one or more strategies to separate a thick GaN layer from the underlying substrate (typically sapphire) using some sort of predefined weak interlayer or interface to promote the separation of the GaN crystal from the underlying substrate. Approaches of this variety include void-assisted separation (VAS) [27] using a nanonetwork of TiN, the use of gallium droplets to promote separation [28], a "pore-assisted separation" (PAS) technique in which an n + GaN interlayer within the starting template is electrochemically etched to generate pores throughout the layer [29], the use of self-assembled micro- or nano-spheres of silica [30], or thermally induced self-separation [31]. The yields of such approaches vary by technique and by manufacturer, and it is not clear that any one of these particular approaches has been uniformly adopted among wafer producers. Unfortunately, the variation in techniques has also resulted in variations in wafer quality (in terms of densities and types of defects present on the wafer surfaces) among wafer manufacturers [32]. This variability has likely further contributed to the slow adoption of bulk GaN wafers by device manufacturers, which in turn has kept the bulk GaN wafer prices relatively high.

The alternative to producing GaN substrates using a single-wafer method is to start with an existing native GaN wafer or "seed" and use HVPE to replicate the underlying crystal structure and grow the native GaN seed out into a thick boule from which GaN wafers can be subsequently cut. Such a method would work in principle but there are several major challenges to such an approach. The first issue is finding a viable seed wafer on which to grow thick GaN. Wafers that have already been prepared using one of the HVPE techniques outlined above tend to have bent lattices as already pointed out, nonzero values of residual strain due to varying density of defects within the wafers, or suffer from nonuniformities across the wafer surface stemming from patterning processes used during wafer preparation. Thick growth on such substrates is therefore limited to only a couple of millimeters in our experience, likely due to issues with the accumulation of stress on an already strained starting seed. The second issue is related to a more rapid buildup of strain near the edges of the seeds which eventually leads to cracking of the boule. This is a more fundamental issue and occurs even if very high-quality seeds with flat lattices produced using a solution-based method are used as seeds for the HVPE growth. The edge effect issue stems from the fact that during HVPE growth, the physical perimeter of the seed presents a different crystalline orientation onto which epitaxy can take place. Growth on these crystal planes present at the wafer edge (which are different than the surface *c*-plane, {0001}) will typically form semipolar facets in the *m*-plane family at the wafer edges, which tend to incorporate impurities at a much higher rate than GaN grown on the *c*-plane [33]. This gives rise to a region of the crystal, termed the "wing" region of the GaN growth [34], which exhibits a different lattice parameter than the GaN grown within the central region of the seed due to the difference in the impurity incorporation rates, Fig. 5.5. These differences in growth

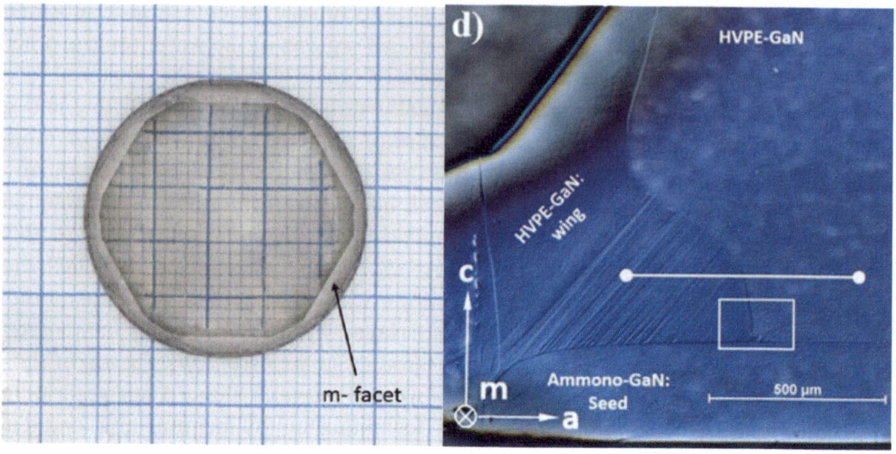

Fig. 5.5 Three-mm thick GaN boule grown by IHPP using HVPE, reproduced from Kucharski et al. [15], with the permission of AIP Publishing. Cross-sectional SEM image of a GaN-on-GaN boule showing the crystal habit with edge faceting. (From Amilusik et al. [37], ©The Japan Society of Applied Physics. Reproduced by permission of IOP Publishing. All rights reserved)

habit and lattice parameter at the edge give rise to strain buildup at the edges of the growing crystal [35], eventually resulting in cracking at the boule edge and a limitation to the ultimate length that one can grow a GaN boule. Despite this, high-quality boules with 3–4 mm thick HVPE layers on ammonothermally grown crystals have been demonstrated by both IHPP and MCC [15, 36], although the ability to reach cm-level boule growth, which would be ideal for driving GaN wafer costs down, appears to be out of reach in light of these edge issues.

Approaches to mitigate the edge effect have been proposed such as using a metal ring to prevent the growth of GaN at the wafer edges [15], the careful profiling of the thermal field around the growing GaN crystal to change the equilibrium crystal habit shape [37], or in Kyma's approach, preparing the initial seeds in a hexagonal shape to present only *a*-planes at the seed edges to suppress the large semipolar facet which forms within the *m*-plane family. Alternatively, if the impurity concentrations of O and Si can be made to be sufficiently low within the HVPE reactor itself, the strain buildup from enhanced impurity incorporation in the wings at the edges of the growing boule could in principle be reduced, allowing thicker boules to be grown before strain-induced cracking occurs.

If one were to simply ignore the edges of the boule and use seeds that are much larger than the eventual wafers being developed, in practice a thick boule could be formed, but developing seed wafers at diameters of 6″ or greater remains challenging. While Sumitomo Electric demonstrated a 6″ wafer in the early days when bulk GaN wafers were beginning to become available [38], the 6″ wafer was never widely available nor widely adopted likely due to either poor yields or high costs. In our experience, even if one utilizes a large-area HVPE reactor, the ability to produce large area wafers (>4″) when growing on non-native seeds is significantly

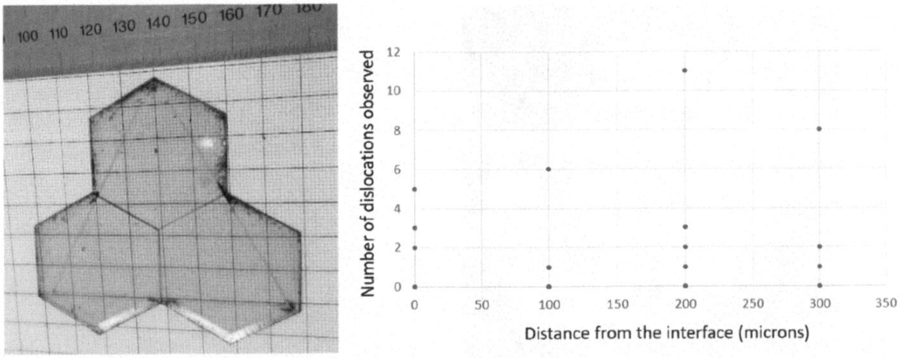

Fig. 5.6 (Left) Image of a trio of tiled GaN wafers from 2-inch seeds grown by Kyma toward the development of large-area GaN substrates. (Right) Numbers of dislocations observed by ECCI at the interface and at distances of 100, 200, and 300 microns from the interfaces

more challenging than producing 2″ or 4″ crystals, likely due to larger off-angle variations across the large wafer area or nonuniformities stemming from the non-native wafers themselves (especially sapphire).

The solution-based techniques appear to be very slow at producing large-diameter bulk GaN seed wafers. Thus, a "tiling" or "mosaic" method has been proposed as one way to quickly realize large area GaN wafers and which could in principle generate large area seeds for subsequent boule growth while mitigating the issue of the large off-angle variation over large areas. In the technique, multiple GaN seeds are prepared which have the same quality, miscut, miscut direction, and thickness, and then are carefully arranged like tiles inside the HVPE reactor and subsequently "stitched" together. The stitching run carried out by HVPE must be conducted initially under conditions that promote lateral growth to ensure that the individual seeds grow together. The resulting crystals could then in principle be processed to make a large area seed wafer for HVPE boule growth. While this approach may seem cumbersome, the general idea was demonstrated in a CVD reactor for the production of single-crystal diamond substrates back in 1994 using *hundreds* of 250-micron-wide cubes [39]. For GaN growth, the technique was utilized to demonstrate a 7″ freestanding GaN substrate [40], and the approach has been replicated using 2″ seed wafers at Kyma. Figure 5.6 shows a photograph of a trio of 2″ seed wafers that were successfully stitched together along with a tally of the numbers of dislocations observed within regions of 26 × 17 μm immediately over the top of and several hundred microns away from the stitch interface using electron channeling contrast imaging (ECCI). The dislocation counts shown in the figure represent densities in the range of $\sim 1 \times 10^5 - 1 \times 10^6$ cm^{-2}, and the lack of an increase in the density of dislocations observed at the stitch interfaces as compared to regions away from the interfaces demonstrates that the stitching technique can be successfully utilized to prepare large-area high-quality seeds from existing high-quality seeds.

Regardless of whether one is producing GaN wafers one at a time or using thick GaN boules that are subsequently sliced, there are several other important

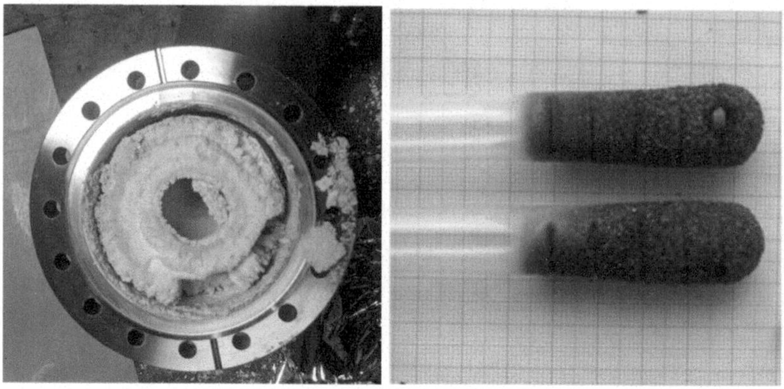

Fig. 5.7 Photographs of (left) buildup of NH_4Cl in the exhaust piping of a GaN HVPE reactor and (right) parasitic GaN buildup on an injector component after long GaN growth. (Image on right from Bockowski et al. [34] ©The Japan Society of Applied Physics. Reproduced by permission of IOP Publishing. All rights reserved)

considerations to be addressed in order to utilize the HVPE technique successfully for bulk GaN growth. The first issue relates to the buildup of NH_4Cl in the reactor's exhaust pathway. This compound is formed in potentially large quantities downstream of the growth zone from the reaction between HCl and NH_3. In HVPE systems that are actively pumped (i.e., under sub-atmospheric growth), this compound must be condensed and collected upstream of the vacuum pump in order to prevent damage to the pumping system. After large amounts are built up, the system will physically clog with NH_4Cl, which must be removed before additional growth can be performed, reducing the productivity of the system (see Fig. 5.7). Systems that are used at atmospheric growth pressures are less sensitive to this issue, assuming that the NH_4Cl can be transported before it condenses inside the HVPE reactor to an external abatement system (a so-called "burn-box") which can pyrolyze the compound.

The second important issue that must be addressed for bulk GaN growth by HVPE relates to the buildup of parasitic GaN within the reactor itself. Parasitic GaN buildup is not only a concern because it represents inefficient conversion of the Ga precursor into productive GaN (i.e., incorporating into the growing wafer or boule) but also because any parasitic buildup that occurs upstream of the growing GaN crystal can become a source of particulate, which will become a defect or a polycrystalline inclusion if it finds its way to the crystal growth front. In the extreme case, the parasitic buildup can also cause reactor failure by clogging the gas injectors, which would in turn require the injectors to be replaced (see Fig. 5.7) [34]. Careful analysis and computational simulations of specific reactor geometries must be performed in order to avoid growth conditions that allow parasitic GaN deposition upstream of the crystal growth front to form [41].

While HVPE systems can grow upwards of 100× faster than solution-based methods, in the long run, it is possible that the HVPE technique would not be the most cost-effective approach for preparing highly n-type GaN substrates since large ammonothermal reactors can potentially process thousands of seeds at once. Taking into account the relatively low growth rate of an individual crystal in the ammonothermal process, the large numbers of boules that could be grown simultaneously in an ammonothermal reactor, and factoring in that an ammonothermal tool may cost 5× – 10× that of an HVPE reactor, researchers at Mitsubishi Chemical suggested that the annual output from their ammonothermal reactor could exceed that of an HVPE by 20–50×, while cost per wafer could reach to 1/5th–1/10th that of HVPE [42]. Eventually, economies of scale, actual system and infrastructure costs, and intellectual property concerns will dictate which technique will win. In either case, the HVPE technique will likely continue to have a place in the world of bulk GaN production as a seed producer to such an ammonothermal growth system, and also since its superior doping control over the ammonothermal technique allows one to produce semi-insulating wafers as well as wafers doped with a particular impurity, for example, RF and mm wave applications, nuclear detector and scintillator applications, as well as other niche applications.

3 HVPE for GaN-Based Epilayers

While HVPE is not typically considered a technique for producing device epilayers within the nitride community, the idea of using HVPE for the growth of heterostructures and full device structures is not new. Early efforts to prepare (In, Ga)As-based LEDs and photodetectors were largely carried out by HVPE, but the high growth rates and the high degree of system "inertia" in HVPE reactors (i.e., the ability to change the chemistry inside the tool so that growth of abrupt interfaces and heterostructures is realized) have meant that more complex device designs are very difficult to realize within HVPE reactors. Having said this, the rise of MOCVD in the 1970s and 1980s and its ability to produce both ternary compounds (particularly Al-containing compounds that are uniquely challenging to carry out in HVPE systems due to a reaction between AlCl and quartz), as well as MOCVD's ability to produce very abrupt interfaces meant that by the 1990s when nitride-based semiconductors were becoming more popular, the use of HVPE for nitride-based devices was largely ignored.

Despite this, there were several early successes in preparing nitride-based devices using HVPE, largely led by the late Vladimir Dmitriev and his colleagues at TDI, including PN diodes [43], UV LEDs [44], and AlGaN/GaN structures [45]. What was unique about these early efforts was the ability of the HVPE reactor to be able to produce relatively complex epilayers with sufficient control of the interfaces to realize functional devices. The challenge of preparing abrupt interfaces in spite of high growth rates and high system inertia was addressed by utilizing a horizonal HVPE growth system in which the substrate was able to be physically pulled from the growth zone for a short period of time while the reactor chemistry was changed, and then reinserted into the growth zone in order to deposit the next layer. The

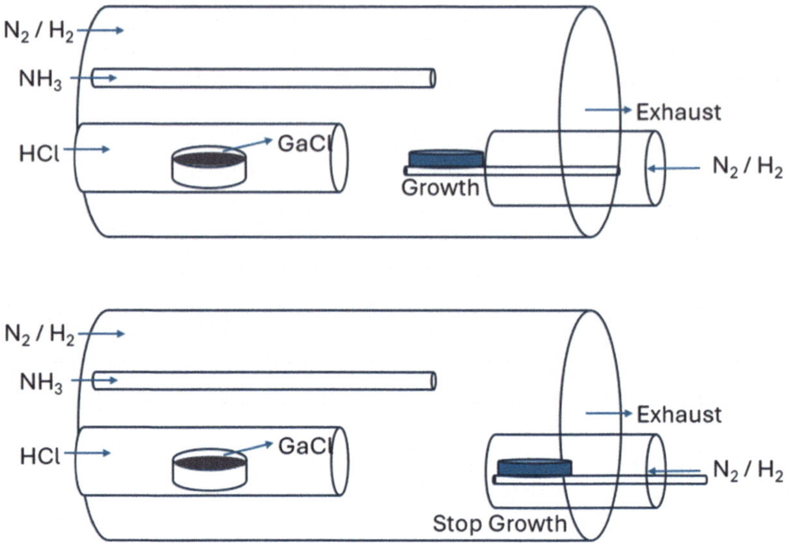

Fig. 5.8 Schematic of an HVPE reactor utilizing a stop-growth region for producing nitride-based epilayers

general idea is shown schematically in Fig. 5.8. Despite these early efforts, the growing use of MOCVD in the commercial production of blue LEDs largely supplanted HVPE from the general discussion around epigrowth for the next decade or so, and so HVPE as it pertained to III-nitrides was largely relegated to the domain of the production of bulk GaN substrates.

The other challenge with utilizing HVPE in GaN-based epilayers stems from the fact that the HVPE reactor itself, which is typically made from quartz, is a source of unintentional donors in GaN, specifically oxygen and silicon [46]. This phenomenon means that the lowest unintended carrier concentrations that are realized in HVPE-grown GaN films are typically around or 1×10^{17} cm^{-3} or higher. Such a background is not an issue for most optoelectronic devices, nor is it a concern for HEMT structures (the buffer layers often being intentionally compensated anyway), but for power electronics applications, drift regions need to have significantly lower doping concentrations in order to support higher voltages. With a background doping concentration of 1×10^{17} cm^{-3}, GaN could only be expected to support about 300 V in the optimum case (Fig. 5.9 shows the calculated free carrier densities and minimum drift layer thicknesses as a function of the desired blocking voltage in GaN), and so HVPE-based drift layers make little sense unless the background O and Si concentrations can be made to be much lower.

While MOCVD-grown GaN layers are able to reach free carrier densities in the low 10^{16} cm^{-3} range, such films suffer from compensation from residual carbon in the films [47], which causes degradation of the mobility of electrons in the lightly doped drift layers [48], as well as premature voltage breakdown [49]. Approaches to reduce the carbon concentrations in the GaN layers tended to result in very low

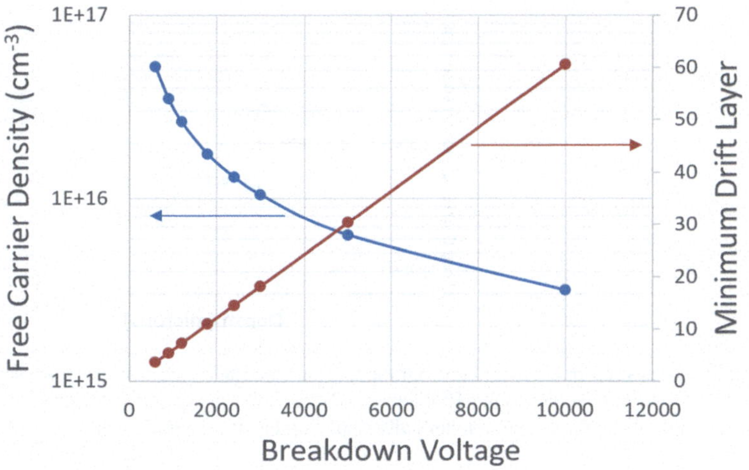

Fig. 5.9 Theoretical ideal breakdown voltage of an abrupt parallel plate junction which can be realized in GaN versus free carrier concentration as well as the minimum epilayer thickness required to realize a given breakdown voltage

growth rates (<2 µm/h), which make the thick epilayer growth times exceedingly long [50]. Furthermore, the thick films required of higher voltage power electronics devices are difficult to reliably grow in MOCVD reactors, which need to be opened up and cleaned after ~50–100 microns of growth to avoid issues with particulate showering onto the films during growth.

For all of these reasons, the HVPE technique, which can be completely free of carbon and whose reactors were often designed with bulk GaN grown in mind (e.g., they can be capable of growing upwards of 1–2 centimeters of GaN before requiring maintenance), has been proposed as a more appropriate technique to serve the epilayer growth needs of high-voltage power electronics devices, if the background doping concentrations could be brought under control [51].

To address the issue of unintended Si and O impurities in HVPE-grown GaN layers, Fujikura et al. [52] removed the quartz components from an HVPE reactor using a technique termed "quartz-free HVPE" or "QF-HVPE" and realized markedly reduced concentrations of Si and O impurities in their films, below the secondary ion mass spectroscopy (SIMS) detection limits. Their results suggested that free electron concentrations could be controlled to levels $<1 \times 10^{15}$ cm^{-3}, with corresponding mobilities >1100 cm^2/V-sec [53]. Coupled with smooth surface morphologies and excellent thickness uniformity, this approach suggested the superiority of HVPE as the ideal growth technique for power device structures aimed at supporting voltages >1 kV.

Kyma's approach to mitigate the background carrier concentrations has been to replace only those quartz components nearest the substrate within the HVPE with materials that do not react with the precursor HCl and NH_3 gasses to form Si-, O-, or C-containing compounds, and to minimize the opportunities for interaction between

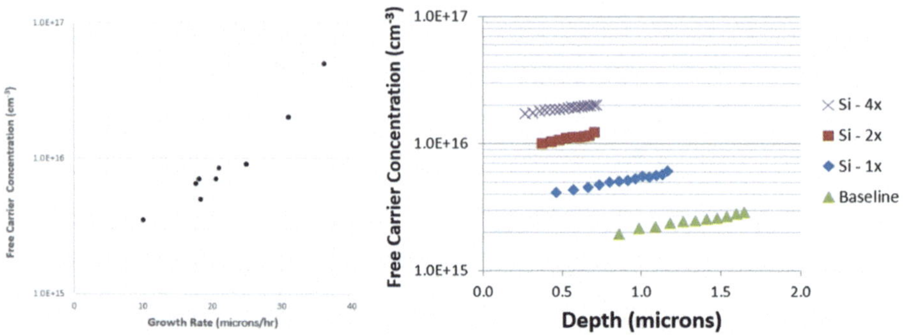

Fig. 5.10 (Left) Measured free carrier concentrations as a function of growth rate for a series of epilayers grown with thicknesses around 10 microns. (Right) Free carrier concentrations measured by CV as a function of depth through intentionally doped epilayers. (From Leach et al. [54])

the remaining quartz components and the HCl and NH_3 precursors by reducing the growth pressure, thereby increasing the gas velocity through the reactor [54]. With this "modified HVPE" approach, the majority of the reactor can still be built from inexpensive quartz components, and sufficiently low background doping concentrations can simultaneously be achieved while maintaining a high growth rate.

As a demonstration of the "modified HVPE" approach, GaN epilayers were grown to a thickness of about 10 μm under a range of V/III ratios from 5 to 17 and at a low growth pressure of 150 torr. Growth rates ranged from 10 to 36 μm/h, and the free carrier concentrations of the resulting films ranged from $\sim 3 \times 10^{15}$ cm^{-3} to 4×10^{16} cm^{-3}, depending on the growth rate (Fig. 5.10). The lowest background carrier concentrations were grown under the lowest growth rates, and the trend between background carrier concentration and growth rate was maintained regardless of the V/III ratio (which was modified by changing both the HCl flow as well as the NH_3 flow). Under moderate growth rates (25 μm/h and lower), the background carrier concentrations were maintained $<1 \times 10^{16}$ cm^{-3}. With the addition of SiH_4 gas into the reactor during GaN growth, the free carrier concentrations could also be controlled from $\sim 3 \times 10^{15}$ cm^{-3} to 2×10^{16} cm^{-3}, Fig. 5.10. The "baseline" data in the figure represent the measured free carrier concentration of a UID GaN layer while each of the other data sets represents GaN grown under various SiH_4 flows with "Si – 4×" and "Si – 2×" representing four times and twice the flow rate as that of "Si – 1×," respectively. The apparent free carrier concentration scales reasonably well with the overall SiH_4 flow rate into the growth space.

Hall measurements of lightly doped or undoped GaN films on insulating Mn-doped bulk GaN substrates grown using the technique were performed both at room temperature as well as at 77 K, and the resulting carrier concentrations and Hall mobilities are given in Fig. 5.11, plotted alongside the room temperature mobilities which were reported for QF-HVPE, as well as MOCVD-grown films. While the modified HVPE technique struggles to realize free carrier concentrations $<1 \times 10^{15}$ cm^{-3}, the mobilities that can be realized are on par with those reported for QF-HVPE, which suggests that either technique would be suitable for the growth

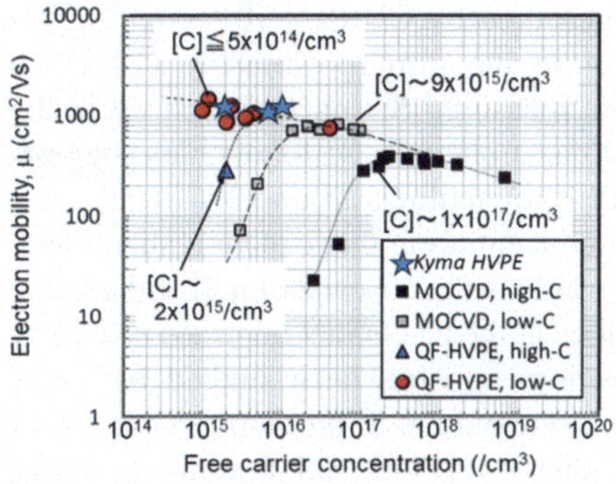

Doping	RT Hall Concentration [carriers cm^{-3}]	RT Hall Mobility [cm^2 V^{-1} s^{-1}]	77 K Hall Concentration [carriers cm^{-3}]	77 K Hall Mobility [cm^2 V^{-1} s^{-1}]
UID	1.8E15	1260	5.4E14	6010
SiH_4-Doped	7.8E15	1070	1.4E15	11 200
SiH_4-Doped	9.5E15	1200	2.3E15	7250

Fig. 5.11 Room temperature Hall mobilities for GaN films grown using the "modified HVPE" (stars) as well as quartz-free HVPE and MOCVD techniques. The table gives the measured RT and 77 K carrier concentrations as well as Hall mobilities for the modified HVPE films. (From Leach et al. [54])

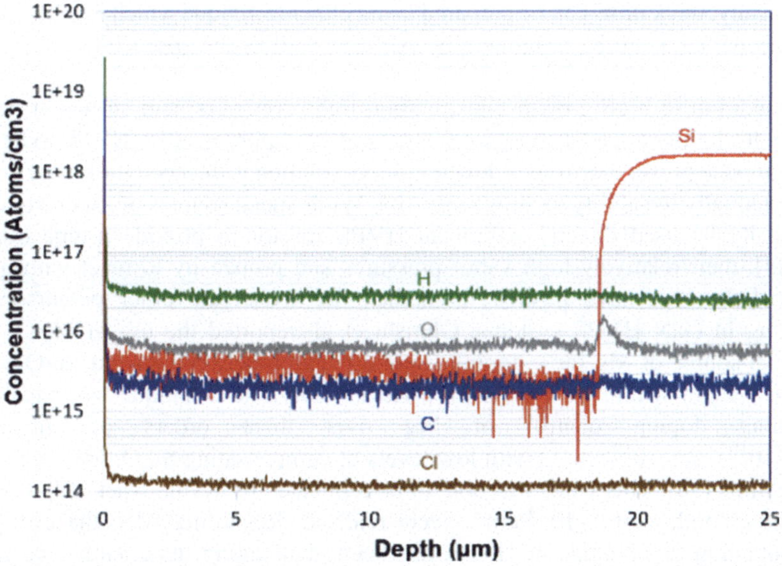

Fig. 5.12 SIMS profile of an intentionally Si-doped GaN-on-GaN film. (From Leach et al. [54])

of GaN films with carrier concentrations over 1×10^{15} cm^{-3}, which correspond to devices with breakdown voltages over 10 kV in the ideal case.

SIMS of an intentionally doped GaN layer grown using the modified HVPE technique was also performed in order to assess the impurity concentrations of Si, O, C, Cl, and H, and the results are given in Fig. 5.12. Concentrations of O, C, and Cl are at the SIMS detection limit (6×10^{15} cm^{-3}, 2×10^{15} cm^{-3}, and 1×10^{14} cm^{-3}, respectively), while the H concentration is slightly above the detection limit (2×10^{15} cm^{-3}), and the Si concentration ($4–5 \times 10^{15}$ cm^{-3}) is slightly lower than the value of the free carrier concentration measured by CV (8×10^{15} cm^{-3}). This implies that an additional donor is present in the film (presumably O) which increases the overall carrier density from the measured Si concentration up to the CV-measured concentration. Assuming full ionization of the Si donors, this suggests that the O concentration in the modified HVPE-grown film is around $3–4 \times 10^{15}$ cm^{-3}, which may represent the minimum carrier concentration that can be realized with the technique. Also notable from the SIMS profile is the absence of a Si spike at the regrowth interface. The presence of a Si "spike" at the GaN/GaN interface is usually detected in GaN homoepitaxy, and the phenomenon has been studied by many groups with no universally accepted solution to avoid the Si buildup reported as of yet [55, 56]. It is believed that the unique chemistry of HVPE and its ability to utilize pregrowth flows of HCl gas at high temperatures is the key to our ability to eliminate this Si spike but further work is needed to clearly understand the mechanisms of its absence in HVPE growth.

It should be noted that for both the "modified HVPE" as well as the "QF-HVPE" techniques, the surface roughness that can be realized is indeed device ready. Specifically, thick films can be prepared using either technique which exhibit smooth surfaces with <1 nm R_a roughness by AFM as well as the absence of any macroscopic texture.

In addition to high growth rates, carbon-free precursors, and sufficient doping control for lightly doped drift regions, another key requirement for HVPE to be used as a complete power electronics device epitaxy solution is the need for p-type doping and potentially the ability to alloy with AlN. While metal sources of Zn or Mg could potentially be used directly within an HVPE reactor to provide p-type dopants [43, 44], their relatively high vapor pressures and propensity to react with quartz (particularly Mg) have pushed researchers to investigate other precursors for p-doping in GaN HVPE systems. Ohnishi et al. reported the use of $MgCl_2$ as a potential source of Mg for p-doping using a reaction between a solid MgO source and HCl vapor within the HVPE [57]. Remarkably, the technique was capable of providing doping control of Mg over three orders of magnitude (~$2 \times 10^{16} – 2 \times 10^{19}$ cm^{-3}) with low levels of compensating Si, O, or C impurities in the films [58]. This work was able to be replicated by Kyma where Mg concentrations as high as 4×10^{20} cm^{-3} were realized. Regarding AlN, the growth of Al-containing compounds in HVPE has been challenging historically due to the propensity for AlCl to react with the quartz in the HVPE reactor, but the use of an external $AlCl_3$ generator has found some success recently for both the growth of III–Vs as well as direct growth of AlN by HVPE [59, 60].

In light of these remarkable results which have been realized by several groups using variations to the traditional HVPE technique, it is clear that HVPE stands as a serious contender to be the growth method of choice for the production of GaN-based power electronics devices, particularly as the voltage blocking values and corresponding drift layer thicknesses are increased.

4 Other Variations on the HVPE Technique for GaN

Tri-halide vapor phase epitaxy (T-HVPE) as the name suggests utilizes a tri-halide species, $GaCl_3$, as a gallium precursor and the general approach dates back to the early 1960s HVPE era when it was utilized for GaAs growth [61]. The $GaCl_3$ precursor has a high vapor pressure, which allows one to use it directly as a gallium source without the need for Cl_2 or HCl, but the material is highly hygroscopic which leads to contamination concerns for GaN growth, particularly O contamination. Accordingly, the technique is not viable using solid $GaCl_3$ sources as precursors. Despite this, the technique became interesting for bulk GaN growth again after Hirasaki et al. [62] suggested the use of a two-stage source zone to produce high-purity $GaCl_3$:

$$\text{Zone 1}: 2Ga + Cl_2 \rightarrow 2GaCl$$

$$\text{Zone 2}: GaCl + Cl_2 \rightarrow GaCl_3$$

While the group was initially focused on growing InGaN using the technique, they subsequently applied it to the growth of thick GaN [63]. The benefit of the technique is a higher growth rate than that which is realized in standard HVPE, its ability to grow at higher temperatures than traditional HVPE which can result in higher GaN crystal quality, and its propensity to grow in the N-polar direction can give rise to expansion of the overall crystal diameter when applied as a boule growth technique to N-polar GaN seeds.

Halide-free vapor phase epitaxy (HF-VPE) is another alternative technique [64] to the general HVPE approach, wherein gallium is directly vaporized from the source and reacts with ammonia:

$$2Ga + 2NH_3 \rightarrow 2GaN + 3H_2$$

The benefits of this approach are that the absence of the halide source makes the technique safer than traditional HVPE, and additionally there are no NH_4Cl by-products generated which makes the exhaust system more manageable. This in principle could make the technique easier to accommodate very long growth runs. The downside of the approach is that the technique tends to have low growth rates unless the gallium source is pushed to relatively high temperatures (up to 1500 °C) and is physically close to the growth region, both of which complicate the overall reactor design.

Oxide vapor phase epitaxy (OVPE) is another variant of the HVPE process, wherein the metal halide precursor is replaced with a gallium suboxide as the group-III source (Ga_2O) [65]. The benefit of this technique is twofold: First, like the HF-VPE, the technique is safer than traditional HVPE and there are also no NH_4Cl byproducts generated, but additionally much higher growth rates can be realized than in the HF-VPE process. The downside of this technique is of course the large concentration of O which invariably incorporates into the growing GaN

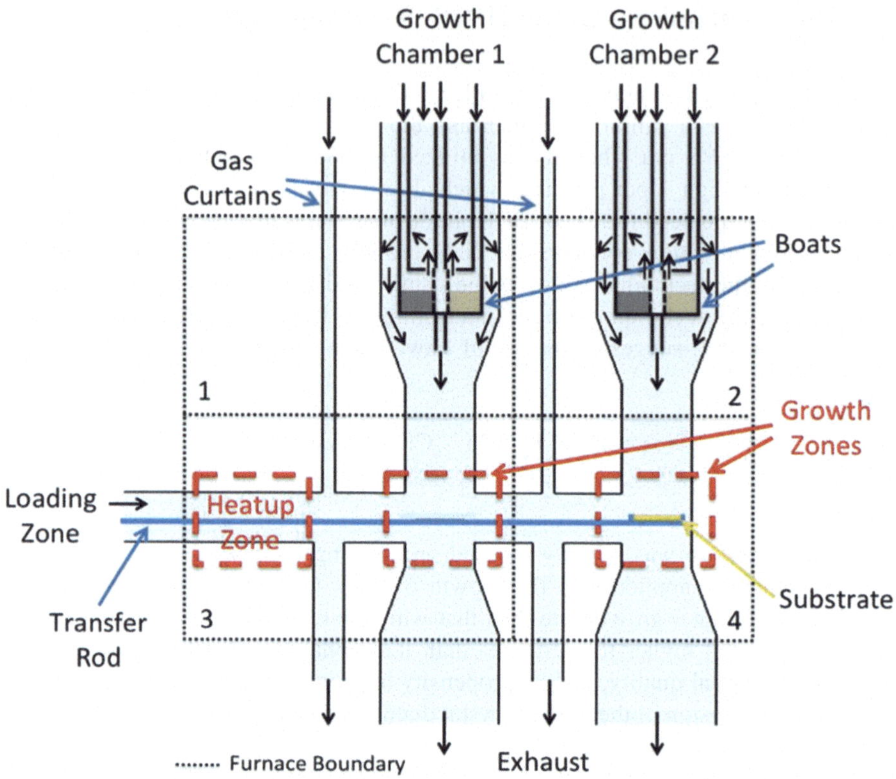

Fig. 5.13 Schematic of the D-HVPE reactor, shown with only two growth chambers. (From Schulte et al. [67])

crystal, making the technique only suitable for the growth of heavily n-type material, and issues related to the formation of polycrystalline Ga_2O_3 during the growth which can degrade the overall GaN crystal quality using this technique [66].

While the approaches mentioned above could primarily impact the use of HVPE for the growth of bulk GaN substrates, another technique called dynamic HVPE or D-HVPE has the potential to improve the ability of HVPE reactors to produce full epilayers at low costs. The technique is somewhat similar in spirit to the "stop growth" reactor mentioned above (Fig. 5.8), but instead of simply stopping growth while the reactor chemistry is changed, the technique relies on a number of HVPE chambers that are physically connected together but chemically independent from one another, and the substrate is shuffled from chamber to chamber to deposit an arbitrary series of layers [67]. A schematic of a simple two-chamber system is given in Fig. 5.13. Key to this technique is the use of gas curtains to prevent cross-contamination from one chamber to the next, and engineering challenges with physically moving hot wafers from one chamber to another. Despite this, the D-HVPE technique has been recently demonstrated to produce high-quality solar cells [68] with remarkably high-throughput and low-cost potential [69]. A pilot production system was recently installed at NREL to demonstrate the applicability of the approach to realizing low-cost III–V devices [70], and while a D-HVPE system that is designed for III-nitride growth has not yet to our knowledge been demonstrated, the concepts in the reactor design are applicable to the use of the technique for III-N devices as well.

5 Concluding Remarks

The halide vapor phase epitaxy of GaN has a long history and a bright future. The technique continues to dominate the commercial GaN substrate market despite inroads being made using solution-based techniques that can produce wafers having remarkably lower concentrations of TDs. Having said this, these types of dislocations are not necessarily "killer" defects in power devices per se, but of course can be problematic for leakage current, premature breakdown, etc. While it is true that the vast majority of GaN substrates that were prepared using the HVPE technique are nucleated on foreign substrates and have correspondingly large TDDs, the HVPE technique can be applied to native GaN crystals themselves, replicating their underlying crystal structure. Thus, if high-quality seeds are made available (e.g., from a solution-based growth approach), and the edge issues discussed in this chapter can be resolved, the technique will continue to play an important role in the production of bulk GaN substrates.

In addition to serving as a bulk GaN crystal growth technique, the HVPE technique is gaining new interest as a solution for overall device epitaxy, particularly for thick drift layers needed in high-voltage power electronics devices. This is due to the high growth rates and chemical purity that can be achieved, particularly when the QF-HVPE or "modified HVPE" techniques are utilized.

References

1. H.P. Maruska, J.J. Tietjen, The preparation and properties of vapor-deposited single crystalline GaN. Appl. Phys. Lett. **15**, 327–329 (1969)
2. J.J. Tietjen, J.A. Amick, The preparation and properties of vapor-deposited epitaxial GaAs1-xPx using arsine and phosphine. J. Electrochem. Soc. **113**, 724 (1966)
3. H.P. Maruska, W.C. Rhines, A modern perspective on the history of semiconductor nitride blue light sources. Solid State Elec. **111**, 32–41 (2015)
4. J.I. Pankove, E.A. Miller, J.E. Berkeyheiser, GaN blue light-emitting diodes. J. Luminescence **5**, 84–86 (1972)
5. T. Kawabata, T. Matsuda, S. Koike, GaN blue light emitting diodes prepared by metalorganic chemical vapor deposition. J. Appl. Phys. **56**, 2367–2368 (1984)
6. S. Strite, H. Morkoç, GaN, AlN, and InN: A review. J. Vac. Sci. Technol. B **10**, 1237–1266 (1992)

7. M. Meneghini, C. De Santi, I. Abid, M. Buffolo, M. Cioni, R.A. Khadar, L. Nela, N. Zagni, A. Chini, F. Medjdoub, G. Meneghesso, G. Verzellesi, E. Zanoni, E. Matioli, GaN-based power devices: Physics, reliability, and perspectives. J. Appl. Phys. **130**, 181101 (2021)
8. T. Hirasaki, K. Hanaoka, H. Murakami, Y. Kumagai, A. Koukitu, Thermodynamic analysis of InGaN HVPE growth using group-III chlorides, bormides, and iodides. Phys. Status Solidi C **210**, 413–416 (2013)
9. I. Akasaki, H. Amano, Y. Koide, K. Hiramatsu, N. Sawaki, Effects of AlN buffer layer on crystallographic structure and on electrical and optical properties of GaN and $Ga_{1-x}Al_xN$ ($0<x\leq0.4$) films grown on sapphire substrate by MOVPE. J. Crys. Growth **98**, 209–219 (1989)
10. S.A. Nikishin, N.N. Faleev, V.G. Antipov, S. Francoeur, L. Grave De Peralta, G.A. Seryogin, H. Temkin, T.I. Prokoyeva, M. Holtz, S.N.G. Chu, High quality GaN grown on Si(111) by gas source molecular beam epitaxy with ammonia. Appl. Phys. Lett. **75**, 2073–2075 (1999)
11. S.-I. Nagahama, N. Iwasa, M. Senoh, T. Matsushita, Y. Sugimoto, H. Kiyoku, T. Kozaki, M. Sano, H. Matsumura, H. Umemoto, K. Chocho, T. Mukai, High-power and long-lifetime InGaN multi-quantum well laser diodes grown on low-dislocation-density GaN substrates. J. J. Appl. Phys. **39**, L647 (2000)
12. A.P. Zhang, J.W. Johnson, B. Luo, F. Ren, S.J. Pearton, S.S. Park, Y.J. Park, J.-I. Chyi, Vertical and lateral GaN rectifiers on free-standing GaN substrates. Appl. Phys. Lett. **79**, 1555 (2001)
13. W. Utsumi, H. Saitoh, H. Kaneko, T. Watanuki, K. Aoki, O. Shimomura, Congruent melting of gallium nitride at 6GPa and its application to single crystal growth. Nat. Mater. **2**, 735–738 (2003)
14. T. Paskova, D.A. Hanser, K.R. Evans, GaN substrates for III-Nitride devices. Proc. IEEE **98**, 1324 (2010)
15. R. Kucharski, T. Sochacki, B. Lucznik, M. Bockowski, Growth of bulk GaN crystals. J. Appl. Phys. **128**, 050902 (2020)
16. S.K. Mathis, A.E. Romanov, L.F. Chen, G.E. Beltz, W. Pompe, J.S. Speck, Modeling of threading dislocation reduction in growing GaN layers. J. Cryst. Growth **231**, 371–390 (2001)
17. J.W.P. Hsu, M.J. Manfra, D.V. Lang, S. Richter, S.N.G. Chu, A.M. Sergent, R.N. Kleiman, L.N. Pfeiffer, R.J. Molnar, Inhomogeneous spatial distribution of reverse bias leakage in GaN Schottky diodes. Appl. Phys. Lett. **78**, 1685 (2001)
18. C. Mion, J.F. Muth, E.A. Preble, D. Hanser, Accurate dependence of gallium nitride thermal conductivity on dislocation density. Appl. Phys. Lett. **89**, 092123 (2006)
19. S.Y. Karpov, Bandgap engineering of III-nitride devices on low-defect substrates, in *III-Nitride Devices and Nanoengineering*, ed. by Z.C. Feng, (Imperial College Press, Ch.13, 2008), pp. 367–398
20. J.H. Leach, Y. Shishkin, K. Udwary, E.A. Preble, K.R. Evans, Large-area bow-free n^+ GaN templates by HVPE for LEDs. Proc. SPIE **8986**, 898604 (2014)
21. T. Suski, G. Staszczak, S. Grzanka, R. Czernecki, E. Litwin-Staszewska, R. Pitrzkowski, L.H. Dmowski, A. Khachapuridze, M. Krysko, P. Perlin, I. Grzegory, Hole carrier concentration and photoluminescence in magnesium doped InGaN and GaN grown on sapphire and GaN misoriented surfaces. J. Appl. Phys. **108**, 023516 (2010)
22. J.A. Grenko, C.L. Reynolds Jr., D.W. Barlage, M.A.L. Johnson, S.E. Lappi, C.W. Ebert, E.A. Preble, T. Paskova, K.R. Evans, Physical properties of AlGaN/GaN heterostructures grown on vicinal substrates. J. Electon. Mat. **39**, 504 (2010)
23. I. Grzegory, M. Bockowski, B. Lucznik, S. Krukowski, M. Wroblewski, S. Porowski, Recent results in the crystal growth of GaN at high N_2 pressure. MRS Internet J. Semicond. Res. **1**, 20 (1996)
24. S.S. Park, I.-W. Park, S.H. Choh, Free-standing GaN substrates by hydride vapor phase epitaxy. Jpn. J. Appl. Phys. **39**, L1141 (2000)
25. K. Motoki, T. Okahisa, S. Nakahata, N. Matsumoto, H. Kimura, H. Kasai, K. Takemoto, K. Uematsu, M. Ueno, Y. Kumagai, A. Koukitu, H. Seki, Growth and characterization of freestanding GaN substrates. J. Cryst. Growth **237-239**, 912–921 (2002)
26. K. Motoki, Development of gallium nitride substrates. SEI Tech. Rev. **70**, 28 (2010)

27. H. Fujikura, T. Inoue, T. Kitamura, T. Konno, T. Suzuki, T. Fujimoto, T. Yoshida, M. Shibata, T. Saito, "Development of GaN single-crystal substrates," SUMITOMO KAGAKU 2018, Report 5. https://www.sumitomo-chem.co.jp/english/rd/report/files/docs/05_2018e.pdf
28. Y.J. Choi, H.-K. Oh, J.-G. Kim, H.-H. Hwang, H.-Y. Lee, W.-J. Lee, B.-C. Shin, J. Hwang, Reducing the warpage and dislocation density of GaN template grown by HVPE with gallium droplet treatment. Phys. Status Solidi C **7**, 1770–1774 (2010)
29. M. Yokoyama, F. Horikiri, H. Mori, T. Konno, H. Fujikura, Fabrication of free-standing GaN substrates using electrochemically formed porous separation layers. Appl. Phys. Express **17**, 055502 (2024)
30. Q. Li, J.J. Figiel, G.T. Wang, Dislocation density reduction in GaN by dislocation filtering through a self-assembled monolayer of silica microspheres. Appl. Phys. Lett. **94**, 231105 (2009)
31. K. Yamane, M. Ueno, H. Furuya, N. Okada, K. Tadatomo, Successful natural stress-induced separation of hydride vapor phase epitaxy-grown GaN layers on sapphire substrates. J. Cryst. Growth **358**, 1–4 (2012)
32. J.K. Hite, T.J. Anderson, L.E. Luna, J.C. Gallagher, M.A. Mastro, J.A. Freitas, C.R. Eddy Jr., Influence of HVPE substrates on homoepitaxy of GaN grown by MOCVD. J. Cryst. Growth **498**, 352–356 (2018)
33. M. Krysko, M. Sarzynski, J. Domagała, I. Grzegory, B. Łucznik, G. Kamler, S. Porowski, M. Leszczyński, The influence of lattice parameter variation on microstructure of GaN single crystals. J. Alloys Comp **401**, 261 (2005)
34. M. Bockowski, M. Iwinska, M. Amilusik, M. Fijalkowski, B. Lucznik, T. Sochacki, Challenges and future perspectives in HVPE-GaN growth on ammonothermal seeds. Semicond. Sci. Technol. **31**, 093005 (2016)
35. J.Z. Domagala, J. Smalc-Koziorowska, M. Iwinska, T. Sochacki, M. Amilusik, B. Lucznik, M. Fijalkowski, G. Kamler, I. Grzegory, R. Kucharski, M. Zajac, M. Bockowski, Influence of edge-grown HVPE GaN on the structural quality of c-plane oriented HVPE-GaN grown on ammonothermal GaN substrates. J. Cryst. Growth **456**, 80–85 (2016)
36. K. Iso, H. Ikeda, T. Mochizuki, Y. Mikawa, S. Izumisawa, High quality GaN crystal grown by hydride vapor-phase epitaxy on SCAAT™. Appl. Phys. Exp. **13**, 085508 (2020)
37. M. Amilusik, D. Wlodarczyk, A. Suchocki, M. Bockowski, Micro-raman studies of strain in bulk GaN grown by hydride vapor phase epitaxy on ammonothermal GaN seeds. J. J. Appl. Phys **58**, SCCB32 (2019)
38. https://www.eetimes.com/sumitomo-claims-first-6-in-gan-wafers/
39. M.W. Geis, N.N. Efremow, R. Susalka, J.C. Twichell, K.A. Snail, C. Spiro, B. Sweeting, S. Holly, Mosaic diamond substrates approaching single-crystal quality using cube-shaped diamond seeds. Diamond Rel. Mater. **4**, 76–82 (1994)
40. T. Yoshida, M. Imanishi, T. Kitamura, K. Otaka, M. Imade, M. Shibata, Y. Mori, Development of a GaN substrate with a large diameter and small orientation deviation. Phys. Status Solidi B **254**, 1600671 (2017)
41. M. Zenk, G. Lukin, D. Bastin, R. Doradziński, F.C. Beyer, E. Meissner, J. Friedrich, Numerical analysis of gas flow instabilities in simplified vertical HVPE GaN reactors. Crystals **12**, 1248 (2022)
42. Y. Mikawa, Y. Kagamitani, T. Ishinabe, H. Ikeda, T. Mochizuki, S. Izumisawa, T. Ohuchi, K. Kurimoto, Q. Bao, Toward mass production of GaN substrates by acidic ammonothermal technology, in *ICNS-14, 14th Conf. Nitride Semi., GR12-5*, 16 November 2023, Fukuoka, Japan
43. A. Nikolaev, Y. Melnik, N. Kuznetsov, A.M. Strelchuk, A.P. Kovarsky, K.V. Vassilevski, V.A. Dmitriev, GaN PN-structures grown by hydride vapor phase epitaxy. MRS Online Proc. Library **482**, 13–18 (1997). https://doi.org/10.1557/PROC-482-251
44. A.S. Usikov, D.V. Tsvetkov, M.A. Mastro, A.I. Pechnikov, V.A. Soukhoveev, Y.V. Shapovalova, O.V. Kovalenkov, G.H. Gainer, S.Y. Karpov, V.A. Dmitriev, B. O'Meara, S.A. Gurevich, E.M. Arakcheeva, A.L. Zakhgeim, H. Helava, Indium-free violet LEDs grown by HVPE. Phys. Status Solidi (c) **7**, 2265–2269 (2003). https://doi.org/10.1002/pssc.200303521

45. M.A. Mastro, D. Tsvetkov, V. Soukhoveev, A. Usikov, V. Dmitriev, B. Luo, F. Ren, K.H. Baik, S.J. Pearton, Hydride vapor phase epitaxy-grown AlGaN/GaN high electron mobility transistors. Solid-State Electron **47**(6), 1075–1079 (2003). https://doi.org/10.1016/S0038-1101(02)00473-2
46. M. Murthy, J.A. Frietas Jr., J. Kim, E.R. Glaser, D. Storm, Residual impurities in GaN substrates and epitaxial layers grown by various techniques. J. Cryst. Growth **305**, 393–398 (2007)
47. N. Sawada, T. Narita, M. Kanechika, T. Uesugi, T. Kachi, M. Horita, T. Kimoto, J. Suda, Sources of carrier compensation in metalorganic vapor phase epitaxy-grown homoepitaxial n-type GaN layers with various doping concentrations. Appl. Phys. Express **11**, 041001 (2008)
48. F. Kaess, S. Mita, J. Xie, P. Reddy, A. Klump, L.H. Hernandez-Balderrama, S. Washiyama, A. Franke, R. Kirste, A. Hoffmann, R. Collazo, Z. Sitar, Correlation between mobility collapse and carbon impurities in Si-doped GaN grown by low pressure metalorganic chemical vapor deposition. J. Appl. Phys. **120**, 105701 (2016)
49. R.P. Tompkins, J.R. Smith, K.W. Kirchner, K.A. Jones, J.H. Leach, K. Udwary, E. Preble, P. Suvarna, J.M. Leathersich, F. Shahedipour-Sandvik, GaN power Schottky diodes with drift layers grown on four substrates. J. Electron. Mater. **43**, 850 (2014)
50. Y. Cao, R. Chu, R. Li, M. Chen, R. Chang, B. Hughes, High-voltage vertical GaN Schottky diode enabled by low-carbon metal-organic chemical vapor deposition. Appl. Phys. Lett. **108**, 062103 (2016)
51. R.P. Tompkins, T.A. Walsh, M.A. Derenge, K.W. Kirchner, S. Zhou, C.B. Nguyen, K.A. Jones, G. Mulholland, R. Metzger, J.H. Leach, P. Suvarna, M. Tungare, F. Shahedipour-Sandvik, HVPE GaN for high power electronic Schottky diodes. Solid-State Electon. **79**, 238–243 (2013)
52. H. Fujikura, T. Konno, T. Yoshida, F. Horikiri, Hydride-vapor-phase epitaxial growth of highly pure GaN layers with smooth as-grown surfaces on freestanding GaN substrates. J. J. Appl. Phys. **56**, 085503 (2017)
53. H. Fujikura, T. Konno, T. Kimura, Y. Narita, F. Horikiri, Homo-epitaxial growth of n-GaN layers free from carbon-induced mobility collapse and off-angle-dependent doping variation by quartz-free hydride vapor phase epitaxy. Appl. Phys. Lett. **117**, 012103 (2020)
54. J.H. Leach, K. Udwary, G. Dodson, T.B. Tran, H.A. Splawn, Low-pressure, modified halide vapor-phase epitaxy for chemically pure GaN epilayers. Phys. Status Solidi B, 2400035 (2024)
55. J.K. Hite, M.A. Mastro, L.E. Luna, T.J. Anderson, Understanding interfaces in homoepitaxial growth. ECS Trans. **86**, 15 (2018)
56. H. Fu, K. Fu, C. Yang, H. Liu, K.A. Hatch, P. Peri, D.H. Mudiyanselage, B. Li, T.-H. Kim, S.R. Alugubelli, P.-Y. Su, D.C. Messina, X. Deng, C.-Y. Cheng, R.V. Meidanshahi, X. Huang, H. Chen, T.-H. Yang, J. Zhou, A.M. Armstrong, A.A. Allerman, E.T. Yu, J. Han, S.M. Goodnick, D.J. Smith, R.J. Nemanich, F.A. Ponce, Y. Zhao, Selective area regrowth and doping for vertical gallium nitride power devices: Materials challenges and recent progress. Mater. Today **49**, 296 (2021)
57. K. Ohnishi, Y. Amano, N. Fujimoto, S. Nitta, Y. Honda, H. Amano, Halide vapor phase epitaxy of p-type Mg-doped GaN utilizing MgO. Appl. Phys. Exp. **13**, 061007 (2020)
58. K. Ohnishi, N. Fujimoto, S. Nitta, H. Watanabe, S. Lu, M. Deki, Y. Honda, H. Amano, Tuning the p-type doping of GaN over three orders of magnitude via efficient Mg doping during halide vapor phase epitaxy. J. Appl. Phys. **132**, 145703 (2022)
59. K.L. Schulte, W. Metaferia, J. Simon, D. Guiling, K. Udwary, G. Dodson, J.H. Leach, A.J. Ptak, Growth of AlGaAs, AlInP, and AlGaInP by hydride vapor phase epitaxy. ACS Appl. Energy Mater. **2**, 8405–8410 (2019)
60. K. Udwary, J.H. Leach, G. Dodson, Low-temperature hydride-assisted HVPE growth of single crystal AlN films, in *62nd EMC Conference, Paper NN04* (2020)
61. M. Rubenstein, E. Meyers, Epitaxial synthesis of GaAs using a flow system. J. Electron. Soc. **113**, 365–367 (1966)
62. T. Hirasaki, K. Asano, M. Banno, M. Ishikawa, F. Sakuma, H. Muakami, Y. Kumagai, Growth of thick InGaN layers by tri-halide vapor phase epitaxy. J. J. Appl. Phys **53**, 05FL02 (2014)

63. H. Murakami, N. Takekawa, A. Shiono, Q.T. Thieu, R. Togashi, Y. Kumagai, K. Matsumoto, A. Koukitu, Tri-halide vapor phase epitaxy of thick GaN using gaseous $GaCl_3$ precursor. J. Cryst. Growth **456**, 140–144 (2016)
64. D. Nakamura, T. Kimura, K. Horibuchi, Halogen-free vapor phase epitaxy for high-rate growth of GaN bulk crystals. Appl. Phys. Express **10**, 045504 (2017)
65. M. Imade, Y. Bu, T. Sumi, A. Kitamoto, M. Yoshimura, T. Sasaki, M. Isemura, Y. Mori, Vapor-phase epitaxial growth of GaN films using Ga_2O vapor and NH_3. J. Cryst. Growth **350**, 56 (2012)
66. A. Kitamoto, J. Takino, T. Sumi, M. Kamiyama, S. Tsuno, K. Ishibashi, Y. Gunji, M. Imanishi, Y. Okayama, M. Nobuoka, M. Isemura, M. Yoshimura, Y. Mori, Effect of methane additive on GaN growth using the OVPE method. Jpn. J. Appl. Phys. **58**, SC1021 (2019)
67. K.L. Schulte, J. Simon, A. Roy, R.C. Reedy, D.L. Young, T.F. Kuech, A.J. Ptak, Computational fluid dynamics-aided analysis of a hydride vapor phase epitaxy reactor. J. Cryst. Growth **434**, 138–147 (2016)
68. K.L. Schulte, J. Simon, M.A. Steiner, A.J. Ptak, Modeling and design of III-V heterojunction solar cells for enhanced performance. Cell Reports Phys. Sci. **4**, 101541 (2023)
69. K.A. Horowitz, T.W. Remo, B. Smith, A.J. Ptak, A techno-economic analysis and cost reduction roadmap for III-V solar cells, in *No. NREL/TP-6A20–72103 United States 10.2172/1484349 NREL English*, (National Renewable Energy Lab. (NREL), Golden)
70. A.J. Ptak, M. Hassanaly, K. Udwary, J.H. Leach, G. Dodson, H. Splawn, K.L. Schulte, J. Simon, Toward high-throughput deposition of III-V materials and devices using halide vapor phase epitaxy. Proc. SPIE **12881**, 1288102 (2024)

Chapter 6
X-Ray Topography and High-Resolution X-Ray Diffraction Characterization of GaN Materials for Power Electronics Applications

Balaji Raghothamachar and Michael Dudley

1 Introduction

Gallium nitride (GaN) [1], along with other wide bandgap semiconductors, especially silicon carbide (SiC) [2], is being developed to replace silicon (Si) in the applications of power electronic devices as the latter has nearly reached its performance limits. Devices based on wide bandgap semiconductors are characterized by large electric breakdown fields, higher current density and thermal conductivity, faster switching, and lower on-resistance. To date, GaN has been successfully applied in optoelectronic devices such as light emitting diodes and laser diodes [3], as well as in lateral power electronic devices such as high electron mobility transistors (HEMT) using non-native substrates such as silicon, GaAs, and sapphire [4]. To achieve the full potential of GaN, vertical power devices on high-quality, bulk GaN substrates are desirable. The availability of bulk GaN substrates prepared from ammonothermal, hydride vapor phase epitaxy (HVPE), and other methods has enabled the fabrication of vertical device architectures, but this also requires the development of reliable selective area doping processes [5]. Common selective area doping (SAD) processes include ion implantation [6], etch and regrowth [7], diffusion [8], and more recently neutron transmutation [9]. The relationship between processing, microstructure, and device performance in selectively doped devices must be understood to overcome the limitations of these doping techniques and obtain reproducible and reliable results. X-ray diffraction-based techniques offer a noninvasive and nondestructive approach to characterize the distribution of defects and strain in single crystal materials at each stage of device processing starting from substrates, through epilayer growth, selective area doping, annealing, and other

B. Raghothamachar (✉) · M. Dudley
Stony Brook University, Stony Brook, NY, USA
e-mail: balaji.raghothamachar@stonybrook.edu; michael.dudley@stonybrook.edu

I. C. Kizilyalli et al. (eds.), *Gallium Nitride and Related Materials*, The Materials Research Society Series, https://doi.org/10.1007/978-3-031-83056-3_6

processes. Analysis of the distribution can provide valuable insights to modify or optimize these processes.

In this review, synchrotron X-ray topography techniques [10] and high-resolution X-ray diffraction (HRXRD) [11] were used to evaluate microstructure in GaN materials at different processing stages of different SAD methods. Synchrotron monochromatic beam X-ray topography (SMBXT) [10] was adopted to give information on dislocation densities and dislocation behavior when the material is subject to different processes. Synchrotron X-ray plane-wave topography (SXPWT), also known as synchrotron X-ray rocking curve topography (SXRCT) [12], was also conducted for strain mapping, giving information about the effect on strain from different SAD processes. HRXRD was used to perform rocking curve measurements and obtain reciprocal space maps. Other characterization techniques such as chemical etching, SEM, TEM, and optical microscopy are employed in a complementary manner.

2 Extended Defects in GaN

Gallium nitride is a direct bandgap semiconductor and typically crystallizes in the hexagonal wurtzite structure (2H-GaN; space group: P6 3 mc; $a = 3.186$ Å; and $c = 5.185$ Å). GaN also crystallizes in the cubic zincblende structure (3C-GaN; $a = 4.5036$ Å) but this phase is metastable and generally, the 2H-GaN polytype with a bandgap of 3.4 eV is developed for power devices. The 2H-GaN structure is highly anisotropic characterized by the basal c-plane and the m- and a-plane perpendicular to the basal planes (Fig. 6.1b).

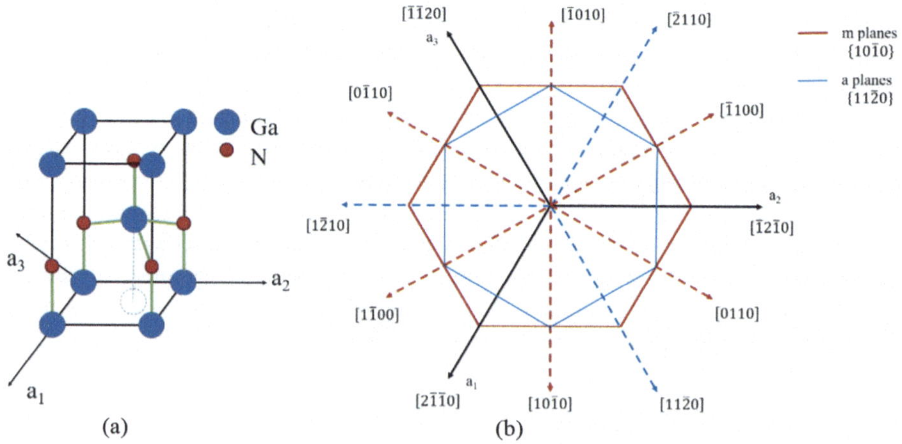

Fig. 6.1 (**a**) Crystal structure of 2H-GaN and (**b**) major crystallographic planes in 2H-GaN. The [0001] or c-plane is parallel to the plane of the page

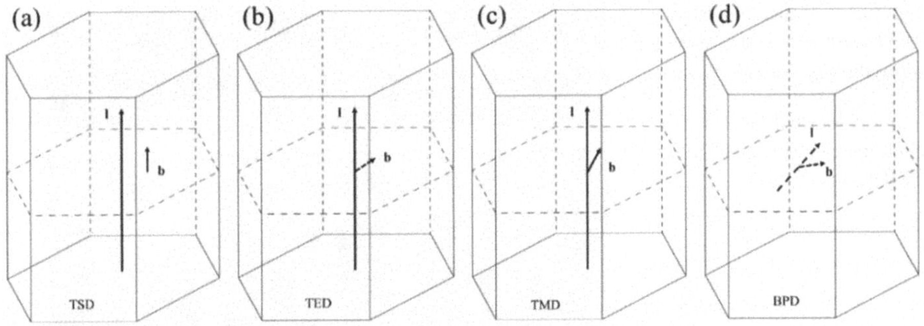

Fig. 6.2 Burgers vectors (b) and line directions (l) of dislocations commonly observed in 2H-GaN. (**a**) Threading screw dislocation (b = [0001]; 1 = [0001]); (**b**) threading edge dislocation ($b = \frac{1}{3}\langle 11\bar{2}0\rangle$; l = [0001]); (**c**) threading mixed dislocation ($b = \frac{1}{3}\langle 11\bar{2}3\rangle$; l = [0001]); and (**d**) basal plane dislocation ($b = \frac{1}{3}\langle 11\bar{2}0\rangle$; l in the basal plane). Note: Line directions of TSDs, TEDs, and TMDs can deviate from [0001]

The primary extended defects of concern in GaN crystals are dislocations broadly characterized into growth and stress-induced dislocations. Threading screw dislocations (TSDs), threading edge dislocations (TEDs), and threading mixed dislocations (TMDs) are growth dislocations and basal plane dislocations (BPDs) are stress-induced dislocations although TEDs can also glide when they lie in the prismatic $\{10\bar{1}0\}$ planes.

Growth dislocations are formed during crystal growth and can originate from (a) the relaxation of stresses arising from the different lattice parameters and thermal expansion coefficients from foreign substrates; (b) the continuation of dislocations present in the native seed; and (c) the relaxation of stresses from coalescence between different misoriented growth centers. The nucleation of TSDs/TMDs at the nucleation sites during growth occurs to accommodate the local elastic strain in the lattice due to the existence of various types of lattice inhomogeneity such as inclusions, precipitates, and voids [13]. TSDs have both their line direction and Burgers vectors along [0001] (Fig. 6.2a) while TMDs run roughly along the c-axis but with Burgers vectors that can be expressed as $\boldsymbol{nc} + \boldsymbol{ma}$, consisting of both c-components and one of the six a-components (Fig. 6.2c). Common types of TMDs are $\boldsymbol{c} + \boldsymbol{a}$ types, in which the screw component can be either $+\boldsymbol{c}$ or $-\boldsymbol{c}$, and the edge component can be either one of the six $\frac{1}{3}\langle 11\bar{2}0\rangle$ types. So, the Burgers vectors of the $\boldsymbol{c} + \boldsymbol{a}$ types of TMDs are $\frac{1}{3}[\bar{1}2\bar{1}3]$, $\frac{1}{3}[\bar{2}113]$, $\frac{1}{3}[\bar{1}\bar{1}23]$, $\frac{1}{3}[1\bar{2}13]$, $\frac{1}{3}[2\bar{1}\bar{1}3]$, $\frac{1}{3}[11\bar{2}3]$, $\frac{1}{3}[\bar{1}2\bar{1}\bar{3}]$, $\frac{1}{3}[\bar{2}11\bar{3}]$, $\frac{1}{3}[\bar{1}\bar{1}2\bar{3}]$, $\frac{1}{3}[1\bar{2}1\bar{3}]$, $\frac{1}{3}[2\bar{1}\bar{1}\bar{3}]$, and $\frac{1}{3}[11\bar{2}\bar{3}]$. Images of TSDs and TMDs observed in an X-ray topograph recorded with the monochromatic X-ray beam are shown in Fig. 6.3. TSDs/TMDs can affect the performance of devices by causing increased reverse leakage and soft breakdown [14].

TEDs are edge type threading dislocations with Burgers vectors of $\frac{1}{3}\langle 11\bar{2}0\rangle$ (Fig. 6.2b) and line directions approximately parallel to the [0001] direction. Six different types of TEDs exist in GaN crystals. The Burgers vectors are $\frac{1}{3}[\bar{1}2\bar{1}0]$,

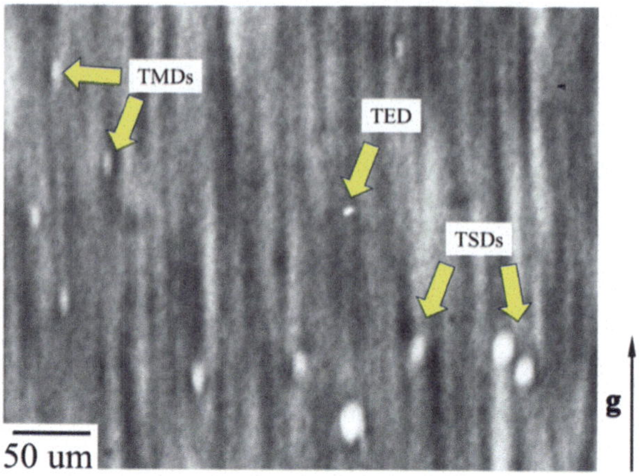

Fig. 6.3 Grazing incidence monochromatic beam x-ray topograph ($g = 11\bar{2}4$; 9.37 keV) from an ammonothermal GaN wafer showing examples of TSDs, TMDs, and TEDs

$\frac{1}{3}[\bar{2}110]$, $\frac{1}{3}[\bar{1}\bar{1}20]$, $\frac{1}{3}[1\bar{2}10]$, $\frac{1}{3}[2\bar{1}\bar{1}0]$, and $\frac{1}{3}[11\bar{2}0]$. While most TEDs are inherited from the native seed, arrays of TEDs can form to accommodate the misorientation between adjacent growth centers [15]. As shown in Fig. 6.3, TEDs usually have smaller contrasts than TSDs due to a smaller Burgers vector.

For 2H-GaN, the primary slip system is $(0001)\langle 11\bar{2}0\rangle$, the basal slip system. BPDs have Burgers vectors of $\frac{1}{3}\langle 11\bar{2}0\rangle$ and glide on the basal plane (Fig. 6.2d). During crystal growth and device fabrication processes, thermal stress is the main source of external force. Therefore, processes with a relatively higher temperature such as annealing or the HVPE method are more likely to generate BPDs than those with a lower temperature such as the ammonothermal method. Additionally, the nucleation of BPDs can occur at various sites of stress concentration such as edges, scratches, inclusions, and the crystal surface. Besides the basal slip system, secondary slip systems exist in the hexagonal system including prismatic slip systems $\{10\bar{1}0\}\langle 11\bar{2}0\rangle$ and the pyramidal slip systems $\{10\bar{1}1\}\langle 11\bar{2}0\rangle$ or $\{11\bar{2}2\}\langle 11\bar{2}0\rangle$. However, observations of prismatic slip or pyramidal slip in GaN crystals have not been reported.

3 X-Ray Characterization Techniques for Extended Defects and Strain

3.1 Synchrotron X-Ray Topography

X-ray topography is an X-ray diffraction imaging technique capable of providing information on the nature and distribution of imperfections such as dislocations, inclusions/precipitates, stacking faults, growth sector boundaries, twins and low angle grain boundaries, and lattice distortions in crystalline materials. Different imperfections are identified through the interpretation of contrast using

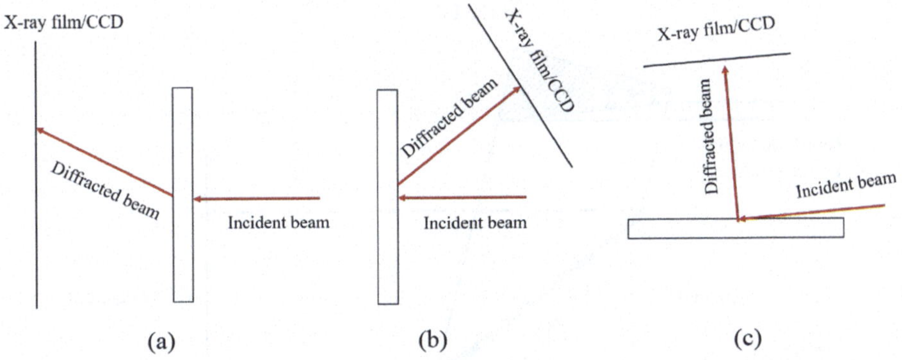

Fig. 6.4 Schematics of diffraction geometries employed for synchrotron X-ray topography. (**a**) Transmission; (**b**) back reflection; and (**c**) grazing incidence

well-established kinematical and dynamical theories of X-ray diffraction. Details of the X-ray topography technique can be found in [2]. Compared with conventional X-ray sources, X-ray topography using the synchrotron X-ray source has many advantages such as high brightness, high natural collimation, low emittance, and wide tunability of energy/wavelength. Synchrotron White Beam X-ray Topography (SWBXT), often used in transmission geometry (Fig. 6.4a) for thin or low absorbing crystals like SiC or AlN, is not suitable for a highly absorbing material such as GaN. Therefore, reflection geometries such as grazing incidence (Fig. 6.4c) and back reflection (Fig. 6.4b) geometries are employed to image defects up to the effective penetration depth of X-rays using a synchrotron monochromatic beam obtained by passing the synchrotron white beam through a Si(111) double crystal monochromator (DCM), i.e., synchrotron monochromatic beam X-ray topography (SMBXT). The penetration depth is determined by the absorption coefficient of the crystal and the path length that X-rays travel inside the sample. For kinematical diffraction, the penetration depth is given by

$$\mathrm{t} = \frac{-\,ln\left(\frac{I}{I_0}\right)}{\mu\left(\frac{1}{\sin \Phi_0} + \frac{1}{\sin \Phi_h}\right)} \tag{6.1}$$

where I/I_0 is the ratio of diffracted beam intensity and incident beam intensity and is taken to be 0.1. μ is the linear absorption coefficient of the material, Φ_0 is the angle between the surface and incident beam, and Φ_h is the angle between the surface and diffracted beam. For the $11\overline{2}4$ reflection of GaN at 9.37KeV, Φ_0 is set as 2° and Φ_H is calculated to be 80.13° for a penetration depth of about 2 μm. Grazing incidence geometry is especially suitable for studying defects in epitaxial thin film layers. Moreover, by adjusting the wavelength of the monochromatic beam, different penetration depths can be achieved.

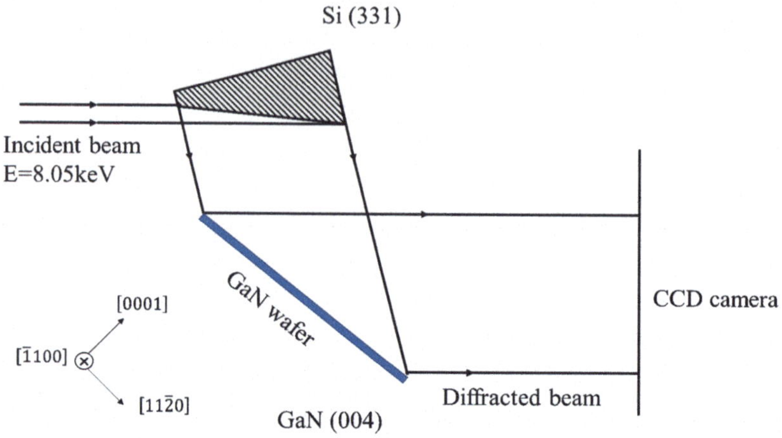

Fig. 6.5 Schematic of the experimental setup for SXPWT for strain/tilt mapping

While SMBXT is suitable for imaging of defects, to obtain quantitative information on the distribution of lattice strains and tilt in a sample, it can be modified by the introduction of a beam conditioner to obtain a highly strain-sensitive setup (Fig. 6.5) described as Synchrotron X-ray Plane Wave Topography (SXPWT) [12]. Recent applications of this technique include SiC [12, 16], diamond [17, 18], and GaN [19]. This is important for GaN, as lattice strain can be introduced during device fabrication processes which can affect homogeneity and thereby performance of devices. The experimental setup for SXPWT (Fig. 6.5) involves the incident beam ($E = 8.05$ keV) selected by the Si(111) DCM conditioned by a highly asymmetric Si (331) beam conditioner with d-spacing matching that of GaN (0004) planes for the highest resolution. The GaN sample acts as the second crystal and diffracts with (0004) planes in reflection geometry.

After recording the 0004 data by rocking the sample about the $\left[\bar{1}100\right]$ axis, the sample is then rotated about the [0001] axis by 180° and the 0004 data are recorded again with opposite diffraction vector. According to Bragg's Law, the deviation of an arbitrary position in the crystal can be expressed as

$$\omega^{x_i(x_j)} = \frac{\partial u_i}{\partial x_i} \tan \theta_B + \frac{\partial u_i}{\partial x_k} \tag{6.2}$$

where $\omega^{x_i(x_j)}$ means that the deviation is from rotating the diffraction plane x_i about the x_j axis and θ_B is the perfect Bragg angle for the crystal. $\frac{\partial u_i}{\partial x_i}$ and $\frac{\partial u_i}{\partial x_k}$ are the terms of local lattice dilation/compression and shear/rotation. To deconvolute these two terms, an opposite diffraction vector is used by rotating the diffraction plane about its normal by 180°. A similar equation is given by

$$\omega^{-x_i(x_j)} = \frac{\partial u_i}{\partial x_i} \tan \theta_B - \frac{\partial u_i}{\partial x_k} \tag{6.3}$$

Combining Eqs. (6.2) and (6.3), we have

$$\frac{\partial u_i}{\partial x_i} = \frac{\omega^{x_i(x_j)} + \omega^{-x_i(x_j)}}{2 \tan \theta_B} \tag{6.4}$$

$$\frac{\partial u_i}{\partial x_k} = \frac{\omega^{x_i(x_j)} - \omega^{-x_i(x_j)}}{2} \tag{6.5}$$

Thus, maps of the strain (dilation/compression) and tilt (shear/rotation) can be plotted from the recorded SXPWT data in the 0° and 180° positions.

3.2 *Ray-tracing Simulations of Dislocations*

The image contrast of dislocations on synchrotron X-ray topographs recorded in back reflection and grazing incidence geometries has been shown to be dominated by the orientation contrast mechanism [20]. Therefore, the ray-tracing method can be effectively used to simulate the images of dislocations of different Burgers vectors (TSDs, TMDs, TEDs, and BPDs) and compared to identify the Burgers vectors of all dislocations on X-ray topographs from GaN materials. This technique has previously been successfully applied in both back reflection and grazing incidence geometries for characterizing dislocations in SiC [21], aluminum nitride (AlN) [15], and zinc oxide (ZnO) [22]. This method has the advantage that Burgers vector characterization of dislocations can be carried out using a single reflection from a commercial (0001) substrate wafer without the need for additional reflections (as required for $\mathbf{g} \cdot \mathbf{b} = 0$ & $\mathbf{g} \cdot \mathbf{b} \times \mathbf{l} = 0$ criteria) or axial-cut wafers to characterize dislocations parallel to the c-axis. Details on ray-tracing simulations of dislocations in GaN can be found elsewhere [23]. Briefly, in ray-tracing simulations, the mosaic region around the dislocation (or other defect) is divided into many cubic diffraction units with their local misorientations coinciding with the long-range displacement field of the dislocation or defect. The key point is to obtain the plane normal after distortion due to the strain fields associated with the defects. The plane normal $\boldsymbol{n}(x, y, z)$ after distortion is given by

$$\boldsymbol{n}(x, y, z) = \boldsymbol{n}^0(x, y, z) - \nabla \left[\boldsymbol{n}^0(x, y, z) \cdot \boldsymbol{u}(x, y, z)\right] \tag{6.6}$$

where $\boldsymbol{n}^0(x, y, z)$ is the plane normal before distortion and $\boldsymbol{u}(x, y, z)$ is the displacement field of the dislocation or other defect. When X-rays are incident (with incidence wave vector $\boldsymbol{s_0}$) on the crystal surface, local distortion on the reflecting planes will result in the variation of diffracted beam directions $\boldsymbol{s_g}$. The diffracted

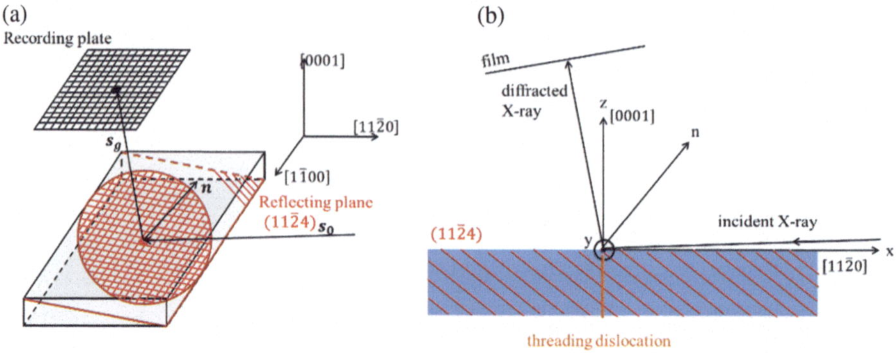

Fig. 6.6 (**a**) Schematic diagram for ray-tracing simulation and (**b**) the coordinate system setup used for ray-tracing simulation in grazing incidence geometry

beam direction is given by $\boldsymbol{s_0} \times \boldsymbol{n} = -\boldsymbol{n} \times \boldsymbol{s_g}$. Diffracted beams with slightly different directions will be collected on the recording plane, and contrast is formed as the diffracted beams overlap or separate. A schematic of the ray-tracing simulation principle is shown in Fig. 6.6a and the coordinate system employed for the simulation is illustrated in Fig. 6.6b. The simulations consider the effects of surface relaxation as well as photoelectric absorption to predict dislocation contrast. Table 6.1 summarizes the results of simulations for different types of dislocations expected to be observed in GaN substrates.

3.3 High-Resolution X-Ray Diffraction (HRXRD)

HRXRD [11] involves the recording of double- and triple-axis X-ray diffraction rocking curves to evaluate the crystalline quality of single crystal materials. Based on Bragg's law

$$2\mathrm{d}\sin\theta = \lambda \tag{6.7}$$

where d is the planar spacing of the diffracting planes, θ is the Bragg angle, and λ is the wavelength of X-rays, it involves recording the diffracted intensity as the sample is rocked through the Bragg condition. The full-width at half-maximum (FWHM) of the rocking curve is compared to that for a perfect crystal to evaluate the crystalline quality of the sample. According to dynamical theory, the FWHM for a perfect crystal is given by [24]

$$2\delta_{\mathrm{hs}} = \frac{2r_e\lambda^2}{\pi V \sin 2\theta_B}\frac{1}{|\gamma|}|C|\sqrt{F_{\mathrm{hkl}}F_{\overline{\mathrm{hkl}}}} \quad \text{Bragg case} \tag{6.8a}$$

Table 6.1 Ray-tracing simulations of different dislocations on grazing incidence topographs in a (0001) GaN wafer ($\mathbf{g} = 11\bar{2}4$, ($\mathbf{g}\uparrow$))

Dislocation type		Simulated image (50 μm × 50 μm)	Burgers vector	Dislocation type	Simulated image (50 μm × 50 μm)	Burgers vector
BPD	Edge	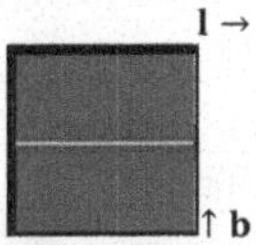	$\frac{1}{3}[11\bar{2}0]$	Edge	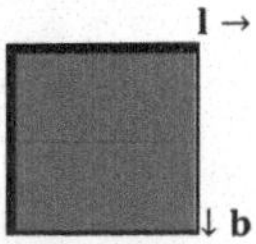	$-\frac{1}{3}[11\bar{2}0]$
	Screw right-handed	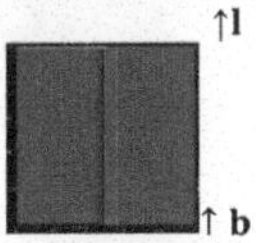	$\frac{1}{3}[11\bar{2}0]$	Screw left-handed	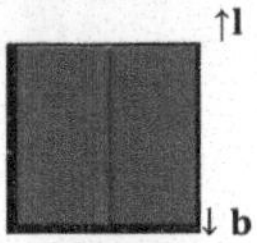	$-\frac{1}{3}[11\bar{2}0]$
TSD	Right-handed		[0001]	Left-handed		$[000\bar{1}]$
TED	I		$\frac{1}{3}[\bar{1}2\bar{1}0]$	II	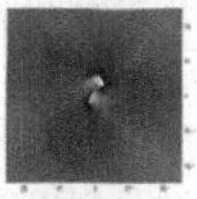	$\frac{1}{3}[\bar{2}110]$
	III	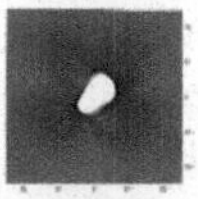	$\frac{1}{3}[\bar{1}\bar{1}20]$	IV		$\frac{1}{3}[1\bar{2}10]$
	V		$\frac{1}{3}[2\bar{1}\bar{1}0]$	VI		$\frac{1}{3}[11\bar{2}0]$
TMD	**c** + 0°**a**		$\frac{1}{3}[11\bar{2}3]$	**c** + 60°**a**		$\frac{1}{3}[\bar{1}2\bar{1}3]$
	c + 120°**a**		$\frac{1}{3}[\bar{2}113]$	**c** + 180°**a**	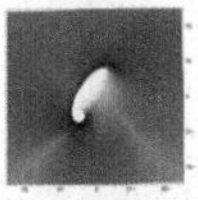	$\frac{1}{3}[\bar{1}\bar{1}23]$
	c + 240°**a**		$\frac{1}{3}[1\bar{2}13]$	**c** + 300°**a**		$\frac{1}{3}[2\bar{1}\bar{1}3]$
	-**c** + 0°**a**	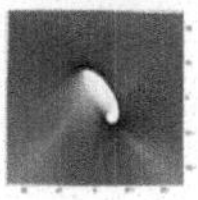	$\frac{1}{3}[11\bar{2}\bar{3}]$	-**c** + 60°**a**		$\frac{1}{3}[\bar{1}2\bar{1}\bar{3}]$

(continued)

Table 6.1 (continued)

Dislocation type	Simulated image (50 μm × 50 μm)	Burgers vector	Dislocation type	Simulated image (50 μm × 50 μm)	Burgers vector
-c + 120°**a**		$\frac{1}{3}[\bar{2}11\bar{3}]$	**-c** + 180°**a**		$\frac{1}{3}[\bar{1}\bar{1}2\bar{3}]$
-c + 240°**a**		$\frac{1}{3}[1\bar{2}1\bar{3}]$	**-c** + 300°**a**		$\frac{1}{3}[2\bar{1}\bar{1}3]$

$$2\delta_{Os} = \frac{2r_e\lambda^2}{\pi V \sin 2\theta_B}|\gamma||C|\sqrt{F_{hkl}F_{\overline{hkl}}} \quad \text{Laue case} \tag{6.8b}$$

where r_e is the classical radius of an electron, V is the unit cell volume, θ_B is the Bragg angle, γ is the asymmetric ratio, C is the polarization factor, and F_{hkl} is the structure factor.

These widths are applicable for an incident plane wave X-ray beam and must be modified for the divergent X-ray beam conditioned by mirrors and monochromators in a diffractometer. This is achieved by using a computer program, HXRD [25] that can simulate rocking curves of perfect crystals recorded with different X-ray sources, mirrors, and monochromators. For the simulation of rocking curves of implanted samples, Rocking Curve Analysis by Dynamical Simulation (RADS) software [26] was adopted, and the depth profile of strain levels in the implanted layers was obtained. In the double-axis configuration, an open detector is used to collect all diffracted beams, while in the triple-axis configuration, an analyzer crystal is placed in front of the detector to narrow down the acceptance angle of the detector, giving a higher sensitivity and capability to separate lattice tilts and strains contributions to the broadening of the double axis rocking curve.

4 Bulk GaN Substrates

The commercial realization of GaN power devices requires low-cost, high-quality GaN substrates wafers. Due to the extremely high melting temperature and low decomposition pressure of GaN, melt growth techniques such as Czochralski and Bridgman methods are precluded and crystallization from the gas phase, solution, and their combinations must be employed. The high-pressure, high-temperature method solution growth method developed in the early 1980s [27] is characterized by slow growth rates and extreme growth conditions that limit the industrial application of the method. Small, high-quality crystals produced by this method can be used as seeds for other growth methods. The physical vapor transport (PVT) method, well developed for SiC bulk growth, requires the use of ammonia in addition to the GaN powder source to overcome the dissociation of GaN under the growth temperature of around 1700 °C and has limited development [28–30]. In the Na-flux method, the poor solubility of N_2 in Ga melt is increased by introducing Na to the Ga melt. Compared with the high-temperature high-pressure method, the growth temperature and pressure are significantly reduced but industrial production has yet to be realized [31–37].

4.1 HVPE Substrates

The most common growth method for manufacturing the GaN substrate is hydride vapor phase epitaxy (HVPE) because of its high growth rate and crystallization of high-purity GaN materials. Using foreign substrates such as sapphire, SiC, or GaAs, free-standing and large diameter GaN wafers have been obtained [38–44]. In HVPE, the chloride gas of metal Ga and NH_3 are reacted on the substrate and a GaN film is grown in an excess ammonia environment. Horizontal and vertical growth systems can be used for HVPE consisting of a source zone and a growth zone. The temperature of the source zone is kept at around 800 °C to 900 °C, while the temperature of the growth zone is about 1000 °C to 1100 °C. Inert gases (N_2, He, Ar, or H_2) are used as carrier gases. Among the substrates used for growing GaN crystals using the HVPE method, sapphire is commonly used due to low cost and availability of large size substrates while 6H-SiC is most favorable based on lattice parameter and thermal expansion coefficient mismatch but high cost. Other substrates with matched lattice parameters ($LiAlO_2$, $LiGaO_2$, etc.) have also been used. However, poor chemical and thermal stability of these substrates require a critical growth condition, limiting the wide application of these substrates. X-ray topographs and double-axis rocking curves from typical GaN substrates from different vendors shown in Fig. 6.7 show a nearly uniform distribution of dislocations with dislocation densities of the order of 10^6 cm^{-2} (the limit for X-ray topography) with FWHM widths ranging from 50 to 100″. Wafers also suffer from a high level of lattice distortion (radius of curvature <20 m) originating from the thermal expansion coefficient mismatch of the newly grown material with the foreign substrate used [45].

4.2 Patterned HVPE Substrates

To improve the quality of HVPE-grown GaN crystal, a patterned growth method has been developed [40, 42, 44]. In this method, the substrate used for HVPE growth is prepared by depositing a patterned layer typically composed of inverted polarized GaN crystals on which the growth rate is lower. During growth, hexagonal inverse

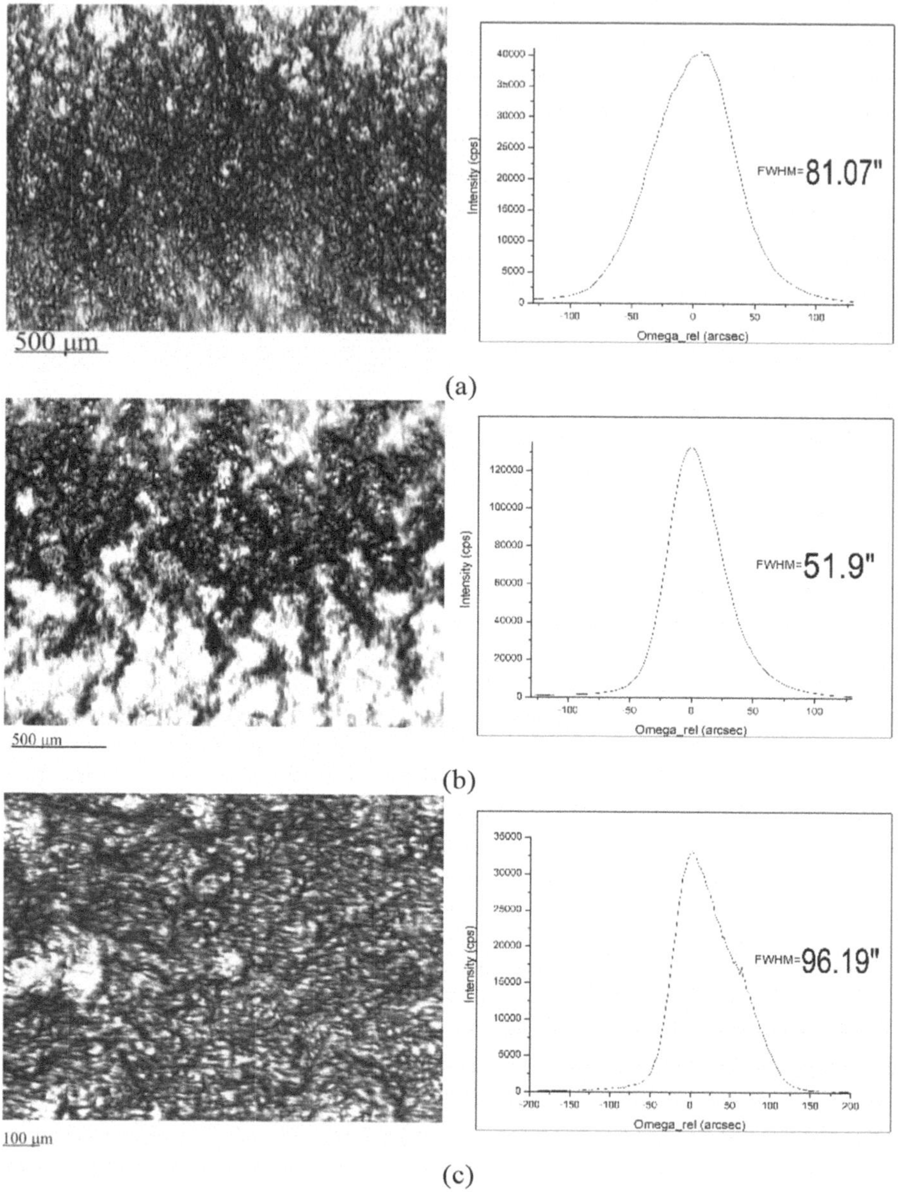

Fig. 6.7 High magnification topographs (**g** = $11\overline{2}4$; 9.37 KeV) showing typical defect distributions and 0002 double-axis rocking curves from HVPE substrates from different vendors. (**a**) Dislocation density ~ 10^6 cm^{-2}, FWHM 81.07″. (**b**) Dislocation density ~ 10^6 cm^{-2}, FWHM 51.9″. (**c**) Dislocation density ~ 10^6 cm^{-2}, FWHM 96.19″ (note: FWHM of the perfect GaN crystal for the diffractometer setup used is 22″)

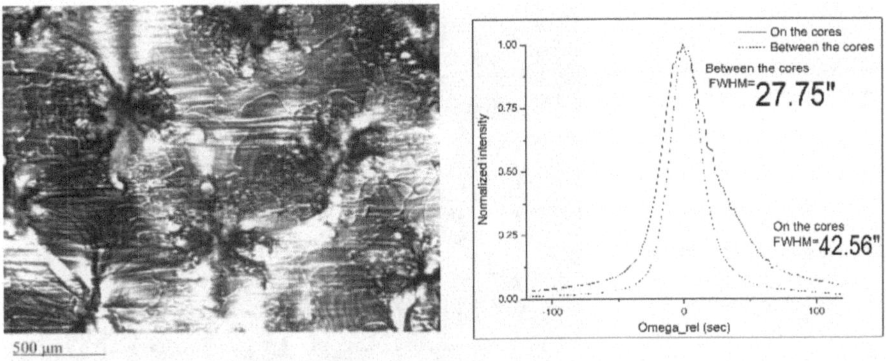

(a)

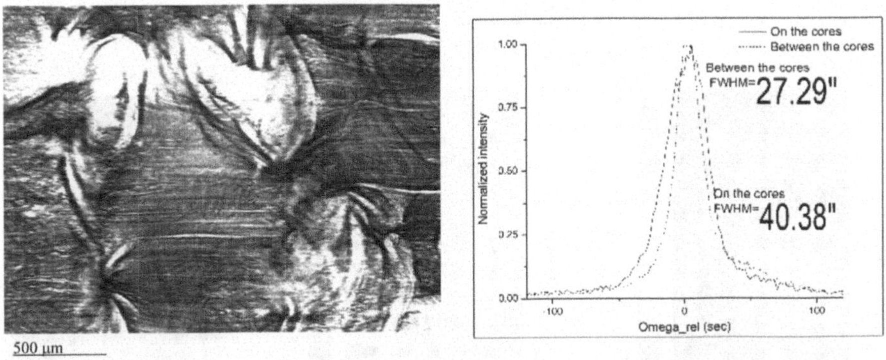

(b)

Fig. 6.8 High-magnification topographs (**g** = $11\overline{2}4$; 9.37 KeV) showing typical defect distributions and 0002 double-axis rocking curves from patterned HVPE substrates from different vendors. (**a**) Near core TD density ~ 10^3 cm^{-2}, BPD density ~ 10^4–10^5 cm^{-2}, FWHM 42.56″. Between cores TD density ~ 10^2 cm^{-2}, BPD density ~ 10^4 cm^{-2}, FWHM 25.77″. (**b**) Near core TD density ~ 10^6 cm^{-2}, BPD density ~ 10^4–10^5 cm^{-2}, FWHM 40.38″. Between cores TD density $<10^2$ cm^{-2}, BPD density ~ 10^4 cm^{-2}, FWHM 27.29″

pyramidal pits are created by $\{11\overline{2}2\}$ facets. Dislocations existing on the surface of the facets move horizontally toward the patterned dots. After growth, strain centers (dot cores) are formed with radial tilt around them [46] and areas between centers are characterized by lower dislocation densities. X-ray topographs from patterned HVPE GaN substrates from different vendors (Fig. 6.8) show dislocations concentrated to small areas with large areas with very low dislocation densities [23]. This is demonstrated by the low FWHM values in areas between the strain centers (or cores), which is around 27″ (22″ for perfect crystal). BPDs emanating from the strain centers sometimes propagate deep into the surrounding regions potentially damaging to device operation but the areas of low dislocation densities are comparable to active device area for 1.2 kV GaN vertical power FinFETs [47] and are thus potentially suitable for the fabrication of vertical power devices.

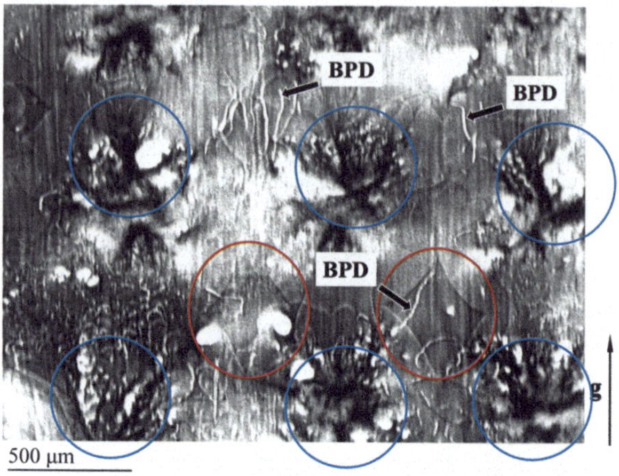

Fig. 6.9 Magnified SMBXT image ($\boldsymbol{g} = 11\bar{2}4$; 9.37 keV) of a patterned HVPE-grown GaN substrate. Low dislocation density can be found between the strain centers marked by blue circles. Red circles mark the enclosed domain shapes. The domain boundaries have linear contrast but are different from BPDs marked by black arrows

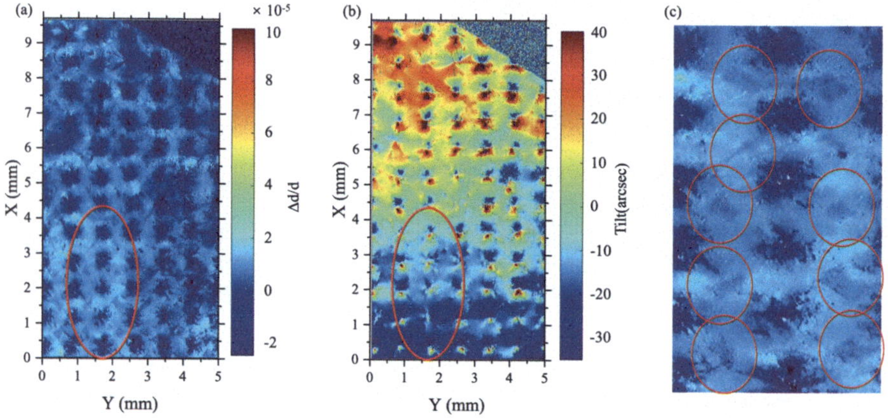

Fig. 6.10 (**a**) Strain map and (**b**) tilt map generated from SXPWT ($\boldsymbol{g}$ = 0008; 8.05 KeV). The domain features are visible in the strain map but not in the tilt map indicating the Bragg peak shift in the domain region is from pure strain. (**c**) Magnified image showing the domain features characterized by compressive strain in the level of 10^{-5}

Another interesting feature of patterned HVPE GaN wafers is the presence of domain boundaries different from dislocations or low-angle grain boundaries (Fig. 6.9). HRTEM analysis reveals no difference in lattice structures on either side of the domain boundaries. Strain and tilt maps generated by deconvoluting the SXPWT data reveal that the regions enclosed by these boundaries are characterized by compressive strain values of the order of magnitude of 10^{-5} (Fig. 6.10).

SIMS analysis revealed that the concentration of oxygen from outside the domain is at the level of 10^{18} atoms/cm^3, while the oxygen concentration level is below the detection limit of 10^{15} atoms/cm^3 inside the domain suggesting that the two types of domains separated by the boundaries are different growth sectors [48]. The origin of such growth sectors lies in the HVPE growth process (Fig. 6.11). In the patterned

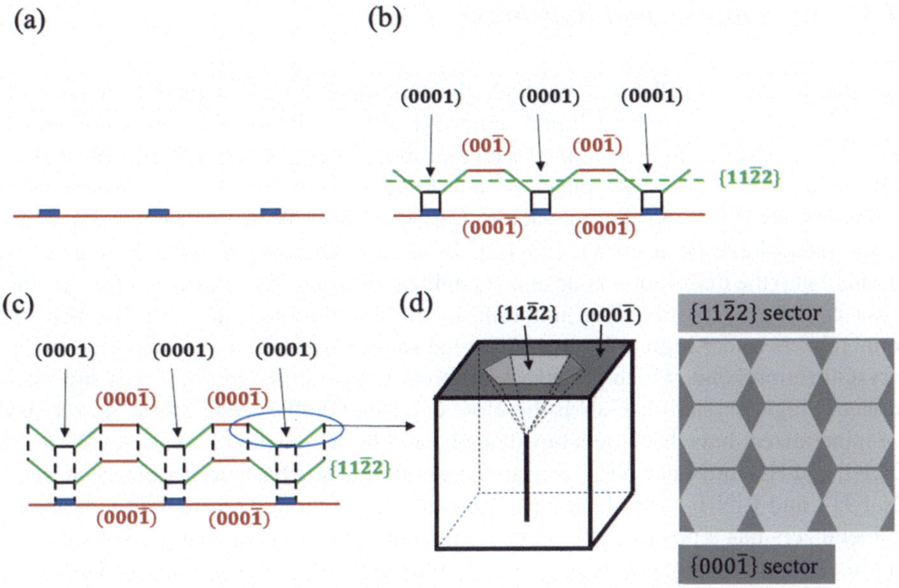

Fig. 6.11 Formation of the growth sectors during patterned HVPE growth. (**a**) A pattern of inverted polarized materials is first deposited on the growth template. (**b**) The growth rate on top of the inverted polarized materials is lower than the other areas. $\{11\bar{2}2\}$ type facets form. (**c**) Growth on both $(000\bar{1})$ and $\{11\bar{2}2\}$ type facets will result in different growth sectors separated by growth sector boundaries. (**d**) The 6 $\{11\bar{2}2\}$ facets have a hexagonal outline when viewed from the top. As the four-fold $\{11\bar{2}2\}$ sectors expand, the $(000\bar{1})$ sectors become diamond shaped

HVPE growth, a pattern of inverted polarized materials is first deposited on the growth template (Fig. 6.11a). The growth rate on top of the inverted polarized materials is lower than the other areas, as shown in Fig. 6.11b. From a study of growth rates along different directions by kinetic Wulff plots [49], $\{11\bar{2}2\}$ type facets are predicted to form. Growth on both $(000\bar{1})$ and $\{11\bar{2}2\}$ type facets will result in different growth sectors separated by growth sector boundaries (Fig. 6.11c). The 6 $\{11\bar{2}2\}$ facets have a hexagonal outline when viewed from the top, as shown in Fig. 6.11d. As the four-fold $\{11\bar{2}2\}$ sectors expand, the $(000\bar{1})$ sectors in between become diamond shaped as shown in Fig. 6.11e. Despite the difference in strain between the growth sectors being small, the growth sector boundaries are mostly located in the low dislocation density regions, target locations for power device fabrication. The small strain difference near the growth sector boundaries can potentially induce lattice defects during epitaxy, ion implantation, and annealing thereby influencing the performance of the devices fabricated in those regions.

4.3 Ammonothermal Substrates

Similar to the hydrothermal method, the ammonothermal method is a kind of solvothermal growth where liquid ammonia is the solvent. The ammonothermal growth method was first developed with the ammonobasic solution in the mid-1990s [50] and later small GaN platelets and needles were successfully grown with ammonobasic [51] and ammonoacidic [52] solutions. With further development, 2 in wafers have been grown [53–62]. In ammonothermal growth, the reactor is divided into the dissolution zone and crystallization zone. The nutrient, which in this case is GaN, is dissolved in ammonia in the dissolution zone, with the help of mineralizers under high pressures. Then the saturated solution is transported to the crystallization zone, which is under a different temperature, and becomes supersaturated. Single crystalline GaN is therefore deposited on the seed surface. Two types of mineralizers have been developed, acidic and basic. Acidic mineralizers (NH_4Cl, NH_4I, and HI) introduce NH_4^+ ions to the solution, while basic mineralizers (KNH_2, $NaNH_2$, and $LiNH_2$) introduce NH_2^- ions to the solution. The solvent with acidic mineralizers has a positive solubility coefficient, which means that the solubility is higher when the temperature is higher. On the contrary, the solubility of GaN in a basic environment is lower when the temperature is higher. As a result, the temperature of the crystallization zone must be lower than that of the dissolution zone if an acidic mineralizer is adopted, while the temperature of the crystallization zone must be higher if a basic mineralizer is adopted. As shown in Fig. 6.12a, compared to GaN substrates obtained by other methods, ammonothermal-grown samples have the lowest dislocation densities and show the lowest FWHM widths and relative flat curvature. Analysis of dislocations indicates that no BPDs are observed and TEDs, TSDs, and TMDs can be identified by comparing the image contrasts with ray-tracing simulation images (Table 6.1). TMDs dominate among the threading dislocations accounting for more than 90% [45]. The absence of BPDs is expected since the low temperatures (<600 °C) and relatively small temperature gradients at which ammonothermal growth is carried out are unlikely to generate sufficiently large thermal gradient stresses to induce deformation. Compared to patterned substrates [23] that contain BPDs, the ammonothermal substrates may thus be less vulnerable to BPD-induced mechanisms of device failure. As shown in Fig. 6.12b, the use of HVPE seeds in ammonothermal growth simply replicates dislocations from the HVPE seed resulting in high dislocation densities. But by using ammonothermal GaN substrates as seeds for HVPE growth, high-quality substrates can be obtained as shown in Fig. 6.12c where the low density of dislocations from the ammonothermal layer is replicated into the HVPE layer with no new defects nucleated and no significant change in strain.

Ammonothermal-grown GaN substrate wafers sometimes contain domain features (Fig. 6.13) that are characterized by compressive strains of the order of 10^{-5} similar to patterned HVPE GaN. However, HRTEM studies show no difference in crystal structure and SIMS analysis reveals no difference in concentrations of common impurity species across the domain boundaries. The origins of these

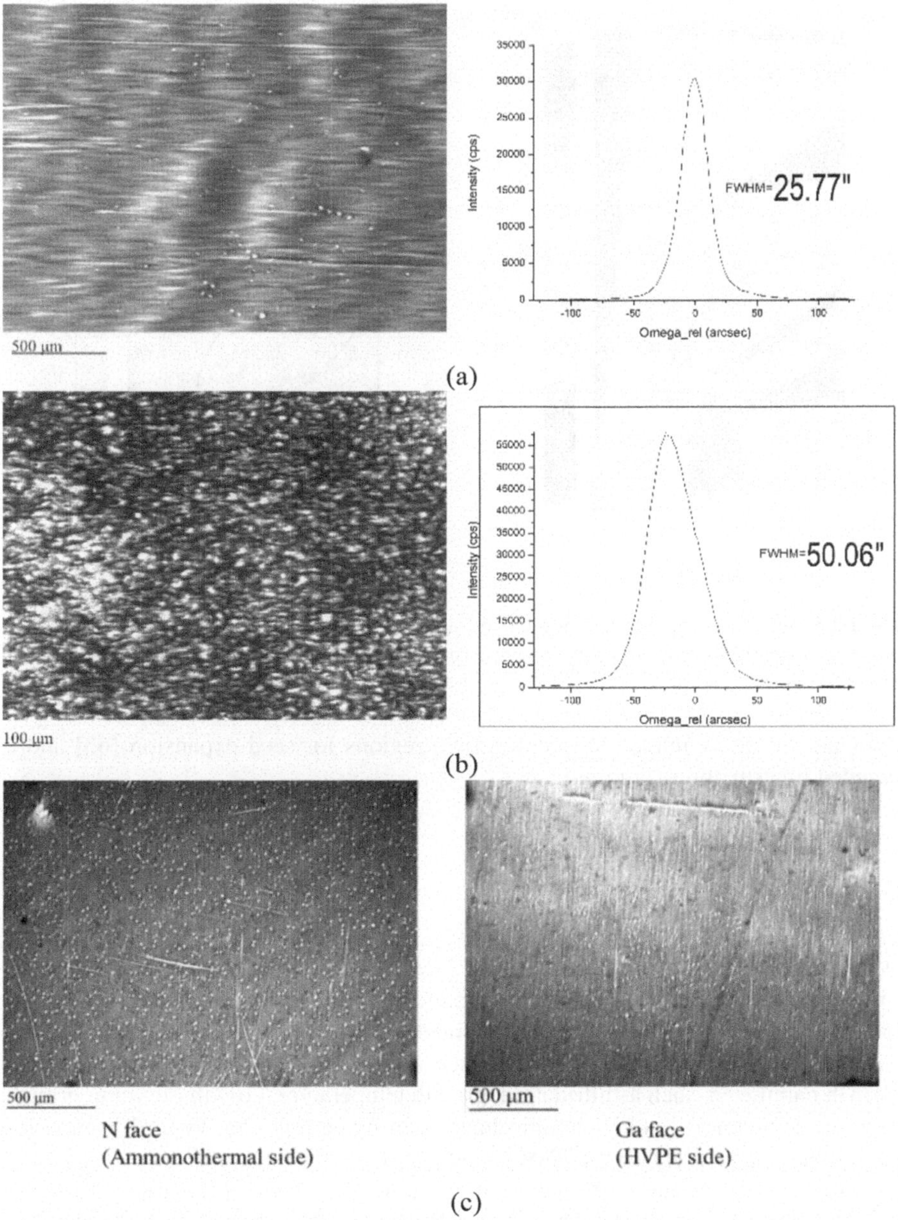

Fig. 6.12 High-magnification X-ray topographs (**g** = $11\bar{2}4$; 9.37 KeV) showing typical defect distributions and double-axis rocking curves from (**a**) the ammonothermal GaN substrate—threading dislocation (TD) density: 6.2×10^3 cm^{-2}; FWHM 25.77″. (**b**) Ammonothermal GaN substrate grown on HVPE seed—dislocation density: ~10^5–10^6 cm^{-2}; FWHM 50.05″. (**c**) HVPE GaN substrate grown on the ammonothermal seed. Both N face (ammonothermal) and Ga face (HVPE) showing threading dislocations of the order of 10^4 cm^{-2} indicating replication of defects with few new defects nucleated

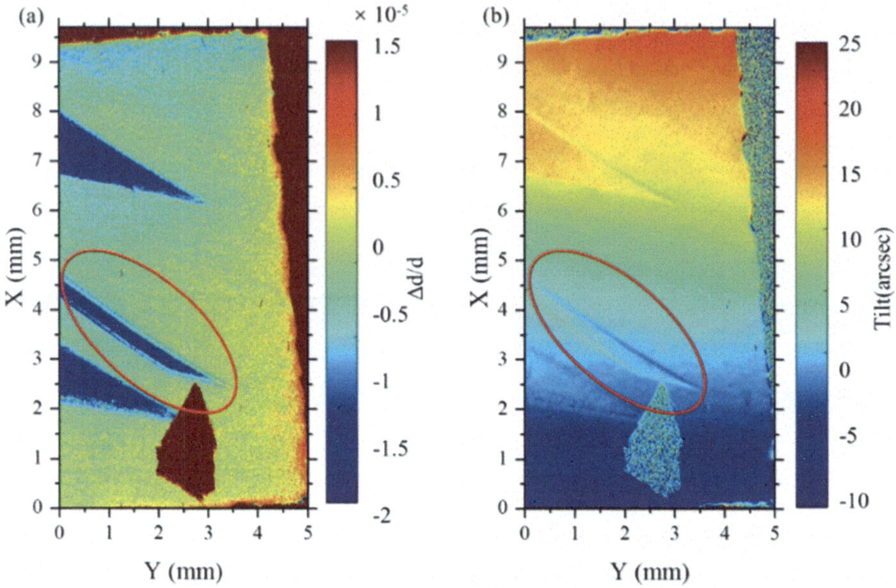

Fig. 6.13 (**a**) Strain map and (**b**) tilt map generated from SXPWT ($\boldsymbol{g}$ = 0008; 8.05 KeV) of an ammonothermal GaN wafer showing domains. The domain features are visible in the strain map but not in the tilt map indicating the difference in strain but not in the tilt

domains are likely related to lateral-grown regions for seed expansion [63] and a detailed investigation is ongoing.

5 GaN Epilayer Growth

After substrate selection, the next stage of device fabrication is homoepitaxial GaN growth. For selective area doping, unintentionally doped homoepitaxial GaN films are typically grown by metal organic chemical vapor deposition (MOCVD) resulting in n-type epilayers. Various efforts on the optimization of buffer layer and other growth parameters such as nitridation, growth temperature, growth pressure, growth rate, layer thickness, and III/V ratio have been made [64–75]. Typical growth for GaN utilizes trimethylgallium (TMGa) and ammonia (NH_3) as precursors at 300 torr pressure at temperatures around 1040 °C [76, 77]. X-ray topography imaging conducted before and after MOCVD growth (Fig. 6.19a & b) reveals that threading dislocations are replicated into the epilayers with no nucleation of new defects at the interface. Strain and tilt maps also reveal no significant changes. However, the contrast of threading dislocations is either enlarged or transformed into dark contrast. This is likely associated with the formation of helical dislocations [78]. The observation of dark contrast is likely related to the edge component of the threading dislocations.

6 Selective Area Doping

For the fabrication of GaN power devices in the vertical architecture, selective area doping methods being developed are discussed with their impact on the microstructures. The unintentionally doped epilayer is typically n-type and p-type doping needs to be achieved usually by the introduction of Mg into the epilayer.

6.1 Ion Implantation

Ion implantation [79] is a common doping technique employed in semiconductor device fabrication where the ions of the dopant element are accelerated to a high energy and bombard the target material to be implanted. Damage to the crystal lattice occurs as the ions with high energy penetrate into a GaN crystal, generating a collision cascade consisting of Ga and N vacancies, Ga and N interstitials, and sometimes an amorphous zone. While the level of damage can be characterized by Rutherford backscattering/channeling (RBS/C) spectrometry [80] and defects identified by cross-sectional transmission electron microscopy (XTEM) [81], X-ray diffraction (XRD) [82], and X-ray topography [83] can be nondestructively used to study the level of strains caused by lattice distortions in the crystal. For GaN materials implanted with ions, high-temperature annealing is necessary to heal the implantation-induced damage and also activate the implanted species by their migration onto energetically favorable lattice sites. As a rule of thumb, a temperature of about two-third of the melting point in Kelvin is required to remove the extended defect. As the melting point of GaN is estimated to be around 2500 °C, the annealing temperature is expected to be around 1650 °C. Unfortunately, at such temperatures, GaN undergoes material decomposition caused by the loss of nitrogen from the surface. To prevent decomposition of GaN at high temperatures, annealing approaches include the capping of the GaN surface with protective layers [65] and annealing under nitrogen overpressure [84]. It is essential to completely cap the GaN surface as improper capping can result in loss of material and increased levels of strain as shown in Fig. 6.14. Changes in microstructures and strain distribution during ion implantation and annealing processes may have an impact on the performance of the devices similar to that observed in 4H-SiC [85].

6.1.1 Mg/N Co-implantation and Gyrotron Microwave Annealing

Mg ion implantation for p-type doping in GaN can give rise to nitrogen vacancies, which limits the free hole concentration and therefore compensates the acceptors. Co-implantation of Mg and N can remove the nitrogen vacancies and give better p-type conductivities [86]. A set of six ammonothermal GaN epiwafers subject to Mg/N co-implantation at different N concentrations at 600 °C were analyzed by

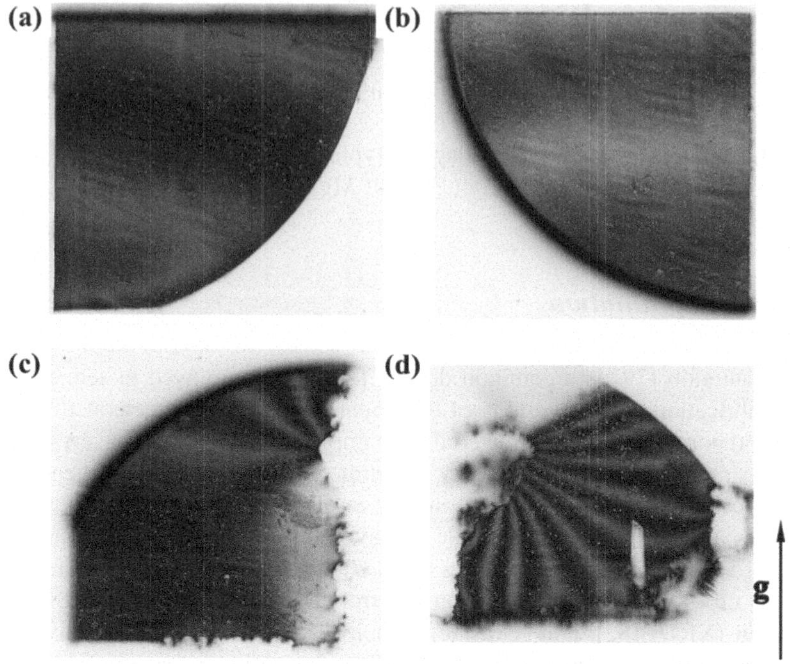

Fig. 6.14 SMBXT ($g = 11\bar{2}4$; 9.37 KeV) images of ammonothermal GaN epiwafers subject to annealing after implantation and capping: (**a**) As-implanted; (**b**) annealed at 1150 °C, (**c**) annealed at 1225 °C, and (**d**) annealed at 1300 °C. Note the loss of material at 1225 °C and 1300 °C and resultant strains indicated by narrower and increased number of contours

HRXRD in conjunction with RADS simulations [87]. The high temperature for implantation was intended to reduce implantation-induced damage by dynamic annealing during the implantation. However, the rocking curves from the as-implanted samples reveal a satellite peak at a lower Bragg angle than the main peak due to a tensile strain gradient in the implanted layer compared to the unimplanted epilayer and substrate. The tensile strain gradient is of the order of 10^{-3} and increases with increasing N concentration. Using RADS, simulated rocking curves were fitted to the recorded rocking curves of the as-implanted samples (Fig. 6.15a, c, e) and the corresponding strain profiles correlated well with the anticipated implantation profiles from the TRIM simulations (Fig. 6.15b, d, f). The strain profiles clearly follow the trend of dopant concentrations along the depth. The N dopant concentration of 5×10^{19} cm^{-2} exceeds stoichiometric levels and results in elevated strains even though the light nitrogen atoms are expected to introduce lower lattice damage.

Implanted samples were capped with AlN and annealed via gyrotron microwave in short thermal cycles at a peak temperature of 1225 °C for a total of 2 min under 3 MPa N_2 ambient to increase the equilibrium thermal decomposition temperature of GaN [77]. Comparison between the 0004 triple-axis rocking curves of as-implanted

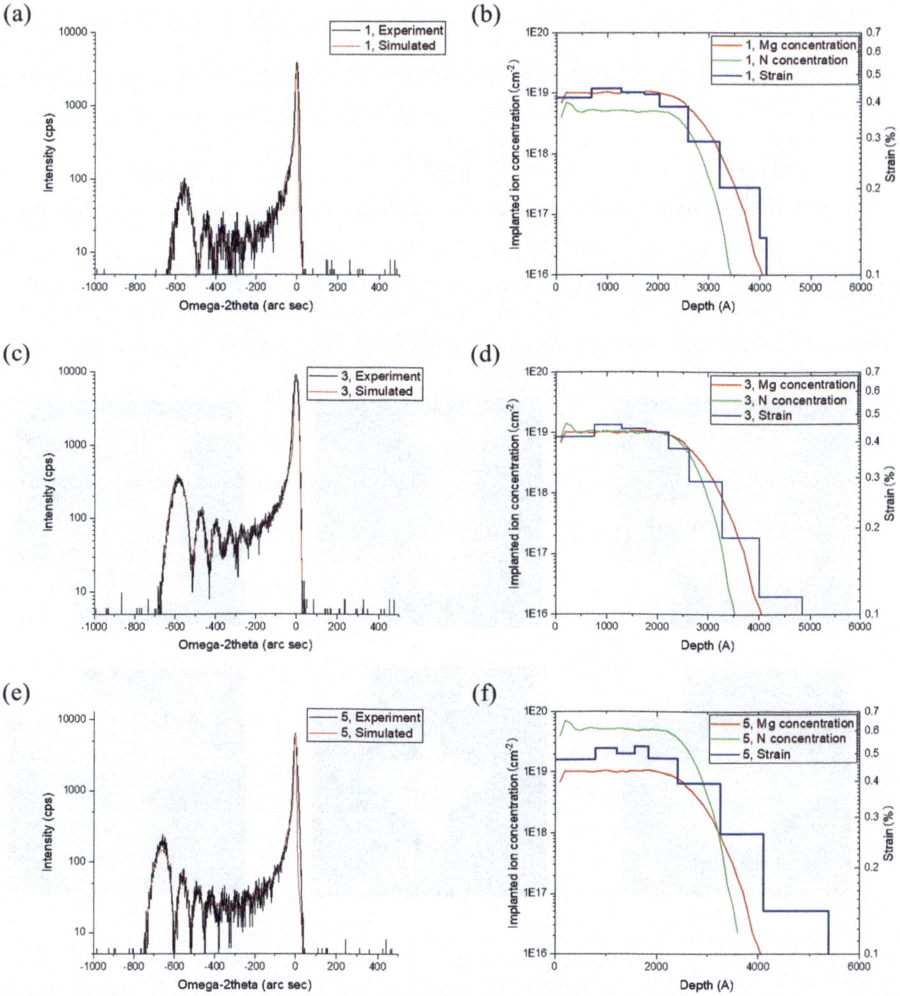

Fig. 6.15 The 0004 rocking curves (black) superimposed on RADS simulated rocking curve (red) of the as-implanted samples: (**a**) #1 (Mg:1 × 10^{19} cm^{-2}; N:5 × 10^{18} cm^{-2}); (**c**) #3 (Mg:1 × 10^{19} cm^{-2}; N:1 × 10^{19} cm^{-2}); and (**e**) #5 (Mg:1 × 10^{19} cm^{-2}; N:5 × 10^{19} cm^{-2}). (**b**) (**d**) (**f**) TRIM simulation and the depth profiles of strain from the RADS simulation of the as-implanted samples #1, #3, and #5

and annealed samples implanted with the same Mg/N concentrations is shown in Fig. 6.16. The rocking curves indicate that samples with lower N concentrations show complete recovery of strain, while the highest N concentration retains a satellite peak adjacent to the main peak indicating a small residual strain and incomplete recovery. A similar assessment can also be derived from the reciprocal space maps shown in Fig. 6.17. Annealing for a longer time and/or at a higher

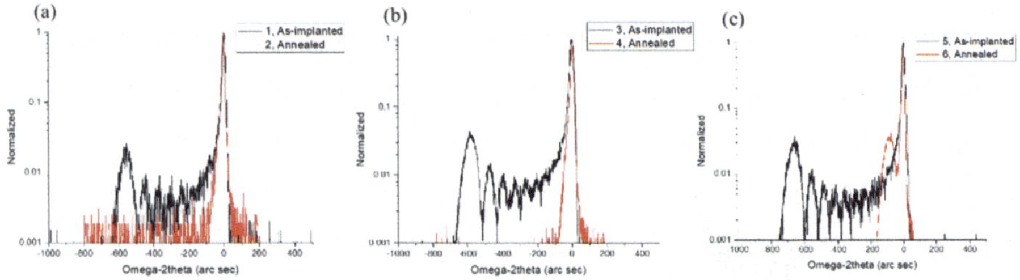

Fig. 6.16 The 0004 rocking curves of as-implanted GaN samples (black) superimposed annealed (red) samples (**a**) #1 and #2 (Mg:1 × 10^{19} cm^{-2}; N:5 × 10^{18} cm^{-2}); (**b**) #3 and #4 (Mg:1 × 10^{19} cm^{-2}; N:1 × 10^{19} cm^{-2}); and (**c**) #5 and #6 (Mg:1 × 10^{19} cm^{-2}; N:5 × 10^{19} cm^{-2})

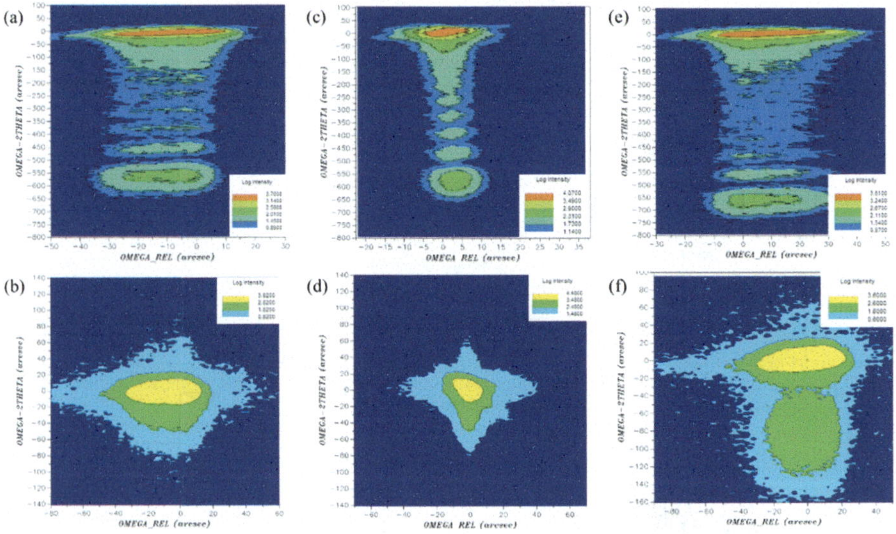

Fig. 6.17 The 0004 reciprocal space maps of Mg/N-implanted GaN samples: As-implanted (**a**) #1 (Mg:1 × 10^{19} cm^{-2}; N:5 × 10^{18} cm^{-2}); (**b**) #3 (Mg:1 × 10^{19} cm^{-2}; N:1 × 10^{19} cm^{-2}); (**c**) #5 (Mg:1 × 10^{19} cm^{-2}; N:5 × 10^{19} cm^{-2}); implanted and annealed (**d**) #2 (Mg:1 × 10^{19} cm^{-2}; N:5 × 10^{18} cm^{-2}); (**e**) #4 (Mg:1 × 10^{19} cm^{-2}; N:1 × 10^{19} cm^{-2}); and (**f**) #6 (Mg:1 × 10^{19} cm^{-2}; N:5 × 10^{19} cm^{-2})

temperature is necessary for complete damage recovery of co-implanted samples with above-stoichiometric nitrogen concentrations.

6.1.2 Mg Ion Implantation and High-Pressure Annealing

While capped gyrotron microwave annealing is capable of removing the strain induced by ion implantation, alloying and etch damage from the capping and capping removal processes can be potential challenges [76, 77]. Annealing under

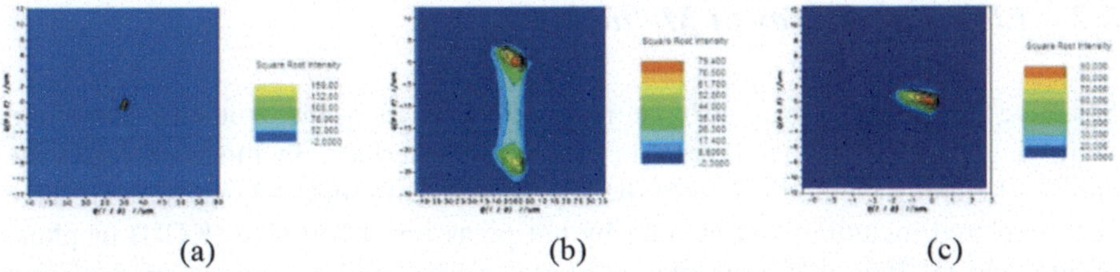

Fig. 6.18 The 0004 reciprocal space maps from an ammonothermal-grown GaN wafer (a) after epilayer growth; (b) implantation with Mg-ions; and (c) annealing under 1GPa at 1400 °C for 10 minutes

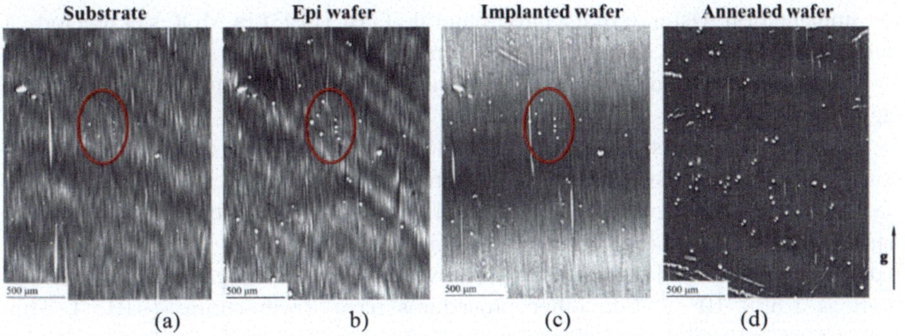

Fig. 6.19 SMBXT images (**g** = $11\bar{2}4$; 9.37 KeV) of the same regions of an ammonothermal GaN wafer: (**a**) substrate, (**b**) after MOCVD epilayer growth, (**c**) after Mg-ion implantation, and (**d**) after high-pressure annealing showing an essentially unchanged dislocation distribution. Note: The region from the annealed wafer is different from the first three stages of processing and it shows similar dislocation distribution

ultra-high pressure without capping has been developed as an alternative to recover the lattice damage after ion implantation [84]. An unintentionally doped epilayer grown on an ammonothermal GaN wafer implanted with Mg-ions then annealed at a pressure of 1 GPa at 1400 °C for 10 minutes was characterized after each processing step [88]. A comparison of the reciprocal space maps of the wafer after epilayer growth, implantation, and annealing (Fig. 6.18) shows the satellite peak from the implantation damaged layers is no longer visible on the reciprocal space map of the annealed wafer indicating complete recovery with no residual strains. Annealing at 1400 °C also eliminates inversion domains containing inactive Mg and improves dopant activation efficiency [89]. X-ray topographs recorded after each processing step (Fig. 6.19) show that no new dislocations are generated during epitaxial growth. While the overall dislocation distribution appears essentially unchanged after implantation and annealing, TEM studies show that ion implantation results in the nucleation of small dislocation loops, stacking faults, and defect clusters whose density is lowered with increasing annealing temperatures and times [89, 90].

6.2 Etch and Regrowth Methods

Selective area doping can also be realized by etching the epitaxial layer and regrowing doped epitaxial layers on the etched interface. In this technique, the choice of etching method is one of the key factors. Different etching methods have been explored including wet etching by hot potassium hydroxide (KOH) or phosphoric acid (H_3PO_4) [91], plasma-based etching like inductively coupled plasma (ICP) [92], and vapor-based etching by tertiary butyl chloride (TBCl) [93]. The wet etching methods can selectively etch the areas around dislocations and give a rough surface with etch pits [94]. Plasma etching can achieve an anisotropic etching profile with nearly vertical sidewalls, but lattice damage occurs by the same mechanism as ion implantation. Using tertiary butyl-chloride (TBCl) [93] in MOCVD reactors, a rate of at least 3 μm/h can be achieved at a relatively low temperature of about 800 ° C. This etching method can potentially remove the damage incurred during plasma etching.

HVPE GaN wafers were subject to ICP etching, TBCl etching as well as combined ICP and TBCl etching, and regrowth procedures were characterized by SMBXT but no changes in dislocation microstructure were observed (not shown here) [87]. SXPWT measurements to obtain strain (Fig. 6.20) and tilt (Fig. 6.21) maps of the etched and regrown samples reveal that ICP etching results in strains as high as 1.6×10^{-4} while other procedures have lower strains ($\sim 10^{-5}$). This nonuniform lattice damage induced by the ICP etching process is similar to damages induced by ion bombardment in the ion implantation process. TBCl etching, on the other hand, is a relatively damage-free vapor-based etching method. As shown by the low strains in the sample with ICP followed by TBCl etching, TBCl can recover some of the damage induced by ICP etching by the removal of severely damaged material on the surface. The lattice tilt is unaffected by the different etching processes.

6.3 Solid State Diffusion Doping

Solid-state diffusion doping of Mg has been demonstrated in GaN materials [8, 95–97] and therefore selective area doping using this approach has been attempted [98, 99]. Using a bi-layer structure, with a solid Mg source (patterned MgF_2, MgO, or Mg_3N_2) in contact with the HVPE GaN epiwafer combined with a metal capping layer such as gold or yttrium to prevent GaN decomposition up to 1000 °C, significant diffusion of Mg into GaN has been achieved at temperatures as low as 800 °C and time on the order of minutes with concentrations exceeding 10^{19} cm^{-2} and Mg extending >100 nm into the GaN film. Efforts are ongoing to improve the concentration of ionized Mg at room temperature to form a PN diode. From a structural point of view, SMBXT and HRXRD analysis of the microstructure of HVPE GaN wafers subject to diffusion doping (Fig. 6.22) reveals no significant

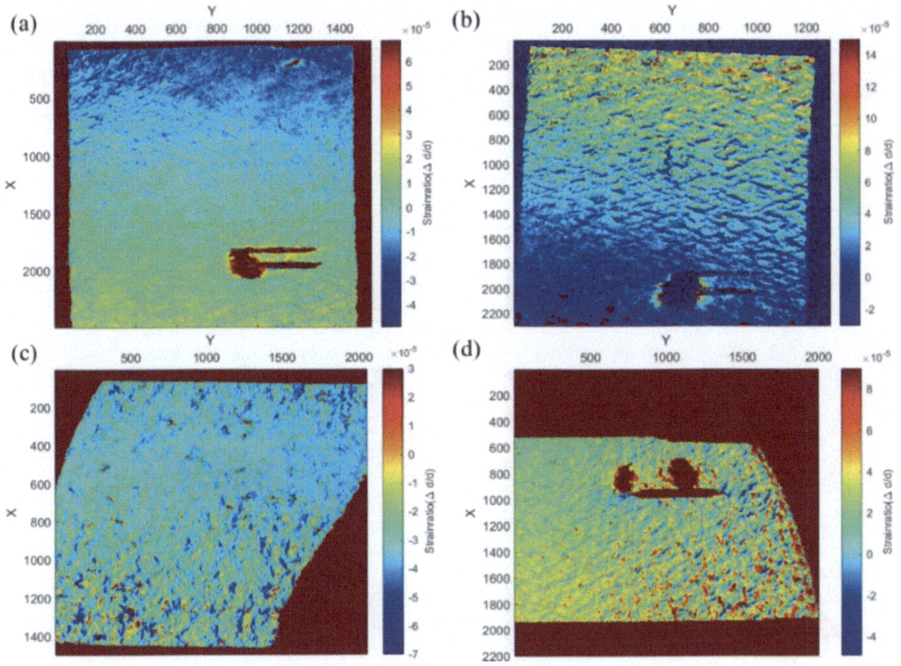

Fig. 6.20 Strain maps of the etched and regrown GaN samples extracted from 0004 SXPWT data. (**a**) Continuously grown without the etching process (strain range: 4×10^{-5}). (**b**) ICP etched before regrowth (strain range: 1.6×10^{-4}). (**c**) TBCl etched before regrowth (strain range: 2.5×10^{-5}). (**d**) ICP etched followed by TBCl etching before regrowth (strain range: 6×10^{-5}). Note: samples in (c) and (d) are smaller but the strain range is not affected

changes in defect distribution and lattice strain/tilt introduced during the diffusion doping process. Employing a + 4 V in situ bias during the diffusion process gave rise to an increase in the surface Mg doping concentration but again did not affect the microstructure (Fig. 6.23) [99].

6.4 *Neutron Transmutation Doping (NTD)*

Different from the conventional doping methods, neutron transmutation doping (NTD) techniques [9] do not require the dopant element to be introduced from outside the material. Nuclear reactions between Ga atoms and thermal neutrons yield Ge atoms, which have been widely used as a favorable n-type dopant for GaN materials. Ge substituting on the Ga site induces far less stress compared to Si because the atom radius of Ge is similar to Ga. During the neutron irradiation process at a site such as the University of Missouri Research Reactor (MURR), nuclear reactions in GaN materials yield $^{70}_{32}$Ge, $^{72}_{32}$Ge and $^{14}_{6}$C. The isotopes of Ge become

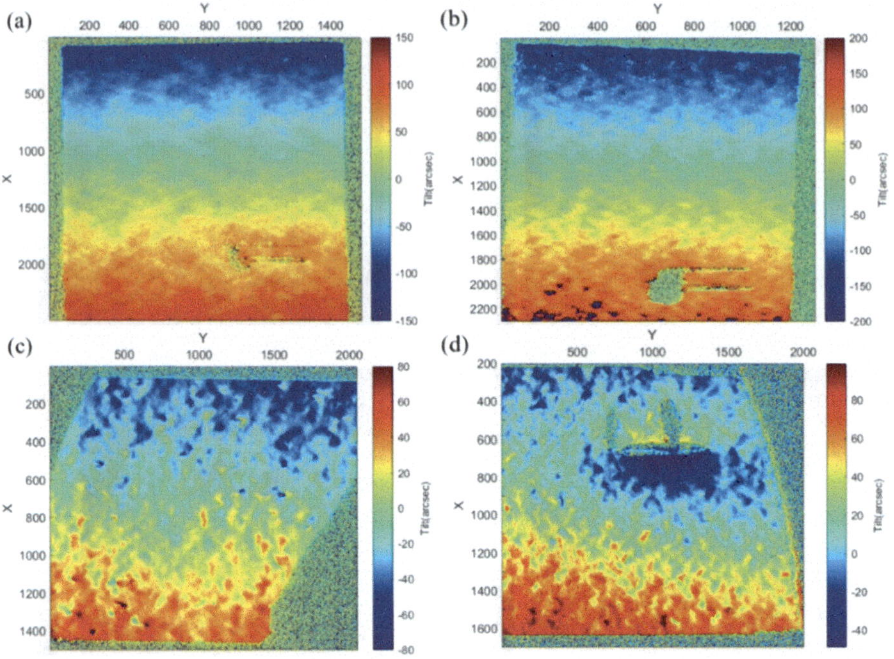

Fig. 6.21 Tilt maps of etched and regrown GaN samples extracted from 0004 SWPWT data. (**a**) Continuously grown without the etching process (tilt range: 230″). (**b**) ICP etched before regrowth (tilt range: 300″). (**c**) TBCl etched before regrowth (tilt range: 160″). (**d**) ICP etched followed by TBCl etching before regrowth (tilt range: 115″). Note samples in (c) and (d) are smaller and therefore tilt ranges are smaller

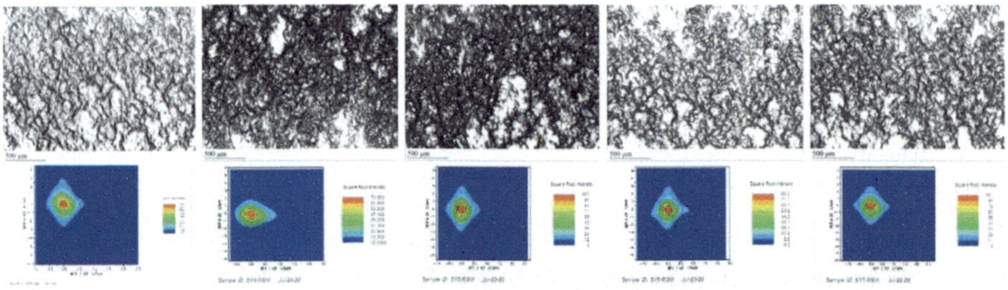

(a) (b) (c) (d) (e)

Fig. 6.22 SMBXT images ($\mathbf{g} = 11\bar{2}4$; 9.37 KeV) and 0002 reciprocal space maps of GaN samples subject to different diffusion doping conditions. (**a**) Control; (**b**) Au/MgF2, 800 °C 5 min; (**c**) Au/MgF2, 900 °C 5 min; (**d**) Y/MgF2, 800 °C 5 min; and (**e**) Y/MgF2, 900 °C 5 min

stable after the decay period and act as the n-type dopant in GaN. The biggest advantages are extremely consistent and uniform doping ranging from 10^{16}Ge atoms/cm^3 to 10^{18} Ge atoms/cm^3 and the realization of selective area doping by the employment of highly neutron absorption masks such as gadolinium

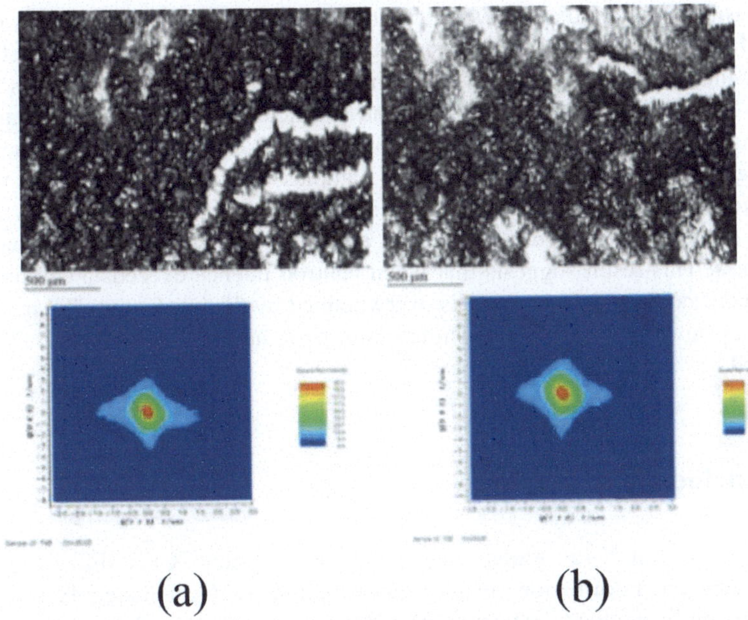

Fig. 6.23 SMBXT images ($\mathbf{g} = 11\bar{2}4$; 9.37 KeV) and 0002 reciprocal space maps of GaN samples subject to different diffusion doping conditions: (**a**) No bias and (**b**) +4v bias

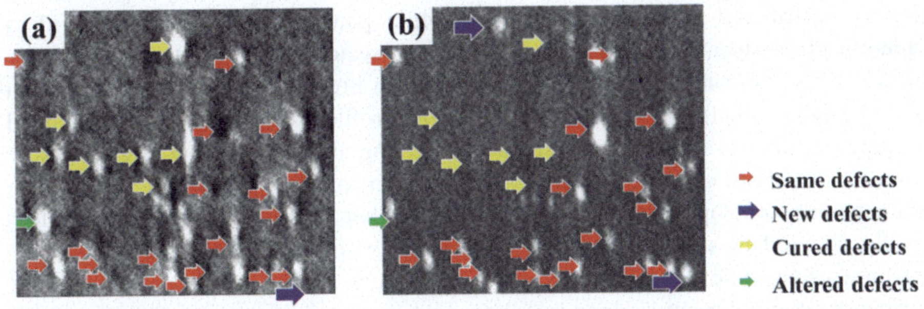

Fig. 6.24 SMBXT images (($\mathbf{g} = 11\bar{2}4$; 9.37 keV) showing the change in defect configurations after neutron irradiation

(Gd) [100]. While the use of thermal neutrons, which have lower energy than fast neutrons, is less likely to damage the crystal structure of the GaN, the radiation process still generates defects such as Frenkel pairs, vacancy-interstitial pairs such as nitrogen vacancies and interstitials, Ga vacancies, and Ga interstitials. An annealing process below 1000 °C is generally carried out after capping the wafer to heal the damage and lower the resistivity. A curing effect of extended defects has been observed to take place during irradiation [101]. In an SMBXT study conducted on a set of ammonothermal-grown GaN wafers, dislocation distribution is slightly changed after irradiation. As marked in Fig. 6.24, contrast from some dislocations is not visible after irradiation while contrast from a few new dislocations can be found. This change in dislocation distribution is attributed to both a curing effect as well as the damage from the neutron irradiation process. This includes lattice damage (formation of interstitial atoms and vacancies) as well as heat damage that can induce interactions between these point defects and the existing dislocations, resulting in the climb of the dislocations. Samples with the lowest neutron fluence were found to contain more cured dislocations than newly generated ones, while the samples with higher neutron fluence had more newly generated dislocations than cured ones. This result suggests that when neutron fluence exceeds a certain level, the effect of neutron defect curing is overwhelmed by the irradiation damage which introduces new defects. This threshold could be related to the type of GaN wafer irradiated.

7 Concluding Remarks

Using synchrotron X-ray topography and high-resolution X-ray diffraction techniques, valuable information on the microstructure of GaN materials for power electronic applications at each stage of different selective area doping processes is obtained. Ammonothermal-grown GaN substrates have the best quality and remain the most promising for vertical device fabrication while HVPE on ammonothermal substrates have similar quality. HVPE wafers suffer from relatively high but uniformly distributed dislocation densities, while patterned HVPE wafers with a nonuniform dislocation distribution and growth sectors can exhibit nonuniformity in device performance. Using HRXRD data from implanted wafers in conjunction with RADS simulation, the induced strain depth profile is correlated to the implanted doping profile. Both gyrotron microwave annealing after capping and high-pressure annealing without capping can effectively recover ion implantation-induced lattice damage but some optimization is required for higher doping concentrations. Inductively coupled plasma (ICP) etching followed by tertiary butyl chloride (TBCl) etching offers a low strain approach for the etch and regrowth processes. The diffusion doping process does not introduce new strains in the GaN material. Neutron transmutation doping by Ge atoms can achieve uniform selective area n-type doping using Gd masks and can lead to the curing of defects at low neutron fluence while higher fluence leads to irradiation damage and nucleation of new defects.

Acknowledgments This work has been supported by the ARPA-E project # DE-AR0001115 under the PNDIODES Program, Dr. Isik C. Kizilyalli, Program Director, and Dr. Eric P. Carlson, Technical Support. Topography data were recorded using the resources of the Advanced Photon Source (Beamline 1-BM), a U.S. DOE Office of Science User Facility operated for the DOE Office of Science by Argonne National Laboratory under contract no. DE-AC02-06CH11357. Joint Photon Sciences Institute at Stony Brook University provided partial support for travel and subsistence at the Advanced Photon Source. The views expressed in the chapter do not necessarily represent the views of the U.S. Department of Energy or the United States Government.

References

1. M. Meneghini, C. de Santi, I. Abid, M. Buffolo, M. Cioni, R.A. Khadar, L. Nela, N. Zagni, A. Chini, F. Medjdoub, G. Meneghesso, G. Verzellesi, E. Zanoni, E. Matiolii, GaN-based power devices: Physics, reliability, and perspectives. J. Appl. Phys. **130**(18), 181101 (2021). https://doi.org/10.1063/5.0061354
2. X. She, A.Q. Huang, Ó. Lucía, B. Ozpineci, Review of silicon carbide power devices and their applications. IEEE Trans. Ind. Electron. **64**(10), 8193–8205 (2017). https://doi.org/10.1109/TIE.2017.2652401
3. H.A.A. Abdul Amir, M.A. Fakhri, A.A. Alwahib, Review of GaN optical device characteristics, applications, and optical analysis technology. Materials Today: Proceedings **42**(5), 2815–2821 (2021). https://doi.org/10.1016/j.matpr.2020.12.727
4. R. Reiner, M. Basler, S. Moench, P. Waltereit, F. Benkhelifa, M. Mikulla, R. Quay, Lateral GaN Power Devices and Integrated GaN Power Circuits: Status and Recent Progress, in *Components of Power Electronics and their Applications 2023*, (ETG Symposium, Bad Nauheim, Germany, 2023), pp. 89–96. https://ieeexplore.ieee.org/document/10227691
5. H. Fu, K. Fu, C. Yang, H. Liu, K.A. Hatch, P. Peri, D.H. Mudiyanselage, B. Li, T.-H. Kim, S.R. Alugubelli, P.-Y. Su, D.C. Messina, X. Deng, C.-Y. Cheng, R.V. Meidanshahi, X. Huang, H. Chen, T.-H. Yang, J. Zhou, A.M. Armstrong, A.A. Allerman, E.T. Yu, J. Han, S.M. Goodnick, D.J. Smith, R.J. Nemanich, F.A. Ponce, Y. Zhao, Selective area regrowth and doping for vertical gallium nitride power devices: Materials challenges and recent progress. Mater. Today **49**, 296–323 (2021). https://doi.org/10.1016/j.mattod.2021.04.011
6. T. Narita, T. Kachi, K. Kataoka, T. Uesugi, P-type doping of GaN (000-1) by magnesium ion implantation. Appl. Phys. Express **10**, 016501 (2017). https://doi.org/10.7567/APEX.10.016501
7. P. Peri, K. Fu, H. Fu, Y. Zhao, D.J. Smith, Characterization of as-grown and regrown GaN-on-GaN structures for vertical p-n power devices. J. Electron. Mater. **50**, 2637–2642 (2021). https://doi.org/10.1007/s11664-021-08769-z
8. M. Rubin, N. Newman, J.S. Chan, T.C. Fu, J. T., Ross; p-type gallium nitride by reactive ion-beam molecular beam epitaxy with ion implantation, diffusion, or coevaporation of Mg. Appl. Phys. Lett. **64**(1), 64–66 (1994). https://doi.org/10.1063/1.110870
9. R. Barber, Q. Nguyen, J. Brockman, J. Gahl, J. Kwon, Thermal neutron transmutation doping of GaN semiconductors. Sci. Rep. **10**, 16295 (2020). https://doi.org/10.1038/s41598-020-72862-2
10. B. Raghothamachar, M. Dudley, G. Dhanaraj, X-ray topography techniques for defect characterization of crystals, in *Springer Handbook of Crystal Growth*, Springer Handbooks, ed. by G. Dhanaraj, K. Byrappa, V. Prasad, M. Dudley, (Springer, Berlin, Heidelberg, 2010), pp. 1425–1451. https://doi.org/10.1007/978-3-540-74761-1_42
11. M. Dudley, X.R. Huang, in *X-Ray Characterization of Epitaxial Layers, in Encyclopedia of Materials: Science and Technology*, ed. by K.J.H. Buschow, R.W. Cahn, M.C. Flemings, B. Ilschner, E.J. Kramer, S. Mahajan, (Elsevier Science, New York, 2001), pp. 2775–2786. https://doi.org/10.1016/B0-08-043152-6/00494-0
12. H. Peng, Z. Chen, Y. Liu, B. Raghothamachar, X. Huang, L. Assoufid, M. Dudley, Quantitative analysis of dislocations in 4H-SiC wafers using synchrotron X-ray topography with ultra-high angular resolution. J. Appl. Cryst. **55**(3), 544–550 (2022). https://doi.org/10.1107/S1600576722004046
13. M. Dudley, X. Huang, W. Huang, A. Powell, S. Wang, P. Neudeck, M. Skowronski, The mechanism of micropipe nucleation at inclusions in silicon carbide. Appl. Phys. Lett. **75**(6), 784–786 (1999). https://doi.org/10.1063/1.124512
14. B. Setera, A. Christou, Threading dislocations in GaN high-voltage switches. Microelectron. Reliab. **124**, 114336 (2021). https://doi.org/10.1016/j.microrel.2021.114336

15. T. Zhou, B. Raghothamachar, F. Wu, R. Dalmau, B. Moody, S. Craft, R. Schlesser, M. Dudley, Z. Sitar, Characterization of threading dislocations in PVT-grown AlN substrates via x-ray topography and ray tracing simulation. J. Electron. Mater. **43**, 838–842 (2014). https://doi.org/10.1007/s11664-013-2968-2
16. J. Guo, Y. Yang, B. Raghothamachar, M. Dudley, S. Stoupin, Mapping of lattice strain in 4H-SiC crystals by synchrotron double-crystal x-ray topography. J. Electron. Mater. **47**(2), 903–909 (2018). https://doi.org/10.1007/s11664-017-5789-x
17. A.T. Macrander, S. Krasnicki, Y. Zhong, J. Maj, Y.S. Chu, Strain mapping with parts-per-million resolution in synthetic type-Ib diamond plates. Appl. Phys. Lett. **87**(19), 194113 (2005). https://doi.org/10.1063/1.2126790
18. G. Yang, R. Jones, F. Klein, K. Finkelstein, K. Livingston, Rocking curve imaging for diamond radiator crystal selection. Diamond Rel. Mater. **19**(7–9), 719–722 (2010). https://doi.org/10.1016/j.diamond.2009.12.017
19. J. Kim, O. Seo, C. Song, S. Hiroi, Y. Chen, Y. Irokawa, T. Nabatame, Y. Koide, O. Sakata, Lattice-plane bending angle modulation of Mg-doped GaN homoepitaxial layer observed by X-ray diffraction topography. Cryst. Eng. Comm. **21**(14), 2281–2285 (2019). https://doi.org/10.1080/27660400.2023.2199130
20. M. Dudley, X.R. Huang, W. Huang, Assessment of orientation and extinction contrast contributions to the direct dislocation image. J. Phys. D. Appl. Phys. **32**, A139–A144 (1999). https://doi.org/10.1088/0022-3727/32/10A/329
21. Q. Cheng, Z. Chen, S. Hu, Y. Liu, B. Raghothamachar, M. Dudley, Analysis of dislocation configurations in SiC crystals through X-ray topography aided by ray tracing simulations. Mater. Sci. Semicond. Proc. **174**, 108207 (2024). https://doi.org/10.1016/j.mssp.2024.108207
22. T. Zhou, B. Raghothamachar, F. Wu, M. Dudley, Grazing incidence X-ray topographic studies of threading dislocations in hydrothermal grown ZnO single crystal substrates. MRS Online Proceedings Library **1494**, 121–126 (2012). https://doi.org/10.1557/opl.2013.261
23. B. Raghothamachar, Y. Liu, H. Peng, T. Ailihumaer, M. Dudley, F.S. Shahedipour-Sandvik, K.A. Jones, A. Armstrong, A.A. Allerman, J. Han, H. Fu, K. Fu, Y. Zhao, X-ray topography characterization of gallium nitride substrates for power device development. J. Cryst. Growth **544**, 125709 (2020). https://doi.org/10.1016/j.jcrysgro.2020.125709
24. A. Authier, Elementary Dynamical Theory, in *Dynamical Theory of X-Ray Diffraction*, (Oxford University Press, 2001), p. 85. https://doi.org/10.1093/acprof:oso/9780198528920.001.0001
25. X. Huang, M. Dudley, A universal computation method for two-beam dynamical X-ray diffraction. Acta Cryst. A **59**, 163–167 (2003). https://doi.org/10.1107/S0108767303000485
26. M. Wormington, C. Panaccione, K.M. Matney, D.K. Bowen, Characterization of structures from X-rayscattering data using genetic algorithms, Phil. Trans.Royal Soc.Lond. Ser. A: Math. Phys. Eng. Sci. **357**, 2827–2848 (1999). https://doi.org/10.1098/rsta.1999.0469
27. S. Porowski, I. Grzegory, Thermodynamical properties of III–V nitrides and crystal growth of GaN at high N_2 pressure. J. Cryst. Growth **178**(1–2), 174–188 (1997). https://doi.org/10.1016/S0022-0248(97)00072-9
28. H. Wu, J. Spinelli, P. Konkapaka, M. Spencer, Rapid growth of bulk GaN crystal using GaN powder as source material. MRS Online Proc. Library **892**, 3001 (2005). https://doi.org/10.1557/PROC-0892-FF30-01
29. D. Siche, D. Gogova, S. Lehmann, T. Fizia, R. Fornari, M. Andrasch, A. Pipa, J. Ehlbeck, PVT growth of GaN bulk crystals. J. Cryst. Growth **318**(1), 406–410 (2011). https://doi.org/10.1016/j.jcrysgro.2010.10.030
30. Y. Naoi, K. Kobatake, S. Kurai, K. Nishino, H. Sato, M. Nozaki, S. Sakai, Y. Shintani, Characterization of bulk GaN grown by sublimation technique. J. Cryst. Growth **189-190**, 163–166 (1998). https://doi.org/10.1016/S0022-0248(98)00218-8
31. M. Aoki, H. Yamane, M. Shimada, S. Sarayama, F.J. DiSalvo, GaN single crystal growth using high-purity Na as a flux. J. Cryst. Growth **242**(1–2), 70–76 (2002). https://doi.org/10.1016/S0022-0248(02)01349-0

32. Z. Si, Z. Liu, H. Gu, Y. Ren, X. Dong, X. Gao, J. Wang, K. Xu, Stress evolution in different growth mechanism of GaN grown by Na-flux method. Jpn. J. Appl. Phys. **59**(11), 110901 (2020). https://doi.org/10.35848/1347-4065/abbd7c
33. P. Von Dollen, S. Pimputkar, M.A. Alreesh, H. Albrithen, S. Suihkonen, S. Nakamura, J.S. Speck, A new system for sodium flux growth of bulk GaN. Part I: System development. J. Cryst. Growth **456**, 58–66 (2016). https://doi.org/10.1016/j.jcrysgro.2016.07.044
34. P. Von Dollen, S. Pimputkar, M.A. Alreesh, H. Albrithen, S. Suihkonen, S. Nakamura, J.S. Speck, A new system for sodium flux growth of bulk GaN. Part II: in situ investigation of growth processes. J. Cryst. Growth **456**, 67–72 (2016). https://doi.org/10.1016/j.jcrysgro.2016.08.018
35. Z. Mingbin, L. Zhenrong, L. Jingsi, F. Shiji, X. Zhuo, Morphology and characterization of GaN crystals grown by Na flux method under 7 MPa nitrogen pressure. Ferroelectrics **454**(1), 145–151 (2013). https://doi.org/10.1080/00150193.2013.842882
36. Y. Song, W. Wang, W. Yuan, X. Wu, X. Chen, Bulk GaN single crystals: Growth conditions by flux method. J. Cryst. Growth **247**(3–4), 275–278 (2003). https://doi.org/10.1016/S0022-0248(02)02014-6
37. F. Kawamura, T. Iwahashi, M. Morishita, K. Omae, M. Yoshimura, Y. Mori, T. Sasaki, Growth of transparent, large size GaN single crystal with low dislocations using Ca-Na flux system. Jpn. J. Appl. Phys. **42**(7A), L729 (2003). https://doi.org/10.1143/JJAP.42.L729
38. K. Motoki, T. Okahisa, S. Nakahata, N. Matsumoto, H. Kimura, H. Kasai, K. Takemoto, K. Uematsu, M. Ueno, Y. Kumagai, Growth and characterization of freestanding GaN substrates. J. Cryst. Growth **237**, 912–921 (2002). https://doi.org/10.1016/S0022-0248(01)02078-4
39. Y. Oshima, T. Eri, M. Shibata, H. Sunakawa, K. Kobayashi, T. Ichihashi, A. Usui, Preparation of freestanding GaN wafers by hydride vapor phase epitaxy with void-assisted separation. Jpn. J. Appl. Phys. **42**(1A), L1 (2003). https://doi.org/10.1143/JJAP.42.L1
40. K. Motoki, T. Okahisa, R. Hirota, S. Nakahata, K. Uematsu, N. Matsumoto, Dislocation reduction in GaN crystal by advanced-DEEP. J. Cryst. Growth **305**(2), 377–383 (2007). https://doi.org/10.1016/j.jcrysgro.2007.03.038
41. K. Fujito, S. Kubo, H. Nagaoka, T. Mochizuki, H. Namita, S. Nagao, Bulk GaN crystals grown by HVPE. J. Cryst. Growth **311**(10), 3011–3014 (2009). https://doi.org/10.1016/j.jcrysgro.2009.01.046
42. K. Motoki, *Development of Gallium Nitride Substrates, SEI Technical Review, Number 70* (Sumitomo Electric Industries, Osaka, Japan, 2010). https://sumitomoelectric.com/rd/technical-reviews/e070
43. H. Geng, H. Sunakawa, N. Sumi, K. Yamamoto, A.A. Yamaguchi, A. Usui, Growth and strain characterization of high quality GaN crystal by HVPE. J. Cryst. Growth **350**(1), 44–49 (2012). https://doi.org/10.1016/j.jcrysgro.2011.12.020
44. T. Nakamura, K. Motoki, GaN substrate technologies for optical devices. Proc. IEEE **101**(10), 2221–2228 (2013). https://doi.org/10.1109/JPROC.2013.2274930
45. Y. Liu, H. Peng, T. Ailihumaer, B. Raghothamachar, M. Dudley, X-ray topography characterization of GaN substrates used for power electronic devices. J. Electron. Mater. **50**, 2981–2989 (2021). https://doi.org/10.1007/s11664-021-08762-6
46. Y. Wang, M.E. Liao, K. Huynh, A.A. Allerman, M.S. Goorsky, Structural characterization of dot-Core GaN substrates with annealing under growth-like conditions using synchrotron monochromatic X-ray topography. ECS J. Solid State Sci. Technol. **10**, 045010 (2021). https://doi.org/10.1149/2162-8777/abf758
47. Y. Zhang, M. Sun, J. Perozek, Z. Liu, A. Zubair, D. Piedra, N. Chowdhury, X. Gao, K. Shepard, T. Palacios, Large-area 1.2-kV GaN vertical power FinFETs with a record switching figure of merit. IEEE Electron Device Lett. **40**(1), 75–78 (2019). https://doi.org/10.1109/LED.2018.2880306

48. Y. Liu, S. Hu, Z. Chen, Q. Cheng, M-K. Cheng, W. Zhao, T. Buyuklimanli, B. Raghothamachar, M. Dudley, Characterization of Growth Sectors in Gallium Nitride Substrate Wafers, Materials Science Forum, Accepted. (2024)
49. Q. Sun, C.D. Yerino, B. Leung, J. Han, M.E. Coltrin, Understanding and controlling heteroepitaxy with the kinetic Wulff plot: A case study with GaN. J. Appl. Phys. **110**(5), 053517 (2011). https://doi.org/10.1063/1.3632073
50. R. Dwiliński, A. Wysmołek, J. Baranowski, M. Kamińska, R. Doradziński, J. Garczyński, L. Sierzputowski, H. Jacobs, GaN synthesis by ammonothermal method. Acta Phys. Pol. A **88**(5), 833–836 (1995). https://doi.org/10.12693/aphyspola.88.833
51. A. Yoshikawa, E. Ohshima, T. Fukuda, H. Tsuji, K. Oshima, Crystal growth of GaN by ammonothermal method. J. Cryst. Growth **260**(1–2), 67–72 (2004). https://doi.org/10.1016/j.jcrysgro.2003.08.031
52. B. Wang, M.J. Callahan, Ammonothermal synthesis of III-nitride crystals. Cryst. Growth Des. **6**(6), 1227–1246 (2006). https://doi.org/10.1021/cg050271r
53. T. Hashimoto, K. Fujito, M. Saito, J.S. Speck, S. Nakamura, Ammonothermal growth of GaN on an over-1-inch seed crystal. Jpn. J. Appl. Phys. **44**(12L), L1570 (2005). https://doi.org/10.1143/JJAP.44.L1570
54. Y. Kagamitani, D. Ehrentraut, A. Yoshikawa, N. Hoshino, T. Fukuda, S. Kawabata, K. Inaba, Ammonothermal epitaxy of thick GaN film using NH4Cl mineralizer. Jpn. J. Appl. Phys. **45**(5R), 4018 (2006). https://doi.org/10.1143/JJAP.45.4018
55. M.P. D'Evelyn, H.C. Hong, D.S. Park, H. Lu, E. Kaminsky, R.R. Melkote, P. Perlin, M. Lesczynski, S. Porowski, R.J. Molnar, Bulk GaN crystal growth by the high-pressure ammonothermal method. J. Cryst. Growth **300**(1), 11–16 (2007). https://doi.org/10.1016/j.jcrysgro.2006.10.232
56. K. Fujii, G. Fujimoto, T. Goto, T. Yao, Y. Kagamitani, N. Hoshino, D. Ehrentraut, T. Fukuda, Photoluminescence from (0 0 0 1) GaN grown by the acidic ammonothermal technique. J. Cryst. Growth **310**(5), 896–899 (2008). https://doi.org/10.1016/j.jcrysgro.2007.11.060
57. D. Bliss, B. Wang, M. Suscavage, R. Lancto, S. Swider, W. Eikenberry, C. Lynch, Ammonothermal GaN: Morphology and properties. J. Cryst. Growth **312**(8), 1069–1073 (2010). https://doi.org/10.1016/j.jcrysgro.2009.12.007
58. R. Dwiliński, R. Doradziński, J. Garczyński, L. Sierzputowski, R. Kucharski, M. Zając, M. Rudziński, R. Kudrawiec, W. Strupiński, J. Misiewicz, Ammonothermal GaN substrates: Growth accomplishments and applications. Phys. Status Solidi (a) **208**(7), 1489–1493 (2011). https://doi.org/10.1002/pssa.201001196
59. D. Tomida, Y. Kagamitani, Q. Bao, K. Hazu, H. Sawayama, S. Chichibu, C. Yokoyama, T. Fukuda, T. Ishiguro, Enhanced growth rate for ammonothermal gallium nitride crystal growth using ammonium iodide mineralizer. J. Cryst. Growth **353**(1), 59–62 (2012). https://doi.org/10.1016/j.jcrysgro.2012.04.042
60. K. Yoshida, K. Aoki, T. Fukuda, High-temperature acidic ammonothermal method for GaN crystal growth. J. Cryst. Growth **393**, 93–97 (2014). https://doi.org/10.1016/j.jcrysgro.2013.09.010
61. S. Pimputkar, S. Kawabata, J. Speck, S. Nakamura, Improved growth rates and purity of basic ammonothermal GaN. J. Cryst. Growth **403**, 7–17 (2014). https://doi.org/10.1016/j.jcrysgro.2014.06.017
62. J. Hertrampf, N. Alt, E. Schlücker, M. Knetzger, E. Meissner, R. Niewa, Ammonothermal synthesis of GaN using Ba (NH2) 2 as mineralizer. J. Cryst. Growth **456**, 2–4 (2016). https://doi.org/10.1016/j.jcrysgro.2016.08.024
63. M. Zajac, R. Kucharski, K. Grabianska, A. Gwardys-Bak, A. Puchalski, D. Wasik, E. Litwin-Staszewska, R. Piotrzkowski, J.Z. Domagala, M. Bockowski, Basic ammonothermal growth of Gallium Nitride – State of the art, challenges, perspectives. Prog. Cryst. Growth Char. Mater. **64**(3), 63–74 (2018). https://doi.org/10.1016/j.pcrysgrow.2018.05.001

64. S. Yoshida, S. Misawa, S. Gonda, Epitaxial growth of GaN/AlN heterostructures. J. Vac. Sci. Technol. B **1**(2), 250–253 (1983). https://doi.org/10.1116/1.582496
65. I. Akasaki, H. Amano, Y. Koide, K. Hiramatsu, N. Sawaki, Effects of AlN buffer layer on crystallographic structure and on electrical and optical properties of GaN and Ga1− xAlxN (0< x≦ 0.4) films grown on sapphire substrate by MOVPE. J. Cryst. Growth **98**(1–2), 209–219 (1989). https://doi.org/10.1016/0022-0248(89)90200-5
66. K. Hiramatsu, S. Itoh, H. Amano, I. Akasaki, N. Kuwano, T. Shiraishi, K. Oki, Growth mechanism of GaN grown on sapphire with A1N buffer layer by MOVPE. J. Cryst. Growth **115**(1–4), 628–633 (1991). https://doi.org/10.1016/0022-0248(91)90816-N
67. S.N.S. Nakamura, GaN growth using GaN buffer layer. Jpn. J. Appl. Phys. **30**(10A), L1705 (1991). https://doi.org/10.1143/JJAP.30.L1705
68. J. Kuznia, M.A. Khan, D. Olson, R. Kaplan, J. Freitas, Influence of buffer layers on the deposition of high quality single crystal GaN over sapphire substrates. J. Appl. Phys. **73**(9), 4700–4702 (1993). https://doi.org/10.1063/1.354069
69. C. Lin, G. Chi, M. Feng, J. Guo, J. Tsang, J.M. Hong, The dependence of the electrical characteristics of the GaN epitaxial layer on the thermal treatment of the GaN buffer layer. Appl. Phys. Lett. **68**(26), 3758–3760 (1996). https://doi.org/10.1063/1.115997
70. O. Briot, J. Alexis, S. Sanchez, B. Gil, R. Aulombard, Influence of the VIII molar ratio on the structural and electronic properties of MOVPE grown GaN. Sol.-Stat. Electron. **41**(2), 315–317 (1997). https://doi.org/10.1016/S0038-1101(96)00235-3
71. C. Wei, J. Edgar, C. Ignatiev, J. Chaudhuri, The role of trimethylgallium flow during nucleation layer deposition in the optimization of epitaxial GaN films. Thin Solid Films **360**(1–2), 34–38 (2000). https://doi.org/10.1016/S0040-6090(99)00875-5
72. H. Schenk, P. Vennégués, O. Tottereau, T. Riemann, J. Christen, Three-dimensionally nucleated growth of gallium nitride by low-pressure metalorganic vapour phase epitaxy. J. Cryst. Growth **258**(3–4), 232–250 (2003). https://doi.org/10.1016/S0022-0248(03)01519-7
73. W. Lundin, E. Zavarin, D. Sizov, M. Sinitsin, A. Tsatsul'nikov, A. Kondratyev, E. Yakovlev, R. Talalaev, Effects of reactor pressure and residence time on GaN MOVPE growth efficiency. J. Cryst. Growth **287**(2), 605–609 (2006). https://doi.org/10.1016/j.jcrysgro.2005.10.084
74. A. Touré, I. Halidou, Z. Benzarti, T. Boufaden, Growth temperature effect on MOVPE Si-doped GaN: Thermodynamic modeling. Microelectron. J. **40**(2), 363–366 (2009). https://doi.org/10.1016/j.mejo.2008.07.059
75. M. Kappers, M. Moram, D.S. Rao, C. McAleese, C. Humphreys, Low dislocation density GaN growth on high-temperature AlN buffer layers on (0 0 0 1) sapphire. J. Cryst. Growth **312**(3), 363–367 (2010). https://doi.org/10.1016/j.jcrysgro.2009.11.014
76. V. Meyers, E. Rocco, T.J. Anderson, J.C. Gallagher, M.A. Ebrish, K. Jones, M. Derenge, M. Shevelev, V. Sklyar, K. Hogan, B. McEwen, F. Shahedipour-Sandvik, p-type conductivity and damage recovery in implanted GaN annealed by rapid gyrotron microwave annealing. J. Appl. Phys. **128**(8), 085701 (2020). https://doi.org/10.1063/5.0016358
77. V. Meyers, E. Rocco, K. Hogan, B. McEwen, M. Shevelev, V. Sklyar, K. Jones, M. Derenge, F. Shahedipour-Sandvik, P-type conductivity and suppression of green luminescence in Mg/N co-implanted GaN by gyrotron microwave annealing. J. Appl. Phys. **130**(8), 085704 (2021). https://doi.org/10.1063/5.0049101
78. K. Horibuchi, S. Yamaguchi, Y. Kimoto, K. Nishikawa, T. Kachi, Semicond. Sci. Tech. **31**, 034002 (2016). https://doi.org/10.1088/0268-1242/31/3/034002
79. S.O. Kucheyev, J.S. Williams, S.J. Pearton, Ion implantation into GaN. Mater. Sci. Eng. R **33**(2–3), 51–108 (2001). https://doi.org/10.1016/S0927-796X(01)00028-6
80. H. Tan, J. Williams, J. Zou, D. Cockayne, S. Pearton, R. Stall, Damage to epitaxial GaN layers by silicon implantation. Appl. Phys. Lett. **69**(16), 2364–2366 (1996). https://doi.org/10.1063/1.117526

81. N. Parikh, A. Suvkhanov, M. Lioubtchenko, E. Carlson, M. Bremser, D. Bray, R. Davis, J. Hunn, Ion implantation of epitaxial GaN films: Damage, doping and activation. Nucl. Instr. Meth. Phys. Res. B **127**, 463–466 (1997). https://doi.org/10.1016/S0168-583X(97)00076-1
82. B. Pong, C. Pan, Y. Teng, G. Chi, W.-H. Li, K. Lee, C.-H. Lee, Structural defects and microstrain in GaN induced by mg ion implantation. J. Appl. Phys. **83**(11), 5992–5996 (1998). https://doi.org/10.1063/1.367465
83. Y. Liu, H. Peng, Z. Chen, T. Ailihumaer, Q. Cheng, S. Hu, B. Raghothamachar, M. Dudley, Microstructure analysis of GaN epitaxial layers during ion implantation using synchrotron X-ray topography. ECS Trans. **104**(7), 113–122 (2021). https://doi.org/10.1149/10407.0113ecst
84. M.H. Breckenridge, J. Tweedie, P. Reddy, Y. Guan, P. Bagheri, D. Szymanski, S. Mita, K. Sierakowski, M. Boćkowski, R. Collazo, Appl. Phys. Lett. **118**, 022101 (2021). https://doi.org/10.1063/5.0038628
85. M. Nagano, H. Tsuchida, T. Suzuki, T. Hatakeyama, J. Senzaki, K. Fukuda, Annealing induced extended defects in as-grown and ion-implanted 4H–SiC epitaxial layers. J. Appl. Phys. **108**(1), 013511 (2010). https://doi.org/10.1063/1.3457840
86. K. Liu, Y.-K. Su, S.-J. Chang, Y. Horikoshi, Magnesium/nitrogen and beryllium/nitrogen coimplantation into GaN. J. Appl. Phys. **98**, 073702 (2005). https://doi.org/10.1063/1.2073969
87. Y. Liu, S. Hu, Z. Chen, Q. Cheng, B. Raghothamachar, M. Dudley, (Invited) Application of synchrotron X-ray topography techniques to the analysis of defect microstructures in bulk GaN substrates and epilayers for power devices. ECS Trans. **111**(2), 117 (2023). https://doi.org/10.1149/11102.0117ecst
88. Y. Liu, H. Peng, Z. Chen, T. Ailihumaer, S. Hu, B. Raghothamachar, M. Dudley, Application of synchrotron X-ray topography to characterization of ion implanted GaN epitaxial layers for the development of vertical power devices. MRS Adv. **6**, 450–455 (2021). https://doi.org/10.1557/s43580-021-00098-x
89. Y. Wang, K. Huynh, M.E. Liao, J. Tweedie, P. Reddy, M.H. Breckenridge, R. Collazo, Z. Sitar, M. Bockowski, X. Huang, M. Wojcik, M.S. Goorsky, Defects Evolution and Mg Segregation in Mg-implanted GaN with Ultra-High-Pressure Annealing, arXiv:2204.10225 [cond-mat.mtrl-sci]
90. Y. Wang, K. Huynh, M.E. Liao, H.-M. Yu, T. Bai, J. Tweedie, M.H. Breckenridge, R. Collazo, Z. Sitar, M. Bockowski, Y. Liu, M.S. Goorsky, Strain recovery and defect characterization in mg-implanted homoepitaxial GaN on high-quality GaN substrates. Phys. Status Solidi B **257**, 1900705 (2020). https://doi.org/10.1002/pssb.201900705
91. D. Stocker, E. Schubert, J. Redwing, Crystallographic wet chemical etching of GaN. Appl. Phys. Lett. **73**(18), 2654–2656 (1998). https://doi.org/10.1063/1.122543
92. R. Shul, G. McClellan, S. Casalnuovo, D. Rieger, S. Pearton, C. Constantine, C. Barratt, R. Karlicek Jr., C. Tran, M. Schurman, Inductively coupled plasma etching of GaN. Appl. Phys. Lett. **69**(8), 1119–1121 (1996). https://doi.org/10.1063/1.117077
93. B. Li, M. Nami, S. Wang, J. Han, In situ and selective area etching of GaN by tertiary butyl chloride (TBCl). Appl. Phys. Lett. **115**, 162101 (2019). https://doi.org/10.1063/1.5120420
94. B. Li, S. Wang, A.S. Chang, L. Lauhon, Y. Liu, B. Raghothamachar, M. Dudley, J. Han, (Invited) Selective area etching and doping of GaN for high-power applications. ECS Trans. **104**(7), 103–112 (2021). https://doi.org/10.1149/10407.0103ecst
95. C.J. Pan, G.C. Chi, The doping of GaN with Mg diffusion. Solid State Electron. **43**(3), 621–623 (1999). https://doi.org/10.1016/S0038-1101(98)00289-5
96. Y.-J. Yang, J.-L. Yen, F.-S. Yang, C.-Y. Lin, P-type GaN formation by Mg diffusion. Jpn. J. Appl. Phys. **39**(5A), L390 (2000). https://doi.org/10.1143/JJAP.39.L390
97. K. Köhler, R. Gutt, J. Wiegert, L. Kirste, Diffusion of Mg dopant in metal-organic vapor-phase epitaxy grown GaN and $Al_xGa_{1-x}N$. J. Appl. Phys. **113**(7), 073514 (2013). https://doi.org/10.1063/1.4792662

98. L.F. Voss, C.D. Frye, N.A. Allen, S.E. Harrison, K. Kweon, J.B. Varley, V. Lordi, R. Nikolic, T.J. Anderson, J.K. Hite, Moderate temperature Mg diffusion doping of GaN, in *ECS Meeting Abstracts*, vol. MA2020-02 1810, (2020). https://doi.org/10.1149/MA2020-02261810mtgabs
99. N. Allen, *Gallium Nitride Superjunction Transistor: Continued Funding Report* (Lawrence Livermore National Lab.(LLNL), Livermore, CA (United States), 2022). https://doi.org/10.2172/1890078
100. J. Kim, M. Frontzek, L. Crow, J.W. Kwon, Selective thermal neutron transmutation doping with Gd masks in GaN semiconductors. Sol.-State Electron. **199**, 108510 (2023). https://doi.org/10.1016/j.sse.2022.108510
101. J. Kim, Y. Liu, B. Raghothamachar, M. Dudley, J.W. Kwon, Defect-curing effects of fast neutrons on n-type GaN. Mater. Chem. Phys. **315**, 128934 (2024). https://doi.org/10.1016/j.matchemphys.2024.128934

Part III
Epitaxial Growth of Gallium Nitride and Processing

Chapter 7
Laser-Assisted MOCVD GaN and the Development of Vertical GaN-on-GaN PN Diodes

Vijay Gopal Thirupakuzi Vangipuram, Yibo Xu, Wu Lu, and Hongping Zhao

1 Introduction

Wide band gap GaN (E_g ~ 3.4 eV) with intrinsic breakdown field E_{br} ~ 3.5 MV/cm and high electron mobility (phonon limited mobility >2000 cm^2/Vs) has a Baliga figure-of-merit (FOM) at least 5× greater than SiC and almost 1000× greater than Si, and can enable significant reduction in conduction and switching losses for power electronics. While significant research has been done on lateral power switching transistors, with significant industry participation, vertical GaN devices are at a relatively less mature stage. Vertical devices are particularly well suited for high current and high voltage applications since they offer better scalability, and are less susceptible to dynamic on-resistance degradation and reliability issues caused by surface or interface states.

The key challenge to achieving high power operation for GaN vertical devices is to realize precise low doping control of the epitaxial layer with a fast enough growth rate. For example, a doping concentration of less than low-10^{15} cm^{-3} and a drift layer thickness of >50 μm are required for a breakdown voltage above 15 kV (Fig. 7.1). Among the various GaN growth techniques, such as molecular beam epitaxy (MBE), hydride vapor phase epitaxy (HVPE), and metal organic chemical

V. G. T. Vangipuram · Y. Xu · W. Lu
Department of Electrical and Computer Engineering, The Ohio State University, Columbus, OH, USA

H. Zhao (✉)
Department of Electrical and Computer Engineering, The Ohio State University, Columbus, OH, USA

Department of Materials Science and Engineering, The Ohio State University, Columbus, OH, USA
e-mail: zhao.2592@osu.edu

I. C. Kizilyalli et al. (eds.), *Gallium Nitride and Related Materials*, The Materials Research Society Series, https://doi.org/10.1007/978-3-031-83056-3_7

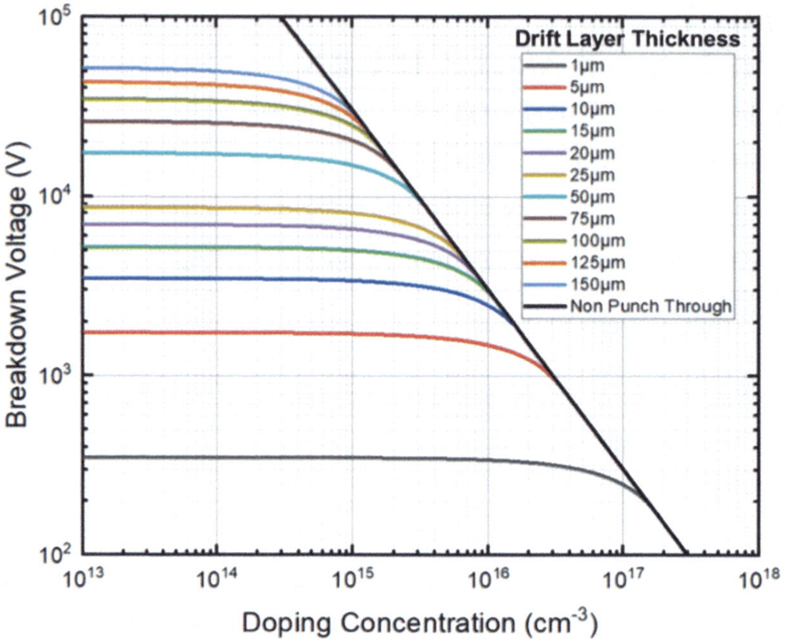

Fig. 7.1 Dependence of V_{br} on the doping concentration in vertical GaN device with different thicknesses of the n-GaN drift region

vapor deposition (MOCVD), HVPE is a method that has been demonstrated with a fast growth rate of GaN (>100 μm/hr). However, HVPE-grown GaN films have not yet shown high breakdown field characteristics successfully. Further investigation is needed to assess the viability of HVPE-grown GaN for high-power vertical power devices. Moreover, the surface morphology of HVPE-grown GaN films is typically rough, requiring chemical mechanical polishing before device fabrication. This rough surface morphology also presents challenges for the in situ growth of GaN PN junctions. MOCVD is an industrially proven and scalable growth method that produces high-quality GaN films with low background doping and a reasonable growth rate (2–3 μm/h). The surface morphology of MOCVD-grown GaN is smooth, making it suitable for homo- and hetero-junction growth and direct device fabrication.

The primary challenge in MOCVD growth of GaN with low controllable doping and fast growth rate (>5 μm/h) for thick drift layer is associated with effectively suppressing impurity incorporation and achieving a low compensation ratio. In this chapter, we will discuss the incorporation of two primary impurities, iron (Fe) and carbon (C), in MOCVD GaN growth. Recently, a novel MOCVD growth technique based on infrared laser-assisted MOCVD (LA-MOCVD) has been demonstrated as an effective technique to suppress C incorporation in GaN, particularly when dealing with fast growth rates [1]. Additionally, using off-axis GaN substrates has shown great promise in suppressing C incorporation in MOCVD GaN [2]. The combination of employing LA-MOCVD and off-axis GaN substrates presents a new path toward achieving high-purity GaN for high-power vertical power devices.

In this chapter, the key elements for developing high-performance vertical GaN-on-GaN PN diodes are introduced. Vertical GaN-on-GaN PN diodes with avalanche breakdown voltages of 1 kV, 1.6 kV, 5 kV, and 7.8 kV are discussed.

2 Impurities (Fe, C) Incorporation in MOCVD GaN

Background silicon (Si), C, and oxygen (O) are considered major impurities in MOCVD-grown GaN that affect the transport properties. Si and O are known as shallow donors in GaN [3]. It is important to minimize the background Si and O in order to achieve low controllable n-doping in the drift layer. C is another impurity that has been extensively studied. Theoretical and experimental studies have proven that C is a deep acceptor in GaN [4–7]. Precursors such as trimethylgallium (TMGa) and triethylgallium (TEGa) are considered to be the main sources of C impurities in MOCVD growth [8, 9], hence C incorporation is strongly influenced by MOCVD growth conditions [10, 11]. Although low net charge concentration can be achieved via compensation, a low compensation ratio (i.e., the ratio of unintentional acceptor concentration $\mathrm{N_A}$ to the intentional donor concentration N_D: N_A/N_D) is necessary to realize high-quality GaN drift layers with high mobility and controllable low doping concentrations. A high compensation ratio could create more scattering centers and lower drift layer mobility [8, 12] resulting in the reduction of carrier lifetime, increased reverse leakage current, and on-resistance (R_{on}).

Despite extensive studies on the incorporation mechanisms of Si, C, O, and H, the unintentionally incorporated impurity of Fe has long been neglected in MOCVD GaN epitaxy. Previously, Fe has been utilized as an intentional dopant in MOCVD to grow semi-insulating GaN using ferrocene as a precursor [13, 14]. Additional studies include the unintentional Fe impurity incorporation from the memory effect and surface segregation of Fe precursor, and Fe-doped GaN substrate [14–18]. Background Fe concentration of 3×10^{15} cm^{-3} was reported in HVPE-grown GaN substrates based on secondary ion mass spectrometry (SIMS) measurement [19]. Furthermore, the existence of background Fe has also been observed in III-V semiconductors [20–23]. Although there are no detailed studies on background Fe in MOCVD GaN, deep-level transient spectroscopy (DLTS) study observed that the concentration of defect state $\mathrm{E_C}$-0.57 eV was positively related to the Fe concentration originating from the Fe-doped GaN substrate during MOCVD AlGaN/GaN growth for high electron mobility transistors (HEMT) [24]. Other studies on deep-level defect centers in AlGaN/GaN HEMT devices suggested that $\mathrm{E_C}$-0.60 eV peak was related to the $\mathrm{Fe_{Ga}}$ defect and $\mathrm{Fe_{Ga}}$-$\mathrm{V_N}$ complexes [24, 25]. Fe impurity was confirmed to induce significant degradation in AlGaN/GaN HEMT device performance such as RF output power [26]. Although the Fe-doped substrate has been identified as an Fe impurity source, this DLTS peak at around ($\mathrm{E_C}$-0.57) eV has been

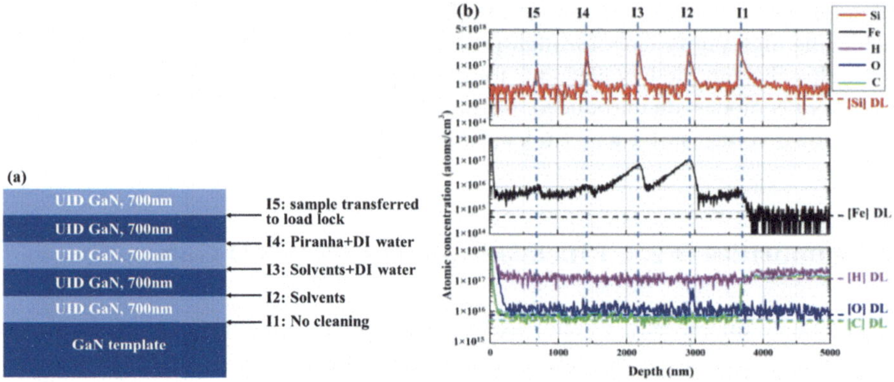

Fig. 7.2 (**a**) Cross-sectional schematic of a UID GaN stack with different pregrowth surface cleaning processes I1-I5. (**b**) Si, Fe, H, C, and O SIMS depth profiles in GaN sub-layers. The vertical dot-dashed lines indicate the regrowth interfaces. The horizontal dashed lines in (**b**) indicate the SIMS DL of each element. Growth conditions in each layer were kept the same. (Reprinted with permission from Ref. [31])

observed in many MOCVD GaN materials in which Fe-doped substrates were not used [27–30].

Recently, the sources of the Fe impurity in the MOCVD-grown GaN were identified and proven by the correlation between SIMS and DLTS [31]. The investigation demonstrated that the Fe impurity level can be minimized by implementing an optimized cleaning process and full susceptor coverage in GaN MOCVD. The studies from this work provide a detailed discussion of the sources and electronic signature of background Fe impurity in MOCVD-grown GaN. Understanding the source and incorporation mechanism of Fe is of great significance toward low unintentional compensation and thus high-performance vertical GaN power devices.

Figure 7.2a illustrates a growth structure that is designed with different cleaning methods at each interface (I1 to I5) [31]. Figure 7.2b shows the Si and Fe SIMS profiles of this sample. Si contamination at the interface is persistent from I1 to I4 sub-layers regardless of the different cleaning processes used, which is between 7×10^{17} and 2×10^{18} cm^{-3}. At the I5 interface, the Si concentration is approximately one order of magnitude lower compared to the other interfaces. This is because the sample is transferred from the growth chamber to the load-lock chamber, without exposing to the atmosphere. Low background Si concentration around $6 \times 10^{15} \sim 1 \times 10^{16}$ cm^{-3} is observed in all epilayers, which enables further lowering of the net charge concentration in the drift layer for higher current blocking capability [31].

Surface cleaning methods appear to have a minimal impact on interfacial and bulk concentrations of H and C. Both concentrations remain near the detection limit (DL) of SIMS. A slight O peak is detected at interface I2, likely attributed to residual organic solvents. However, apart from the peak at I2, O levels closely align with the SIMS detection limit DL across all other interfaces and epilayers. The consistently

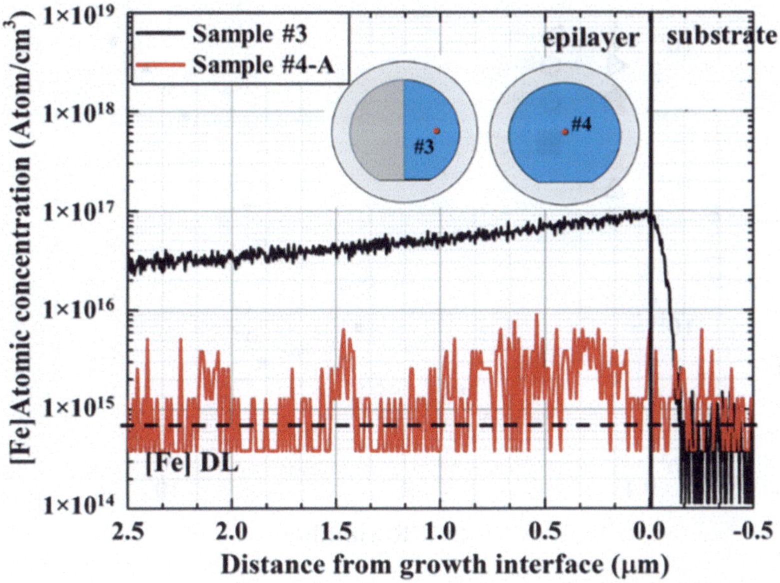

Fig. 7.3 Fe SIMS profiles of Samples #3 and #4-A grown on GaN template. The inset figure illustrates the size of the samples loaded during growth and the SIMS detection point of each sample. The blue area represents the GaN template and the gray area represents the wafer susceptor. Both samples were grown in the same growth conditions. (Reprinted with permission from Ref. [31])

low background levels of Si, H, C, and O suggest minimal dependence on wafer-cleaning processes. In contrast, the Fe SIMS profile shown in Fig. 7.2b has significant changes due to surface treatment. Without any surface cleaning process, the Fe concentration is ~6 × 10^{15} cm^{-3} at the interface I1, subsequently decreasing to ~4 × 10^{15} cm^{-3} as the epilayer growth progressed. However, at interfaces I2 and I3, both subjected to solvent cleaning, Fe contamination levels increase dramatically. The interface Fe concentration at I2 exceeds 1 × 10^{17} cm^{-3} before decreasing to ~1 × 10^{16} cm^{-3} during epilayer growth. Interface I3 exhibits a slightly lower Fe concentration (9 × 10^{16} cm^{-3}), declining to 8 × 10^{15} cm^{-3} within the layer thickness. Interfaces I4 (cleaned with piranha and DI water) and I5 (transferred to load lock) show roughly an order of magnitude lower interface Fe concentration, measuring at 1 × 10^{16} cm^{-3} before decreasing to 5 × 10^{15} cm^{-3} within the epilayer. This indicates that solvent cleaning likely serves as the source of Fe contamination on the regrowth surface [31].

From Fig. 7.2b, it is evident that a background Fe level persists in the epi-GaN layers, ranging from mid- to high-10^{15} cm^{-3}, regardless of the cleaning process employed. This suggests the presence of other potential Fe sources during MOCVD GaN growth. The heated wafer susceptor emerges as a possible impurity source. Figure 7.3a illustrates a quantitative comparison of Fe concentrations between two samples: Sample #3, grown on half of a 2″ GaN template, and Sample #4-A, grown

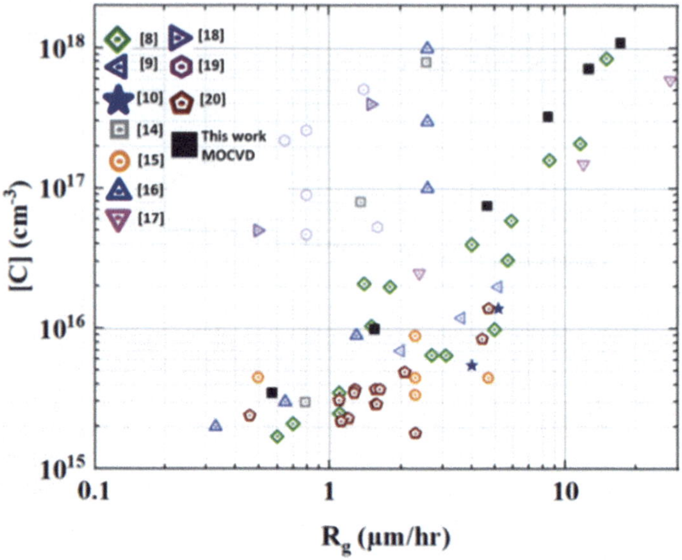

Fig. 7.4 [C] as a function of the GaN growth rate R_g in MOCVD-grown layers from the literature. (Reprinted with permission from Ref. [1])

on a whole 2″ GaN template. Both samples are cleaned with piranha and have identical growth structures of 2.5 μm epilayer growth. The only distinction between the two samples lies in the exposed area to the susceptor [31].

As shown in Fig. 7.3, the Fe concentration at the interface of Sample #3 measures ~9 × 10^{16} cm^{-3}, and steadily decreases to ~3 × 10^{16} cm^{-3}. The decay in Fe concentration does not reach saturation within the 2 μm GaN growth. The decline in Fe concentration in Sample #3 is attributed to a greater influx of Fe from the susceptor, continuously released into the growth environment due to the larger exposed area of the susceptor. As growth progresses, GaN deposition covers the exposed area on the susceptor, likely diminishing Fe diffusion from the susceptor, thus resulting in a gradual decay of Fe concentration. In contrast, the Fe concentration in Sample #4-A falls below the detection limit of SIMS in the epilayer, strongly indicating the susceptor as another crucial source of Fe incorporation in MOCVD GaN. The origin of Fe impurity from the susceptor likely comes from Fe impurities present in the coating material (e.g., SiC) [31].

As previously noted, carbon stands out as another significant impurity within GaN MOCVD growth. Within the GaN crystal structure, C is known to occupy N sites (C_N) and acts as a deep-level acceptor. The metal-organic (MO) precursors utilized in GaN growth, such as TMGa or TEGa, inherently contain C. Consequently, during the MOCVD process of GaN, carbon emerges as a dominant impurity species that readily incorporates into the growing crystal. The incorporation of C impurities in MOCVD GaN becomes more severe when GaN is grown at fast growth rates that require a high flow of TMGa. Figure 7.4 illustrates the C concentration as a function of growth rate. It is evident from this figure that C concentration increases monotonically as the growth rates increases. This presents a significant challenge to achieve low controllable doping in MOCVD GaN with fast growth rates.

3 Laser-Assisted MOCVD GaN Growth

Carbon incorporation in MOCVD growth of III-nitrides is significantly influenced by the stoichiometric balance, which relies on the effective V/III ratio at the growth surface. The input V/III ratio is determined by the molar flow of group V and group III precursors, while the effective V/III ratio refers to the active group V and group III species participating during the growth process. Theoretical studies have suggested that maintaining a high effective V/III ratio is an effective strategy to reduce carbon incorporation [32, 33]. Conventional methods to suppress carbon incorporation in MOCVD GaN growth include (1) increasing the input V/III ratio by increasing the NH_3 flow or reducing the TMGa flow [8–11, 34, 35]; (2) increasing the growth pressure [9, 11, 35]; and (3) increasing the growth temperature [11, 34]. However, these approaches often result in GaN with lower growth rates, increased gas-phase reactions, or increased thermal decomposition. Fundamentally, the low decomposition efficiency of NH_3 in MOCVD GaN growth represents a key factor leading to relatively high carbon incorporation.

NH_3 molecules consist of one nitrogen and three hydrogen atoms. The vibration and rotation of N-H bonds can couple with low-energy photons, allowing the molecule to be excited by these photons. The vibrational and rotational energy states are quantized, meaning that ammonia molecules can only absorb photons with specific energies. Theoretical calculations [36] and experimental measurements [37, 38] have identified several strong absorption peaks corresponding to NH-wagging modes at 932.51 cm^{-1}, 968.32 cm^{-1}, as well as NH rotational-vibrational transition mode at 1084.63 cm^{-1}.

LA-MOCVD employs an infrared laser source to enhance NH_3 decomposition, thereby increasing the effective V/III ratio. Figure 7.5 illustrates a schematic of the LA-MOCVD system [39]. A 1500-watt tunable CO_2 laser (PRC Laser Corp.) serves as the laser source. The CO_2 laser beam is generated and directed into the vertical rotating-disk MOCVD reactor. Positioned parallel to the wafer surface and susceptor, the laser beam facilitates efficient NH_3 decomposition. A beam dump is positioned at the front viewport to absorb any residual laser power. Gas species are uniformly distributed via high-speed rotation of wafer carriers [39].

The absorption of CO_2 laser power is investigated under various controlled chamber conditions [39]. In Fig. 7.6a, the laser power absorption is plotted as a function of laser wavelength under typical reactor conditions for GaN MOCVD growth (965 °C, 200 torr, with NH_3 flow at 4 standard liters per minute (slm) and H_2 flow at 4 slm), with a fixed input laser power of 200 W. Prominently, the most significant absorption occurs at 9.219 μm (1084.71 cm^{-1}), where approximately

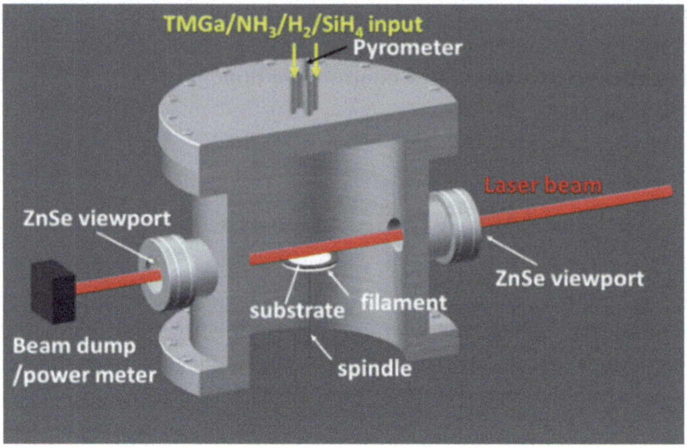

Fig. 7.5 Schematic illustration of LA-MOCVD system. (Reprinted with permission from Ref. [39])

100% of the input laser power is absorbed. This wavelength corresponds to the rotational-vibrational transition of the N-H wagging mode of NH_3 molecules at 1084.63 cm^{-1} [40–42]. Hence, a 9.219 μm laser wavelength has been chosen for LA-MOCVD GaN growths [39]. Figure 7.6b illustrates the laser power absorption (%) as a function of NH_3 flow rate under a chamber pressure of 200 Torr, with a fixed laser wavelength of 9.219 μm and a laser power of 200 W. The absorption increases with the NH_3 flow rate, reaching saturation at 0.75 slm, beyond which further increases result in full absorption of the incident laser power. This study also reveals a strong correlation between laser power absorption and chamber pressure. As shown in Fig. 7.6c, with a fixed NH_3 flow rate of 4 slm, laser absorption is measured as a function of input laser power under different chamber pressures at 75, 100, 150, and 200 Torr. At 75 Torr, full absorption occurs at input powers below 150 W, with absorption gradually decreasing as input power exceeds this threshold. Similar trends are observed for all chamber pressures, although the input power threshold increases with higher chamber pressures. At 200 Torr, the threshold input power is measured at 350 W. Figure 7.6d shows that the threshold input laser power exhibits a linear relationship with chamber pressure. According to the ideal gas law ($PV = nRT$), with fixed temperature (T), reactor volume (V), and gas composition, the amount of NH_3 molecules (n) is directly proportional to chamber pressure (P). Therefore, the threshold input power is proportionate to the amount of NH_3 molecules along the beam pathway [39].

Typical MOCVD GaN growth is performed under nitrogen-rich conditions, with input V/III ratios reaching several thousand. Consequently, the GaN growth rate is governed by the mass transport of Ga species, specifically the flow of TMGa, resulting in a linear relationship between growth rate and TMGa flow rate. However, as TMGa flow rate increases, the effective V/III ratio decreases, leading to a nonlinear relationship between growth rate and TMGa flow due to limited N species

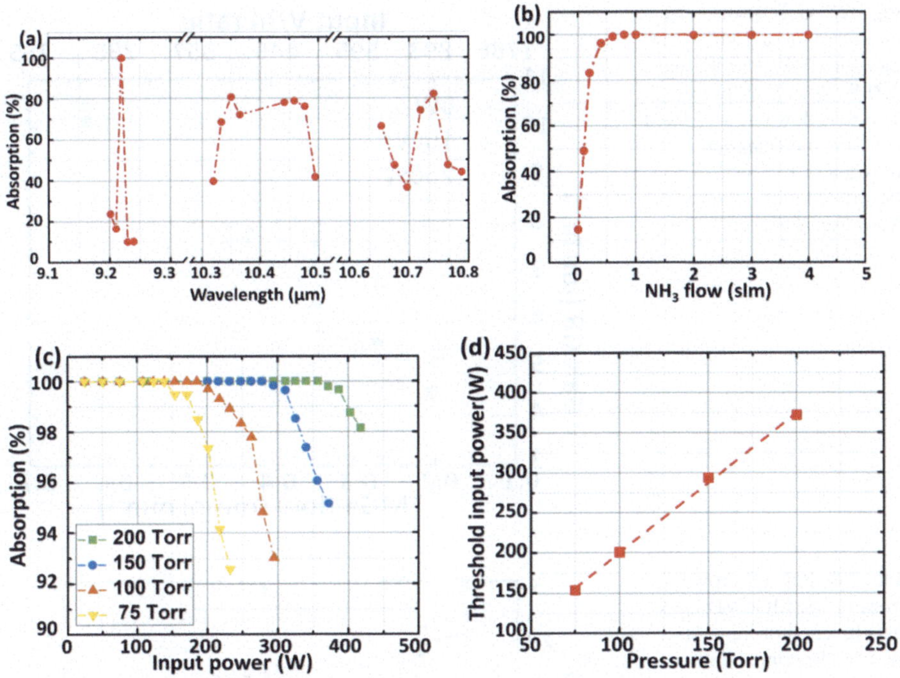

Fig. 7.6 (**a**) Absorption spectrum of CO_2 laser power through MOCVD reactor with 4 slm NH_3 flow rate and 150 torr chamber pressure. (**b**) Laser power absorption as a function of NH_3 flow rate with 200 W input power and 9.219 μm wavelength. The chamber pressure was kept at 150 torr. (**c**) Laser power absorption as a function of input laser power under different chamber pressures with 200 W input power and 9.219 μm wavelength. The NH_3 flow rate was kept at 4 slm. (**d**) The penetration threshold of input laser power as a function of chamber pressure extracted from (**c**). (Reprinted with permission from Ref. [39])

availability. Figure 7.7 shows the GaN growth rate plotted against TMGa flow rate under varying input laser powers: 0 W (no laser), 150 W, and 250 W [39]. For these experiments, the chamber pressure was maintained at 150 Torr, with NH_3 flow constant at 4 slm. In the absence of laser excitation (0 W), the GaN growth rate shows a trend of saturation as the TMGa flow rate increases (>0.4 mmol min^{-1}), indicating the reduction of the effective V/III ratio. Conversely, with laser excitation (150 W and 250 W), the GaN growth rate increased linearly with the TMGa flow rate, indicating that growths were still under N-rich conditions. Consequently, introducing a laser beam during GaN MOCVD growth enhances NH_3 decomposition efficiency. It is worth noting a slight reduction in GaN growth rates with LA-MOCVD compared to conventional growth (0 W), likely due to increased parasitic gas phase reactions facilitated by N radicals (NH_2), which form adducts and particles, thus consuming the precursor supply to the growth surface. Nonetheless, LA-MOCVD remains a promising technique for achieving rapid growth rates of MOCVD GaN with relatively high effective V/III ratios.

Fig. 7.7 GaN growth rate as a function of TMGa flow rate with 0 W, 150 W, and 250 W input laser power. The NH_3 flow rate was fixed at 4 slm and chamber pressure at 150 torr. (Reprinted with permission from Ref. [39])

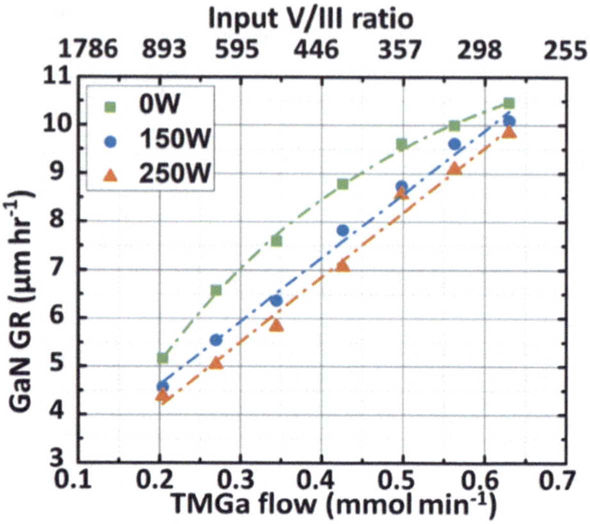

Fig. 7.8 [C] and growth rates of LA-MOCVD GaN as a function of input laser power. (Reprinted with permission from Ref. [39])

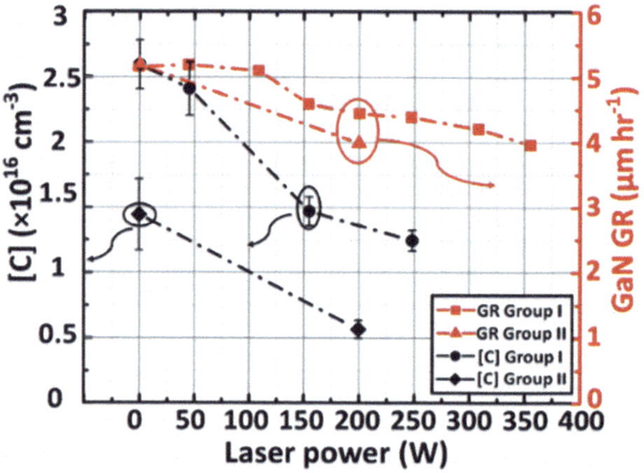

A series of GaN sub-layers were grown under various growth conditions for quantitative SIMS characterization to assess C impurity concentration [39]. The growth conditions are categorized into two groups: Group I (growth pressure 150 Torr) and Group II (growth pressure 200 Torr). Within each group, the input laser power was varied to explore its impact on C incorporation and GaN growth rate. Figure 7.8 illustrates the C concentration and GaN growth rate as a function of laser power. For Group I cases, it is observed that [C] in GaN layers decreases with increasing laser power. Correspondingly, the GaN growth rates in Group I show a slight reduction as the laser power increases. This is attributed to enhanced parasitic gas phase reactions, as discussed previously. A more significant reduction in [C] is observed in LA-MOCVD GaN growth within Group II, where the growth pressure was set at 200 Torr. This set of data directly highlights the effective reduction of

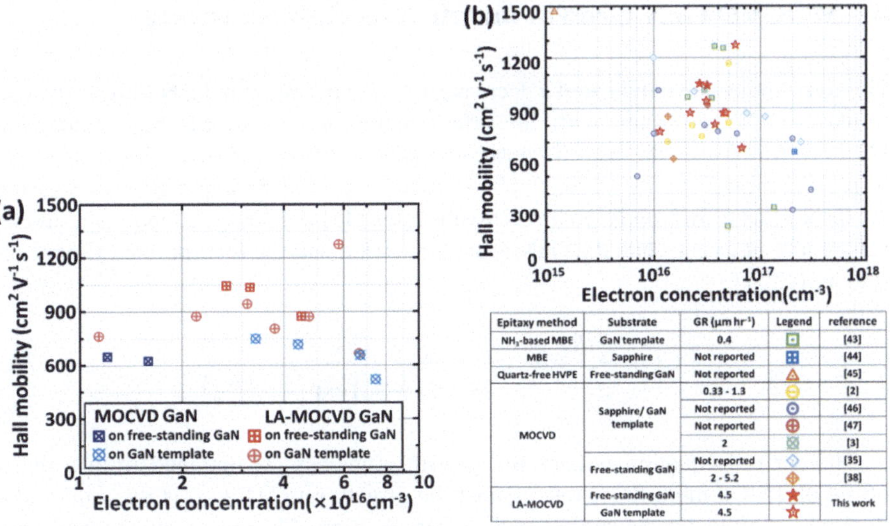

Epitaxy method	Substrate	GR (μm hr^{-1})	Legend	reference
NH_3-based MBE	GaN template	0.4		[43]
MBE	Sapphire	Not reported		[44]
Quartz-free HVPE	Free-standing GaN	Not reported		[45]
MOCVD	Sapphire/ GaN template	0.33 - 1.3		[2]
		Not reported		[46]
		Not reported		[47]
		2		[3]
	Free-standing GaN	Not reported		[35]
		2 - 5.2		[38]
LA-MOCVD	Free-standing GaN	4.5		This work
	GaN template	4.5		

Fig. 7.9 (**a**) Hall mobility of GaN samples grown with conventional MOCVD and LA-MOCVD with electron concentration between 10^{16} and 10^{17} cm^{-3}. The growth rates of conventional MOCVD and LA-MOCVD samples were 5.2 and 4.5 μm hr.$^{-1}$, respectively. (**b**) Hall mobility versus electron concentration of GaN grown with HVPE, MBE, and MOCVD on various substrates from the literature, and LA-MOCVD samples from this study. The reported growth rates from the literature are listed in the table. (Reprinted with permission from Ref. [39])

[C] in LA-MOCVD GaN growth due to enhanced NH_3 decomposition efficiency [39].

Transport properties of samples grown using LA-MOCVD and conventional MOCVD have been compared, through Hall measurements of samples with n$^-$ GaN layers grown on insulating GaN layers containing high C concentrations (~10^{18} cm^{-3}) [39]. The trend reveals consistently higher electron mobilities achieved in samples grown via LA-MOCVD compared to conventional MOCVD growth. In conventional MOCVD growth without laser excitation, room temperature electron mobilities range between 600 and 750 cm^2/Vs. In contrast, LA-MOCVD-grown samples exhibit electron mobilities between 750 and 1275 cm^2/Vs (Fig. 7.9a). This increase in electron mobilities with LA-MOCVD-grown films is likely attributable to reduced impurity scattering resulting from lower C concentrations in the films. Additionally, Fig. 7.9b provides a comparison of electron mobilities for LA-MOCVD-grown films with those of other growth methods and representative MOCVD GaN growth conditions. The comparisons suggest that LA-MOCVD facilitates high mobilities at a relatively fast growth rate, comparable to values achieved through alternate techniques such as MBE, which typically have much slower growth rates.

4 MOCVD GaN Growth on Off-Axis GaN Substrate

Previous studies have indicated a decrease in C incorporation in GaN films grown on m-plane GaN substrates with an offcut angle of 5° toward the -c direction [43, 44]. Moreover, the use of substrates with increased offcut angles toward the m-plane, ranging from 0° to 6.3°, has shown a reduction in impurity incorporation for GaN films grown on N-polar GaN substrates [45]. Recently, a theoretical model was proposed to correlate the GaN growth rate (R_g) and the incorporated C concentration, $[C]$ [1].

$$[C] = \frac{3nR_g}{bf_{\text{total}}}\rho_{\text{GaN}}\frac{k_0}{k_0 + (1 - k_0)\exp\left(-\frac{R_g\delta}{D\sin\theta}\right)} \tag{7.1}$$

The derived equation utilizes the growth efficiency, b, obtained from a linear fitting of TMGa molar flow rate against the growth rate [1]. f_{total} is the total molar flow rate of all incoming species and carrier gas flow. δ is considered the boundary layer thickness, and ρ_{GaN} is the molecular density of GaN per unit volume. n, k_0, and D are the extracted parameters obtained through the fitting. n represents the C adsorption probability, while D represents the bulk diffusion coefficient of the impurity atoms. k_0 is defined as the ratio of the impurity concentration in the solid and on the adsorption layer. The offcut angle is given by θ. In this equation, the resultant C concentration, [C], is thus correlated with the offcut angle as well. The developed equation indicates a reduction in C concentration is possible with an increase in the substrate offcut angle [1].

The combination of employing high-offcut angled substrates and LA-MOCVD growth presents an opportunity to further suppress C concentrations as growth rates increase. The subsequent paragraphs provide insights into C incorporation in Ga-polar GaN-on-GaN growth on substrates with high offcut angles toward the m-plane and its impact on impurity incorporation and epitaxial quality, especially at rapid growth rates.

Ammonothermal GaN substrates with offcut angles of 0.3°, 1°, and 4° toward the m-plane were used for conventional MOCVD (C-MOCVD) and LA-MOCVD growth under various growth conditions corresponding to different growth rates [2]. In this work, Condition C refers to a TMGa flow rate of 409 μmol min^{-1}, resulting in a growth rate of ~8.48 μm/h at an offcut angle of 0.3° for C-MOCVD growth. At the same condition, the growth rate decreased to ~8.25 μm/h for C-MOCVD growth on a substrate with a high offcut angle of 4°. Condition E corresponds to a TMGa flow rate of 911 μmol min^{-1}, leading to a growth rate of ~17.26 μm/h on a sample with an offcut angle of 0.3°. However, the growth rate decreases to ~15.25 μm/h for GaN films grown via C-MOCVD on a substrate with an offcut angle of 4°. Surface morphology comparisons are shown in Fig. 7.10 for corresponding samples [2].

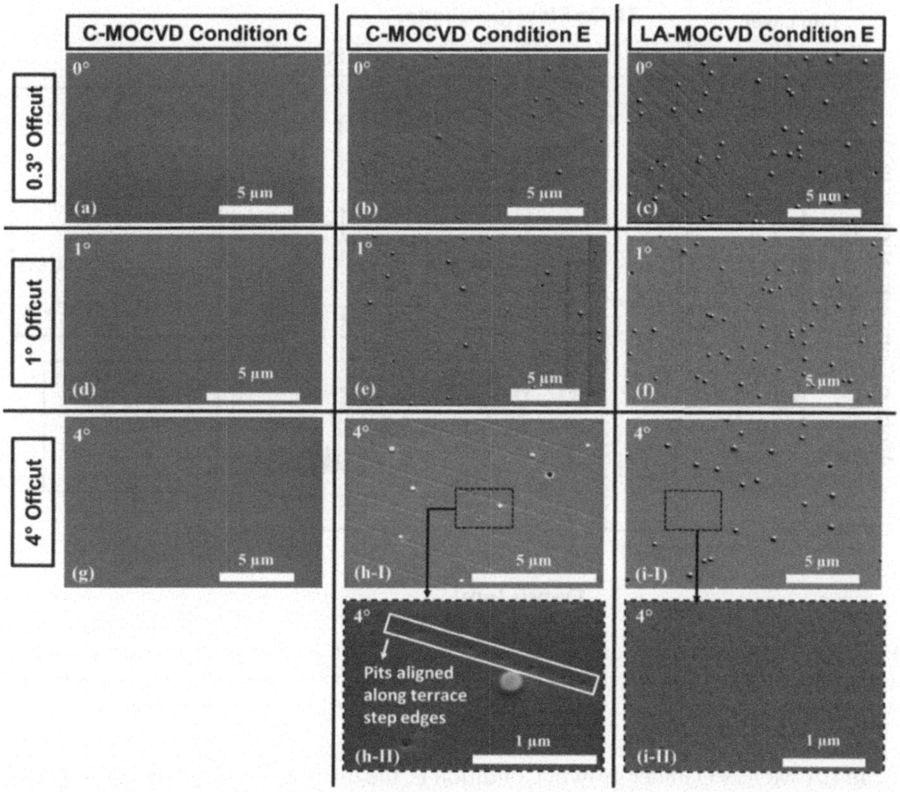

Fig. 7.10 SEM images of GaN samples grown on bulk GaN substrates at different growth conditions. (**a**, **d**, **g**) show images for C-MOCVD GaN grown under Condition C on substrates with 0.3°, 1°, and 4° offcut angles, respectively. (**b**, **e**), (h-I,II) show images for C-MOCVD GaN grown under Condition E on substrates with 0.3°, 1°, and 4° offcut angles, respectively. (**c**, **f**), (i-I,II) show images for LA-MOCVD GaN grown under Condition E on substrates with 0.3°, 1°, and 4° offcut angles, respectively. (Reprinted with permission from Ref [2])

GaN films grown under Condition C via C-MOCVD (Fig. 7.10 a,d,g) exhibit smooth surface morphologies without observable surface microstructures [2]. However, with faster growth rates under Condition E, both C-MOCVD and LA-MOCVD-grown GaN films show similar surface morphologies characterized by micro-steps and the presence of metallic Ga droplets and pit defects, attributed to the high TMGa molar flow rate and hence, low V/III ratio.

In C-MOCVD-grown GaN films (Fig. 7.10b, e, h-I, II), the distribution of pit defects shifts with increasing offcut angle of the GaN substrate [2]. As shown in Fig. 7.10h-II, the pit defects tend to align with the edges of the terrace steps. On surfaces with offcut angles, Ga adatoms exhibit a preference to diffuse and adsorb along these terrace step edges. This preference leads to lower effective V/III ratios at the step edges where Ga adatoms adsorb, consequently resulting in the formation of a high density of pit defects.

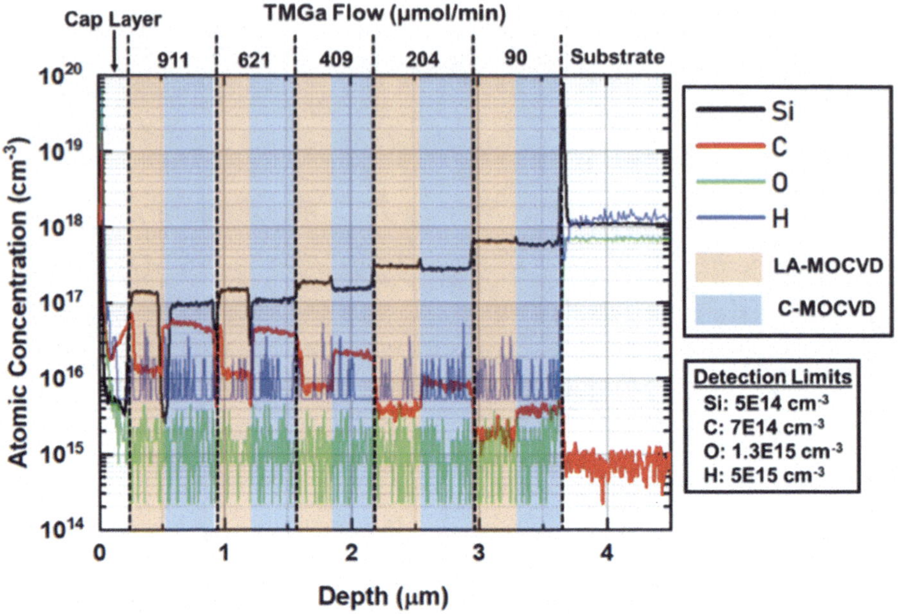

Fig. 7.11 SIMS depth profiles for [Si], [C], [O], and [H] in GaN sublayers grown on a 4° offcut angle GaN substrate. The blue regions indicate sub-layers grown via C-MOCVD, while the red regions indicate sub-layers grown via LA-MOCVD. (Reprinted with permission from Ref [2])

In LA-MOCVD under growth Condition E, the distribution of pit defects on GaN films grown on substrates with a 4° offcut angle (Fig. 7.10i-I, II) appears uniform across the surface in distinct contrast with Fig. 7.10h-I, II [2]. The impact of CO_2 laser excitation in LA-MOCVD growth is twofold. Firstly, the interaction between the laser beam and NH_3 enhances NH_3 decomposition efficiency, thereby increasing the effective V/III ratio and consequently reducing pit density. Secondly, laser-assisted NH_3 decomposition imparts additional kinetic energy, facilitating adatom diffusion on the growth surface and promoting more uniform adatom nucleation. Consequently, this results in a distinct appearance of the surface pit distribution for GaN films grown via LA-MOCVD compared to conventional MOCVD [2].

To quantitatively assess impurities incorporated into GaN films grown on high offcut angles, an SIMS sample with multiple sublayers was grown on a GaN substrate with a 4° offcut angle [2]. The sample comprised sublayers grown at various TMGa molar flow rates ranging from 90 to 911 μmol min^{-1}, corresponding to GaN growth rates between 2.04 and 15.25 μm/h, as measured for C-MOCVD growth. For each sublayer growth condition, both C-MOCVD and LA-MOCVD growth methods were employed, with a constant silane flow maintained throughout the growth stack. As shown in Fig. 7.11, the C concentration increased with increasing TMGa flow rate, which also directly corresponds to the growth rate. However, for each sublayer grown at a specific TMGa flow condition, a reduction in C concentration is observed with LA-MOCVD growth compared to C-MOCVD

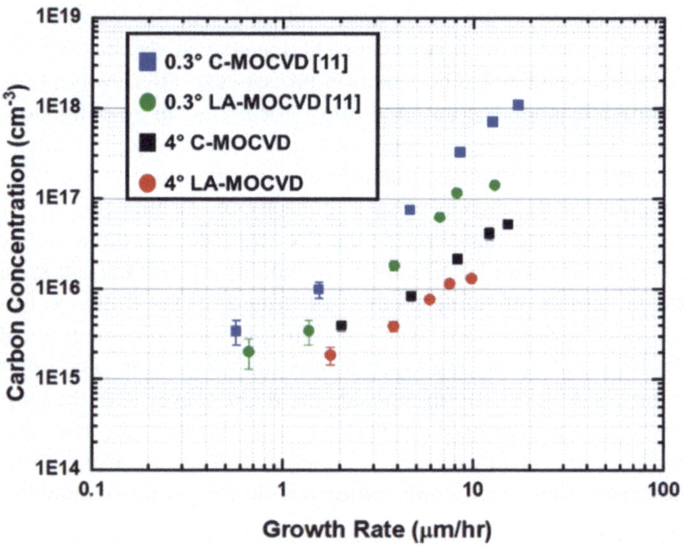

Fig. 7.12 Carbon concentration [C] extracted from SIMS results for C-MOCVD and LA-MOCVD conditions with 0.3° and 4° offcut angles. (Reprinted with permission from Ref [2])

growth. The C concentration within the grown films reduces by 52.4–74.3%, depending on the growth rate, with the utilization of LA-MOCVD when compared with C-MOCVD growth [2].

Figure 7.12 presents a comparison of C concentration against growth rate for films grown via C-MOCVD and LA-MOCVD at various growth rates on substrates with offcut angles of 0.3° and 4°, respectively [2]. It is evident that LA-MOCVD consistently results in a reduction in C concentration compared to C-MOCVD growth. Moreover, an overall decrease in C concentration is observed when substrates with a higher offcut angle are used. By employing a combination of a high-offcut angled substrate for epitaxial GaN growth and the developed LA-MOCVD growth technique, a significant reduction in C concentration was achieved [2]. This reduction in C concentration appears to be more effective as TMGa flow rates increase (i.e., growth rate increases).

5 Device Development and Design of Vertical GaN-on-GaN PN Diodes

5.1 Isolation of Vertical PN Diode

Isolation is one of the critical steps for vertical PN diode fabrication, designed to limit the device's active area. In a vertical PN diode, electrical current flows vertically between the anode and cathode areas. Without proper isolation, the current can spread out underneath the anode area. Not only will the leakage current under reverse bias be higher, but the forward current density can also be misleading if only the anode area is considered. The isolation process can effectively prevent current spreading, reduce leakage current, and allow for the accurate extraction of device characteristics. Typical isolation processes used effectively in vertical GaN PN diodes are etched mesa and ion-implanted isolation regions.

The formation for the device mesa is done by an etching process. By etching away the top p-GaN layer and through the PN interface into the n-GaN drift layer, the device's active area can be effectively limited to prevent current spreading. Ion implantation is another effective method of providing good device isolation. By injecting ions into the top p-GaN layer, e.g., nitrogen ions, to compensate the carriers in p-GaN and form an insulating layer around the anode area. The implanted ions will follow the Gaussian distribution inside the p-GaN layer with the corresponding energy. Therefore, considering the p-GaN layer thickness and carrier density, multiple ion implantations with different energies and doses are usually performed to realize a uniformly distributed implantation profile for good isolation.

5.2 Contacts

Formation of Ohmic contacts on WBG and UWBG materials is one of the major limitations for power electronic applications due to the wide bandgap. Due to the connection between the external circuit and the semiconductor devices, the contact resistance has a critical impact on device performance. Compared with the drift layer resistance, the contact resistance needs to be orders of magnitude lower to reduce the power loss and ensure device stability.

Typically, n-type WBG semiconductors need the contact resistivity in the range of $10^{-5} - 10^{-6}\ \Omega \cdot \text{cm}^2$ to be negligible for device performance [46]. For GaN, in general, to ensure good adhesion and low specific contact resistance, a metal stack starting with Ti is desired for n-GaN ohmic contact. N-type GaN has a work function of about 4.1–4.2 eV, while Ti has a work function of ~4 eV, which are very similar. Naturally formed ohmic contact sometimes can be achieved without thermal annealing. Moreover, the contact metal should have limited diffusion onto the top surface of GaN after annealing. The reaction of Ti, if annealed in N_2 ambient could form TiN. The out-diffusion of N-atoms from GaN leads to the formation of nitrogen vacancies that act as donors and increase the net carrier concentration below the interface [46]. The second layer of metal should have a lower work function to further match the ideal ohmic contact for GaN. Thus, Al is suitable for this purpose, which is indeed generally used [46]. The third and fourth layers of metals generally work to thermally stabilize the contact scheme and antioxidation, with metals such as Ni and Au generally being candidates for the third and fourth layers. After the choice of metal deposition, rapid thermal annealing (RTA) is always desired to form better ohmic contact. The annealing temperature for n-GaN ohmic contact is typically in the range of 600–900 °C. A good specific contact resistivity can reach the range of $10^{-6} - 10^{-7}\ \Omega\ cm^2$ with good quality bulk GaN substrate [46].

P-type ohmic contact presents a major challenge for vertical GaN power device applications. P-type GaN has been suffering from the suitable acceptors (even with Mg) that can properly dope GaN to be p-type material, due to the high activation energy and low activation rate. The low hole mobility and high sheet resistance are the primary issues for p-contact and voltage drop across the p-layer. The typical p-GaN thickness is kept thin (~500 nm) to minimize the voltage drop in p-GaN. Contact metals such as Ni, Pt, and Pd are generally used for potentially lower contact resistance. Ni can potentially form a good ohmic contact with a good p-GaN surface and proper annealing conditions. But the uncertainty of material quality makes the contact to be Schottky. Further, both Pt and Pd have drawbacks as contact metals. Pt-based contacts on GaN are fragile and can easily peel off due to poor adhesion, leading to unstable metal contacts and potential issues with device reliability. Pd, on the other hand, has a better adhesion but is extremely thermally unstable. Even a small variation in annealing temperature can alter the contact properties, causing issues during the device fabrication process. For typical Pt and Pd-based contacts, contact resistivities fall in the range of $10^{-4} - 10^{-5}\ \Omega\cdot cm^2$ with optimized annealing conditions and metal schemes [46].

5.3 *Electrical Field Management*

Vertical GaN PN diodes represent a significant advancement in semiconductor technology, offering remarkable benefits for power electronics due to their high breakdown field and ability to handle high power density [47, 48]. These devices are structured in such a way that current flows vertically through the substrate, as opposed to lateral devices where current flows parallel to the substrate surface. The vertical structure enables the fabrication of devices with larger current handling capabilities and lower on-resistance, making them particularly well suited for applications in power conversion and management. Despite these advantages, vertical GaN PN diodes face unique challenges related to electric field management, which are critical to their performance and reliability.

The high critical electric field of GaN allows for thinner device structures without sacrificing device breakdown. However, this also means that the electric field within the device can become highly concentrated, particularly at junction interfaces underneath the device edge of metal contacts [49]. The electric field profile within the diode without any field management is unevenly distributed, which leads to the device efficiency being far from its theoretical limit. Thus, the reliability and performance of vertical GaN PN diodes are directly influenced by how well the electric field is managed within the device. Without proper management, these high electric field regions can lead to premature and destructive breakdown, reducing the device's breakdown voltage below its theoretical maximum. Efficient electric field management is therefore essential for leveraging the inherent material advantages of

GaN for high-performance power electronics. To address these challenges, various electric field management techniques have been developed and applied to vertical GaN PN diodes such as beveled mesa etching, junction termination extension (JTE), guard rings (GRs), field plate (FP), surface passivation, and high dielectric constant (high-k) material passivation.

Beveled mesa etching reduces field concentration at the edges of the device by shaping the device mesa edge into a beveled or sloped sidewall [50]. It has been verified that a smaller tapered angle can lead to a higher parallel-plane breakdown field which means a higher design efficiency [51]. As an alternative to beveled mesa etching, JTE can spread the electric field by step etching with different depths at the edge of the device [52]. With a properly designed structure and processing multiple etching with different depth toward the drift layer, the electric field can be spread out more evenly. GRs are concentric, lightly doped, isolated box-like structures that surround the active area to provide a gradual change in potential, reducing and mitigating field intensity at the device edges [51]. By introducing ion implantation to inject ions into the semiconductor surface for compensation to create isolation rings around the periphery of the main active area, GRs create a controlled path for the electric field. FPs are metal layers placed over the device contact to attract electric field lines to modify the electric field distribution [50, 51]. This technique requires the strategic placement of metal layers over parts of the semiconductor device to influence the distribution of the electric field. Surface passivation plays an important role in enhancing the performance and reliability of semiconductor devices, especially in suppressing leakage current and electric field management [53]. This method involves the application of an insulating layer on the surface of the semiconductor device to stabilize and protect it from environmental factors, reduce surface states that can trap carriers, and manage the surface electric field, which is crucial for high-voltage applications. Similarly, high-k materials employ insulating layers to protect the device surface and manage the surface electric field. These materials are used to enhance the device breakdown performance, particularly in high-frequency and high-voltage applications [54].

Designing electric field management in vertical GaN PN diodes is not just a matter of enhancing performance, but it is also essential for the practical application and long-term reliability of these devices. The strategies employed to manage the electric field directly impact the device's ability to operate at high voltages and handle high power densities. As GaN technology continues to evolve, innovative electric field management techniques are playing a critical role in unlocking the full potential of vertical GaN PN diodes to make these devices more efficient, reliable, and compact power electronic systems.

The design of vertical GaN power PN diodes for different voltage applications requires different considerations for the implementation of field management methods mentioned above. In the following section, we will briefly discuss several designs of GaN power PN diodes at the voltage rating of 1 kV, 1.6 kV, 5 kV, and 7.8 kV and the applied electric field management [53–56].

5.4 Vertical GaN PN Diode with 1 kV Breakdown Voltage

The targeted voltage rating for low-voltage applications generally falls in the range of <3000 V in general. The design starts with theoretical calculations assuming the critical electric field for GaN to be 3.5 MV/cm [57]. To achieve 1 kV breakdown voltage with considerations of simulation and fabrication efficiency, a punch-through design with a drift layer thickness of 8 μm thick and a doping density of $1 \sim 2 \times 10^{16}$ cm^{-3} was used [55]. A theoretical breakdown voltage of 1.72 kV was calculated accordingly without considering the electrical field crowding.

To reduce the electric field intensity, a technique known as beveled mesa etching was used, as shown in Fig. 7.13a. A mesa profile with a 25° angle was employed to shift the critical electric field from the mesa edge [55]. Figure 7.13b illustrates the profile of the electric field across the device with a reverse bias of 1.32 kV. The peak electric field is at the p-n junction underneath the p-contact edge and penetrates through the drift layer to the n-GaN/n + GaN interface. The peak electric field reaches 3.5 MV/cm at the PN junction before tapering off near the mesa edge, and a smaller peak is observed at the edge of the mesa edge as shown in Fig. 7.13c.

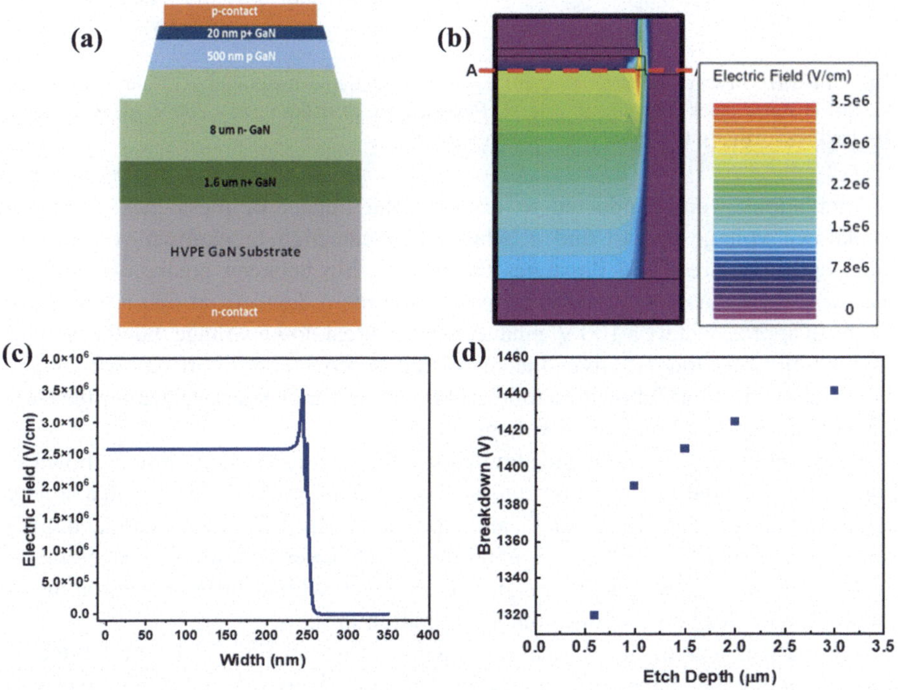

Fig. 7.13 (**a**) 1 kV vertical GaN diode schematic and (**b**) electric field distribution under −1.32 kV. (**c**) Electric field profile along the cutline (AA'). (**d**) Breakdown voltage vs. mesa etching depth. (Reprinted with permission from [55])

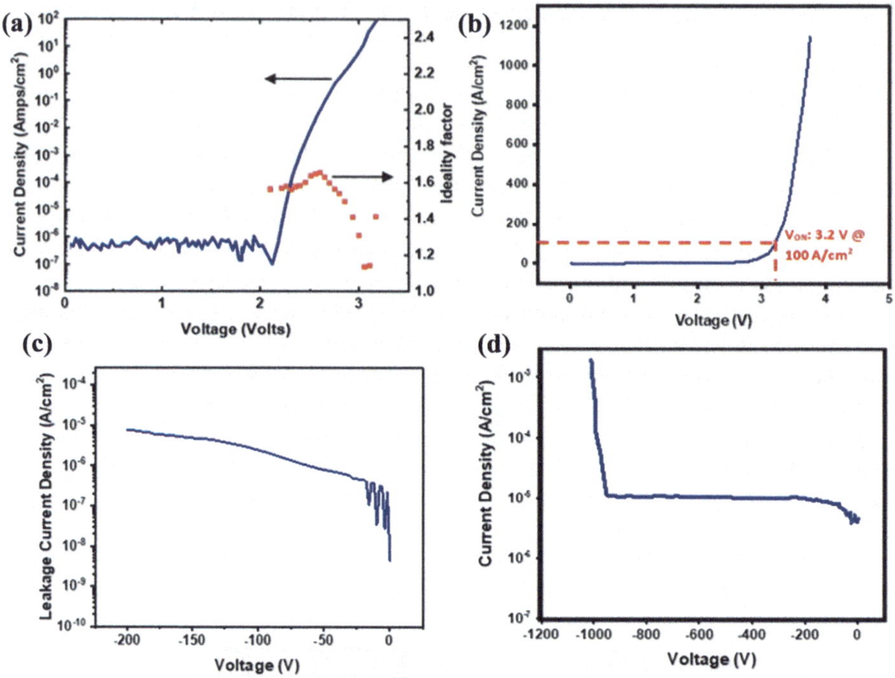

Fig. 7.14 (**a**) Forward characteristics in log scale and the corresponding ideality factor. (**b**) Forward characteristics in the linear scale. (**c**) Reverse characteristics up to—200 V. (**d**) Breakdown characteristics. (Reprinted with permission from [55])

Simulations were conducted to examine the impact of mesa etch depth on breakdown voltage and to find a balance between high breakdown voltage and etch depths. Figure 7.13d illustrates the relationship between breakdown voltage and mesa depth. As can be seen, increasing the mesa depth from 600 nm to 2 μm results in approximately a 100 V enhancement in breakdown voltage, but the etching time is four times longer. Given the significant improvement in breakdown voltage with a relatively short fabrication time, a 600 nm mesa etch was selected for the GaN diode, resulting in a breakdown voltage of 1 kV.

Based on the optimization and simulations, the final device structure is shown in Fig. 7.13a. The finalized structure consisted of a 1.6 μm n + GaN layer with a doping concentration of 2×10^{18} cm^{-3}, an 8 μm n- GaN drift layer with a doping concentration of 1.5×10^{16} cm^{-3}, a 500 nm p-GaN layer with a hole concentration of 1×10^{17} cm^{-3}, and a 20 nm p + GaN layer with a hole concentration of 5×10^{17} cm^{-3} for better ohmic contact formation [55].

The performance of the fabricated device is shown in Fig. 7.14. The fabricated device is in circular shape with a diameter of 650 μm. The extracted ideality factor as a function of bias is shown in Fig. 7.14a where the lowest point is 1.15 as the voltage approaches 3.1 V. The device exhibited a low turn-on voltage of 3.2 V at a current density of 100 A/cm^2 as shown in Fig. 7.14b. The device shows a low leakage current density of 5×10^{-6} A/cm^2 at—200 V as shown in Fig. 7.14c. Figure 7.14d shows the breakdown voltage of the fabricated vertical GaN PN diodes to be 1.028 kV considering a current density of 1 mA/cm^2. The measured breakdown voltage is in great agreement with the numerical simulations with 78% efficiency [55].

5.5 *Vertical GaN PN Diode with 1.6 kV Breakdown Voltage*

With a similar drift layer and doping concentration, GRs and surface passivation were implemented into the design to further enhance the device's design efficiency. A 1.5 kV diode was designed and fabricated, incorporating improved electric field management techniques. Considering an overall efficiency of 70% between fabricated device and analytical calculations, a drift layer thickness of 8 μm with an electron concentration of 9×10^{15} cm^{-3} was chosen. To mitigate field crowding at the PN interface under the edge of the contact, GR structures were implemented for edge termination. Numerical simulations were conducted to optimize the design.

The layer structure is shown in Fig. 7.15a as a starting point of the simulation. Figure 7.15b shows the electric field profile of the structure under a reverse bias of 1.57 kV without any guard rings or passivation. The electric field peaks at the edge

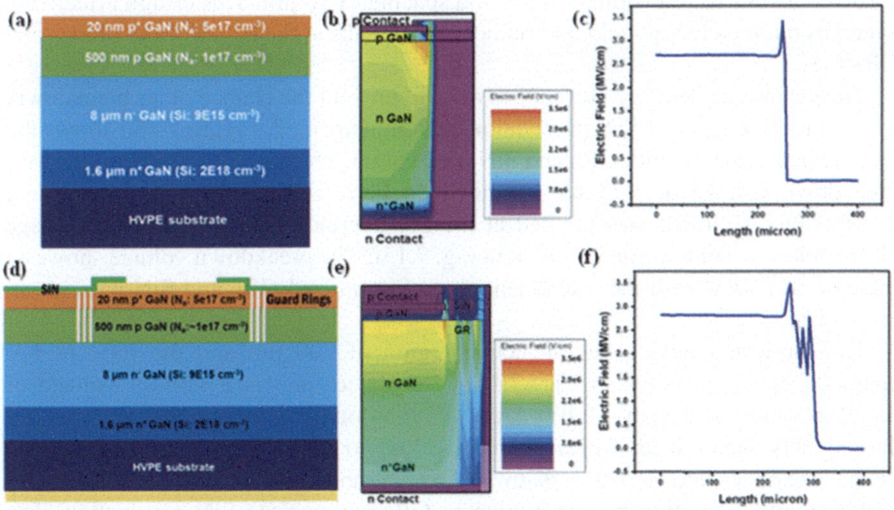

Fig. 7.15 (**a**) Initial layer structure for 1.5 kV vertical GaN PN diode design. (**b**) Electric field distribution under a reverse bias of 1.57 kV without electric field management. (**c**) Electric field profile along the cutline under PN interface at −1.57 kV. (**d**) Finalized device layer structure with GRs and SiN passivation implemented. (**e**) Electric field distribution under a reverse bias of 1.73 kV with optimized electric field management. (**f**) Electric field profile along the cutline under PN interface at −1.73 kV. (Reprinted with permission from [53])

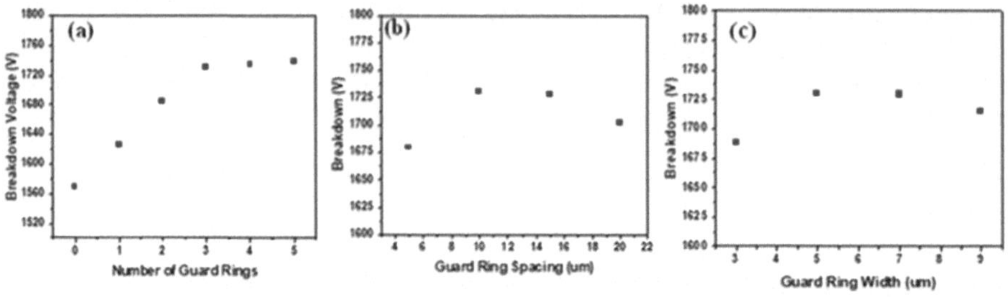

Fig. 7.16 (**a**) Device breakdown voltage vs. GR number at a fixed GR spacing and GR width. (**b**) Device breakdown voltage vs. GR spacing at a fixed GR number and width. (**c**) Device breakdown voltage vs. GR width with fixed GR number and spacing. (Reprinted with permission from [53])

of the contact and has reached the breakdown field of GaN of 3.5 MV/cm as shown in Fig. 7.15c. To reduce the field crowding at the contact edges for enhanced breakdown voltage, ion-implanted GRs were designed. A fine-tuning process was performed to optimize the GR structure, considering the number, width, and spacing of the GRs. First, the GR width and spacing were fixed at 5 μm and 10 μm, respectively, while varying the number of guard rings. As shown in Fig. 7.16a, the breakdown voltage increments saturated at 3 GRs. Next, the GR spacing was varied from 5 μm to 20 μm with a fixed number and width of GRs. The breakdown voltage was highest at a spacing of 10 μm, as shown in Fig. 7.16b. Finally, the GR width was optimized from 3 μm to 9 μm, as shown in Fig. 7.16c. The optimized solution involved 3 GRs with a width of 5 μm and spacing of 10 μm. This design effectively averages out the electric field and reduces field peaks at the edges of the diodes and GRs [53].

To decrease the leakage current density and prevent the chance of air breakdown, a 500 nm thick layer of SiN passivation was implemented. Figure 7.15d shows the final device structure with GRs and SiN passivation. The designed structure shows a breakdown voltage of 1.73 kV as shown in Fig. 7.15e of the 2D electric field distribution. A cutline was applied at the PN interface under breakdown voltage and the electric field profile is shown in Fig. 7.15f. The breakdown voltage shows an increase of 260 V with the implementation of optimized GRs and SiN passivation layer [53].

The fabricated device has an active region of 200 μm and the forward I-V characteristics are shown in Fig. 7.17a. The device shows a turn-on voltage of 4.5 V at a current density of 100 A/cm^2 with an on-resistance of 1.5 mOhm-cm^2. The ideality factor in this bias regime is close to 1.89. Figure 7.17b shows the device's reverse characteristics. A low leakage current of 7×10^{-8} A/cm^2 at −200 V indicates good isolation by ion-implanted GRs and surface SiN passivation. The breakdown voltage was measured to be 1575 V at 2.5×10^{-2} A/cm^2. The measured breakdown voltage shows an efficiency of 72% compared with the theoretical value and 91% with the simulation [53].

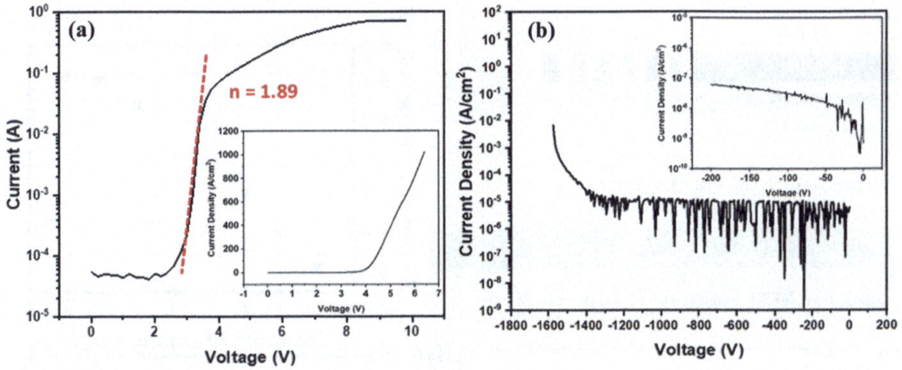

Fig. 7.17 (**a**) Fabricated device forward I-V characteristics and (**b**) device reverse characteristics. (Reprinted with permission from [53])

5.6 *Vertical GaN PN Diode with 5 kV Breakdown Voltage*

More recently, the availability of high-quality free-standing GaN substrates has supported the development of higher breakdown voltage vertical GaN PN power diodes [56]. MOCVD-grown GaN with a fast growth rate has been investigated, enabling the possibility of a thicker drift layer within a reasonable growth period. This advancement has further expanded the design potential to achieve high breakdown voltage for vertical GaN power PN diodes.

A 5 kV breakdown voltage GaN PN diode was targeted, thanks to the high-quality material from MOCVD. Based on the simulation and considering a fabrication efficiency of 65%, a 28 μm n^- drift layer with a doping concentration of 1×10^{15} cm^{-3} is employed. The considerations for electric field management include GRs and a layer spin-on-glass (SoG) for surface passivation. Figure 7.18a shows the targeted device layer structure. The optimization of GRs follows the same strategy as described before. By varying the GR numbers, width, and spacing of the GRs, the optimized GR structure is finalized with 6 GRs with a width of 5 μm and a spacing of 10 μm as shown in Fig. 7.18b–d. With the optimized GR design and a 3 μm thick SoG layer implemented, the device shows a breakdown voltage of 6.4 kV, considering the GaN critical electric field to be 3.5 MV/cm. The 2D electric field distribution under a reverse bias of 6.4 kV is shown in Fig. 7.18e. Figure 7.18f shows the electric field profile along the applied cutline AA′at the PN interface [56].

The fabricated device shows a turn-on voltage of 4.8 V at a current density of 100 A/cm^2 with a specific resistance of 0.9 mΩcm^2 as shown in Fig. 7.19a and b. The device exhibits a low leakage current density of 1×10^{-7} A/cm^2 at −200 V and 1.5×10^{-5} A/cm^2 at −3 kV as shown in Fig. 7.19c and d. Finally, Fig. 7.19d shows the breakdown characteristics of the device to be 4.92 kV considering the current density at 2 mA/cm^2. The fabricated device shows an efficiency of 77% compared with numerical simulation. Based on the extracted on-resistance and the measured

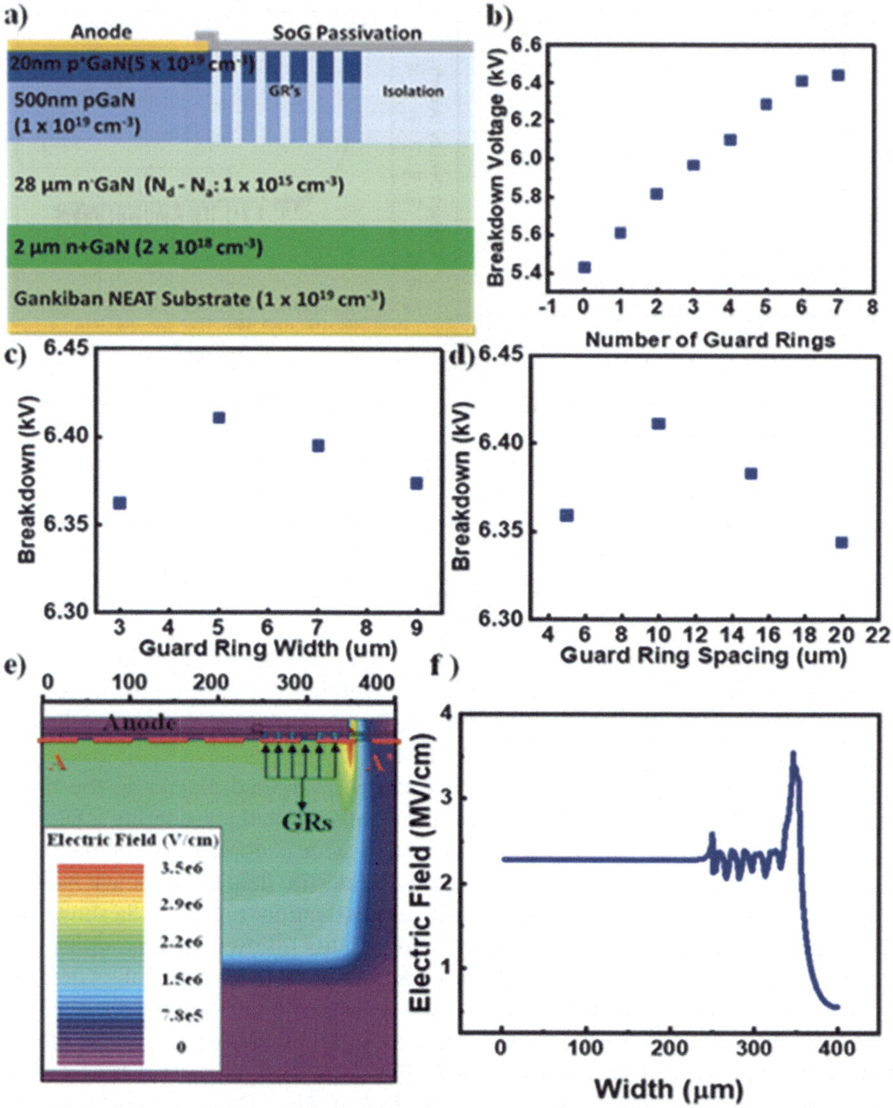

Fig. 7.18 (**a**) Schematic of the GaN p-n junction diode. (**b–d**) GR optimization is based on GR numbers, width, and spacing. (**e**) 2D electric field distribution of the designed structure with optimized GRs structure under a reverse bias of 6.4 kV. (**f**) Electric field profile along the cutline AA' at the on interface under a reverse bias of 6.4 kV. (Reprinted with permission from [56])

breakdown voltage, the device has a remarkable Baliga Figure of Merit (BFOM) of 27 GW/cm^2 [56].

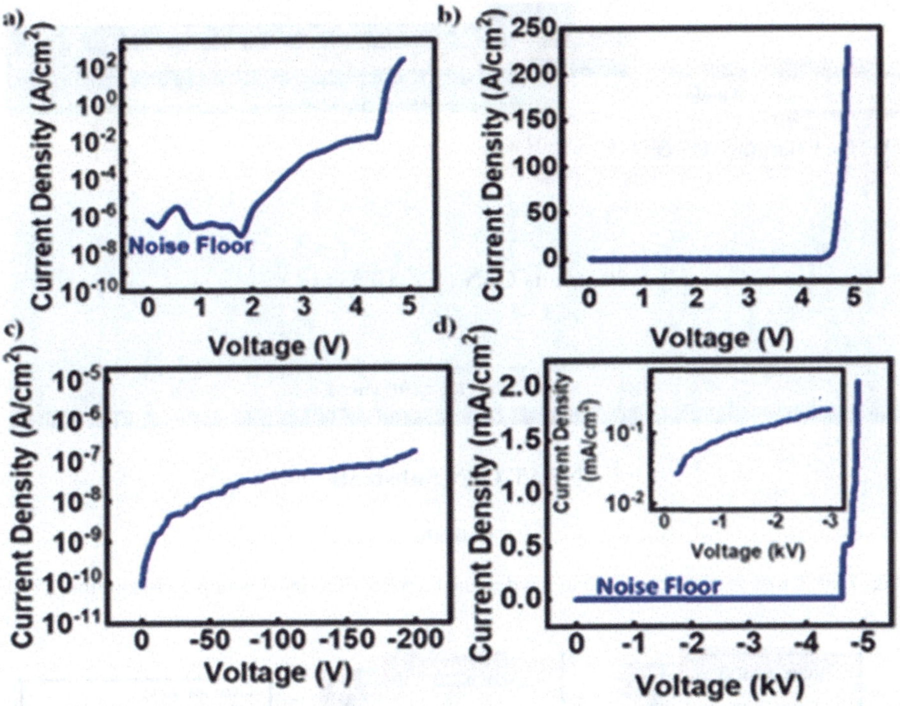

Fig. 7.19 (**a**) Forward I–V characteristics in log scale. (**b**) Forward IV in linear scale. (**c**) Reverse characteristics under a reverse bias of 200 V. (**d**) The device has a breakdown voltage of 4920 V at 2 mA/cm^2. The inset shows the device's reverse leakage current up to 3 kV. (Reprinted with permission from [56])

5.7 *Vertical GaN PN Diode with 7.86 kV Breakdown Voltage*

The most recent work has reported a vertical GaN power PN diode with a breadown voltage of 7.86 kV with a 57 μm thick drift layer and doping of 1×10^{15} cm^{-3}. In addition to GRs, field plate, and surface passivation, one of the key innovations of this work is the implementation of high-k dielectric, perovskite oxide $BaTiO_3$ (BTO) for electric field mitigation. For device design, the goal of this work is to demonstrate GaN PN diodes with a breakdown voltage >7.5 kV. Considering the breakdown efficiency of 80% compared with the numerical simulation, the simulated layer structure was designed to have a 50 μm drift layer with a doping concentration of 1×10^{15} cm^{-3}. The final designed GaN-on-GaN PN diode structure is shown in Fig. 7.20 [54].

Numerical simulations were conducted for electric field management, considering the critical electric field to be 3.5 MV/cm. Similarly, as discussed above, GR structures were optimized and implemented first. Six GRs with a width of 8 μm and a spacing of 14 μm were implemented. Then, high-k dielectrics, SoG, and FP were

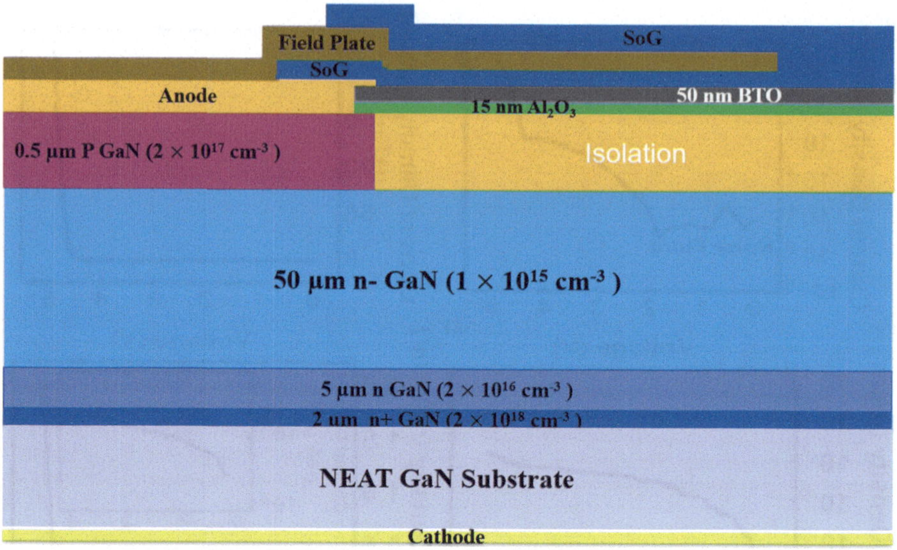

Fig. 7.20 Cross-section schematic of the designed device. (Reprinted with permission from [54])

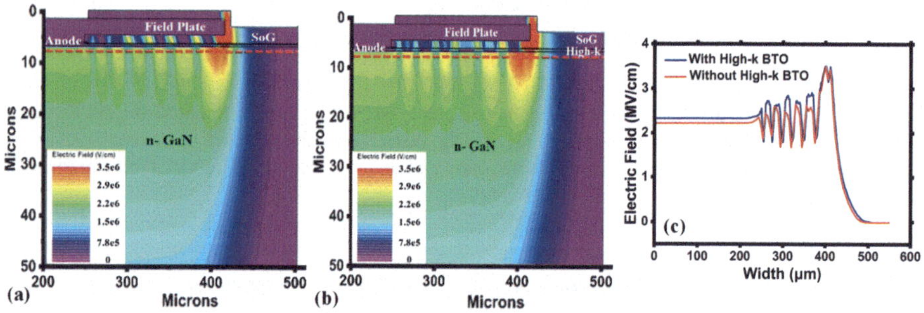

Fig. 7.21 (**a**) The enlarged simulated device electric field profile under 9.05 kV reverse bias with GRs, SoG, and field plate. (**b**) The enlarged simulated electric field profile under a reverse bias of 9.65 kV with GRs, high-k dielectrics, field plate, and SoG. (**c**) Electric field profiles comparison for structures with (**a**) and without high-k BTO (**b**), along the red cutline starting from the center of the anode (0 micron). (Reprinted with permission from [54])

applied on top of the structure. This design made the electric field at the edges of each GR more uniform and shifted the peak electrical field away from the anode area to the edge of the field plate. Figure 7.21a and b shows the electric field distribution of the device structure with and without BTO implementation, respectively, for direct comparison. Without BTO, the electrical peaks at the PN interface indicate a breakdown voltage of 9.05 kV as shown in Fig. 7.21c (the red curve). With BTO implemented, the device breakdown voltage is 9.65 kV as shown in Fig. 7.21c (the

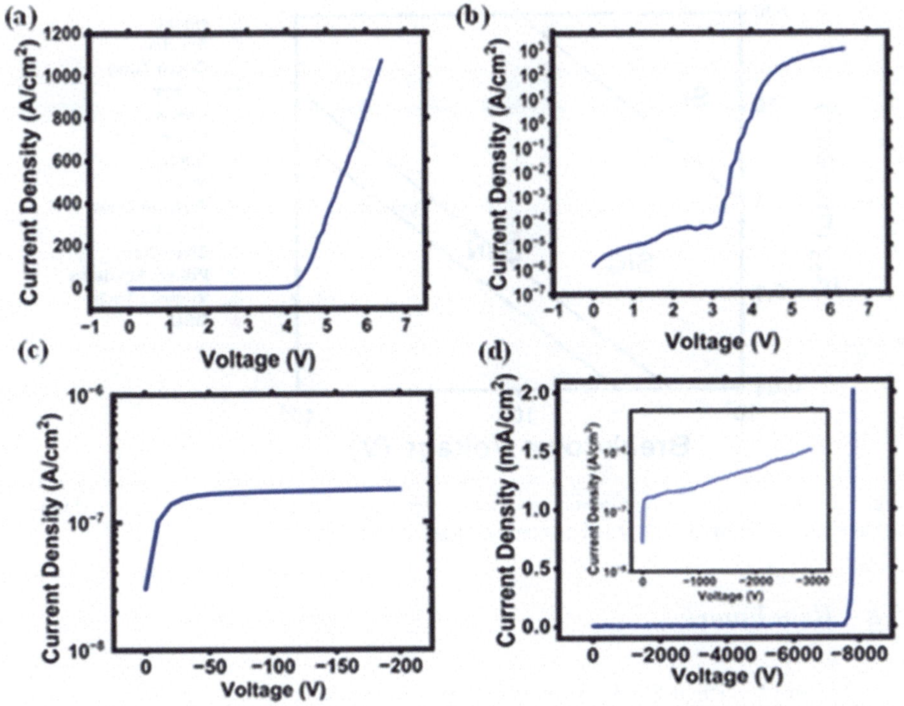

Fig. 7.22 (**a**) Device forward characteristics in linear scale. (**b**) Device forward characteristics in log scale. (**c**) Reverse leakage current density under −200 V. (**e**) Reverse leakage current density under −3000 V. (**f**) The device shows a breakdown behavior under −7.86 kV. (Reprinted with permission from [54])

blue curve). The high-k layers lead to a 600 V improvement in breakdown voltage based on the simulation [54].

The forward characteristic of the fabricated device is shown in Figs. 7.22a and b. The device shows a turn-on voltage of 4.55 V at 100 A/cm^2 and an on-resistance of 2.8 mΩ·cm^2. Figure 7.22b shows the forward characteristic in the log scale, an ideality factor of 1.92 was extracted accordingly. As shown in Figs. 7.22c and d, the device exhibited a low leakage current density of 1.7×10^{-7} A/cm^2 under −200 V and 1.2×10^{-6} A/cm^2 at the reverse bias of −3000 V. The breakdown characteristics are presented in Fig. 7.22d. A breakdown voltage of 7.86 kV was measured at a current density of 2 mA/cm^2. With the extracted on-resistance and breakdown voltage, the device shows a BFOM of 22 GW/cm^2 [54].

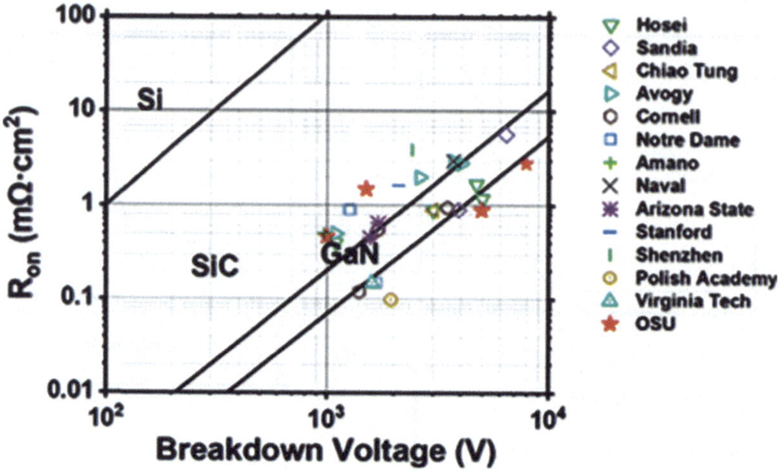

Fig. 7.23 Benchmark of representative reported vertical GaN-on-GaN PN diodes at different voltage ratings. Works discussed above are highlighted as red stars

5.8 Benchmark

The design of vertical GaN power PN diodes for various voltage-level applications has been thoroughly demonstrated. By implementing effective electric field management and optimization techniques in combination with the exceptional material properties of GaN, the potential of vertical GaN-on-GaN power device technology can be sufficiently realized. Figure 7.23 benchmarked the devices introduced above at different voltage ratings together with other reported vertical GaN PN power diodes [14, 16, 50, 52–54, 56, 58–69].

6 Concluding Remarks

This chapter introduced the essential components of developing high-performance vertical GaN power PN diodes, spanning from MOCVD epitaxy to device design and demonstration. It discusses the incorporation mechanisms of two primary impurities (Fe, C) in MOCVD GaN. The novel LA-MOCVD growth technique is demonstrated as an effective method to enhance the effective V/III ratio through efficient NH_3 decomposition, and thus suppress C incorporation, particularly beneficial for MOCVD GaN with rapid growth rates. Furthermore, the use of off-axis GaN substrates is demonstrated as another effective approach to suppress C incorporation. These methodologies offer new avenues for producing high-purity GaN with controllable low doping and thick films necessary for vertical GaN-on-GaN device technologies. Additionally, this chapter introduced the important device elements for vertical GaN-on-GaN PN diodes, including isolation, contacts, and field management. Vertical GaN PN diodes with four voltage ratings (1 kV, 1.6 kV, 5 kV, and 7.8 kV) were presented. With the current device demonstrations, vertical GaN-on-GaN device technology exhibits substantial potential to scale up to even higher voltage ratings exceeding 10 kV with high current ratings. Broadly, the discussed methodologies and demonstrations serve as pivotal building blocks for the ongoing advancement of GaN device technologies.

References

1. Y. Zhang, V.G. Thirupakuzi Vangipuram, K. Zhang, H. Zhao, Investigation of carbon incorporation in laser-assisted MOCVD of GaN. Appl. Phys. Lett. **122**(16), 162101 (2023)
2. V.G. Thirupakuzi Vangipuram, K. Zhang, H. Zhao, Suppressing carbon incorporation in metal–organic chemical vapor deposition GaN using high-offcut-angled substrates. Phys. Status Solidi RRL—Rapid Res. Lett. **18**(3), 2300318 (2024)
3. W.J. Moore, J.A. Freitas Jr., G.C.B. Braga, R.J. Molnar, S.K. Lee, K.Y. Lee, I.J. Song, Identification of Si and O donors in hydride-vapor-phase epitaxial GaN. Appl. Phys. Lett. **79**(16), 2570–2572 (2001)
4. J.L. Lyons, A. Janotti, C.G. Van de Walle, Carbon impurities and the yellow luminescence in GaN. Appl. Phys. Lett. **97**(15), 152108 (2010)
5. A. Armstrong, A.R. Arehart, D. Green, U.K. Mishra, J.S. Speck, S.A. Ringel, Impact of deep levels on the electrical conductivity and luminescence of gallium nitride codoped with carbon and silicon. J. Appl. Phys. **98**(5), 053704 (2005)
6. A. Armstrong, A.R. Arehart, B. Moran, S.P. DenBaars, U.K. Mishra, J.S. Speck, S.A. Ringel, Impact of carbon on trap states in n-type GaN grown by metalorganic chemical vapor deposition. Appl. Phys. Lett. **84**(3), 374–376 (2004)
7. A. Armstrong, A.R. Arehart, B. Moran, S.P. DenBaars, U.K. Mishra, J.S. Speck, S.A. Ringel, Impact of growth pressure on defects in GaN grown by metalorganic chemical vapor deposition, in *2003 Int. Symp. Compd. Semicond. Post-Conf. Proc. IEEE Cat No03TH8767*, (2003), pp. 42–48
8. F. Kaess, S. Mita, J. Xie, P. Reddy, A. Klump, L.H. Hernandez-Balderrama, S. Washiyama, A. Franke, R. Kirste, A. Hoffmann, R. Collazo, Z. Sitar, Correlation between mobility collapse and carbon impurities in Si-doped GaN grown by low pressure metalorganic chemical vapor deposition. J. Appl. Phys. **120**(10), 105701 (2016)
9. Y. Cao, R. Chu, R. Li, M. Chen, R. Chang, B. Hughes, High-voltage vertical GaN Schottky diode enabled by low-carbon metal-organic chemical vapor deposition growth. Appl. Phys. Lett. **108**(6), 062103 (2016)
10. G. Piao, K. Ikenaga, Y. Yano, H. Tokunaga, A. Mishima, Y. Ban, T. Tabuchi, K. Matsumoto, Study of carbon concentration in GaN grown by metalorganic chemical vapor deposition, in *Proceeding 9th Int. Workshop Bulk Nitride Semicond*, vol. 456, (2016), pp. 137–139
11. D.D. Koleske, A.E. Wickenden, R.L. Henry, M.E. Twigg, Influence of MOVPE growth conditions on carbon and silicon concentrations in GaN. J. Cryst. Growth **242**(1), 55–69 (2002)
12. N. Sawada, T. Narita, M. Kanechika, T. Uesugi, T. Kachi, M. Horita, T. Kimoto, J. Suda, Sources of carrier compensation in metalorganic vapor phase epitaxy-grown homoepitaxial n-type GaN layers with various doping concentrations. Appl. Phys. Express **11**(4), 041001 (2018)
13. S. Heikman, S. Keller, T. Mates, S.P. DenBaars, U.K. Mishra, Growth and characteristics of Fe-doped GaN. Proc. Elev. Int. Conf. Met. Nic Vap. Phase Epitaxy **248**, 513–517 (2003)

14. S. Heikman, S. Keller, S.P. DenBaars, U.K. Mishra, Growth of Fe doped semi-insulating GaN by metalorganic chemical vapor deposition. Appl. Phys. Lett. **81**(3), 439–441 (2002)
15. S. Leone, F. Benkhelifa, L. Kirste, C. Manz, S. Mueller, R. Quay, T. Stadelmann, Suppression of iron memory effect in GaN epitaxial layers. Phys. Status Solidi B **255**(5), 1700377 (2018)
16. W. Lee, J.-H. Ryou, D. Yoo, J. Limb, R.D. Dupuis, D. Hanser, E. Preble, N.M. Williams, K. Evans, Optimization of Fe doping at the regrowth interface of GaN for applications to III-nitride-based heterostructure field-effect transistors. Appl. Phys. Lett. **90**(9), 093509 (2007)
17. T. Ishiguro, A. Yamada, J. Kotani, N. Nakamura, T. Kikkawa, K. Watanabe, K. Imanishi, New model of Fe diffusion in highly resistive Fe-doped buffer layer for GaN high-electron-mobility transistor. Jpn. J. Appl. Phys. **52**(8S), 08JB17 (2013)
18. Y. Oshimura, K. Takeda, T. Sugiyama, M. Iwaya, S. Kamiyama, H. Amano, I. Akasaki, A. Bandoh, T. Udagawa, AlGaN/GaN HFETs on Fe-doped GaN substrates. Phys. Status Solidi C **7**(7–8), 1974–1976 (2010)
19. J.A. Freitas, J.C. Culbertson, E.R. Glaser, E. Richter, M. Weyers, A.C. Oliveira, V.K. Garg, Efficient iron doping of HVPE GaN. J. Cryst. Growth **500**, 111–116 (2018)
20. J.V. DiLorenzo, Analysis of impurity distribution in homoepitaxial n on n+ films of GaAs. J. Electrochem. Soc. **118**(10), 1645 (1971)
21. J.V. DiLorenzo, R.B. Marcus, R. Lewis, Analysis of impurity distribution in homoepitaxial n on n+ films of GaAs which contain high-resistivity regions. J. Appl. Phys. **42**(2), 729–739 (1971)
22. Q.-M. Fan, Y.-W. Liu, C.-L. Wei, Depth profiling of iron impurities on GaAs surfaces using total reflection X-ray fluorescence. Fresenius J. Anal. Chem. **354**(2), 193–194 (1996)
23. F. Hasegawa, T. Saito, Occurrence of a high resistance layer at vapor epitaxial GaAs film-substrate interface. Jpn. J. Appl. Phys. **7**(9), 1125 (1968)
24. D.W. Cardwell, A. Sasikumar, A.R. Arehart, S.W. Kaun, J. Lu, S. Keller, J.S. Speck, U.-K. Mishra, S.A. Ringel, J.P. Pelz, Spatially-resolved spectroscopic measurements of Ec − 0.57 eV traps in AlGaN/GaN high electron mobility transistors. Appl. Phys. Lett. **102**(19), 193509 (2013)
25. S. Mukherjee, Y. Puzyrev, J. Chen, D.M. Fleetwood, R.D. Schrimpf, S.T. Pantelides, Hot-carrier degradation in GaN HEMTs due to substitutional iron and its complexes. IEEE Trans. Electron Devices **63**(4), 1486–1494 (2016)
26. A.R. Arehart, A. Sasikumar, S. Rajan, G.D. Via, B. Poling, B. Winningham, E.R. Heller, D. Brown, Y. Pei, F. Recht, U.K. Mishra, S.A. Ringel, Direct observation of 0.57eV trap-related RF output power reduction in AlGaN/GaN high electron mobility transistors. Solid State Electron. **80**, 19–22 (2013)
27. Y. Tokuda, Y. Matsuoka, H. Ueda, O. Ishiguro, N. Soejima, T. Kachi, DLTS study of n-type GaN grown by MOCVD on GaN substrates. E-MRS 2006 Symp. Mater. Sci. Technol. Wide Bandgap Semicond **40**(4), 268–273 (2006)
28. A. Sasikumar, D.W. Cardwell, A.R. Arehart, J. Lu, S.W. Kaun, S. Keller, U.K. Mishra, J.S. Speck, J.P. Pelz, S.A. Ringel, Toward a physical understanding of the reliability-limiting EC-0.57 eV trap in GaN HEMTs, in *2014 IEEE Int. Reliab. Phys. Symp.*, (2014), pp. 2C.1.1–2C.1.6
29. A. Hierro, D. Kwon, S.A. Ringel, M. Hansen, J.S. Speck, U.K. Mishra, S.P. DenBaars, Optically and thermally detected deep levels in n-type Schottky and p+-n GaN diodes. Appl. Phys. Lett. **76**(21), 3064–3066 (2000)
30. S. Chen, U. Honda, T. Shibata, T. Matsumura, Y. Tokuda, K. Ishikawa, M. Hori, H. Ueda, T. Uesugi, T. Kachi, As-grown deep-level defects in n-GaN grown by metal–organic chemical vapor deposition on freestanding GaN. J. Appl. Phys. **112**(5), 053513 (2012)
31. Y. Zhang, Z. Chen, W. Li, H. Lee, M.R. Karim, A.R. Arehart, S.A. Ringel, S. Rajan, H. Zhao, Probing unintentional Fe impurity incorporation in MOCVD homoepitaxy GaN: toward GaN vertical power devices. J. Appl. Phys. **127**(21), 215707 (2020)
32. Y. Inatomi, Y. Kangawa, A. Pimpinelli, T.L. Einstein, Kinetic-thermodynamic model for carbon incorporation during step-flow growth of GaN by metalorganic vapor phase epitaxy. Phys. Rev. Mater. **3**(1), 013401 (2019)

33. D. Yosho, Y. Inatomi, Y. Kangawa, Modeling carbon coverage on polar GaN surfaces during MOVPE. Jpn. J. Appl. Phys. **59**(4), 048002 (2020)
34. T. Ciarkowski, N. Allen, E. Carlson, R. McCarthy, C. Youtsey, J. Wang, P. Fay, J. Xie, L. Guido, Connection between carbon incorporation and growth rate for GaN epitaxial layers prepared by OMVPE. Materials **12**(15), 2455 (2019)
35. N.A. Fichtenbaum, T.E. Mates, S. Keller, S.P. DenBaars, U.K. Mishra, Impurity incorporation in heteroepitaxial N-face and Ga-face GaN films grown by metalorganic chemical vapor deposition. J. Cryst. Growth **310**(6), 1124–1131 (2008)
36. NIST Chemistry WebBook, SRD 69: https://webbook.nist.gov/cgi/cbook.cgi?ID=C7664417&Type=IR-SPEC&Index=1
37. L.S. Fan, Z.Q. Xie, J.B. Park, X.N. He, Y.S. Zhou, L. Jiang, Y.F. Lu, Synthesis of nitrogen-doped diamond films using vibrational excitation of ammonia molecules in laser-assisted combustion flames. J. Laser Appl. **24**(2), 022001 (2012)
38. C.W. David, IR vibration-rotation spectra of the ammonia molecule. J. Chem. Educ. **73**(1), 46 (1996)
39. Y. Zhang, Z. Chen, K. Zhang, Z. Feng, H. Zhao, Laser-assisted metal–organic chemical vapor deposition of gallium nitride. Phys. Status Solidi RRL—Rapid Res. Lett. **15**(6), 2100202 (2021)
40. H.R. Golgir, Y.S. Zhou, D. Li, K. Keramatnejad, W. Xiong, M. Wang, L.J. Jiang, X. Huang, L. Jiang, J.F. Silvain, Y.F. Lu, Resonant and nonresonant vibrational excitation of ammonia molecules in the growth of gallium nitride using laser-assisted metal organic chemical vapour deposition. J. Appl. Phys. **120**(10), 105303 (2016)
41. H. Rabiee Golgir, Y. Gao, Y.S. Zhou, L. Fan, P. Thirugnanam, K. Keramatnejad, L. Jiang, J.-F. Silvain, Y.F. Lu, Low-temperature growth of crystalline gallium nitride films using vibrational excitation of ammonia molecules in laser-assisted metalorganic chemical vapor deposition. Cryst. Growth Des. **14**(12), 6248–6253 (2014)
42. J.O.P. McBride, R.W. Nicholls, The vibration-rotation spectrum of ammonia gas. I. J. Phys. B At. Mol. Phys. **5**(2), 408 (1972)
43. H. Yamada, H. Chonan, T. Takahashi, M. Shimizu, Impact of substrate off-angle on the m-plane GaN Schottky diodes. Jpn. J. Appl. Phys. **57**(4S), 04FG01 (2018)
44. A. Tanaka, Y. Ando, K. Nagamatsu, M. Deki, H. Cheong, B. Ousmane, M. Kushimoto, S. Nitta, Y. Honda, H. Amano, m-Plane GaN Schottky barrier diodes fabricated with MOVPE layer on several off-angle m-plane GaN substrates. Phys. Status Solidi A **215**(9), 1700645 (2018)
45. K. Nagamatsu, Y. Ando, T. Kono, H. Cheong, S. Nitta, Y. Honda, M. Pristovsek, H. Amano, Effect of substrate misorientation on the concentration of impurities and surface morphology of an epitaxial GaN layer on N-polar GaN substrate by MOVPE. J. Cryst. Growth **512**, 78–83 (2019)
46. G. Greco, F. Lucolano, F. Roccaforte, Ohmic contacts to gallium nitride materials. Appl. Surf. Sci. **383**, 324–345 (2016)
47. V.A. Dmitriev, K.G. Irvine, C.H. Carter Jr., N.I. Kuznetsov, E.V. Kalinina, Electric breakdown in GaN p-n junctions. Appl. Phys. Lett. **68**(2), 229–231 (1996)
48. I.C. Kizilyalli, A.P. Edwards, H. Nie, D. Disney, D. Bour, High voltage vertical GaN p-n diodes with avalanche capability. Trans. Elec. Dev. **60**(10), 3067–3070 (2013)
49. Y. Liu, S. Yang, K. Sheng, Design and optimization of vertical GaN PiN diodes with fluorine-implanted termination. IEEE J Electron Devices Soc. **8**, 241–250 (2020)
50. Z. Hu, K. Nomoto, B. Song, M. Zhu, M. Qi, M. Pan, X. Gao, V. Protasenko, D. Jena, H.G. Xing, Near unity ideality factor and Shockley-Read-Hall lifetime in GaN-on-GaN p-n diodes with avalanche breakdown. Appl. Phys. Lett. **107**(24), 243501 (2015)
51. T. Pu, U. Younis, H. Chiu, K. Xu, H. Kuo, X. Liu, Review of recent progress on vertical GaN-based PN diodes. Nanoscale Res. Lett. **16**(101), 101 (2021)
52. L. Yates, B.P. Gunning, M.H. Crawford, J. Steinfeldt, M.L. Smith, V.M. Abate, J.R. Dickerson, A.M. Armstrong, A. Binder, A.A. Allerman, R.J. Kaplar, Demonstration of >6.0-kV breakdown voltage in large area vertical GaN p-n diodes with step-etched junction termination extensions. Trans. Elec. Dev. **69**(4), 1931–1937 (2022)

53. V. Talesara, Y. Zhang, Z. Chen, H. Zhao, W. Lu, Design and development of 1.5 kV vertical GaN pn diodes on HVPE substrate. J. Mater. Res. **36**, 4919–4929 (2021)
54. Y. Xu, V.G.T. Vangipuram, V. Talesara, J. Cheng, Y. Zhang, T. Hashimoto, E. Letts, D. Key, H. Zhao, W. Lu, 7.86 kV GaN-on-GaN PN power diode with $BaTiO_3$ for electrical field management. Appl. Phys. Lett. **1233**(14), 142105 (2023)
55. V. Talesara, Y. Zhang, Z. Chen, H. Zhao, W. Lu, GaN power p-n diodes on hybride vapor epitaxy GaN substrates with near-unity ideality factor and <0.5 mΩ cm^2 specific on-resistance. Phys. Status Solidi RRL **16**(4), 2100599 (2022)
56. V. Talesara, Y. Zhang, V.G.T. Vangipuram, H. Zhao, W. Lu, Vertical GaN-on-GaN pn power diodes with Baliga figure of merit of 27 GW/cm^2. Appl. Phys. Lett. **122**(12), 123501 (2023)
57. T. Maeda, T. Narita, H. Ueda, M. Kanechika, T. Uesugi, T. Kachi, T. Kimoto, M. Horita, J. Suda, Parallel-plane breakdown fields of 2.8–3.5 MV/cm in GaN-on-GaN p-n junction diodes with double-side-depleted shallow bevel termination. IEEE Int. Electron Devices Meet. (IEDM) (2018)
58. H. Ohta, N. Kaneda, F. Horikiri, Y. Narita, T. Yoshida, T. Mishima, T. Nakamura, Vertical GaN p-n junction diodes with high breakdown voltage over 4 kV. Electron Device Lett. **36**(11), 1180–1182 (2015)
59. H. Ohta, K. Hayashi, F. Horikiri, M. Yoshino, T. Nakamura, T. Mishima, 5.0 kV breakdown-voltage vertical GaN p-n junction diodes. Jpn. J. Appl. Phys. **57**(4s), 04FG09 (2018)
60. S.H. Chen, H. Wang, C. Hu, Y. Chen, H. Wang, J. Wang, W. He, X. Sun, H. Chiu, H. Kuo, W. Wang, K. Xu, D. Li, X. Liu, Vertical GaN-on-GaN PIN diodes fabricated on free-standing GaN wafer using an ammonothermal method. J. Alloys Compd. **804**(5), 435–440 (2019)
61. I.C. Kizilyalli, T. Prunty, O. Aktas, 4-kV and 2.8 mΩ -cm^2 vertical GaN p-n diodes with low leakage currents. IEEE Electron Device Lett. **36**(10), 1073–1075 (2015)
62. I.C. Kizilyalli, A.P. Edwards, O. Aktas, T. Prunty, D. Bour, Vertical power p-n diodes based on bulk GaN. IEEE Trans. Electron Devices **62**(2), 414–422 (2014)
63. K. Nomoto, B. Song, Z. Hu, M. Zhu, M. Qi, N. Kaneda, T. Mishima, T. Nakamura, D. Jena, H.G. Xing, 1.7 -kV and 0.55- m$\Omega \cdot$ cm^2 GaN p-n diodes on bulk GaN substrates with avalanche capability. IEEE Electron Device Lett. **37**(2), 161–164 (2015)
64. K. Nomoto, Z. Hu, B. Song, M. Zhu, M. Qi, R. Yan, V. Protasenko, E. Imhoff, J. Kuo, N. Kaneda, T. Mishima, T. Nakamura, D. Jena, H.G. Xing, GaN-on-GaN p-n power diodes with 3.48 kV and 0.95 mΩ-cm^2: a record high figure-of-merit of 12.8 GW.cm^2. IEEE Int Electron Devices Meet. (IEDM) (2015)
65. H. Fukushima, S. Usami, M. Ogura, Y. Ando, A. Tanaka, M. Deki, M. Kushimoto, S. Nitta, Y. Honda, H. Amano, Vertical GaN p-n diode with deeply etched mesa and the capability of avalanche breakdown. Appl. Phys. Express **12**(2), 026502
66. K. Fu, H. Fu, H. Liu, S.R. Alugubelli, T. Yang, X. Huang, H. Chen, I. Baranowski, J. Montes, F.A. Ponce, Y. Zhao, Investigation of GaN-on-GaN vertical p-n diode with regrown p-GaN by metalorganic chemical vapor deposition. Appl. Phys. Lett. **113**(23), 233502 (2018)
67. X. Liu, H. Chiu, H. Wang, C. Hu, H. Wang, H. Kao, F. Chien, 2.4 kV vertical GaN PN diodes on free standing GaN wafer using CMOS-compatible contact materials. IEEE J. Electron Devices Soc. **6**, 825–829 (2018)
68. A. Taube, M. Kaminski, J. Tarenko, O. Sadowski, M. Ekielski, A. Szerling, P. Prystawko, M. Bockowski, I. Grzegory, High breakdown voltage and high current injection vertical GaN-on-GaN p-n diodes with extremely low on-resistance fabricated on ammonothermally grown bulk GaN substrates. IEEE Trans. Elect. Dev. **69**(11), 6244–6259 (2022)
69. J. Liu, R. Zhang, M. Xiao, S. Pidaparthi, H. Cui, A. Edwards, L. Baubutr, C. Drowley, Y. Zhang, Surge current and avalanche ruggedness of 1.2-kV vertical GaN p-n diodes. IEEE Trans. Power Electron. **36**(10), 10959–10964 (2021)

Chapter 8
MOCVD Growth of GaN Drift Layers on Bulk GaN Substrates for Power Electronic Devices

Robert Kaplar, Anthony Rice, Luke Yates, Andrew Binder, Andrew M. Armstrong, and Andrew A. Allerman

1 Introduction

In this chapter, we consider the growth of GaN epitaxial material by metal-organic chemical vapor deposition (MOCVD). The emphasis is on the growth and characterization of epitaxial material on GaN substrates for vertical pn diodes (Fig. 8.1) with n-drift layers from 6 to 50 μm thick. Two commercial MOCVD growth tools, a Veeco D125 and a Taiyo Nippon Sanso (TNSC) SR4000, were used to grow the epitaxial layers. Trimethylgallium and ammonia were the precursors in a mixed N_2/H_2 carrier gas. Silane and bis(cyclopentadienyl) magnesium were the n-type and p-type dopants, respectively. The choice of GaN substrates was critical for the growth of the layers, the ability to control the thickness uniformity and doping, and electrically compensating defects introduced during the growth process. The growth of the epitaxial layers was done at Sandia National Laboratories on two-inch diameter GaN wafers. Characterization of the epitaxial layers and device processing was done both at Sandia and the Naval Research Laboratory.

2 GaN Substrates

The quality of the bulk GaN substrate is critical for the growth of high-quality epitaxial GaN layers. Substrates were characterized by optical microscopy, Raman spectroscopy, photoluminescence (PL), atomic force microscopy, and X-ray topography (XRT). Figure 8.2 shows the optical microscopy and PL characterization for

R. Kaplar (✉) · A. Rice · L. Yates · A. Binder · A. M. Armstrong · A. A. Allerman
Sandia National Laboratories, Albuquerque, NM, USA
e-mail: rjkapla@sandia.gov

I. C. Kizilyalli et al. (eds.), *Gallium Nitride and Related Materials*, The Materials Research Society Series, https://doi.org/10.1007/978-3-031-83056-3_8

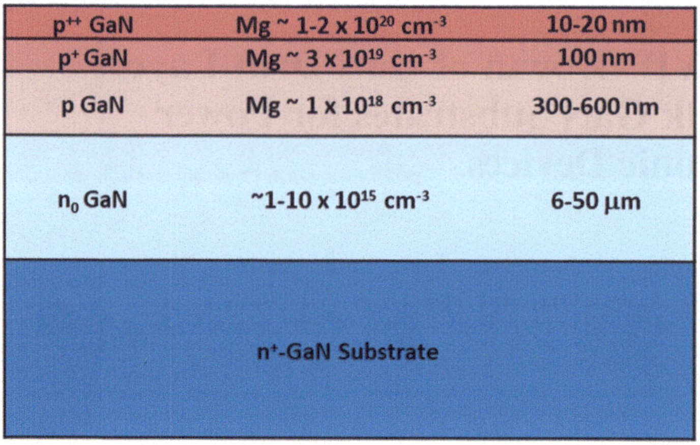

Fig. 8.1 Schematic representing typical GaN pn diode structures fabricated during the course of this work

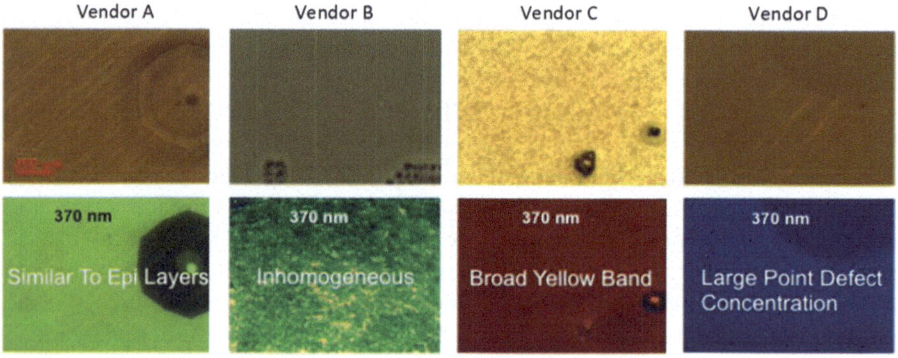

Fig. 8.2 Example of evaluation of GaN substrates from four vendors, using optical microscopy (top row) and PL (bottom row). (Courtesy of NRL)

substrates from four vendors, identified as A, B, C, and D. Substrates can be classified into two main types: Type-I (uniform dislocation density, high cost) and Type-II (nonuniform dislocation density, typically associated with lower cost "dot-core" manufacturing).

Dot-core substrates have been developed as an effective approach to reduce the dislocation density during growth by hydride vapor phase epitaxy (HVPE). This approach was historically targeted at the development of laser diodes, allowing for placement of the laser strip in regions of low dislocation density. Generally, the approach involves growing freestanding GaN substrates with periodically positioned inversion domains known as "cores" [1]; these are visible as black dots and thus the name "dot-core" HVPE substrates. The core positions are fixed using a 2D grid pattern "seed" substrate with spacing up to ~1 mm. During HVPE growth, threading dislocations congregate and concentrate at the centers of the cores, thereby reducing dislocations in regions between the cores. The dislocation density between the cores has been reported to be as low as 1×10^3 cm^{-2}. In comparison, the dislocation density at the center of the cores is usually reported to be well above 1×10^6 cm^{-2} [1–3]. High-performance devices have been demonstrated in the literature using both types of substrates [4]. However, the yield may suffer for devices fabricated on Type-II substrates due to the necessity of aligning the devices to the low-defect regions.

Type-II substrates can generally be categorized as inhomogeneous wafers with further classification into Types-IIa and -IIb, corresponding to a uniform distribution of dot-core structures (Type-IIa) or a nonuniform distribution of dot-core structures (Type IIb). This is a result of the position of the wafer in the boule, where wafers close to the seed reproduce the initial pattern while wafers close to the tail of the boule exhibit an irregular structure due to the movement and consumption of some dot-core features during boule growth. This information is reflected in the wafer ID, so wafers with either of these properties can be selected. A regular dot-core structure pattern can have devices aligned to it but may be dominated by lower-quality material at dot-core structure boundaries, whereas an irregular pattern cannot have devices aligned but will statistically exhibit higher-quality material. Figure 8.3 shows Raman spectroscopy scans of Type-II wafers, illustrating the progression from uniformly distributed dot-core structures to the random distribution of those structures as the wafers move from the seed to the tail. By comparison, Fig. 8.4 shows the Raman spectroscopy scans for Type-I wafers from two different vendors. X-ray topography can distinguish between defects due to the substrate and those due to epitaxial growth which will be discussed later in this chapter.

3 Epitaxial Growth and Characterization

GaN pn diodes with 6 μm drift layers with n-type doping of 1.3×10^{16} cm^{-3} are designed to have a 1.5 kV breakdown. Sandia provided 48 wafers with this basic structure to NRL for wafer characterization, diode processing, and testing. The goal was to determine the yield and any challenges with scaling the fabrication of pn diodes. Figure 8.5 indicates the thickness uniformity that was achieved on a two-inch diameter GaN substrate for a 6 μm drift layer.

The challenge with growing n-type GaN drift layers by MOCVD with low free carrier concentration is that the background carbon concentration incorporated during growth can be comparable to the intentional donor concentration, in this case, silicon. This requires the net carrier concentration to be notably smaller than the donor concentration and the density of compensating acceptors [5]. With the background carbon concentration in the drift layer typically between 2 and 5×10^{15} cm^{-3} depending on growth conditions, the silicon flow must be slightly higher than this to achieve a lower electron concentration. While this type of growth is challenging, it is controllable and reproducible as shown in Fig. 8.6 where a linear relationship is observed between the net carrier concentration (as determined by CV measurements

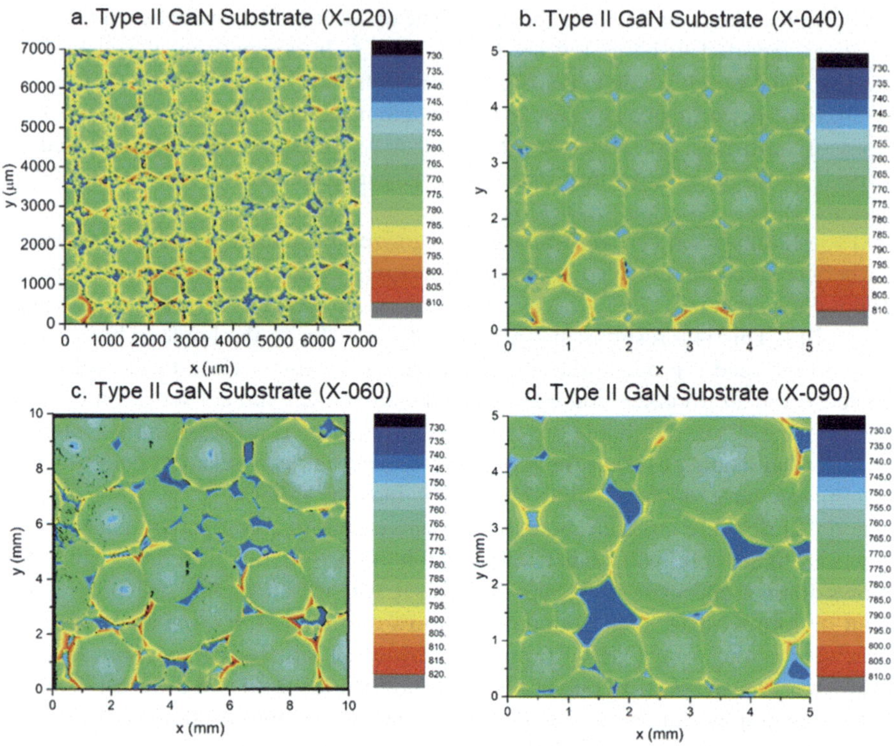

Fig. 8.3 Raman maps of Type-II GaN substrates, going from the boule seed (**a**) to the seed tail (**d**), showing the change from uniformly distributed dot-cores to randomly distributed dot-cores. The Type-IIa substrate is represented by (**a**) and the Type IIb substrate is represented by (**d**). The scales in (**a**) are in μm, while those for (**b–d**) are in mm. The color bar represents the A1 (LO) peak position of the Raman shift (cm^{-1}). (Courtesy of NRL)

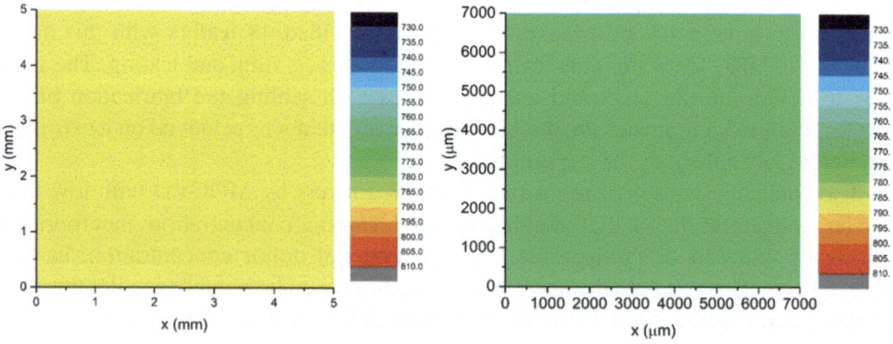

Fig. 8.4 Raman maps of Type-I substrates from different vendors. The scales in the left image are in mm, while those for the right image are in μm. The color bar represents the A1 (LO) peak position of the Raman shift (cm^{-1}). (Courtesy of NRL)

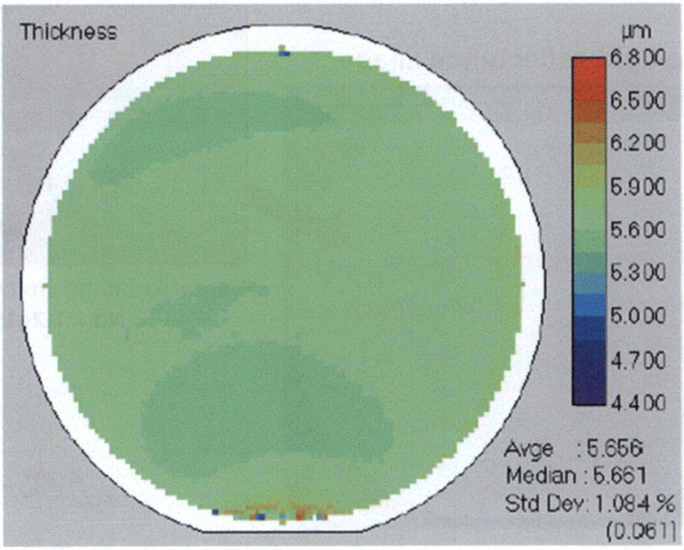

Fig. 8.5 Thickness map of 6 µm epi grown on a two-inch diameter GaN wafer. The thickness variation (standard deviation) is <2%

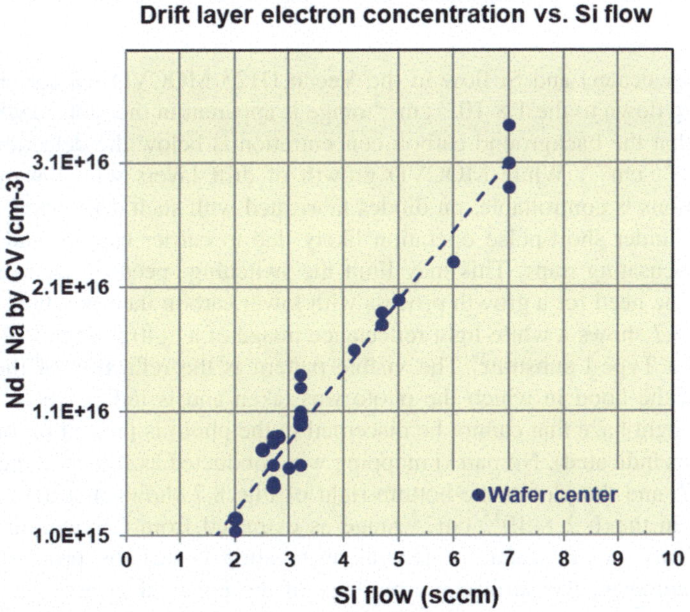

Fig. 8.6 Net doping $N_0 = N_d - N_a$ as determined by capacitance-voltage measurements, plotted against Si flow in the MOCVD reactor

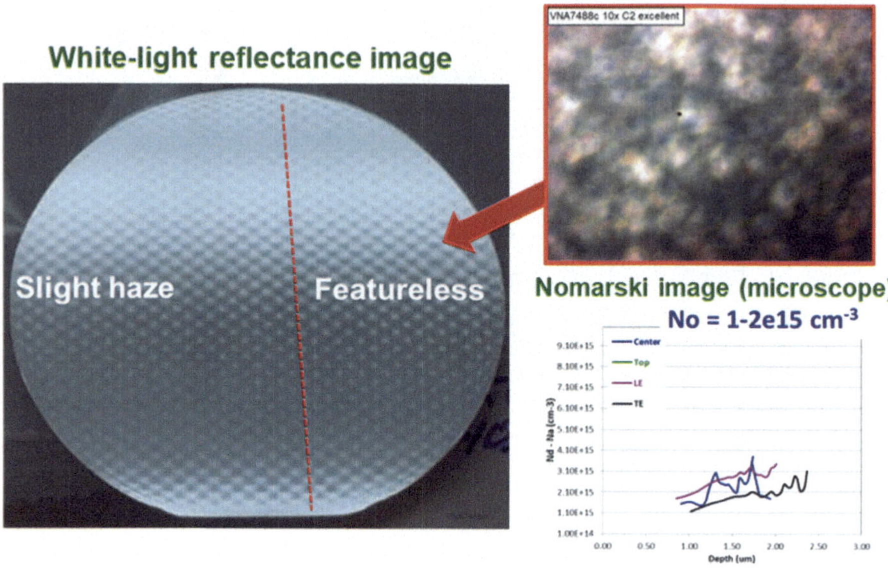

Fig. 8.7 (left) White-light reflectance image of >40 μm GaN drift region grown on Type-I GaN substrate. (upper right) Nomarski microscope image of a portion of the wafer. (bottom right) Net doping density extracted from capacitance-voltage measurements at different points on the wafer

at the wafer center) and Si flow in the Veeco D125 MOCVD reactor. The linear relationship down to the 1×10^{15} cm^{-3} range is apparent in the plot. SIMS analysis indicates that the background carbon concentration is below the detection limit of ~5–7 × 10^{15} cm^{-3}. While MOCVD growth of drift layers with low net carrier concentrations is controllable, pn diodes fabricated with such drift layers exhibited dispersion under short-pulse operation likely due to carrier capture and emission from compensating traps. This may limit the switching speed of such diodes and illustrates the need for a growth process with lower carbon incorporation.

Figure 8.7 shows a white-light reflectance image of a >40 μm thick drift region grown on a Type-I substrate. The visible pattern is the reflection of the lighting element in the hood in which the photo was taken and is indicative of a smooth surface (a light haze that cannot be discerned in the photo is present on one half of the wafer as indicated). Nomarski mapping was conducted as shown at the top right of Fig. 8.7, and the plot at the bottom right of Fig. 8.7 shows that the net doping density is in the 1–2 × 10^{15} cm^{-3} range as extracted from CV measurements at several points on the wafer. A general observation is that the properties of the substrate dominate the surface morphology of the epitaxial growth, as shown in Fig. 8.8 for a pn diode structure with a 45 μm thick drift layer grown on a Type-II substrate (left) and a Type-I substrate (right). Improvement in the consistency of the substrates is necessary to achieve the reproducible morphology of thick drift layers.

Figure 8.9 shows the Nomarski images and the net carrier density as determined by CV measurements for a 60 μm drift layer, the thickest layer grown during this

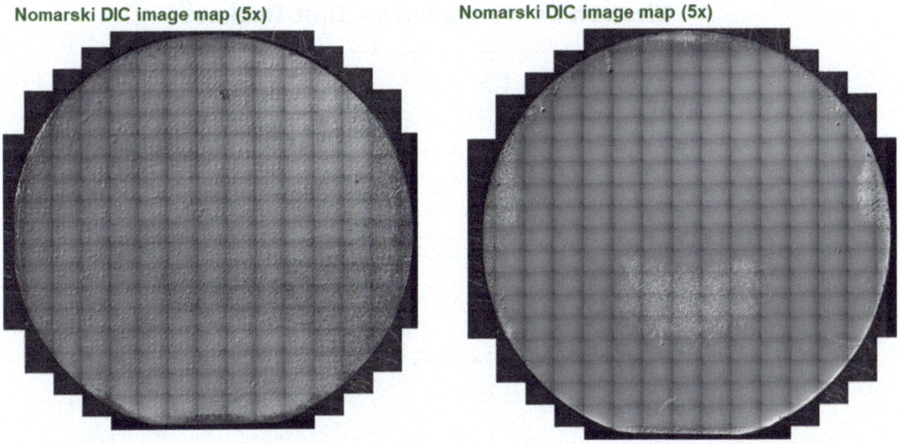

Fig. 8.8 Nomarski images of pn diode structures with 45 μm thick drift regions grown on (left) Type-II GaN substrate where the dot-core grid pattern is visible and (right) Type-I GaN substrate

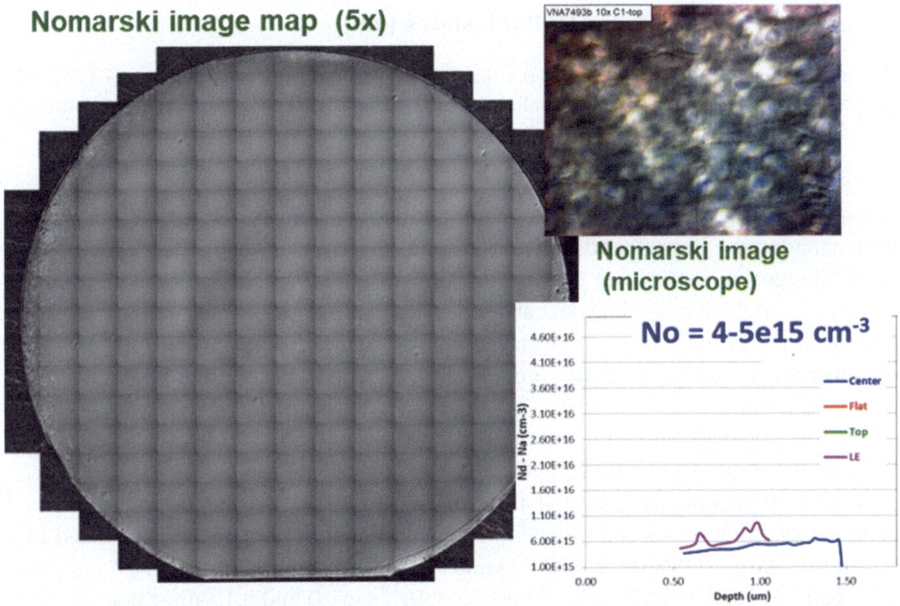

Fig. 8.9 (left and upper right) Nomarski images of 60 μm thick GaN drift layer grown on a Type-I GaN substrate. (bottom right) Net doping density extracted from capacitance-voltage measurements at various points on the wafer

work. The epitaxial layers were grown in the Veeco D125 MOCVD tool on a Type-I GaN substrate. The results indicate the ability to control the net carrier density for a thick drift layer. This wafer featured some surface morphology like that which we have observed in much thinner epilayers (<10 μm) and is related to variations in the

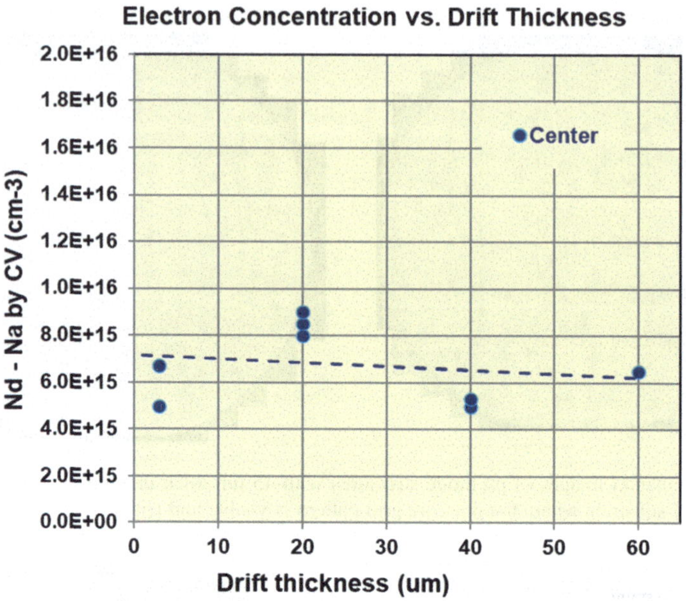

Fig. 8.10 Net doping density extracted from capacitance-voltage measurements for GaN drift regions up to 60 μm thick grown in Sandia's Veeco D125 MOCVD reactor. Multiple growth runs were done to the target thickness, with multiple wafers in each growth run

starting substrate. Figure 8.10 shows the net doping density for n-type drift layers grown under the same conditions to various thicknesses using the Veeco D125 MOCVD reactor. A separate growth run was made for each thickness, with multiple wafers in 3 of the 4 runs. The net carrier density was extracted by CV at the center of each wafer. The net carrier concentration was stable during growth up to 60 μm demonstrating a stable level of compensating acceptors despite the large change in epilayer thickness.

Another example showing that doping at the mid-10^{15} cm^{-3} level was repeatable for growths in the Veeco D125 MOCVD tool is presented in Fig. 8.11.

The pn diode epi structure consisted of a 45-μm-thick drift layer with a targeted carrier concentration of 2–5 × 10^{15} cm^{-3}. Additionally, the p-anode consisted of a 0.5-μm-thick (p^-) GaN layer with a magnesium concentration of ~2 × 10^{18} cm^{-3} but with 100-nm-thick p-GaN layer ([Mg] ~2 × 10^{19} cm^{-3}) and a 10-nm-thick p^{++} GaN contact layer ([Mg] ~1–2 × 10^{20} cm^{-3}). This structure was grown on a Type-I substrate. CV measurements of test diodes from the center of the wafer to the major flat showed carrier concentration decreasing from 5 × 10^{15} cm^{-3} to ~3 × 10^{15} cm^{-3}. This is in good agreement with Hg-probe CV measurements (4–8 × 10^{15} cm^{-3}) of the pn diode wafer prior to activation and device processing. Additionally, Hg-probe CV measurements of a growth calibration layer showed a specially varying carrier concentration ranging from 7 × 10^{15} cm^{-3} to 4 × 10^{15} cm^{-3}. These results provide

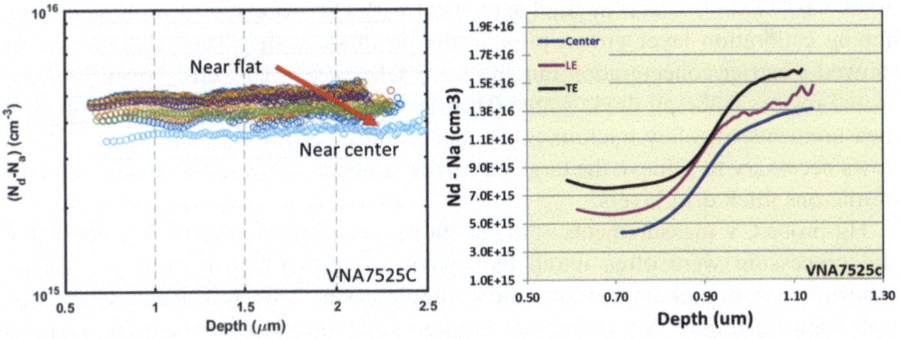

Fig. 8.11 CV measurements of processed GaN pn diodes with a 45-μm-thick drift layer show a carrier concentration ranging between 3 and 5 × 10^{15} cm^{-3}. Further, good agreement is found between the carrier concentration measured on devices with an n-GaN drift layer in a pn diode with the carrier concentration of an n-GaN doping calibration layer measured by Hg-probe CV

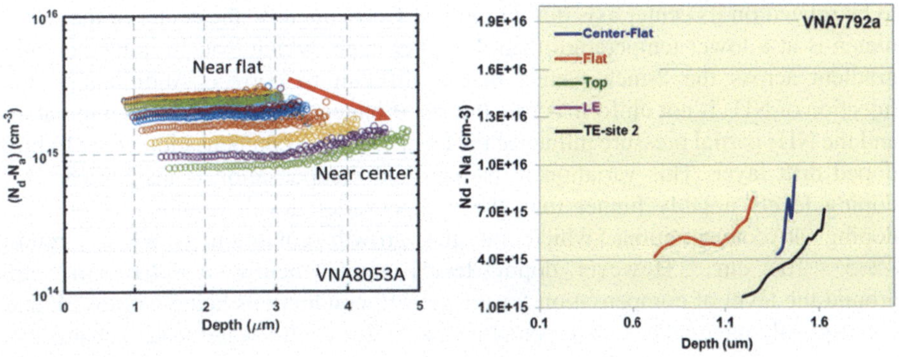

Fig. 8.12 CV measurements of processed GaN pn diodes with a 50-μm-thick drift layer show carrier concentration ranging from 0.8–3.0 × 10^{15} cm^{-3}. Good agreement is found between the carrier concentration measured on devices for an n-GaN drift layer in a pn diode with the carrier concentration of an n-GaN doping calibration layer measured by Hg-probe CV

further evidence of controllable and reproducible carrier concentrations in 45-μm-thick drift layers in the low-to-mid-10^{15} cm^{-3} doping range.

While run-to-run doping control by MOCVD growth was demonstrated, at lower carrier concentrations a variability was observed across a wafer for growths done in the Veeco D125 MOCVD tool. Figure 8.12 shows the pn diode epi structure consisting of a 50 μm thick drift layer with a target carrier concentration of 1–3 × 10^{15} cm^{-3} and a p-anode consisting of a 0.6 μm thick (p^-) GaN layer with a magnesium concentration of ~2.5 × 10^{17} cm^{-3} followed by a 0.1 μm thick p-GaN layer with a magnesium concentration of 3 × 10^{19} cm^{-3}. A thin p^{++} GaN contact layer with a magnesium concentration of 2 × 10^{20} cm^{-3} completes the epi structure. Following diode fabrication, CV measurements of test diodes from the center of the wafer to the major flat showed carrier concentration decreasing from 3 × 10^{15} cm^{-3}

to ~8 × 10^{14} cm^{-3}. This is in good agreement with Hg-probe CV measurements of a doping calibration layer grown prior to the pn diode wafer. The calibration layer showed a carrier concentration ranging from 4.1 to 1.6 × 10^{15} cm^{-3} over the same spatial region as the pn diode wafer. This demonstrated the utility of Hg-probe CV measurements of a few-microns-thick calibration layer for determining the silicon flows necessary to achieve the targeted carrier concentrations in pn diodes with tens of microns thick drift layers.

Hg-probe CV measurements of the pn diode wafer prior to magnesium activation and processing were often unreliable, making it difficult to confirm the carrier concentration in full pn epi structures after growth. This was the case for this wafer, but Hg-probe CV measurements suggest that the carrier concentration ranged from 1.3 × 10^{15} cm^{-3} to ~6 × 10^{14} cm^{-3} moving from the wafer center to the major flat. The nonuniformity in the net carrier concentration observed in these wafers is most likely a consequence of the design of the Veeco D125 MOCVD tool being used. The reactor utilizes a wafer platen that holds three 2-inch diameter wafers. It has heating elements positioned around a central spindle that allows the wafer platen to be rotated on its center axis during growth. Consequently, the center of the wafer platen is at a lower temperature than the outer edge, which leads to a temperature gradient across the 2-inch wafers that is difficult to correct. Additionally, the injection of NH_3 is not uniform across the wafer platen. Both the growth temperature and the NH_3 partial pressure influence the level of carrier compensation of the lightly doped drift layer. This variation in background compensation is insignificant for doping levels notably higher than the compensation level, resulting in uniform doping at concentrations which for the growth conditions used is around ~2–5 × 10^{16} cm^{-3}. However, doping levels targeted below ~5 × 10^{15} cm^{-3} are around the level of compensation for the growth conductions being employed, and consequently the gradients in temperature and NH_3 over the wafer lead to changes in the carrier concentration. Hence, nonuniform carrier concentrations are not necessarily fundamental, but depend on the uniformity of growth conditions over the wafer and the magnitude of the carrier concentration. While carrier concentrations at the low 10^{15} cm^{-3} level were not uniform across a wafer, they were repeatable from run to run and over time, and the concentration was controllable to achieve the device doping targets. Additionally, Veeco reactors with larger platens avoid the temperature nonuniformity caused by the center spindle and offer more uniform NH_3 injection over the wafers. We observed little problem with particles, even for the thickest drift layers grown in our Veeco D125 MOCVD reactor due to its vertical gas-flow geometry.

The growth of a thick epi (~50 μm) n-type drift layer pn diode was also done using the TNSC SR4000 MOCVD tool. Known-good growth conditions for thinner drift layers were used to grow the thicker drift layer. To evaluate the performance of different substrates from different vendors, 2-inch diameter Type-I substrates from different vendors (termed X and Y) were used in the same growth run. The entire pn junction epi structure was grown at atmospheric pressure and required ~37 hours to complete. There were two major concerns during the growth run: controlling the net carrier concentration in the low-10^{15} cm^{-3} range, and the build-up of material in the

Fig. 8.13 Nomarski microscope maps of 50 μm thick epi grown in the TNSC SR4000 MOCVD tool on substrates from two different vendors (vendor X left, vendor Y right)

Fig. 8.14 Nomarski microscope images of chevron features (left) and step bunching (right) after thick epi growth on 2-inch diameter vendor Y substrate

reactor chamber during the long epi growth run, and the subsequent deposition of particulates on the wafers. Figure 8.13 shows the Nomarski pictures of the two wafers. There were 10–20 large particles that were deposited onto each wafer either before or during the epi run. However, the cleaning procedures and reactor preparation that were implemented for the TNSC SR4000 MOCVD tool were effective in reducing the small, high-density particles that are deleterious for yield. Thus, while large particles are still a target for improvement, the particle issue overall was not prohibitive even for this very thick epi.

Other than the large particles, the growth on the vendor X substrate was almost entirely featureless except near the edge and at a "bad" spot near the right side. Figure 8.14 shows a closer view of the surface morphology of the growth on the vendor Y substrate. There were far more surface features including areas of obvious "step meandering" and chevron/arrow-shaped features that were routinely seen on these wafers. While these were ultra-low dislocation density wafers, the nonuniform

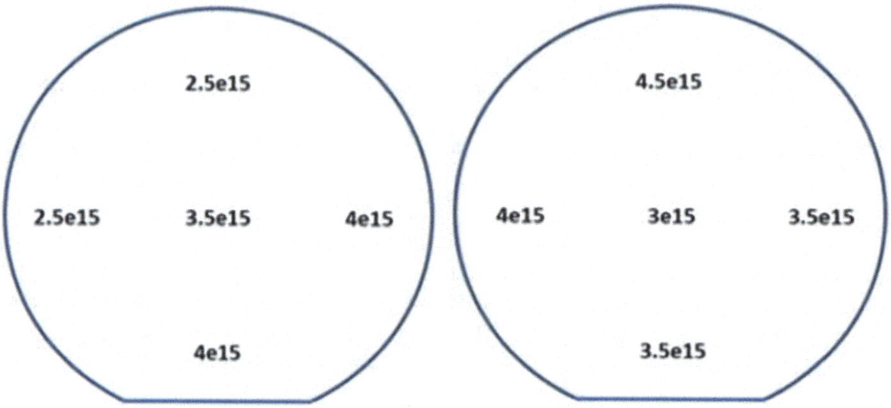

Fig. 8.15 As-grown Hg-probe CV measurements for 50 μm thick pn junctions on vendor X substrate (left) and vendor Y substrate (right)

epi morphology (including significant portions of the surface that were featureless) suggested that the quality of the substrates was still a potential challenge.

The other challenge with the thick epi growths in the TNSC SR4000 MOCVD tool was control of the doping level in the low-10^{15} cm^{-3} range. If the doping was too high, much of the epi thickness would simply add diode on-resistance with no benefit in breakdown voltage. Fortunately, a stable growth condition was achieved. Fig. 8.15 shows the 1 MHz Hg-probe CV measurements taken on the as-grown wafers (i.e., full pn junctions but without the p-GaN activation anneal). The CV measurements suggested a uniform net carrier concentration of ~2–5 × 10^{15} cm^{-3} across each wafer, consistent with the target of ~3 × 10^{15} cm^{-3}.

X-ray topography (XRT) is a powerful tool for characterizing the epitaxial layers because a clear separation can be made between the epitaxial layers and the substrate, depending on the measurement conditions, and is thus capable of differentiating the source of defects. Figure 8.16 shows the XRT results for a 25 μm epitaxial layer grown on a vendor X Type-I substrate. Four different types of defects are noted. First, hexagonal defects are present in the substrate but did not propagate to the epitaxial layers (Region 1). These defects are likely the result of V-pits that fill in during epitaxial crystal growth. Second, threading dislocations can be resolved and are estimated to have a density of ~2 × 10^4 cm^{-2}, though the algorithm for defect detection and counting is still being refined (Region 2). Third, a defect (Region 3) was identified that is present only in the epitaxial layer, which has varying contrasts depending on the beam condition. Fourth, a different type of defect was identified that is also only in the epitaxial layer (Region 4). These defects all manifest as "pits" in the optical profilometry analysis, so the XRT technique helps distinguish the defect types and their impact on devices.

XRT imaging was also done on epitaxial layers from vendor Y. Threading edge dislocations were identified using a sectioning technique. These dislocations have not been previously observed in a lab-grade XRT tool, and are extremely difficult to

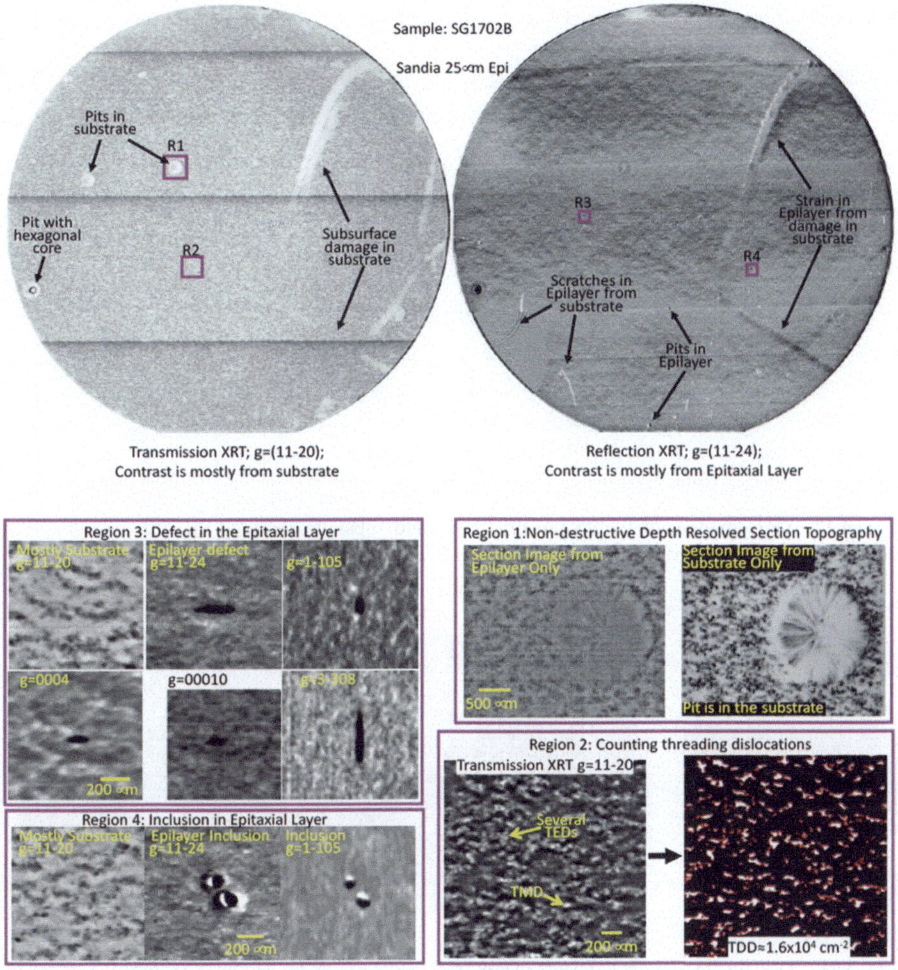

Fig. 8.16 XRT imaging and analysis of 25 μm epi on Type-I vendor X substrate. (Courtesy of NRL)

resolve even at a beamline. Dislocations were counted and are estimated to be ~7×10^3 cm^{-2}. Clear clusters of dislocations propagating from the tiling seam were observed, manifested in Fig. 8.17 as a series of vertical lines around the center of the wafer.

Further XRT analysis of thick epitaxial layers on vendor Y wafers identified a new feature, circular regions containing defect clusters as shown in Fig. 8.18. Given the diffraction conditions where these features were observed, they only resided in the epitaxial layer and were not present in the substrate. The nature of these clusters, how they evolve during epitaxial growth, and what impact they have on device performance need to be better understood.

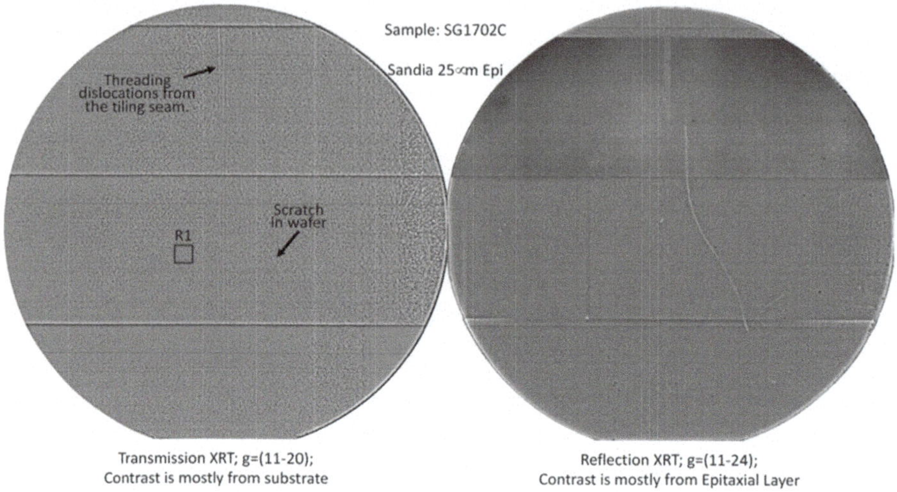

Fig. 8.17 XRT imaging of the 25 μm drift layer and substrate showing the difference in XRT between the epitaxial surface (left) and the Type I vendor Y substrate (right). (Courtesy of NRL)

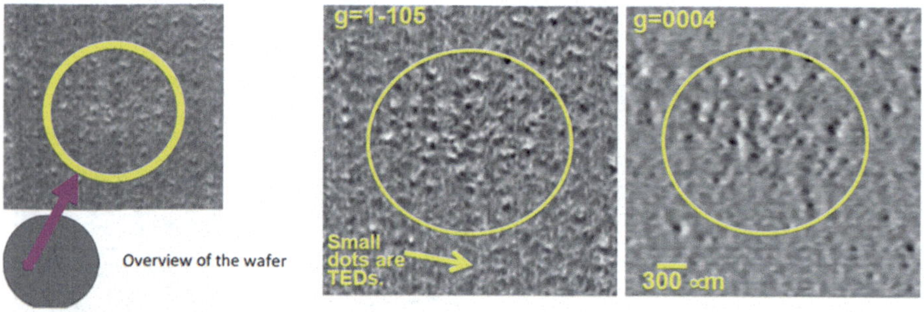

Fig. 8.18 XRT image of defect clusters in epitaxial layers grown on vendor Y wafers. (Courtesy of NRL)

4 Device Results and Yield Analysis

Material from the 50 μm thick drift layer growth was processed into devices at Sandia, the details of which are given in [6]. Figure 8.19 shows the reverse-bias current-voltage (IV) curves and breakdown for two of the best-performing devices. The 1 mm^2 device had a breakdown voltage of 6.0 kV, and the 0.063 mm^2 device had a breakdown of 6.4 kV. The noise floor for the current density is different for the two devices due to different anode areas. Figure 8.20 has the forward-bias IV curves for the same two devices, along with the specific on-resistance. For the smaller

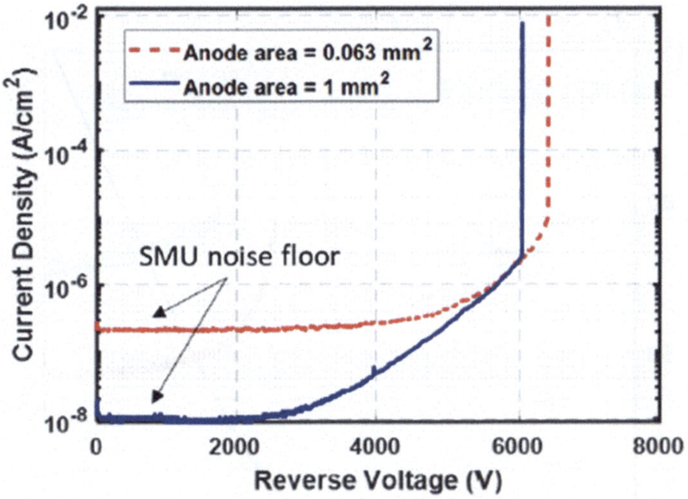

Fig. 8.19 Reverse-bias current-voltage curves for two pn diodes processed from the 50 μm drift layer epitaxial material. (From Ref. [6])

device, the specific on-resistance is 5.7 mΩ-cm^2, assuming an anode area of 0.63 mm^2, and 11.1 mΩ-cm^2 for the larger device assuming an anode area of 1 mm^2.

One critical aspect of growth that correlated with device yield was particle density. NRL used an optical profilometry tool to characterize epitaxial layers and then compared that data to device yield (Fig. 8.21). The yield of the 1 mm^2 devices improved by 50% following the cleaning and maintenance of the TNSC SR4000 MOCVD growth reactor. Further discussion about device yield and epitaxial wafer characterization is discussed in Chap. 10. Figure 8.22 shows another analysis of wafer optical profilometry and device yield. At a qualitative level, the device yield increases with reduced particle count on the epitaxial wafers.

5 Concluding Remarks

This chapter has examined the growth of thick GaN epitaxial layers by MOCVD on GaN substrates. A variety of commercial substrate types, both with and without high-dislocation-density dot-core regions, were studied. Two different MOCVD reactors at Sandia were utilized to grow pn diode structures with drift layer thicknesses varying from 6 to 50 μm. The ability to control both the thickness and doping uniformity, as well as the defect types of concentrations introduced during the growth, were studied using a variety of characterization techniques including Raman, CV profiling, Nomarski microscopy, optical profilometry, and x-ray

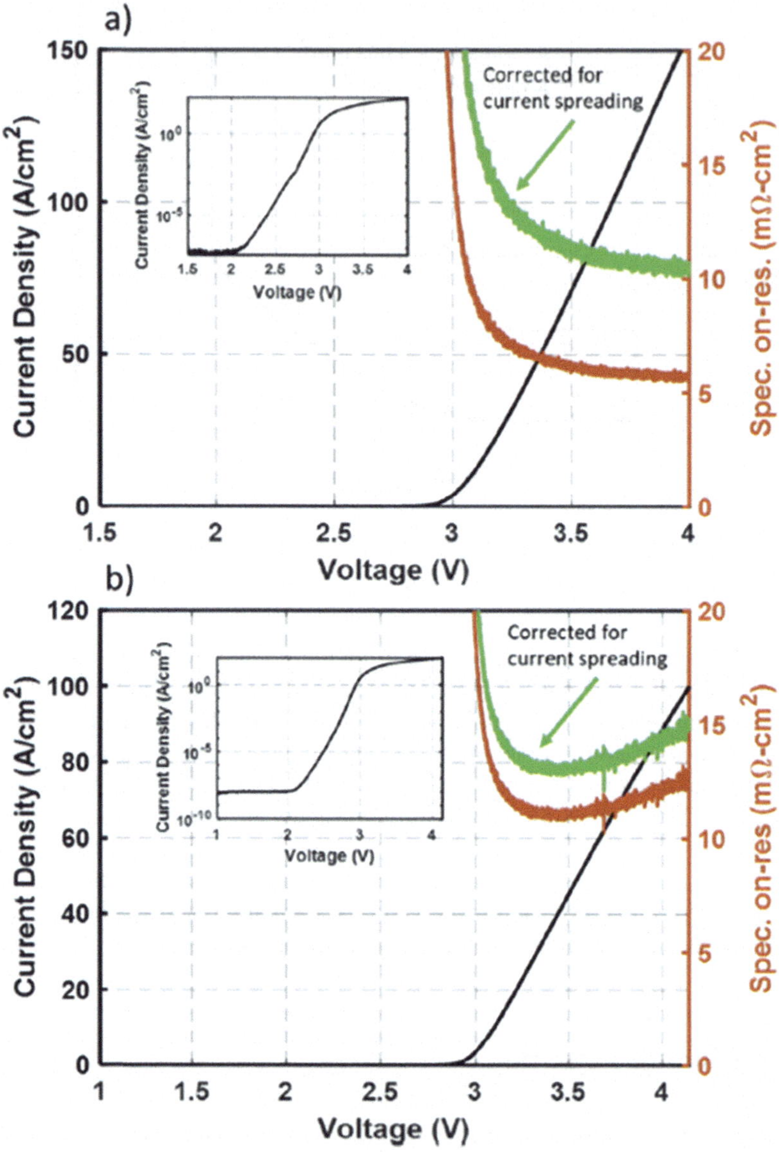

Fig. 8.20 Forward-bias current-voltage and specific on-resistance curves for (**a**) the 0.063 cm^2 area diode and (b) the 1 mm^2 area diode. The insets for each diode show the semi-logarithmic IV curves. (From Ref. [6])

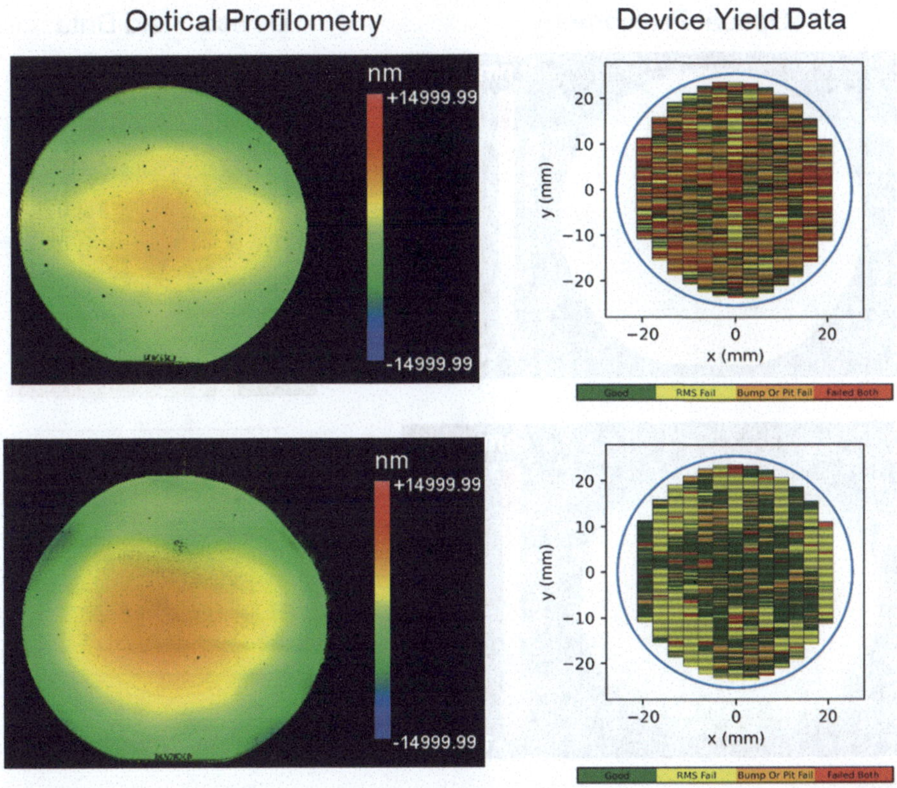

Fig. 8.21 Optical profilometry and device yield analysis of representative 8 μm drift layer wafers grown at Sandia, correlating device yield with epitaxial layer surface morphology. Critical particles that affect device yield are the black spots. For the device yield, green is a device meeting the performance specifications, while yellow, orange, and red are failures. The upper row has a higher level of critical particles that affect the device yield. (Courtesy of NRL)

topography. Epitaxial layers were processed into pn diodes at both Sandia and NRL. Breakdown voltage up to 6.4 kV was observed for 50 μm thick drift layer structures processed at Sandia, and device yield was correlated to epiwafer profilometry maps for devices processed at NRL. This work provides the foundation needed for the maturation of vertical GaN power devices in the future.

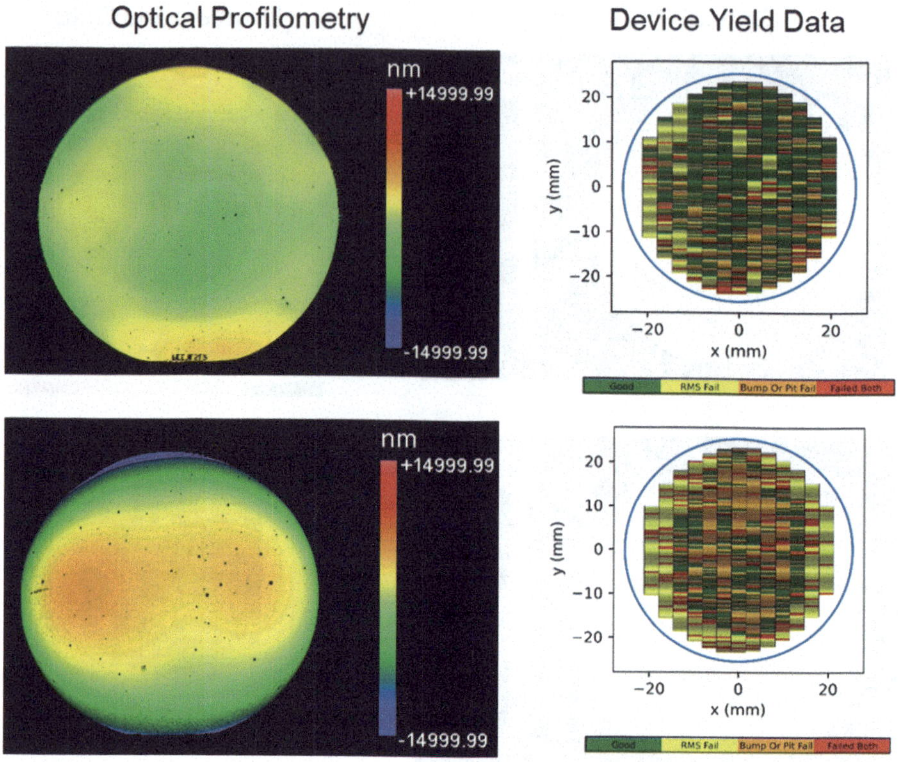

Fig. 8.22 Optical profilometry and device yield data analysis for representative 8 μm drift layer wafers grown at Sandia. Critical particles that affect device yield are the black spots. For the device yield, green is a device meeting the performance specifications, while yellow, orange, and red are failures. (Courtesy of NRL)

Acknowledgments The authors acknowledge Paul Sharps of Sandia for assisting in the preparation of the manuscript; Brendan Gunning, Mary Crawford, Jeffrey Steinfeldt, and Michael Smith of Sandia for technical contributions; and Travis Anderson, Alan G. Jacobs, Jennifer Hite, James Gallagher, Mona Ebrish, Nademullah Mahadik, James Spencer Lundh, and Karl Hobart of NRL for collaboration. The work described herein was performed under the ARPA-E OPEN+ Kilovolt Devices Cohort directed by Dr. Isik Kizilyalli. Sandia National Laboratories is a multi-mission laboratory managed and operated by National Technology & Engineering Solutions of Sandia, LLC, a wholly owned subsidiary of Honeywell International Inc., for the U.S. Department of Energy's National Nuclear Security Administration under contract DE-NA0003525. The views expressed in the chapter do not necessarily represent the views of the U.S. Department of Energy or the U.S. Government.

References

1. T. Nakamura, K. Motoki, GaN substrate technologies for optical devices. Proc. IEEE **101**(10), 2221 (2013)
2. K. Motoki, Development of gallium nitride substrates. SEI Tech. Rev. **70**, 28 (2010)

3. K. Motoki, T. Okahisa, R. Hirota, S. Nakahata, K. Uematsu, N. Matsumoto, Dislocation reduction in GaN crystal by advanced-DEEP. J. Cryst. Growth **305**, 377 (2007)
4. Y. Wang, M.E. Liao, K. Huynh, A.A. Allerman, M.S. Goorsky, Structural characterization of dot-core GaN substrates with annealing under growth-like conditions using synchrotron monochromatic x-ray topography. ECS J. Solid-State Sci. Tech. **10**, 045010 (2021)
5. M.P. King, R.J. Kaplar, J.R. Dickerson, S.R. Lee, A.A. Allerman, M.H. Crawford, A.J. Fischer, M.J. Marinella, J.D. Flicker, R.M. Fleming, I.C. Kizilyalli, O. Aktas, A.M. Armstrong, Identification of the primary compensating defect level responsible for determining blocking voltage of vertical GaN power diodes. Appl. Phys. Lett. **109**, 183503 (2016)
6. L. Yates, B.P. Gunning, M.H. Crawford, J. Steinfeldt, M.L. Smith, V.M. Abate, J.R. Dickerson, A.M. Armstrong, A. Binder, A.A. Allerman, R.J. Kaplar, Demonstration of >6.0-kV breakdown voltage in large-area vertical pn diodes with step-etched junction termination extensions. IEEE Trans. Elec. Dev. **69**(4), 1931 (2022)

Chapter 9
MBE Grown GaN P-N Diodes

Zachary J. Biegler, Esmat Farzana, and James S. Speck

1 Introduction

Molecular beam epitaxy (MBE) has played a major role in the development of compound semiconductors materials, heterostructures and devices. In this chapter, we review work on the development of MBE in the context of GaN vertical pn junctions.

The concept of a critical breakdown electric field is somewhat vague (please see a detailed discussion of the relevant issues in Chap. 24 by Jun Suda) but it is still a widely used material parameter. For GaN, the critical field strength F_c is ~3 MV/cm. De-rating the field by 50% as a reasonable safety margin and another 50% for one-sided p^+-n junctions with a triangular electric field, the average field in a depletion region is <1 MV/cm or <100 V/μm. Thus, for kV class devices the n-drift regions must be >10 μm thick. In the context of growth, such thick epitaxial layers require reasonably high growth rates while sustaining controlled low net donor doping $N_D - N_A$ in the drift regions where N_D is the intentional donor concentration and N_A is the unintentional compensating acceptor concentration. We will return to the topic of growth rate and net doping control later in this chapter.

For thick epitaxial layer growth, hydride vapor phase epitaxy (HVPE) offers outstanding material quality and high growth rate. Typically, HVPE is not used for forming junctions or structures with complicated doping profiles. Metalorganic chemical vapor deposition (MOCVD) is commonly used for devices and the progress in controlled doping and outstanding junctions as described in Chaps. 7 and 8.

Z. J. Biegler · J. S. Speck (✉)
University of California, Santa Barbara, CA, USA
e-mail: speck@mrl.ucsb.edu

E. Farzana
University of California, Santa Barbara, CA, USA

Iowa State University, Ames, IA, USA

I. C. Kizilyalli et al. (eds.), *Gallium Nitride and Related Materials*, The Materials Research Society Series, https://doi.org/10.1007/978-3-031-83056-3_9

Molecular beam epitaxy (MBE) of the group III-nitrides has been a valuable tool for basic research and development of advanced electron devices—particularly GaN-based high electron mobility transistors (HEMTs) [1]. There are two commonly employed techniques for nitride MBE: plasma-assisted MBE (PAMBE) and NH_3 MBE.

2 Plasma-Assisted MBE

In PAMBE growth of nitrides, the group III flux is provided by evaporation from liquid elemental sources Al, Ga, and In. The N flux is provided by a rf plasma source that cracks N_2 into atomic nitrogen N with a ~1% cracking efficiency [2]. Modern rf plasma sources are based on rf cavities to crack the molecular species while containing nearly all ions and electrons within the plasma source. The growth flux is typically "leaked" from the plasma cavity through an array of small apertures, usually on the order of 0.008–0.02 in diameter. Silicon is the most widely used donor in GaN MBE and is typically provided by a high-temperature effusion cell. For p-type doping, Mg is the only viable acceptor for GaN, and it is typically evaporated from the solid in a low-temperature effusion cell.

Typically, in the MBE growth of III-V compounds, the group V/group III flux ratios (e.g., ratio of As_4 or As_2 flux to Ga flux for GaAs—'V/III' ratio) are quite high—often on the order of 1000 or more. In PAMBE growth of GaN, the highest quality growth is realized for slightly group III-rich conditions—often referred to as metal-rich growth (V/III ratio < 1). The metal-rich growth of GaN on the (0001) metal face in PAMBE growth promotes the formation of a chemisorbed group III adlayer. The saturated adlayer is two monolayers thick of group III element—typically the gallium adsorbed layer in GaN PAMBE growth. The adlayer serves as an autosurfactant in the growth and promotes the diffusion of active nitrogen within the adlayer to step edges [3]. Excessively group III-rich growth results in group III droplet formation on the growth surface, which results in locally poor morphology in the region. Growth with lower group III flux results in the intermediate growth region with a partial wetting layer or the N-rich regime without any measurable group III adlayer.

Growth diagrams provide an easy way to understand the dependence of growth modes on fluxes and temperature. For PAMBE growth of GaN, the growth is typically on the (0001) c-plane and the nitrogen plasma source is usually operating at constant conditions that have been optimized for growth rate and material quality. Figure 9.1 shows a PAMBE GaN growth diagram that was developed using both reflection high energy diffraction (RHEED) and line-of-sight desorption mass spectroscopy. RHEED provides information on the growth mode and the desorption mass spectroscopy can be used to determine the gallium adlayer coverage θ^{Ga} on the growth surface [4]. For an active nitrogen flux f_N greater than the gallium flux f_{Ga}, the growth surface is usually either 3D or rough at lower temperatures (e.g., $T < \sim 725$ °C), whereas for higher T, the growth can be layer-by-layer. For metal-rich growth,

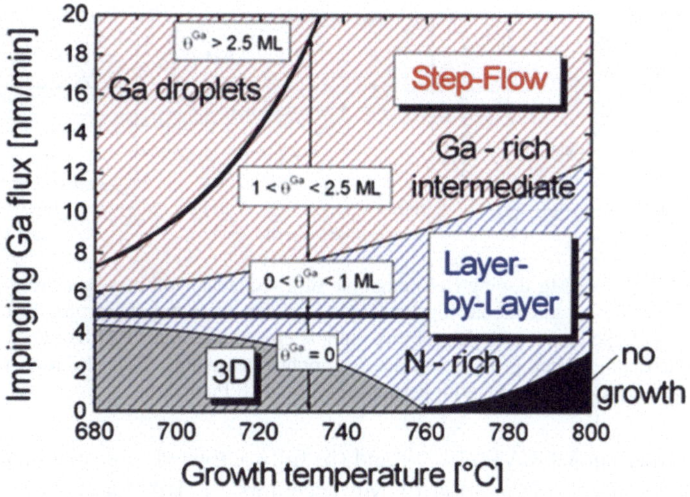

Fig. 9.1 Growth diagram for PAMBE GaN layers. The flux is in units of nm/min. This is a flux such that if there was sufficient active nitrogen, the growth rate of a Ga growth would be the given value. These values were obtained using a quadrupole mass spectrometer to measure the post-growth Ga desorption to determine the Ga adlayer coverage. (Reprinted from [4], with permission from AIP Publishing)

the excess gallium forms an adlayer on the surface that increases in coverage with an increasing flux ratio f_{Ga}/f_N. For (0001) growth, Ga adlayer coverages >1 ML result in step flow growth. The best morphologies are realized under conditions with a saturated Ga adlayer (contracted Ga bilayer) [4]. Excessively high Ga fluxes lead to the formation of metal droplets on the surface and poor morphology.

Realization of uniform morphology and properties is a challenge in PAMBE growth, as any lateral variation in metal flux or temperature will result in either droplet formation or growth without a saturated Ga-wetting layer. Uniform morphology can be realized by modulating the growth between the droplet regime and intermediate regime in a way such that the time average growth is within the droplet regime. The modulated growth was first demonstrated by Poblenz et al. [5].

Highly controlled donor and acceptor doping has been well demonstrated in PAMBE. Intentional doping with silicon, the most commonly used donor in GaN, is usually realized by a high-temperature effusion cell. Early work on optimizing the electron mobility in GaN by Heying et al., using mesa-defined junction isolated test structures, yielded record room temperature bulk electron mobility value of 1191 cm^2/Vs for growth in the intermediate regime near the cross-over to droplet formation [6]. Later work by Koblmüeller et al. with n-GaN layers grown at higher temperatures in the N-rich intermediate growth regime resulted in layers with electron mobilities in excess of 1100 cm^2/Vs [7].

Magnesium is the only viable acceptor for realizing p-type behavior in GaN. MBE, in comparison with MOCVD, offers two key advantages for Mg doping: (i) very abrupt turn-on and turn-off [Mg] profiles (absence of the well-known Mg

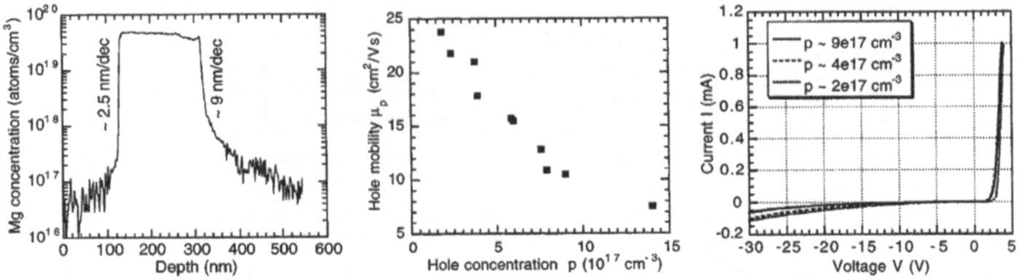

Fig. 9.2 PAMBE GaN Mg doping (left) example of a secondary ion mass spectroscopy (SIMS) Mg depth profile for an intentional top hat doping profile. Note the sharp turn-on and turn-off profiles. (center) Dependence of hole mobility on hole concentration for PAMBE GaN:Mg. (right) I-V for PAMBE pin diodes grown on GaN-on-sapphire templates. Note the high reverse bias leakage. (Reprinted from Ref. [8], with permission from AIP Publishing)

"memory effect" in MOCVD) [8, 9]; and (ii) fully activated as grown material– i.e., absence of hydrogen passivation of Mg acceptors [8, 10]. Early work at UCSB showed measured hole concentrations from ~10^{17} cm^{-3} to >10^{18} cm^{-3} with room temperature mobilities ranging from 24 cm^2/Vs at low doping to 7 cm^2/Vs at high doping—these values are approximately the same as today's start-of-the-art mobilities in MOCVD Mg-doped GaN [8]. The early work on Mg doping also resulted in among the earlier pn diodes by MBE. However, as a major drawback for PAMBE on dislocated GaN-on-sapphire MOCVD templates, threading dislocations typically have metal-rich cores that act as very strong leakage pathways [11]. The abrupt doping profiles, mobility of holes as a function of carrier concentration, and early GaN pn diode IV curves can be seen in Fig. 9.2.

In the early 2000s, bulk GaN substrates hardly existed, and thus, PAMBE seemed unviable for vertical structures due to threading dislocation-related leakage. However, for lateral structures, namely AlGaN/GaN high electron mobility transistors (HEMTs), the threading dislocations are perpendicular to channel and thus have fairly minor effect on both leakage and channel mobility. Many of the key early developments in GaN-based HEMTs were pioneered by PAMBE and then later transitioned to MOCVD.

3 NH_3-MBE

Early work in the plasma-based MBE GaN growth showed potential limitations in the growth rate due to inefficient cracking of the nitrogen molecule and potential damage due to ions created in the plasma. NH_3, another source of active nitrogen used by MOCVD reactors, circumvents the need for a plasma source through thermal decomposition, providing the active nitrogen species [12]. This process enables a significant increase in the active nitrogen flux during growth, as compared to PAMBE, enabling growth in the far N-rich regime [13]. Additionally, the lack of plasma completely removes the potential ion damage.

UCSB [14, 15] and CRHEA-CNRS [12] were early adopters of NH_3-MBE growth of III-nitrides. As with PAMBE systems, the group III source is provided through standard effusion cells. The NH_3, however, is introduced through a showerhead system to obtain uniform gas flow. This line includes gas purifiers and VCR fittings to minimize risks of contamination of the gas source.

NH_3-MBE growth of GaN presents unique challenges and benefits over PAMBE growth. The large gas flow rates of NH_3 require large throughput pumps, such as turbopumps, for operation. Traditional MBEs include cryo-panels, shielding inside the MBE chamber cooled with liquid N_2 (LN2) to obtain lower vacuum base pressures by freezing gaseous species to the walls of the chamber. Naturally, the NH_3 also freezes to the cryo-panel, which results in a potential overpressure situation with loss of liquid nitrogen (LN2). This presents an added safety risk, which can be mitigated by a regular "boil off" procedure, or recovery, where the LN2 is carefully removed from the cryo-panel and the additional NH_3 is pumped out of the system as it is boiled off. Additionally, although it is standard for Al to creep up the sides of the effusion cells, the NH_3 environment seems to exacerbate this issue, resulting in either more frequent openings for Al cell replacement, the need to have specially designed cells, or loading of less Al to mitigate this problem.

However, the benefits of NH_3 usage far outweigh these added risks. NH_3 systems allow growth in the far nitrogen-rich regime, eliminating the necessity of excess metal adlayer desorption cycles typically noted in PAMBE growth. Additionally, after thermally cracking the NH_3, excess hydrogen remains in the growth environment. Normally, these unintended impurities can cause detrimental effects on the grown crystals, but in GaN grown on dislocated template layers on foreign substrates, the excess hydrogen acts as a means of passivating the dangling bonds in the threading dislocations [13, 16]. This results in decreased reverse bias leakage currents in NH_3-MBE-grown GaN layers as compared to those grown in PAMBE. For vertical or quasi-vertical devices, these passivated threading dislocations result in a lower need to migrate toward significantly more expensive freestanding GaN substrates.

NH_3-MBE GaN growth has a wide range of conditions that result in stable growth in the nitrogen-rich regime. A growth diagram from Corrian et al., shown in Fig. 9.3, shows the wide range of NH_3 flow and substrate temperatures enabling step flow growth [15].

The growth diagram in Fig. 9.3 details the regimes where GaN growth is stable under given beam equivalent pressures (BEP). Given that the Ga BEP is held constant at 3.4×10^{-7} Torr, the NH_3 BEP can be directly converted to V/III ratios, ranging from 200 to 1000, which were obtained with gas flow rates on the order of 20–200 sccm of NH_3. At low temperatures, and low gas flow rates, the growth begins in a region similar to PAMBE growth, which is Ga-rich. This is typically marked by Ga droplet formation on the surface of the sample and spiral hillock growth around threading dislocations. Increasing the gas flow, but keeping the temperature the same (moving from sample vii to sample i), moves the growth out of the Ga-rich regime into the N-rich regime. However, the unstable layer-by-layer regime results in roughening due to the lower adatom mobility from the loss of the

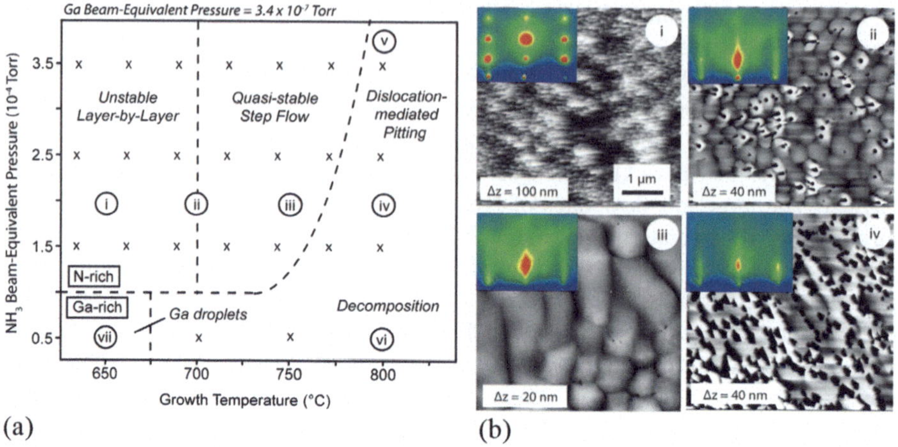

Fig. 9.3 (**a**) Growth diagram of NH_3-MBE grown GaN layers. Layers were grown at conditions marked with "x" and with numerals i-vii. Further details are available for the samples marked with the numerals in the referenced work. (**b**) AFM and RHEED images for layers grown in various regimes. (Reprinted from [15], with permission from AIP Publishing)

Ga wetting layer in this intermediate regime between island growth and step flow. RHEED often shows a 3D surface for films grown in this growth regime. Increasing the temperature (moving from i to ii) results in two beneficial processes. First, the adatoms on the surface gain more energy, which helps to smooth out the surface and prevent the 3D growth, but also increases the active nitrogen obtained from the NH_3, further suppressing any desorption of Ga adatoms. Continuing to increase the temperature (moving from ii to iii) results in a transition to the quasi-stable step flow regime. Films grown in NH_3-MBE are often grown in this regime as it facilitates a 2D, smooth, surface with minimal pitting. As the temperature continues to increase (moving from iii to iv), the surface begins to form significant pitting, with the surface forming pits at threading dislocations. RHEED in this region often includes chevrons corresponding to the faceting associated with the $\{10\bar{1}3\}$ planes. In this higher temperature regime (at around 800 °C), there are two competing processes: growth and decomposition. With enough NH_3 flux, it is possible to suppress the decomposition of the layer, but without active growth, the GaN layer will decompose at these temperatures. This process is extremely noticeable when decreasing the flux (moving from sample iv to vi) [15].

Similar to PAMBE GaN, NH_3-GaN maintains the use of Si and Mg as the n and p-type dopants, respectively. Both dopants are solid sources sublimated with effusion cells. The vapor pressure of Si, being relatively low, necessitates the use of a high-temperature dopant cell capable of reaching ~1400 °C. There are minimal impacts to the Si charge resulting from the NH_3 environment. The Mg charge, however, is impacted by the NH_3, and it is recommended to have a valved Mg cell to prevent the contamination of the Mg charge with either the active nitrogen or the hydrogen, as this will reduce the amount of volatile Mg. Additionally, to prevent the buildup of Mg at the valve, there should be dual zone heating on the cell.

One of the major advantages of doping from NH_3-MBE is the ability to easily include large amounts of Mg in the layers. Typically, PAMBE growth operates in the metal-rich regime, limiting the amount of Mg that is incorporated at higher temperatures. This requires lower substrate temperatures and higher Mg cell temperatures (thereby increasing Mg flux) to incorporate higher ($> 5 \times 10^{19}$ cm^{-3}) Mg concentrations. This often results in effusion cell temperatures on the order of >500 °C for PAMBE growth. When operating in the N-rich regime, growth in NH_3 systems can easily incorporate $>1 \times 10^{20}$ cm^{-3} at cell temperatures of ~250 °C. Although, these layers are still grown at somewhat lower substrate growth temperatures (740–760 ° C) to facilitate Mg incorporation and improve hole generation [17]. As with PAMBE, the p-GaN layers in NH_3-MBE are active as grown, enabling buried p-type layers. Traditionally grown p-GaN in NH_3-MBE show hole and mobility levels on the order of 1×10^{18} cm^{-3} and 7–9 cm^2/Vs for layers with Mg concentrations in the mid-10^{19} cm^{-3} range [17]. This results in resistivities on the order of 0.7 Ω cm. However, it is possible to improve the resistivity of the layer by the use of an In surfactant during growth, thereby improving hole generation and reducing the layer resistivities to 0.59 Ω cm for similar Mg concentrations [18].

For the n-type dopant in a pn junction, Si is the dopant of choice for MBE-grown GaN. In NH_3-MBE, these layers are grown around 820 °C with a Si either from silane [19] or from a silicon effusion cell [20]. Layers grown on nucleation layers on Al_2O_3 with electron concentrations in the low to high 10^{17} cm^{-3} range show mobilities varying from 547 to 356 cm^2/Vs, respectively. Layers grown directly on Al_2O_3, even with nucleation layers in MBE, typically have very high threading dislocation densities (TDDs). By growing on commercially available GaN/Al_2O_3 templates, it is possible to lower the TDD to ~5×10^8 cm^{-3}. By doing this, one can obtain higher electron mobilities due to less scattering from the TDD [20]. This improvement can be seen in Fig. 9.4.

Further reduction of the TDD by moving to freestanding GaN layers (which often have TDDs $<10^6$ cm^{-3} for HVPE-grown layers and ~10^4 cm^{-3} for layers that are grown via ammonothermal evaporation or on ammonothermal templates) can dramatically increase the room temperature mobilities of the carriers. In the paper by Kyle et al., layers grown on FS GaN with threading dislocation densities on the order of 2×10^6 cm^{-3} show room temperature mobilities up to 1265 cm^2/Vs with carrier concentrations in the mid 10^{16} cm^{-3} range [20]. The highest mobility in this study (achieved upon cooling the sample for the temperature-dependent Hall effect measurement) was 3327 cm^2/Vs. This highlights, again, the need for native substrates for GaN devices, as even NH_3-MBE material sees a dramatic increase in material performance through the reduction of the TDD.

Although the vast majority of n-type doping in NH_3-GaN is achieved with Si, Ge has also been explored as an n-type dopant due to the potentially reduced stress in the films [21]. Although able to obtain reasonable mobilities at ~400 to 100 cm^2/Vs for carrier densities of ~4×10^{17} cm^{-3} to 2×10^{20} cm^{-3}, the films are somewhat plagued by morphological challenges [21].

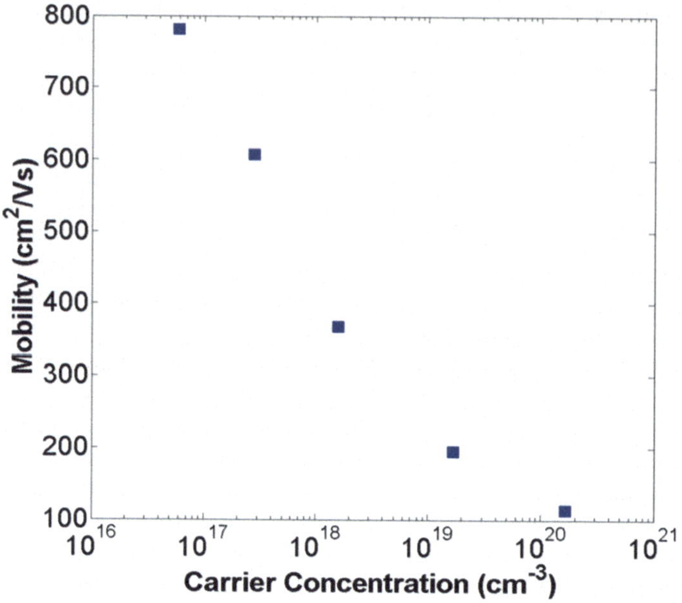

Fig. 9.4 Mobility of GaN:Si layers for different carrier concentrations on GaN/Al_2O_3 templates. (Reprinted from [20], with permission from AIP Publishing)

Although most of the discussion has focused around GaN layers, ternary alloys are possible in NH_3-MBE as well. $Al_xGa_{1-x}N$ (and AlN) result in higher desorption temperatures than GaN, enabling growth at higher substrate temperatures. This, in turn, increases the amount of available active nitrogen. As such, it is trivial to add Al into the growth layer. Growth of n-type AlGaN via NH_3-MBE has been reported since the early 2000s, with carrier concentrations in the 10^{19} cm^{-3} range up to about 85% Al [22]. For AlGaN layers at about 60%, carrier concentrations in the mid-10^{19} cm^{-3} with mobility of 48 cm^2/Vs have been reported [23]. These layers show the ability to grow and effectively dope the AlGaN layers for devices. However, there is still significant research to be done with full AlN layers via NH_3-MBE.

InGaN, on the other hand, faces some challenges for growth in the NH_3 environment. InN and GaN have dramatically different thermal stabilities, resulting in the need to grow InGaN at significantly reduced temperatures than GaN [14, 24]. Typical growths of InGaN result in maximum incorporations on the order of 15% of In at temperatures as low as 550 °C [14, 25]. Lower temperatures than this result in decreased active nitrogen due to the poor cracking efficiency of NH_3 at these temperatures [25].

NH_3-MBE has also been employed in work relating to high electron mobility transistors (HEMTs) [26]. In work by Poblenz et al., $Al_{0.3}Ga_{0.6}N$ layers were grown on 4H-SiC on top of GaN nucleation layers. The channel mobility at room temperature was 1643 cm^2/Vs for a sheet charge density of 1.0×10^{13} cm^{-2}. These HEMTs resulted in an output power density of 11.2 W/mm and PAE of 58% at 10 GHz for a V_{DS} of 48 V.

By far, the most relevant work for NH_3-MBE material is in vertical power electronics. Development of high-quality vertical structures dramatically reduces the size, and thereby the cost, of each device. NH_3-MBE is exceptionally well suited for this task due to its ability to achieve high-quality growth in the nitrogen-rich regime [15] and the inclusion of hydrogen in the growth environment, which helps suppress the reverse bias leakage through the threading dislocations [11, 16]. This allows NH_3-MBE devices on highly dislocated substrates (such as GaN on Al_2O_3 templates) to compete with PAMBE devices on more expensive freestanding GaN substrates. Diodes grown using NH_3-MBE by Hurni et al. show exceptionally low ideality factors of about 1.3 for $n/n^-/p^+$ junctions [13]. They also showed that the n^-layer played a role in both the reverse leakage and the diode ideality, where increasing the Si concentration in the n^-layer increased both reverse leakage and the ideality factor.

4 MBE P-N Diodes

P-N diodes, themselves, are some of the more basic device structures that can utilize the full bandgap of the material. As such, they typically offer significantly reduced reverse bias leakage as compared to Schottky diodes made of the same material. Even more complicated device structures (such as Bipolar Junction Transistors and Field Effect Transistors) often rely on local pn junctions for the devices to operate properly. Here, MBE GaN has a distinct advantage over MOCVD GaN. Namely, p-GaN by MBE is active as grown and does not need an additional post-growth anneal [18]. This enables buried p-GaN layers, which can facilitate more complicated device structures, such as bipolar junction transistors, without complicated device architecture to allow for the hydrogen escape necessary to activate MOCVD p-GaN layers. In addition, MOCVD suffers from a "memory effect" for the Mg dopants, in stark contrast to the abrupt junctions possible in MBE GaN [9].

An additional advantage MBE GaN has over MOCVD GaN is the (generally) cleaner growth environment. By growing at significantly reduced pressures, typical unintentional doping concentrations for MBE GaN are on the order of 5×10^{15} cm^{-3} to 2×10^{16} cm^{-3} [27]. This enables significantly reduced UID concentrations in drift layers, thereby increasing the maximum reverse bias voltage available. In addition, the reduction in drift layer concentration typically leads to improved diode ideality factors [13]. This can be seen in the work done by Hurni et al. for diodes grown on Lumilog GaN:Fe MOCVD templates (TDD around 5×10^8 cm^{-3}) with a 300 nm n^+ layer, a 450 nm n^- layer (of varying doping shown in Table 9.1), and a 100 nm p^+ layer. The improvement in the ideality factor from higher doped (sample D) to lower doped (sample A) is shown in Fig. 9.5.

Despite these advantages, however, the lowest reported ideality factor comes from MOCVD grown material on freestanding GaN substrates, about 1.1 [28].

Table 9.1 Sample number designation for Hurni et al., 2010

Sample number	300 nm n^+ layer	450 nm n^- layer	100 nm p^+ layer
A	$n \sim 3 \times 10^{18}$ cm^{-3}	$n \sim 3.5 \times 10^{17}$ cm^{-3}	[Mg] ~ 10^{20} cm^{-3}
B	$n \sim 3 \times 10^{18}$ cm^{-3}	$n \sim 7 \times 10^{17}$ cm^{-3}	[Mg] ~ 10^{20} cm^{-3}
C	$n \sim 3 \times 10^{18}$ cm^{-3}	$n \sim 3 \times 10^{18}$ cm^{-3}	[Mg] ~ 10^{20} cm^{-3}
D	$n \sim 4 \times 10^{18}$ cm^{-3}	$n \sim 4 \times 10^{18}$ cm^{-3}	[Mg] ~ 10^{20} cm^{-3}

Reprinted from [13], with permission from AIP Publishing

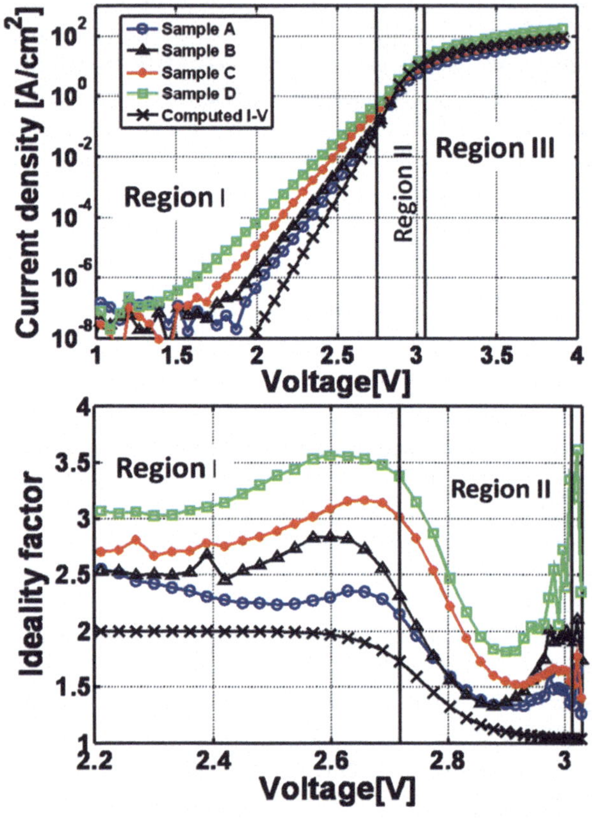

Fig. 9.5 Top: Log scale J-V characteristics for each diode, including a computed curve fitted to sample (**b**). Bottom, calculated ideality factor for samples (**a**–**d**), including the ideality factor for the computed J-V curve. (Reprinted from [13], with permission from AIP Publishing)

When examining vertical structures for high power, there are multiple metrics that are significant. First, increasing the depletion region width increases the voltage allowed across the device in reverse bias. Therefore, it is necessary to have thick draft regions with exceptionally low doping values to hold this depletion region. This results in the necessity to obtain low UID doping levels in drift layers. However, the decrease in carriers in these regions results in dramatic increases in specific on-resistance. Therefore, it is necessary to obtain drift regions with exceptionally high mobility of the carriers, to offset the increase in the specific on-resistance. GaN layers by NH_3-MBE and PAMBE have shown low UID doping

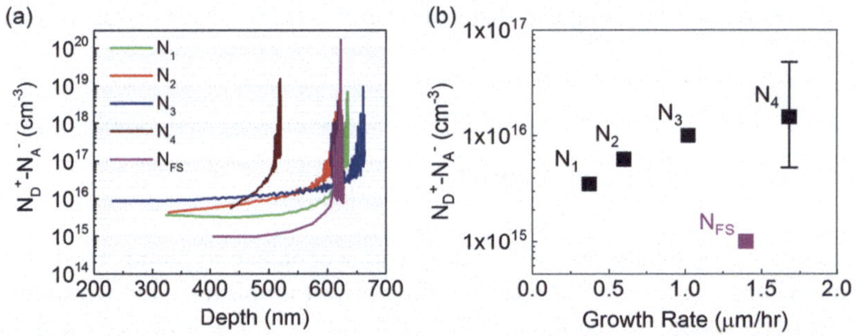

Fig 9.6 (**a**) Capacitance-Voltage measurements on MOS-CV structures for UID structures grown at increasing growth rates. (**b**) The obtained net dopant concentration from the C-V measurements in (**a**). N_1-N_4 were grown on template layers, but N_{FS} was grown on a freestanding GaN substrate [27]

Fig. 9.7 Simulated forward J-V curves for an ideal diode without dislocations included and a diode with 10^8 cm^{-2} dislocations. (Reprinted from [29], with permission from AIP Publishing)

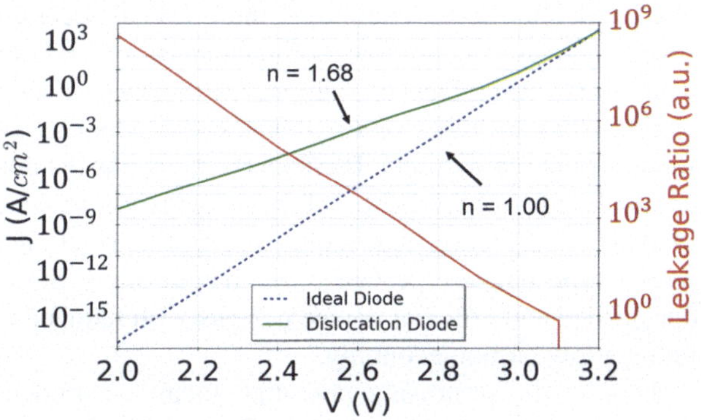

in the range of 5×10^{15} cm^{-3} to 2×10^{16} cm^{-3} [27]. Wang et al. showed that for NH_3-MBE GaN epilayer on Saphhire template, increasing growth rate from 0.37 μm/hr. to 1 μm/hr. resulted in an increased UID doping concentration from 3.5×10^{15} to 1×10^{16} cm^{-3} [23]. However, growing on freestanding GaN substrates allows for significantly lower background doping of ~1×10^{15} cm^{-3}, even at faster growth rate of 1.4 μm/hr., likely due to low donor-like impurities and dislocation complexes (Fig. 9.6) [23].

Additionally, the mobilities of these layers are exceptionally high. PAMBE-grown layers have shown mobilities as high as 1191 cm^2/Vs on highly dislocated (TDD ~ 5×10^8 cm^{-2}) MOCVD templates [6]. For similar TDDs, NH_3-MBE shows room temperature mobilities at about 966 cm^2/Vs. But, this is improved to 1265 cm^2/Vs on freestanding GaN templates with low threading dislocation densities (~ 2×10^6 cm^{-2}) [20].

Substrates with lower threading dislocation densities would not only improve electron mobility but also reduce both forward [29] and reverse bias leakage [30]. In the simulation work done by Robertson et al., the inclusion of threading dislocations dramatically increased the forward leakage current and increased the diode ideality factor (as compared to a simulated dislocation-free diode). These results can be seen in Fig. 9.7 [29].

5 Vertical GaN Power Diodes

GaN devices have already received widespread acceptance in commercial electronics in many areas, such as solid-state lighting, power amplifiers, and RF transistors, with operating voltage in the range of 15–650 V [31, 32]. For devices with higher operating voltage range beyond 650 V, the power electronics market is still dominated by conventional silicon technology due to their lower cost, easier availability of bulk crystals, and decades of development history. However, in grid-scale applications, these silicon-based power converters require multilevel bulky series connection to reach the desired multi-kV rating due to the inherent material limits of low bandgap silicon. These lead the power devices to suffer from slow switching speed, large system volume, complex circuitry, and enormous energy wastage from extensive heating. Hence, to improve efficiency, switching speed, functionality, and compactness in high-voltage devices, the adoption of WBG semiconductors is a fundamental requirement. Particularly, GaN has the potential to dramatically change the landscape of high-power electronics taking the advantage of its high breakdown field (~3.3 MV/cm), high mobility, good thermal conductivity, and availability of both p-type and n-type GaN, which will allow higher voltage rating in compact device size and efficient device designs with reduced energy wastage toward developing a low carbon economy.

Despite the promising potential, today's commercial GaN power devices are mostly limited to 650 V, which is realized with lateral device topologies. To expand application space of GaN devices beyond the 650 V, vertical devices can be effective pathways to support superior power handling capability compared to lateral ones due to their (i) high-voltage sustainably with better electric field confinement in the WBG GaN drift layer, (ii) high current capability enabled by entire backside ohmic, (iii) scaling feasibility, and (iv) better thermal management offered by large heat dissipation volume [31, 33]. However, compared to lateral counterparts, demonstration of vertical GaN power devices remained limited due to lack of high-quality and cost-effective GaN substrates in prior years. Although foreign substrates, such as sapphire and silicon, were utilized for vertical GaN devices, these were not suitable to achieve a high-voltage rating due to dislocation leakage current, degraded transport properties, as well as limited thickness of high-quality GaN epitaxial layer growth.

In recent years, the advances in freestanding (FS) GaN substrates with low dislocation density ($<10^6$ cm^{-2}) have spurred expansive interest to overcome these challenges of high-power vertical GaN devices. Moreover, the continued progress in high-quality thick homoepitaxial GaN epilayer growth with low-background impurity, as described previously, has created excellent opportunities for vertical GaN power devices. The FS GaN substrates bring exciting potential to obtain thicker homoepitaxial drift layer and reduced dislocation-assisted leakage while the low-background impurities in epitaxial layers will contribute to low on-resistance.

6 Vertical GaN Power Devices: Drift Layer, Surface Morphology, and Surfactants

The previous sections demonstrated homoepitaxial GaN epilayer growth by NH_3-MBE at UCSB with low unintentional background impurity. In addition to the low-background doping, good surface morphology is also a fundamental requirement for high-performance devices. However, this is often a challenge for high-voltage vertical devices as the target thick drift layer synthesis requires fast growth rate (≥ 1 μm/hr) to complete the growth within a feasible time frame. Growth rates on the order of 1–8 μm/hr. have been demonstrated in PAMBE films by growing with a high flux plasma source [34, 35] and NH_3-MBE [36]. Such fast growth rate, in turn, leads to rough surface with steps and pits. To maintain improved surface morphology, surfactants are typically used for III-nitrides that can reduce the diffusion barrier of adatoms and thereby, promote their smoother incorporation to fill in kinks and vacancies available in the surface as shown in Fig. 9.8.

A widely used surfactant for III-nitrides is indium, which can reduce the adatom diffusion barrier of both the constituent Ga and N atoms, decreasing from 0.7 to 0.12 eV for Ga and from 1.3 to 0.5 eV for N [38].

As such, indium surfactants were utilized in the NH_3-MBE GaN-on-GaN growth at UCSB for thick drift layers, up to 10 μm, at fast growth rates (~1 μm/hr) [36]. The reduction of RMS surface roughness was evident from 1.1 nm to 0.90 nm when indium flux was introduced with BEP of 5×10^{-8} Torr during GaN epilayer growth on GaN-on-sapphire template (Fig. 9.9a and b). Further reduction of surface roughness was achieved by having GaN epilayer grown on FS GaN substrates, which enabled substantially low RMS roughness of 0.21 nm (Fig. 9.9c and d). The lower RMS in the homoepitaxial GaN layer was contributed by the orders of magnitude

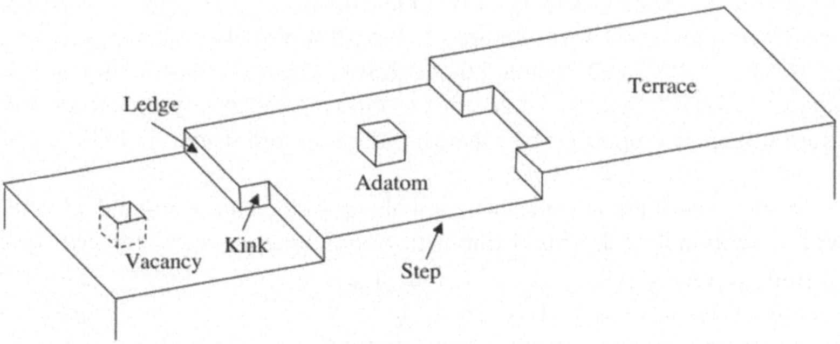

Fig. 9.8 Schematic of a sample surface showing the adatom, kinks, and surface vacancies. (Reprinted from [37], with permission from Elsevier)

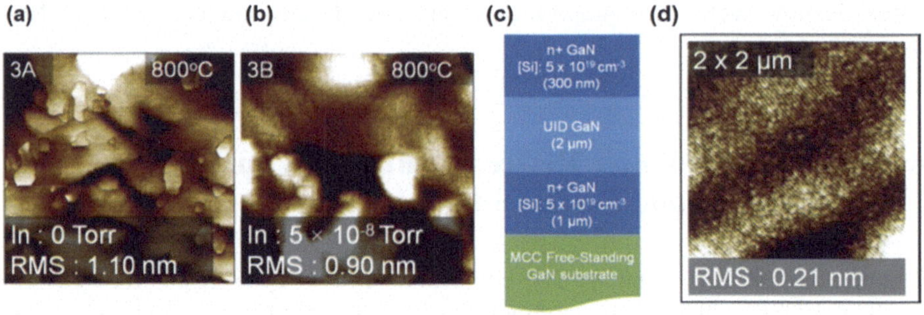

Fig. 9.9 Effect on Indium surfactant to reduce surface roughness in GaN epilayer grown on GaN-on-sapphire template. The $2 \times 2\ \mu m^2$ AFM scans are for indium flux BEPs of (**a**) 0 and (**b**) 5×10^{-8} Torr (**c**) Homoepitaxial GaN layer grown on MCC FS GaN substrate using indium flux BEP of 5×10^{-8} Torr and (**d**) further enhancement in smooth surface morphology with lower RMS of 0.21 nm was achieved with the applied indium surfactant in homoepitaxial GaN [36]

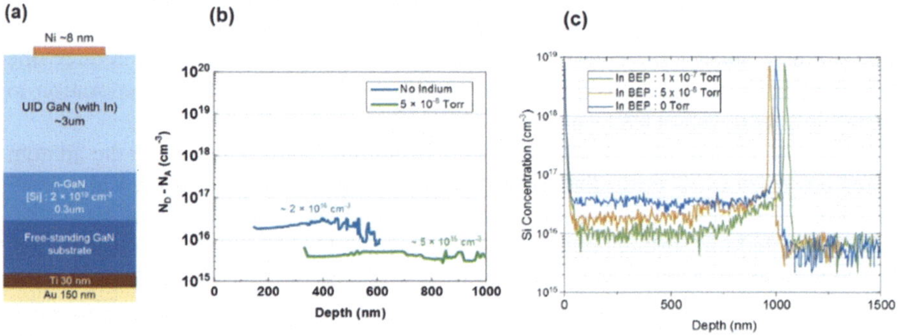

Fig. 9.10 (**a**) Fabricated Schottky diode with homoepitaxial UID GaN grown by Indium surfactant using indium flux BEPs of 5×10^{-8} Torr (**b**) Reduced background unintentional doping achieved by using indium as surfactant (**c**) Lower background doping was consistent with the trend of reduced silicon incorporation from SIMS in the GaN epilayer with increased indium flux BEP [36]

lower threading dislocation density of FS GaN substrates ($\sim 10^6\ cm^{-2}$) by Mitsubishi Chemical Corporation (MCC) compared to the MOCVD GaN-on-sapphire template layer ($\sim 10^8\ cm^{-2}$). It should be noted that careful optimization of indium surfactants is needed to make the best use of adatom mobility since excessive indium flux can deteriorate surface morphology by hindering the required reaction of Ga atoms with NH_3.

The introduction of an indium surfactant also provided an additional advantage of reduced unintentional background impurities, which is necessary for drift layers of vertical high-voltage devices. A reduction of unintentional doping concentration, measured by CV, from $2 \times 10^{16}\ cm^{-3}$ to $5 \times 10^{15}\ cm^{-3}$ was evident in the homoepitaxial GaN films with introduced indium flux BEPs of 5×10^{-8} Torr as compared to the reference without indium flux (Fig. 9.10a and b). This was also commensurate with the trend of reduced silicon impurity concentration from SIMS with increased indium flux, as shown in Fig. 9.10c. A possible underlying mechanism can be the propensity of indium atoms to bond at step edges which may inhibit silicon incorporation on the GaN surface. Thus, this combination of smooth surface morphology and lower unintentional background doping, enabled by indium surfactant, can further improve drift layer quality for high-performance vertical GaN devices.

7 Vertical GaN Power Devices from NH_3-MBE GaN

For vertical devices rated for high voltage, it is necessary to take advantage of the critical breakdown field of WBG GaN using thick drift layers with very low unintentional doping as well as extremely clean background with minimal compensating defects, as described before. However, in terms of device design, several fundamental requirements also need to be addressed including processing techniques, passivation, surface and interface quality, and edge termination, as follows:

(i) Effective edge termination is a vital requirement for managing high electric fields in power devices. At high reverse voltage, the electric field often gets concentrated at junction peripheries, which introduces non-uniform high field at edges and leads to early device failure. To redistribute the localized high fields, efficient edge termination techniques need to be implemented, such as field plate, guard ring, deep etch, ion implantation, and junction termination extension.

(ii) To achieve the full utilization of the GaN drift layer with a uniform electric field, it is also of great importance to obtain a punch-through design taking into account the drift layer doping concentration, thickness, and thereby, adopt necessary field management. The field management can be employed using one or more edge termination methods based on the device geometry and epilayer properties to minimize all concentrated high field locations and achieve a uniform field profile.

(iii) The dielectric and interface quality needs to be carefully optimized with high-quality deposition methods and surface preparation techniques. If the field-plate dielectric is of poor quality or the dielectric has an intrinsic breakdown field less than that of the GaN drift layer, then the diode will suffer from early parasitic breakdown within dielectric rather than utilizing the material benefits of GaN.

(iv) Process-induced damage also needs to be critically addressed since such damage can degrade device performance by introducing unintentional leakage paths. For example, device processing often involves plasma-based dielectric deposition, dry etch, or ion implantation that can create physical damage in device layers from energetic ion bombardment. Hence, the fabrication process needs to be coordinated with necessary damage mitigation techniques, such as wet etch, annealing, or passivation techniques.

(v) Finally, electrical contacts and relevant contact layers need to be well developed so that low contact resistance can be obtained to ensure high on-state current.

7.1 Device Structure

To realize high-performance vertical GaN p^+n power diodes using the NH_3-MBE-grown GaN layers, the above-mentioned strategies were rigorously adopted. The epilayers were grown on a Lumilog FS GaN substrate with dislocation density of high 10^6 cm ^{2}. The epilayer growth was commenced with a ~0.25 μm n^+ GaN ([Si]: 1×10^{19} cm^{-3}) buffer layer to mitigate any defect incorporation from the substrate during the subsequent device layer growth. The successive device layers included a ~4 μm thick UID GaN for the n^- drift layer and ~0.4 μm p^+ GaN ([Mg]: 3×10^{19} cm^{-3}) to complete the p^+n junction (Fig. 9.11a). To facilitate a high-quality anode ohmic contact layer, a 10 nm highly Mg-doped ([Mg]: 3×10^{20} cm^{-3}) p^{++} GaN cap layer was grown. It is to be mentioned that the entire epilayer structure, consisting of p^{++} cap/p^+-n^-diode/n^+ buffer, was synthesized with an uninterrupted NH_3-MBE growth in Ga-limited growth condition using a substrate temperature of 800 °C for n^-/n^+ layers and 750 °C for the p^{++}/p^+ layers. A growth rate of 0.32, 0.6, 0.32, and 0.076 μm/hr. was used for the n^+, n^-, p^+, and p^{++} layers, respectively [33].

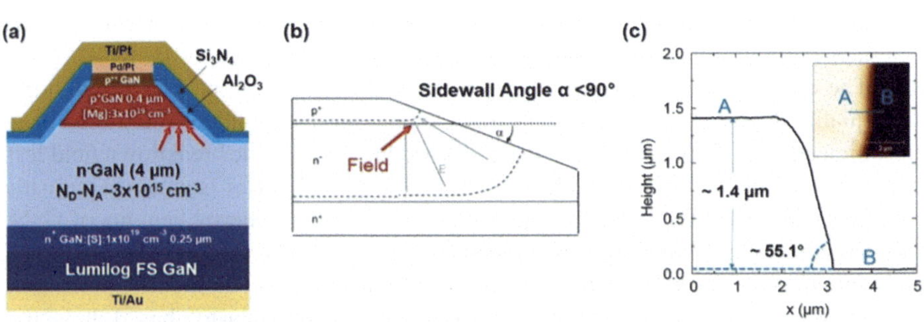

Fig. 9.11 (**a**) Schematic of the fabricated NH_3MBE vertical GaN-on-GaN p^+n diode with a sidewall angle field-plate formed with dielectric stack of Si_3N_4/Al_2O_3. (**b**) Field management to spread out concentrated electric field at p^+n junction by using etched edge with a less than 90° slanted angle [39] (**c**) Measured etch depth and sidewall angle using AFM (inset) along cutline AB that demonstrates an etched depth of 1.4 μm and a sidewall angle of ~55°. (Reprinted from [33] with permission)

7.2 Field Management

The p^+n diodes were fabricated with circular anode contacts using Pd/Pt metals on the p^{++} GaN cap layer with diameters of 80–100 μm. The backside cathode ohmic contact was formed by Ti/ Au deposition on the conductive FS GaN substrate. For efficient field management, a combination of deep etch, subsequent wet etch, dielectric passivation, and field plate was implemented. The deep etch was performed with Cl_2-based inductively coupled plasma (ICP) with a depth of 1.4 μm. Subsequently, wet etch treatment was performed with KOH to reduce dry etch-induced damages [33, 40]. Further passivation of sidewall damage was performed with a thin Al_2O_3 (~26 nm) deposited by atomic layer deposition (ALD) at 300 °C. The Al_2O_3 also served as a protective layer in the dielectric field-plate stack of Si_3N_4/Al_2O_3. The field-plate dielectric Si_3N_4 (~205 nm) was formed by plasma-enhanced chemical vapor deposition (PECVD) where the prior deposited ALD Al_2O_3 protected the surface from plasma damage during PECVD Si_3N_4 growth.

The field plate was also designed with a sidewall angle of 55° instead of a steep 90°angle, as shown in Fig. 9.11. Such slanted angle design allows gradual smoothing of the electric field and better uniformity compared to the abrupt steep 90° angle field plates (Fig. 9.11b) [39, 41]. It is to be noted that the combination of optimized dry etch and wet etch process contributed to this sidewall angle of 55°. Further modification or reduction of sidewall angle is also possible with photoresist reflow and photomasking techniques with different etch selectivity [41]. In the sample, some diodes were also kept without field plates to serve as baseline devices to extract epilayer doping properties as well as pursue comparative evaluation of the diode performance with respect to field plate ones.

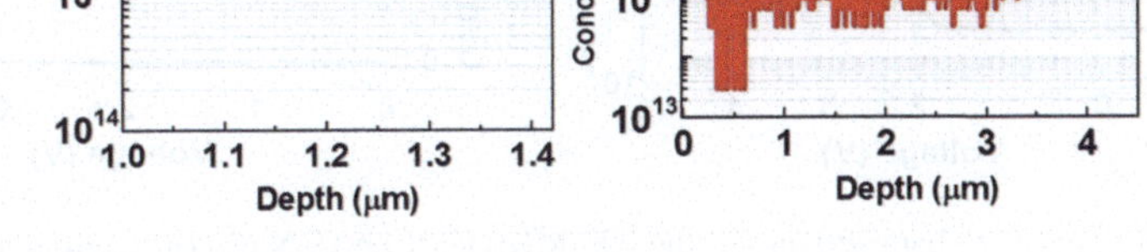

Fig. 9.12 (**a**) C-V extracted net doping concentration (N_D^+-N_A^-) showing a low unintentional background doping of ~3 × 10^{15} cm^{-3} (**b**) SIMS profile of Si, Mg, and O in the NH_3MBE p^{++} /p^+/ n−/n^+ GaN sample showing background silicon impurities of ~2 × 10^{15} cm^{-3}. The silicon and oxygen likely acted as unintentional doping source. (Reprinted from [33] with permission)

8 Device Characteristics

The electronic properties of the NH_3-MBE GaN p^+n diodes were systematically characterized with current density-voltage (J-V) and capacitance-voltage (C-V) measurements. The C-V measurements, performed at 1 MHz frequency, exhibited remarkably low net background doping of ~3×10^{15} cm^{-3} in the drift layer, seen in Fig. 9.12a. This was almost in line with the very low background silicon concentration (~2×10^{15} cm^{-3}) from SIMS, shown in Fig. 9.12b. SIMS also detected a background oxygen concentration (~2×10^{16} cm^{-3}), which may have contributed to the remaining background doping by partial ionization. Moreover, the compensating carbon concentration in these samples was below the detection limit, indicating high-purity GaN with minimal background compensation. It is to be noted that this unintentional doping is about order of magnitude lower than most conventional MOCVD reports in vertical GaN p^+n diodes, revealing an extremely clean growth environment facilitated by the NH_3-MBE method [33].

8.1 *Forward Transport Properties*

In the forward current transport property of pn diodes, the current is typically dominated by generation-recombination mechanism at low biases. Later, at higher forward biases, the diffusion process provides the main current transport, leading to an overall equation expressed as:

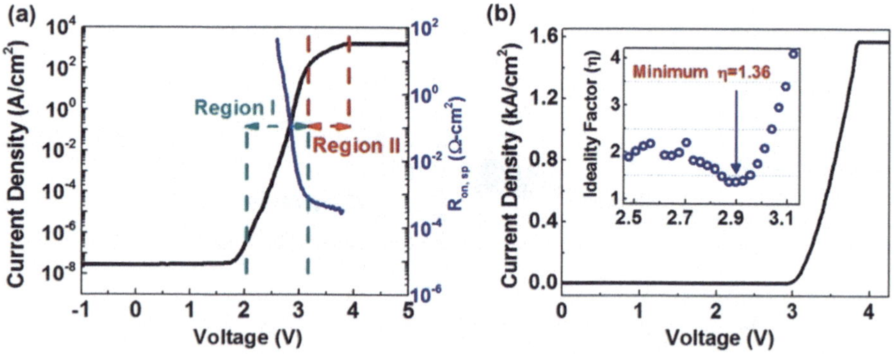

Fig. 9.13 Forward J-V characteristics of the NH_3MBE GaN-on-GaN p^+n (**a**) in log scale showing the different regions based on dominated current transport mechanism where Region I is generation-recombination dominated and Region II is diffusion dominated. The differential specific on-resistance is shown in right axis (blue) (**b**) in linear scale where the inset shows the extracted ideality factor with a minimum value of 1.36. (Reprinted from [33] with permission)

$$J = J_o^{GR}\left(e^{\frac{qV}{2kT}} - 1\right) + J_o\left(e^{\frac{qV}{kT}} - 1\right) \tag{9.1}$$

where the first term in Eq. (9.1) implies a generation-recombination-dominated transport with ideality factor 2 and the second term indicates the diffusion-dominated transport with unity ideality factor 1. The terms $J_o$and J_o^{GR} indicate the reverse saturation current density for respective cases, T refers to the temperature, and k is the Boltzmann constant (Fig. 9.13).

The fabricated vertical NH_3-MBE GaN p^+n diodes also exhibited an initial generation-recombination dominated current transport for low forward bias (V_f) of $2 < V_f < 2.8$ V (Region I) with an ideality factor ~2. At higher applied biases of $V_f > 2.8$ V in Region II, the diffusion process starts to dominate. Here, the vertical p^+n GaN diodes revealed outstanding transport characteristics, reaching a low minimum ideality factor of 1.36. The turn-on voltage of the p^+n GaN diodes was obtained as 3.15 V extracted at a forward current density of 100 A/cm^2, which is expected based on the GaN bandgap. The diodes also exhibited a low minimum differential specific on-resistance ($R_{on,sp}$) of 0.28 mΩ-cm^2. A high forward current density was achieved with a maximum of ~1.6 kA/cm^2 at 3.85 V before the measurement reached the compliance limit of the equipment.

8.2 *Reverse Breakdown Properties*

The NH_3-MBE GaN p^+n diodes also exhibited excellent reverse breakdown properties of greater than 1 kV which was the equipment limit (Fig. 9.14a). Some diodes without field plates had breakdown voltages of 890 V. The breakdown performance was analyzed using Silvaco TCAD software. The TCAD modeling indicated that the

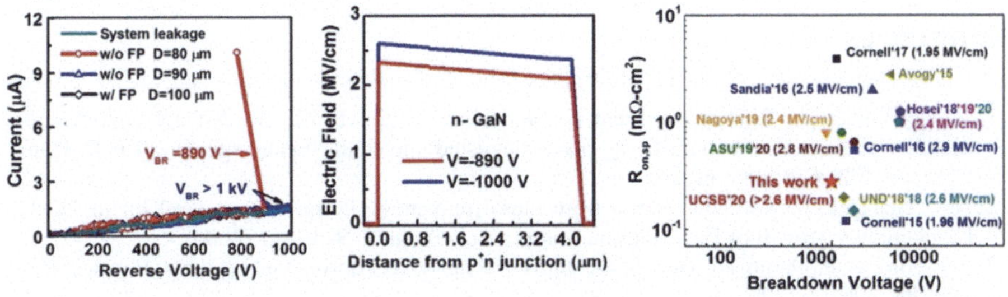

Fig. 9.14 (**a**) Reverse J-V characteristics of the NH_3MBE GaN-on-GaN p^+n diodes with or without field plate. The best diodes did not show breakdown up to 1 kV (equipment limit), whereas some diodes without field-plate revealed breakdown at 890 V. Note that the system leakage obscured to measure the actual reverse leakage current of the p^+n diodes (**b**) Simulated electric field profile in Silvaco TCAD showing a punch-through electric field profile for both 890 V and 1 kV cases with peak fields of 2.3 MV/cm and 2.6 MV/cm, respectively. (**c**) Benchmark plot of minimum $R_{on,sp}$ and breakdown voltage for this work and prior reports. (Reprinted from [33] with permission)

diodes reached a punch-through electric field profile where the drift region was entirely depleted. The peak electric field (E_C) at 890 V and 1 kV reverse bias was obtained as 2.3 MV/cm and 2.6 MV/cm, respectively, at the abrupt p^+n junction interface (Fig. 9.14b).

It is to be noted that the measurement system leakage and maximum measurement capability of 1 kV of the equipment limited our ability to determine the actual reverse current density and breakdown voltage of the best diodes. Yet, the combination of low minimum ideality factor (1.36), low specific on-resistance (0.28 mΩ-cm^2), and breakdown voltage of more than 1 kV are among the best-achieved properties compared to the reported homoepitaxial p^+n GaN diodes which required thicker MOCVD drift layer to achieve similar breakdown voltage (Fig. 9.14c). The high-quality NH_3-MBE GaN drift layer with an order of magnitude lower unintentional doping and minimal compensating impurities as well as synergistic device development strategies, contributed to the superior performance of the NH_3-MBE GaN p^+n diodes compared to the available reports.

9 Conclusion

In summary, the MBE GaN brings enormous potential for vertical GaN power switches, allowing for an excellent combination of thick and clean epilayers, controlled doping, fast growth rate, and smooth surface morphology. Coordinated efforts on epilayer growth and device development, through the use of edge termination and minimal process-induced damage, kilovolt range GaN p-n diodes have been demonstrated. Further studies on expanding MBE GaN to new device designs can enable its widespread adoption for various multi-kilovolt power devices and achieve the desired avalanche capabilities in compact vertical GaN devices.

References

1. A. Corrion et al., Review of recent developments in growth of AlGaN/GaN high-electron mobility transistors on 4H-SiC by plasma-assisted molecular beam epitaxy. IEICE Trans. Electron. **E89-C**(7), 906 (2006)
2. A.V. Blant et al., Nitrogen species from radio frequency plasma sources used for molecular beam epitaxy growth of GaN. Plasma Sources Sci. Technol. **9**, 12–17 (2000)
3. S.T. King et al., Atomistic view of the autosurfactant effect during GaN epitaxy. Physical Rev. Lett. **98**, 206106 (2007)
4. G. Koblmüller et al., *In situ* investigation of growth modes during plasma-assisted molecular beam epitaxy of (0001) GaN. App. Phys. Lett. **91**, 161904 (2007)
5. C. Poblenz et al., Uniformity and control of surface morphology during growth of GaN by molecular beam epitaxy. J. Vac. Sci. Technol. B **23**, 1379–1385 (2005)
6. B. Heying, Optimization of the surface morphologies and electron mobilities in GaN grown by plasma-assisted molecular beam epitaxy. Appl. Phys. Lett. **77**, 2885–2887 (2000)

7. G. Koblmüller et al., High electron mobility GaN grown under N-rich conditions by plasma-assisted molecular beam epitaxy. Appl. Phys. Lett. **91**, 221905 (2007)
8. I.P. Smorchkova et al., Mg doping of GaN layers grown by plasma-assisted molecular-beam epitaxy. Appl. Phys. Lett. **76**, 718–720 (2000)
9. H. Xing et al., Memory effect and redistribution of mg into sequentially regrown GaN layer by metalorganic chemical vapor deposition. Jpn. J. Appl. Phys. **42**, 50 (2003)
10. S. Nakamura et al., High-power GaN P-N junction blue-light-emitting diodes. Jpn. J. Appl. Phys., Part 2 **30**, L1998 (1991)
11. J.E. Northrup, Screw dislocations in GaN: The Ga-filled core model. Appl. Phys. Lett. **78**, 2288–2290 (2001)
12. N. Grandjean et al., Group-III nitride quantum heterostructures grown by molecular beam epitaxy. J. Phys. Condens. Matter **13**, 6946–6960 (2001)
13. C.A. Hurni et al., p-n junctions on Ga-face GaN grown by NH3 molecular beam epitaxy with low ideality factors and low reverse currents. Appl. Phys. Lett. **97**, 222113 (2010)
14. D.A. Browne et al., Indium and impurity incorporation in InGaN films on polar, nonpolar, and semipolar GaN orientations grown by ammonia molecular beam epitaxy. J. Vac. Sci. Tech. A. **30**, 041513 (2012)
15. A.L. Corrion et al., Growth regimes during homoepitaxial growth of GaN by ammonia molecular beam epitaxy. J. Appl. Phys. **112**, 054903 (2012)
16. J.E. Northrup, Theory of intrinsic and H-passivated screw dislocations in GaN. Phys. Rev. B **66**, 045204 (2002)
17. C.A. Hurni et al., Effects of growth temperature on Mg-doped GaN grown by ammonia molecular beam epitaxy. Appl. Phys. Lett. **101**, 102106 (2012)
18. E.C.H. Kyle et al., Increased p-type conductivity through use of an indium surfactant in the growth of Mg-doped GaN. Appl. Phys. Lett. **106**, 222103 (2015)
19. H. Tang, J.B. Webb, Growth of high mobility GaN by ammonia-molecular beam epitaxy. Appl. Phys. Lett. **74**, 2373–2374 (1999)
20. E.C.H. Kyle et al., High-electron-mobility GaN grown on free-standing GaN templates by ammonia-based molecular beam epitaxy. J. Appl. Phys. **115**, 193702 (2014)
21. M.N. Fireman et al., High germanium doping of GaN films by ammonia molecular beam epitaxy. J. Crys. Growth **508**, 19–23 (2019)
22. B. Borisov et al., Si-doped AlxGa1-xN ($0.56 \leq x \leq 1$) layers grown by molecular beam epitaxy with ammonia. Appl. Phys. Lett. **87**, 132106 (2005)
23. J. Wang et al., High conductivity n-Al0.6Ga0.4N by ammonia-assisted molecular beam epitaxy for buried tunnel junctions in UV emitters. Opt. Express **29**(25) (2021)
24. O. Ambacher et al., Thermal stability and desorption of Group III nitrides prepared by metal organic chemical vapor deposition. J. Vac. Sci. Tech. B. **14**, 3532–3542 (1996)
25. M.N. Fireman et al., Isotype InGaN/GaN heterobarrier diodes by ammonia molecular beam epitaxy. J. Appl. Phys. **119**, 055709 (2016)
26. C. Poblenz et al., Power performance of AlGaN/GaN HEMTs grown on SiC by ammonia-MBE at 4 and 10 GHz. IEEE Electron Device Lett. **28**(11), 945 (2007)
27. J. Wang et al., Impact of growth parameters on the background doping of GaN films grown by ammonia and plasma-assisted molecular beam epitaxy for high-voltage vertical power switches. APL Mater. **9**, 081118 (2021)
28. Z. Hu et al., Near unity ideality factor and Schockley-Read-Hall lifetime in GaN-on-GaN p-n diodes with avalanche breakdown. Appl. Phys. Lett. **107**, 243501 (2015)
29. C.A. Robertson et al., Modeling dislocation-related leakage currents in GaN p-n diodes. J. Appl. Phys. **126**, 245705 (2019)
30. K.S. Qwah et al., Modeling dislocation-related reverse bias leakage in GaN p-n diodes. Semicond. Sci. Technol. **36**, 075001 (2021)
31. Y. Zhang et al., Gallium nitride vertical power devices on foreign substrates: A review and outlook. J. Phys. D. Appl. Phys. **51**, 273001 (2018)

32. H. Teo et al., Emerging GaN technologies for power, RF, digital, and quantum computing applications: Recent advances and prospects. J. Appl. Phys. **130**(16), 160902 (2021)
33. E. Farzana et al., Over 1 kV vertical GaN-on-GaN p-n diodes with low on-resistance using ammonia molecular beam epitaxy. IEEE Electron Device Lett. **41**(12), 1806 (2020)
34. B.M. McSkimming et al., Plasma assisted molecular beam epitaxy of GaN with growth rates > 2.6 μm/hr. J. Crystal Growth. **386**, 168–175 (2014)
35. B.M. McSkimming et al., High active nitrogen flux growth of GaN by plasma assisted molecular beam epitaxy. J. Vac. Sci. Technol. A **33**, 05E128 (2015)
36. K.S. Qwah et al., Indium as a surfactant: Effects on growth morphology and background impurity in GaN films grown by ammonia-assisted molecular beam epitaxy. APL Mater. **10**, 081107 (2022)
37. V. Loddo et al., Chapter 7 - Kinetic aspects of heterogeneous catalytic versus photocatalytic reactions, in *Heterogeneous photocatalysis*, (Elsevier, 2019), p. 2D15
38. J. Neugebauer et al., Adatom kinetics on and below the surface: The existence of a new diffusion channel. Phys. Rev. Lett. **90**(5), 056101 (2003)
39. J. Lutz, H. Schlangenotto, U. Scheuermann, R. De Doncker, Chapter 4-Short introduction to power device technology, in *Semiconductor power devices: Physics, characteristics, reliability*, (Springer, 2011), p. 142
40. R.T. Ley et al., Revealing the importance of light extraction efficiency in InGaN/GaN microLEDs via chemical treatment and dielectric passivation. Appl. Phys. Lett. **116**, 2511041–2511045 (2020)
41. N. Allen et al., Vertical Ga_2O_3 Schottky barrier diodes with small-angle beveled field plates: A Baliga's figure-of-merit of 0.6 GW/cm^2. IEEE Electron Device Lett. **40**(9), 1399 (2019)

Chapter 10
Gallium Nitride Wafer Metrology and Device Processing

Travis J. Anderson, Alan G. Jacobs, James C. Gallagher, Nadeemullah A. Mahadik, James Spencer Lundh, and Karl D. Hobart

1 Introduction

Vertical GaN power switching devices are of significant interest for high-efficiency power conversion circuits in medium voltage applications. Vertical device architectures are critical to fully exploit the superior material properties of GaN over conventional power device technologies such as Si and SiC. A vertical architecture offers higher blocking voltage, reduced die size, lower on-resistance and lower inductance for a given voltage node compared to a lateral counterpart such as a high electron mobility transistor (HEMT). Bulk GaN device technology is rapidly maturing with the widespread availability of 50 mm diameter free-standing GaN wafers by both hydride vapor phase epitaxy (HVPE) and ammonothermal techniques. In addition, recent advances in development and commercial production have enabled substrates with 10^3 defect/cm^2 and 100 mm diameter. While pioneering work has produced promising device results on small area structures, the realization of a robust device technology base has been limited by a poor understanding of the performance limiting issues in GaN epitaxial layers as well as a relatively immature processing technology for vertical devices compared to lateral HEMTs, where low resistance contacts, field plates, recess etching, gate dielectrics, and scaling are well understood. In addition, device reliability is relatively poorly understood due to a lack of large data sets from electrical stressing. Reports in the literature have demonstrated that devices fabricated on substrates from different vendors can have vastly different reliability under electrical stressing conditions despite appearing identical after fabrication and screening. Also, it is

T. J. Anderson (✉)
University of Florida, Gainesville, FL, USA
e-mail: TJAnderson@che.ufl.edu

A. G. Jacobs · J. C. Gallagher · N. A. Mahadik · J. S. Lundh · K. D. Hobart
U.S. Naval Research Laboratory, Washington, DC, USA

I. C. Kizilyalli et al. (eds.), *Gallium Nitride and Related Materials*, The Materials Research Society Series, https://doi.org/10.1007/978-3-031-83056-3_10

known in the literature that even adjacent devices on a single wafer can have vastly variable leakage current and breakdown voltage despite appearing similar optically. In this chapter, we will systematically address the fabrication challenges associated with vertical GaN devices by systematically isolating and addressing the critical process modules and demonstrating appropriate solutions, thus describing the fundamental building blocks of vertical GaN technology [1–5].

2 Metrology

2.1 Overview

Before device fabrication or even epitaxial growth, incoming wafers and epitaxial layers must be evaluated to determine viability for further steps. While vendors typically provide data about resistivity, offcut, defect density, and wafer shape, there are additional critical factors that can impact device performance. Most notably, uniformity of properties across the wafer will impact yield. Typical metrology techniques include electrical screening, probing by optical techniques, and probing by X-ray techniques. We have focused on methods that can be expanded to wafer-scale mapping using step-and-repeat measurements rather than methods that yield detailed information from extremely limited regions, such as electron microscopy-based techniques such as cathodoluminescence (CL), electron channeling contrast imaging (ECCI), or transmission electron microscopy (TEM). This is particularly important when considering device manufacturing, as the incoming metrology must be relatively fast and scalable to production levels. Substantial research and investment in metrology of SiC wafers several decades ago enabled rapid maturing of device technology. A series of studies correlating the wafer properties to electrical performance under DC and bias stressing conditions led to the discovery of the basal plane dislocation as the critical "killer" defect. This enabled the development of tools such as ultraviolet photoluminescence (UVPL) mapping to rapidly screen wafers and develop techniques for mitigating this dislocation during the growth processes. While UVPL imaging yields less information for GaN than SiC due the direct bandgap nature of the material, other techniques are highly relevant as described below [6, 7].

2.2 Mercury Probe

The mercury probe is a well-known tool for evaluating the electrical parameters of many semiconductor materials by making non-destructive electrical contact to a sample prior to processing. As applied to vertical GaN P-i-N epitaxial layers grown on a substrate, the mercury-sample structures are used to measure the capacitance-voltage properties of the drift layer to extract carrier concentration. The analysis of this is somewhat similar to a mercury-oxide-semiconductor structure because the as-grown p-GaN layers are not electrically active, thus Eq. (10.1) can be applied to the depletion region of the C-V curve at high frequency.

$$N_D = \frac{2}{q\varepsilon A^2 \left(\frac{d\frac{1}{C^2}}{dV}\right)} \tag{10.1}$$

The analysis of a bare drift layers on a substrate is more complicated, as the mercury forms a poor contact to the lightly doped GaN layer. Since the mercury is typically a relatively large dot (~0.7 mm), the measurement can also be contaminated by pinholes in the substrate or epitaxial layers that cause leakage paths. This technique cannot resolve highly localized variations in carrier concentration that might occur to due to local miscut variation. However, the advantage of the technique is that any residual mercury contamination can be easily etched using HCl treatments to recover a pristine GaN wafer. While a simple lab system only measures a single point, advances in commercial tool development have realized automated mapping systems that can readily perform probing and analysis at the wafer scale. An example data set of 32 curves measured on a 50 mm wafer and the corresponding analyzed map is shown in Fig. 10.1.

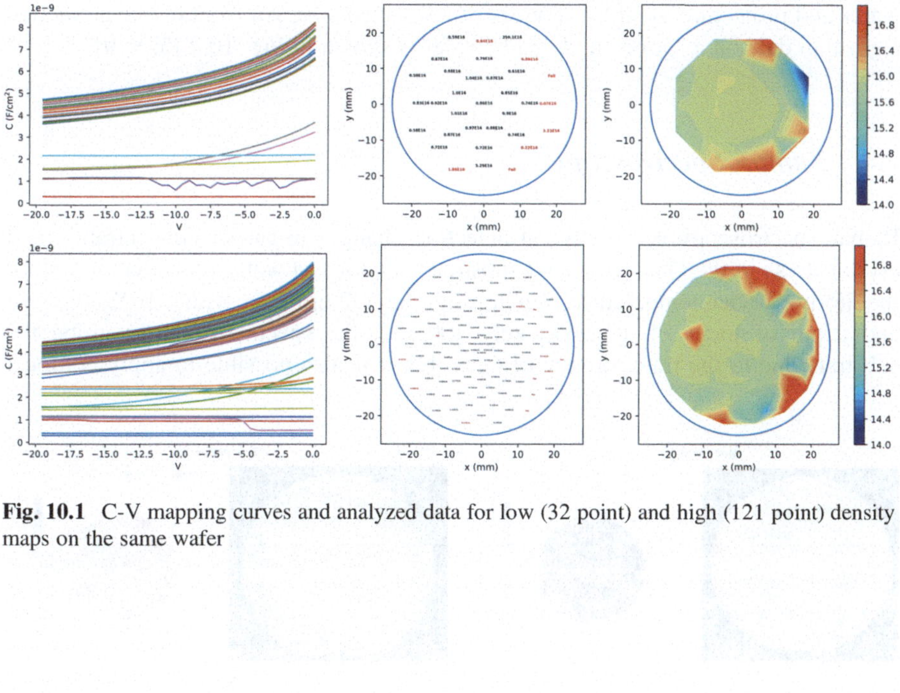

Fig. 10.1 C-V mapping curves and analyzed data for low (32 point) and high (121 point) density maps on the same wafer

2.3 *Optical Profilometry*

Optical profilometry has proven to be an extremely valuable tool to monitor GaN wafers and epitaxial layers. Mapping systems are readily available that are capable of acquiring surface height and roughness data with micron-scale resolution across a full wafer. While a raw image will yield information about macroscopic defects, the availability of a large array of surface height data lends itself to the development of algorithms for in-depth analysis using machine learning models. Starting with the raw data of surface height versus position, an algorithm was developed to bin the data to regions corresponding roughly to the size of a device, thus creating a new array. Statistical analysis is performed on each region, looking at the distribution of surface height. A relatively flat or even Gaussian distribution of height corresponds to a uniform part of the wafer, while a large portion of outlier data is used to identify the presence of a bump (hillock) or pit in that region based on an extreme studentized deviation (ESD) test. These regions are flagged as anomalous and expected to have low yield. Compiling this information, we developed a map of both RMS roughness and outlier area on the wafer. We have then correlated the presence of such defects to leakage currents in devices and thus are able to statistically predict the yield of devices fabricated on wafer. Experimental data is ~90% accurate compared to modeled yield, which is acceptable given the presence of fab defects and yield as well. Continuing to train a machine learning model to the experimental data sets, we have developed a true predictive algorithm that is capable of assessing an incoming wafer and projecting yield for devices of a selected size. An example of predictive algorithm data on a "good" and "bad" wafer is shown in Fig. 10.2 [5, 8, 9].

2.4 *Raman Spectroscopy*

Raman spectroscopy is effective at detecting changes in carrier concentration and crystal stress in the GaN films and substrates, and it is a well-known technique for monitoring device temperature. When using the 532 nm laser with a 10X objective and depth of focus of approximately 100 μm, the spectra primarily probe the substrate due to the transparency of GaN, but it is also possible to apply confocal

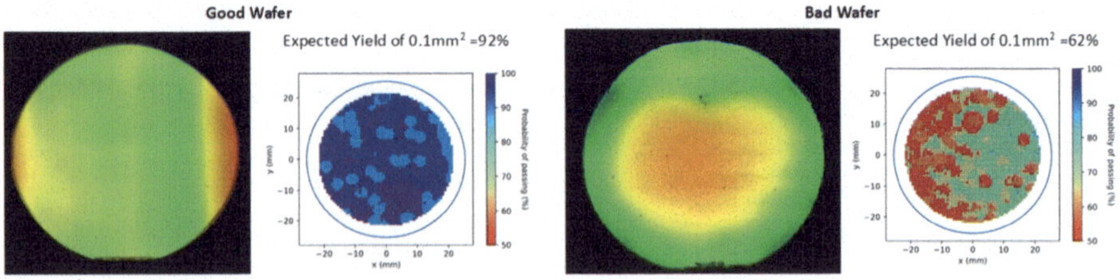

Fig. 10.2 Optical profilometry image and predicted yield for multiple wafers

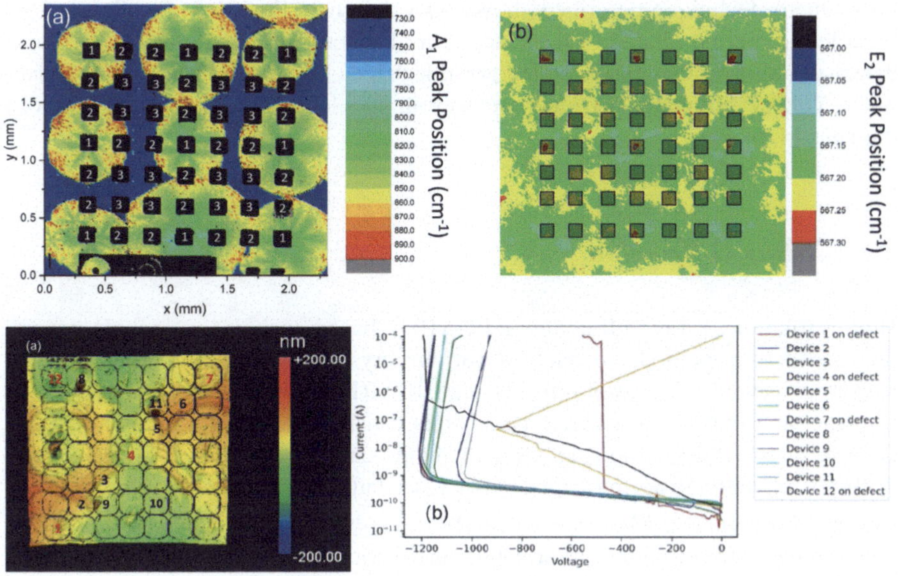

Fig. 10.3 Raman maps of a region of a dot-core wafer, A_1 position relating to substrate n-type carrier density (top left), and E_2 relating to crystal stress (top right), correlated to optical profilometry image (bottom left) and device I-V behavior

techniques to resolve epitaxial layers grown on substrates. This technique can map a full 2-inch wafer with ~20 μm resolution in 1–3 days, depending on the acceptable noise level. Some GaN wafers, such as those grown using the dot-core technique, have a regular pattern of defects, while others appear quite uniform. The A1 (LO) peak is sensitive to carrier concentration in the 10^{17}–10^{19} range using the following equation [10]:

$$n = 1.1 \times 10^{17} \Delta\omega^{0.764} \tag{10.2}$$

Similarly, the E2 peak position is correlated to stress. An example of a GaN wafer mapped with the technique is shown in Fig. 10.3. The wafer used in this study has periodic points of crystal stress associated with dot-core type of manufacturing. These domains also exhibit increased carrier concentration in a hexagonal structure detected by the shifting of the A1 (LO) peak, with regions of relatively low carrier concentration in between. This structure is interesting for devices, because the "core" regions are known to have high dislocation density, which is what causes the crystal stress, but also high carrier concentration. The region between domains, while relatively defect-free, also exhibits reduced carrier concentration. This is likely caused by the unintentional incorporation of compensating impurities on the facets of the domains as they grow together. Using the Raman map, we were able to fabricate diodes that were aligned in 7 × 7 grids such that the corner points were aligned to the domains. The same regions were also mapped using the optical profilometry technique described above, also shown in Fig. 10.3. Finally, devices were tested and correlated to the maps. The devices fabricated directly on top of defective regions often exhibited increased leakage current and/or premature breakdown voltage compared to other devices in the less defective region between domains [4, 5, 11].

2.5 X-Ray Diffraction

High-resolution X-ray diffraction (HR-XRD) based techniques include several non-destructive methods to analyze various crystallographic properties of GaN substrates as well as epitaxial films. Using HR-XRD in triple-axis geometry, precise in-plane, and out-of-plane lattice parameters can be obtained for ammonothermal GaN substrates. Furthermore, by utilizing reciprocal space mapping of the GaN($11\overline{2}4$) Bragg reflection in asymmetric geometry, the lattice parameters of epitaxial layers can also be obtained. More advanced HR-XRD-based techniques can be used to obtain novel microstructural information as described below:

2.6 Whole-Wafer Mapping of Lattice Distortion Variation

Lattice distortions such as lattice plane curvature (LPC) and lattice strain ($\frac{\Delta d}{d}$) in GaN substrates can cause degradation in both device yield as well as reliability. Higher lattice strain can lead to generation of added defects such as misfit dislocations, which could propagate into the epitaxial layers and cause higher reverse bias currents and other reliability issues. Higher LPC can lead to wafer shape change during device fabrication during the several thermal process steps. Higher variation in wafer shape can cause problems during lithography steps and will lead to poor device yield. The lattice strain and LPC are convolved together in the rocking curve as: $\Delta w = \tan\theta_B \frac{\Delta d}{d} + (\widehat{n_r} \cdot \widehat{n_m})\Delta\rho$, where θ_B is the Bragg angle, d is the lattice spacing, $\widehat{n_r}$:normal vector, $\widehat{n_m}$: misorientation/tilt vector, and $\Delta\rho$: measured angle [U. Bonse, Z. Phys. 153, 278 (1958)]. In order to deconvolve the lattice strain and LPC along the two orthogonal directions with respect to the wafer flat (typically [$11\overline{2}0$]), whole-wafer rocking curve mapping (RCM) technique can be implemented by collecting point-wise rocking curve measurements taken from three azimuthal directions with respect to the wafer flat: Φ=0°, 90°, and −90°, where the vicinal offcut in the wafer is along the m-plane. The resulting equations obtained that relate the variation in rocking curves to lattice strain and LPC are: $\Delta w_0 = \tan\theta_B \frac{\Delta d}{d} + LPC_{[11\overline{2}0]}$, $\Delta w_{90} = \tan\theta_B \frac{\Delta d}{d} + LPC_{[1\overline{1}00]}$, and $\Delta w_{-90} = \tan\theta_B \frac{\Delta d}{d} - LPC_{[1\overline{1}00]}$. Due to the full rotation, Δw_{90} and Δw_{-90} will have opposite sign tilts. From these three equations, the variation in lattice strain as well as LPC can be precisely obtained. This technique

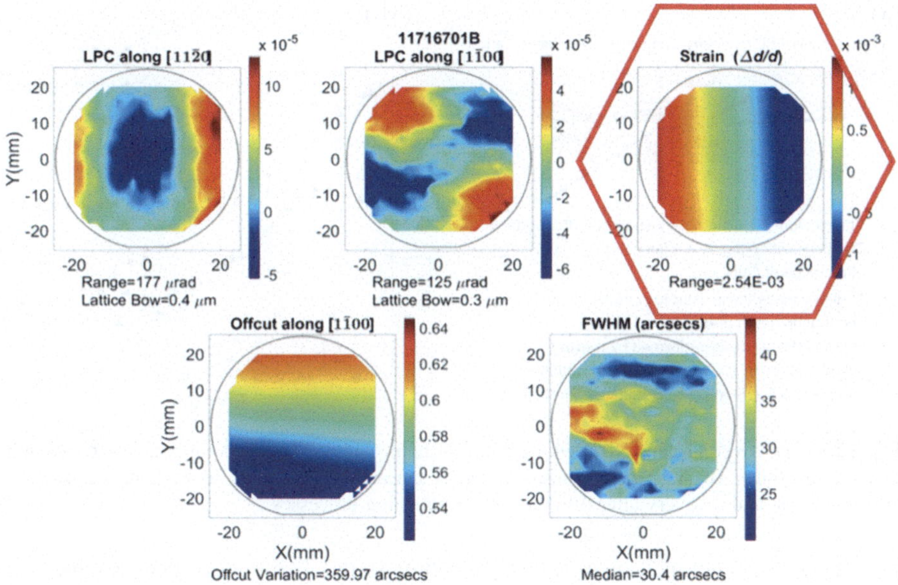

Fig. 10.4 Whole-wafer maps of variation in lattice plane curvature along two orthogonal directions, lattice strain, offcut, and rocking curve FWHM

was implemented on an ammonothermal GaN substrate that contained several microcracks. Figure 10.4 shows the whole-wafer maps of variation in LPC, lattice strain, offcut, and rocking curve full width at half maximum (FWHM) for the wafer. The average lattice bow is also calculated. The LPC variations as well as lattice bow in both directions in this wafer, are in the range of a typical ammonothermal GaN wafer. However, the lattice strain map indicated a variation that is in the 10^{-3} range, which is ~100× higher than typical ammonothermal GaN wafers. Hence, one of the causes for the microcracks in this wafer was due to high lattice strain. The offcut map shows ~0.1° variation with higher offcuts toward the top and lower toward the bottom of the wafer. The FWHM map is an indication of overall crystalline quality, and shows higher FWHM near the center toward the left side of the wafer.

2.7 *Depth-Resolved Determination of Sub-surface Lattice Damage and Recovery*

Sub-surface lattice damage is critical for the growth of epitaxial layers and can significantly influence the formation of defects. The GaN substrate can be polished to obtain a very smooth surface. However, the crystal lattice below the surface up to ~1 μm plays a significant role in the crystalline quality of the epitaxial layers grown on it. The GaN substrate, after slicing from the boule undergoes grinding, lapping,

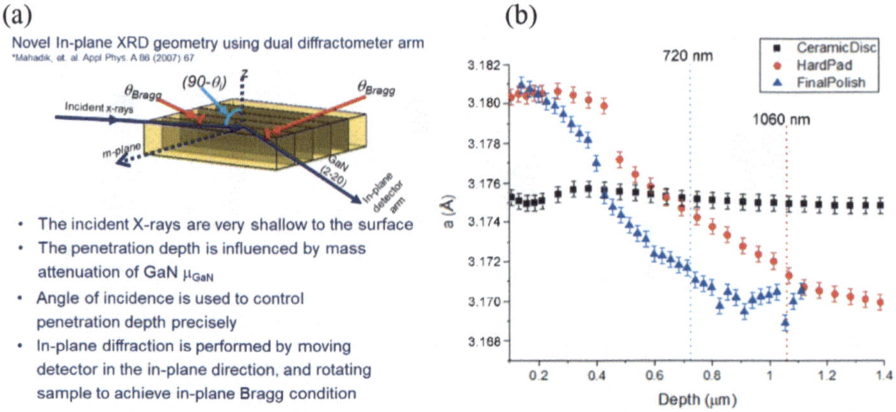

Fig. 10.5 (**a**) Novel in-plane XRD with grazing incidence angle x-ray to control penetration depth in GaN substrates. (**b**) Variation in in-plane lattice parameter with depth to evaluate sub-surface lattice damage and recovery after different wafer polishing processes

polishing, and chemical-mechanical polishing steps. These steps are designed to remove the damage associated with the slicing process. However, these mechanical processes can also introduce sub-surface damage, and there is a trade-off between damage removal and generation.

To evaluate sub-surface damage, non-destructively, a novel technique using in-plane XRD with grazing incidence angle was developed. In this technique, the incidence x-rays are kept at a shallow angle (~0.5°–3°), and the penetration depth of x-rays is influenced by the X-ray mass attenuation of GaN. In this manner, the penetration depth can be precisely controlled to a certain depth. In the out-of-plane direction, the GaN samples cannot be aligned to a Bragg condition. However, in the in-plane direction, the sample can be rotated such that a Bragg condition can be satisfied, for instance GaN ($2\bar{2}00$) can be achieved by rotating the GaN sample ~33.9°. With a dual-arm 2theta goniometer, the detector can also be positioned to observe the Bragg reflection as shown in Fig. 10.5a. Using this setup, depth-dependent lattice parameters can be measured for GaN samples after various stages of grinding and polishing as shown the Fig. 10.5b. From this figure, it can be observed that after the grinding process with a ceramic disc, the sub-surface damage does not recover even beyond 1 μm from the surface. Upon subsequent lapping, there is an increase in the lattice parameter very close to the surface but the sub-surface damage is recovered ~1 μm below the surface. After final polishing process, there is an improvement in the lattice recovery within ~700 nm below the sample surface. Hence, using this non-destructive technique, sub-surface lattice properties of GaN and other samples can be measured.

2.8 X-Ray Topography

High-resolution X-ray topography (XRT) is a non-destructive imaging technique that can be utilized to perform detailed microstructural analysis of various extended defects in both substrates and epitaxial layers. XRT studies have been performed on ammonothermally grown bulk GaN substrates to reveal and catalog several extended defects in the substrate. Others have also employed XRT to structurally characterize different types of GaN substrates, such as the so-called advanced dislocation elimination by the epitaxial growth with inverse pyramidal pits (A-DEEPs) and void-assisted separation (VAS). While XRT has been primarily used to examine the defects in the substrate, we characterize both the substrate and epitaxial layer (with an emphasis on the latter) in this work. By employing both transmission and reflection geometries, the contrast in XRT images can be predominately from the substrate or from within a thick (tens of micrometers) epitaxial layer. A typical pair of images is shown in Fig. 10.6. The first feature clearly visible is a vertical line perpendicular to the flat. This is a tiling seam formed from the merging of multiple smaller seed crystals to form the 50 mm wafer, which produces a cluster of threading dislocations along the length of the wafer. Since it is present in both transmission and reflection geometry, these dislocations extend into the epitaxial layer. Given the discussion above, these will certainly impact device behavior thus creating a large exclusion zone at the center of the wafer. Further analysis of different diffraction conditions yields small clusters of threading dislocations throughout the image, specifically in the reflection geometry. This indicates that new dislocations are formed during epitaxial layer growth, which could be the result of difference in lattice parameter between the N+ substrate and lightly doped epitaxial layer, dislocation conversion and/or glide, or thermal variations in the growth reactor [12].

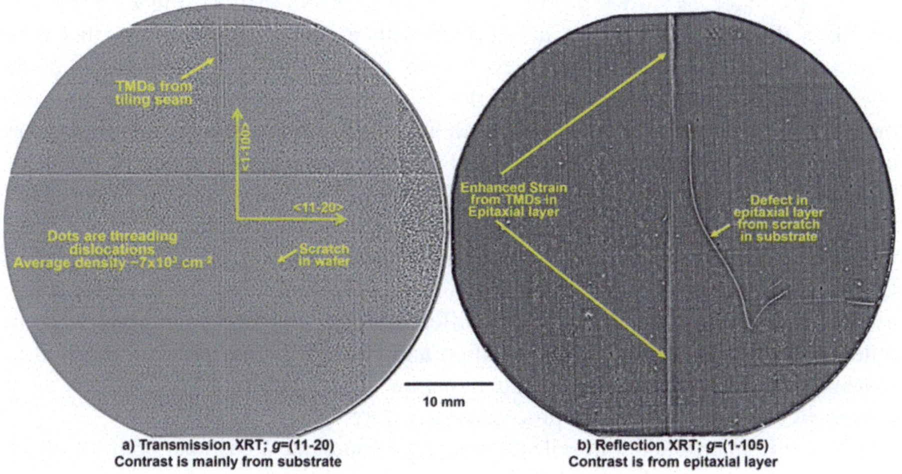

Fig. 10.6 XRT images of ammonothermal GaN wafer with epitaxial layers

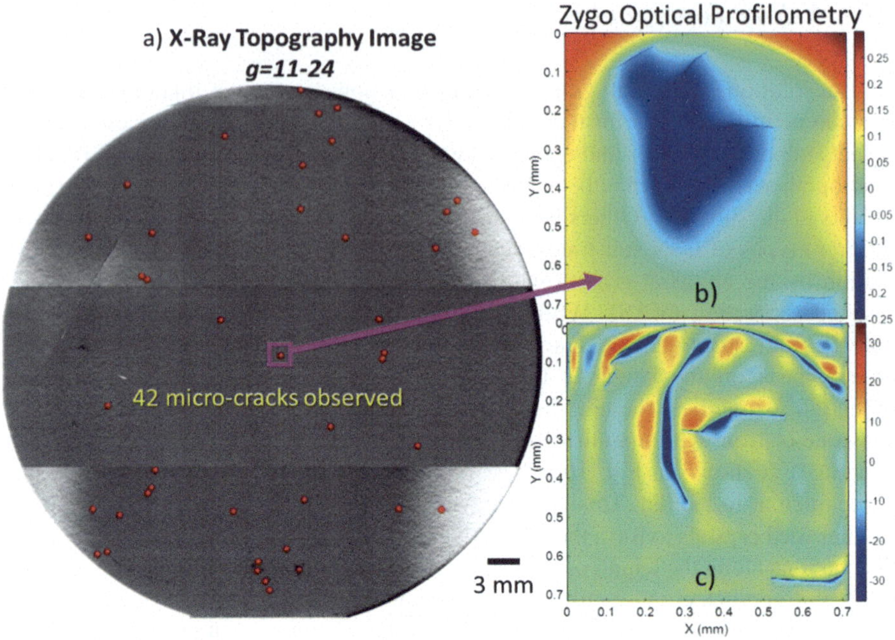

Fig. 10.7 (**a**) Full wafer XRT image of GaN wafer with the presence of 42 microcracks. (**b**) Zygo image of the micro-crack showing an overall divoted structure, and (**c**) Zygo image of the micro-crack after removal of low spatial frequencies to observe fine structure

Additionally, XRT analysis of ammonothermal GaN wafers from different vendors was performed to investigate microstructure of micro-crack formation on those wafers. There were 42 microcracks observed from the XRT image ($g=\left[11\overline{2}4\right]$) of the entire GaN wafer, as shown in Fig. 10.7a. Surface profilometry using Zygo system reveals a divoted structure on the surface with a central crack region that dips ~0.4 μm below the GaN surface, as shown in Fig. 10.7b. Upon removing the low-frequency spatial components of the Zygo image, which effectively approximates the divoted pit to a fifth-order polynomial and removes the overall pit, we can observe bands of surface undulations that radiating horizontally from the center of this micro-crack, as shown in Fig. 10.7c. This suggests that during the micro-crack formation, the localized surface of the GaN has a radial undulation formed from mechanical stress, which could have an additional impact on epitaxial quality. Detailed analysis of this micro-crack was also performed using multi-vector XRT, as shown below in Fig. 10.8. Three different XRT imaging conditions were in reflection geometry using Cu kα radiation and two XRT images were acquired in transmission geometry using Mo kα radiation. The measured XRT images were processed using custom image processing software.

The XRT images taken in reflection geometry indicate a stronger contrast of the micro-crack, which indicates that these were likely initiated and present near the surface. The crack appears as lobes that branch out from a central point that has a

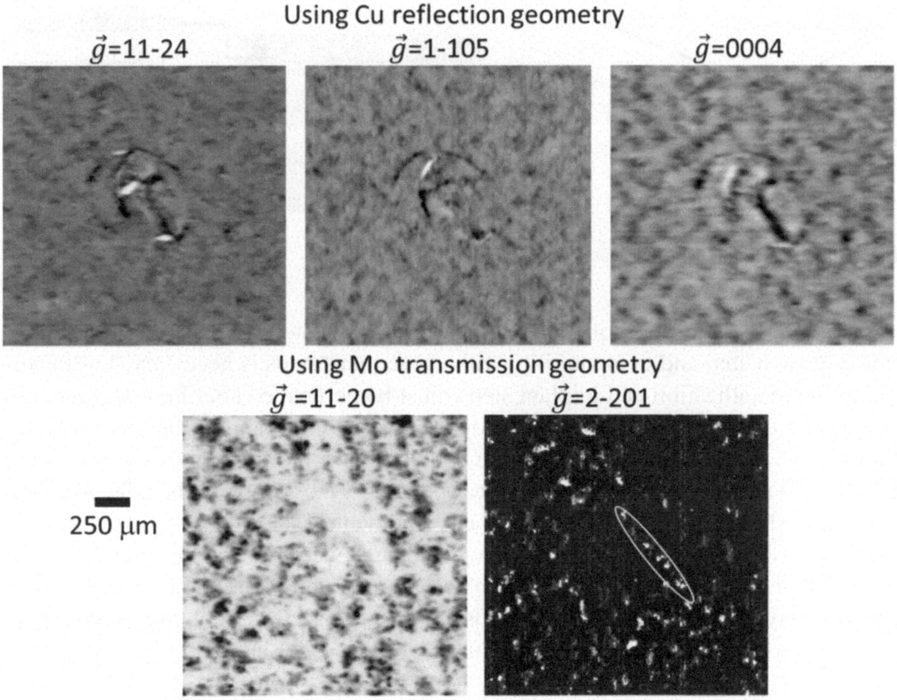

Fig. 10.8 XRT images using various g-vectors showing differing contrast of features within the structure of the micro-crack

high-stress region. This was confirmed using X-ray rocking curve mapping technique, which shows the presence of high strain in this sample. It is likely that the micro-crack could have been eventuated during the wafering process, which locally relieves the high residual stress. Analyzing the various XRT images, from the $g = 2–201$ transmission image, a dot contrast appears segregated along the central region of the crack. This is not observed in the $g = 11–20$ XRT image. Using the $g.b$ criterion, these dots can be determined to be threading screw-type dislocations that segregate along the central part of the micro-crack.

3 Device Processing

3.1 Overview

The process module development for a vertical GaN P-i-N diode includes optimization of epitaxial layer stack to include an appropriate substrate vendor, drift layer and anode design, p-ohmic metallization, n-ohmic metallization, planar implant and edge termination processes utilizing nitrogen ion implantation, and surface

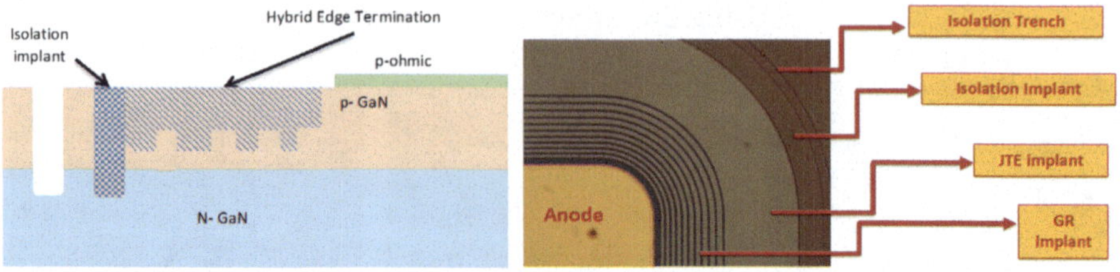

Fig. 10.9 Schematic of device architecture

passivation. In addition, sequencing of the fabrication steps to ensure proper cleanliness at each step and compatibility with subsequent steps is necessary. For example, some metallization and implant steps must be performed after high-temperature processing to avoid unintentionally alloying the metal or altering the ion implantation profile. Here, we will focus on the details of specific process modules that are critical to achieving appropriate electrical behavior, most notably the edge termination step. The key consideration in the development of all process modules here is manufacturing robustness. The goal is to design a device architecture and fabrication process that is relatively insensitive to inherent minor variations in the process and/or epitaxial layers that will inherently arise in a large-scale production process. An example cross section and plan-view image of devices is shown in Fig. 10.9.

3.2 *Epitaxial Layer Design*

Appropriate design of the epitaxial layers to achieve target performance, as well as validation of incoming wafers to ensure they meet the design specifications, is a key step before processing begins. The metrology and validation techniques are discussed in the previous sections, but the most notable one is the mercury probe evaluation to validate drift layer carrier concentration, which is typically an N-layer formed by intentional Si doping, not a truly intrinsic layer. Device simulation packages are extremely helpful to facilitate the development of appropriate designs, but even a simple calculation based on theoretical parallel plane limits is useful for an initial design. There is inherently a trade-off between drift layer thickness, doping, specific on-resistance, and breakdown voltage, demonstrated by the following equation:

$$\mathrm{BV} = E_{\mathrm{crit}}\, t - \frac{qN_d t^2}{2\epsilon_s} \tag{10.3}$$

Where the breakdown voltage is dictated by the critical electric field as a function of doping, the drift layer thickness, and drift layer doping. While small variations in the literature values for impact ionization coefficients, thus for the critical electric field in GaN yields variation in the calculated punch-through breakdown voltage, it

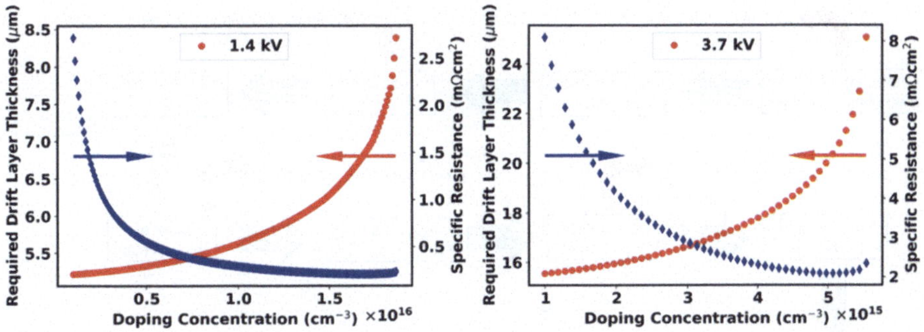

Fig. 10.10 Example design calculations for 1.2 and 3.3 kV-class diodes

can be estimated from the relationship shown in Eq. (10.3). It is clear that for a 1.2 kV-class diode, the drift layer thickness needs to be ~8 μm if the carrier concentration is ~1 × 10^{16} cm^{-2}. Similarly, a 3.3 kV-class diode design would have a drift layer thickness of ~25 μm if the carrier concentration is ~4 × 10^{15} cm^{-2}. While the target could be achieved with thinner drift layers, it is desirable to design for a ~20% higher target than necessary to allow for non-ideality in growth and edge termination processes. An example of the full design options for these technology nodes calculated in TCAD software is shown in Fig. 10.10 [13–16].

A further consideration is the anode thickness and doping levels. These designs are primarily guided by the specifics of the edge termination process, used to manage electric fields at the device periphery. Regardless of the doping target, it is necessary to include a highly doped p++ layer to facilitate ohmic contact to the anode. This is typically Mg-doped at ~1 × 10^{20} cm^{-2} or higher. Preferably, this should be present only under the ohmic contact, as the additional charge introduced by this layer can impact passivation and edge termination processes.

3.3 Process Flow

An overview of an example process flow with early metalization for process evaluation is shown in Fig. 10.11. Following incoming metrology and evaluation, the first step is the activation of the p-GaN layer. Various processes exist in the literature, but in this work, activation was achieved by annealing at 700 °C for 20 min in a nitrogen ambient. The placement of global alignment marks is the next step. In this process, marks etched by Cl_2 ICP were preferred over deposited metal typically used, as the etched marks are more robust to the aggressive cleaning steps used later in the process, which could attack some metallization. The next step in the process was deposition of both the front-side and back-side ohmic metallization. The n-GaN ohmic is a typical Ti/Al/Ni/Au stack commonly reported in the literature, while the p-GaN ohmic consists of a Pd/Pt/Au stack. This is well known to form an ohmic contact to Mg-doped GaN, though alternative approaches including Ni-based

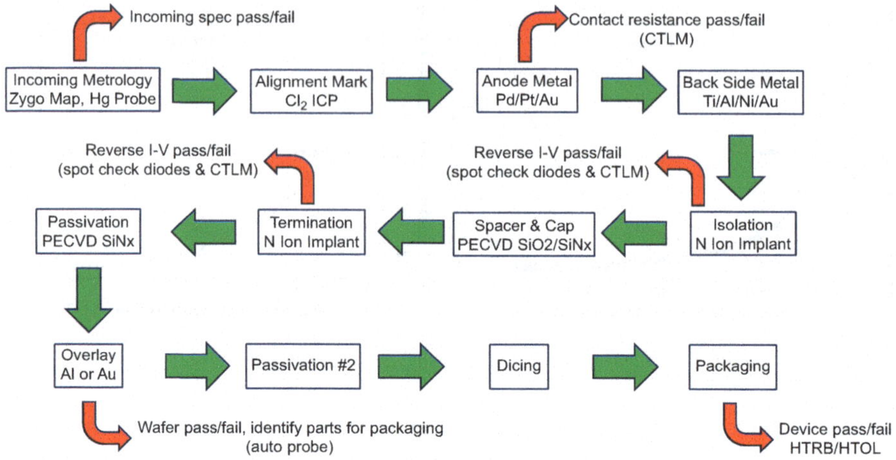

Fig. 10.11 Example process flow for vertical GaN device

and Ag-based stacks have been reported in the literature. The metals are deposited in an e-beam evaporator and patterned using lift-off. Depositing the metals here enables testing of contacts to ensure contact resistance is on-spec before committing to further processing. We have achieved contact resistance as low as 2×10^{-4} cm^{-2} in this process. The next steps are the ion implantation modules. Here, conductivity in the p-GaN layer is controlled utilizing N ion implantation. The mechanism is that the ion implantation creates damage in the p-GaN, which makes the film semi-insulating. This behavior was confirmed experimentally, with measured resistance $>10^8$ Ω/sq. in implanted regions. Isolation is patterned in a ring around the device using a box profile extending deeper than the p-GaN layer. This is typically accomplished using as little as three energies, but a more abrupt box profile can be formed with more energies. A second ion implantation step is performed for edge termination in a junction termination extension profile again using a box profile with as little as two energies. Guard ring structures can be superimposed on this pattern by using a spacer layer in a material such as SiO_2 which can easily be patterned and etched. In this type of structure, the spacer layer shifts the profile shallower, but still penetrating the p-GaN. This results in the castellated profile shown above which is referred to as the "hybrid edge termination". After these steps, implant test structures should be evaluated to confirm appropriate conductivity in the implanted regions. At this point, forward and reverse I-V curves can be measured either in vacuum or fluorinert to validate the on-resistance and breakdown voltage. If acceptable, the wafers should go forward for passivation. We have successfully evaluated typical passivation layers used for GaN devices such as SiN_X and SiO_2, though this topic requires further investigation. In a true production process, a thick overlay metallization would be used to facilitate packaging. In addition, back-side thinning would be performed to reduce on-resistance. However, these steps have been omitted from this demonstration in order to streamline the process. Finally, after probing the devices would be singulated and packaged, which will be discussed later.

3.4 Process Validation and Device Testing

Following fabrication, the wafers are tested under DC and pulsed forward I-V to quantify performance. In the process described here we have observed a maximum current of 18A for a 1mm^2 device and on-resistance of 1.2 mOhm-cm^2 under pulsed I-V conditions. The maximum current capability was relatively consistent, though testing was later limited to 10A to avoid damaging diodes during testing due to current localization at the probe tips. We performed many systematic evaluations to assess the appropriate probe configuration and metallization thickness appropriate for large area devices. We found that testing with a single probe on a 100 nm thick anode layer introduced an artificially high on-resistance due to substantial spreading resistance effects, but adjusting to a four probe configuration (3× force and 1× sense) and increasing thickness to 500 nm was sufficient to mitigate this effect. Typical I-V results from a "good" wafer are shown in Fig. 10.12. We were able to clearly achieve avalanche breakdown at 1.4 kV using an optimized termination design (discussed below).

The uniformity was quantified using automated probing techniques. We observed a spread in ON-resistance from 1 to 4 mΩ-cm^2, which we attribute to both variations in the anode p++ cap process due to epitaxial layer variability as well as instability in the metallization process due to the inherent nature of a multi-user facility. We also observed a spread in breakdown voltage from 600 to 1400 V, which we have determined is a function of the drift layer doping, anode doping, anode thickness, and edge termination design. An example of this is shown in Fig. 10.13. This is an example of a wafer where we observed significant variation within wafer for reverse I-V performance only. It is important to note that the drift layer doping and forward I-V behavior were generally relatively constant. Also note that generally, the devices

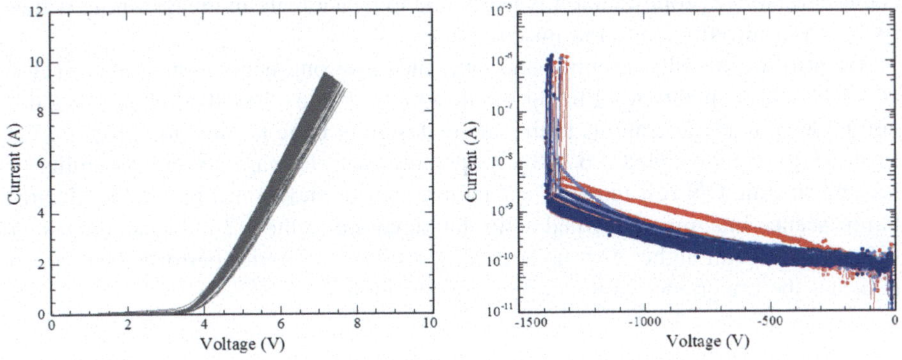

Fig. 10.12 Typical I-V behavior of 1 mm^2 1.2 kV-class Foundry diodes

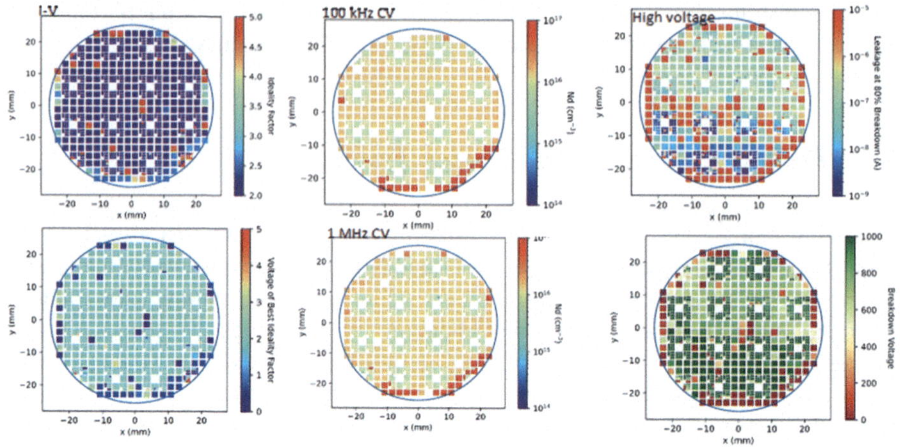

Fig. 10.13 Typical auto probe map of Foundry wafer

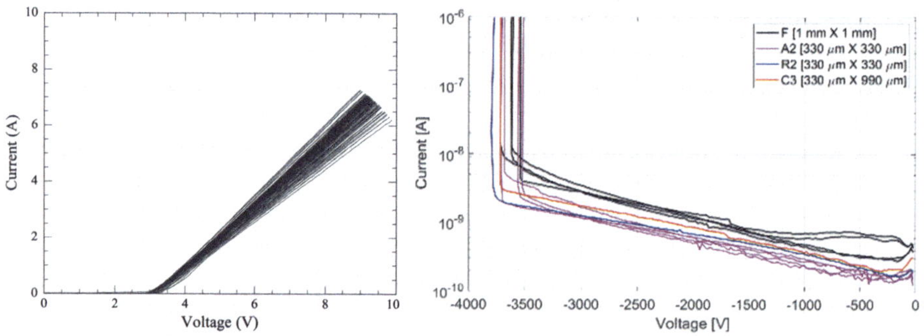

Fig. 10.14 Typical I-V behavior of 1mm^2 3.3 kV-class Foundry diodes

with the highest breakdown voltage also had the highest leakage current, while the devices with low leakage current also had low breakdown voltage. The cause of this is unclear, but we propose that it is likely due to an anomaly in the incoming wafer, such as a localized variation in miscut angle.

We also successfully designed and fabricated a second-generation diode targeted at 3.3 kV-class operation, utilizing the design guidelines described above. We did not implement any alterations to the anode design or edge termination process. We successfully demonstrated 3.6–3.8 kV operation with forward current capability of 8A and specific ON-resistance of <7 mOhm-cm^2 as shown in Fig. 10.14. Electroluminescence imaging confirmed a similar signature to the 1.2 kV-class diodes. A second wafer with higher drift layer doping resulted in inferior performance, which indicates the importance of the doping specification.

3.5 Edge Termination

In conventional power semiconductor device technology such as Si and SiC, there are two commonly used edge termination approaches—junction termination extension (JTE), and guard rings (GR), also called Floating Field Rings (FFR). The former utilizes a region of lighter doping extending from the edge of the anode to the edge of an isolation region. The dose in this region is optimized to properly balance and shape the electric field. The latter approach utilizes floating regions of heavily doped material, spaced in rings of varying width and spacing to distribute the electric field. Many years of technology development have realized variants on these designs for further field shaping, such as multi-zone JTE and grayscale JTE, both of which implement a dose that is laterally varying through the width of the edge termination region. The III-N material system presents several unique challenges, making the optimization of known JTE and GR-type structures challenging and limited to compensation doping approach described here. Due to the wide bandgap nature of the material the dose window for optimal edge termination efficiency is extremely small. As discussed above, we have utilized a "hybrid" structure which is based on JTE architecture but superimposes an additional layer of guard ring (GR) structures to further shape the charge profile in the anode. This produces a discretized charge profile that approximates the charge profile obtained from using a smoothly tapered (e.g., linear, convex, or concave) termination scheme but in a much more streamlined and reproducible manner. The true "bevel" or grayscale lithography process is extremely challenging in the III-N system because the bevel angle must be very shallow ($<1°$) and anode doping extremely light ($<5 \times 10^{17}$ cm^{-2}) to achieve high voltage operation [17–20].

To effectively implement the termination process, we also must design an appropriate anode thickness and doping profile. To study this, we developed several controlled experiments to implement in the process flow, based on a fixed implant profile. By varying the anode thickness, we were able to systematically vary the thickness of the remaining p-GaN layer, since the implant depth is constant and the wafers are processed in parallel. Since the target implant depth is ~300 nm, we varied the anode thickness from 500 to 300 nm. In doing so, we observed that by making the anode thin, and thus the underlying p-GaN layer constituting the termination region <10 nm, we were able to achieve $>100\times$ reduction in leakage current under reverse bias conditions. In addition, temperature dependence of the reverse I-V behavior changed to show an increased breakdown voltage at elevated temperature. Finally, we also observed the electroluminescence spot moving from the edge of the isolation implant to the edge of the anode. The latter two observations are consistent with avalanche breakdown. We performed a similar experiment of holding the anode thickness constant at 400 nm while varying the Mg doping level from 2E19 down to 5×10^{17} cm^{-2}. In this case, the optimal Mg concentration was 1×10^{18} cm^{-2}, which also resulted in an abrupt avalanche breakdown, confirmed by temperature dependence and electroluminescence imaging. These results are all

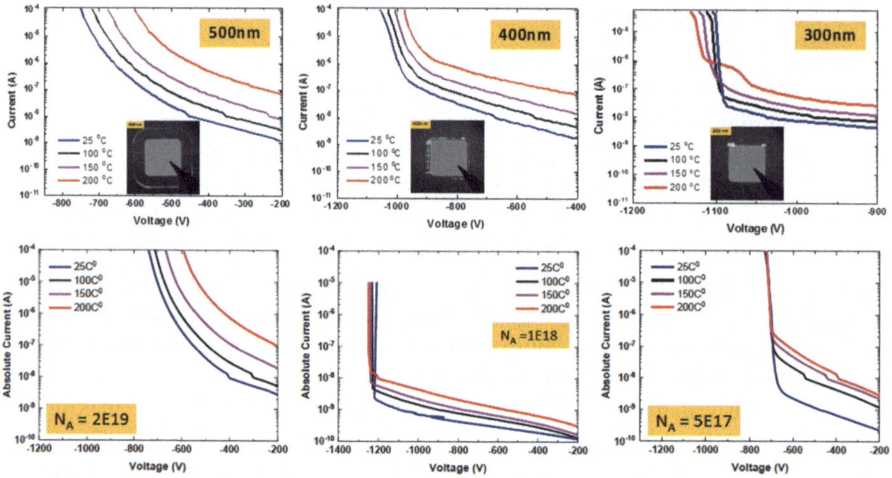

Fig. 10.15 Temperature-dependent reverse I-V behavior of six different anode designs

consistent with simulation results reported in the literature. A summary of the I-V behavior is shown in Fig. 10.15 [21–24].

In addition to the temperature-dependent reverse breakdown characterization, electroluminescence (EL) imaging is a useful technique for confirming avalanche breakdown as well as spatially identifying leakage current paths in devices and evaluating the effectiveness of edge termination designs. Exemplary EL images are shown in Fig. 10.16a–c for vertical GaN p-i-n diodes with different edge termination designs. TCAD simulation (Fig. 10.12d) shows that if implemented properly, the peak electric field at the junction should shift from the start of the isolation implant to the start of the "hybrid" edge termination (guard rings superimposed on a JTE) which is directly outside the anode metal edge. Accordingly, EL imaging of the diodes with the different edge termination designs shows this shift in the location of EL from the start of the isolation implant for the ISO and JTE structures (Figs. 10.16a and 10.1b) to the anode edge for the "hybrid" 12R edge termination design (Fig. 10.16c). It should be noted that ideally the avalanche process would result in uniform EL under the anode and may be seen as a uniform ring around the anode. The observation of EL at a single point at the edge of the anode for the 12R structure (Fig. 10.16c) indicates there is local avalanching due to non-idealities.

Figure 10.17 shows EL and reverse breakdown characterization of two vertical GaN p-i-n diodes incorporating two different JTE implant profiles. The unoptimized JTE design shows local avalanching at the outer edge of the JTE and start of the isolation implant, whereas EL from the improved JTE design is not readily visible as it is located underneath the anode. The change in the EL signature also corresponds to significant changes in the breakdown characteristics of the diodes (Fig. 10.15c). The unoptimized JTE has a "soft" breakdown, and the leakage current steadily increases from ~10^{-7} to >10^{-5} A/cm^2 until breakdown is realized at a voltage of −1021 V. In contrast, the improved JTE has a "hard" breakdown, where the leakage

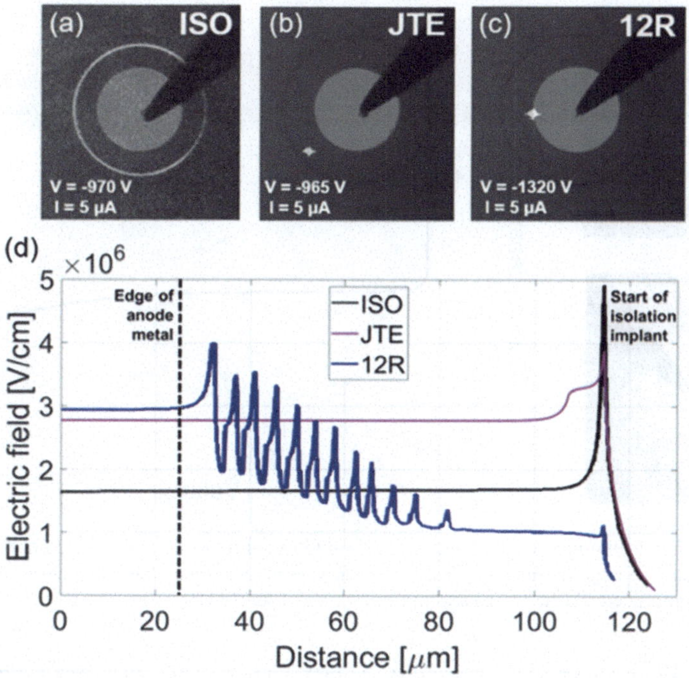

Fig. 10.16 Electroluminescence images of vertical GaN p-i-n diodes with (**a**) only an isolation implant (ISO), (**b**) a junction termination extension (JTE), and (**c**) the "hybrid" edge termination consisting of 12 guard rings superimposed on a JTE (12R). All diodes were imaged with 5 μA of leakage current. (**d**) Electric field profile along the junction of the ISO, JTE, and 12R structured vertical GaN diodes

current remains low (~10^{-6} A/cm^2) and relatively constant until its breakdown voltage of −1454 V is reached. The observed changes in the leakage current before breakdown were also accompanied by a 42% increase in the breakdown voltage from the improved JTE design. As demonstrated, the combination of EL imaging and reverse breakdown characterization helps provide a compelling assessment of edge termination design in vertical GaN diodes.

3.6 Lifetime Testing of 3.3 kV Devices

Application of the above process to 3.3 kV class epi, as shown in Fig. 10.18, but with a slightly higher drift layer doping, ~6×10^{14} cm^{-3}, yields equivalent devices with 3.3 kV breakdown voltage rather than the 3.8 kV shown above. On wafer testing indicates several classes of reverse bias behavior, (i) low leakage (<5 nA) until a sharp breakdown (black), (ii) increased leakage before a sharp breakdown (blue dashed), and (iii) a soft breakdown behavior (red), as shown in Fig. 10.18. Testing

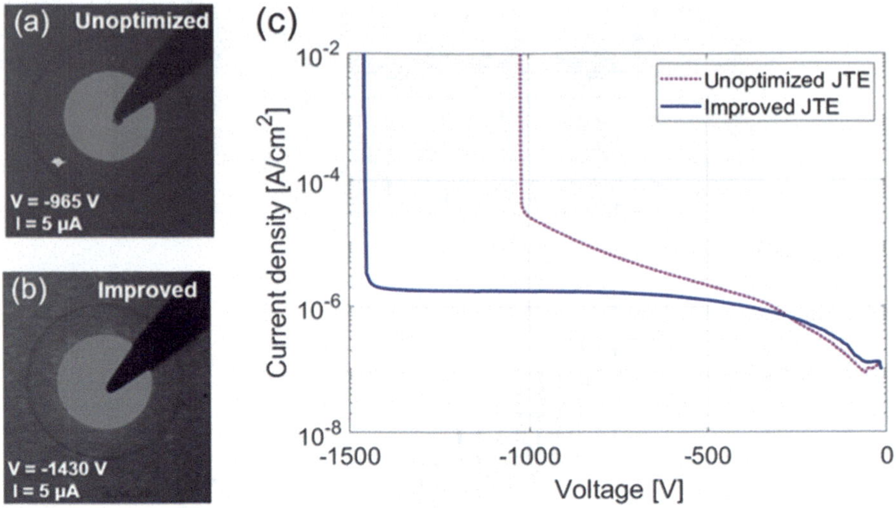

Fig. 10.17 Electroluminescence images of vertical GaN p-i-n diodes with an (**a**) unoptimized and (**b**) optimized JTE edge termination designs. (**c**) Reverse breakdown characterization of the diodes with the unoptimized (dashed line) and optimized (solid line) JTE

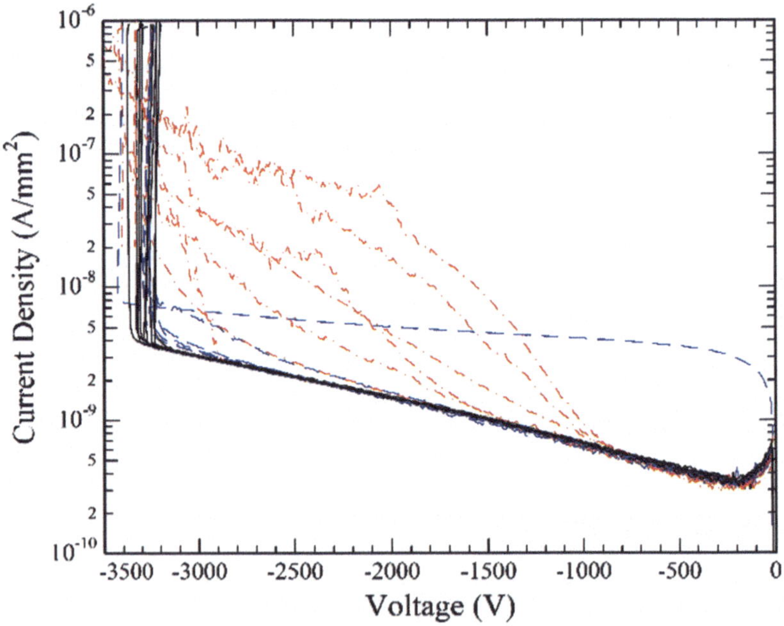

Fig. 10.18 Breakdown curves exhibiting three families of 3.3 kV device breakdown

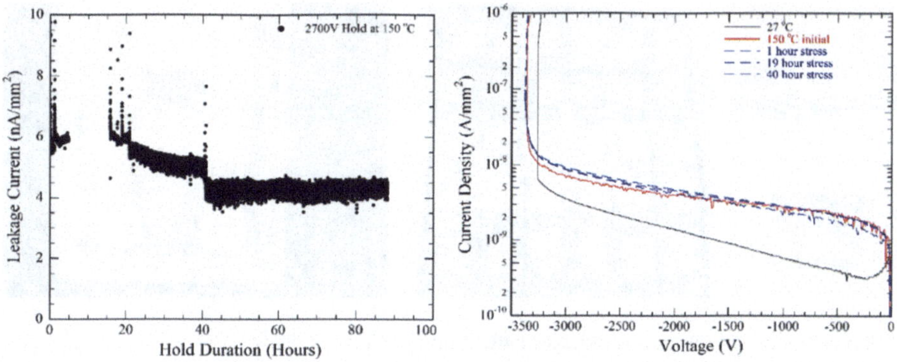

Fig. 10.19 Leakage current at 2.7 kV as a function of HTRB stress duration and breakdown curves before and after stressing

after singulation into individual die yields no change. Forward EL of low leakage devices shows uniform emission around the periphery of the anode metal while reverse EL is dark. At elevated temperature, diodes exhibit increased breakdown voltage consistent with avalanche behavior while still exhibiting a low but slightly elevated leakage (<10 nA at 3 kV) as expected.

While significant study remains to be completed, initial high temperature reverse bias (HTRB) stressing at 150 °C and ~80% of the breakdown voltage (2.7 kV) elicits no significant change in leakage current with time over 100 h as seen in Fig. 10.19. After each interrupt for reverse bias testing, the device stabilizes to ~6 nA of leakage at 2.7 kV. Reverse IV testing after short duration testing yielded no significant change in this singular device through 40 hours of stress. Forward EL exhibits no changes while reverse EL remained dark at this low current. Long duration testing is ongoing but will require further statistical treatment of multiple devices to elucidate the behavior of device cohorts with stress and failure modes of various classes of device behavior.

3.7 Dicing and Packaging

The final steps of any device process are singulation and packaging. While it is relatively straightforward to dice GaN wafers with standard dicing blades and conditions, care must be taken to avoid damaging the edges of the die. In particular, we have observed that a dicing edge too close to the isolation edge can damage the device, particularly with thinned wafers. The reason for this is that the dicing process can cause microcracks that propagate through the isolation region. Alternatives such as laser dicing are possible. Commercial packaging and processes using standard die attach and wire bonds have proven to be successful. In Fig. 10.20, we show an example of a device packaged using a surface mount package at a commercial

Fig. 10.20 Optical images of packaged parts from 1.2 kV-class diode wafer

facility. Following wire bonding, the cavities are potted with Hysol encapsulation. The supplier was able to place ~16 wire bonds on a 0.1 mm^2 device, and ~64 wire bonds on the 1 mm^2 device. Forward and reverse I-V testing on die before and after packaging has identified no obvious degradation in performance following singulation and packaging.

4 Concluding Remarks

A scalable, planar process for vertical GaN P-iN diode fabrication has been demonstrated. We have developed process modules for incoming metrology and device evaluation, in addition to device processing. We have identified appropriate edge termination designs for the given anode specifications. Yield is highly wafer-dependent but correlated to incoming metrology. Preliminary work is ongoing to establish packaging processes and develop electrical stress protocols to identify fundamental failure mechanisms.

References

1. I.C. Kizilyalli, P. Bui-Quang, D. Disney, H. Bhatia, O. Aktas, Reliability studies of vertical GaN devices based on bulk GaN substrates. Microelectron. Reliab. **55**, 1654–1661 (2015)
2. T. Narita, H. Yoshida, K. Tomita, K. Kataoka, H. Sakurai, M. Horita, M. Bockowski, N. Ikarashi, J. Suda, T. Kachi, Y. Tokuda, Progress on and challenges of p-type formation for GaN power devices. J. Appl. Phys. **128**, 090901 (2020)
3. T.J. Anderson, J.D. Greenlee, B.N. Feigelson, J.K. Hite, K.D. Hobart, F.J. Kub, Improvements in the annealing of mg ion implanted GaN and related devices. IEEE Trans. Semicond. Manuf. **29**, 343–348 (2016)
4. J.K. Hite, M.A. Mastro, T.J. Anderson, J.C. Gallagher, M. Ebrish, J.A. Freitas Jr., GaN homoepitaxial growth and substrate-dependent effects for vertical power devices. ECS Trans. **98**, 63–67 (2020)

5. J.C. Gallagher, M.A. Ebrish, M.A. Porter, A.G. Jacobs, B.P. Gunning, R.J. Kaplar, K.D. Hobart, T.J. Anderson, Optimizing performance and yield of vertical GaN diodes using wafer scale optical techniques. Sci. Rep. **12**, 1 (2022)
6. K.X. Liu, R.E. Stahlbush, M.E. Twigg, J.D. Caldwell, E.R. Glaser, K.D. Hobart, F.J. Kub, Photoluminescence and electroluminescence imaging of carrot defect in 4H-SiC epitaxy. J. Electron. Mater. **36**, 297 (2007)
7. J.D. Caldwell, R.E. Stahlbush, M.G. Ancona, O.J. Glembocki, K.D. Hobart, On the driving force for recombination-induced stacking fault motion in 4H-SiC. J. Appl. Phys. **108**, 044503 (2010)
8. J.C. Gallagher, M.A. Mastro, M.A. Ebrish, A.G. Jacobs, B.P. Gunning, R.J. Kaplar, K.D. Hobart, T.J. Anderson, Using machine learning with optical profilometry for GaN wafer screening. Sci. Rep. **13**, 3352 (2023)
9. J.C. Gallagher, M.A. Mastro, A.G. Jacobs, R.J. Kaplar, K.D. Hobart, T.J. Anderson, Detecting defects that reduce breakdown voltage using machine learning and optical profilometry. Sci. Rep. **14**, 7740 (2024)
10. M. Kuball, Raman spectroscopy of GaN, AlGaN, and AlN for process and growth monitoring/control. Surf. Interface Anal. **31**, 987–999 (2001)
11. J.C. Gallagher, T.J. Anderson, A.D. Koehler, M.A. Ebrish, G.M. Foster, M.A. Mastro, J.K. Hite, B.P. Gunning, R.J. Kaplar, K.D. Hobart, F.J. Kub, Effect of GaN substrate properties on vertical GaN PiN diode electrical performance. J. Electron. Mater. **50**, 3013 (2021)
12. M.E. Liao, N.A. Mahadik, J.C. Gallagher, B.P. Gunning, R.J. Kaplar, T.J. Anderson, Microstrucutral evolution of extended defects in 25um thick GaN homoepitaxial layers. Appl. Phys. Lett. **122**, 242101 (2023)
13. T. Maeda et al., Impact ionization coefficients and critical electric field in GaN. J. Appl. Phys. **129**(18) (2021)
14. L. Cao et al., Experimental characterization of impact ionization coefficients for electrons and holes in GaN grown on bulk GaN substrates. Appl. Phys. Lett. **112**(26) (2018)
15. D. Ji, B. Ercan, S. Chowdhury, Experimental determination of impact ionization coefficients of electrons and holes in gallium nitride using homojunction structures. Appl. Phys. Lett. **115**(7) (2019)
16. B.J. Baliga, Gallium nitride devices for power electronic applications. Semiconductor Sci. Technol. **28**(7), 074011 (2013)
17. J. Baliga, *Fundamentals of Power Semiconductor Devices* (Springer, 2008), pp. 107–132
18. P. Pandey, T.M. Nelson, W.M. Collings, M.R. Hontz, D.G. Georgiev, A.D. Koehler, T.J. Anderson, J.C. Gallagher, G.M. Foster, A. Jacobs, M.A. Ebrish, B.P. Gunning, R.J. Kaplar, K.D. Hobart, R. Khanna, A simple edge termination design for vertical GaN P-N diodes. IEEE Trans. Electron Devices **69**, 5096 (2022)
19. T. Nelson, P. Pandey, D.G. Georgiev, M.R. Hontz, A.D. Koehler, K.D. Hobart, T.J. Anderson, A. Ildefonso, R. Khanna, Hybrid edge termination in vertical GaN: Approximating beveled edge termination via discrete implantations. IEEE Trans. Electron Devices **69**, 6940 (2022)
20. P. Pandey, T.M. Nelson, M.R. Hontz, D.G. Georgiev, R. Khanna, A.G. Jacobs, J.S. Lundh, J.C. Gallagher, A.D. Koehler, K.D. Hobart, T.J. Anderson, Hybrid edge termination for high voltage vertical GaN devices: Empirical validation and robust processing tolerance. IEEE Trans. Electron Devices (2024) (in press)
21. M.A. Ebrish, M.A. Porter, A.G. Jacobs, J.C. Gallagher, R.J. Kaplar, B.P. Gunning, K.D. Hobart, T.J. Anderson, Impact of anode thickness on breakdown mechanisms in vertical GaN PiN diodes with planar edge termination. Crystals **12**, 623 (2022)
22. M.A. Ebrish, M. Porter, A. Jacobs, J.C. Gallagher, R.J. Kaplar, B.P. Gunning, K.D. Hobart, T.J. Anderson, Study of anode doping and avalanche in foundry-compatible 1.2kV vertical GaN PiN diodes. Appl. Phys. Express **16**, 116501 (2023)

23. M.A. Ebrish, T.J. Anderson, A.D. Koehler, G.M. Foster, J.C. Gallagher, R.J. Kaplar, B.P. Gunning, K.D. Hobart, A study on the impact of mid-gap defects on vertical GaN diodes. IEEE Trans. Semicond. Manuf. **33**, 546 (2020)
24. J.S. Lundh, A.G. Jacobs, P. Pandey, T. Nelson, D. Georgiev, A.D. Koehler, R. Khanna, M.J. Tadjer, K.D. Hobart, T.J. Anderson, Experimental validation of robust hybrid edge termination structures in vertical GaN PiN diodes with avalanche capability. IEEE Electron Device Lett. **45**, 873 (2024)

Chapter 11
Layer Transfer of Gallium Nitride by Controlled Spalling

Stephen W. Bedell

1 Introduction

When a layer possessing tensile stress is adhered to the surface of a brittle substrate, a peculiar mode of fracture is possible whereby a crack tip can propagate below, and parallel to, the layer-substrate interface. This mode of fracture, referred to as *substrate spalling*, results from the edge load created by the tensile stressor layer, giving rise to a mixture of type I stress (opening mode) and type II stress (shear), which guides the crack to an equilibrium depth below the interface. The stress field that is transferred to a crack tip in a material body through external loads is typically described using stress intensity factors (K), that are decomposed into pure type I or opening mode (K_I), type II or in-plane shear mode (K_{II}) and type III or out-of-plane shear (K_{III}) [1]. Figure 11.1 is an illustration (adapted from [2]) of the effect of in-plane shear on an advancing crack tip in a substrate. The effect of the shear stress (K_{II}) is to deflect the crack in the direction that minimizes shear ($K_{II} = 0$). For a surface layer under compression, the substrate crack will deflect up and result in delamination or layer cracking. For a surface layer in tension, the crack will deflect into the substrate until it reaches a depth where $K_{II} = 0$, and it will propagate, provided that the K_I value at that depth exceeds the fracture toughness K_{IC} of the substrate. The crack trajectory is stable because K_{II} is *corrective*. An analytical model of spalling was developed in the late 1980s [2, 3], which offered a straight-forward means of calculating critical loading conditions (film stress and thickness values) for which spalling fracture is possible, as well as computing the equilibrium crack depth (where $K_{II} = 0$) for a given substrate material. A thorough review of mixed-mode fracture is given in [4] for the interested reader.

S. W. Bedell (✉)
IBM T. J. Watson Research Center, Yorktown Heights, NY, USA
e-mail: bedells@us.ibm.com

I. C. Kizilyalli et al. (eds.), *Gallium Nitride and Related Materials*, The Materials Research Society Series, https://doi.org/10.1007/978-3-031-83056-3_11

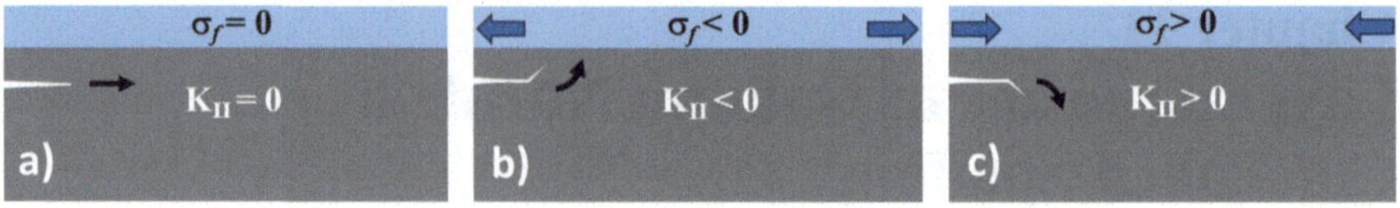

Fig. 11.1 The effect of in-plane shear stress on crack trajectory (Adapted from Ref. [2] with permission from Elsevier). In (**a**), a film with zero stress (blue layer) is on the surface of a brittle substrate (gray) with a subsurface crack. An opening mode stress will result in a straight crack path. In (**b**), the surface film is under compressive stress, resulting in a negative value of K_{II} at the crack tip, which deflects the crack toward the surface. In (**c**), a film in tension results in a positive K_{II} value deflecting the crack tip into the substrate. The crack will continue propagating into the substrate until $K_{II} = 0$. This mode of fracture is called substrate spalling

Substrate spalling is generally considered a material failure mode and, in the specific area of semiconductor technology, has posed challenges to under-bump metallization (UBM) schemes used in packaging [5]. However, there have also been historical attempts to exploit this unique fracture mode for useful purposes. One of the earliest practical applications of substrate spalling was the texturing of glass in a process called "glass chipping" [6]. By applying glue to the surface of glass, discontinuous spalling of the glass surface could be induced as the glue layer dried and contracted, creating a decorative fracture pattern. More recently, there have been attempts to use spalling as a method for removing continuous thin layers in semiconductor technology using metal layers as the stressed surface layer [7, 8]. There were several issues that limited the success of these early attempts at using substrate spalling for surface layer transfer from substrates. First, fracture of the substrate occurred spontaneously. Specifically, the tensile stress in the surface film (Nichrome in [7] and Ag in [8]) was sufficiently high to permit both crack initiation and propagation. When fracture initiation is spontaneous, it generally results in multiple fracture fronts that lead to fracture instability and layer cracking where these different fracture paths meet. Additionally, if a fracture is to be used as a method of layer transfer, it must occur at a temperature well below the brittle-to-ductile transition temperature to ensure dislocations are not generated at the crack tip (*c.f.* Figure 6.22 in [1]). This limits the usefulness of thermal expansion originated stress used in [8].

2 Controlled Spalling

By recognizing that a stressed film could be deposited on a substrate with sufficient loading conditions (thickness and stress) to propagate fracture but insufficient to spontaneously initiate the crack, a process was developed [9, 10] that exploited this metastability and allowed one to manually control the fracture *after* the deposition of the stressor layer using a flexible handle layer. The process was called Controlled Spalling and it eliminated the need for high-temperature process steps, and the occurrence of spontaneous fracture observed in previous work. By controlling the

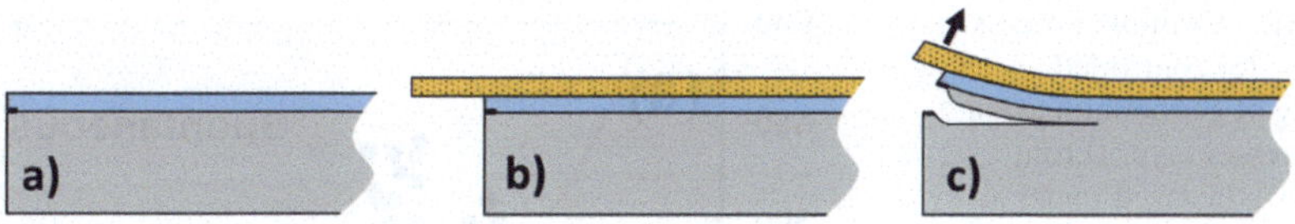

Fig. 11.2 Schematic depiction of the controlled spalling process. In (**a**) a tensile stressed layer (blue) is deposited on a brittle substrate (gray). The combination of stress and thickness is below the values that result in spontaneous fracture but sufficient to sustain fracture once initiated. In (**b**) a flexible handling layer (e.g., tape) is attached to the surface of the stressor layer and (**c**) shows the use of the handle layer to initiate and guide a single fracture front. Also depicted in the image is the adhesion demoter region used for reliable fracture initiation (described in text)

magnitude of the tensile stress and the thickness of the metal stressor layer independently, the stressor/substrate combination could be made metastable against fracture, and the spalling depth could be controlled (spalling depth is proportional to the stressor thickness). The first demonstration of controlling the spalled layer thickness using this approach was shown for Ge [10]. The flexible handle layer is applied to the metal stressor surface to help initiate and mechanically guide fracture propagation in a controllable manner. By creating a well-defined fracture initiation region and propagating a single fracture front across the surface, large-area, crack-free layer transfer was demonstrated on a variety of substrates. Figure 11.2 shows a schematic illustration of the controlled spalling process, which enables low-temperature layer transfer and a simplified film-handling scheme.

As depicted in Fig. 11.2, the general steps of controlled spalling are (a) deposition of a stressor layer, (b) application of a flexible handle layer, and (c) fracture initiation and propagation using the handle layer. The stressor layer can be any material that can strongly adhere to the substrate, possess sufficient elastic energy to induce and propagate fracture in the substrate without yielding or fracturing itself (high yield strength and fracture toughness), and is preferably cost-effective and simple to apply. Although many options exist, Ni metal was found to be the easiest to implement and satisfied the above requirements. For spalling in low fracture toughness substrate materials (Si, Ge, or GaAs), Ni deposited by magnetron sputtering is typically used. The stress is controlled very well by regulating the Ar pressure during sputtering. For harder materials (such as GaN), the required Ni thickness is too great for sputtering, and electroplating was used (described later). The role of the handle layer is to provide a means of both initiating fracture in a prescribed region of the substrate, and to guide a single fracture front across the substrate. The handle layer must also permit the load from the stressor layer to be transferred to the crack tip to maintain the spalling condition ($K_{II} = 0$ inside the substrate). If the handle layer is too thick or has too high of an elastic modulus (or combinations thereof), then the shear stress cannot be transferred to the crack tip and spalling mode fracture will not be possible. Handle layer materials that work well are typically roll-applied thin polyimide, or low-modulus PVC-based tapes. Each substrate material will have its own "process window" for controlled spalling that depends on the fracture toughness of the material and cleavage planes. Figure 11.3 shows an example of the

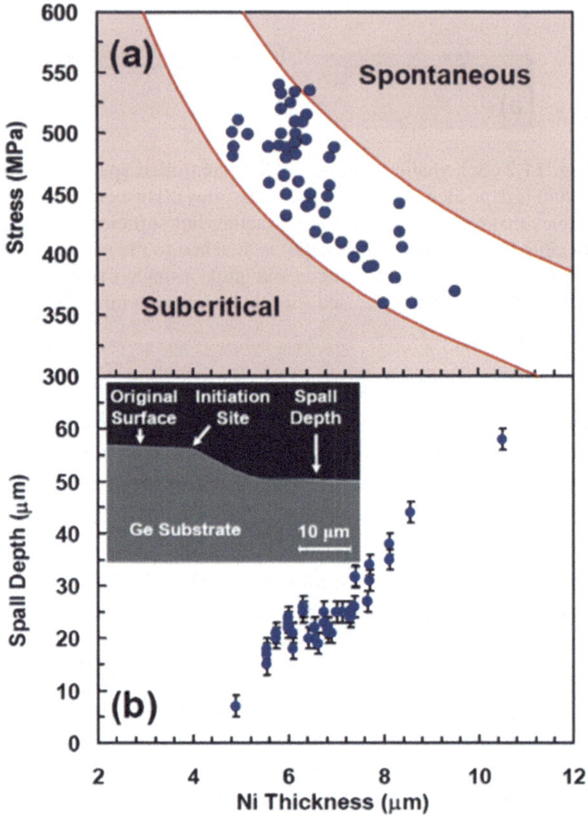

Fig. 11.3 Example process window for controlled spalling of germanium using a Ni stressor layer (From Ref. [10], licensed under a Creative Commons Attribution (CC BY 3.0) license). The fracture metastability is shown in (**a**), and the resulting fracture depth control is shown in (**b**)

process window for spalling 4″ <001> Ge substrates using a Ni stressor. The metastable region between the sub-critical and spontaneous spalling areas in the stress versus thickness plot defines the controlled spalling process window. Also shown is the measured spalling depth resulting from the varying Ni thickness.

There are a few practical considerations when attempting to use substrate spalling for consistent, routine layer transfer. Achieving consistent fracture initiation after application of the handle layer is one of the more challenging aspects of controlled spalling. Through trial and error, it was found that the most reliable method of initiating a crack into the substrate was by having the stressor layer terminate abruptly at the edge of the substrate. This sudden termination creates a stress discontinuity localized at the edge of the stressor layer and serves as the source of the initial crack. If the edge of the stressor layer is sufficiently abrupt, a small mechanical perturbation using the handle layer is enough to initiate fracture into the substrate. Many attempts were made to deposit the stressor layer through a physical mask to realize a sharp boundary, but the most successful strategy was to create a thin boundary around the circumference of the substrate before stressor deposition that compromised the adhesion of the stressor layer to the substrate in that region. Any material that could withstand the stressor deposition process and remain weakly adhered to the substrate worked well. Initially, photoresist was used, but it was found that simple ink from inexpensive permanent markers worked equally well. The materials used in these regions were called "adhesion demoters" [11] and served the dual purpose of naturally creating an almost ideal stress singularity as well as ensuring that fracture initiation was consistent around the boundary during spalling. If, at any point during the spalling process, fracture along the boundary was arrested (e.g., due to a stressor layer defect or missing adhesion demoter), a perturbation in the fracture path would occur due to the change in loading conditions and create deviations in the spalling depth, or even film cracking. Additionally, because the handle layer was usually an adhesive tape, special precautions were taken to ensure that the adhesive was only attached to the area where fracture and layer transfer were to occur. If the handle layer was accidentally "stuck" to a region outside the spalling area, that would be sufficient to cause a change in loading and subsequently affect the quality of the spalled layer.

Once the handle/stressor/spalled layer stack is peeled away from the substrate, it will tend to curl cylindrically to minimize strain energy. A successful strategy was adapted from wafer dicing technology whereby the spalled layer is vacuum chucked flat, and a metal frame is attached to the exposed adhesive of the handle layer. This kept the transferred layer flat and permitted additional processing such as device fabrication steps or wafer bonding.

Subsequent to the early work on controlled spalling, there have been numerous publications expanding the understanding and usefulness of this technique [12].

3 Controlled Spalling of Epitaxial GaN on Sapphire

The first published work on controlled spalling of GaN demonstrated wafer-scale removal of epitaxial $In_xGa_{(1-x)}N$/GaN multi-quantum well (MQW) LED structures from a 2″ sapphire substrate [13]. This situation differs from the "classic" substrate spalling picture [3] in that the depth of spalling is not determined by the $K_{II} = 0$ criteria due to the very high fracture toughness of sapphire suppressing fracture at the equilibrium depth. Because of this, the critical loading condition using a 400 MPa Ni layer required 25 μm of stressor thickness. A calculation of the K_I and K_{II} values as a function of depth into the substrate is shown in Fig. 11.4 for the case of a 25 μm thick 400 MPa Ni layer on a sapphire wafer and helps qualitatively to understand the observed behavior. The computed depth where $K_{II} = 0$ is approximately 28 μm below the surface and represents the equilibrium spalling depth. The corresponding value for K_I at that depth, however, is only 1.61 MPa*m$^{1/2}$, which is well below the reported fracture toughness of sapphire $K_{IC} = 2.5$ MPa*m$^{1/2}$ [14]. The computed value of K_I in the epitaxial GaN region (upper 5 μm) is between 1 and 1.5 MPa*m$^{1/2}$ which is in the range of the reported K_{IC} of GaN [15, 16]. The result is that controlled spalling can be used to selectively remove heteroepitaxial layers from hard substrates such as sapphire (or SiC). In the case of GaN grown on patterned sapphire substrates

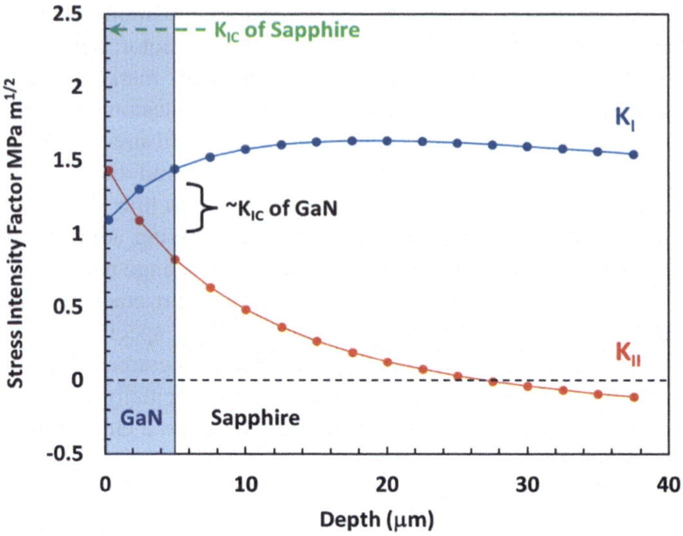

Fig. 11.4 Calculation of K_I and K_{II} as a function of depth in a sapphire substrate using the procedure described in Ref. [3]. As seen in the image, at a depth corresponding to $K_{II} = 0$, the value of K_I is around 1.6 $MPa{*}m^{1/2}$ and insufficient to fracture the sapphire. The value of K_I near the surface where the epitaxial GaN is grown is at or above the K_{IC} of GaN, so fracture occurs there. This is an example of $K_{II} \neq 0$ spalling

(PSS), spalling leaves a layer of epitaxial GaN on the patterned Al_2O_3 surface (Fig. 11.5 left), whereas the entire GaN epitaxial layer is removed from planar GaN leaving an atomically smooth transferred GaN layer (Fig. 11.5 right).

An important lesson from the early work on GaN spalling in [13] indicated that electroplating was the only practical method of Ni stressor deposition. Naturally, nearly all modern plating bath formulations have as their goal low intrinsic stress and high luster (low roughness). To achieve these goals, additives such as surfactants are included, and the bath chemistry and plating conditions (such as heating) become complicated. The unstated issue with many additives is that they compromise the fracture toughness of the plated Ni itself, which leads to cracking of the Ni film under stress. Controlled spalling benefits from the fact that the simplest plating bath is nearly ideal with minimal modifications, provided an adequate thin "seed" layer is first deposited by PVD (usually sputtered Ni or Ti/Ni). The plating bath chemistry that was found to work extremely well was 300 g/l of nickel chloride hexahydrate ($NiCl_2{\bullet}6H_2O$) and 30 g/l of boric acid (to limit H_2 evolution) at room temperature. The plating rate is typically 1 μm/min using a current density of 50 mA/cm^2. Another benefit of using this chemistry is that the pure Ni anode is efficiently etched during electroplating, which preserves the Ni content of the bath. This results in very consistent characteristics of the plated Ni without the need for bath maintenance and monitoring. At room temperature, using the above plating conditions, the stress in the Ni is typically between 450 and 500 MPa. To reduce the stress in the plated Ni, the bath can be heated, and to increase the Ni stress, NH_3Cl (10 g/l) can be added to

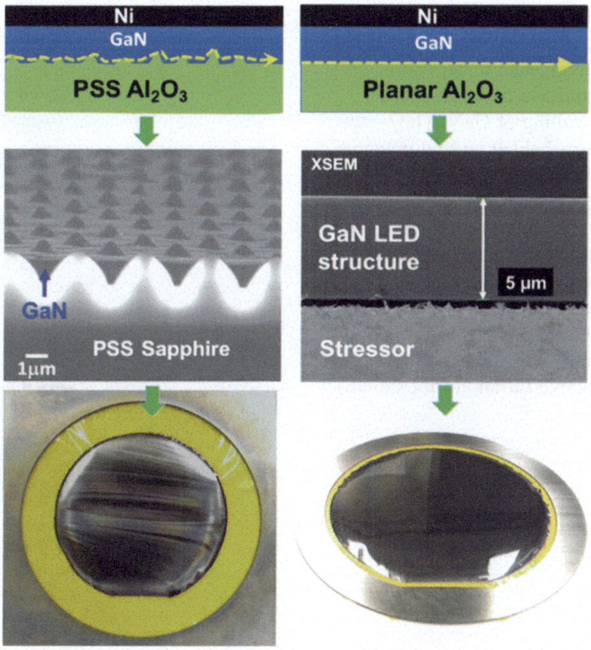

Fig. 11.5 $K_{II} \neq 0$ spalling of epitaxial GaN grown on patterned sapphire substrates (PSS) (left) and planar sapphire (right). The fracture path when spalling GaN/PSS is such that approximately 1 μm of GaN remains between the patterned features, whereas spalling removes GaN at the sapphire/GaN interface, leaving atomically smooth surfaces. The lower images show a 2″ spalled PSS GaN layer (left) and a 4″ spalled GaN from planar sapphire (right) mounted to handling frames

the bath. This gives a practical stress range of 250 to 800 MPa for this plating chemistry.

Although the plating chemistry is very forgiving, Ni thickness uniformity and particle inclusion must be controlled extremely well for producing high-quality spalled films. For example, if the plating rate near the wafer edges is too high (which is common), the critical loading conditions for spontaneous spalling can be met at the edge, while the Ni thickness is sub-critical near the wafer center and can result in a half-spalled wafer. Because fracture is initiated at the wafer edge, it is better to have the Ni at the edges slightly thinner than the center, thereby guaranteeing that the entire wafer area is ready to spall when the edges exceed the critical loading condition. Custom plating equipment was developed to enable independent control of the Ni plating rate at the wafer edge and the 2″ wafer plating system is shown in Fig. 11.6. The sacrificial cathode that surrounds the wafer is called a thief (or current thief) and serves to limit current crowding at the wafer edge that otherwise leads to high edge plating rate. When the thief is held at the same potential as the wafer (ground), the plating rate at the very edge of the wafer is reduced compared to the rest of the wafer. Therefore, by varying the potential of the thief electrode during plating, the thickness profile of the Ni can be controlled to optimize fracture initiation and uniform spalling.

Figure 11.7 shows the Ni thickness uniformity for the case of both a grounded and floating thief and illustrates the control of the Ni thickness distribution using this approach. Measuring the Ni thickness to within 10's of μm from the edge was accomplished by masking and etching a portion of the Ni using $FeCl_3$ and measuring

Fig. 11.6 Picture of a custom 2″ electroplating setup used to deposit uniform Ni layers for the spalling of GaN. The wafer was held at ground potential through a conductive elastomer that also doubled as a vacuum chuck seal. The thief was used to tune the edge thickness profile as described in the text. Particles were controlled using an anode bag and active filtration (not shown)

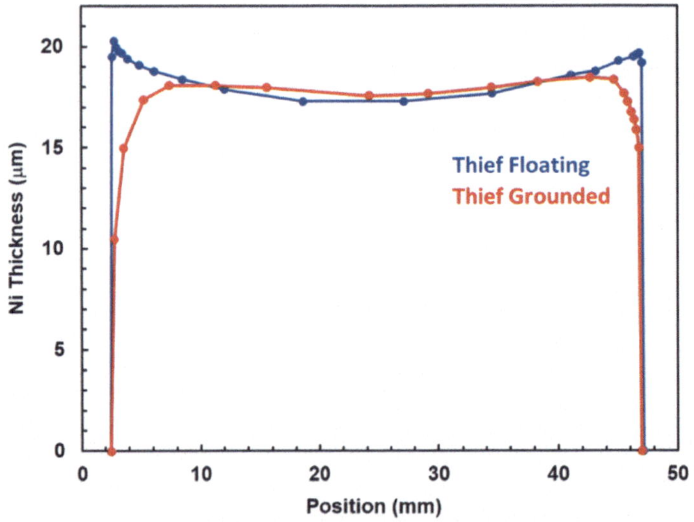

Fig. 11.7 Ni thickness as a function of position for 2″ electroplating using the setup shown in Fig. 11.6. By controlling the potential of the thief cathode during plating the thickness of the Ni at the edge could be optimized for controlled spalling of GaN

the local step height using profilometry. The thief is fabricated using Au-coated Ti and enables infinite reuse by selectively etching the plated Ni after each run. Also seen in Fig. 11.6 is the vacuum chuck base used to hold the wafer down through a conductive elastomer seal and the removable vessel wall with inlet and outlet ports for bath circulation. The vertical plating apparatus shown in Fig. 11.6 has superior Ni uniformity than first-generation horizontal plating setups, but managing particles in such a geometry requires the use of an anode bag and bath circulation/filtration (not shown).

4 Controlled Spalling of Bulk GaN

The preservation of crystal quality after controlled spalling of GaN from sapphire substrates was an important demonstration [13] in addition to the increasing list of useful materials compatible with this technology. There are, however, commercially available systems already in existence that use laser lift-off (LLO) [17] to transfer GaN (LEDs) from sapphire to metal substrates. Also, heteroepitaxial GaN MQW LED structures grown on sapphire typically have very high crystal defect densities (~10^8 cm^{-2}), so the preservation of lattice perfection is less of a concern in that material system. For high-current density devices such as laser diodes or power electronics, crystal defects in GaN have been shown to impact performance and device longevity [18, 19]. For these types of devices, bulk GaN is the substrate of choice due to the lower defect density (~10^6 cm^{-2}). The very high cost, as well as the lower thermal conductivity (compared to SiC), make layer transfer from a bulk GaN substrate an attractive technology. Other attempts to develop layer transfer methods compatible with bulk GaN crystals generally compromised the high crystal quality [20–22], which makes controlled spalling a truly differentiating technology. An excellent review of the various approaches to GaN layer transfer can be found in Youtsey et al. [23].

Controlled spalling of bulk GaN is unlike spalling epitaxial layers from sapphire in that the $K_{II} = 0$ condition is obeyed and follows the Suo and Hutchinson (S&H) model described in [3]. To determine the critical loading conditions, and controlled spalling process window for bulk GaN, a range of Ni thickness regions was electroplated on a 2″ bulk GaN substrate. First a uniform layer of Ni was deposited over the entire wafer surface, then a small rectangular piece of polyimide tape was applied to exclude that region from further plating. The process was repeated until spontaneous spalling occurred at the wafer edge. Spalling proceeded until it reached a masked region where the Ni was too thin to sustain fracture propagation, as shown in Fig. 11.8. These data combined with other data from 2″ bulk. GaN spalling experiments allow one to construct the controlled spalling process window for bulk GaN similar to that for Ge shown in Fig. 11.2 (from Ref. [11]). Due to the paucity of material data available for bulk GaN, reliable fracture toughness K_{IC} was unavailable except for nanoindentation measurements [15]. The issue with relying on the K_{IC} for bulk GaN measured using indentation is that the measured value of 0.79 MPa*$m^{1/2}$ most likely represents fracture on the GaN habit plane, rather than the [0001] c-planes fractured during spalling. However, with the measured critical loading conditions for spalling, the K_{IC} can be inferred by inverting the S&H model to fit the measured data. The best fit K_{IC} value to the critical loading conditions measured for bulk GaN is around 1.7 MPa*$m^{1/2}$ and is shown in Fig. 11.9.

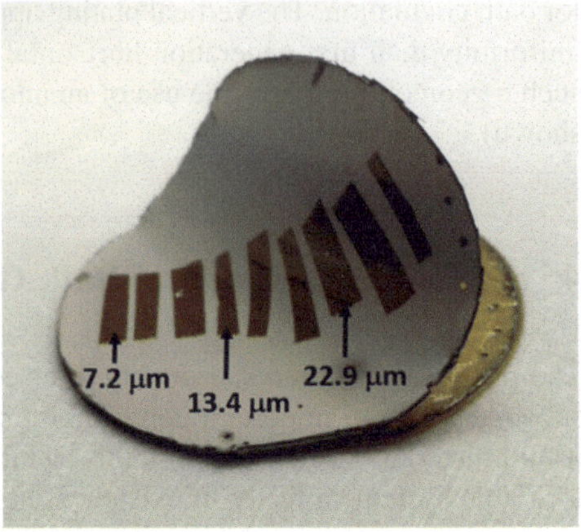

Fig. 11.8 An image of a 2″ bulk GaN wafer with multiple Ni thickness values used to efficiently map out the controlled spalling process window. Spalling was allowed to occur spontaneously to ensure the full metastability window was captured. (From Ref. [16] with permission from AIP publishing)

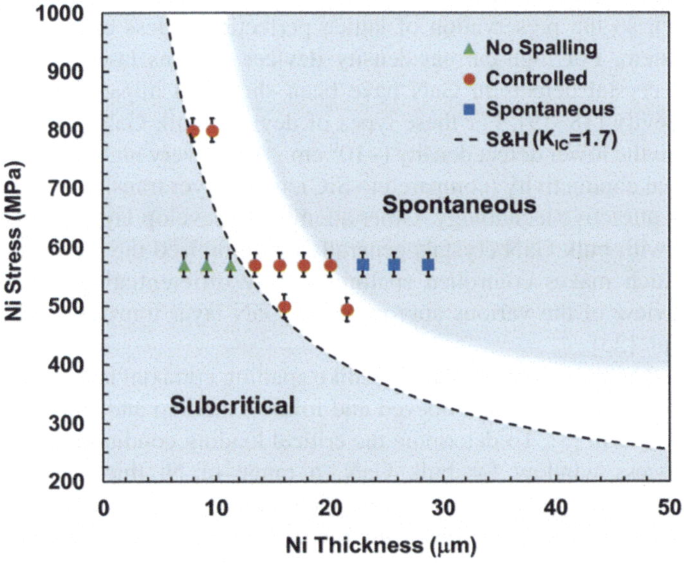

Fig. 11.9 Process window for controlled spalling of bulk GaN using a Ni stressor layer (From Ref. [16] with permission from AIP publishing). By inverting the spalling equations from Ref. [3], a fracture toughness for the GaN [0001] plane was estimated to be 1.7 MPa*m$^{1/2}$

5 Characteristics of Spalled Bulk GaN Layers

Figure 11.10 shows an image of a layer that was removed from a 2″ bulk GaN surface that has been mounted to a stainless steel frame for the sake of film handing, as described above. The GaN thickness values are included in the image (in units of μm) and were measured using calibrated X-ray fluorescence (XRF). Analysis using the S&H model also provides a calculation of spalling depth. Applying the

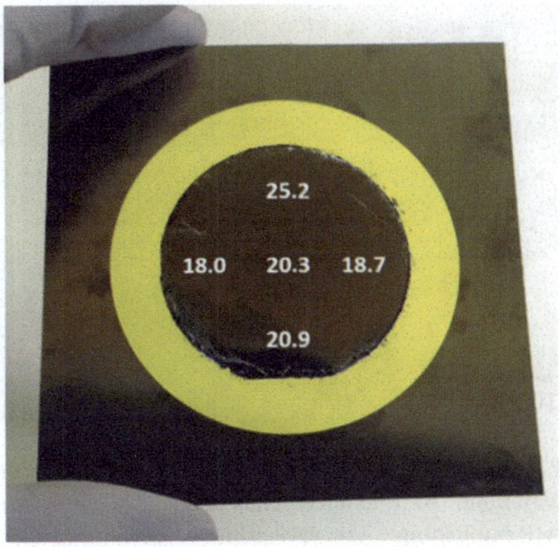

Fig. 11.10 Successful crack-free controlled spalling and mounting of a 2″ bulk GaN layer (From Ref. [16] with permission from AIP publishing). The thickness values measured using calibrated X-ray fluorescence are shown on the sample in units of microns

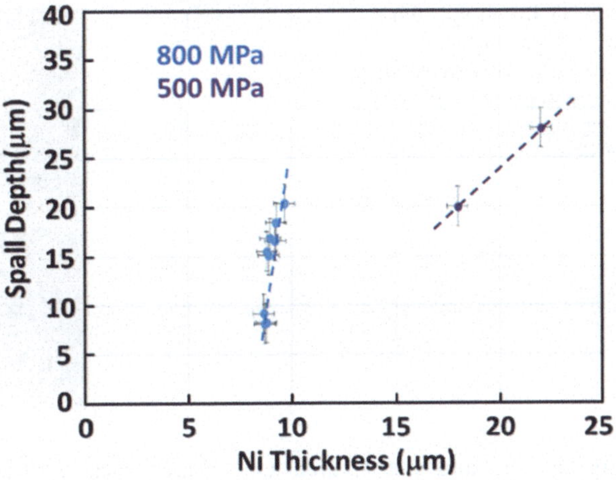

Fig. 11.11 Spalling depth as a function of Ni thickness for two different Ni stress values. The stress-independent spalling depth observed in germanium (Fig. 11.3) is not observed in the bulk GaN system

conditions used for the layer transfer shown in Fig. 11.10 (500 MPa, 22 μm Ni) yields a calculated spalling depth of 22.1 μm, in reasonable agreement with the thickness values measured on the sample. Attempts to reduce the GaN spalling depth by increasing the electroplated Ni stress and decreasing the thickness resulted in the relationship shown in Fig. 11.11. The dashed lines are added to guide the eye. Although the S&H model predicts that the spalling depth is independent of Ni stress, the experimentally measured relationships between Ni thickness and spalling depth for the two Ni stress values shown suggests there is a stress-dependence. The large slope observed for the 800 MPa data indicates that small variations in Ni thickness will translate into large spalling depth variation and should be avoided if uniform film removal is the goal.

Figure 11.12 shows a high-magnification and low-magnification cross-section transmission electron micrograph (XTEM) image for both the spalled bulk GaN

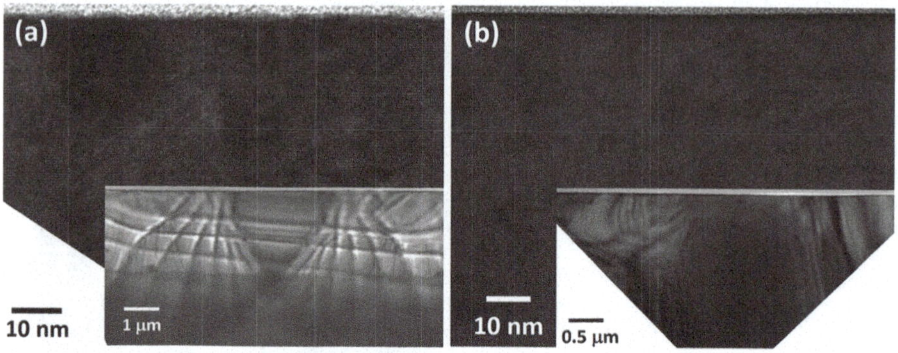

Fig. 11.12 High and low (inset) magnification cross-section TEM images of (**a**) residual bulk GaN wafer and (**b**) transferred GaN layer after controlled spalling (From Ref. [16] with permission from AIP publishing). The crystalline defectivity of the bulk GaN is preserved during layer transfer

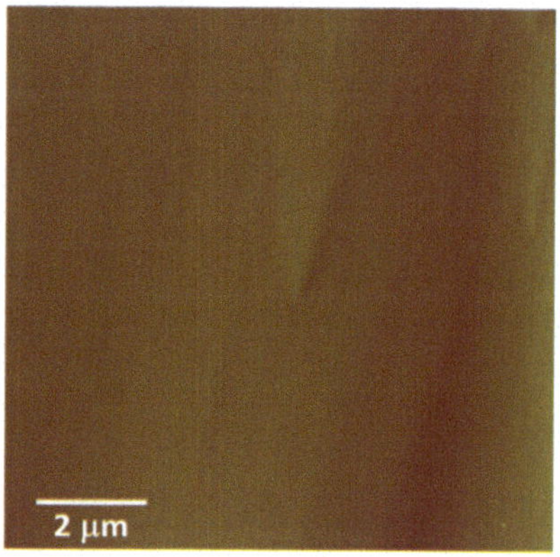

Fig. 11.13 Atomic force microscope image (10 × 10 μm) showing that the as-fractured surface is very smooth *locally* (1.3 nm RMS)

surface (a) and the spalled GaN layer surface (b), demonstrating that room-temperature fracture does not introduce defects in the GaN crystal lattice.

Atomic force microscopy (AFM) taken on a 10 × 10 μm region of the spalled surface shows a locally smooth fractured surface of 1.3 nm root mean squared (RMS) roughness and is shown in Fig. 11.13. The macroscopic scale of roughness is, in general, much larger than the local roughness values.

At a length scale of ~mm and greater, the surface characteristics are determined by the dynamics of fracture propagation. These dynamics are almost exclusively determined by phenomena involving the boundary or edge effects. Figure 11.14 shows the surface morphology in two different regions of the residual bulk GaN wafer after spalling. For surface regions where fracture was uninterrupted and occurred at a constant velocity, the morphology is smooth and uniform, with an oscillation amplitude of approximately 500 nm over a length scale of several millimeters. When fracture propagation is interrupted, a line feature is created on the surface, orthogonal to the direction of propagation. These line features are referred to as "start/stop" artifacts and are local depth perturbations with an

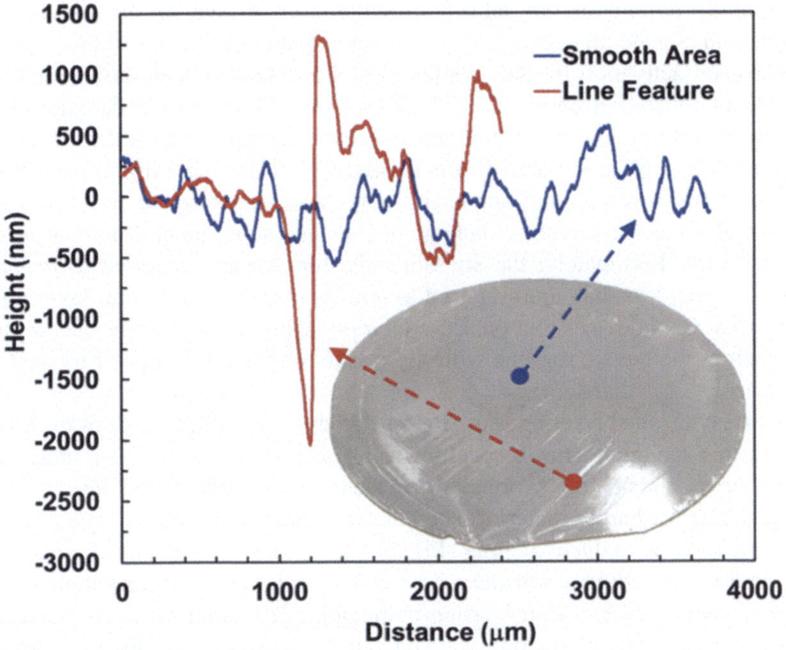

Fig. 11.14 Wafer-scale surface roughness measured on a spalled bulk GaN surface by profilometry (From Ref. [16] with permission from AIP publishing). There are large regions of "typical" roughness that have an amplitude of about +/− 500 nm and line artifacts with localized deviations in depth of up to 2 μm. These "start/stop" features can be traced back to imperfections along the wafer edge

amplitude between 1 and 2 μm. It is these perturbations, largely due to imperfections near the wafer edge, that limit the ability to reuse bulk GaN wafers multiple times. Because the [0001] planes are not native cleavage planes for the wurtzite GaN system, spalling of semipolar and non-polar GaN wafers may have much lower as-spalled roughness than the c-plane spalling shown here.

6 Spalling of Bulk GaN Devices

One important advantage that controlled spalling offers compared to other GaN layer transfer techniques is that spalling permits layer separation at any point in the device fabrication process. In the case of flexible lightweight solar cells [24], the simplest approach was to first spall the multijunction III-V layers from the Ge substrate, then fabricate the photovoltaic devices directly on the spalled and framed thin films. Conversely, when demonstrating ultra-flexible CMOS circuits [25], spalling was performed as the last step of the fabrication process. There are three physical phenonmena that conspire to limit this versatility in the GaN material system.

First, spalling requires strong adhesion of the stressor layer to the surface of the substrate and is generally accomplished using a sputtered Ti/Ni seed layer. Second, p-type dopant activation in GaN results in low acceptor concentrations due to the high dopant ionization energy (~200 eV) of Mg. Third, lattice damage in GaN generally leads to excess donor states that can easily compensate for the low acceptor concentrations in p-GaN. Early efforts to apply controlled spalling to prefabricated diodes on GaN revealed that sputtering conditions that resulted in good stressor adhesion, also led to subsurface damage in GaN and subsequent deactivation of the p-GaN surface. Fortunately, the stressor adhesion strength required to *propagate* fracture is much less than that required to *initiate* fracture. Successful layer transfer of functional GaN devices that preserved the p-GaN regions could be accomplished by protecting the device regions with the p-GaN ohmic contact prior to seed layer deposition.

Because controlled spalling uses stress to produce thin films, most of the effort in applying this technique for a specific device application is spent managing the residual stress and unconventional form factor of the transferred layers. This is especially true for harder materials such as GaN that require much higher Ni stress and/or thickness to induce spalling. Figure 11.15 shows an example process flow using controlled spalling to transfer a p-i-n diode device layer fabricated on a 2″ bulk GaN wafer onto a 2" Si wafer. Starting in the upper left and moving clockwise, a 2" GaN wafer is shown with the ohmic contact of the p-i-n diodes visible (a). Next, the wafer has a layer of ink deposited around the bevel of the wafer to act as an adhesion demoter (not shown) followed by sputter deposition of a Ti/Ni bilayer seed and then 20 μm thick ~500 MPa electroplated Ni layer (b). After Ni deposition a layer of 25 μm thick polyimide tape is roll-applied to the Ni surface (c) and used to initiate fracture at the wafer edge (d). The handle layer is used to continue spalling ~25 μm GaN from the substrate and the tape is mounted to a stainless steel frame to permit subsequent processing (e). Rear contact metallization is deposited on the as-spalled n-GaN surface while it is framed (not shown). The metallized n-GaN film side is then bonded to a Si wafer either permanently (usually using electroplated Sn vacuum bonding) or temporarily using a low-temperature mounting adhesive. Once bonded to the Si wafer, the polyimide tape can be removed, and the Ni stressor can be etched away to reveal the original device surface (f).

An alternative post-spalling process flow is shown in Fig. 11.16. A similar device structure is spalled and frame-mounted using the same process flow shown in Fig. 11.15. Instead of bonding the spalled GaN device layer to a rigid substrate, a ~20 μm thick layer of Cu is electroplated directly on the rear contact metallization (a) and provides mechanical support, electrical contact, and thermal sinking for the devices. Next, the handling tape can be removed (b) and the Ni is selectively removed using Transene Nickel Etchant TFG (c). This process has the advantage of being extremely simple and very high yielding as there are no bonding defects associated with direct plating. The one drawback is the high curvature associated with the residual compressive stress in the GaN after Ni removal. This can be controlled by plating the Cu thicker (at the expense of increasing compressive strain in GaN) or by removing a portion of the Ni stressor prior to Cu electroplating.

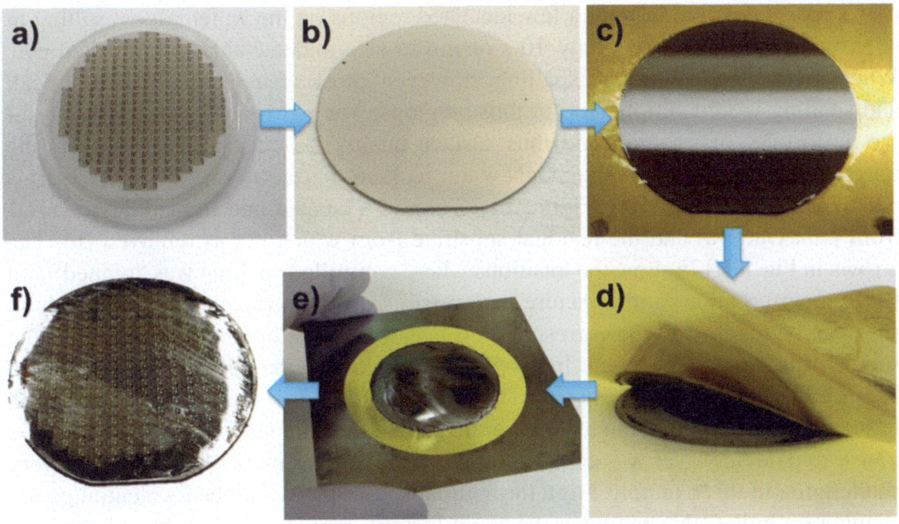

Fig. 11.15 Example process flow for spalling and mounting bulk GaN p-i-n diode devices. The ohmic contacts seen in (**a**) protect the underlying p-GaN regions from sputtering damage during seed layer formation. The spalled device layer (**e**) is then bonded to a Si wafer, and the original surface is recovered after handle layer and Ni removal (**f**)

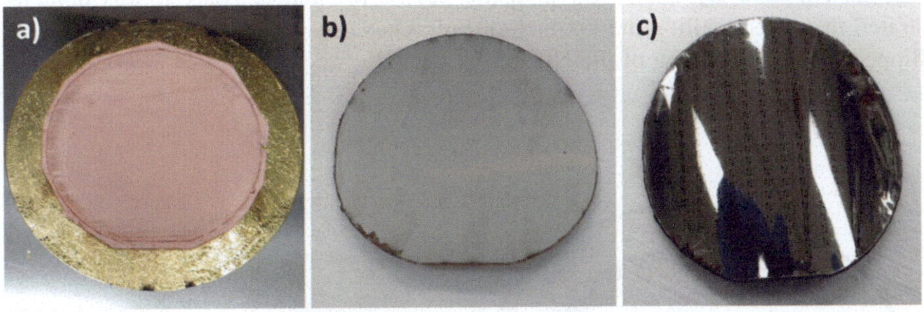

Fig. 11.16 A bonding-free approach for spalled GaN device handling. Instead of bonding the spalled layer to a substrate, a thick Cu layer is directly electroplated on the back n-GaN contacts (**a**). This permits the removal of the handle layer (**b**) and the Ni stressor (**c**), revealing the original device surface

7 Challenges with Bulk GaN Spalling

One of the motivating factors in pursuing a quality-preserving layer transfer process for bulk GaN was to enable substrate reuse, thereby dramatically altering the cost structure of this expensive material. Controlled spalling of bulk GaN as currently practiced would render the residual bulk GaN surface difficult to recover. The custom electroplating equipment described above is very simple, leading to subtle issues like H_2 evolution from a few localized regions on the wafer edge resulting in slightly thinner Ni regions only 10's of μm in size undetectable to the naked eye. When performing controlled spalling, however, those areas could be sub-critical, and spalling would simply stop in those regions. The extremely high sensitivity of the uniformity of fracture (and thus surface quality) to edge imperfections is the biggest challenge with enabling GaN wafer reclaim. Furthermore, bulk GaN itself is not free of bulk defectivity. Figure 11.17 shows a spalled and framed device layer from bulk GaN (a) and the residual substrate from which it was removed (b). The arrows in Fig. 11.17a show the locations where controlled spalling was "pinned" and led to a perturbation in the fracture path (visible on the surface). Each one of these sites can be mapped to the corresponding location on the bulk wafer to identify the source of the fracture pinning. In this case, inclusions in the bulk GaN crystal could be traced back to each pinning site. The inset in Fig. 11.17b is an optical micrograph of the inclusion found in the location indicated on the bulk GaN wafer. Although the transferred device layer was continuous and crack-free, wafer reclaim would be very challenging in this example. High throughput sputtering would be a potential game-changer for bulk GaN spalling as the Ni uniformity could be made extremely high up to the wafer edge without any of the aforementioned electroplating issues.

The ability to transfer bulk GaN device layers while preserving crystalline quality does provide value outside of any wafer reclaim potential. Devices on thin (~25 μm) GaN would have superior performance over their bulk counterparts due to the higher thermal conductivity and lower parasitic device resistance. Also, it was demonstrated that after spalling the Ga-face of bulk GaN, the residual substrate could be flipped over and spalling of the N-face performed.

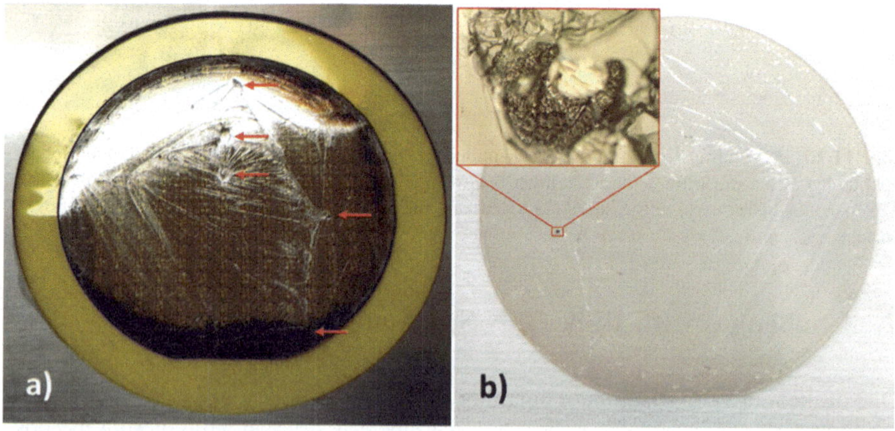

Fig. 11.17 Example of bulk GaN substrate defects resulting in poor surface quality after spalling. The transferred layer (**a**) was crack-free but fracture perturbations can be seen. The source wafer (**b**) was used to identify the origin of these fracture pinning sites. As seen in the inset of (**b**) polycrystalline inclusions in the bulk GaN wafer, they were observed at the pinning sites

8 Concluding Remarks

Layer transfer by controlled spalling was briefly reviewed with particular attention to bulk GaN. Although the general procedure of controlled spalling is simple, the requirement of a very thick Ni stressor layer due to the hardness of GaN necessitated the fabrication of custom equipment and film-handling schemes. The details of the equipment and procedures were discussed as well as some of the shortcomings and remaining issues. Despite these challenges, wafer-scale layer transfer from bulk GaN wafers can be carried out and the material quality can be preserved. If wafer reclaim is not an issue, and one seeks only the performance advantages inherent to thin film GaN, then controlled spalling is likely the most cost-effective means of achieving such a device at a large-scale.

Acknowledgments This work was supported, in part, by the ARPA-E SWITCHES program. I would also like to thank Paul Lauro, Keith Fogel, Devendra Sadana, and the IBM management team for their support.

References

1. B. Lawn, *Fracture of Brittle Solids*, 2nd edn. (Cambridge Univ. Press, New York, 1993)
2. M.D. Thouless et al., The edge cracking and spalling of brittle plates. Acta Metall. **35**, 1333 (1987)
3. Z. Suo, J.W. Hutchinson, Steady-state cracking in brittle substrates beneath adherent films. Int. J. Solids Struct. **25**, 1337 (1989)
4. J.W. Hutchinson, Z. Suo, Mixed mode cracking in layered materials. Adv. Appl. Mech. **29**, 63 (1991)
5. C.Y. Liu, S.J. Wang, Prevention of spalling by the self-formed reaction barrier layer on controlled collapse chip connections under bump metallizations. J. Electron. Mater. **32**, L1 (2003)
6. E.A. Savary, Process of chipping or ornamenting glass. US Patent 495,375 (1893)
7. M. Tanielian et al., A new technique of forming thin free-standing single-crystal films. J. Electrochem. Soc. **132**, 507 (1985)
8. F. Dross et al., Stress-induced large-area lift-off of crystalline Si films. Appl. Phys. A Mater. Sci. Process. **89**, 149 (2007)
9. S.W. Bedell et al., Kerf-less removal of Si, Ge, and III–V layers by controlled spalling to enable low-cost PV technologies. IEEE J. Photovoltaics **2**, 141 (2012)
10. S.W. Bedell et al., Layer transfer by controlled spalling. J. Phys. D. Appl. Phys. **46**, 152002 (2013)
11. Bedell et al., Edge-exclusion spalling method for improving substrate reusability. US Patent 8,748,296 B2 (2014)
12. J. Chen, C.E. Packard, Controlled spalling-based mechanical substrate exfoliation for III-V solar cells: A review. Sol. Energy Mater. Sol. Cells **225**, 111018 (2021)
13. S.W. Bedell et al., Vertical light-emitting diode fabrication by controlled spalling. Appl. Phys. Express **6**, 112301 (2013)
14. W.W. Gerberich et al., A crack extension force correlation for hard materials. Int. J. Fract. **148**, 109 (2007)

15. M.D. Drory et al., Hardness and fracture toughness of bulk single crystal gallium nitride. Appl. Phys. Lett. **69**, 4044 (1996)
16. Bedell et al., Layer transfer of bulk gallium nitride by controlled spalling. J. Appl. Phys. **122**, 025103 (2017)
17. W.S. Wong et al., Damage-free separation of GaN thin films from sapphire substrates. Appl. Phys. Lett. **72**, 599 (1998)
18. S.J. Pearton et al., New applications for gallium nitride. Mater. Today **5**, 24 (2002)
19. I.C. Kizilyalli et al., Reliability studies of vertical GaN devices based on bulk GaN substrates. Microelectron. Reliab. **55**, 1654 (2015)
20. D. Iida et al., Laser lift-off technique for freestanding GaN substrate using an In droplet formed by thermal decomposition of GaInN and its application to light-emitting diodes. Appl. Phys. Lett. **105**, 072101 (2014)
21. O. Moutanabbir et al., Stress adjustment and bonding of H-implanted 2 in freestanding GaN wafer: The concept of double-sided splitting. Electrochem. Solid State Lett. **12**, H105 (2009)
22. D.J. Rogers et al., Use of ZnO thin films as sacrificial templates for metal organic vapor phase epitaxy and chemical lift-off of GaN. Appl. Phys. Lett. **91**, 071120 (2007)
23. C. Youtsey et al., Epitaxial lift-off for III-nitride devices. Semiconduct. Semimet. **102**, 467 (2019)
24. D. Shahrjerdi et al., Ultralight high-efficiency flexible InGaP/(In)GaAs tandem solar cells on plastic. Adv. Energy Mater. **3**, 566 (2013)
25. D. Shahrjerdi, S.W. Bedell, Extremely flexible nanoscale ultrathin body silicon integrated circuits on plastic. Nano Lett. **13**, 315 (2013)

Part IV
Selective Area Doping, Processes, and Characterization

Chapter 12
In Situ TBCl Etching and Selective-Area Growth and Doping of GaN

Bingjun Li, Alexander S. Chang, Lincoln J. Lauhon, and Jung Han

1 Introduction

Gallium nitride (GaN) has a great potential in high-power and high-frequency applications due to its wide bandgap and good transport property. GaN-based high electron mobility transistors (HEMTs) have been deployed for microwave power amplifiers for both commercial and military applications. To take advantage of the merits of GaN' properties, more sophisticated device configurations such as current-aperture vertical electron transistors (CAVETs), junction-barrier Schottky (JBS) diodes, and superjunction (SJ) devices require the ability to form in-plane, lateral PN junctions by selective-area doping (SAD) [1–3]. In mature Si and SiC material systems, SAD is achieved through dopant diffusion or ion implantation. Toward this goal for GaN, great progress has been made with the implantation of Mg to achieve selective p-type doping. By employing ultra-high-pressure annealing (UHPA), defects have been minimized, and an activation ratio of 70% was reported. A PN junction formed through implantation followed by UHPA has shown a breakdown voltage of 900 V and a negligible leakage current [4–7].

An alternative approach to realize selective-area doping of GaN involves creating vertical trenches in n^-GaN through selective-area etching (SAE), followed by trench filling with p-GaN through selective-area regrowth. In this process, realizing a clean and damage-free selective-area etching process of GaN is considered to be the most challenging and important step. Due to the chemical inertness of c-plane GaN, wet etching of c-plane GaN to create trenches with high aspect ratio is extremely difficult. Instead, typically Cl-based plasma etching or dry etching is

B. Li (✉) · J. Han
Department of Electrical Engineering, Yale University, New Haven, CT, USA
e-mail: bingjun.li@aya.yale.edu; jung.han@yale.edu

A. S. Chang · L. J. Lauhon
Department of Material Science and Engineering, Northwestern University, Evanston, IL, USA

I. C. Kizilyalli et al. (eds.), *Gallium Nitride and Related Materials*, The Materials Research Society Series, https://doi.org/10.1007/978-3-031-83056-3_12

used to achieve anisotropic profile in GaN [8, 9]. However, dry etching process involves high-energy ions, which is known to produce damage [10, 11]. In addition, impurities, such as silicon (Si), are introduced to the as-etched surface. An imperfect etched-and-regrown interface, located near the active region of a power device, typically leads to a high ideality factor in the forward current, high reverse leakage current, and premature breakdown. Attempts have been reported to mitigate the plasma-induced damage, including thermal annealing of dry-etched surface [12], chemical treatment of as-etched surface [13], and a multistep dry etching process [14]. Using these methods, planar diodes with a low leakage current have been demonstrated. However, when it came to selective-area doping, leakage current remained much higher than expected, probably due to sidewall etch damage [15]. Therefore, it is necessary to develop an alternative etching method that introduces less damage.

Recent studies by Liu et al. and Su et al. have unveiled unexpected magnesium (Mg, p-type dopant) doping distribution during the p-GaN selective regrowth in the trench. With spatially resolved cathodoluminescence (CL), they observed a lower doping concentration near the trench sidewall compared to the center of the trench [20, 21]. To control the p-doping profile during SAD, a comprehensive understanding of the origin of the observed nonuniformity and the evolution of growth fronts during nonplanar regrowth is crucial. Nonplanar regrowth in SAD, however, is intrinsically different from the well-known selective-area growth of GaN for nanorods and epitaxial-lateral overgrowth (ELO) [16–19]. It will be discussed in a later section that the growth of nanorods and ELO are dominated by the convex growth front, while p-GaN filling in the trench is instead a result of concave growth.

In this chapter, we will discuss an alternative etching technique for GaN using a chlorine-based precursor, tertiarybutylchloride (TBCl), within an MOCVD reactor. We have conducted electrical and material characterizations on both the as-etched surfaces and etched-and-regrown interfaces. The results will be discussed to confirm that TBCl etching is an intrinsically clean and low-damage etching process. We will also delve into the limitations of TBCl etching with regard to impurity removal, which causes surface roughening. Furthermore, we will explain the cause of the nonuniform Mg doping profile during the selective-area growth of p-GaN in the trench. The growth evolution in this process will be analyzed and elucidated through the application of the kinetic Wulff diagram and Wulff construction.

2 In Situ Etching of GaN by TBCl

Vapor phase chemical etching of GaN is a promising alternative to plasma-based dry etching process which tends to create lattice damages and defects. At elevated temperatures, vapor phase chemical etching exhibits a significantly enhanced etch rate of c-plane GaN when compared to the room temperature wet etching process [22–24]. Additionally, vapor phase etching is more likely to achieve selective-area etching with minimal damages, given the absence of high-energy ions and UV photons.

Koleske et al. studied the high-temperature H_2 etching of GaN comprehensively and identified the corresponding rate-limiting factors [22]. A major limitation with GaN etching by H_2 is the formation of nonstoichiometric surface region since N atoms or molecules tend to preferentially desorb. The resulting surface suffers from nitrogen vacancies and possibly the formation of Ga droplets and surface roughening in the more severe case [23]. In order to facilitate the removal of excess Ga by desorption, Fahle et al. tested the use of HCl in an MOCVD reactor [24]. A considerable etch rate at a temperature of as low as 625 °C was reported. However, HCl is known to be highly corrosive and is not considered compatible with MOCVD hardware. Identifying a vapor phase etchant compatible with standard MOCVD equipment is crucial and has the potential to expand opportunities for in situ etching of GaN.

Tertiarybutly-chloride (TBCl) is a metalorganic precursor that has been used in MOCVD or chemical beam epitaxy (CBE) for in situ etching and cleaning of InP and GaAs [25–29]. In contrast to HCl, TBCl does not decompose to form Cl or HCl at room temperature and is reported to be compatible with MOCVD tools [27]. To assess the viability of TBCl for etching GaN, we introduced this precursor into our MOCVD system and performed in situ TBCl etching. In the following section, we will discuss the etching chemistry of applying TBCl to GaN, along with material and electrical characterizations of the as-etched GaN.

2.1 Chemistry of TBCl Etching of GaN

TBCl etching of GaN was investigated in a wide range of temperature, TBCl flow rate, and NH_3 flow rate [30]. The etch rate under each condition was measured by in situ reflectometry. Figure 12.1 shows an example of in situ reflectance trace collected during MOCVD growth and etching of GaN, in which we colored the background into different segments for growth (gray sections) and etching (white sections) under a constant reactor pressure of 200 mbar with 2 slm of NH_3. To prevent any etching-induced surface roughness from hindering the accuracy of the measurement, we adopted a cyclic etching-growth process in the experiment. For example, we chose to start each etching experiment after the reflectance curve passed the valley in Fig. 12.1. The curve traced itself in the *opposite* direction as the film thickness decreased during etching cycles, thus providing the initial evidence of in situ etching. We let TBCl etch approximately 60 nm of GaN ($>$ 1/4 λ, wavelength in GaN) so clear troughs and peaks can be recorded, and stopped flowing TBCl after the reflectance passes the peak. The etch rate was determined based on the duration of a half period. Following the etching, more than 100 nm of high-temperature GaN (~1000 °C) was grown to recover/smoothen the surface in most of the cases.

Figure 12.2a summarizes the measured etch rates as a function of temperatures and TBCl flow rates. The etch rates clearly follow a linear pattern at a constant

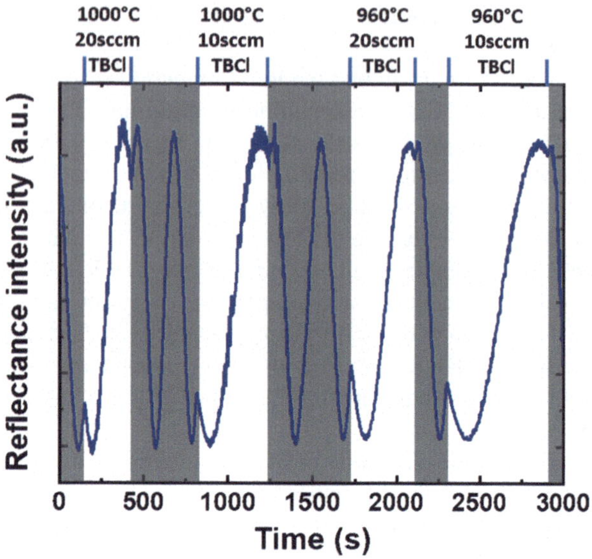

Fig. 12.1 In situ optical reflectance curve to measure the growth (gray region) and etch (white region) rate of GaN grown on a sapphire substrate. The reactor pressure and NH_3 flow rate were kept at 200 mbar and 2 slm, respectively. (Reprinted from ref. [30], with the permission of AIP Publishing)

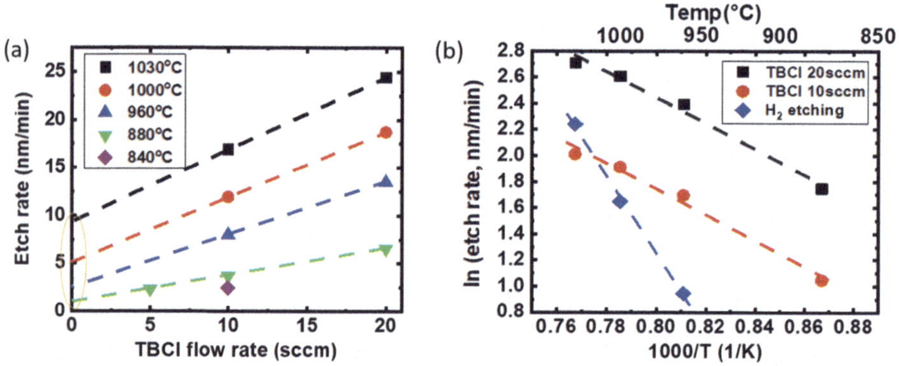

Fig. 12.2 (**a**) Measured planar etch rates of GaN vs TBCl flow rates, The reactor pressure was at 200 mbar, the flow rates of NH_3 and H_2 were held constant at 2.0 and 5.5 slm, respectively; (**b**) Arrhenius plot of etch rate of GaN vs temperature, and the activation energy of H_2 and TBCl etching are 2.57 eV and 0.85 eV, respectively. (Reprinted from ref. [30], with the permission of AIP Publishing)

temperature, indicating the effect of adding TBCl. This linear dependence has also been reported for TBCl etching of GaAs and InP [27, 29]. Extrapolation of the straight lines to the y-axis (zero TBCl flow) gives a measurement of the background etching or desorption rates from H_2-related etching of GaN (circled by a yellow solid oval curve). These measured values agree with a separate H_2-etching experiment within the same temperature range [22]. Assuming that H_2-etching and TBCl etching takes place concurrently yet independently, the net etch rate from TBCl was extracted based on the difference between the total-etch rate measured by reflectance

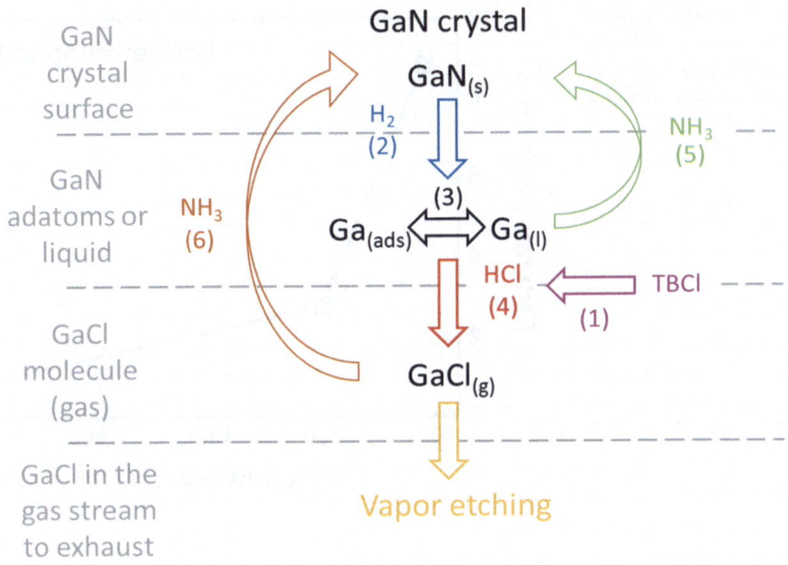

Fig. 12.3 Schematic drawing of the reactions involved during TBCl etching of GaN

and the offsets at zero TBCl flow. The rate of two etching mechanisms in the form of the Arrhenius plot is shown in Fig. 12.2b. An activation energy of 2.57 eV was extracted for H_2 etching, indicating that the decomposition rate was limited by N_2 formation and desorption [31]. In contrast, the activation energy derived from TBCl etching, over the temperature range of 880 to 1030 °C, is estimated to be between 0.8 and 0.9 eV for the two different TBCl flow rates (10 and 20 sccm), indicative of a different etching mechanism from H_2 etching. Such a reduced activation energy associated with TBCl etching also suggests that it is possible to achieve a reasonable etch rate by TBCl at low temperatures, which is often desirable for controllability.

To visualize the reactions involved in the TBCl etching of GaN, Fig. 12.3 schematically shows both forward and backward reactions at or near GaN surfaces. Detail of each chemical reaction is listed in (1)–(6) below. They were summarized based on the literature reports regarding GaN decomposition in H_2 environments [22], HCl etching of GaN [24, 36], and TBCl etching of conventional III-V materials (GaAs and InP) [27, 32–35].

As shown in Fig. 12.3, reactions (12.1)–(12.4) drive the etching of GaN forward. In the gas phase, TBCl fully pyrolyzes above a temperature of 550 °C, and the Cl-radicals should bind with H (supplied from the co-flowed NH_3) to form HCl in gas phase, following Reaction 12.1. In Reaction 12.2, GaN decomposes into Ga and N-related species in H_2 environment. The exact form of N molecule depends strongly on the etching condition, such as temperature and pressure [22]. Ga atoms, having a low vapor pressure, tend to aggregate and form Ga droplets on the surface in the absence of NH_3 and/or HCl, as indicated by Reaction 12.3. In the presence of HCl, Ga adatoms are converted to GaCl immediately and desorb

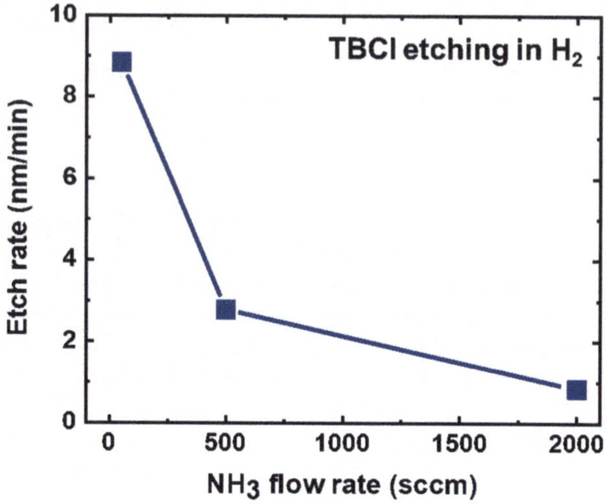

Fig. 12.4 Measured planar etch rate of GaN vs. NH_3 flow rate using hydrogen carrier gas (5.5slm) at 800 °C and 200 mbar with 5 sccm of TBCl. (Reprinted from ref. [30], with the permission of AIP Publishing)

afterward (Reaction 12.4) [24]. According to the reports of Rebey et al. and Jani et al., the activation energy of the GaCl-desorption process is 0.8–0.9 eV [32, 35]. This value correlates well with the one extracted from Fig. 12.2b, indicating that the rate-limiting factor of TBCl etching under a high NH_3 flow rate of 2 slm is the desorption of GaCl in Reaction 12.4.

$$\mathrm{TBCl} + \mathrm{H} \rightarrow (\mathrm{CH_3})_3\mathrm{C} + \mathrm{HCl}_{(g)} \tag{12.1}$$

$$\mathrm{GaN} + \mathrm{H}_{2(g)} \rightarrow \mathrm{Ga}_{(ads)} + \mathrm{N} - \text{related molecule}_{(g)} \tag{12.2}$$

$$\mathrm{Ga}_{(ads)} \rightarrow \mathrm{Ga}_{(l)} \tag{12.3}$$

$$2\mathrm{Ga}_{(ads)} + 2\mathrm{HCl}_{(g)} \rightarrow 2\mathrm{GaCl}_{(g)} + \mathrm{H}_{2(g)} \tag{12.4}$$

$$\mathrm{Ga}_{(ads)} + \mathrm{NH}_{3(g)} \rightarrow \mathrm{GaN} + \frac{3}{2}\mathrm{H}_{2(g)} \tag{12.5}$$

$$\mathrm{GaCl}_{(g)} + \mathrm{NH}_{3(g)} \rightarrow \mathrm{GaN} + \mathrm{HCl}_{(g)} + \mathrm{H}_{2(g)} \tag{12.6}$$

Besides the TBCl flow rate and reactor temperature, the etching process depends sensitively on the NH_3 flow rate. The role of NH_3 in the etching process is indicated by Reactions 12.5 and 12.6, as shown in Fig. 12.3. In Reaction 12.5, NH_3 converts Ga adatoms back into GaN. In addition, NH_3 also reacts with the etching product, GaCl, to form GaN (Reaction 12.6), which is the reaction that happens during the hydride vapor phase epitaxy (HVPE) growth of GaN [37]. An increase in the partial pressure of NH_3 will drive the reactions (Fig. 12.3) upward or backward in terms of etching, thus greatly inhibiting the in situ etching.

Recognizing as the NH_3 flow rate was decreased from the typical 2000 sccm to around 50 sccm, a 10 × increase in the etch rate was observed with rates as high as 0.5 μm/hr. at 800 °C. Figure 12.4 shows the measured etch rate versus the flow rate of NH_3 at a relatively low temperature of 800 °C under a constant TBCl flow rate (5 sccm), and pressure (200 mbar). More importantly, a decrease in NH_3 partial pressure is likely to facilitate the desorption of GaCl and drive Reaction 12.4 forward, which has been reported in the TBCl etching of InP [29]. It is important to mention that after decreasing the NH_3 flow rate to 50 sccm, the etching reaction transitions from GaCl-desorption limited to a diffusion-limited regime. The reason is that knowing the molar flow rate of TBCl, the etch rate (~0.5 μm/hr) is comparable to the growth rate of GaN under a similar molar flow rate of TMGa.

2.2 *Morphology GaN Surface After TBCl Etching*

An optimal regrowth surface should be atomically smooth. A rough surface tends to attract impurities during the subsequent regrowth [38]. Should this roughness extend to the PN junction interface, it can result in a locally high electric field [39]. Atomic force microscopy (AFM) was employed to reveal the surface morphology of as-etched GaN samples.

During the TBCL etching, we maintained a relatively low MOCVD reactor pressure of 50 mbar and a low NH_3 flow of 14 sccm to enhance the desorption of the etching product based on Figs. 12.3 and 12.4. As will be discussed later in Sect. 4.1, this condition resulted in a smooth trench profile in masked selective-area etching. A constant flow rate of 2.5 sccm was used for TBCl. To avoid any adverse effect of ex-situ impurities to the morphology of as-etched surface, TBCl etching was performed on as-grown GaN templates on GaN substrates without breaking the chamber vacuum.

AFM images of as-etched surfaces are presented in Fig. 12.5, where the temperature was set to be at 800 °C (Fig. 12.5a) and 860 °C (Fig. 12.5b). For the sample etched at 800 °C, pronounced step bunching with a step height of 2–3 nm is observed in Fig. 12.5a, while Fig. 12.5b (at 860 °C) exhibits step-flow patterns. These two morphologies closely resemble the simulations of the thermal desorption of GaN

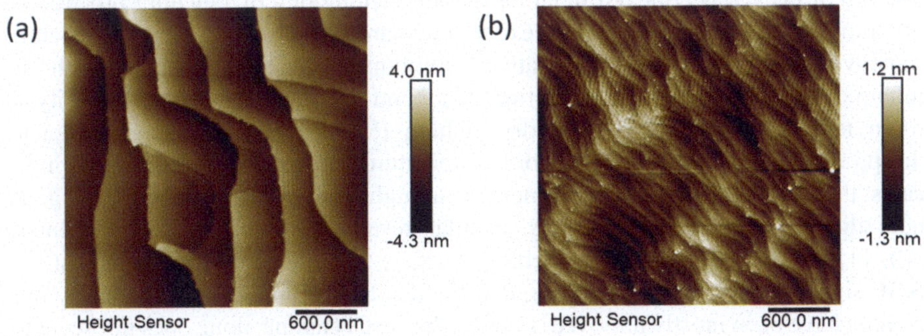

Fig. 12.5 AFM images of GaN surface etched by TBCl at (**a**) 800 °C and (**b**) 860 °C. (Reprinted with permission from ref. 47. Copyright 2021 American Chemical Society)

using Monte Carlo (MC) method [40], where the key surface mechanisms include the detachment of atoms from specific sites on step edges, diffusion of atoms on terraces, desorption of atoms into the vapor phase, and the re-incorporation of atoms to adjacent step edges. The presence or absence of step bunching in etching can be determined by the degree of "coupling" among adjacent atomic steps. When adatoms (GaCl in this case) have a diffusion length that is less than the width of the terrace, the atomic steps are considered as noninteracting (or decoupled) and each will move independently, thus maintaining a steady-state step-flow surface morphology. Conversely, when the adatom diffusion length exceeds the terrace width, the adjacent steps are coupled and step bunching becomes statistically favorable. Here, the diffusion length (x_s) is defined as $x_s = \sqrt{\frac{D}{p_d}}$, where p_d is the desorption probability in units of inverse time, and D is the diffusion coefficient. At an elevated temperature, the desorption of GaCl is enhanced. As a result, step-flow patterns were formed on the surface at 860 °C, in contrast to a morphology with step bunching at 800 °C.

2.3 *Optical, Elemental, and Electrical Characterizations of As-Etched GaN Surface*

Next, the property of the TBCl-etched surfaces was examined by both material and electrical techniques. Four GaN samples (prepared on bulk GaN substrates) were studied. Surface A has an as-grown surface by MOCVD to serve as a reference with an ideal surface property. Surfaces B and D started with MOCVD-grown GaN (like surface A) but were etched by conventional dry etching and in situ TBCl etching, respectively. Surface C was subject to dry etching (like Surface B), followed by additional TBCl etching. Unless otherwise mentioned, TBCl etching was performed at 860 °C at a pressure of 50 mbar with 2.5 sccm of TBCl and 14 sccm of NH_3.

Photoluminescence (PL) and X-ray photoelectron spectroscopy (XPS, Cl 2p) spectra of these four surfaces are presented in Fig. 12.6a, b, respectively. Here, 300 nm of GaN was removed in either the dry etching or TBCl etching process. A 325 nm HeCd laser was used as an excitation source for PL measurement, which is sensitive to the presence of nonradiative recombination centers. Compared with the reference Surface A, which has a distinct near band edge (NBE) emission intensity at room temperature in Fig. 12.6a, dry etching (Surface B) leads to a complete disappearance of NBE signal at room temperature. The missing NBE signal indicates that dry etching process generates a high density of defects, serving as nonradiative recombination centers. Similar results were reported in references [10, 11]. TBCl etching after dry etching (Surface C) leads to a full recovery of the NBE signal in Fig. 12.6a, indicating that the damage or impurities incurred by dry etching have been mostly removed. Finally, the TBCl etching alone does not seem to introduce any adverse effects, or nonradiative defects (Surface D). XPS of Cl 2p (Fig. 12.6b) shows that dry etching causes an increase in the Cl signal (red curve),

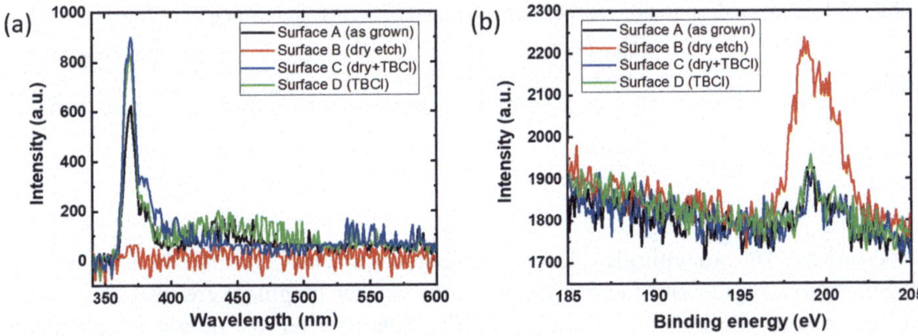

Fig. 12.6 (**a**) PL and (**b**) Cl 2p XPS spectra measured on Surface A–D. (Reprinted from ref. 41, with permission from Elsevier)

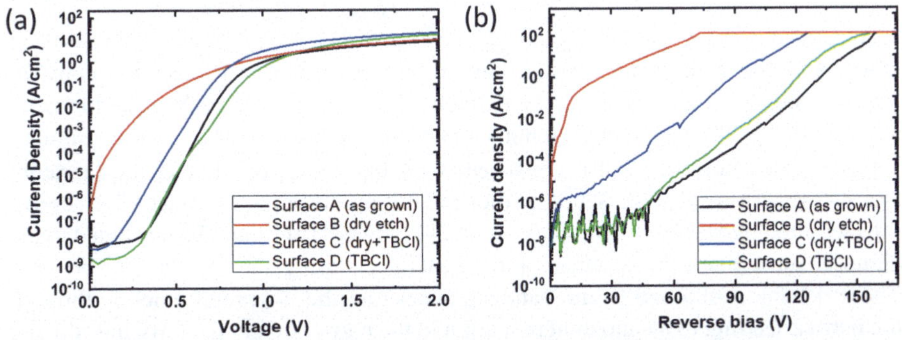

Fig. 12.7 (**a**) Forward and (**b**) reverse I-V characteristics of Schottky diodes built on Surface A–D

which can be removed by the TBCl etching (blue curve) [41]. Other than PL and XPS measurements, ion-beam channeling measurement was also conducted on these surfaces, yielding results consistent with PL measurements. Details of the ion-beam measurement and findings are elaborated in a separate chapter (Chap. 20) in this book by R.S. Goldman et al.

More importantly, we also placed Ni/Au Schottky contacts on these surfaces and investigated their electrical properties. In this study, 150 nm of GaN was removed in either the dry etching or TBCl etching process. Prior to the deposition of Schottky contacts, samples were cleaned in acid (1:1 HCl 60 sec + dilute HF 60 sec). The n-metal stacks on the bottom of substrates consisted of Ti/Al/Ni/Au (20/200/20/50 nm). No mesa isolation and special field management techniques were employed.

In Fig. 12.7a, the representative forward current-voltage (I-V) properties of Schottky diodes with an 80 μm diameter anode are presented. Dry-etched surface (Surface B) shows pronounced leakage current under low bias voltage, which is three to four orders of magnitude higher than the other three. Conversely, TBCl-etched sample (Surface D) exhibits I-V relations similar to that of the as-grown one (Surface A), with an I_{on}/I_{off} ratio of above 10^9. Notably, Surface C demonstrates

Table 12.1 Extracted saturation current density J_s and Schottky barrier height of Surface A–D

	Ideality factor (n)	Schottky barrier height (eV)
Surface A	1.07	1.11
Surface B	1.12	0.66
Surface C	1.79	0.96
Surface D	1.18	1.11

three-orders of magnitude lower leakage current compared to Surface B. Subsequently, ideality factors and Schottky barrier heights were also calculated, and the results are listed in Table 12.1. The Schottky barrier height is calculated based on the following equation [42].

$$\phi_B = \frac{kT}{q} \ln\left(\frac{A^*T^2}{J_s}\right) \tag{12.7}$$

where k is the Boltzmann constant and A* is the effective Richardson constant (26.64 A/cm^{2}K^2) [42, 43]. J_s can be extracted by fitting forward I-V behavior using $J = J_s \exp\left(\frac{qV}{nkT}\right)$. The extracted Schottky barriers for Surface A and D are similar, 1.11 eV. This number is also consistent with the value reported in the literature [42, 44, 45]. However, for Schottky diodes built on dry-etched surface (Surface B), the measured barrier height, 0.66 eV, is notably lower. This could be attributed to the damage, most likely N vacancies [46], generated during the dry etching process. These defects positioned in the bandgap accelerate the hopping/tunneling through the barrier, leading to an apparent reduction in the barrier height [42]. By introducing additional TBCl etching, Surface C exhibits a substantial recovery in the barrier height, reaching 0.96 eV. This suggests less defects near the surface compared to Surface B.

Leakage currents under reverse bias were also measured and are shown in Fig. 12.7b. As expected, dry-etched sample (Surface B) shows a high leakage due to the damage near the surface. Consistent with the findings in the forward bias, Surface C has a more than 1000× reduction in the leakage current compared to Surface B. More importantly, the TBCl-etched sample (Surface D) exhibits a similar current blocking capability as that of as-grown one (Surface A).

In conclusion, multi-modal characterizations (PL, XPS, ion-beam channeling, and metal-GaN diodes) produce a consistent picture that the in situ TBCl etching is a clean and low-damage etching process and it is capable of removing pre-existing damage from prior dry etching.

3 Characterization of Etched-and-Regrown Diodes

In the previous section, we demonstrated a controlled and low-damage in situ etching process for GaN by TBCl. The etching process involves Cl-chemistry at elevated temperatures. Effective Cl-chemistry to etch GaN circumvents the need of using energetic ions and photons and essentially eliminates the undesirable effects of etch damage. However, as it turns out, the Cl-chemistry also imposes limitations to the removal of elements and impurities whose nitride bond strengths exceed GaN, specifically in the presence of impurities (Si and Al) on the surface of GaN [47]. In this section, we will first discuss the origin of impurities and their effects on the morphology of TBCl-etched surfaces. This investigation will guide the design of planar etched-and-regrown diodes in the latter part of this section and assist in interpreting the results.

3.1 1The Origin of Impurities (Si and Al) and Their Effects in Mediating TBCl Etching

Both Si and Al are known to be elements with very high earth abundance. Observations of Si contaminants and other impurities on the surface of GaN have been reported by numerous investigators [13, 14, 48–50]. In particular, the source of Si was found to be ubiquitous. As the presence of Si impurities can interfere with the chemical etching of GaN by TBCl, we have conducted a series of experiments of preparing GaN surfaces under different conditions (with different doses of silicon contaminants), followed by performing regrowth of GaN on the exposed surface and characterizing the two-dimensional (2D) density of Si contaminants. Figure 12.8 summarizes the measured (2D) density of Si impurity positioned at the regrowth interface subject to a variety of processes. Here, the 2D density (cm^{-2}) was obtained by integrating the secondary ion mass spectrometry (SIMS) profile of Si (cm^{-3})

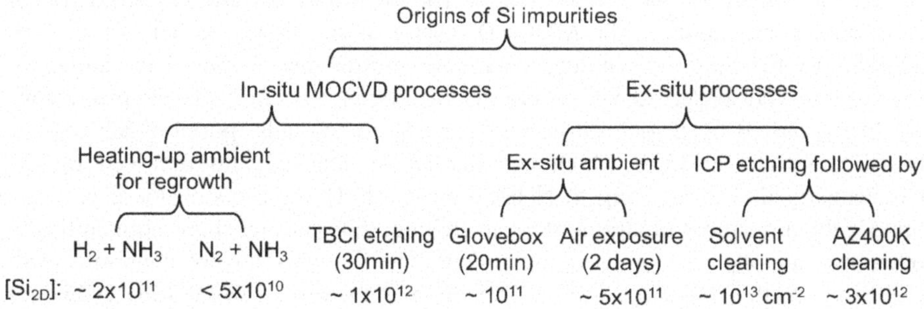

Fig. 12.8 Summary of 2D density of Si impurities ([Si_{2D}] (cm^{-2})) at the regrowth interface introduced through different processes. (Reprinted with permission from ref. [47]. Copyright 2021 American Chemical Society)

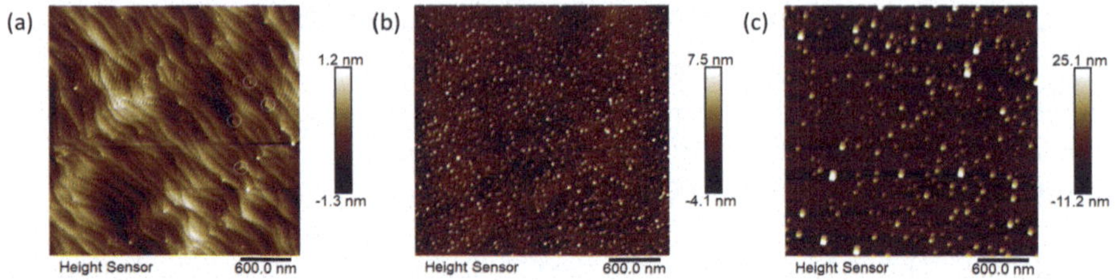

Fig. 12.9 AFM images of TBCl-etched surface (**a**) without intentional Si doping and the density of Si impurities is ~1×10^{12} cm^{-2} from parasitic etching of reactor parts; (**b**) with a much higher density of Si ~2×10^{13} cm^{-2} to illustrate the mediating effect of Si in TBCl etching. Reprinted with permission from ref. [47]. Copyright 2021 American Chemical Society; and (**c**) with Si ~1.1×10^{13}/cm^2 and Al on the surface, introduced from dry etching. Reprinted from ref. [41], with permission from Elsevier

across the interface. On one hand, in the course of analyzing in situ processes within the MOCVD reactor (Aixtron 200/4 HT), it was observed that silicon (Si) tends to accumulate on the GaN surface as the temperature ramps up in a hydrogen (H_2) ambient before regrowth. However, switching to N_2 as the carrier gas can mitigate this problem to some extent. The source of Si is traced back to the Si containing components in the reactor, such as the quartz liner, quartz susceptor holder, and SiC-coated susceptor. Additionally, during the TBCl etching, Si further accumulates on the GaN surface due to the parasitic etching of Si-containing parts and remains resistant to the subsequent TBCl etching. A quantitative analysis of the limitations of TBCl in removing Si could be found in [47]. On the other hand, ex-situ sources of Si include exposure to glovebox, to cleanroom environment, and most seriously, after the dry etching. The use of wet chemical treatment, especially AZ400K, can reduce the silicon contamination introduced by dry etching by more than 60%.

A direct consequence of surface Si during the TBCl etching is the interference of etching kinetics and morphology. Figure 12.9a, b shows two AFM images of GaN surfaces (grown on bulk GaN substrates) after 30 min of TBCl etching with different amount of residual Si impurity. Figure 12.9a is replicated from Fig. 12.5b, with a surface Si density of ~1×10^{12} cm^{-2}. The Si originated entirely from the Si containing components in the MOCVD reactor as mentioned earlier [51, 52]. In addition to the step-flow mode, nanoscale protrusions (white dots circled in Fig. 12.9a) were observed with a density of around 10^9 cm^{-2}. These protrusions are likely caused by Si-rich clusters or particle compounds (possibly SiN) in the N-rich environment [53], which can hardly be attacked and removed by the Cl-chemistry due to a strong Si-N bond energy [54]. As a result, these particles impede the movement of atomic steps in etching, thus causing sharp pointed tips in otherwise smooth contours of atomic steps. Our findings do not contradict with earlier reports that TBCl was used to remove Si on the surface of InP [28], as Si does not form strong chemical bond with P [54]. In addition, the sample with a much higher surface density of Si (2×10^{13} cm^{-2}) is presented in Fig. 12.9b, showing a completely different morphology after etching. The surface is dominated by a high density of protrusions and is significantly roughened. It is worth mentioning that fluorine (F)-based vapor phase etching is a possible solution to remove Si contaminant on the GaN surface, knowing that Si forms a much stronger bond with fluorine than nitrogen does [54].

Besides silicon, we have studied the effect of aluminum (Al) as an impurity of GaN. Al is introduced to the GaN surface unintentionally during the plasma dry etching process, as many dry etching tools are constructed using aluminum parts [55]. Similar to Si, Al forms a very strong chemical bond with nitrogen [24], which cannot be easily attacked by the TBCl chemistry. The consequence of the abundant Si (1×10^{13} cm^{-2}) and Al from dry etching is that subsequent TBCl etching induces an even rougher surface consisting of 3D islands, as shown in Fig. 12.9c. During the subsequent regrowth, surface roughening and micro-faceting inevitably lead to an enhanced incorporation of impurities [38].

3.2 Electrical Characterization of Etched-and-Regrown Interfaces

With the knowledge of the sources and roles of impurities, four PN diodes, designed to ascertain the intrinsic merit of TBCl etching, were grown and fabricated. Figure 12.10 shows the key processes in preparing the etched-and-regrown diodes (PND B, C, and D), with a continuous-grown diode (PND A) as a reference. When finished, these four diodes have nominally the same p-n diode design, having approximately 15 nm of p^+-GaN and 0.4 µm of p-GaN on 1.5 µm of UID-GaN, which was prepared on a n^+-GaN layer on GaN substrates. For all the etched-and-

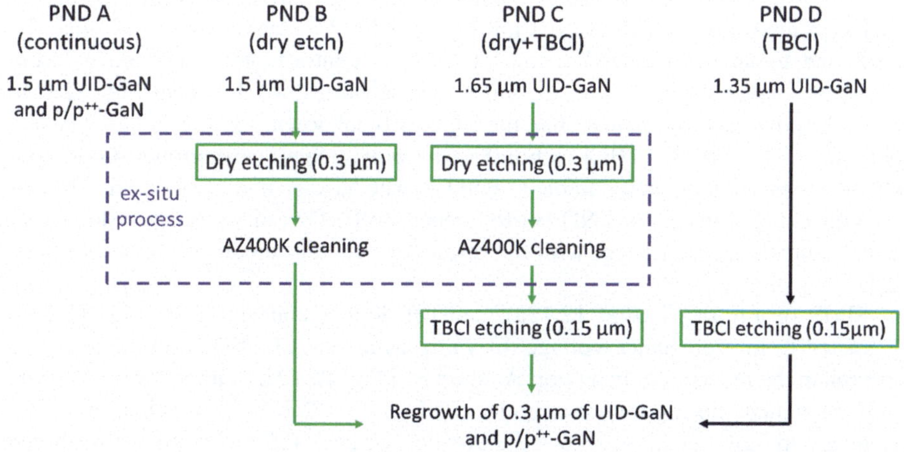

Fig. 12.10 Detailed process flows on PND A–D. Processes that introduce Si are indicated by either green arrows or outlines, and ex-situ processes are enclosed by purple dashed lines. (Reprinted with permission from ref. [47]. Copyright 2021 American Chemical Society)

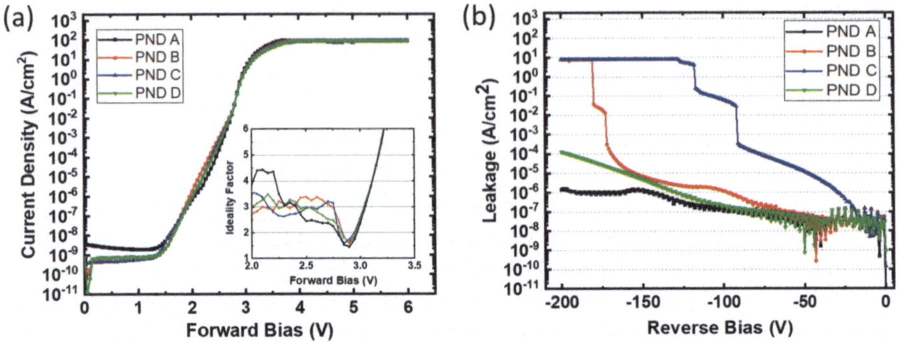

Fig. 12.11 (**a**) Forward I-V characteristics, with calculated ideality factor (inset); (**b**) reverse bias I-V characteristics of PND A–D (see Fig. 12.10 for a description of the processes for these diodes). (Reprinted with permission from ref. [47]. Copyright 2021 American Chemical Society)

regrown devices (PND B, C, and D), a 0.3 μm of UID GaN layer was first grown on the etched surface before the growth of the p-layers such that the quality of the regrowth interface will not be masked by the p-type doping at the p-n junction. It is worth mentioning that TBCl-etched sample (PND D) was prepared entirely in the MOCVD reactor to minimize the presence of Si impurities and to study the intrinsic properties of the TBCl etching.

Figure 12.11a, b shows the representative I-V characteristics under forward and reverse bias, respectively. The diameters of the measured devices were 400 μm. All four devices exhibit similar forward performance with a differential specific on-resistance $R_{on,sp} \sim 0.3 m\Omega \cdot cm^2$, and the ideality factors (inset of Fig. 12.11a) for all of them are in the range of 1.5–1.7. The similarity in forward bias characteristics reflects the fact that the PN junction is hardly affected by the quality of regrowth interfaces, which are 0.3 μm underneath. An ideality factor > 1 may be attributed to an unoptimized p-contact [56]. In contrast, great differences were observed among the four devices under reverse biases, as shown in Fig. 12.11b. PND A (continuously grown) has the highest breakdown and the lowest leakage current, while PND D (TBCl etching) is the only etched-and-regrown diode with characteristics that are comparable to PND A. The leakage and breakdown behavior are noticeably worse from PND B (dry etched) to PND C (dry etching followed by TBCl), which is inconsistent with the results of Schottky diodes presented in Sect. 2.3.

SIMS profile of Si (Fig. 12.12a), and net doping concentration (Fig. 12.12b) extracted from capacitance-voltage (C-V) measurement of PND A–D are analyzed to explain the reverse I-V behavior. As shown in Fig. 12.12a, Si impurity is observed in three etched-and-regrown p-n diodes (PND B–D). PND C, a combination of dry etching and TBCl etching, shows the broadest Si peak (25 nm) with an integrated density twice as high as PND B ($[Si_{2D}]$ of PND C: 3.8×10^{12} cm^{-2}). The elevated Si concentration in PND C arose from a combined effect of Si incorporation during TBCl etching and the subsequent regrowth on a rough surface. The net doping

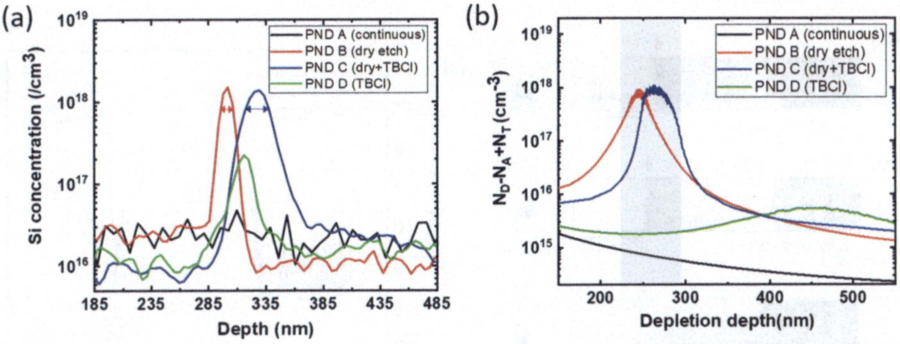

Fig. 12.12 (**a**) SIMS profile of Si near the regrowth interface and (**b**) net charge profiles calculated from C-V profiles. The light blue region indicates the designed position of regrowth interface. (Reprinted with permission from ref. [47]. Copyright 2021 American Chemical Society)

profile (Fig. 12.12b) of PND B and C shows clearly peak donor concentration of around 8×10^{17} cm^{-3} at the regrowth interface, while the full-width at half-maximum (FWHM) of PND C is larger than PND B. The broadened distribution of net donors in PND C is consistent with the SIMS result. High donor concentration at the regrowth interface typically induces field crowding near the PN junction. Consequently, it results in an elevated leakage current and premature breakdown, particularly for the PND C in Fig. 12.11b.

From SIMS measurement, the TBCl-etched PND D ([Si_{2D}]: 7.0×10^{11} cm^{-2}) has the lowest density of Si at the interface among the regrown diodes as the sample was not exposed to air. However, there are no obvious donor-related features at the regrowth interface (at around 250–300 nm) for PND D in Fig. 12.12b. The apparent disagreement between SIMS measurement of Si impurity and net doping concentration at the regrowth interface is likely due to two reasons [47]: (1) When the doping profile abruptly changes in a scale comparable to the Debye length (L_D), its variation cannot be resolved accurately [57, 58]; (2) Si shallow donors is likely to be compensated by deep acceptors. Instead, there is a broad peak with a net donor concentration below 1×10^{16} cm^{-3} at an apparent depth of around 450 nm below the PN junction, as shown in Fig. 12.12b. Further investigation reveals that this peak only exists when the C-V scan was conducted from the forward-biased condition (start from 2.7 V) to the reverse-biased condition. If PND D is allowed to "rest," followed by C-V scans from reverse bias to the forward bias, the broad peak at around 450 nm can no longer be revealed, which indicates that this broad peak is related to the hole traps [59, 60].

Moreover, regrown diodes (PND B–D) were investigated using a cross-sectional scanning spreading resistance microscope (SSRM), the result of which is presented in Fig. 12.13. Figure 12.13a–c are the measured resistance maps of PND B–D. Darker color represents lower resistance. Taking PND B as an example, the resistance map covered n-type GaN substrate, UID-GaN, regrowth interface, and regrown layers from left to right. Figure 12.13d summarizes the resistance profile

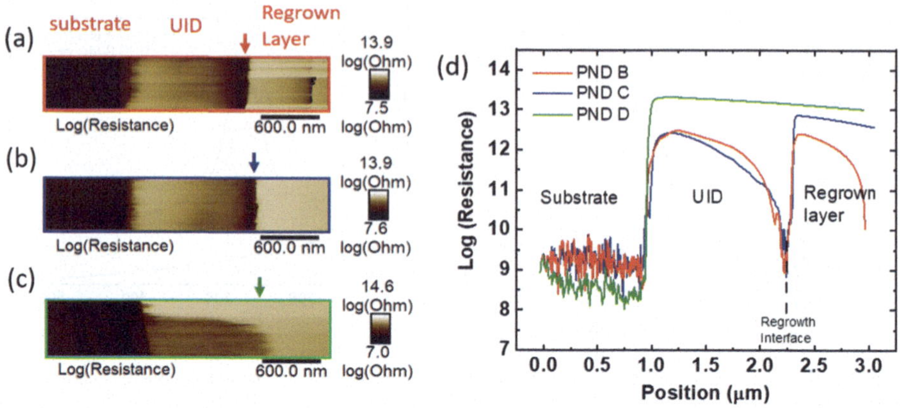

Fig. 12.13 SSRM resistance maps of sample cross-sections right after cleaving (**a**) PND B (dry etch); (**b**) PND C (dry+TBCl); and (**c**) PND D (TBCl). The arrows indicate the positions of regrowth interfaces. (**d**) Resistance profile of PND B–D. Setting for the SSRM measurement: -1 V bias with 24 nm of tip force using Bruker DDESP-V2 probe

of three samples, obtained by performing a horizontal cut across resistance maps. Electrically conductive n-type layers exist at the regrowth interface of PND B and C, while no detectable donor concentration was observed in the case of TBCl-etched diode (PND D). The result from SSRM further corroborates with the net doping profile from C-V scan (Fig. 12.12b).

In conclusion, although defects are possibly observed around the etch-and-regrowth interface, the device comparison indicates that TBCl etching is a very promising alternative to the conventional ICP etching. By minimizing pre-existing silicon impurities and avoiding ICP etching, advanced device architectures involving lateral PN junctions could be potentially achieved with high performance.

4 Selective-Area Etching and Doping of GaN with MOCVD

Up to now we have discussed the mechanisms and basic properties of in situ etching of GaN by TBCl. The studies were carried out over the entire GaN surface in the absence of any pattering or masks that are required for selective etching. With the foundational understanding of planar TBCl etching, we will proceed in Sect. 4 to summarize investigations of selective-area etching (SAE) of GaN by TBCl and selective-area doping (SAD) through regrowth.

4.1 Selective-Area Etching of GaN by TBCl

Selective-area etching process involves two main steps: (1) creating a masking layer with designed patterns and (2) etching within the exposed region to create vertical trenches. As was discussed in Sect. 3.1, TBCl etching depends sensitively on the presence of impurities. The common step of patterning dielectric masks through lithography and dry etching turns out to be a source of introducing impurities to the exposed GaN, which in turn interfered with the TBCl etching of GaN and causes deterioration of surface morphology. This subsection will explore how, through the optimization of etching conditions, one can achieve selective-area etching with a smooth morphology, even in the presence of impurities.

Selective-area etching by in situ TBCl was conducted on GaN-on-sapphire samples with patterned SiO_2 masks. Here, the SiO_2 was deposited using plasma-enhanced chemical vapor deposition (PECVD) and patterned using photolithography and F-based dry etching. Varied reactor pressures and NH_3 flow rates were explored at a reactor temperature of approximately 800 °C. Secondary electron microscope (SEM) images of trenches etched under these conditions are presented in Fig. 12.14. The upper right panel of Fig. 12.14 reveals the formation of large pyramids within the trench (~3 μm width) after 40 min etching, using 2 slm of NH_3 under a reactor pressure of 200 mbar. By successively reducing the NH_3 flow rate and the reactor pressure, the surface pyramids disappeared and a smooth morphology was achieved at a flow rate of 14 sccm of NH_3 and a reactor pressure of 50 mbar, a condition we subsequently referred to as the optimized condition. A mechanistic discussion of the morphology-dependence on the processing conditions will be

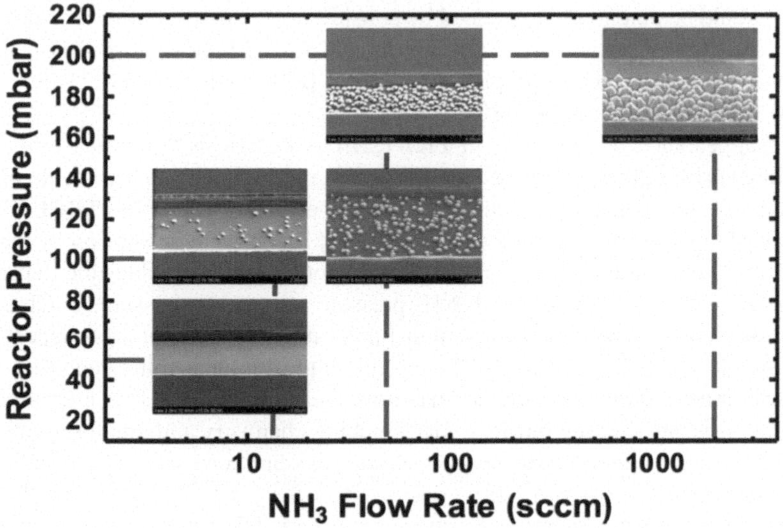

Fig. 12.14 30-Degree tilted SEM image of SAE with different NH_3 flow rate and reactor pressure at 800 °C and 2.5 sccm of TBCl. (Reprinted from ref. [30], with the permission of AIP Publishing)

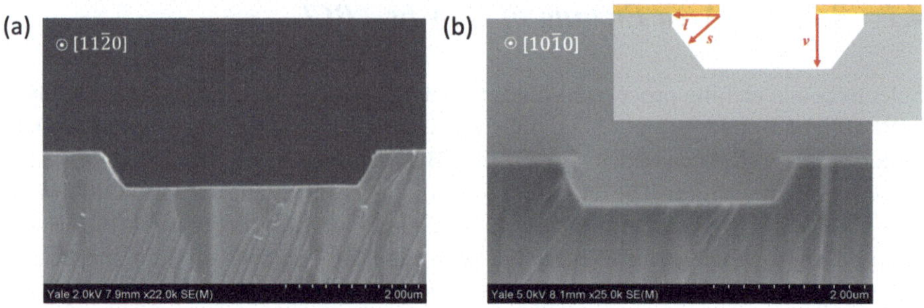

Fig. 12.15 Cross-section SEM images of stripes along (**a**) [11$\bar{2}$0] and (**b**) [10$\bar{1}$0], etched under the optimized condition. Etching length of different planes (10$\bar{1}$1) in (**a**) and (11$\bar{2}$2) in (**b**) could be determined by the method shown in the inset of (**b**) [61]. (Reprinted from ref. 30, with the permission of AIP Publishing)

given later in the section; examining the cross-sectional trench profile obtained in the optimized condition using SEM, we found that the sidewall of a- and m-axis stripes are defined by the {10$\bar{1}$1} and {11$\bar{2}$2} facets, respectively, as shown in Fig. 12.15a, b. The overhang from the edge of the SiO_2 mask gives important information regarding the crystallographic anisotropy of TBCl etch; the quantitative analysis in turns aided the interpretation of Fig. 12.14. We adopted the methods mentioned in [61] to extract the etch rates of {10$\bar{1}$1} and {11$\bar{2}$2} facets and c-plane. The etch rate can be determined by dividing the linear dimension from the edge of mask to the surface of interest (*l*, *s*, and *v* in the inset of Fig. 12.15b) by the time duration. The etch rate of {10$\bar{1}$1} facet was nearly zero, while {11$\bar{2}$2} surface exhibited an etching rate equal to one-third of the c-plane. The difference in etch rate between {10$\bar{1}$1} and {11$\bar{2}$2} could be potentially explained by considering the density of surface dangling bonds and thus the stability of surface atoms. {11$\bar{2}$2 } facet possesses a higher density of dangling bonds compared to {10$\bar{1}$1 } [62], rendering {11$\bar{2}$2} less stable and more reactive during the TBCl etching.

The appearance of pyramids in Fig. 12.14 is known as "micromasking" [25] when the surface during etching is partially covered by micro- or nanoscale objects that impede etching. In the study of in situ etching of InP by TBCl, similar micromasking was reported and attributed to the incomplete desorption of reaction products from etching ($InCl_3$) [25, 30]. This hypothesis seemed plausible in the case of GaN as under conditions of high NH_3 pressure and reactor pressure, $GaCl_{x(s)}$ may not desorb readily from this surface, thus hindering the etching process beneath and resulting in pyramids. Conversely, this morphological issue is mitigated when GaCl is rapidly removed by reducing the NH_3 flow rate and pressure. This interpretation, though consistent with the intrinsic TBCl etching chemistry and the observations in Fig. 12.14, was subsequently revised. Given the limitations of TBCl etching in removing impurities (Si and Al) as discussed in Sect. 3.1, the pyramids observed in SAE might not be an intrinsic characteristic of TBCl etching but a result of extrinsic impurities, such as from patterning the SiO_2 mask. An alternative model is provided as follows to explain the result presented in Fig. 12.14.

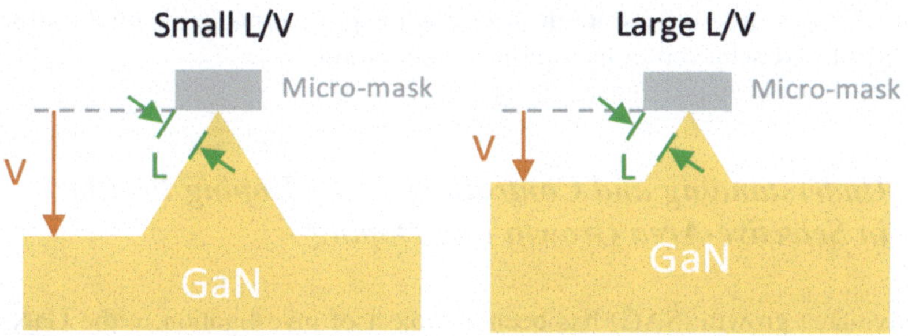

Fig. 12.16 Schematic drawings of TBCl-etched surface under a condition with small L/V ratio and large L/V ratio. In order to present different L/V ratios, the etching distance of the sidewall facet is fixed here, while the vertical depth varies

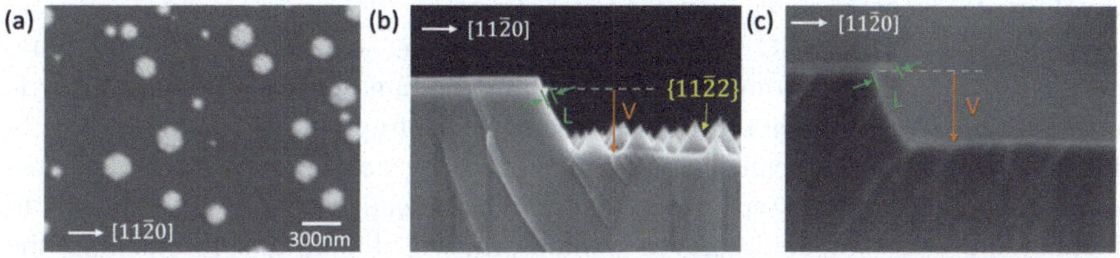

Fig. 12.17 (**a**) A zoom-in top-view SEM image of pyramids from an unoptimized condition. These hexagonal pyramids have their sidewalls consisting of $\{11\bar{2}m\}$ planes; Cross-sectional SEM images of trenches (m-direction) from (**b**) unoptimized condition; (**c**) optimized condition. Green arrows indicate the etching distance of trench sidewall ($\{11\bar{2}2\}$), and orange arrows indicate the vertical etching depth of c-plane GaN

Figure 12.16 illustrates two scenarios after etching under different conditions with the same duration (t). We presume a constant etching distance of sidewall facet (L), while the vertical etching depth (V) in the left panel is greater than the one in the right panel of Fig. 12.16. Given a constant mask size on the surface, a larger L/V ratio produces smaller pyramids (right panel of Fig. 12.16) and possibly complete absence of observable surface features. In this model, it is important to first identify the group of facets bounding these pyramids. Figure 12.17a, b presents, respectively, a top-view and a cross-sectional SEM image of trenches etched under unoptimized conditions. Figure 12.17c is the cross-section of a trench obtained under the optimized condition. Both trenches in Fig. 12.17b, c are along the m-direction. It is evident from Fig. 12.17a that pyramids are bounded by six equivalent $\{11\bar{2}m\}$ facets. The sidewalls of pyramids, as indicated by the yellow arrow in Fig. 12.17b, are in parallel with the sidewall of the trench, and can be further identified as $\{11\bar{2}2\}$. Consequently, we can determine the aforementioned L/V ratio under each condition by the etching distance of trench sidewall (green arrows in Fig. 12.17b, c) and that of c-plane GaN (orange arrows in Fig. 12.17b, c). The L/V ratio in the optimized condition (Fig. 12.17c) is found to be three times larger than the unoptimized one

(Fig. 12.17b). As a result, a smooth surface with no discernable pyramidal structures under SEM was achieved in the optimized condition.

4.2 Understanding and Controlling of 3D Doping Profile in Selective-Area Growth and Doping

Selective-area growth (SAG) has been a subject of investigation in the GaN community for three decades. The main research interest in this field was motivated by dislocation reduction through epitaxial-lateral overgrowth (ELO) and controlled preparation of nanowires, nanorods, and pyramids for optoelectronic applications [16–19]. A clear distinction between these SAG experiments is the growth evolution involving facets having a convex characteristic.

In the context of SAG of p-GaN in the trench, selective growth involving concave-shape surfaces dominates the growth evolution. Growth rate anisotropy is characterized by a different set of rules. Such an anisotropy in growth rate is directly linked to the observed spatial inhomogeneity in Mg doping during selective-area doping, as will be discussed in Sect. 4.2.2. The objective of this subsection is to identify the origin of the observed nonuniform Mg doping and to elucidate the growth evolution in nonplanar growth. These insights into both aspects will guide the design of future experiments aimed at controlling the 3D p-doping profile.

4.2.1 Origin of Nonuniform Mg Doping Profile

To provide insight into the origin of the observed nonuniformity in Mg doping during selective-area regrowth, we employed atom probe tomography (APT) which has a sub-nanometer spatial resolution and concurrently near part-per-million sensitivity in elemental analysis, thus rendering the possibility of 3D reconstruction of spatial doping profile. This study unveiled a correlation between the local Mg concentration and the local growth rate [63]. The SAG experiment for p-GaN doping was designed in a particular way (as illustrated in Fig. 12.18a): seven nominally 3 nm thick $Al_{0.06}Ga_{0.94}N$ marker layers were incorporated into the regrown p-GaN within the trench at intervals equidistant in time. The marker layers were designed for two purposes: to extract the local growth rate from the spacing between two marker layers, and to delineate the evolution of growth fronts. Following the growth of the first marker layer, Mg dopants were introduced. Subsequently, three lift-outs (Fig. 12.18b) were performed from three locations in the trench, denoted as sites 1–3 in Fig. 12.18a. As observed from Al markers at sites 1 and 2, semi-polar facets developed, becoming the dominant growth front near the edge of the trench, whereas the c-plane dominated the growth at the center of the trench (site 3). Semi-polar facets in site 2 were found to evolve >three times faster than the c-plane at the trench center (site 3). Doping profiles of Mg and extracted local growth rates in sites 2 and

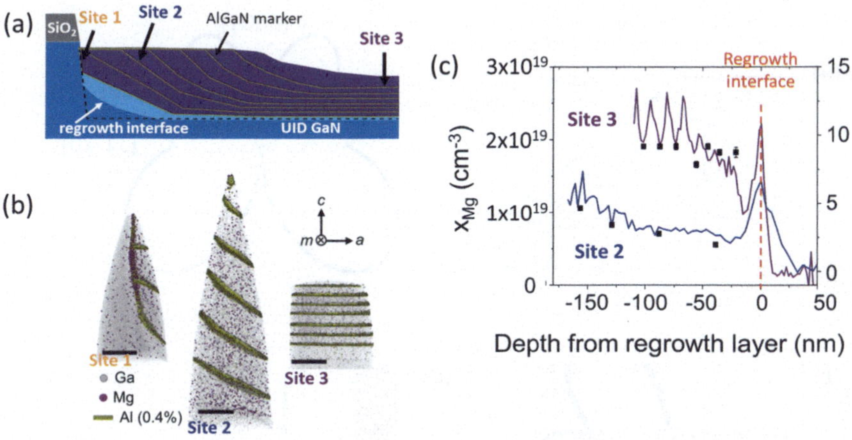

Fig. 12.18 (**a**) Schematic of designed structure for APT analysis with seven Al marker layers. Purple region is Mg-doped. (**b**) APT reconstructions of regrown p-GaN from three sites. Ga atoms are shown as gray dots, and Mg atoms are shown and purple spheres. Regions in which $x_{Al} > 0.4$ at. % are encapsulated in dark-yellow isosurfaces. Scale bars are 50 nm. (**c**) Mg concentration (lines) and inverse growth rate (black squares) versus distance from the regrowth interface for trenches along m-axis for site 2 and 3. (Reprinted with permission from ref. [63]. Copyright 2021 American Chemical Society)

3 were plotted together in Fig. 12.18c. An unambiguous inverse proportionality between the local growth rate and Mg doping concentration was observed in Fig. 12.18c. It indicates that the reduced Mg concentration observed near the trench edge is due to a locally enhanced growth rate of GaN near the trench edge. This observation provides a crucial link to the understanding and control of Mg doping through the control of local growth rate of GaN. The observed inverse proportionality also suggests that, unlike Ga atoms that diffuse freely within the trench [64, 65] through surface diffusion, Mg doping atoms hardly diffuse on the GaN surface, and tend to incorporate into GaN crystal upon adsorbing.

4.2.2 Understanding the Growth Evolution and Anisotropy in Nonplanar Selective-Area Epitaxy

As discussed in the preceding section, the lower Mg doping concentration near the sidewall of trench is a result of an enhanced growth of a semi-polar facet in comparison to the center region of the trench with primarily c-plane surface. The semi-polar facet in Fig. 12.18b (site 2) has an inclination angle of approximately 30°, which differs from the typical "stable" facets reported in the ELO of GaN [16, 17, 61]. This discrepancy is a result of different surface curvature during the epitaxy. Convex growth fronts dominate the ELO process, while concave surface governs the process of p-GaN filling in the trench. The intrinsic difference between convex and concave growths is given by the Wulff model illustrated in Fig. 12.19a [61]. It

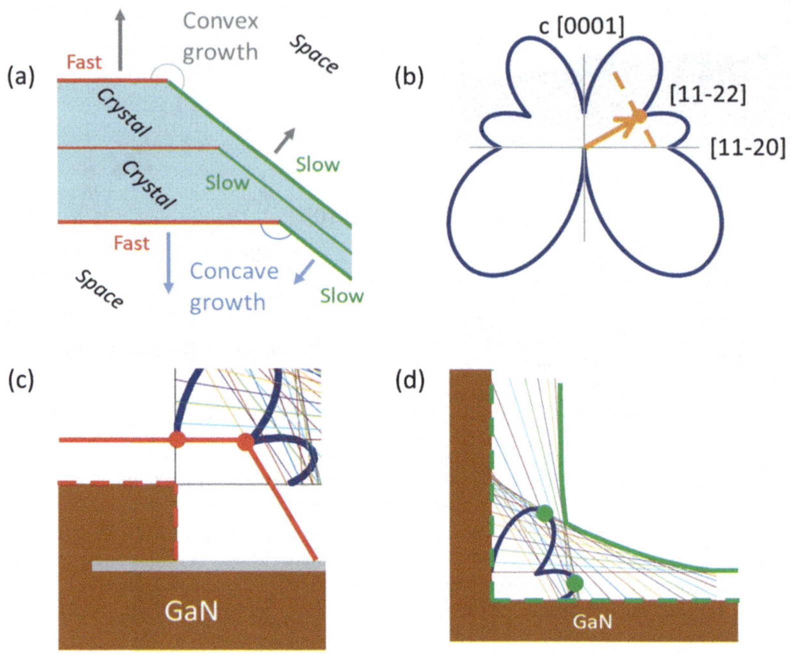

Fig. 12.19 (**a**) Schematic diagram showing the principle of morphological evolution for convex (growing outwards) and concave (growing inwards) growth fronts. The convex growth front will be dominated by the slow (green) facets while the concave growth front will be by the fast (red) facets. (**b**) A representative 2D Wulff diagram. The direction of the red arrow is [11–22]. The red dash line, which is normal to the red arrow and passing through the red point, represents the (11–22) plane. The relative growth rate of an individual plane is determined by the relative length of these arrows. (**c**) Using the Wulff construction to predict the shape in a convex growth case. The red dash line is the contour of the starting surface. The red solid line is the steady-state shape. (**d**) Using the Wulff construction to predict the shape in the concave growth. The green dash line is the contour of the starting surface. The green solid line is the steady-state shape

considers the advances of two facets with different growth rates with either a convex (growing outwards) or a concave (growing inwards) nature. In the convex growth, the rapid growing (red) facet tends to diminish, while the size of slower (green) one expands. Conversely, in the case of concave growth, the faster-growing (red) facet consumes the slower (green) one.

To achieve a more accurate prediction of crystal shape during the epitaxy, it is necessary to consider the anisotropy of growth rates across all orientations. The Wulff construction was first introduced to determine the shape of crystal under thermal equilibrium, based on a polar plot of surface energy (Wulff diagram) [66, 67]. With crystal growth and shape evolution under nonequilibrium conditions, Wulff diagram is substituted by the kinetic Wulff diagram (ν-plot). Figure 12.19b gives an example of a kinetic Wulff diagram (ν-plot) for GaN growth, projected onto the crystallographic c-a plane, with local minimum points at (0001), $(000\overline{1})$, $\{11\overline{2}2\}$, and $\{11\overline{2}0\}$ [61, 68, 69]. Every point on the ν-plot represents the relative growth rate

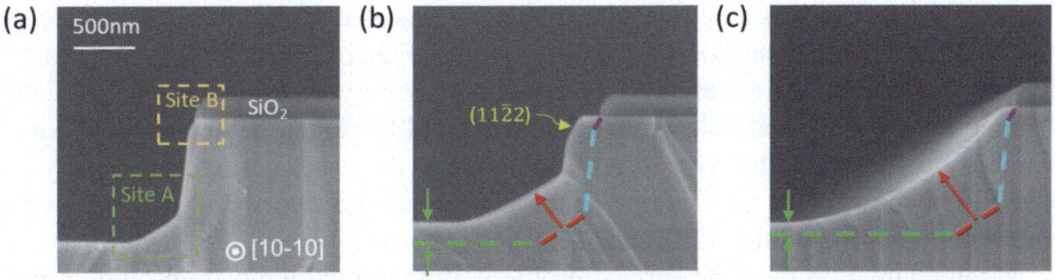

Fig. 12.20 Cross-section SEM images of (**a**) template; (**b**) growth on template under condition 1; (**c**) growth on template under condition 2. The dashed lines in (**b**) and (**c**) represent the contour of template profile approximately

of a plane (orange dash line in Fig. 12.19b) perpendicular to the radius vector (orange arrow in Fig. 12.19b) from the origin. The relative growth rate of the plane is proportional to the length of the vector.

Figure 12.19c illustrates the evolution of convex growth on a mesa stripe (bound by red dashed line) along m-direction and the procedure to predict the shape of growth front based on the ν-plot in Fig. 12.19b. The shape evolution is constrained between the initial c- and a-planes (red dash lines in Fig. 12.19c), and only a section of the ν-plot from [0001] to [11$\bar{2}$0] was considered. Subsequently, we placed the origin of the ν-plot at the intersection point of two nearby facets, and constructed normal plane in Fig. 12.19c at the end of *every* surface energy radius vector from the origin. As was stated earlier, slow growing facets dominate the convex growth. In this case, the steady-state configuration of the crystal is outlined in solid red in Fig. 12.19c, which is the inner most set of the normal segments [61, 70]. A similar procedure can be applied to concave growths as well, where fast-growing facets consume slower ones, and the outermost set of normal segments is selected (green contour in Fig. 12.19d) [70].

By comparing convex and concave growths (Fig. 12.19c, d) demonstrates that convex growth is mainly governed by facets with the growth rate of local minimum (red points in Fig. 12.19c), while concave growth is determined by the regions near the local maximum in the ν-plot (green points in Fig. 12.19d). This distinction leads to a rarely-seen semi-polar plane with a small inclination angle observed at site 2 in Fig. 12.18b. Additionally, the continuously curved surface, as shown in Fig. 12.19d (green contour), instead of a well-defined facet can be observed at the concave growth front, if the growth rate around the local maximum does not change abruptly.

Subsequently, two nonplanar growth fronts of UID-GaN were analyzed using the aforementioned model. These growth fronts were prepared on "templates" under two distinct conditions, with the flow rate of TMGa and growth time held constant. To generate templates, trenches along m-direction were initially patterned through dry etching. Subsequently, ~100 nm of GaN was deposited in the trench to bury the high-damage interface underneath and ensure a consistent starting surface condition for two SAG conditions. The cross-sectional SEM images of template and growth fronts under conditions 12.1 and 12.2 are displayed in Fig. 12.20a–c, respectively. The initial template profiles were outlined approximately as dashed lines in

Fig. 12.20b, c. Analysis is separated into two regions, site A and site B, as labeled in Fig. 12.20a. Taking condition 12.1 as an example (Fig. 12.20b), site A is dominated by concave growth. The presence of a continuously curved surface instead of well-defined facets suggested that vicinal planes around the local maximum have comparable growth rates in the kinetic Wulff diagram as discussed earlier. Conversely, at site B, convex growth governs the shape evolution, and two slow-growing facets were observed, including $(11\overline{2}2)$ and (0001).

Upon comparing the final surface profile with the initial template profile, it was observed that the two conditions resulted in distinct growth anisotropy. By comparing the length of red arrows in Fig. 12.20b, c, the low-angle surface (red) evolved much faster in condition 12.2 than in condition 12.1, given a constant TMGa molar flow rate and growth time. The ratio of growth rate between low-angle surface (red) and c-plane (green) in condition 12.1 is approximately half of the ratio in condition 12.2. As discussed in the preceding section, the nonuniform Mg doping profile in the trench is a consequence of anisotropic growth rate between mainly the low-angle surface and c-plane. The comparison between conditions 12.1 and 12.2 suggests that growth anisotropy and the consequent 3D Mg doping profile can be manipulated and engineered by tuning the MOCVD growth conditions.

4.2.3 Correlated Mapping of Doping and Conductivity

The Mg doping profiles arising from the complex growth kinetics described above have important consequences for the electrical characteristics of junctions formed by nonplanar regrowth. In this context, we note that APT provides knowledge of the chemical doping, but not the dopant activity that ultimately determines the carrier type, concentration, and equilibrium band profiles. Indeed, electron and hole concentrations may be affected by unintentional dopants and defects, as well as dopant clusters and complexes formed between Mg and defects. To optimize growth processes for optimal device performance, it is important to correlate spatial variations in composition with spatial variations in conductivity, and develop insights into the electrical nature of intentional and unintentional impurities. Toward that end, correlated APT and SSRM investigations were carried out on regrown *p*-GaN junctions in trenches defined by SiO_2 masks (Fig. 12.21a) [71]. APT analysis was carried out as described in Sect. 4.2.1, and cross-sectional samples for SSRM analysis were created by cleaving perpendicular to the trench direction. The hole-dominated and electron-dominated conductivities are shown in Fig. 12.21b, c, respectively. The dark (electron insulating) region in Fig. 12.21c corresponds to the region of hole-dominated conductivity, as seen in the corresponding bright (hole conducting) region of Fig. 12.21b. We note that the brightest regions in each SSRM map occur at the regrowth interface, where the overlapping electron and hole conductivity is suggestive of defect-mediated tunneling current associated with leakage pathways [72].

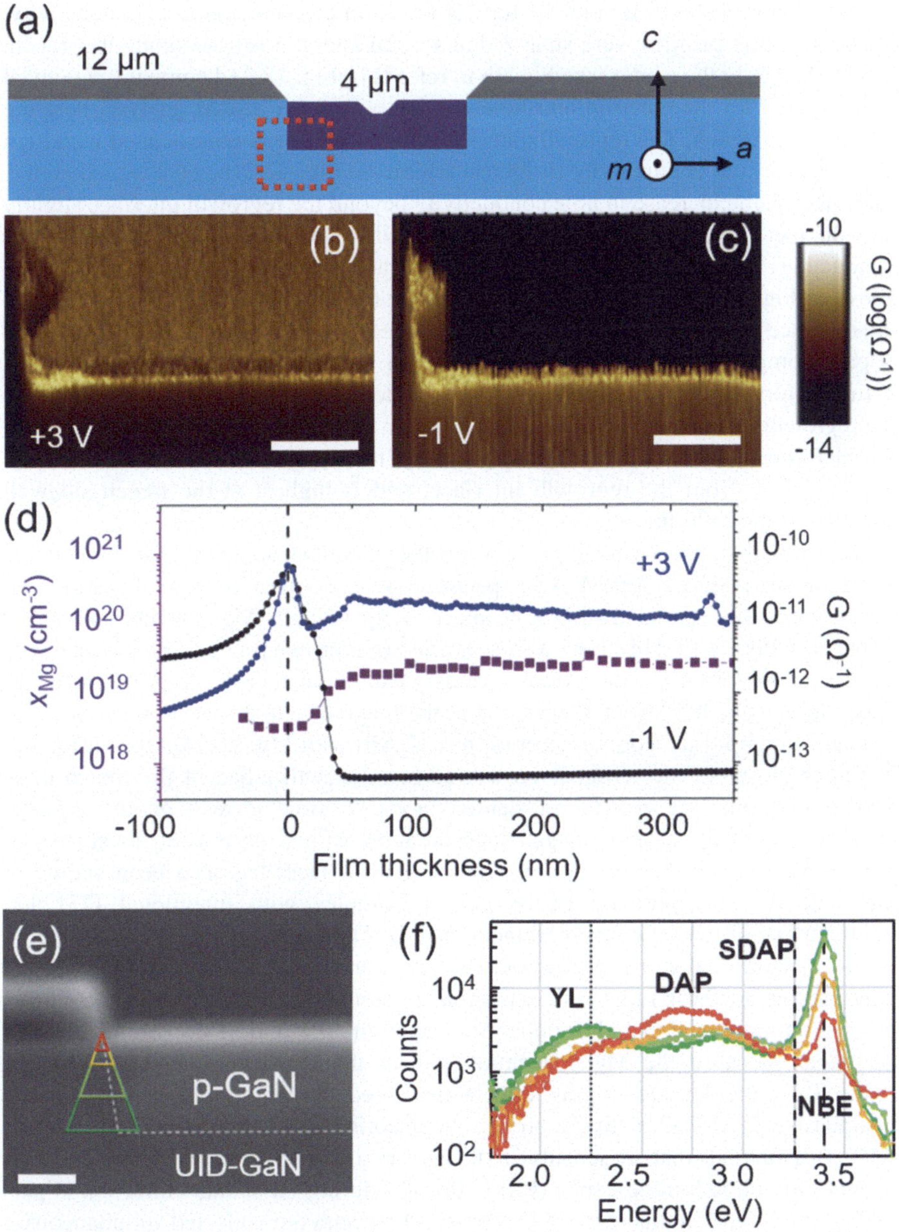

Fig. 12.21 (**a**) Cross-section schematic of p-GaN (purple) grown in an n-GaN trench (blue) with an oxide mask (gray). The dashed red line indicates the area of SSRM analysis. (**b, c**) SSRM maps of (**b**) hole and (**c**) electron conductivity taken at +3 V and − 1 V sample bias, respectively. Scale bars are 200 nm. (**d**) Plots of hole (blue) and electron (black) conductivity along the growth direction and through the regrowth interface defined as zero thickness. The APT-measured Mg concentration is plotted in purple. (**e**) Cross-sectional SEM of the trench edge. The red, orange, light green, and green regions indicate the approximate interaction volumes in CL measurements at primary beam energies of 4, 6, 9, and 12 keV, respectively. Scale bar is 250 nm. (**f**) Depth-dependent CL spectra acquired in plan-view. YL yellow luminescence; DAP Mg-O donor-acceptor pair; SDAP Mg-Si shallow donor-acceptor pair; NBE near band-edge emission

APT analysis sheds light on spatial variations in electron and hole conductivity. Vertical doping profiles were analyzed at several lateral positions along the trench, from the edge to the center (see Fig. 3b in ref. [71]); Fig. 12.21d compares a vertical Mg profile with the electron and hole conductivity from SSRM analysis near the center of the trench. The regrowth interface at zero thickness is associated with high electron and hole conductivity (brightest regions in Fig. 12.21b, c), as noted above. Importantly, the increase in hole conductivity beyond the regrowth interface slightly lags the increase in Mg doping; this can be readily observed in dark (reduced hole conducting) region between regrowth interface and the uniform conductivity region above in Fig. 12.21b. In other words, the decrease in hole conduction approaching the interface from above arises not just from a decreasing Mg concentration, but also dopant compensation. This claim is based on: (1) the lack of evidence for Mg complex formation in the APT analysis; (2) detection of Si in SIMS analysis of the regrowth interface [47]; and (3) the detection of oxygen at the regrowth interface by APT (Figs. 5c, 7a of ref [71]). Specifically, the oxygen concentration decreases moving away from the regrowth interface, and is highest at the trench sidewall adjacent to the SiO_2 mask.

To strengthen the correlation between the chemical analysis of APT and the electrical analysis of SSRM, CL spectroscopy was used to provide additional insights into the spatial distribution and activity of both Mg dopants and donor impurities. Figure 12.21e shows a cross-section of regrown p-GaN that was analyzed by in plan-view by CL with different primary excitation energies to probe the depth dependence (Fig. 12.21e). CL emission in the range of 2.6–2.9 eV was attributed to oxygen, which forms a donor-acceptor pair (DAP) with Mg. The DAP contribution increases closer to the terminal growth surface; growth adjacent the trench edge transitions from fast growth on slanted facets to slow growth on the c-plane, consistent with increasing oxygen incorporation with a decreasing local growth rate. We also note that the polarity of the growth facet has also been shown to influence the incorporation of oxygen in samples with intentional [73] and unintentional inversion domain boundaries [74]. The peak at 3.25 eV was attributed to Si-Mg shallow donor-acceptor pair (SDAP) transitions, which compete with the near band-edge (NBE) excitonic emission. In summary, the CL analysis confirms that the unintentional O and Si dopants detected in APT and SIMS compensate Mg acceptors. Because the Mg concentration near the regrowth interface is further reduced by the kinetic doping effects described above, the electrical junction mapped by SSRM is relatively broad, underscoring the importance of correlating APT with methods that are sensitive to the electrical state of dopants. The integration of real-time photoluminescence (PL) collection during APT data collection is particularly powerful in this regard [75]. When PL spectra are collected simultaneously with field evaporation, that is, as regions of the sample are removed, the deconvolution of overlapping dopant-related emission peaks is facilitated.

5 Conclusion and Future Directions

This chapter discusses scientific topics toward selective-area doping of GaN with in situ TBCl etching during MOCVD operation. The TBCl etching proceeds in a way similar to the HCl etching through a gas-phase exchange reaction. The etching depends sensitively on the ammonia partial pressure and can be carried out over a wide temperature range. Through both electrical and material characterizations, it has been demonstrated that TBCl etching is a low-damage and clean process, leading to close-to-ideal as-etched surface and etched-and-regrown interface. However, the presence of impurities, especially Si and Al which form strong bonding with nitrogen, can interfere and at times adversely affect the surface morphology after TBCl etching. The possible sources of silicon were identified and quantified. The 3D doping profiling with selective-area growth and doping was examined using atom probe tomography (APT) with a high spatial resolution, unveiling an inverse proportionality between Mg doping concentration and local growth rate. Kinetic Wulff diagram and Wulff construction were introduced to elucidate the growth evolution, offering guidance for controlling the 3D doping profile in p-GaN SAG for future applications.

Going forward, it is important to improve the impurity control in the lithographic patterning processing and perhaps in employing more potent in situ chemistry (than the Cl-chemistry), such as using fluorine-based vapor phase etching, to remove pre-existing Si impurity. Microfabrication with Si-free masking layers will be an important direction to enable TBCl etching. Patterning of sputtered AlN masks through liftoff is a potential approach to reduce impurities in the opening area prior to the TBCl etching. Employing AlN mask also minimizes unintentional doping during subsequent selective-area growth.

Acknowledgments This work is supported by the Advanced Research Projects Agency-Energy (ARPA-E), U.S. Department of Energy, under Award Number DE-AR0000871 as part of the PNDIODES program managed by Dr. Isik Kizilyalli. We acknowledge fruitful collaborations with Professors Rachel Goldman (U Michigan), Christian Wetzel (RPI), Rebecca Anthony (Michigan State U) and their group members whose names appear in related references. Effort from Dr. Sizhen Wang (Yale) is also acknowledged.

References

1. I. Ben-Yaacov, Y.-K. Seck, U.K. Mishra, S.P. DenBaars, AlGaN/GaN current aperture vertical electron transistors with regrown channels. J. Appl. Phys. **95**(4), 2073–2078 (2004)
2. S. Chowdhury, B.L. Swenson, U.K. Mishra, Enhancement and depletion mode AlGaN/GaN CAVET with mg-ion-implanted GaN as current blocking layer. IEEE Electron Device Lett. **29**(6), 543–545 (2008)
3. F. Udrea, G. Deboy, T. Fujihira, Superjunction power devices, history, development, and future prospects. IEEE Trans. Electron Devices **64**(3), 713–727 (2017)

4. J.D. Greenlee, B.N. Feigelson, T.J. Anderson, J.K. Hite, K.D. Hobart, F.J. Kub, Symmetric multicycle rapid thermal annealing: Enhanced activation of implanted dopants in GaN. ECS J. Solid State Sci. Technol. **4**(9), 382–386 (2015)
5. A. Uedono et al., Annealing behavior of vacancy-type defects in Mg- and H-implanted GaN studied using monoenergetic positron beams. Phys. Status Solidi B **256**(10), 1900104 (2019)
6. H. Sakurai et al., Highly effective activation of mg-implanted p-type GaN by ultra-high-pressure annealing. Appl. Phys. Lett. **115**(14), 142104 (2019)
7. H. Sakurai et al., Effects of the sequential implantation of mg and N ions into GaN for p-type doping. Appl. Phys. Express **14**(11), 111001 (2021)
8. R.J. Shul et al., Inductively coupled plasma etching of GaN. Appl. Phys. Lett. **69**(8), 1119–1121 (1996)
9. S.J. Pearton, R.J. Shul, F. Ren, A review of dry etching of GaN and related materials. MRS Int. J. Nitride Semicond. Res. **5**(1), e11 (2000)
10. Z. Liu et al., Investigation of effects of ion energies on both plasma-induced damage and surface morphologies and optimization of high-temperature Cl_2 plasma etching of GaN. Jpn. J. Appl. Phys. **56**(2), 026502 (2017)
11. Z. Liu et al., Suppression of plasma-induced damage on GaN etched by a Cl_2 plasma at high temperatures. Jpn. J. Appl. Phys. **54**(6S2), 06GB04 (2015)
12. S.H. Chan et al., High electron mobility recovery in AlGaN/GaN 2DEG channels regrown on etched surfaces. Semicond. Sci. Technol. **31**(6), 065008 (2016)
13. G.W. Pickrell et al., Investigation of dry-etch-induced defects in >600 V regrown, vertical, GaN, p-n diodes using deep-level optical spectroscopy. J. Appl. Phys. **126**(14), 145703 (2019)
14. K. Fu et al., Demonstration of 1.27 kV etch-then-regrow GaN p-n junctions with low leakage for GaN power electronics. IEEE Electron Device Lett. **40**(11), 1728–1731 (2019)
15. C. Yang et al., GaN vertical-channel junction field-effect transistors with regrown p-GaN by MOCVD. IEEE Trans. Electron Devices **67**(10), 3972–3977 (2020)
16. T.S. Zheleva, O.-H. Nam, M.D. Bremser, R.F. Davis, Dislocation density reduction via lateral epitaxy in selectively grown GaN structures. Appl. Phys. Lett. **71**(17), 2472–2474 (1997)
17. H. Marchand et al., Mechanisms of lateral epitaxial overgrowth of gallium nitride by metalorganic chemical vapor deposition. J. Cryst. Growth **195**(1), 328–332 (1998)
18. Y.-H. Ra, R.T. Rashid, X. Liu, S.M. Sadaf, K. Mashooq, Z. Mi, An electrically pumped surface-emitting semiconductor green laser. Sci Adv **6 (1), eaav7523** (2020)
19. S.D. Hersee, X. Sun, X. Wang, The controlled growth of GaN nanowires. Nano Lett. **6**(8), 1808–1811 (2006)
20. P.-Y. Su et al., Lateral and vertical growth of Mg-doped GaN on trench-patterned GaN films. Appl. Phys. Lett. **117**(10), 102110 (2020)
21. H. Liu et al., Non-uniform Mg distribution in GaN epilayers grown on mesa structures for applications in GaN power electronics. Appl. Phys. Lett. **114**(8), 082102 (2019)
22. D.D. Koleske, A.E. Wickenden, R.L. Henry, J.C. Culbertson, M.E. Twigg, GaN decomposition in H_2 and N_2 at MOVPE temperatures and pressures. J. Cryst. Growth **223**(4), 466–483 (2001)
23. T.H. Myers et al., Use of high temperature hydrogen annealing to remove sub-surface damage in bulk GaN. J. Cryst. Growth **246**(3), 244–251 (2002)
24. D. Fahle, T. Kruecken, M. Dauelsberg, H. Kalisch, M. Heuken, A. Vescan, In-situ decomposition and etching of AlN and GaN in the presence of HCl. J. Cryst. Growth **393**, 89–92 (2014)
25. D. Franke, N. Sabelfeld, W. Ebert, P. Harde, P. Wolfram, N. Grote, Surface quality of InP etched with tertiarybutylchloride in an MOVPE reactor. J. Cryst. Growth **248**, 421–425 (2003)
26. R. Gessner, A. Dobbinson, A. Miler, J. Rieger, E. Veuhoff, Fabrication of AlGaInAs and GaInAsP buried heterostructure lasers by in situ etching. J. Cryst. Growth **248**, 426–430 (2003)
27. M. Kondow, B. Shi, C.W. Tu, Chemical Beam etching of GaAs using a novel precursor of Tertiarybutylchloride (TBCl). Jpn. J. Appl. Phys. **38**, L617–L619 (1999)
28. K. Naniwae, M. Ohya, K. Hamamoto, K. Nishi, T. Sasaki, Cleaning of residual silicon on InP regrowth interface in MOVPE reactor, in 16th IPRM. 2004 International Conference on Indium Phosphide and Related Materials (2004)

29. P. Wolfram, W. Ebert, J. Kreissl, N. Grote, MOVPE-based in situ etching of in(GaAs)P/InP using tertiarybutylchloride. J. Cryst. Growth **221**(1), 177–182 (2000)
30. B. Li, M. Nami, S. Wang, J. Han, *In-situ* and selective area etching of GaN by Tertiarybutylchloride (TBCl). Appl. Phys. Lett. **115**, 162101 (2019)
31. I. Gorczyca, A. Svane, N.E. Christensen, Theory of point defects in GaN, AlN, and BN: Relaxation and pressure effects. Phys. Rev. B **60**(11), 8147–8157 (1999)
32. A. Rebey, A. Bchetnia, B. El Jani, Etching of GaAs by CCl_4 and VCl_4 in a metalorganic vapor-phase epitaxy reactor. J. Cryst. Growth **194**(3), 286–291 (1998)
33. S. Chaochin, X. Ming, D. Zi-Guo, M.F. Vernon, B.E. Bent, Dry etching of GaAs with Cl2: Correlation between the surface cl coverage and the etching rate at steady state. Surf. Sci. **282**(3), 357–370 (1993)
34. C. Su, Z.-G. Dai, W. Luo, D.-H. Sun, M.F. Vernon, B.E. Bent, Chemical dry etching of GaAs (100) by HCl: Products, rate, and a kinetic model. Surf. Sci. **312**(1), 181–197 (1994)
35. B. El Jani, J.-C. Grenet, M. Guittard, B. Senouci, In situ etching of GaAs using AsCl3 in MOVPE. I. J. Cryst. Growth **58**(2), 381–386 (1982)
36. Dirk Fahle, Ph.D. thesis, RWTH Aachen University (2014)
37. A. Koukitu, S.-i. Hama, T. Taki, and H. Seki, "Thermodynamic analysis of hydride vapor phase epitaxy of GaN," Jpn. J. Appl. Phys., 37, 762–765 (1998)
38. S.C. Cruz, S. Keller, T.E. Mates, U.K. Mishra, S.P. DenBaars, Crystallographic orientation dependence of dopant and impurity incorporation in GaN films grown by metalorganic chemical vapor deposition. J. Cryst. Growth **311**(15), 3817–3823 (2009)
39. I.C. Kizilyalli, P. Bui-Quang, D. Disney, H. Bhatia, O. Aktas, Reliability studies of vertical GaN devices based on bulk GaN substrates. Microelectron. Reliab. **55**(9), 1654–1661 (2015)
40. M.A. Załuska-Kotur, F. Krzyżewski, Step bunching process induced by the flow of steps at the sublimated crystal surface. J. Appl. Phys. **111**(11), 114311 (2012)
41. B. Li, S. Wang, M. Nami, J. Han, A study of *in-situ* etching of GaN in metalorganic chemical vapor deposition (MOCVD) by tertiarybutylchloride (TBCl). J. Cryst. Growth **534**, 125492 (2020)
42. K. Fu et al., Reverse leakage analysis for as-grown and regrown vertical GaN-on-GaN Schottky barrier diodes. IEEE J. Electron Devices Soc. **8**, 74–83 (2020)
43. A. Kumar, S. Arafin, M.C. Amann, R. Singh, Temperature dependence of electrical characteristics of Pt/GaN Schottky diode fabricated by UHV e-beam evaporation. Nanoscale Res. Lett. **8**(1), 481 (2013)
44. A.C. Schmitz, A.T. Ping, M.A. Khan, Q. Chen, J.W. Yang, I. Adesida, Schottky barrier properties of various metals on n-type GaN. Semicond. Sci. Technol. **11**(10), 1464–1467 (1996)
45. P. Hacke, T. Detchprohm, K. Hiramatsu, N. Sawaki, Schottky barrier on n-type GaN grown by hydride vapor phase epitaxy. Appl. Phys. Lett. **63**(19), 2676–2678 (1993)
46. H.K. Cho, F.A. Khan, I. Adesida, Z.Q. Fang, D.C. Look, Deep level characteristics in n-GaN with inductively coupled plasma damage. J. Phys. D. Appl. Phys. **41**(15), 155314 (2008)
47. B. Li, S. Wang, M. Nami, A.M. Armstrong, J. Han, Etched-and-regrown GaN P-N diodes with low-defect interfaces prepared by in-situ TBCl etching. ACS Appl. Mater. Interfaces **13**(44), 53220–53226 (2021)
48. A. Aragon et al., Interfacial impurities and their electronic signatures in high-voltage regrown nonpolar m-plane GaN vertical p-n diodes. Phys. Status Solidi A **217**(7), 1900757 (2020)
49. G. Koblmüller, R.M. Chu, A. Raman, U.K. Mishra, J.S. Speck, High-temperature molecular beam epitaxial growth of AlGaN/GaN on GaN templates with reduced interface impurity levels. J. Appl. Phys. **107**(4), 043527 (2010)
50. K. Fu et al., Investigation of GaN-on-GaN vertical p-n diode with regrown p-GaN by metalorganic chemical vapor deposition. Appl. Phys. Lett. **113**(23), 233502 (2018)
51. Z.Y. Xie, C.H. Wei, L.Y. Li, Q.M. Yu, J.H. Edgar, Gaseous etching of 6H–SiC at relatively low temperatures. J. Cryst. Growth **217**(1), 115–124 (2000)

52. S.-i. Nakamura, T. Kimoto, H. Matsunami, S. Tanaka, N. Teraguchi, and A. Suzuki, "Formation of periodic steps with a unit-cell height on 6H–SiC (0001) surface by HCl etching," Appl. Phys. Lett., 76 (23), 3412–3414 (2000)
53. A.L. Rosa, J. Neugebauer, Understanding Si adsorption on GaN(0001) surfaces using first-principles calculations. Phys. Rev. B **73**(20), 205314 (2006)
54. D.R. Lide, CRC Handbook of Chemistry and Physics, Internet Version 2005. CRC Press
55. K.J. Nordheden, B.A. Pathak, J.L. Alexander, ICP etching of ZnO in BCl3/SF6 gas mixtures, in Proc. SPIE 7217, Zinc Oxide Materials and Devices IV, 72170Q (2009)
56. J.M. Shah, Y.L. Li, T. Gessmann, E.F. Schubert, Experimental analysis and theoretical model for anomalously high ideality factors ($n \gg 2.0$) in AlGaN/GaN p-n junction diodes. J. Appl. Phys. **94**(4), 2627–2630 (2003)
57. L.C. Kimerling, Influence of deep traps on the measurement of free-carrier distributions in semiconductors by junction capacitance techniques. J. Appl. Phys. **45**(4), 1839–1845 (1974)
58. W.C. Johnson, P.T. Panousis, The influence of debye length on the C-V measurement of doping profiles. IEEE Trans. Electron Devices **18**(10), 965–973 (1971)
59. A. Hierro et al., Optically and thermally detected deep levels in n-type Schottky and p^+-n GaN diodes. Appl. Phys. Lett. **76**(21), 3064–3066 (2000)
60. E. Calleja et al., Yellow luminescence and related deep states in undoped GaN. Phys. Rev. B **55**(7), 4689–4694 (1997)
61. Q. Sun, C.D. Yerino, B. Leung, J. Han, M.E. Coltrin, Understanding and controlling heteroepitaxy with the kinetic Wulff plot: A case study with GaN. J. Appl. Phys. **110**(5), 053517 (2011)
62. K. Hiramatsu et al., Recent progress in selective area growth and epitaxial lateral overgrowth of III-nitrides. Effects of reactor pressure in MOVPE growth. Phys. Stat. Sol (a) **176**(1), 535–543 (1999)
63. A.S. Chang et al., Selective area regrowth produces nonuniform mg doping profiles in nonplanar GaN p–n junctions. ACS Appl. Electron. Mater. **3**(2), 704–710 (2021)
64. M.E. Coltrin, C.C. Mitchell, Mass transport and kinetic limitations in MOCVD selective-area growth. J. Cryst. Growth **254**, 35–45 (2003)
65. D. Du et al., Systematic prediction of kinetically limited crystal growth morphologies. PRL **95**, 155503 (2005)
66. G. Wulff, XXV. Zur Frage der Geschwindigkeit des Wachsthums und der Auflösung der Krystallflächen. Zeitschrift für Kristallographie **34**(1–6), 449–530 (1901)
67. C. Herring, Some theorems on the free energies of crystal surfaces. Phys. Rev. **82**(1), 87–93 (1951)
68. Q. Sun et al., Understanding nonpolar GaN growth through kinetic Wulff plots. **104**(9), 093523 (2008)
69. Q. Sun et al., Morphological and microstructural evolution in the two-step growth of nonpolar a-plane GaN on r-plane sapphire. J. Appl. Phys. **106**(12), 123519 (2009)
70. S.H. Jones, L.K. Seidel, K.M. Lau, M. Harold, Patterned substrate epitaxy surface shapes. J. Cryst. Growth **108**(1), 73–88 (1991)
71. A.S. Chang, B. Li, S. Wang, S. Frisone, R.S. Goldman, J. Han, L.J. Lauhon, Unveiling the influence of selective-area-regrowth interfaces on local electronic properties of GaN p-n junctions for efficient power devices. Nano Energy **102**, 107689 (2022)
72. M. Xiao, X. Yan, J. Xie, E. Beam, Y. Cao, H. Wang, Y. Zhang, Origin of leakage current in vertical GaN devices with nonplanar regrown p-GaN. Appl. Phys. Lett. **117**(18), 183502 (2020)
73. D. Szymanski, D. Khachariya, T.B. Eldred, P. Bagheri, S. Washiyama, A. Chang, S. Pavlidis, R. Kirste, P. Reddy, E. Kohn, L. Lauhon, R. Collazo, Z. Sitar, GaN lateral polar junction arrays with 3D control of doping by supersaturation modulated growth: A path toward III-nitride superjunctions. J. Appl. Phys. **131**(1), 015703 (2022)

74. P. Blanchard, M. Brubaker, T. Harvey, A. Roshko, N. Sanford, J. Weber, K.A. Bertness, Characterization of sub-monolayer contaminants at the regrowth Interface in GaN nanowires grown by selective-area molecular beam epitaxy. Crystals **8**(4), 178 (2018)
75. E. Weikum, A. Diaz-Damian, J. Houard, G.D. Costa, F. Delaroche, A. Vella, G. Muzioł, H. Turski, L. Rigutti, Microscopic correlation of doping distribution and luminescence in a nitride laser junction by Photonic Atom Probe. Phys. Rev. Mater. **8**, 074603 (2024)

Chapter 13
Selective Area Regrowth and Doping for Vertical GaN Power Devices

Houqiang Fu, Kai Fu, Dinusha Herath Mudiyanselage, Cheng Chang, Imteaz Rahaman, and Yuji Zhao

1 Introduction

Gallium nitride (GaN) is a wide bandgap semiconductor that has been extensively studied for next-generation efficient power electronic devices and systems due to its larger bandgap, higher critical electric field, higher frequency switching, and higher Baliga's figure of merit (BFOM) than the incumbent Si [1–12]. To fully harness GaN's potential, vertical GaN power devices on bulk GaN substrates have been extensively explored in recent years. Various vertical GaN diodes and transistors have been demonstrated, such as Schottky barrier diodes (SBDs) [13–21], p-n diodes [22–42], advanced power rectifiers [43–47], and current aperture vertical electron transistors (CAVETs), trench metal-oxide-semiconductor field-effect transistors (MOSFETs), and fin field-effect transistors (FinFETs) [48–58]. Compared with lateral GaN devices, such as GaN high electron mobility transistors (HEMTs) typically on foreign substrates, vertical GaN devices offer significant advantages, including higher voltages, higher currents, smaller chip areas, better thermal management, absence of surface-related issues, and avalanche capability [1, 2, 7]. Nevertheless, a critical component still absent for vertical GaN power devices is laterally patterned GaN p-n junctions (Fig. 13.1a). These laterally patterned p-n junctions are

---

Houqiang Fu and Kai Fu contributed equally.

H. Fu · D. H. Mudiyanselage
Arizona State University, Tempe, AZ, USA
e-mail: houqiang@asu.edu

K. Fu · I. Rahaman
The University of Utah, Salt Lake City, UT, USA
e-mail: kai.fu@utah.edu

C. Chang · Y. Zhao (✉)
Rice University, Houston, TX, USA
e-mail: yuji.zhao@rice.edu

I. C. Kizilyalli et al. (eds.), *Gallium Nitride and Related Materials*, The Materials Research Society Series, https://doi.org/10.1007/978-3-031-83056-3_13

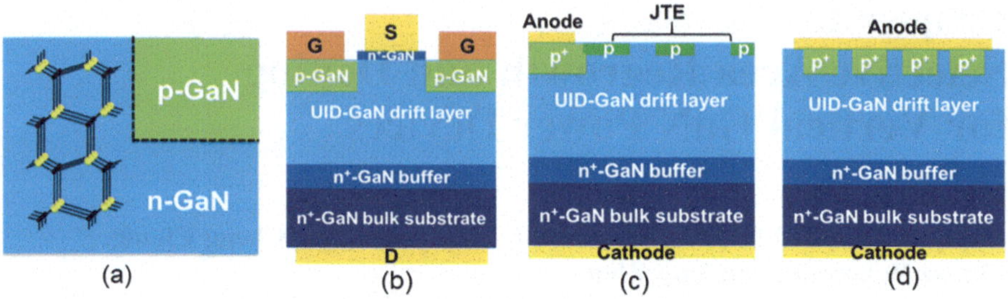

Fig. 13.1 (**a**) Schematic of laterally patterned p-n junctions via selective area regrowth and doping. Schematics of several devices that require laterally patterned p-n junctions, including (**b**) vertical JFETs, (**c**) JTE, and (**d**) JBS diodes. (Reprinted from [59], with the permission of Elsevier)

the building blocks for numerous advanced GaN power devices and edge terminations, such as vertical junction field-effect transistors (JFETs) (Fig. 13.1b), junction termination extension (JTE) (Fig. 13.1c), and junction barrier Schottky (JBS) diodes (Fig. 13.1d).

Laterally patterned GaN p-n junctions can be achieved through selective area doping using ion implantation and epitaxial regrowth. Despite its mature in Si and SiC power devices [60–62], ion implantation technology remains a significant challenge for GaN power devices [16]. Ion bombardment during implantation can induce substantial material damage, necessitating high-temperature annealing (>1000 °C) for damage recovery. Moreover, some implanted atoms may occupy interstitial sites that require thermal annealing to restore them to their proper substitutional sites for dopant activation [62, 63]. However, the high-temperature annealing can decompose GaN when above 900 °C and cause killer defects and surface degradation [64, 65]. Various techniques, such as capping layers, multicycle rapid thermal annealing, and ultra-high-pressure annealing [66–68], have been proposed to mitigate GaN decomposition at high temperatures. Furthermore, the conductivity of implanted p-GaN remains lower than that of epitaxially grown p-GaN [26, 43], negatively impacting device performance.

Selective area regrowth and doping offers a promising alternative for realizing laterally patterned GaN p-n junctions. However, regrown GaN p-n junctions often suffer from high leakage currents and low breakdown voltages [69, 70]. This chapter highlights the challenges and recent advancements in selective area regrowth and doping by metalorganic chemical vapor deposition (MOCVD) for vertical GaN power devices. It will first discuss the general MOCVD regrowth process and the regrowth dynamics and doping in trenches. The important material characterization techniques for GaN selective area regrowth and doping will be detailed, including dopant profiling via cathodoluminescence (CL) and secondary electrons (SEs) in scanning electron microscopy (SEM), and regrowth interface characterizations using electron holography, secondary ion mass spectrometry (SIMS), capacitance-voltage (*C-V*), and transmission electron microscopy (TEM). Finally, this chapter will discuss two novel etching technologies aimed at reducing etching damage, including low-power dry etching, and atomic layer etching (ALE), followed by the successful demonstration of regrown GaN p-n junctions and vertical GaN JFETs with regrown p-GaN gate.

2 Selective Area Regrowth and Doping

2.1 Regrowth Process in MOCVD

To achieve selective area doping through regrowth, dry etching is often employed to create selective regions or trenches. Nonetheless, this etch-then-regrow process poses considerable challenges, one of which is introducing impurities at the interface. Even without etching, regrowth itself introduces a notable concentration of impurities at the interface, including silicon (Si), carbon (C), and oxygen (O). Thermal desorption of impurities via annealing in MOCVD or molecular beam epitaxy (MBE) has been utilized for in situ cleaning. Annealing in NH_3 at 700–800 °C or employing multi-step alternating growth and annealing under vacuum at 780 °C have been proven efficient in removing C and O [71, 72]. However, the residual Si could not be removed by these methods. While Si and O are shallow donors for n-type doping and may not pose a problem for regrowing n-type GaN, they are highly undesirable for regrowing p-GaN as they can compensate acceptors and reduce hole concentrations. Additionally, the dry etching process can introduce etching damage at the regrowth interface, leading to the formation of deep-level defects or trap states. These issues contribute to high leakage and premature breakdown in regrown GaN p-n junctions.

Figure 13.2a depicts a typical regrowth process in MOCVD, comprising three stages: pre-regrowth, temperature and gas ramp-up, and regrowth. Initially, an n^+-GaN buffer layer and an unintentionally doped (UID) GaN were grown on GaN substrates. Subsequently, the sample underwent processing outside the reactor before being reloaded. To explore the impact of growth conditions on regrowth, four different combinations of growth gases were utilized during the ramp-up phase, including $N_2/NH_3/H_2$, N_2, N_2/trimethylgallium (TMGa, 6 sccm), and N_2/TMGa (12 sccm). A UID-GaN layer was then regrown on top of the sample. Figure 13.2b displays the SIMS profiles of the regrown UID-GaN under different gas conditions during the ramp-up phase. Si peaks were detected in all samples. While a Ga-rich environment aids in reducing Si incorporation [73], introducing TMGa during the ramp-up phase notably increased Si, O, and C concentrations. Among the conditions used, the sample exposed to $N_2/NH_3/H_2$ exhibited the lowest Si concentration at the regrowth interface. Studies have shown that minimizing the exposure to air can mitigate residual Si, such as storing the sample in a nitrogen box during processing [74]. Additionally, surface treatment and etching conditions before regrowth play crucial roles in influencing interfacial impurity concentrations, as discussed in subsequent sections. Overall, residual Si at the interface remains a persistent challenge in regrowth processes, with its origin yet to be fully elucidated.

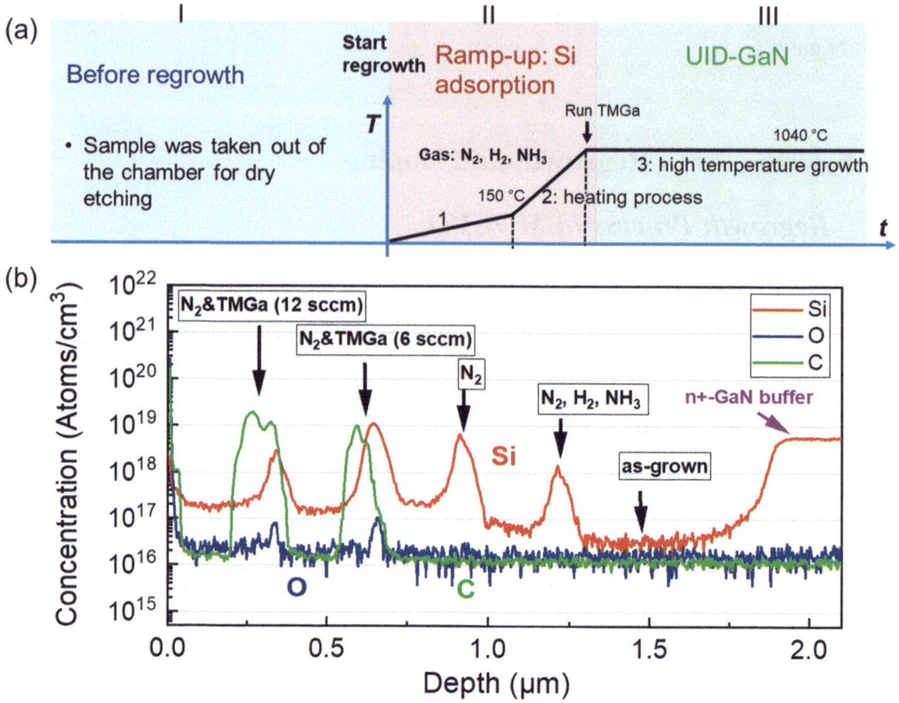

Fig. 13.2 (**a**) Typical regrowth process for undoped GaN in MOCVD. (**b**) SIMS profile of the regrown undoped GaN under different ambient gases during the ramp-up process. (Reprinted from [59], with the permission of Elsevier)

2.2 *Regrowth and Doping in Trenches*

The formation of laterally patterned p-n junctions on n^+-GaN substrates was achieved through a multi-step process using MOCVD. A UID-GaN layer was first grown on the substrate. Subsequently, the samples were removed from the MOCVD reactor, and trenches were created using inductively coupled plasma (ICP) etching. Following this, the samples were reloaded into the MOCVD reactor for the regrowth of p-GaN. This process employed a maskless regrowth approach, as previous studies indicated that regrowth with SiO_2 or Si_3N_4 masks introduced unintentional impurities like Si or O [75–77]. SEM can be used to characterize regrown p-GaN within a trench because the p-GaN exhibited a bright appearance, while the n-GaN appeared darker [78]. However, the regrowth of p-GaN within the trench exhibited non-uniformity. Specifically, the p-GaN layer on the trench sidewall appeared much thicker compared to other regions due to faster lateral growth rates relative to basal plane growth. This discrepancy in growth rates could potentially lead to lower concentrations of Mg acceptors in the sidewall region.

To understand the growth dynamics, we conducted selective area regrowth experiments with trenches of varying widths and depths. It was observed that as

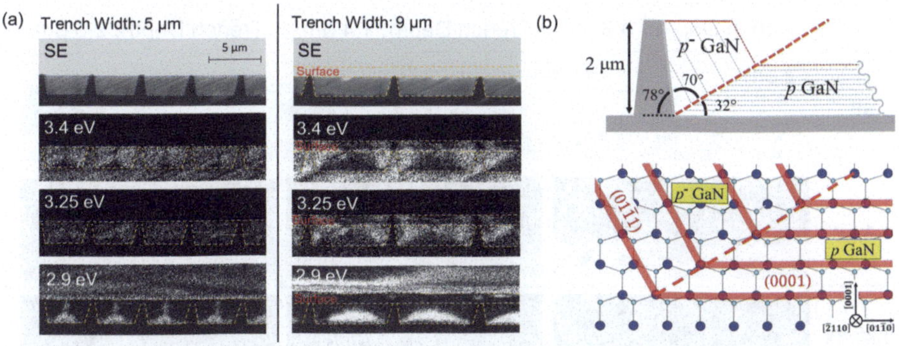

Fig. 13.3 (**a**) SE images and CL mappings at 3.4 eV, 3.25 eV, and 2.9 eV for selective area regrowth in trenches with trench widths of 5 µm and 9 µm. Reprinted from [59], with the permission of Elsevier. (**b**) p-GaN regrowth in trenches with both lateral and vertical growth. Reprinted from [80], with the permission of AIP Publishing

the trench width decreased, the thickness of the regrown p-GaN layer became more uniform across the trench, as depicted in Fig. 13.3a. Notably, for trench widths less than 9 µm, the p-GaN region within the trench exhibited a relatively flat profile. This phenomenon arises from the fact that in narrower trenches, the growth fronts originating from the sidewalls and the basal plane can merge more rapidly, resulting in the formation of a unified growth front and uniform p-GaN growth. To analyze the Mg distribution within the trenches, CL imaging was employed. It is worth noting that strong CL emission at 2.9 eV serves as a reliable indicator for high acceptor concentrations [79]; further elaboration on this topic will be provided later. As illustrated in the 2.9-eV CL mapping (Fig. 13.3a), the central region of the trench exhibited significantly higher brightness compared to the sidewalls [80], signifying a higher acceptor concentration in the former. Moreover, with an increase in trench width, the brightness of the central region of the trench intensified, indicating a shift toward a more dominant basal plane growth mechanism.

To comprehend the underlying physics, we analyzed the lateral growth from the sidewall and the vertical growth from the basal plane of the trench [80]. Surface steps of the growth front, as well as their aggregation into supersteps, instigate a competition between step flow from growth on microfacets and vertical growth from basal planes [81]. Monochromatic CL images revealed that growth on basal planes exhibited strong 2.9 eV emission, whereas growth from the sidewall facets did not exhibit such emission [82]. The interface between these regions formed an angle of approximately 32° with the basal plane, as illustrated in the lattice structure in Fig. 13.3b [80]. The $\{01\bar{1}1\}$ and {0001} facets are the most stable in GaN and have been observed in epitaxial lateral overgrowth studies [83]. The sidewall facet can be conceptualized as a highly tilted facet comprising a high density of basal plane steps, where each step acts as an effective site for adatom capture and facilitates lateral growth. The presence of Mg on the $\{01\bar{1}1\}$ facet can enhance the adsorption of Ga and facilitate the accumulation of Ga adatoms near the sidewall.

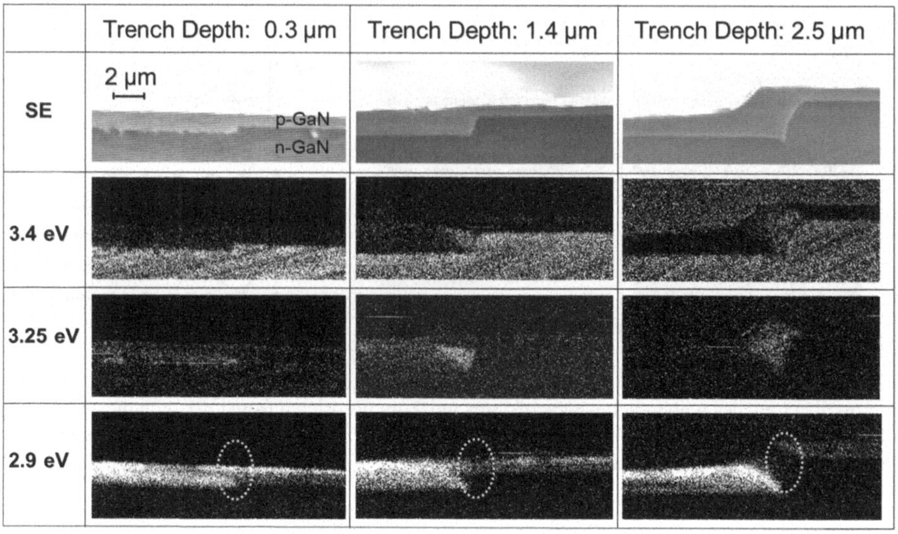

Fig. 13.4 SE images and CL mappings at 3.4 eV, 3.25 eV, and 2.9 eV for selective area regrowth of p-GaN in trenches with different trench depths. The sidewall regions are indicated using circles. (Reprinted from [59], with the permission of Elsevier)

It was determined that the lateral growth rate was twice as large as the vertical growth rate. Consequently, with the same quantity of Mg acceptors available during growth, the acceptor concentration near the sidewall would be lower than that in the central region of the trench. This was confirmed by CL analysis, wherein the vertical growth region exhibited p-type behavior, while the lateral growth region showed semi-insulating characteristics with a low acceptor concentration. Furthermore, previous studies indicated that Mg incorporation efficiency varied across different facets [79, 82]. Therefore, optimizing the trench width (neither too wide nor too narrow) is crucial to ensure uniform p-GaN thickness within the trenches and maximize basal plane growth to achieve the highest acceptor concentrations in the regrown p-GaN trenches.

The influence of trench depth on selective area regrowth was examined, as shown in Fig. 13.4. In shallow trenches with a depth of 0.3 μm, no conclusive evidence of lateral growth from the sidewall was observed through CL mapping. However, with increasing trench depth, the sidewall surface area expanded, leading to an increase in the lateral growth region with lower acceptor concentrations. This was corroborated by the diminished CL emission intensity at 2.9 eV near the sidewall. Furthermore, Fig. 13.5a illustrates that the reverse current leakage of the selectively doped p-n junctions increased with escalating trench depth. To elucidate this trend, the model depicted in Fig. 13.5b was proposed. According to the CL findings, the sidewalls in deeper trenches exhibited lower acceptor concentrations, potentially resulting in a p^-GaN or even intrinsic region. Consequently, the junction barrier near the sidewall may be diminished, creating a leakage path. This issue could potentially be

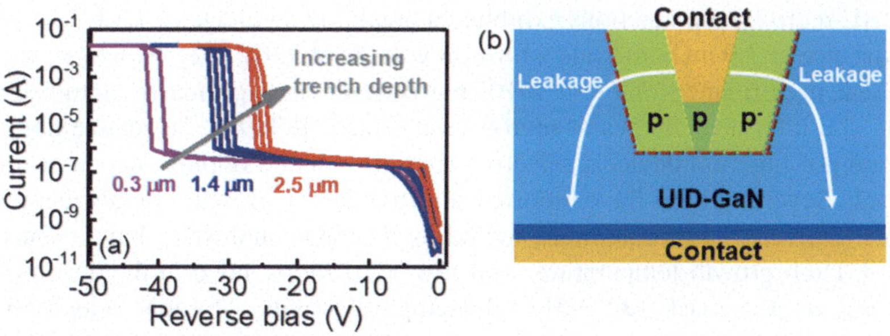

Fig. 13.5 (**a**) Reverse leakage characteristics for selective-area doped GaN p-n junctions with different trench depths. (**b**) Proposed mechanism for the reverse leakage of selective-area doped GaN p-n junctions. (Reprinted from [59], with the permission of Elsevier)

addressed by adjusting the growth mode or enhancing acceptor doping or activation efficiency along the sidewall. In essence, utilizing shallow trenches for selective area growth might prove advantageous in mitigating the reverse leakage of selectively doped p-n junctions.

2.3 MOCVD and MBE Regrowth Comparison

MOCVD growth of GaN typically operates at high temperatures (> 950 °C) using NH_3 as the nitrogen precursor. However, significant challenges arise when the device structure involves burying p-GaN beneath subsequent n-GaN growth, as seen in current aperture vertical electron transistors [1, 2], trench-MOSFETs [48–53], and tunnel junctions (TJs) [84]. The p-type conductivity of buried p-GaN is often hindered because active hydrogen (H) atoms from NH_3 or H_2 decomposition during n-GaN growth can diffuse backward and passivate magnesium (Mg) dopants in p-GaN under high temperatures [85]. Activating buried p-GaN post-growth becomes difficult due to the high diffusion barrier of H atoms in n-GaN [86]. Efforts have been made to activate buried p-GaN laterally through mesa sidewalls [87]. Additionally, Mg from buried p-GaN can diffuse into overlayers, potentially affecting the properties of the two-dimensional electron gas (2DEG) in CAVET-like structures. Similarly, defects generated during ion implantation to form the current blocking layer (CBL) in CAVET structures were reported to diffuse to the active region, causing current collapse during operation [2].

MBE growth of GaN, conducted in a hydrogen-free environment, offers potential solutions to these challenges. MBE overgrowth on MOCVD-grown Mg-doped GaN preserves the activated state of p-GaN [88]. MBE-regrown trench-MOSFETs [89, 90] and PolarMOS have also been demonstrated [91]. Moreover, MBE operates at lower temperatures (< 800 °C), reducing Mg atom diffusion and point defect migration to active regions. Chowdhury et al. achieved dispersion-free CAVET structures by combining Mg implantation and MBE-regrown channels

[2]. MBE-regrown p-n junctions exhibited a breakdown voltage of 1.1 kV, a specific on-resistance of 3.9 mΩ·cm^2, and a turn-on voltage of 3.9 V [92]. However, the less-than-ideal performance of these MBE-regrown devices primarily stemmed from defects and interface charges at regrowth interfaces [89–92]. To realize nonplanar or lateral p-n junction devices, selective area growth and doping offer a promising approach. Several groups have reported selective area regrowth of GaN using MBE reactors. Different mask materials, including Ti, SiO_2, and SiN_x, have been tested [93–95]. High growth temperatures and low V/III ratios aided in the desorption of Ga atoms on the mask [94, 95], enhancing selectivity. Optimal selectivity and surface morphology were achieved on periodic arrays of nanocolumn structures, while large-area devices exhibited a porous surface morphology under low V/III conditions [93]. Additionally, spike-like structures at mask edges were observed due to the short diffusion length of adatoms [91, 96].

3 Material Characterization

Due to the complicated nature of growth, defects, and structures involved in selective area growth and doping, it is critical to develop a proper material characterization toolkit to provide information on the growth dynamics, impurity elimination/reduction, and regrown devices. This section delves into various material characterization techniques that are useful for selective area regrowth and doping. CL is used to understand the acceptor distribution in nonplanar structures after p-GaN regrowth, and SEs are capable of doping profiling for both lateral and vertical structures at the sub-micron scale. The regrowth interfaces are also characterized by electron holography for electrostatic potential profiles at the regrowth interfaces, SIMS, TEM, and *C-V*.

3.1 Cathodoluminescence

Traditional SIMS cannot provide reliable chemical analysis for complex structures due to its limited lateral spatial resolution, while CL has the potential to overcome this shortcoming. During the regrowth on a mesa or trench structure, there were basal plane growth on the upper and lower mesa regions and lateral growth on the mesa sidewall. These directions could significantly affect acceptor incorporation efficiency for their different crystal orientations [97–99]. Liu et al. [79] used CL to investigate the optical properties of *p*-GaN in a mesa structure. An empirical correlation has been established between the Mg concentration, the optical properties, and the electronic properties of *p*-GaN. The SE image of the epitaxially regrown structure is shown in Fig. 13.6a, and Fig. 13.6b, c show the CL mappings of the structure at 3.25 and 2.9 eV, respectively. The CL mapping at 3.25 eV showed that the *p*-GaN layer exhibited a uniform emission, while the CL mapping at 2.9 eV

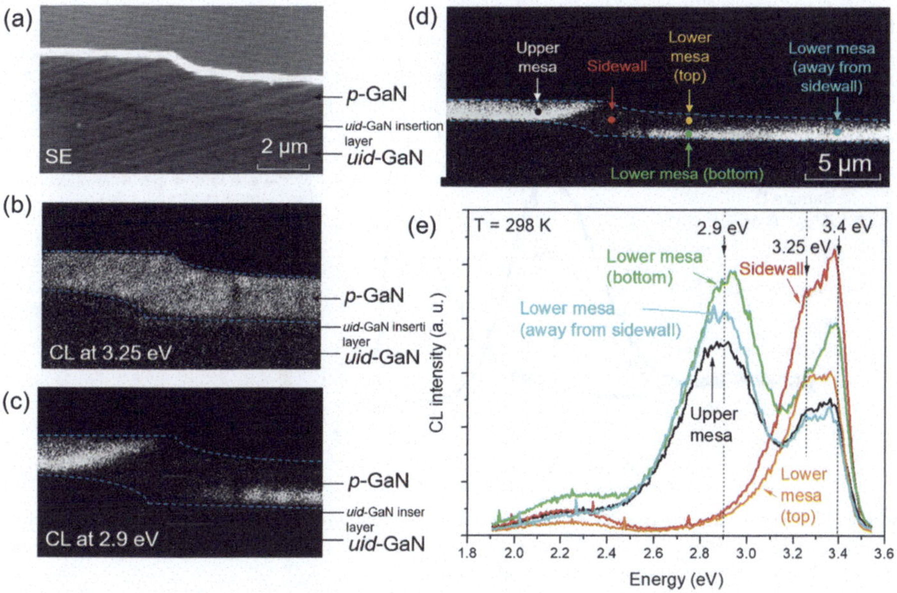

Fig. 13.6 (**a**) SE image of the regrown p-GaN/UID-GaN mesa structures mesa structure. (**b**, **c**) CL images at 3.25 eV and 2.9 eV with lines showing the boundaries between the regrown p-GaN and UID-GaN. (**d**) CL mapping at 2.9 eV for the regrown p-GaN/UID-GaN mesa structure. (**e**) Spot CL spectra from different locations in (**d**). (Reprinted from [79], with the permission of AIP Publishing)

exhibited a weaker emission in the vicinity of the sidewall. To further understand the CL mappings, the spot mode CL was used to study the CL emissions at specific locations, as shown in Fig. 13.6d, and the extracted spectra are shown in Fig. 13.6e. Three common peaks at 3.4, 3.25, and 2.9 eV were observed at all locations. The CL spectral shape was similar for the upper mesa, the bottom of the lower mesa, and the lower mesa away from the sidewall, where emission was dominated by 2.9 eV peak. The emissions at the sidewall and the top of the lower mesa were dominated by 3.4 and 3.25 eV peaks, while the 2.9 eV peak was very weak.

To establish a correlation between the doping concentration and the optical and electronic properties of *p*-GaN films, three samples with different Mg concentrations were grown and characterized in Fig. 13.7a. The intensity of the 2.9 eV peak increased with increasing acceptor concentration, while the intensities of the other two high-energy peaks decreased. The ionized shallow and deep donors are responsible for the 3.25 eV and 2.9 eV transitions, respectively. At low Mg concentration (Fig. 13.7b), only shallow donors are partially ionized due to self-compensation, leading to dominant 3.4 eV and 3.25 eV emissions. At high Mg concentration (Fig. 13.7c), the electrons in the conduction band would first relax to the deep donor state and then transition to the acceptor state. With this information, it was found that the sidewall and the lower mesa top suffered from acceptor deficiency, which is attributed to the non-uniform Mg concentration due to different incorporation efficiencies around the mesa structure having different crystal planes.

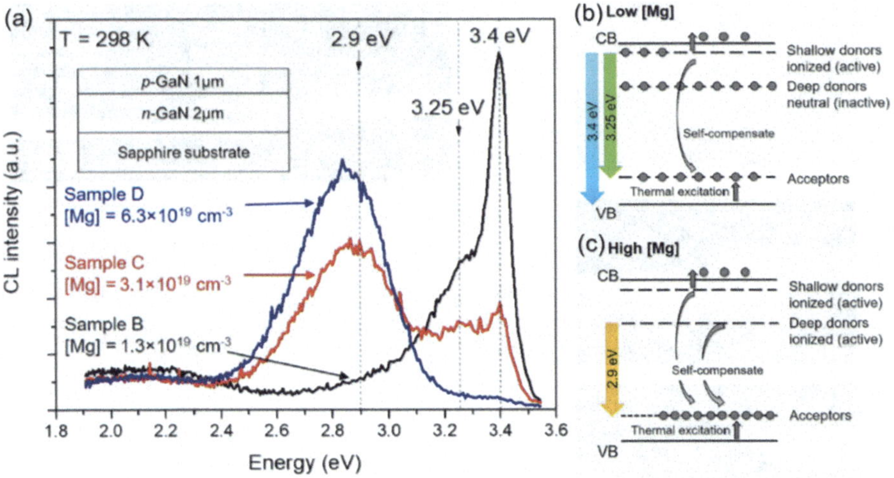

Fig. 13.7 (**a**) CL spectra of p-GaN samples with different Mg concentrations. Proposed mechanisms to explain different CL spectra of p-GaN with (**b**) low [Mg] and (**c**) high [Mg]. (Reprinted from [79], with the permission of AIP Publishing)

3.2 Secondary Electrons

The spatial dopant distribution in structures containing *p*-GaN, UID-GaN, and *n*-GaN is important for selective area regrowth and doping and associated devices. SEs in a SEM can also be used to map out accepter distributions. In contrast to the poor lateral resolution of SIMS, SEs can realize dopant profiling at the sub-micron scale. The process of SE emission in a material involves three main steps: generation of SEs within the bulk by primary electrons, their transport toward the surface, and their emission into the vacuum that depends on doping levels and sample preparation [100, 101]. In Alugubelli's work [78], they reported on a quantitative approach using SEs for determining both the lateral and vertical dopant distribution profiles in *p-i-n* GaN structures. In Fig. 13.8a, the incoming electrons are primary electrons, and the outgoing electrons consist of the SEs from the beam-specimen interaction and backscattered electrons (BSEs) stemming from the specimen and the chamber wall [100–102]. Because the BSE yields η (I_{BSE}/I_b *where* I_{BSE} is the backscattered electron current, I_b is the primary electron beam current) are low for GaN and relatively constant as a function of primary electron beam accelerating voltage [103], the measured total electron emission yields reflected the variations of the SE yield δ (I_{SE}/I_b *where* I_{SE} is the secondary electron current) with accelerating voltage. Figure 13.8b shows an SE image of a GaN *p-i-n* structure at 2 kV beam voltage, where the SE image contrast between different layers was visible. The emission intensity was highest for the *p*-layer, followed by the *n*-layer, and it was lowest for the *i*-layer, as shown in Fig. 13.8c.

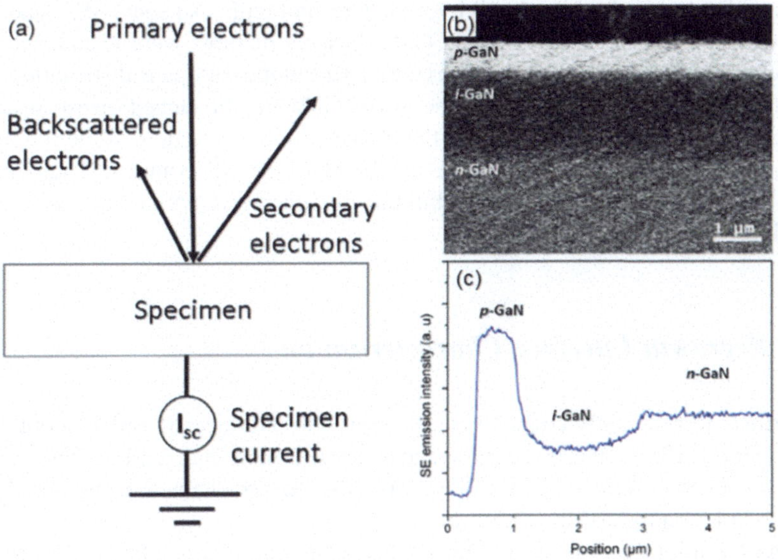

Fig. 13.8 (**a**) Schematics of different electrons flowing in and out of a sample in an SEM. (**b**) SE image and the (**c**) SE emission intensity profile of a GaN *p-i-n* structure. (Reprinted from [78], with the permission of AIP Publishing)

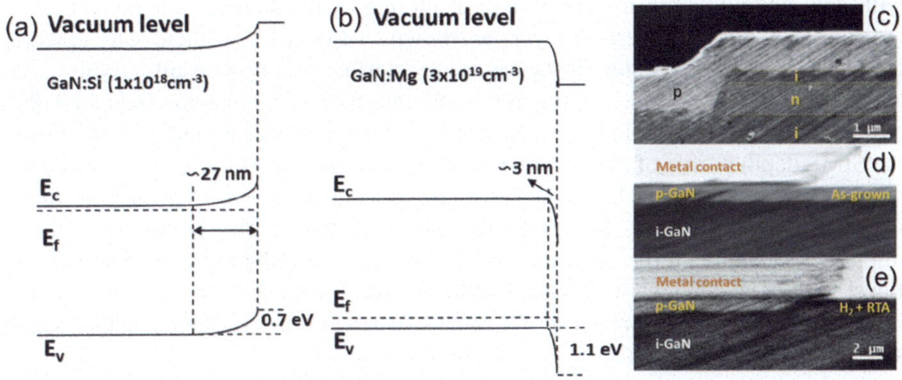

Fig. 13.9 Band diagram and band bending near the surface of (**a**) n-GaN and (**b**) p-GaN. SE images of (**c**) etch-then-regrow *p-i-n* mesa structure, (**d**) as-grown p-GaN, and (**e**) hydrogen passivated p-GaN. (Reprinted from [78], with the permission of AIP Publishing)

Alugubelli et al. proposed a mechanism to explain the observed different emission intensities between *p*-GaN, *n*-GaN, and *i*-GaN. *p*-GaN usually has a downward band bending near the surface (Fig. 13.9a), and *n*-GaN usually exhibits an upward band bending near the surface (Fig. 13.9b) that can deplete electrons from the surface and reduce the SE emission [104, 105]. However, the downward band bending of the *p*-GaN surface can attract electrons toward the surface and thus increase its SE emission. Due to the low conductivity, *i*-GaN cannot replenish some SEs, leading to the lowest SE emission [106]. Some GaN device structures were visualized using this technique. In Fig. 13.9c, the horizontal and vertical dopant distribution were identified from the SE image of the *p-i-n* GaN on an etched mesa structure. Figure 13.9d shows the SE image of the planar *p-i*-GaN junction, while the hydrogenated *p*-GaN region in Fig. 13.9e exhibited *i*-type SE contrast. These results indicate the effectiveness of this technique for imaging GaN devices with layers with different doping profiles.

3.3 Regrowth Interface Characterization

It is critical to accurately characterize the regrowth interfaces in order to obtain the best device performance through selective area growth and doping. This section discusses several useful techniques for regrowth interface characterization, including electron holography, SIMS, TEM, and *C-V*.

Electron holography is an interferometric technique that can be used to retrieve the phase and the amplitude shifts of the electron wave within the material concerning vacuum, where the phase shift is proportional to a product of electrostatic potential energy and the sample thickness. Alugubelli et al. [107] used electron holography to map out the electrostatic potential profile of regrown GaN p-i-n diodes with and without a thin insertion layer at the regrown interface. For the regrown structure without the insertion layer, as shown in Fig. 13.10a, there was upward curvature on the *p*-GaN side of the regrowth interface and downward curvature on the *i*-GaN or *n*-GaN side of the regrowth interface. This feature indicated the electronic charge accumulation on each side of the interface. Figure 13.10b shows the SIMS depth profile of the device, where the regrowth interface region showed an overlap of Mg, Si, and O peaks. The high concentrations of Si at the regrowth interface may also affect Mg gettering and precipitates. The presence of these impurities can explain the potential profile curvature near the regrowth interface in Fig. 13.10a. The downward curvature of the potential profile on the *i*-GaN side of the interface is related to the presence of Si and O, which may act as donors on the *i*-GaN side of the regrowth interface,

turning it into *n*-type GaN. The overlap of Mg, Si, and O can lead to an enhancement in hole concentration by up to two orders of magnitude due to co-doping effect [108]. The increase in hole concentration would result in an upward curvature in the potential profile on the *p*-GaN side of the etched interface. A thin *i*-GaN layer was used as the insertion layer before regrowing *p*-GaN layer. A dip was observed in Fig. 13.10c, which indicated a positive sheet charge that was surrounded by negatively charged regions. The SIMS analysis in Fig. 13.10d shows that the insertion layer moved the Si peak away from Mg. It should be noted that the broadening of the signals at the regrowth interface is an artifact due to the SIMS sputtering process. The presence of Si impurities at the regrowth interface acts as shallow donors, resulting in the formation of a thin n-type layer sandwiched between

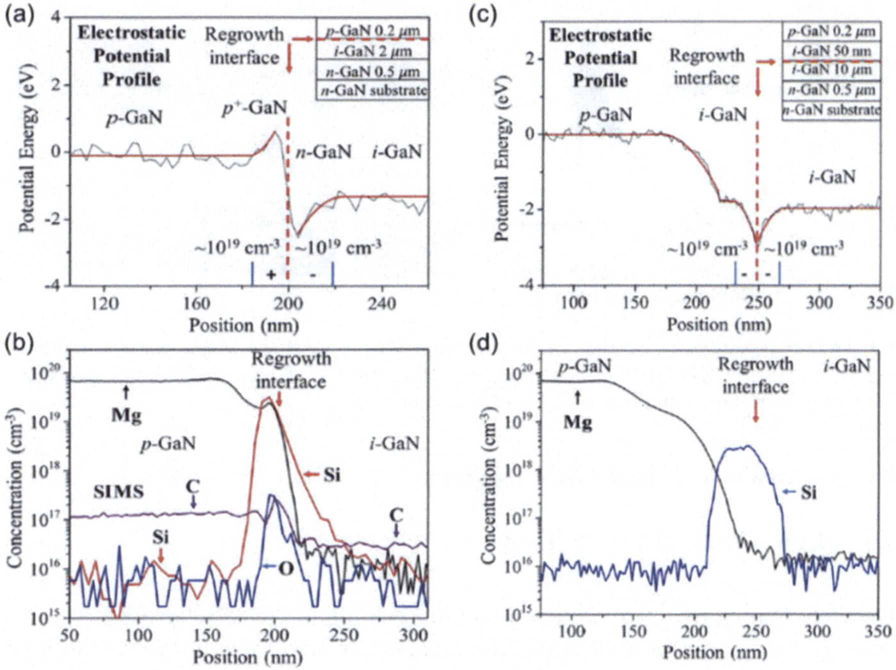

Fig. 13.10 (**a**) Electrostatic potential profile and (**b**) SIMS profile of the p-i interface in a regrown GaN p-i-n structure by electron holography. (**c**) The electrostatic potential profile and (**d**) SIMS profile of the regrown GaN *p-i-n* structure with a 50 nm undoped GaN insertion layer. (Reprinted from [107], with the permission of AIP Publishing)

the 50 nm insertion layer and the *i*-GaN layer beneath. Adding an undoped layer on top of the etched surface helps move the defective regrowth interface away from the *p-i* junction and prevents co-doping, leading to no upward curvature at the *p*-GaN side.

Fu et al. [69] reported frequency dispersion in *C-V* measurements of regrown GaN p-n junctions, suggesting the presence of defects and charges. The net charge distribution in the regrown p-n junction in Fig. 13.11a, b showed a high donor charge concentration near the regrowth interface. As the previous SIMS results indicate, these charges may stem from the high concentrations of impurities at the regrowth interface, especially Si. Fu et al. [109] and Stricklin et al. [110] indicated that reduced air exposure can help reduce interfacial impurities. In addition, TEM was also used to visualize the regrowth interface in Fig. 13.11c. Based on these results, the regrowth interface of a regrown GaN p-n diode may behave like a p+/n + tunneling junction due to the presence of high concentrations of donor and acceptor, which is consistent with the observed tunneling-dominated reverse leakage current [69]. In addition, these charges at the regrowth interface led to higher local electrical fields, higher reverse leakage, and premature device breakdown, which are big hurdles for selective area regrowth and doping.

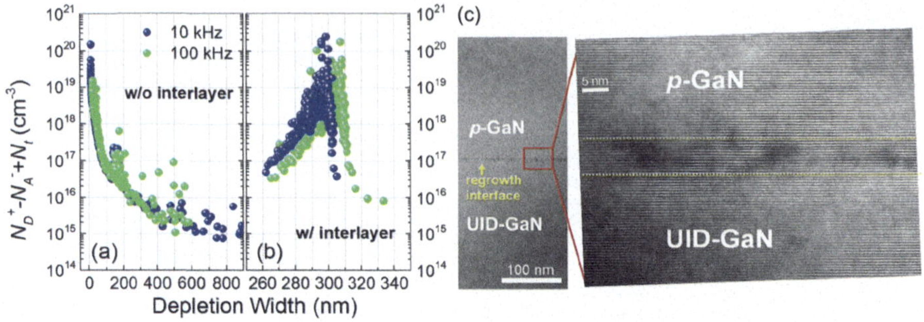

Fig. 13.11 (**a**) Net charge distribution for regrown GaN p-n junctions without an insertion layer and (**b**) with an UID-GaN insertion layer. (**c**) TEM images of regrown GaN p-n junctions. (Reprinted from [69], with the permission of AIP Publishing)

4 Regrown Vertical GaN Devices

4.1 Low-Power Dry Etching

The etch-then-regrow process, involving dry etching to create selectively regrown areas, poses significant challenges. The process can lead to an interface with high concentrations of contaminants [69, 107, 109] like Si, C, and O, even without etching. The etch-then-regrow process can also cause surface damage during dry etching process [109]. Consequently, very high reverse leakage currents have emerged as a major challenge for regrown p-n junctions. Hence, exploring the impacts of surface treatment and ICP etching on the electrical characteristics of regrown GaN p-n junctions is crucial. Planar regrown p-n junctions were used as a test structure to obtain essential knowledge about the etch-then-regrow process. The details of the vertical diode fabrication can be found in Ref. [111]. Additionally, hydrogen plasma treatment was employed for the edge termination since the approach was proven effective in reducing electric field crowding and enabled avalanche breakdown [59, 112–114].

Two samples were co-loaded into the MOCVD reactor for p-GaN regrowth without any surface treatment. One of the samples was without ICP etching, and the other sample was with ICP etching at a power of 70 W. The ICP power of 70 W resulted in an etching rate of ~280 nm/min, a reasonable value for GaN device fabrication. However, both samples suffered from very high reverse leakage current and low breakdown voltage, as shown in Fig. 13.12a, b. This result indicates that both non-etched and etched surfaces are not optimal for regrowth. The two samples were then treated in sequence with UV-ozone, hydrofluoric acid (HF), and hydrochloric acid (HCl) acid treatments prior to the regrowth. The UV-ozone treatment can oxidize the surface and any organic residue contaminant, and the acids can remove oxidized materials and further clean the surface. The non-etched sample showed a significant reduction in the reverse leakage current (Fig. 13.12a). However, the etched sample had little improvement (Fig. 13.12b). It is hypothesized

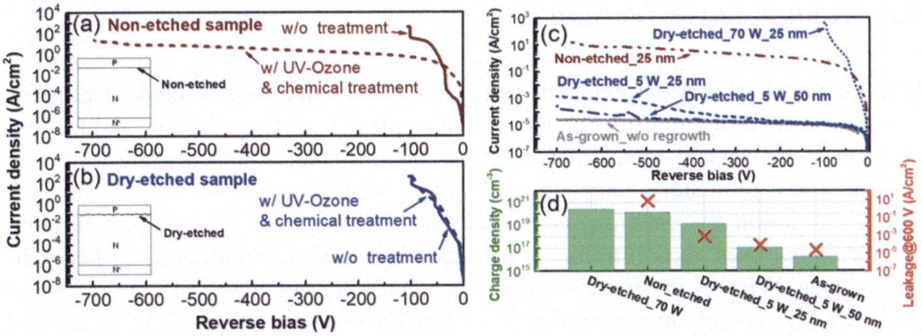

Fig. 13.12 Reverse *I-V* curves of (**a**) non-etched samples and (**b**) etched samples with and without UV-chemical treatment. (**c**) Reverse *I-V* characteristics of etched samples with varying etching powers and insertion layer thicknesses with the as-grown sample as a reference. (**d**) The leakage current and interface charge density for regrown and as-grown p-n junctions [111]. (© [2019] IEEE. Reprinted with permission from [111])

that the etching damage in the etched sample was too severe to be recovered by any surface treatment. To validate this hypothesis, a significantly lower ICP etching power of 5 W (with an etching rate of ~20 nm/min) was employed. As depicted in Fig. 13.12c, this led to a reduction in leakage current by over 4–5 orders of magnitude [111]. Additionally, it is worth noting that a thin UID-GaN insertion layer is advantageous for further reducing the leakage current. This layer helps to shift the regrowth interface (insertion-UID-GaN/drift layer) away from the peak electric field in the p-n junction depletion region, thereby decreasing trap-assisted reverse leakage by mitigating the electric field at the regrowth interface where numerous interface states are present. The thickness of the insertion layer needs to be optimized, as a layer that is too thin may not be effective, while one that is too thick may not be practical.

Figure 13.12d illustrates a comparison of the charge concentrations at the regrowth interfaces and the reverse leakage currents of the samples depicted in Fig. 13.12c. A higher charge concentration at the regrowth interface was found to correspond to a larger reverse leakage current. The regrown samples exhibited significant concentration peaks ranging from 10^{17} to 10^{20} cm^{-3} at the regrowth interface, whereas the as-grown sample displayed a low and uniform charge distribution of approximately 10^{16} cm^{-3}. These substantial charge accumulations at the regrowth interface can result in a larger electric field and a thinner effective barrier, enhancing carrier tunneling and increasing leakage current. The etched sample subjected to 5 W ICP etching and featuring a 50-nanometer insertion layer exhibited very low interfacial charge density and reverse leakage current, comparable to that of the as-grown sample. These findings suggest that low-power ICP etching, in conjunction with appropriate surface treatment and insertion layer thickness, can not only mitigate dry etching damage but also eliminate interfacial charges, both of which are highly advantageous for selective area regrowth.

4.2 Atomic Layer Etching

Progress in enhancing regrown GaN p-n junctions and their associated devices relies on effectively mitigating the challenges posed by ICP etching damage induced by photons, radicals, and ions. These factors can introduce defects, impurities, and surface states, ultimately compromising device performance [59, 109, 111]. ALE offers a solution by enabling controlled material removal through self-limiting chemical processes, thus eliminating damaged layers without inducing further etching damage. With its limited ion component, ALE presents the best opportunity to achieve a high-quality interface. Figure 13.13a illustrates the ALE processes tailored to GaN [59, 112]. Unlike several other materials where ALE has been applied, direct fluorination of GaN using HF is not feasible. Therefore, the proposed ALE process initiates with surface modification via plasma-enhanced atomic layer oxidation, converting the surface to Ga_2O_3. Subsequently, exposure to HF gas reduces the oxide, forming GaF_3 at the surface. TMGa or trimethylaluminum (TMAl) then undergoes ligand exchange with the fluoride surface, yielding volatile $GaF(CH_3)_2$ and removing the converted surface layer. The ALE of Ga_2O_3 is a fully thermal process and is integral to the GaN etching. These methods align with previously reported thermal ALE of Al_2O_3 and TiN [113, 114].

The initial step of the ALE process involved oxidizing the GaN surface using O_2 plasma exposure (Fig. 13.13b). Saturation of the oxide thickness at 0.5 nm was achieved within 20 seconds of exposure at 100 W RF power. Subsequently, HF exposure reduced the oxide, yielding GaF_3 (Fig. 13.13c). The introduction of TMA or TMG facilitated methyl group exchange at the surface, generating volatile GaF $(CH_3)_2$ and $AlF(CH_3)_2$ (Fig. 13.13d). X-ray photoelectron spectroscopy (XPS) revealed incomplete removal of the surface oxide during HF exposure, necessitating

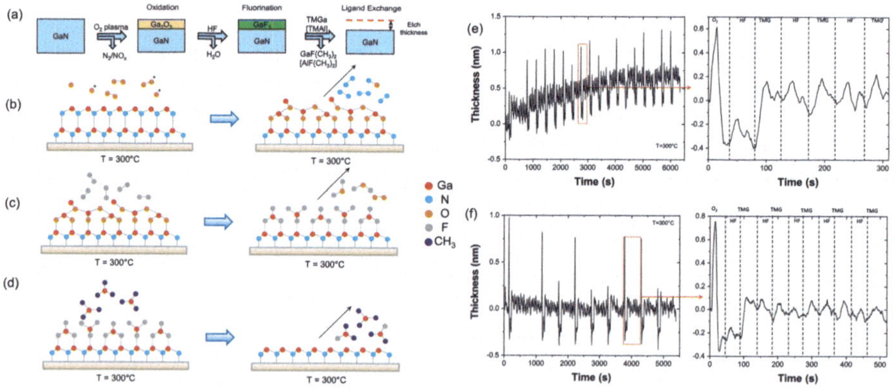

Fig. 13.13 (**a**) Method for ALE processing of GaN. (**b-d**) Schematics of reactions occurring at the surface during each cycle [59]. Measurement of the GaN thickness using in situ ellipsometry during ALE process with (**e**) 20 supercycles with three alternating exposures of HF and TMG per O_2 plasma, and (**f**) 10 supercycles with five alternating exposures of HF and TMG per O_2 plasma [59]. (Reprinted from [59], with the permission of Elsevier)

multiple cycles of Ga_2O_3 etching after oxidation. Given the high oxidation rate, several HF and TMG/TMA exposures were necessary to eliminate the surface oxide. While three alternating exposures of HF and TMG led to oxide deposition observed via in situ ellipsometry (Fig. 13.13e), five alternating exposures achieved complete oxide removal (Fig. 13.13f), yielding an etching rate of approximately 0.5 nm/cycle. These findings underscore the efficacy of the ALE process for GaN surfaces, offering valuable preparation of interfaces before regrowth in selective area doping.

4.3 Vertical GaN Diodes with Regrowth

Regrown p-n junctions play a pivotal role in selective area regrowth and doping. However, both regrown p-GaN and UID n-GaN encounter challenges associated with high-concentration donor-like contamination at the regrowth interface (Fig. 13.14a, b). These interfacial charges, as characterized in Schottky barrier diodes, create a heavily doped interlayer, leading to a reduced depletion region and lower breakdown voltage (Fig. 13.14c, d) [69, 115]. Furthermore, both regrown p-GaN and UID-GaN exhibit increased edge threading dislocations (Fig. 13.14e, f) [69, 115]. Studies indicate that Schottky barrier diodes on both regrown and as-grown samples showed similar forward *I-V* curves [126] and underwent the

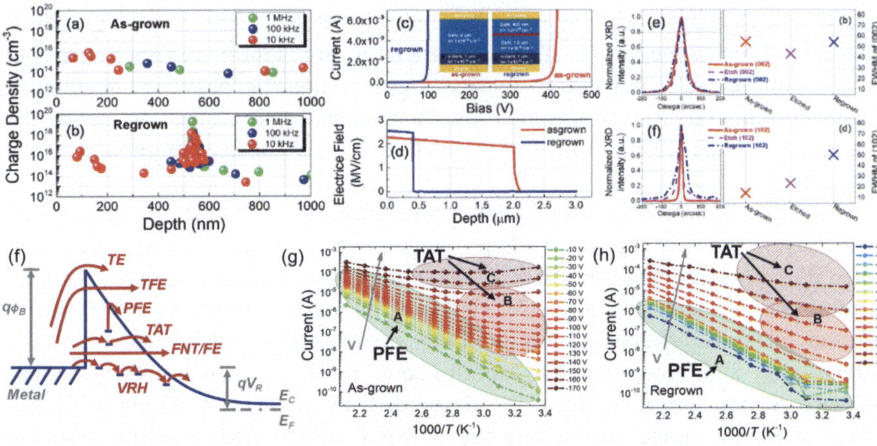

Fig. 13.14 Charge density of (**a**) as-grown and (**b**) regrown samples extracted from *C-V* results. Simulations of (**c**) reverse *I-V* and (**d**) electric field at breakdown voltage (BV), for as-grown and regrown samples. The simulated BV matches the experiments well. XRD omega rocking curves for (**e**) (002) and (102) reflections for both samples. The full-width-half-maximum (FWHM) for (002) and (102) reflections are typically related to screw and edge threading dislocations, respectively. (**f**) Possible leakage processes for the Schottky barrier diodes. Current vs. temperature for (**g**) as-grown and (**h**) regrown didoes at different voltages. Leakage in region "A" is due to Poole-Frenkel emission (PFE) process while the trap-assisted tunneling (TAT) is for regions "B" and "C". (Reprinted with permission from Ref. [115]. Copyright 2020 under the terms of CC BY 4.0)

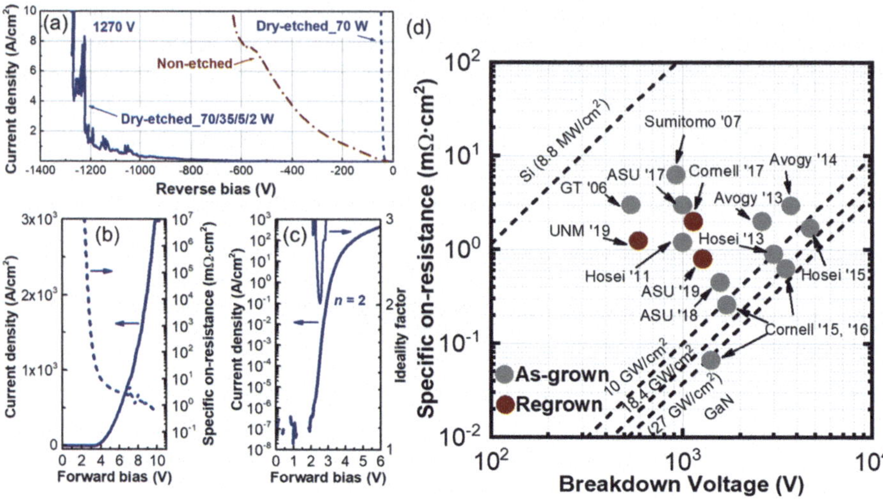

Fig. 13.15 (**a**) Comparison of reverse *I-V* curves for regrown p-n diodes on the dry-etched samples by multi-step ICP etching and non-etched sample. (**b**) Forward *I-V* curve in linear scale and on-resistance in semi-log scale for regrown p-n junction by multi-step ICP etching. (**c**) Forward I-V curve and ideality factor in semi-log scale. (**d**) Specific on-resistance vs. breakdown voltage for as-grown and regrown GaN-on-GaN vertical *p-n* diodes [111]. © [2019] IEEE. (Reprinted with permission from [111])

same reverse leakage transport processes, namely, Poole-Frenkel emission (PFE) and trap-assisted tunneling (TAT), as illustrated in Fig. 13.14f, g. Given that both processes are linked to defect/trap density, regrown diodes experienced heightened leakage currents due to their elevated interfacial state density. Thus, mitigating the interface charge density holds the key to reducing reverse leakage current in regrown devices.

While high-performance lateral GaN p-n junctions are still in the developmental stages, several vertical regrown GaN p-n junctions have been successfully demonstrated. For instance, diodes in Hu et al. [92] exhibited a specific on-resistance of 3.9 mΩ·cm^2 and an ideality factor of 2.5. However, compared to the as-grown device with an identical structure, the regrown device displayed elevated specific on-resistance and recombination currents due to the regrowth interface. The increased reverse leakage current in the regrown device was attributed to defect-assisted hopping conduction within the regrown p-GaN and/or at the regrowth interface. For various applications, the etch-then-regrow process remains indispensable. Fu et al. [69, 109, 111] conducted a systematic investigation into the origins and impacts of interfacial contaminants and the etch-then-grow process. They showcased high-performance MOCVD regrown GaN p-n junctions, achieving the highest breakdown voltages of ~1.3 kV using a novel multi-step ICP etching approach (Fig. 13.15a). As previously discussed, the low-power ICP dry etching can effectively reduce the interface charge density and lower the reverse leakage current of regrown p-n junctions [111]. However, the low etching rate of this method may not always be practical, especially for device structures with deep trenches or mesas. To address this challenge, a multi-step etching protocol was developed, gradually reducing ICP etching powers from 70 W to 35 W to 5 W to 2 W. This strategy not only ensures decent etching rates but also significantly reduces etching damage, resulting in improved regrowth surfaces. The resulting devices exhibited an on-resistance of 0.8 mΩ·cm^2, an on/off ratio of 10^{10}, and an ideality factor of ~2, closely resembling values for the as-grown samples (Fig. 13.15b, c). Additionally, Yamada et al. [116] demonstrated that subsequent thermal annealing after low-power ICP etching could further enhance device performance. Figure 13.15d illustrates a benchmark plot for both as-grown and regrown vertical GaN-on-GaN p-n diodes [59]. The Baliga's figure of merit for the regrown p-n diodes by Fu et al. reached 2.0 GW/cm^2, ranking among the best performance in regrown devices and even comparable to some of the as-grown devices. With ongoing advancements in etching technology, surface treatment, and device fabrication, the performance of regrown GaN p-n junctions is expected to approach the material limit of GaN.

4.4 Vertical GaN Junction Field-Effect Transistors

Initially proposed by Kizilyalli et al. [117], and subsequently refined through device simulations and optimization by Ji et al. [118], the development of vertical GaN JFETs encountered early hurdles in achieving optimal performance on bulk GaN substrates. Kotzea et al. demonstrated these devices on sapphire substrates, achieving a low on/off ratio of less than 2 through selective area regrowth [119]. Yang et al. reported the first experimental demonstration of high-performance vertical GaN JFETs on bulk GaN substrates, achieving a notable on/off ratio of approximately 100 [120]. The successfully tested vertical GaN JFETs featured six fins, each measuring 160 μm in length and 1 μm in width, with current density uniformly distributed across the total active area of the fins. The detailed fabrication process can be found elsewhere [120]. Additionally, the development process focused on realizing normally-off vertical GaN JFETs. Initial attempts revealed challenges such as the formation of a slanted channel sidewall, resulting in variable channel width and a tendency toward normally-on operation.

In the pursuit of normally-off vertical GaN JFETs, two sets of experiments were conducted to optimize device growth and structures. In the first-generation vertical GaN JFETs, no TMAH treatment was applied before p-GaN gate regrowth. A cross-sectional SEM image of the vertical GaN JFETs (Fig. 13.16a) revealed a slanted sidewall, resulting in a channel region with a widening width from top to bottom. Furthermore, an n$^+$-GaN layer served as the source. Transfer characteristics and gate leakage of the vertical GaN JFETs are depicted in Fig. 13.16b. While the device exhibited gate modulation, it operated in a normally-on mode, likely due to the slanted channel. Additionally, the I_{on}/I_{off} ratio was ~6, and gate leakage was relatively high. The transconductance gm displayed two peaks (Fig. 13.16c), indicating the presence of interface charges. While the first-generation vertical GaN JFETs

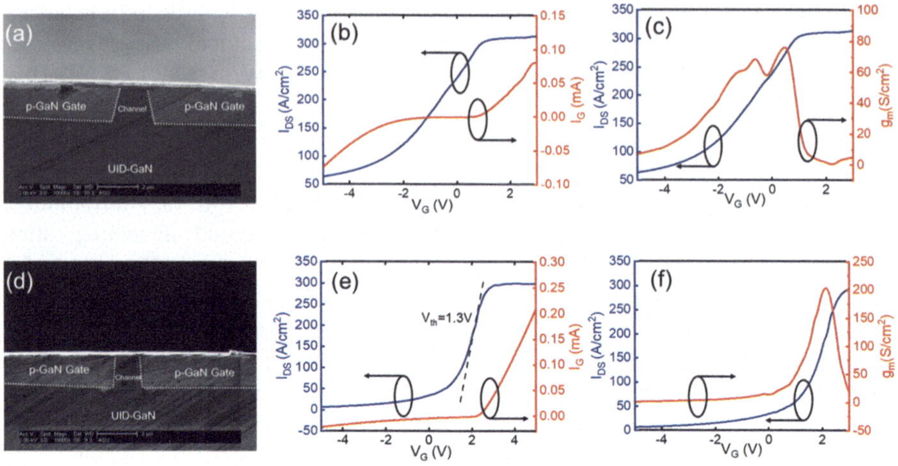

Fig. 13.16 (**a–c**) Cross-sectional SEM image and electrical characteristics of the first-generation vertical GaN JFETs. (**d–f**) Cross-sectional SEM image and electrical characteristics of the second-generation vertical GaN JFETs [120]. © [2020] IEEE. (Reprinted with permission [120])

exhibited transistor characteristics, their relatively poor performance may stem from various factors, including interfacial defects, impurities, etching damage, a slanted channel, and source-to-gate tunneling. To enhance the performance of second-generation vertical GaN JFETs and achieve normally-off operation, TMAH wet etching, UV-ozone, and HF treatments were implemented prior to regrowth to mitigate potential etching damage, interfacial charges, and impurities. TMAH etching also facilitated the formation of a more vertical sidewall aligned to the *m*-plane (Fig. 13.16d). Moreover, the absence of an n^+-GaN layer as the source was ensured to prevent the formation of Zener junctions between the source and gate, which could lead to substantial gate leakage. Instead, the source region underwent ICP etching to create donor-like surface defects for improved contacts. Transfer characteristics and gate leakage of the second-generation vertical GaN JFETs are illustrated in Fig. 13.16e. A threshold voltage (V_{th}) of ~1.3 V was achieved, with a drain current density reaching ~300 A/cm^2 at $V_{GS} = 3.4$ V and $V_{DS} = 10$ V. Subsequent increases in V_{GS} resulted in forward biasing of the gate-to-source junctions and a rise in gate current. The I_{on}/I_{off} was significantly enhanced to approximately 100, while gate leakage was reduced compared to first-generation vertical GaN JFETs. Transconductance curves of the devices (Fig. 13.16f) displayed a single peak with an impressive peak value of about 200 S/cm^2, among the best reported for vertical GaN JFETs. However, the breakdown voltages of the devices remained relatively low (<100 V) and require further improvement. Significant advancements have been achieved recently [121, 122]. Liu et al. [122] showcased fin JFETs with avalanche capability, reaching 1.2 kV-class voltages.

5 Concluding Remarks

This chapter discusses the recent progress of selective area regrowth and doping using MOCVD for vertical GaN power devices. Both lateral and vertical growth modes were observed in GaN regrowth in trenches, where the trench width and depth need to be optimized for maskless selective area regrowth. The uneven growth rate in the two growth modes also resulted in non-uniform distribution of dopants in the regrown p-GaN in a mesa or trench structure. Comprehensive material characterization techniques have been used to map out the dopant distribution in both lateral and vertical directions at a sub-micron scale via CL and SEs and to analyze the regrowth interfaces via electron holography, SIMS, and *C-V*. It was found that the regrowth interface had high concentrations of impurities and charges and large band bending, leading to characteristics similar to a p^+ -n^+ tunneling junction and large leakage in regrown GaN p-n junctions. Two novel etching techniques have been developed to alleviate the etching damage and high charge density at the regrowth interface, including low-power ICP dry etching and ALE. With these advancements in selective area regrowth and doping, high-voltage low-leakage regrown GaN p-n junctions and the normally-off vertical GaN JFETs with regrown p-GaN gate were successfully demonstrated. This work shows that selective area regrowth and doping via MOCVD are promising for the future development of novel advanced vertical GaN power devices.

Acknowledgments This work was supported in part by ARPA-E PNDIODES Program under Grant No. DE-AR0000868 and in part by the National Science Foundation (NSF) under Award No. ECCS-2302696. We thank Dawei Wang, Hunter D. Ellis, Dr. Stephen M. Goodnick, Dr. David J. Smith, Dr. Robert J. Nemanich, and Dr. Fernando A. Ponce for their contributions to the work.

References

1. S. Chowdhury et al., Lateral and vertical transistors using the AlGaN/GaN heterostructure. IEEE Trans. Electron Devices **60**, 3060 (2013)
2. S. Chowdhury et al., Current status and scope of gallium nitride-based vertical transistors for high-power electronics application*. Semicond. Sci. Technol. **28**, 074014 (2013)
3. H. Amano et al., The 2018 GaN power electronics roadmap. J. Phys. D Appl. Phys. (2018)
4. Y. Zhang et al., Gallium nitride vertical power devices on foreign substrates: A review and outlook. J. Phys. D Appl. Phys. **51**, 273001 (2018)
5. J.Y. Tsao et al., Ultrawide-bandgap semiconductors: Research opportunities and challenges. Adv. Electron. Mater. **4** (2018)
6. J. Wei et al., GaN power IC technology on P-GaN gate HEMT platform. Jpn. J. Appl. Phys. **59**, SG0801 (2020)
7. H. Fu et al., *Vertical GaN-on-GaN Power Devices, in Wide Bandgap Semiconductor-Based Electronics* (IOP Publishing, 2020)
8. K.J. Chen et al., GaN-on-Si power technology: Devices and applications. IEEE Trans. Electron Devices **64**, 779 (2017)

9. R. Sun et al., GaN power integration for high frequency and high efficiency power applications: A review. IEEE Access **8**, 15529 (2020)
10. T. Ueda, GaN power devices: Current status and future challenges. Jpn. J. Appl. **58**, SC0804 (2019)
11. T. Oka, Recent development of vertical GaN power devices. Jpn. J. Appl. Phys. **58**, SB0805 (2019)
12. I.C. Kizilyalli et al., Vertical power P-n diodes based on bulk GaN. IEEE Trans. Electron Devices **62**, 414 (2014)
13. Y. Cao et al., High-voltage vertical GaN Schottky diode enabled by low-carbon metal-organic chemical vapor deposition growth. Appl. Phys. Lett. **108** (2016)
14. H. Fu et al., Ultra-low turn-on voltage and on-resistance vertical GaN-on-GaN Schottky power diodes with high mobility double drift layers. Appl. Phys. Lett. **111** (2017)
15. S. Han et al., Fluorine-implanted termination for vertical GaN Schottky rectifier with high blocking voltage and low forward voltage drop. IEEE Electron Device Lett. **40**, 1040 (2019)
16. S. Han et al., High-voltage and high-I_ON/I_OFF vertical GaN-ON-GaN Schottky barrier diode with nitridation-based termination. IEEE Electron Device Lett. **39**, 572 (2018)
17. Y. Cao et al., Improved performance in vertical GaN Schottky diode assisted by AlGaN tunneling barrier. Appl. Phys. Lett. **108** (2016)
18. L. Sang et al., Initial leakage current paths in the vertical-type GaN-on-GaN Schottky barrier diodes. Appl. Phys. Lett. **111** (2017)
19. B. Ren et al., Nearly ideal vertical GaN Schottky barrier diodes with ultralow turn-on voltage and on-resistance. Appl. Phys. Express **10**, 051001 (2017)
20. N. Tanaka et al., 50 a vertical GaN Schottky barrier diode on a free-standing GaN substrate with blocking voltage of 790 V. Appl. Phys. Express **8**, 071001 (2015)
21. Y. Saitoh et al., Extremely low on-resistance and high breakdown voltage observed in vertical GaN Schottky barrier diodes with high-mobility drift layers on low-dislocation-density GaN substrates. Appl. Phys. Express **3**, 081001 (2010)
22. H. Fu et al., High voltage vertical GaN P-n diodes with hydrogen-plasma based guard rings. IEEE Electron Device Lett. **41**, 127 (2019)
23. I.C. Kizilyalli et al., 4-KV and 2.8- MΩ-Cm^2 vertical GaN p-n diodes with low leakage currents. IEEE Electron Device Lett. **36**, 1073 (2015)
24. H. Fu et al., Effect of buffer layer design on vertical GaN-on-GaN p-n and Schottky power diodes. IEEE Electron Device Lett. **38**, 763 (2017)
25. I.C. Kizilyalli et al., High voltage vertical GaN P-n diodes with avalanche capability. IEEE Trans. Electron Devices **60**, 3067 (2013)
26. H. Fu et al., Nonpolar vertical GaN-on-GaN p–n diodes grown on free-standing (1010) m-plane GaN substrates. Appl. Phys. Express **11**, 111003 (2018)
27. I.C. Kizilyalli et al., 3.7 KV vertical GaN PN diodes. IEEE Electron Device Lett. **35**, 247 (2013)
28. J. Wang et al., High voltage, high current GaN-on-GaN p-n diodes with partially compensated edge termination. Appl. Phys. Lett. **113** (2018)
29. K. Nomoto et al., GaN-on-GaN p-n Power Diodes with 3.48 KV and 0.95 MΩ-Cm^2: A Record High Figure-of-Merit of 12.8 GW/Cm^2, in *2015 IEEE International Electron Devices Meeting (IEDM)*, December 2015
30. J. Wang et al., High Voltage Vertical P-n Diodes with Ion-Implanted Edge Termination and Sputtered SiNx Passivation on GaN Substrates, in *2017 IEEE International Electron Devices Meeting (IEDM)*, December 2017
31. Z. Hu et al., Near Unity ideality factor and Shockley-read-hall lifetime in GaN-on-GaN p-n diodes with avalanche breakdown. Appl. Phys. Lett. **107** (2015)
32. K. Nomoto et al., 1.7-KV and 0.55- $m\omega$. cm^2 GaN p-n diodes on bulk GaN substrates with avalanche capability. IEEE Electron Device Lett. **37**, 161 (2015)
33. H. Fukushima et al., Deeply and vertically etched Butte structure of vertical GaN p–n diode with avalanche capability. Jpn. J. Appl. Phys. **58**, SCCD25 (2019)

34. H. Fukushima et al., Vertical GaN p–n diode with deeply etched Mesa and the capability of avalanche breakdown. Appl. Phys. Express **12**, 026502 (2019)
35. T. Maeda et al., Parallel-Plane Breakdown Fields of 2.8–3.5 MV/Cm in GaN-on-GaN p-n Junction Diodes with Double-Side-Depleted Shallow Bevel Termination, in *2018 IEEE International Electron Devices Meeting (IEDM)*, December 2018
36. H. Fu et al., High performance vertical GaN-on-GaN p-n power diodes with hydrogen-plasma-based edge termination. IEEE Electron Device Lett. **39**, 1018 (2018)
37. T. Maeda et al., Design and fabrication of GaN P-n junction diodes with negative beveled-Mesa termination. IEEE Electron Device Lett. **40**, 941 (2019)
38. H. Fu et al., Implantation-and etching-free high voltage vertical GaN p–n diodes terminated by plasma-hydrogenated p-GaN: Revealing the role of thermal annealing. Appl. Phys. Express **12**, 051015 (2019)
39. S. Usami et al., Correlation between dislocations and leakage current of P-n diodes on a free-standing GaN substrate. Appl. Phys. Lett. **112** (2018)
40. B. Rackauskas et al., Leakage mechanisms in GaN-on-GaN vertical Pn diodes. Appl. Phys. Lett. **112** (2018)
41. D. Ji et al., Experimental determination of impact ionization coefficients of electrons and holes in gallium nitride using homojunction structures. Appl. Phys. Lett. **115** (2019)
42. L. Cao et al., Experimental characterization of impact ionization coefficients for electrons and holes in GaN grown on bulk GaN substrates. Appl. Phys. Lett. **112** (2018)
43. Y. Zhang et al., Vertical GaN junction barrier Schottky rectifiers by selective ion implantation. IEEE Electron Device Lett. **38**, 1097 (2017)
44. Y. Zhang et al., Novel GaN Trench MIS Barrier Schottky Rectifiers with Implanted Field Rings, in *2016 IEEE International Electron Devices Meeting (IEDM)*, December 2016
45. Y. Zhang et al., Trench formation and corner rounding in vertical GaN power devices. Appl. Phys. Lett. **110** (2017)
46. T. Hayashida et al., Vertical GaN merged PiN Schottky diode with a breakdown voltage of 2 KV. Appl. Phys. Express **10**, 061003 (2017)
47. W. Li et al., Design and realization of GaN trench junction-barrier-Schottky-diodes. IEEE Trans. Electron Devices **64**, 1635 (2017)
48. C. Gupta et al., *In situ oxide,* GaN interlayer-based vertical trench MOSFET (OG-FET) on bulk GaN substrates. IEEE Electron Device Lett. **38**, 353 (2017)
49. D. Ji et al., Demonstrating >1.4 KV OG-FET Performance with a Novel Double Field-Plated Geometry and the Successful Scaling of Large-Area Devices, in *2017 IEEE International Electron Devices Meeting (IEDM)*, December 2017
50. D. Ji et al., Large-area in-situ oxide, GaN interlayer-based vertical trench MOSFET (OG-FET). IEEE Electron Device Lett. **39**, 711 (2018)
51. T. Oka et al., 1.8 MΩ·cm^2 vertical GaN-based trench metal–oxide–semiconductor field-effect transistors on a free-standing GaN substrate for 1.2-KV-class operation. Appl. Phys. Express **8**, 054101 (2015)
52. T. Oka et al., Vertical GaN-based trench metal oxide semiconductor field-effect transistors on a free-standing GaN substrate with blocking voltage of 1.6 KV. Appl. Phys. Express **7**, 021002 (2014)
53. R. Li et al., 600 V/1.7 - normally-off Gan vertical trench metal-oxide-semiconductor field-effect transistor. IEEE Electron Device Lett. **37**, 1466 (2016)
54. H. Nie et al., 1.5-KV and 2.2-m Ω-Cm^2 vertical GaN transistors on bulk-GaN substrates. IEEE Electron Device Lett. **35**, 939 (2014)
55. D. Shibata et al., 1.7 KV/1.0 MΩcm^2 Normally-off Vertical GaN Transistor on GaN Substrate with Regrown p-GaN/AlGaN/GaN Semipolar Gate Structure, in *2016 IEEE International Electron Devices Meeting (IEDM)*, December 2016
56. D. Ji et al., Normally OFF trench CAVET with active mg-doped GaN as current blocking layer. IEEE Trans. Electron Devices **64**, 805 (2016)

57. M. Sun et al., High-performance GaN vertical fin power transistors on bulk GaN substrates. IEEE Electron Device Lett. **38**, 509 (2017)
58. Y. Zhang et al., 1200 V GaN Vertical Fin Power Field-Effect Transistors, in *2017 IEEE International Electron Devices Meeting (IEDM)*, December 2017
59. H. Fu et al., Selective area regrowth and doping for vertical gallium nitride power devices: Materials challenges and recent Progress. Mater. Today **49**, 296 (2021)
60. B.J. Baliga, *Fundamentals of Power Semiconductor Devices* (Springer US, Boston, 2008)
61. B.J. Baliga, *Silicon Carbide Power Devices* (World Scientific, 2006)
62. W. Sung et al., A new edge termination technique for high-voltage devices in 4H-SiC–multiple-floating-zone junction termination extension. IEEE Electron Device Lett. **32**, 880 (2011)
63. A.V. Bolotnikov et al., Junction termination extension implementing drive-in diffusion of boron for high-voltage SiC devices. IEEE Trans. Electron Devices **57**, 1930 (2010)
64. R. Ghandi et al., High-voltage (2.8 KV) implantation-free 4H-SiC BJTs with Long-term stability of the current gain. IEEE Trans. Electron Devices **58**, 2665 (2011)
65. A. Salemi et al., 15 KV-class implantation-free 4H-SiC BJTs with record high current gain. IEEE Electron Device Lett. **39**, 63 (2017)
66. B.N. Feigelson et al., Multicycle rapid thermal annealing technique and its application for the electrical activation of mg implanted in GaN. J. Cryst. Growth **350**, 21 (2012)
67. J.D. Greenlee et al., Symmetric multicycle rapid thermal annealing: Enhanced activation of implanted dopants in GaN. ECS J. Solid State Sci. Tech. **4**, P382 (2015)
68. H. Sakurai et al., Highly effective activation of mg-implanted p-type GaN by ultra-high-pressure annealing. Appl. Phys. Lett. **115** (2019)
69. K. Fu et al., Investigation of GaN-on-GaN vertical p-n diode with regrown p-GaN by metalorganic chemical vapor deposition. Appl. Phys. Lett. **113** (2018)
70. M. Monavarian et al., High-voltage regrown nonpolar m -plane vertical p-n diodes: A step toward future selective-area-doped power switches. IEEE Electron Device Lett. **40**, 387 (2019)
71. S.W. King et al., Cleaning of AlN and GaN surfaces. J. Appl. Phys. **84**, 5248 (1998)
72. G. Koblmüller et al., High-temperature molecular beam epitaxial growth of AlGaN/GaN on GaN templates with reduced Interface impurity levels. J. Appl. Phys. **107** (2010)
73. A.L. Rosa et al., Understanding Si adsorption on GaN (0001) surfaces using first-principles calculations. Phys. Rev. B **73** (2006)
74. A. Aragon et al., Interfacial impurities and their electronic signatures in high-voltage regrown nonpolar m-plane GaN vertical p–n diodes. Phys. Status Solidi A **217** (2020)
75. M. Yang et al., Maskless selective growth of semi-polar (11$\bar{2}$2) GaN on Si (311) substrate by metal organic vapor phase epitaxy. J. Cryst. Growth **311**, 2914 (2009)
76. Z. Zheng et al., Nonalloyed Ohmic contact of AlGaN/GaN HEMTs by selective area growth of single-crystal N+-GaN using plasma assisted molecular beam epitaxy. Phys. Status Solidi A **208**, 951 (2011)
77. F. Yang et al., The suppression of background doping in selective area growth technique for high performance normally off AlGaN/GaN MOSFET. J. Mater. Sci. **26**, 9753 (2015)
78. S.R. Alugubelli et al., Dopant profiling in P-i-n GaN structures using secondary electrons. J. Appl. Phys. **126** (2019)
79. H. Liu et al., Non-uniform mg distribution in GaN Epilayers grown on Mesa structures for applications in GaN power electronics. Appl. Phys. Lett. **114** (2019)
80. P.-Y. Su et al., Lateral and vertical growth of Mg-doped GaN on Trench-patterned GaN films. Appl. Phys. Lett. **117** (2020)
81. P.-Y. Su et al., The effect of low-angle off-Axis GaN substrate orientation on the surface morphology of Mg-doped GaN Epilayers. J. Appl. Phys. **128** (2020)
82. H. Liu et al., Influence of substrate Misorientation on the optical properties of mg-doped GaN. J. Appl. Phys. **127** (2020)
83. K. Hiramatsu et al., Fabrication and characterization of low defect density GaN using facet-controlled epitaxial lateral overgrowth (FACELO). J. Cryst. Growth **221**, 316 (2000)

84. J. Wang et al., III-nitride blue light-emitting diodes utilizing hybrid tunnel junction with low excess voltage. Semicond. Sci. Tech. **35**, 125026 (2020)
85. S. Nakamura et al., Hole compensation mechanism of P-type GaN films. Jpn. J. Appl. Phys. **31**, 1258 (1992)
86. S.M. Myers et al., Diffusion, release, and uptake of hydrogen in magnesium-doped gallium nitride: Theory and experiment. J. Appl. Phys. **89**, 3195 (2001)
87. Y. Kuwano et al., Lateral hydrogen diffusion at P-GaN layers in nitride-based light emitting diodes with tunnel junctions. Jpn. J. Appl. Phys. **52**, 08JK12 (2013)
88. W. Li et al., Activation of buried P-GaN in MOCVD-regrown vertical structures. Appl. Phys. Lett. **113** (2018)
89. W. Li et al., Development of GaN vertical trench-MOSFET with MBE regrown channel. IEEE Trans. Electron Devices **65**, 2558 (2018)
90. W. Li et al., 600 V GaN Vertical V-Trench MOSFET with MBE Regrown Channel, in 2017 75th Annual Device Research Conference (DRC), June 2017
91. W. Li et al., Realization of GaN PolarMOS using selective-area regrowth by MBE and its breakdown mechanisms. Jpn. J. Appl. Phys. **58**, SCCD15 (2019)
92. Z. Hu et al., 1.1-KV vertical GaN p-n diodes with p-GaN regrown by molecular beam epitaxy. IEEE Electron Device Lett. **38**, 1071 (2017)
93. S. Guha et al., Selective area metalorganic molecular-beam epitaxy of GaN and the growth of luminescent microcolumns on Si/SiO_2. Appl. Phys. Lett. **75**, 463 (1999)
94. H. Sekiguchi et al., Ti-mask selective-area growth of GaN by RF-plasma-assisted molecular-beam epitaxy for fabricating regularly arranged InGaN/GaN nanocolumns. Appl. Phys. Express **1**, 124002 (2008)
95. A. Bengoechea-Encabo et al., Understanding the selective area growth of GaN nanocolumns by MBE using Ti nanomasks. J. Cryst. Growth **325**, 89 (2011)
96. R. Yeluri et al., Design, fabrication, and performance analysis of GaN vertical electron transistors with a buried p/n junction. Appl. Phys. Lett. **106** (2015)
97. S.C. Cruz et al., Crystallographic orientation dependence of dopant and impurity incorporation in GaN films grown by metalorganic chemical vapor deposition. J. Cryst. Growth **311**, 3817 (2009)
98. A. Chakraborty et al., Nonpolar A-plane p-type GaN and p-n junction diodes. J. Appl. Phys. **96**, 4494 (2004)
99. J. Wang et al., Mg doping of 3D semipolar InGaN/GaN-based light emitting diodes. Phys. Status Solidi A **211**, 2645 (2014)
100. J. Cazaux, From the physics of secondary electron emission to image contrasts in scanning electron microscopy. J. Electron Microsc. **61**, 261 (2012)
101. J.R. Dennison et al., Evolution of the electron yield curves of insulators as a function of impinging electron Fluence and energy. IEEE Trans. Plasma Sci. **34**, 2204 (2006)
102. H. Seiler et al., Secondary electron emission in the scanning electron microscope. J. Appl. Phys. **54**, R1 (1983)
103. L. Reimer et al., Measuring the backscattering coefficient and secondary electron yield inside a scanning electron microscope. Scanning **3**, 35 (1980)
104. J.P. Long et al., Band bending and photoemission-induced surface photovoltages on clean n - and p -GaN (0001) surfaces. Phys. Rev. B **66** (2002)
105. K.M. Tracy et al., Preparation and characterization of atomically clean, stoichiometric surfaces of n- and p-type GaN (0001). J. Appl. Phys. **94**, 3163 (2003)
106. A. Shih et al., Secondary electron emission from diamond surfaces. J. Appl. Phys. **82**, 1860 (1997)
107. S.R. Alugubelli et al., Determination of electronic band structure by electron holography of etched-and-regrown interfaces in GaN p-i-n diodes. Appl. Phys. Lett. **115** (2019)
108. H. Katayama-Yoshida et al., Codoping method for the fabrication of low-resistivity wide band-gap semiconductors in p-type GaN, p-type AlN and n-type diamond: Prediction versus experiment. J. Phys. Condens. Matter **13**, 8901 (2001)

109. K. Fu et al., The impact of interfacial Si contamination on GaN-on-GaN regrowth for high power vertical devices. Appl. Phys. Lett. **118** (2021)
110. I. Stricklin et al., Investigation of interfacial impurities in M-plane GaN regrown p-n junctions for high-power vertical electronic devices, in *Wide Bandgap Power and Energy Devices and Applications III*, edited by M. Matin, A. K. Dutta, and S. Chowdhury, 2018
111. K. Fu et al., Demonstration of 1.27 KV etch-then-regrow GaN p-n junctions with low leakage for GaN power electronics. IEEE Electron Device Lett. **40**, 1728 (2019)
112. K.A. Hatch et al., External charge compensation in etched gallium nitride measured by X-ray photoelectron spectroscopy. J. Appl. Phys. **131** (2022)
113. Y. Lee et al., Thermal atomic layer etching of titanium nitride using sequential, self-limiting reactions: Oxidation to TiO_2 and fluorination to volatile TiF_4. Chem. Mater. **29**, 8202 (2017)
114. Y. Lee et al., Trimethylaluminum as the metal precursor for the atomic layer etching of Al_2O_3 using sequential, self-limiting thermal reactions. Chem. Mater. **28**, 7657 (2016)
115. K. Fu et al., Reverse leakage analysis for as-grown and regrown vertical GaN-on-GaN Schottky barrier diodes. IEEE J. Electron Devices Soc. **8**, 74 (2020)
116. S. Yamada et al., Reduction of plasma-induced damage in n-type GaN by multistep-bias etching in inductively coupled plasma reactive ion etching. Appl. Phys. Express **13**, 016505 (2019)
117. I.C. Kizilyalli et al., Characterization of vertical GaN p–n diodes and junction field-effect transistors on bulk GaN down to cryogenic temperatures. Semicond. Sci. Technol. **30**, 124001 (2015)
118. D. Ji et al., Design of 1.2 KV power switches with low R_{ON} using GaN-based vertical JFET. IEEE Trans. Electron Devices (2015)
119. S. Kotzea et al., Demonstration of a GaN-based vertical-channel JFET fabricated by selective-area regrowth. IEEE Trans. Electron Devices **65**, 5329 (2018)
120. C. Yang et al., GaN vertical-channel junction field-effect transistors with regrown p-GaN by MOCVD. IEEE Trans. Electron Devices **67**, 3972 (2020)
121. R. Zhang et al., Robust through-fin avalanche in vertical GaN fin-JFET with soft failure mode. IEEE Electron Device Lett. **43**, 366 (2022)
122. J. Liu et al., 1.2-KV vertical GaN fin-JFETs: High-temperature characteristics and avalanche capability. IEEE Trans. Electron Devices **68**, 2025 (2021)

Chapter 14
Acceptor Dopant Ion Implant and Gyrotron Rapid Thermal Annealing of GaN

F. Shadi Shahedipour-Sandvik, Vincent Meyers, Emma Rocco, and Benjamin McEwen

1 Introduction

GaN grown by metalorganic chemical vapor deposition (MOCVD) and doped in situ with Mg is passivated by H and generally not compensated during growth. Post-growth annealing at a modest temperature as low as 650 °C dissociates Mg-H clusters formed during MOCVD growth and activates the material *p*-type [1]. Mg-implanted GaN, by contrast, is highly resistive after implantation due to high degree of compensation. The implantation process generates large numbers of charged point defects and defect clusters. These are thermally stable at temperatures of at least 1100 °C [2]. These defects are predominantly donor-like and compensate for the holes contributed by activated Mg [3, 4]. *P*-conductivity in implanted GaN has therefore only been achieved at high temperatures—generally above 1200 °C [5]. This is well beyond thermodynamic equilibrium near ambient N_2 overpressure. The equilibrium partial pressure of N_2 in GaN increases superlinearly from 0.1 MPa at 845 °C to 100 MPa at 1200 °C [6].

To achieve p-type Mg-implanted material, annealing must therefore be performed using a technique like ultra high-pressure annealing (UHPA), in which an autoclave achieves N_2 overpressure of 1 GPa and permits equilibrium heating of GaN to about 1500 °C. This method has proved effective at activating implanted Mg and removing implant-induced defects [7]. The alternative is metastable annealing, wherein some combination of GaN surface protection (typically through deposition of a cap),

F. S. Shahedipour-Sandvik (✉) · B. McEwen
State University of New York, University at Albany, Albany, NY, USA
e-mail: sshahedipour-sandvik@albany.edu

V. Meyers
Sandia National Laboratory, Albuquerque, NM, USA

E. Rocco
U.S. Naval Research Laboratory, Washington, DC, USA

I. C. Kizilyalli et al. (eds.), *Gallium Nitride and Related Materials*, The Materials Research Society Series, https://doi.org/10.1007/978-3-031-83056-3_14

moderately elevated N2 overpressure, and short-duration pulses are performed sequentially to enable nonequilibrium heating with arbitrary total soak times.

The first demonstration of this was by the Naval Research Laboratory, which developed multicycle rapid thermal annealing (MRTA). This method heats by inductively coupling with a graphite crucible in an N_2 atmosphere at around 3 MPa, heating samples to temperatures in excess of 1500 °C with a few-second pulse duration. Feigelson et al. report achieving over 8% Mg activation efficiency through a nonequilibrium MRTA [8, 9].

Microwave annealing is capable of similar metastable heating via a high thermal ramp-rate by coupling a microwave beam directly with a semiconductor. Through this method, heating at ramp rates >1000 °C/s [10] is possible. In other materials (Si and SiC), scalability to production-scale wafer sizes has been demonstrated [11, 12]. Annealing using a gyrotron microwave source recovers implant-induced lattice dilation as effectively as other heating methods and achieves effective p-type conductivity in Mg-implanted GaN [13, 14]. As shown in Sect. 2, the selection of an appropriate microwave source has the potential to increase the efficiency of GaN heating by microwave even further.

In devices where the electric field shape is a critical design parameter, the ability to determine the distribution of charge carriers and ionized dopants is therefore also critical. These devices include power electronics [15–17], solid-state charge collection devices [18, 19], and optoelectronic sensors [20]. The example of power electronics illustrates this, where achievement of performance at the Baliga FOM requires the formation of selectively doped regions to hold off breakdown at high reverse bias [21]. The junction termination extension (JTE) [22, 23] is frequently used for this purpose. JTEs operate efficiently in Si [24] and SiC [25] but have only recently been demonstrated in GaN [26]. While this is due partially to the challenges of achieving low-resistivity *p*-type GaN, it is also due to the delicate dependence of device breakdown field on selective-area doped (SAD) region (including JTE) conductivity/geometry. Diffusion (at rate D_B) is rapid at the temperatures needed for efficient activation of Mg-implanted *p*-GaN due to its Arrhenius relationship to temperature T (i.e., it is of the form $D(T) = e^{\frac{-E_M}{k_B T}}$ where E_M is the migration energy). This E_M, critically, reflects the most energetically favorable diffusion mechanism. GaN implanted with Mg and annealed at 1400 °C exhibits rapid diffusion at rates of around 10^{-12} cm^{-2}/s. Such levels of diffusion rates are observed in both metastable microwave annealing [27] and equilibrium thermal annealing [28], indicating the D_B is attributable to the defectivity of the implanted material rather than the heating method. At this rate, an initially box-like profile of implanted dopant such as Mg in-diffuses to a graded doping profile after activation-annealing nearly twice the original thickness. By contrast, the lower starting concentration of point defects in in situ doped GaN:Mg results in diffusivity up to two orders of magnitude lower [29].

While increased defectivity of implanted material evidently leads to greater dopant diffusivity, the defects responsible must be investigated in order to understand the effect of annealing on them. Defect formation energy predicts the relative populations of impurities in different lattice locations, such that a higher concentration of a defect state results from a lower formation energy. Upon implantation, the potential GaN acceptor species Be readily forms Be_i [4, 30]. Mg forms almost exclusively Mg_{Ga} states when implanted into GaN when implant is performed at elevated temperatures [31], potentially because the Mg_i formation energy is high when the Fermi level is above the valence band [32]. Ion implant also displaces large numbers of lattice atoms, produces high concentrations of point and cluster native defects [33, 34]. It is shown that Mg diffuses more readily in GaN when V_N and V_{Ga} are present [35, 36], suggesting that the presence of high defect concentrations should enable significantly faster diffusion of Mg than in low-defectivity (e.g. in situ doped material). As described above, however, implant-induced defect concentrations are significantly reduced during annealing. This suggests a general method (as is demonstrated in this chapter) for achieving benefits that might appear mutually exclusive: achieving high *p*-conductivity by application of very high-temperature annealing while retaining the implant dopant profile characteristic of lower-temperature annealing. Annealing ion-implanted material in two thermodynamic regimes, first where implant-induced defects agglomerate and dopant diffuses slowly, then where defects annihilate and diffusion can occur more quickly, can reduce defect-mediated diffusion pathways while enabling high conductivity and full crystalline recovery from implant-induced damage.

2 Criteria for Selecting Microwave Annealing Source

Microwave heating of a semiconductor can be modeled as energy dissipation in a lossy dielectric. Therefore, the principles that enable minimization of energy dissipation by an antenna in a dielectric also enable maximization of energy dissipation (heating efficiency) of a semiconductor by a microwave source. Heating is primarily a bulk process based on the sample electrical properties and microwave frequency. Attenuation of the oscillating microwave field intensity is exponential with depth, and the power dissipated, P_d, in the semiconductor is calculated by integration of the attenuation of the incident field E_o over the sample volume v [37]:

$$P_d = \iiint \sigma \vec{E}^2 \, dv = \frac{1}{2} \iiint \sigma E_0^2 e^{-2\alpha z_0} dv \tag{14.1}$$

where σ is the sample conductivity and the attenuation constant α is a function of σ, microwave frequency ω and relative real permittivity ε'. For a given material relative permittivity and thickness, the frequency at which microwave/RF is most evenly dissipated is $\omega \sim \frac{2c}{d\epsilon''}$, where c is the speed of light in a vacuum and ϵ'' is the imaginary component of permittivity [37]. Maximizing the efficiency of energy absorption in a given semiconductor (and therefore maximizing the efficiency of the overall annealing process) can therefore be accomplished primarily through selection of an appropriate microwave output frequency. For example, a 1 μm doped region and typical GaN $\epsilon'' = 20$ [38], the frequency that should most efficiently absorb is in the high-terahertz regime at 4×10^{12} Hz.

An exceptionally flexible type of microwave source, and the one utilized for the advances described below, is the gyrotron. Output frequency is controlled by the source cavity geometry and is widely variable, ranging from 20 GHz [39] to the low-terahertz range at 527 GHz [40]. This makes gyrotron processing of thin (<1 mm) wafers with little energy dissipation outside the semiconductor sample volume possible.

Gyrotron output of a quasi-Gaussian intensity profile can be focused using relatively simple beam optics, enabling a large amount of power (often 10 s of kW) to be focused on a small area for rapid heating. Microwave frequency can be preferably selected such that the surface ($\alpha \leq d$) rather than bulk ($\alpha \geq d$) absorbs the majority of the energy and therefore heats more rapidly.

The mature emission source technology [41] and widespread prevalence of 2.45 GHz sources, however, mean that a large portion of the microwave annealing research applied to semiconductors has utilized this relatively low frequency [42–45]. The effects of higher frequencies, such as those in the gyrotron range, have demonstrated effects not achievable at lower frequency [46, 47]. These effects are often observed at low temperatures relative to the temperature required by a thermal anneal method [46, 48]. High-power, high-frequency emission utilizing other RF/microwave sources may be applicable, and has not to-date been explored [49].

The potential advantages of microwave annealing over thermal counterparts derive from the mechanism of energy absorption in the semiconductor. Thermal annealing either heats a semiconductor holder, which in turn heats the semiconductor via conduction or convection, or heats the semiconductor and holder together. Microwave energy can interact selectively with the semiconducting medium, transmitting through or reflecting from the annealing vessel/holder. This is because it is sensitive to the bulk electrical properties of the semiconductor. In microwave annealing, an oscillating field interacts with dipolar structures within the medium (e.g. charged defects states) as well as the bulk. The effect of a microwave anneal at a given temperature may therefore differ significantly from a thermal anneal [50].

The interaction of a microwave beam (e.g. a gyrotron operating at 60 GHz in the experimentation described below) with a semiconductor is best modeled as absorption of energy in a lossy dielectric [37]. In this sense the basic parameters of microwave heating are well-understood. The magnitude of an electromagnetic wave $\vec{E}$ proceeding in $\hat{z}$ through a lossy medium attenuates as

$$\vec{E} = E_0 e^{-\alpha z} \tag{14.2}$$

where E_0 is the incident wave amplitude, z is the distance within the semiconductor from the face on which the microwave beam is incident, and α [cm^{-1}] is the attenuation constant. This is defined as

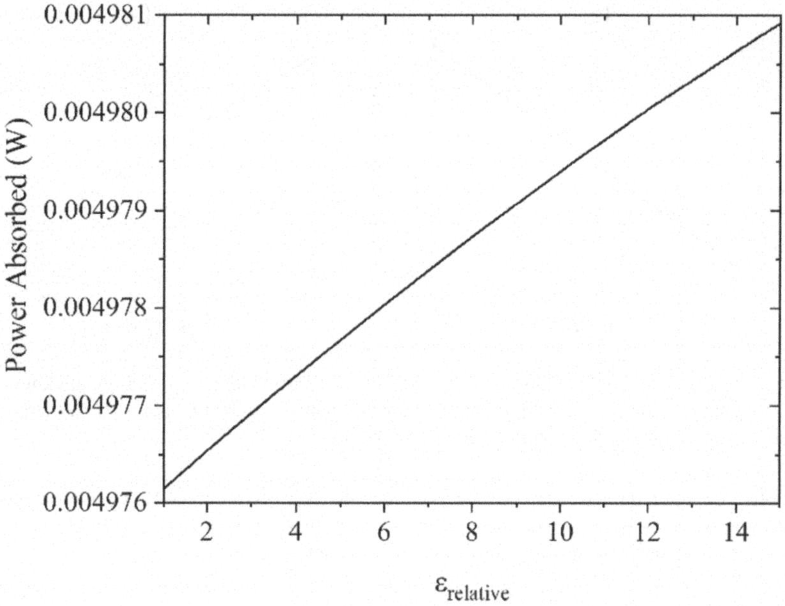

Fig. 14.1 Power dissipated in a generic solid at a frequency ω = 2π(60 GHz), using an incident field of E_0 = 1 kV/m, with an area of 1 cm^2, thickness 1 μm, and bulk resistivity ρ = 1 Ω cm, varying ε from 1–15

$$\alpha = \omega\sqrt{\epsilon\mu_0} \times \sqrt{\frac{1-\cos(\delta)}{2\cos(\delta)}} \tag{14.3}$$

where ε is the relative permittivity, ω the angular frequency, μ_0 the permeability of free space and the dielectric loss angle is $\delta = \frac{\sigma}{\omega\epsilon}$. The attenuation constant also defines the skin depth of a lossy dielectric as $d_{skin} = \alpha^{-1}$ [cm]. Power dissipated P_d in a layer of lossy dielectric can thus be calculated. Ohm's law $\vec{J} = \sigma \times \vec{E}$ allows the definition of power as $\vec{J} \times \vec{E} = \sigma \times \vec{E}^2$ which, when integrated over a volume v yields Eq. (14.1). The indefinite integral over a volume with area A and depth z_0 yields the expression for the total power dissipated in a volume as a function of the incident electric field magnitude E_0

$$P_d[W] = \frac{A}{4\alpha}\sigma\vec{E}^2\left(1 - e^{-2\alpha z_0}\right) \tag{14.4}$$

The parameter dependence of dissipated power is well-illustrated in univariate plots of P_d holding other parameters at typical values, in this case typical values for GaN. Using Eqs. (14.1) and (14.4), these dependencies (within selected material property ranges) are plotted in Figs. 14.1 and 14.2.

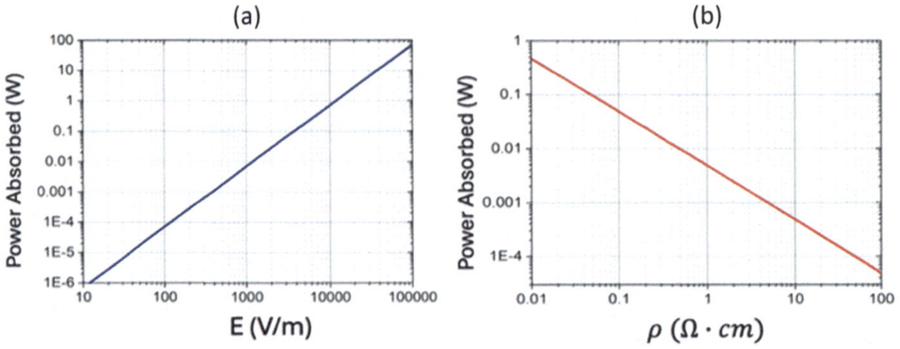

Fig. 14.2 Power dissipated in GaN at a frequency of $\omega = 2\pi(60\ \text{GHz})$, $\varepsilon_{HF} = 5.35$, thickness 1 μm, and (**a**) varying electric field amplitude with a constant resistivity of p-GaN $\rho = 0.7\ \Omega$ cm, and (**b**) varying resistivity with a constant field of $E_0 = 1$ kV/m

Table 14.1 Computed microwave volumetric power dissipation for a 60 GHz gyrotron microwave at 1 kV/m propagating in material of the specified conductivity, permittivity, and thickness denoted as P(σ,ε,t). Skin depth d for these materials is also computed

Material	Thickness [μm]	Typical conductivity [S/cm]	Real permittivity [unitless]	Dissipated power [W/cm^3]	Skin depth [mm]
Semi-insulating GaN	0.25	10–6	5.35	5×10^{-5}	122,000
p-GaN	0.25	1.5	5.35	71.3	0.183
n-GaN	330	100	5.35	156	20.6
n-Si, p-Si	700	100	11.7	73.5	20.6
Intrinsic Si	700	0.01	11.7	0.502	18.2
n-SiC	300	50	6.52	121	29.1

From these figures the exponential dependence on microwave field strength and semiconductor (or any workpiece) resistivity is illustrated. Typical electrical parameters conductivity (σ) and permittivity (ε), absorbed power and skin depth for a frequency of 60 GHz and an arbitrary field strength of 1 kV/m can be calculated using Eqs. (14.3) and (14.4), as shown in Table 14.1. These values are presented to convey the dependence of absorbed microwave power to the electrical properties of the material absorbing power (being annealed, in this case).

The microwave absorption behavior described above can be applied to any microwave source and semiconductor. In the annealing processing described here, a number of observations may be made about the expected dynamics of gyrotron microwave annealing of GaN. As-implanted layers have high resistivity and therefore absorb energy inefficiently. They are instead heated by conduction from lower resistivity (e.g. *n*-doped) layers of the semiconductor. An effective anneal process activates the doped regions, decreasing their resistivity and cause them to absorb with increasing efficiency. The skin depth is shallower than a typical GaN wafer thickness (several hundred μm), and the sample is placed on a highly conductive substrate that reflects normally incident microwave beam. As a result of this, a GaN sample can be heated to 1500 °C and removed by hand ~1 minute afterward without any active cooling system.

The gyrotron microwave source used in this study has an operating frequency of 60 GHz with maximum power output of 40 kW. The gyrotron output is conditioned to a circular profile with an approximately Gaussian intensity distribution. Microwave output is steered and focused by lenses and mirrors into a pressure vessel with a microwave- and optically transparent quartz top cover. The microwave mirror is rastered during annealing to increase the uniformity of heating on-wafer. The pressure vessel is made of stainless steel with a 1.25-inch thick quartz window and can contain ambient up to 4 MPa. GaN described here is annealed under 3 MPa high purity N_2. During annealing, GaN samples are placed on a substrate of conductive SiC-coated graphite. The high electrical conductivity of this substrate means it (like the steel portion of the pressure vessel) acts as an efficient reflector of microwave power. Sample temperature is measured by optical pyrometer. Annealing is performed in a "pulsed" heating scheme, where power is applied to heat the sample at 180 °C/s to a specified maximum temperature and held for a set period of time. Then power is cut off and the sample permitted to cool to below 1000 °C. This is repeated until the desired total anneal soak time is attained (typically 1 minute).

3 Effectiveness of Metastable Annealing via Gyrotron

Comprehensive experimentation and analysis of GaN:Mg annealing by gyrotron has been conducted and a summary is presented here. Greater detail can be found elsewhere [27, 51–54]. Full recovery of lattice structure in the implant region, substantial reduction of point defects, and *p*-type conductivity are achieved with a thermal budget of 1450 °C annealing for approximately 1 minute. Three factors cumulatively enable this heating far outside the thermodynamic stability of GaN. The first is the deposition of an AlN cap on both sides of the sample—which provides a high-thermal-stability intimate contact with the GaN to help prevent N desorption. The second is a modest N_2 overpressure, which raises the equilibrium thermal stability of GaN to ~1000 °C. The third is carefully controlled anneal pulse length, which in combination with a cap and N_2 overpressure, is found to be up to 8 s at 1450 °C. This approach to inhibiting thermal degradation is thus, to some degree, parametric, enabling a large possible space of damage-inhibiting factors depending on the microwave-semiconductor annealing system design.

The anneal conditions described enable *p*-type conductivity to be achieved in GaN implanted with Mg and Mg + N to a depth of 250 nm. A useful simple model (demonstration) device is performing ion implantation in unintentionally doped (*u*-) GaN grown on bulk *n*-GaN. Effective annealing then yields *p-i-n* structures which can be tested as lateral (for Ohmic contact formation) or vertical (as a diode) devices. Ion implantation results in a large amount of N vacancies which compensate for the acceptors. A complete removal of these defects is very difficult even with a post-implant high-temperature annealing. The use of the above-stoichiometric [N] introduced by co-implanting it in the Mg-implanted region is believed to help remove V_N during the post-implant anneal step, thereby increasing net hole concentration [4, 55].

Our studies show that annealing of material implanted with Mg alone (to 10^{19} cm^{-3}) results in relatively resistive *p*-type material and low hole concentration, whereas co-implantation of Mg with N improves acceptor activation, achieving free hole concentrations up to 10^{17} cm^{-3}. This free carrier concentration compares favorably even with in situ doped *p*-GaN with comparable Mg concentrations. Ohmic conduction in transfer length method (TLM) arrays using *p*-GaN contacts (Ni/Au) on implant-annealed material (both Mg and Mg + N) exhibit linear I-V, confirming both the conductivity type and the high quality of the surface after annealing and cap removal. Vertical diode devices with Ohmic *p* and *n* contacts on the front and back side of the material exhibit clear *p-n* junction characteristics [52, 56].

Utilization of co-located photoluminescence (PL) measurements and diode measurements of implanted and annealed GaN show that the low-temperature PL spectrum can be used independently as a method to indicate *p*-type conductivity. This is useful given that PL is a nondestructive, on-wafer measurement with relatively rapid feedback characteristics. PL band signatures of interest include the UVL_{Mg}, the GL2, and the YL, and their relative intensities (I_{UVL}, etc.), which can provide indication of conductivity type. In GaN:Mg, the UVL_{Mg} arises at 3.27 eV from the Mg_{Ga} acceptor state as a relatively efficient band with a narrow FWHM [56, 57]. The YL1 is a broad band centered at 2.2 eV that dominates much of the GaN PL spectrum at or below room temperature. It is attributed to the C_N state (typically unintentionally incorporated) and is therefore ubiquitous in MOCVD GaN regardless of doping type [58]. Ion implantation creates large concentrations of V_N and V_{Ga} that are otherwise not present in large numbers in MOCVD-grown GaN. The isolated V_{Ga} is not associated with a luminescence band. The donor-like V_N is present as the GL2 band at about 2.33 eV [59]. Finally, the BL_{Mg} band arises due to a deep Mg_{Ga} when [Mg] is high—typically above 10^{19} cm^{-3} [60, 61].

Increasing annealing temperature in the range from 1200 °C to 1450 °C (as described above, for ~1 minute) induces complex behavior in the evolution of the PL spectrum. Implantation generates large densities of nonradiative centers in the implant region, leading to low photoluminescence quantum efficiency (and intensity here) from the implanted region [62], which remains low even after annealing of up to 1100 °C. Annealing at 1250 °C causes sufficient removal of nonradiative centers to result in a clear PL spectrum dominated by the UVL_{Mg} and GL2. As shown in Fig. 14.3, further increases in the annealing temperature result in monotonic increases in the intensity ratio of UVL_{Mg} and GL2. Because these intensities are related to the concentration of Mg_{Ga} and V_N, they can serve as a noncontact, nondestructive indicator of electrical conductivity type. For the implant conditions described above, when the acceptor-related UVL_{Mg} band intensity is much higher

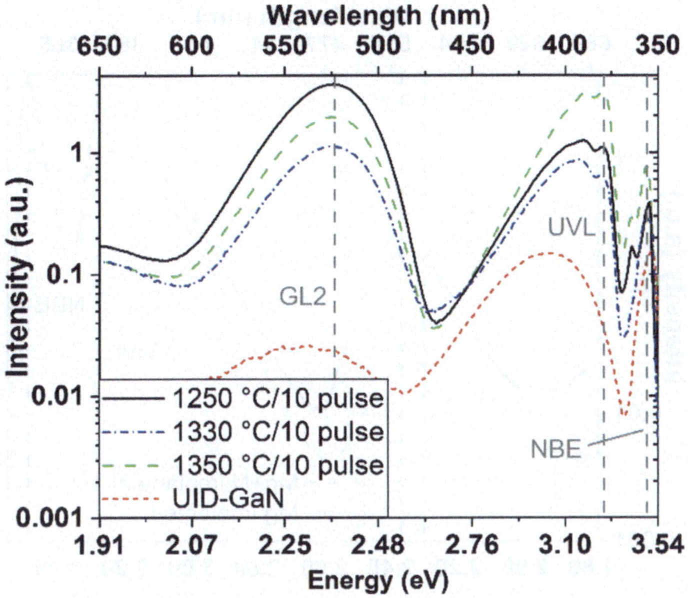

Fig. 14.3 Low-temperature photoluminescence spectra of the as-grown reference u-GaN film as well as Mg-implanted samples C, D, and E after gyrotron anneal at 1250 °C, 1330 °C, and 1350 °C, respectively. Luminescence bands GL2 at 2.3–2.4 eV, UVL at 3.27 eV, and NBE at 3.48 eV are visible in each implanted/annealed sample, while the n^- reference is dominated by NBE, BL, and YL. (Reprinted from [51], with the permission of AIP Publishing)

than the intensity of the V_N-related GL2 band (typically ~5×), the material within the excitation volume is found to be *p*-type.

For Mg-only implantation at the implant conditions considered here, *p*-type material is observed where the magnitude of the UVL band significantly exceeds the magnitude of the GL2 ($I_{UVL} > I_{GL2}$). Where $I_{UVL} < I_{GL2,}$ the material exhibits semi-insulating electrical behavior. This demonstrates that suppression of V_N is critical to obtaining *p*-conductivity in the material. A comparison of the post-anneal PL spectra of singly and co-implanted material shows that co-implant is effective in suppressing the GL2 band, indicating substantially reduced V_N. Figure 14.4 shows luminescence spectra of material implanted with [Mg] $= 10^{19}$ cm^{-3} to a depth of 250 nm, or with Mg and N to a depth of 250 nm in a 1:1 concentration ratio in *u*-GaN, followed by a 1400 °C gyrotron anneal. This suppression of the GL2 band in implanted material has been observed only recently but holds significant promise as a mechanism for controlling conductivity in implanted *p*-GaN [36, 52, 63].

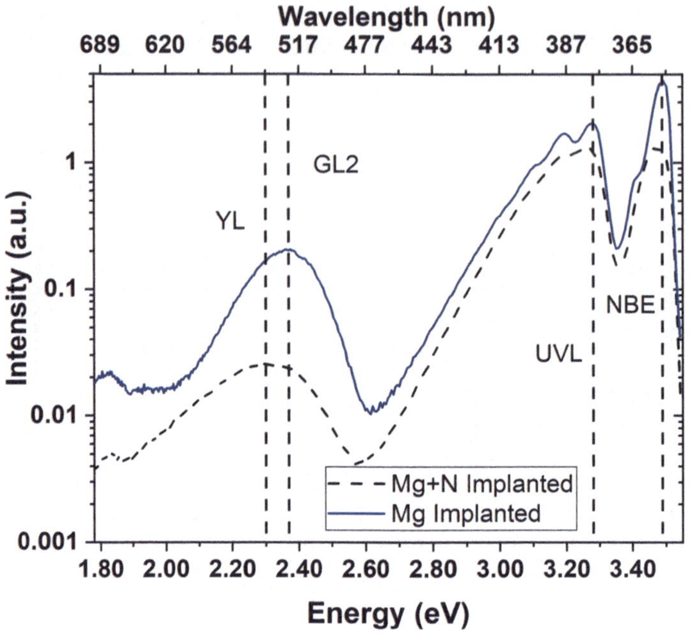

Fig. 14.4 Low-temperature photoluminescence spectra of samples implanted with Mg and Mg + N after annealing at 1400 °C for approximately 1 minute. (Reprinted from [52], with the permission of AIP Publishing)

4 Suppression of Mg Diffusion Due to Ultra High-Temperature Annealing

It is highly desirable to find experimental conditions where reduction in both implant-induced defects and Mg dopant diffusion can be achieved simultaneously. Reduction in the concentration of point defects and healing of implant-induced lattice distortion at temperatures below 1000 °C has been reported previously [64, 65]. However, annealing at this temperature is insufficient to induce *p*-type conductivity in Mg-implanted GaN. As shown above, it can be achieved at annealing temperatures around 1400 °C; however, annealing implant-damaged material at this temperature induces rapid defect-mediated diffusion. Considering the need for these separate thermodynamic regimes, a promising approach to achieving *p*-type activation in implanted regions while suppressing diffusion may be conceived: an initial anneal at a moderate temperature to reduce implant-induced diffusion-mediating defects followed by a high-temperature anneal to activate the implanted regions and create *p*-type material.

To test this proposition, it is helpful to compare diffusion due to these anneal conditions separately. Samples were prepared with Mg + N co-implanted box profiles [Mg] = [N] = 10^{19} cm^{-3} to a depth of 250 nm and capped with AlN. Complete details of the growth and implantation conditions are described elsewhere

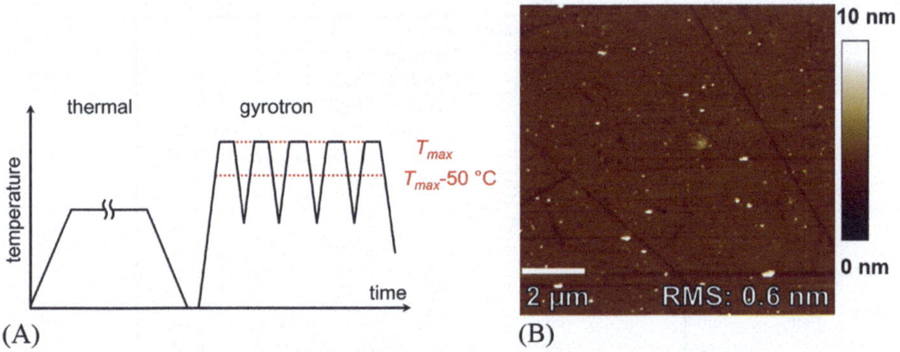

Fig. 14.5 (**a**) schematic of the thermal profiles of continuous thermal and pulsed gyrotron-microwave annealing steps, and (**b**) representative AFM micrograph of the GaN surface after implantation, implant anneal, and AlN cap removal process. (Reprinted from [53], with the permission of AIP Publishing)

Table 14.2 Five annealing conditions performed for this study to compare the effect of thermal annealing only, gyrotron-microwave annealing only, and the two methods performed sequentially

Anneal experiment	Thermal anneal	Gyrotron anneal T_{max}
A	Yes	n/a
B	No	1420 °C
C	No	1500 °C
D	Yes	1420 °C
E	Yes	1500 °C

Reprinted from [53], with the permission of AIP Publishing

[51, 52]. Two methods of annealing were used, as shown schematically in Fig. 14.5a and detailed in Table 14.2. Method one is a conventional tube furnace anneal at 1000 °C for 30 min under flowing nitrogen ambient at atmospheric pressure. As with gyrotron anneals, anneal time in the furnace is specified as the time that the sample was within 50 °C of the nominal maximum temperature. Method two is pulsed gyrotron annealing performed in heating cycles ramping to 1420 or 1500 °C, holding temperature constant for 5–8 s, and cooling to 775 °C between pulses for a total soak time of <60 s. Experiments consist of method 1 only, method 2 only, or method 1 followed by method 2, as summarized in Table 2. Diffusion of Mg was analyzed by secondary ion mass spectrometry (SIMS) after confirmation of a high-quality surface in Fig. 14.5b.

Some level of Mg diffusion is observed after all annealing conditions except anneal A. However, significantly greater diffusion is observed after 1-step annealing conditions B and C than after 2-step annealing conditions D and E. Among the 1-step annealing conditions B and C, [Mg] after anneal C is diffused into the GaN to a greater extent than after anneal B, as expected due to the higher annealing temperature (Fig. 14.6).

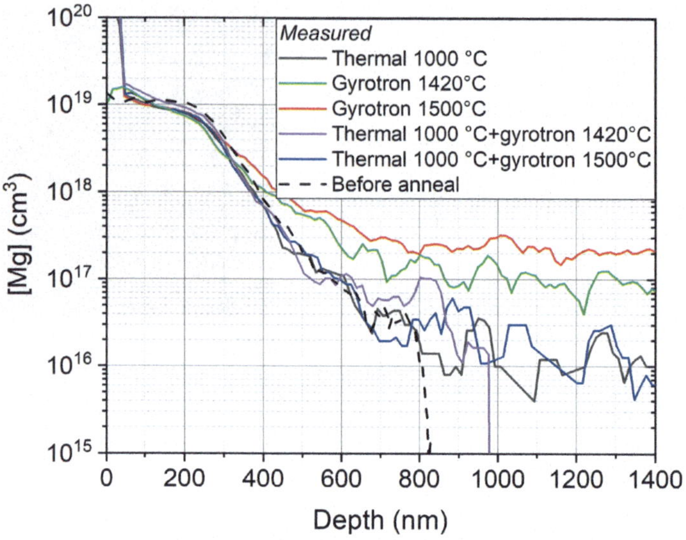

Fig. 14.6 SIMS depth profiles before and after one-step or two-step annealing at moderate temperature and high temperature, as described in Table 14.2. (Reprinted from [53], with the permission of AIP Publishing)

Table 14.3 $\tilde{D}(C)$ at and below the Matano plane extracted from each Mg profile after annealing

Anneal	$\tilde{D}(C)$ [cm^2 × s^{-1}] at		
	x_M	x_M + 100 nm	x_M + 500 nm
A	n/a[a]	n/a[a]	n/a[a]
B	1.2×10^{-12}	3.4×10^{-12}	8.1×10^{-13}
C	2.0×10^{-12}	2.5×10^{-12}	1.2×10^{-12}
D	2.7×10^{-13}	3.4×10^{-13}	1.4×10^{-14}
E	4.3×10^{-13}	6.9×10^{-13}	5.9×10^{-14}

Reprinted from [53], with the permission of AIP Publishing
[a]No significant diffusion after anneal

Impurity diffusion in GaN is most often modeled under the first-order Gaussian solution of Fick's second law that assumes a homogeneous medium [29]. In this case, homogeneity is an incorrect assumption because implanted Mg is diffusing from a (relatively) disordered implant region to a more pristine region below it. In this case, diffusion using a Boltzmann-Matano approach is appropriate. This approach has recently proved to be a powerful tool for understanding implant-annealed GaN diffusion dynamics [53, 66, 67]. In this method, diffusivity $\tilde{D}(C)$ is defined as a function of concentration, and dopant diffuses from one region (the implant volume) to another (pristine GaN below the implant volume) at a depth termed the Matano plane x_M.

Applying the Boltzmann-Matano method to the samples under study, $\tilde{D}(C)$ was calculated between x_M and x_M + 500 nm for each annealing condition, given in

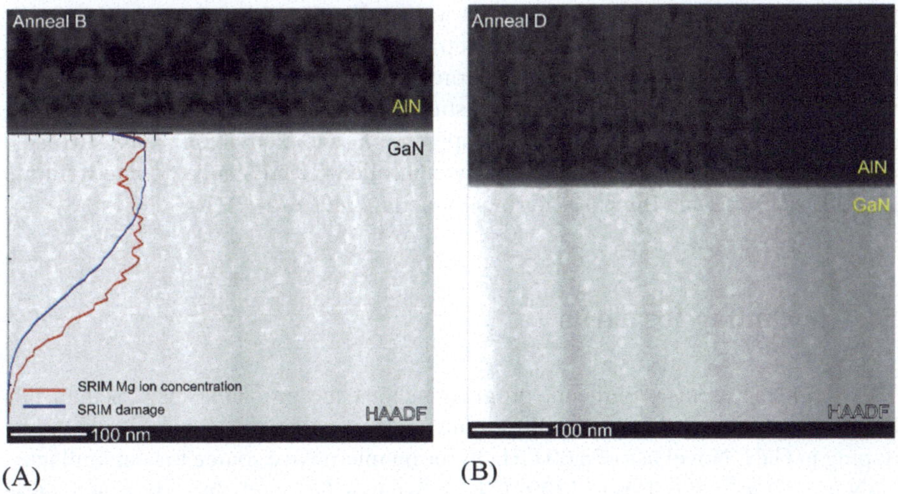

Fig. 14.7 Mg-implanted GaN imaged by TEM prior to removal of protective AlN cap and after (**a**) gyrotron-only anneal B with inset line profiles indicating SRIM-simulated [Mg] and displaced lattice atoms as a function of depth and (**b**) thermal and gyrotron anneal D. (Reprinted from [53], with the permission of AIP Publishing)

Table 14.3. A decrease in Mg diffusivity of approximately 10× is achieved through the use of a two-step anneal (D, E) compared to one-step anneal (B, C).

The decrease in diffusivity $\tilde{D}(C)$ observed when a 1000 °C anneal is performed prior to the high-temperature microwave gyrotron anneal of implanted samples indicates that the 1000 °C step increases migration energy E_M of Mg. This is in turn evidence that the dominant diffusion pathway is altered by the addition of the 1000 °C pre-anneal. Assuming a typical $E_M = 2$ eV [68], diffusivity is larger at 1450 °C than at 1000 °C by a factor of 100, meaning that diffusion of Mg occurs almost entirely in the higher-temperature pulsed gyrotron annealing, rather than in the thermal anneal step. However, damage reduction performed by the moderate-temperature step influences the Mg diffusivity during the high-temperature step.

For each anneal condition, $\tilde{D}(C)$ is comparable at x_M and 100 nm below x_M (where C is typically an order of magnitude lower). Both experimental [69] and theoretical studies [35] have found that the V_{Ga} is at least as diffusive as Mg in GaN. This suggests that V_{Ga} may diffuse with Mg and mediate diffusion below the original implant profile. At a depth of x_M + 500 nm, however, Mg diffusivity decreases by ~10x relative to the implanted region. At this depth, SRIM predicts a negligible concentration of vacancies and TEM imaging indicates no defect clustering, as shown in Fig. 14.7.

This supports the assumption of vacancy-mediated higher diffusivity in/near the implanted region and lower diffusivity at greater depth in the material where vacancy defects are considerably lower. This two-step annealing approach for implanted GaN reduces redistribution of dopant but does not fully suppress it. Even the very short annealing times explored here cause distortion of the implant profile. Related multi-step annealing work indicates that a three-step anneal improves recovery of ion-implanted GaN lattice over a two-step anneal without compromising electrical characteristics [70]. This illustrates the promise of greater controllability using multi-step approaches and the need for further study of the interacting effects of annealing conditions and methods on defect suppression and activation. This two-step approach provides a path for more tightly controlled selective-area doping in future power electronics and dimension-critical junction devices.

5 Concluding Remarks

This work represents significant progress against the problems discussed in this book—those of creating, controlling, and understanding the dynamics of *p*-type doping in GaN. Novel use of a 60 GHz gyrotron microwave source to heat implanted GaN in <10 second pulses to 1450 °C is shown to achieve reliable *p*-type activation in Mg-implanted GaN with minimal degradation of the surface. Co-implant of Mg with N in a 1:1 ratio is shown to achieve $[h^+]$ up to 10^{17} cm^{-3} through the mechanism of removing V_N created by the implant. A 3-pronged approach is demonstrated to enable extremely nonequilibrium heating: a protective cap of AlN, a moderate N_2 overpressure, and rapid, pulsed heating. It is also shown that the rapid diffusion induced by this ultra-high temperature annealing can be significantly suppressed by using a 1000 °C pre-anneal step to reduce the implant damage that facilitates Mg diffusion. Microwave annealing, despite its proven utility for semiconductor annealing, remains almost entirely unexplored, especially using the high frequencies that promise the most efficient heating. The work presented in this chapter shows that focused research efforts on the use of high-frequency microwave annealing of implanted (Al)GaN:Mg,N is of great promise to achieve conductive *p*-AlGaN.

References

1. D.H. Youn, M. Lachab, M. Hao, T. Sugahara, H. Takenaka, Y. Naoi, S. Sakai, Investigation on the p-type activation mechanism in Mg-doped GaN films grown by metalorganic chemical vapor deposition. Jpn. J. Appl. Phys. **38**(2R), 631 (1999)
2. A. Uedono, H. Iguchi, T. Narita, K. Kataoka, W. Egger, T. Koschine, C. Hugenschmidt, M. Dickmann, K. Shima, K. Kojima, S. Chichibu, Annealing behavior of vacancy-type defects in mg-and H-implanted GaN studied using monoenergetic positron beams. Phys. Status Solidi B **256**(10), 1900104 (2019)
3. A. Wenzel, C. Liu, B. Rauschenbach, Effect of implantation-parameters on the structural properties of Mg-ion implanted GaN. Mater. Sci. Eng. B **59**(1–3), 191–194 (1999)
4. K.T. Liu, Y.K. Su, S.J. Chang, Y. Horikoshi, Magnesium/nitrogen and beryllium/nitrogen coimplantation into GaN. J. Appl. Phys. **98**(7) (2005)
5. T. Niwa, T. Fuji, T. Oka, High carrier activation of Mg ion-implanted GaN by conventional rapid thermal annealing. Appl. Phys. Expr. **10**(9) (2017)

6. J. Unland, B. Onderka, A. Davydov, R. Schmid-Fetzer, Thermodynamics and phase stability in the Ga–N system. J. Cryst. Growth **256**(1–2), 33–51 (2003)
7. H. Sakurai, M. Omori, S. Yamada, Y. Furukawa, H. Suzuki, T. Narita, K. Kataoka, M. Horita, M. Bockowski, J. Suda, T. Kachi, Highly effective activation of Mg-implanted p-type GaN by ultra-high-pressure annealing. Appl. Phys. Lett. **115**(14), 142104 (2019)
8. B.N. Feigelson, T.J. Anderson, M. Abraham, J.A. Freitas, J.K. Hite, C.R. Eddy, F.J. Kub, Multicycle rapid thermal annealing technique and its application for the electrical activation of Mg implanted in GaN. J. Cryst. Growth **350**(1), 21–26 (2012)
9. J.D. Greenlee, B.N. Feigelson, T.J. Anderson, J.K. Hite, K.D. Hobart, F.J. Kub, Symmetric multicycle rapid thermal annealing: Enhanced activation of implanted dopants in GaN. ECS J. Solid State Sci Technol. **4**(9), 382–386 (2015)
10. S. Grundas, Advances in induction and microwave heating of mineral and organic materials. BoD–Books on Demand, 2011, pp. 459–481
11. T.L. Alford, D.C. Thompson, J.W. Mayer, N.D. Theodore, Dopant activation in ion implanted silicon by microwave annealing. J. Appl. Phys. **106**(11), 114902 (2009)
12. S.G. Sundaresan, M.V. Rao, Y.L. Tian, M.C. Ridgway, J.A. Schreifels, J.J. Kopanski, Ultrahigh-temperature microwave annealing of Al+-and P+-implanted 4H-SiC. J. Appl. Phys. **101**(7), 073708 (2007)
13. K. Hogan, S. Tozie, E. R. I. Mahaboob, V. Meyers, B. McEwen, F. Shahedipour-Sandvik, R. Tompkins, M. Derenge, K. Jones, M. Shevelev, Novel Gyrotron Beam Annealing Method for Mg-Implanted Bulk GaN, IRPS IEEE Special Issue, 2019
14. K. Hogan, E. Rocco, S. Tozier, V. Meyers, B. McEwen, I. Mahaboob, M. Shevelev, V. Sklyar, R. Tompkins, M. Derenge, K. Jones, M. Taheri, J. Hart, W. Sung, F. Shahedipour-Sandvik, *Toward Selective Activation of Mg Doped GaN: Impact of Pulsed and Continuous Gyrotron Annealing* (Bellevue, 2019)
15. M. Xiao, R. Zhang, D. Dong, H. Wang, Y. Zhang, Design and simulation of GaN superjunction transistors with 2-DEG channels and fin channels. IEEE J. Emerg. Select. Top. Pwr Electron. **7**(3), 1475–1484 (2019)
16. G. Feng, J. Suda, T. Kimoto, Space-modulated junction termination extension for ultrahigh-voltage pin diodes in 4H-SiC. IEEE Trans. Electron Devices **59**(2), 414–418 (2011)
17. I. Mahaboob, M. Yakimov, E. Rocco, K. Hogan, F. Shahedipour-Sandvik, Drain-voltage-induced secondary effects in AlGaN/GaN HEMTs with integrated body-diode. IEEE Trans. Electron Devices **67**(10), 3983–3987 (2020)
18. K. Hogan, M. Rodriguez, E. Rocco, V. Meyers, B. McEwen, F. Shahedipour-Sandvik, In operando investigation of GaN PIN device characteristics under electron irradiation energies comparable to Pm-147 source for betavoltaic application. AIP Adv. **10**(8), 085110 (2020)
19. R. Zheng, J. Lu, Y. Liu, X. Li, X. Xu, R. He, Z. Tao, Y. Gao, Comparative study of GaN betavoltaic battery based on pn junction and Schottky barrier diode. Radiat. Phys. Chem. **168**, 108595 (2020)
20. E. Rocco, I. Mahaboob, K. Hogan, V. Meyers, B. McEwen, L.D. Bell, F. Shahedipour-Sandvik, Impurity incorporation and diffusion from regrowth interfaces in N-polar GaN photocathodes and the impact on quantum efficiency. J. Appl. Phys. **129**(19), 195701 (2021)
21. X. She, A. Huang, O. Lucia, B. Ozpineci, Review of silicon carbide power devices and their applications. IEEE Trans. Ind. Electron. **64**(10), 8193–8205 (2017)
22. J. Wierer, J. Dickerson, A. Allerman, A. Armstrong, M. Crawford, R. Kaplar, Simulations of junction termination extensions in vertical GaN power diodes. IEEE Trans. Electron Devices **64**(5), 2291–2297 (2017)
23. T. Anderson, J. Greenlee, B. Feigelson, J. Hite, F. Kub, K. Hobart, Improved vertical GaN Schottky diodes with ion implanted junction termination extension. ECS J. Solid State Sci. Technol. **5**(6), Q176 (2016)
24. V. Temple, W. Tantraporn, Junction termination extension for near-ideal breakdown voltage in pn junctions. IEEE Trans. Electron Devices **33**(10), 1601–1608 (1986)

25. A. Bolotnikov, P. Muzykov, Q. Zhang, A. Agarwal, T. Sudarshan, Junction termination extension implementing drive-in diffusion of boron for high-voltage SiC devices. IEEE Trans. Electron Devices **57**(8), 1930–1935 (2010)
26. H. Lee, Y. Zhang, Z. Chen, M. Rahman, H. Zhao, S. Rajan, Design and fabrication of vertical GaN pn diode with step-etched triple-zone junction termination extension. IEEE Trans. Electron Devices **67**(9), 3553–3557 (2020)
27. V. Meyers, *Advancing Selective-Area Doping and Annealing Strategies for High-Conductivity P-Type AlGaN System (PhD Diss.)* (State University of New York at Albany, Albany, 2023)
28. H. Sakurai, T. Narita, M. Omori, S. Yamada, A. Koura, M. Iwinska, K. Kataoka, M. Horita, N. Ikarashi, M. Bockowski, J. Suda, Redistribution of Mg and H atoms in Mg-implanted GaN through ultra-high-pressure annealing. Appl. Phys. Express **13**(8), 086501 (2020)
29. Z. Benzarti, I. Halidou, Z. Bougrioua, T. Boufaden, B. El Jani, Magnesium diffusion profile in GaN grown by MOVPE. J. Cryst. Growth **310**(14), 3274–3277 (2008)
30. C. Ronning, E.P. Carlson, D.B. Thomson, R.F. Davis, Optical activation of be implanted into GaN. Appl. Phys. Lett. **73**(12), 1622–1624 (1998)
31. U. Wahl, L. Amorim, V. Augustyns, A. Costa, E. David-Bosne, T. Lima, G. Lippertz, J. Correia, M. da Silva, M. Kappers, K. Temst, Lattice location of Mg in GaN: A fresh look at doping limitations. Phys. Rev. Lett. **118**(9), 095501 (2017)
32. G. Miceli, A. Pasquarello, Self-compensation due to point defects in Mg-doped GaN. Phys. Rev. B **93**(16), 165207 (2016)
33. K. Iwata, H. Sakurai, S. Arai, T. Nakashima, T. Narita, K. Kataoka, M. Bockowski, M. Nagao, J. Suda, T. Kachi, N. Ikarashi, Defect evolution in Mg ions implanted GaN upon high temperature and ultrahigh N2 partial pressure annealing: Transmission electron microscopy analysis. J. Appl. Phys. **127**(10), 105106 (2020)
34. K. Lorenz, E. Wendler, A. Redondo-Cubero, N. Catarino, M. Chauvat, S. Schwaiger, F. Scholz, E. Alves, P. Ruterana, Implantation damage formation in a-, c-and m-plane GaN. Acta Mater. **123**, 177–187 (2017)
35. S. Limpijumnong, C.G. Van de Walle, Diffusivity of native defects in GaN. Phys. Rev. B **69**(3), 035207 (2004)
36. K. Shima, R. Tanaka, S. Takashima, K. Ueno, M. Edo, K. Kojima, A. Uedono, S. Ishibashi, S. Chichibu, Improved minority carrier lifetime in p-type GaN segments prepared by vacancy-guided redistribution of Mg. Appl. Phys. Lett. **119**(18), 182106 (2021)
37. C. Balanis, *Advanced Engineering Electromagnetics* (Wiley, Hoboken, 2012)
38. T. Kim, S. Hwang, J. Byun, N. Barange, H. Park, Y. Dong Kim, Temperature dependent dielectric function and the E critical points of hexagonal GaN from 30 to 690 K. AIP Adv. **4**(2) (2014)
39. Y. Bykov, A. Eremeev, M. Glyavin, V. Kholoptsev, A. Luchinin, I. Plotnikov, G. Denisov, A. Bogdashev, G. Kalynova, V. Semenov, N. Zharova, 24-84-GHz gyrotron systems for technological microwave applications. IEEE Trans. Plasma Sci. **32**(1), 67–72 (2004)
40. M. Blank, P. Borchard, S. Cauffman, K. Felch, M. Rosay, L. Tometich, Experimental demonstration of a 527 GHz gyrotron for dynamic nuclear polarization, in IEEE International Conference on Plasma Science, 2013
41. R.K. Jeyachitra, R. Sukanesh, Highly selective and tunable microwave photonic filter using parallel Fabry- Perot filters for universal mobile telecommunications system noise and interference suppression. IET Commun. **5**, 1123–1128 (2011)
42. R. Nipoti, A. Nath, S. Qadri, Y. Tian, C. Albonetti, A. Carnera, M. Rao, High-dose phosphorus-implanted 4H-SiC: Microwave and conventional post-implantation annealing at temperatures$\geq$ 1700° C. J. Electron. Mater. **41**, 457–465 (2012)
43. R. Nipoti, A. Nath, M. Rao, A. Hallén, A. Carnera, Y. Tian, Microwave annealing of very high dose aluminum-implanted 4H-SiC. Appl. Phys. Express **4**(11), 111301 (2011)
44. M. Rao, Ultra-fast microwave heating for large bandgap semiconductor processing, in *Advances in Induction and Microwave Heating of Mineral and Organic Materials*, (Intech Open, 2011), pp. 459–481

45. H. Li, B. Li, L. Deng, P. Xu, Y. Du, S. Ouyang, Z. Liu, Evidence for non-thermal microwave effect in processing of tailing-based glass-ceramics. J. Eur. Ceram. Soc. **39**(4), 1389–1396 (2019)
46. M. Janney, H. Kimrey, W. Allen, J. Kiggans, Enhanced diffusion in sapphire during microwave heating. J. Mater. Sci. **32**, 1347–1355 (1997)
47. J. Olea, G. González-Díaz, D. Pastor, E. García-Hemme, D. Caudevilla, S. Algaidy, F. Pérez-Zenteno, S. Duarte-Cano, R. García-Hernansanz, A. del Prado, E. San Andrés, Ti supersaturated Si by microwave annealing processes. Semicond. Sci. Technol. **38**(2), 024004 (2023)
48. P. Kohli, S. Ganguly, T. Kirichenko, H. Li, S. Banerjee, E. Graetz, M. Shevelev, Microwave annealing for ultra-shallow junction formation. J. Electron. Mater. **31**, 214–219 (2002)
49. J. Bragg, J. Dickens, A. Neuber, Ferrimagnetic nonlinear transmission lines as high-power microwave sources. IEEE Trans. Plasma Sci. **41**(1), 232–237 (2012)
50. C. Fu, Y. Wang, P. Xu, L. Yue, F. Sun, D. Zhang, S.-L. Zhang, J. Luo, C. Zhao, D. Wu, Understanding the microwave annealing of silicon. AIP Adv. **7**(3), 035214 (2017)
51. V. Meyers, E. Rocco, T.J. Anderson, J.C. Gallagher, M.A. Ebrish, K. Jones, M. Derenge, M. Shevelev, V. Sklyar, K. Hogan, B. McEwen, F. Shahedipour-Sandvik, P-type conductivity and damage recovery in implanted GaN annealed by rapid gyrotron microwave annealing. J. Appl. Phys. **128**(8), 085701 (2020)
52. V. Meyers, E. Rocco, K. Hogan, B. McEwen, K. Jones, M. Derenge, M. Shevelev, V. Sklyar, F. Shahedipour-Sandvik, P-type conductivity and suppression of green luminescence in Mg/N co-implanted GaN by Gyrotron microwave annealing. J. Appl. Phys. **130**(8), 085704 (2021)
53. V. Meyers, E. Rocco, B. McEwen, M. Shevelev, V. Sklyar, F. Shahedipour-Sandvik, Defect-mediated diffusion of implanted Mg in GaN: Suppressing dopant redistribution by sequential thermal and microwave annealing. J. Appl. Phys. **133**(15) (2022)
54. B. Raghothamachar, Y. Liu, H. Peng, T. Ailihumaer, M. Dudley, F.S. Shahedipour-Sandvik, K.A. Jones, A. Armstrong, A.A. Allerman, J. Han, H. Fu, X-ray topography characterization of gallium nitride substrates for power device development. J. Cryst. Growth **544**, 125709 (2020)
55. R. Tanaka, S. Takashima, K. Ueno, H. Matsuyama, M. Edo, Demonstration of 1200 V/1.4 mΩ cm2 vertical GaN planar MOSFET fabricated by an all ion implantation process. Jpn. J. Appl. Phys. **59**(SG), SGGD02 (2020)
56. B. Monemar, P. Paskov, G. Pozina, C. Hemmingsson, J. Bergman, S. Khromov, V. Izyumskaya, V. Avrutin, X. Li, H. Morkoç, H. Amano, Properties of the main Mg-related acceptors in GaN from optical and structural studies. J. Appl. Phys. **115**(5) (2014)
57. M.A. Reshchikov, P. Ghimire, D.O. Demchenko, Magnesium acceptor in gallium nitride. I. Photoluminescence from Mg-doped GaN. Phys. Rev. B **97**(20), 205204 (2018)
58. M. Reshchikov, On the origin of the Yellow luminescence band in GaN. Phys. Status Solidi (B), 2200488, 2022
59. M.A. Reshchikov, D.O. Demchenko, J.D. McNamara, S. Fernández-Garrido, R. Calarco, Green luminescence in Mg-doped GaN. Phys. Rev. B **90**(3), 035207 (2014)
60. L. Eckey, U. Von Gfug, J. Holst, A. Hoffmann, A. Kaschner, H. Siegle, C. Thomsen, B. Schineller, K. Heime, M. Heuken, O. Schön, Photoluminescence and Raman study of compensation effects in mg-doped GaN epilayers. J. Appl. Phys. **84**(10), 5828–5830 (1998)
61. U. Kaufmann, M. Kunzer, M. Maier, H. Obloh, A. Ramakrishnan, B. Santic, P. Schlotter, Nature of the 2.8 eV photoluminescence band in Mg doped GaN. Appl. Phys. Lett. **72**(11), 1326–1328 (1998)
62. K. Kojima, S. Takashima, M. Edo, K. Ueno, M. Shimizu, T. Takahashi, S. Ishibashi, A. Uedono, S. Chichibu, Nitrogen vacancies as a common element of the green luminescence and nonradiative recombination centers in Mg-implanted GaN layers formed on a GaN substrate. Appl. Phys. Express **10**(6), 061002 (2017)
63. H. Sakurai, T. Narita, K. Kataoka, K. Hirukawa, K. Sumida, S. Yamada, K. Sierakowski, M. Horita, N. Ikarashi, M. Bockowski, J. Suda, Effects of the sequential implantation of Mg and N ions into GaN for p-type doping. Appl. Phys. Express **14**(11), 111001 (2021)

64. A. Uedono, S. Takashima, M. Edo, K. Ueno, H. Matsuyama, H. Kudo, H. Naramoto, S. Ishibashi, Vacancy-type defects and their annealing behaviors in Mg-implanted GaN studied by a monoenergetic positron beam. Phys. Status Solidi B **12**(252), 2794–2801 (2015)
65. Y. Wang, K. Huynh, M. Liao, H. Yu, T. Bai, J. Tweedie, M. Breckenridge, R. Collazo, Z. Sitar, M. Bockowski, Y. Liu, Strain recovery and defect characterization in Mg-implanted homoepitaxial GaN on high-quality GaN substrates. Phys. Status Solidi B **257**(4), 1900705 (2020)
66. R. Jakiela, K. Sierakowski, T. Sochacki, M. Iwinska, M. Fijalkowski, A. Barcz, M. Bockowski, Investigation of diffusion mechanism of beryllium in GaN. Phys. B Condens. Matter **594**, 412316 (2020)
67. K. Sierakowski, P. Jaroszynski, R. Jakiela, M. Fijalkowski, T. Sochacki, M. Iwinska, M. Turek, K. Lorenz, M. Bockowski, Lateral and vertical diffusion of magnesium in ion-implanted Halide Vapor Phase Epitaxy gallium nitride. Mater. Sci. Semicond. Process. **171**, 108022 (2024)
68. L.M. Amorim, *Lattice Site Location of Electrical Dopant Impurities in Group-III Nitrides (Doctoral Dissertation)* (KU Leuven, Leuven, 2016)
69. A. Uedono, S. Takashima, M. Edo, K. Ueno, H. Matsuyama, W. Egger, T. Koschine, C. Hugenschmidt, M. Dickmann, K. Kojima, S. Chichibu, Carrier trapping by vacancy-type defects in Mg-implanted GaN studied using monoenergetic positron beams. Phys. Status Solidi B **255**(4), 1700521 (2018)
70. J. Greenlee, B. Feigelson, T. Anderson, J. Hite, K. Hobart, F. Kub, From MRTA to SMRTA: Improvements in activating implanted dopants in GaN. ECS Trans. **69**(14), 97 (2015)

Chapter 15
Ion Implantation for p-Type Conductivity in GaN

Pramod Reddy, Michal Bockowski, Ramón Collazo, and Zlatko Sitar

1 Introduction

GaN-based high-power switches represent a significant advancement toward the development of energy-efficient power grids [1, 2]. These switches are known for their potential to significantly reduce power losses and accommodate higher currents, thereby enhancing overall efficiency. Reduced power losses and higher currents leading to higher efficiencies are expected in devices such as vertical GaN CAVET and junction field effect transistors (JFET) power switches [3–5]. Controlling the selective area doping of both n- and p-regions is necessary to realize these device structures. Although controlled p-type carrier concentrations and conductivities in GaN have been reproducibly demonstrated by Mg doping during epitaxy, achieving p-type conductivity with selective area doping (SAD) remained a challenge.

In this section, we briefly present different approaches that were employed to achieve p-type SAD in GaN and their challenges including diffusion, regrowth, and implantation. In the following sections, we focus on implantation-based SAD, post-implantation damage recovery and dopant activation. Finally, we present implementation of implantation-based SAD to achieve kV-class GaN JBS diodes.

Mg diffusion into GaN films from a Mg source at the surface as an approach has been attempted with only limited success. Issues of high thermal diffusion

P. Reddy
Adroit Materials Inc., Cary, NC, USA

M. Bockowski
Institute of High-Pressure Physics, Warsaw, Poland

R. Collazo (✉) · Z. Sitar
Department of Materials Science and Engineering, North Carolina State University, Raleigh, NC, USA
e-mail: rcollaz@ncsu.edu

I. C. Kizilyalli et al. (eds.), *Gallium Nitride and Related Materials*, The Materials Research Society Series, https://doi.org/10.1007/978-3-031-83056-3_15

temperatures, degradation of the GaN surface, and a non-uniform Mg doping profile have inhibited the adoption of this technique for selective area p-doping in GaN [6–8]. A second approach is the etch and regrowth. This would involve selective area etching of underlying GaN epitaxial layers to produce trenches followed by selective area regrowth of p-type GaN in the trenches. Progress in this direction, including a p-n junction diode, has been shown by several groups [9–11]; however, the regrowth interface is plagued by incorporation of donor impurities such as the silicon "spike" and interface states, leading to large reverse leakage currents. Consequently, numerous technical challenges exist, preventing it from gaining widespread adoption.

2 Mg Ion Implantation in GaN for Selective Area p-Type Doping

The primary technique of choice for selective area doping in Si- and SiC-based technology has been selective area ion implantation [12, 13]. Mg is the primary p-type dopant in GaN and has been the acceptor of choice for implantation studies. However, despite early success in n-type doping of GaN via ion implantation [14], p-type doping in the III-nitrides via ion implantation remained a challenge due to the induced damage. Mg implantation damage is assumed to consist of displaced Ga and N matrix atoms from their lattice sites due to the momentum transfer during implantation. The damage manifests itself as a significant strain profile in the implantation direction. For example, as shown in Fig. 15.1 (left), after implantation, a series of XRD peaks appeared at low angles in the 2θ-ω scan of the symmetric (0 0 0 2) reflection [15]. These peaks correspond to the induced strain profile in the implanted region [16]. Further, ion implantation introduces compensating defects as part of implantation damage. For example, Fig. 15.1 (right) shows the SRIM calculated vacancy concentrations generated during Ar and N implantation in

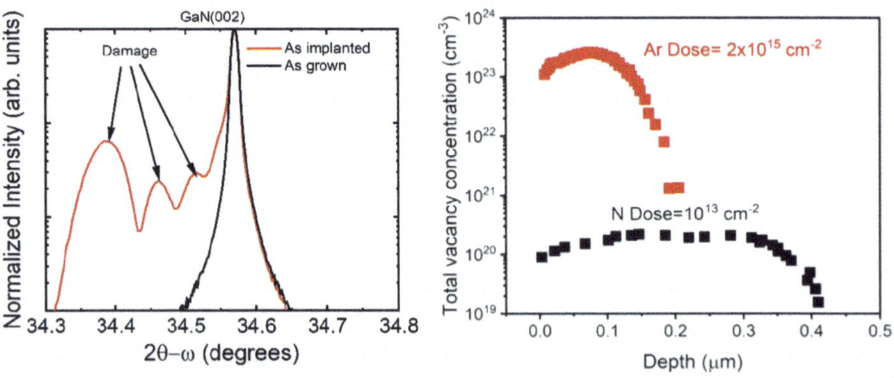

Fig. 15.1 (left) XRD profiles of implanted and as-grown GaN and (right) SRIM simulations calculating the generated vacancy profile for low dose of N and high dose of Ar in GaN

GaN. It is clear that even at a lower dose of 10^{13} cm^{-2}, vacancy concentrations in excess of 10^{20} cm^{-3} are generated, resulting in a highly defective and insulating region. Thus, it is standard practice to anneal the samples after implantation in order to recover crystallinity and to eliminate the resulting strain, as well as to mitigate the concentration of point defects. In the subsequent sections, we will explore recent efforts and advances aimed at achieving Mg-implanted GaN with low defect densities and high carrier concentrations through the process of implantation followed by annealing.

3 Post-implantation Annealing for Damage Recovery and Defects Mitigation

The post-implantation annealing (PIA) process is challenging for GaN. Damage recovery for GaN occurs at temperatures >1200 °C [16], yet the material begins to degrade when temperatures reach ~850 °C [17–19]. The standard free energy of formation of GaN is shown in Fig. 15.2 (left), showing decomposition for temperatures >850 °C at 1 atm of N_2. Annealing at temperatures >1100 °C results in surface decomposition, as shown in Fig. 15.3, posing significant challenges for the commercial adoption of implantation-based doping [20]. In order to prevent surface decomposition during the annealing at elevated temperatures, initial efforts have focused on the use of capping layers (e.g., AlN, Si_3N_4, SiO_2). However, these resulted in unintentional introduction of impurities or alloying elements during the annealing process and potential etch damage to the p-GaN when removing the sacrificial capping layers [11, 21]. This added layers of complexity and potential failure points in the fabrication process and thus required a different solution to

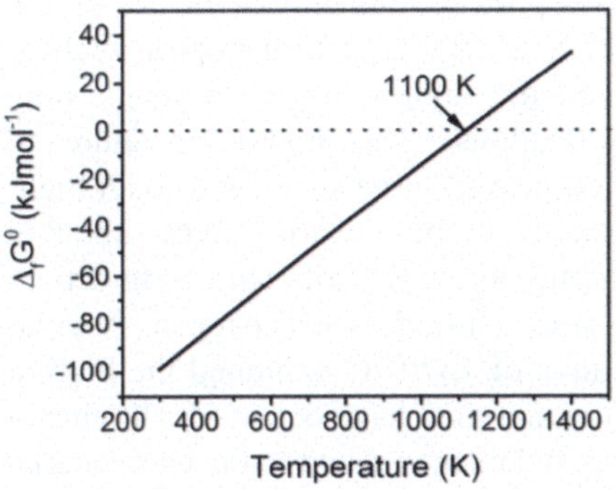

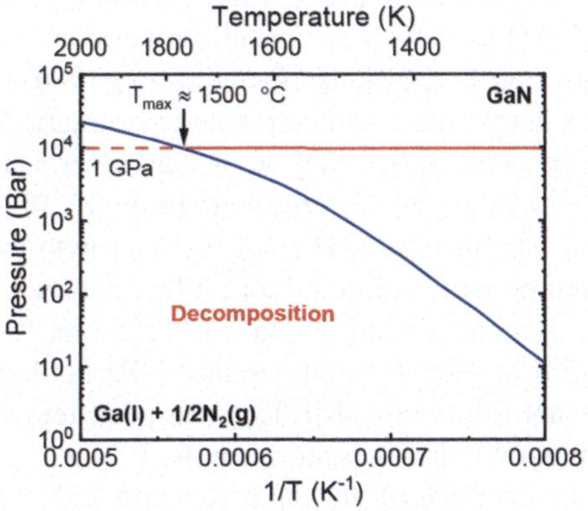

Fig. 15.2 The (left) temperature dependence of standard free energy of GaN and (right) the stabilizing pressure of N_2 as a function of temperature [31, 32]. The standard free energies are obtained from Jacob and Rajitha [33]

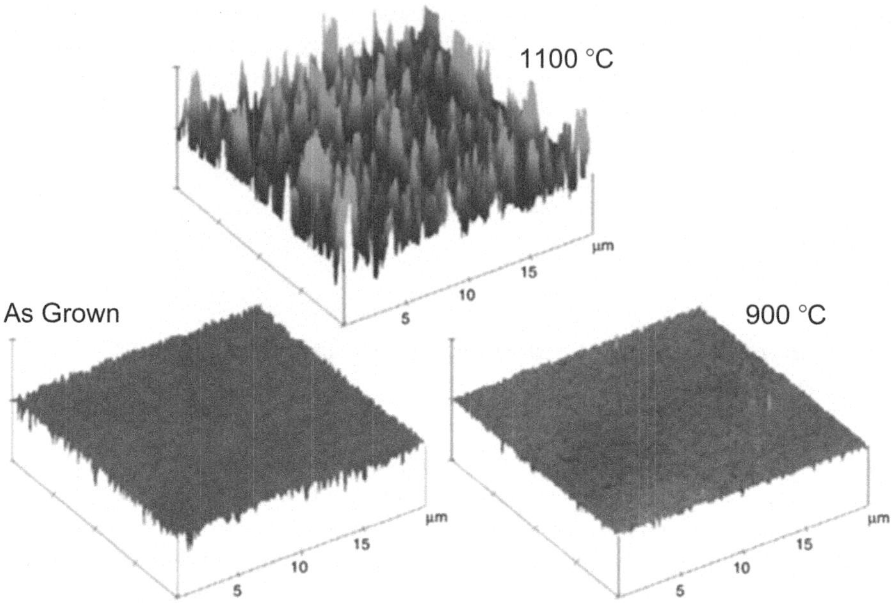

Fig. 15.3 Surface decomposition occurring at higher annealing temperatures as evidenced by AFM micrography. (Reproduced from [20])

p-type Mg implantation in GaN. Consequently, in search of a solution, various innovative PIA methods have been developed, including rapid thermal annealing (RTA) [22], multicycle RTA (MRTA) [23–25], laser thermal annealing (LTA) [26], and microwave annealing [27, 28]. The MRTA technique, originally proposed by Feigelson et al., involves a typical annealing phase at 1050 °C followed by rapid heating and cooling cycles [25]. This technique employed elevated N_2 pressure (20–40 atm) to prevent the decomposition of GaN at 1050 °C due to the pressure-dependent decomposition temperature, which will be discussed in the following section [25, 29]. Jacobs et al. explored various protective materials for use during both traditional annealing and symmetric MRTA (SMRTA) to further optimize these processes [30]. AlN as a cap layer achieved the highest levels of magnesium activation as determined through PL evaluations. The high-temperature to low-temperature (HT–LT) AlN growth method proved to be superior to both high-temperature sputtered and fully sputtered methods. In the scenarios involving other cap materials, all displayed ultraviolet (UV) emission was attributed to Mg activation at elevated temperatures. Particularly, silicon nitride (SiN)-capped, N-polar GaN following SMRTA with peak temperatures of 1370 °C exhibited the highest favorable luminescence ratios. While Mg activation may have occurred, alloying at the interface of the cap layers and GaN, and etch damage related to the cap material removal limited the practical application of the aforementioned procedures.

4 Ultra-High Pressure Annealing for Effective Activation of Mg-Implanted GaN

The surface of GaN may be stabilized by employing an overpressure of N_2. The required pressure to stabilize GaN as a function of annealing temperature is shown in Fig. 15.2 (right). Prior studies with protective cap layers indicated high annealing temperatures >1300 °C were necessary for effective Mg activation in implanted GaN. From Fig. 15.2 (right), it is clear that while an overpressure of 1–2 MPa (10–20 bar) was sufficient to stabilize GaN at 1050 °C, which was the base annealing temperature employed in MRTA, overpressures of several 100 MPa (1000 bar) to 1 GPa (10,000 bar) are necessary to stabilize GaN at annealing temperatures of 1300–1500 °C, required for Mg activation, damage recovery, and point defect reduction.

Accordingly, ultra-high pressure annealing conducted at temperatures ranging from 1300 °C to 1480 °C and under nitrogen pressures of 0.3–1 GPa for the annealing of Mg-implanted GaN samples showed promising results [34–36]. Samples annealed using ultra-high pressure demonstrated significantly greater near-band-edge (NBE) and donor-acceptor pair band emission in cathodoluminescence (CL) measurements, outperforming those treated with conventional RTA by more than an order of magnitude. Concurrently, green luminescence (GL) was substantially reduced. Remarkably, an activation ratio exceeding 70% was achieved, as verified by temperature-dependent Hall and secondary ion mass spectrometry (SIMS) measurements [36]. In this section, we will discuss the ultra-high pressure annealing (UHPA) of Mg-implanted GaN.

4.1 High-Temperature Surface Stabilization

A schematic of the UHPA annealing is shown in Fig. 15.4. To prevent the decomposition of GaN into Ga(g) and N_2, in addition to the N_2 overpressure, the gallium vapor pressure has to be maintained as well. Thus, implanted GaN is annealed under an overpressure of N_2 in contact with polycrystalline GaN and buried in GaN powder, which both act as the source of Ga(g), maintaining the equilibrium Ga pressure and preventing GaN decomposition [32]. Consequently, surface degradation is prevented, and no additional roughening is observed after the annealing, even at 1400 °C, as evidenced by AFM micrography (Fig. 15.5) [15]. Note that no capping layer was employed to stabilize the surface. The surface of the implanted GaN films had the same smooth bilayer step flow structure before and after annealing. There was no evidence of decomposition.

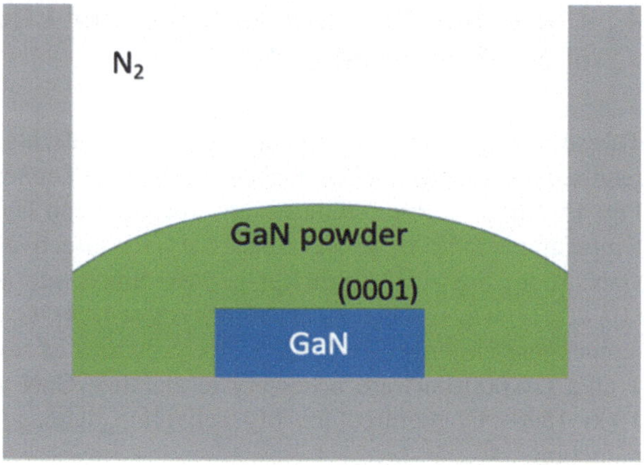

Fig. 15.4 A schematic of the UHPA annealing of GaN [32]. The blue GaN represents the implanted wafer

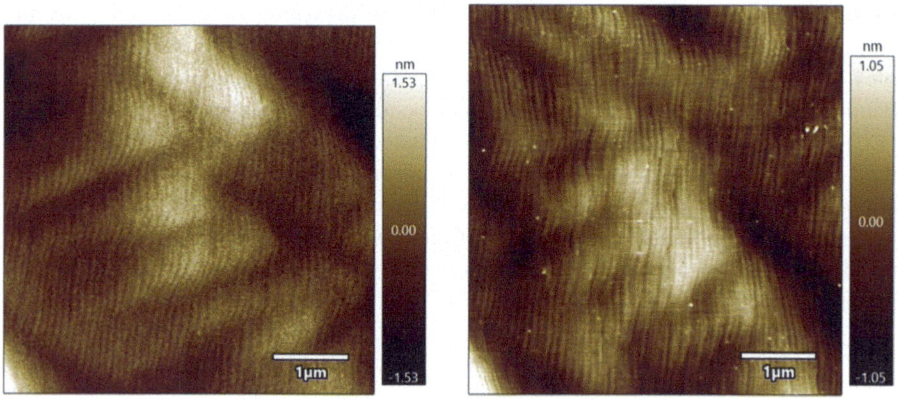

Fig. 15.5 (left) AFM of an as implanted homoepitaxial GaN film; (right) GaN surface after annealing at 1400 °C and 1 GPa N_2, for 10 minutes. (Reproduced from [15])

4.2 Implantation Crystal Damage Recovery

As discussed before, Mg implantation damage is assumed to consist of displaced Ga and N matrix atoms from their lattice sites due to momentum transfer during implantation and the damage manifested itself as a strain profile in the implantation direction. Thus, it was necessary to anneal the samples after implantation in order to recover crystallinity and eliminate the generated strain. The XRD analysis of the implanted GaN as a function of the annealing temperature and time is shown in Fig. 15.6. It is clear that the identified strain peaks and, hence, implantation damage are reduced with an increase in the annealing temperature and time. A majority of the strain relief occurred by annealing at 1200 °C, with the residual damage exhibited as an increase in the FWHM of the (002) peak as compared to the as-grown sample. Increasing the annealing time to 100 minutes or increasing the annealing temperature

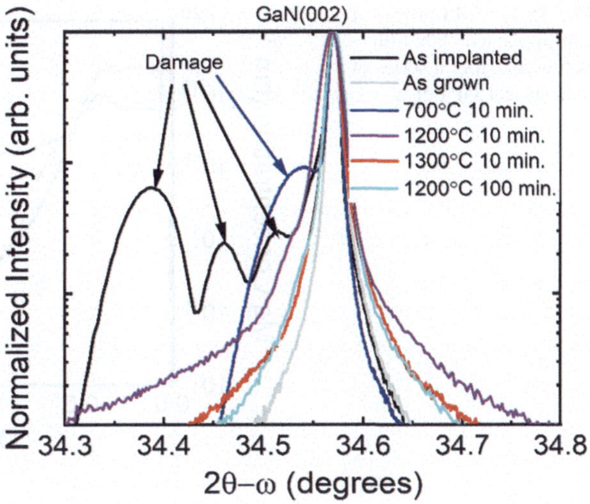

Fig. 15.6 XRD 2θ-ω profiles for a series of Mg-Implanted and annealed homoepitaxial GaN films. (Reproduced from [15])

to 1300 °C, resulted in full recovery of the crystal lattice damage as measured by XRD.

Moreover, following the annealing process, various additional defects, including dislocation loops, inversion domains, and stacking faults, may form within the implanted area [37–39]. These defects are thought to originate from point defects induced during the implantation phase [16]. Wang et al. further investigated the effects of UHPA in Mg-implanted GaN, revealing annealing at 1300 °C for 10 minutes effectively removes implantation-induced strain while leaving behind extended defects such as pyramidal and trapezoidal inversion domains and prismatic dislocation loops [16]. These defects lead to magnesium segregation and reduced electrical activity. Crucially, increasing the annealing temperature above 1400 °C not only reduced this defect density but also eliminated magnesium segregation, further boosting dopant activation efficiency and significantly improving p-type conductivity in GaN [40].

4.3 Dopant Diffusion

It should be highlighted that during the high-temperature annealing, Mg atoms diffuse into the underlying layers and form a diffusion "tail." The influence of annealing on the implanted Mg profile has been investigated by secondary ion mass spectroscopy (SIMS) by Sakurai et al. [41] and Breckenridge et al. [15] by studying the as implanted and annealed samples. As an example, diffusion profiles of "box" implanted GaN with an initial implant profile of 2×10^{19} cm^{-3} for 400 nm are shown as a function of UHPA performed at different temperatures for 40 minutes in Fig. 15.7. Some minor diffusion occurred at 1200 °C, which coincided with full

Fig. 15.7 SIMS profile of Mg-implanted GaN films before and after a series of anneals at 1 GPa of N_2

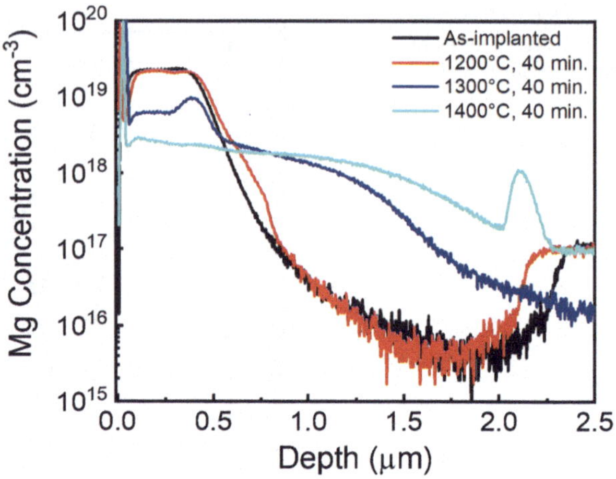

crystal damage recovery (Fig. 15.6). Annealing at higher temperatures (>1300 °C) resulted in significant bulk diffusion of Mg over length scales of 1 μm. The Mg concentration in the implanted region near the surface was accordingly reduced by about an order of magnitude, which is in agreement with the total implant dose conservation.

Interestingly, the XRD and SIMS analyses demonstrated a clear contrast between crystal damage recovery occurring for temperatures ≤1200 °C at 10 minutes and the bulk diffusion occurring at higher annealing temperatures and longer annealing times. This indicated that the mechanism of damage recovery was related to a fast and short-range diffusion with lower activation energy that occurs before the onset of bulk diffusion.

Due to the reduction in Mg concentration in the implanted region and the formation of a diffusion tail, as suggested by Sakurai et al., the UHPA sequence must be meticulously planned, taking into account the diffusion profiles of the implanted species [36].

5 p-Type Conductivity and Diffusion Budget

In this section, we report on studies of Mg activation by the UHPA process and associated p-type conductivity and their dependence on a thermal diffusion budget. Mg implantation was performed at room temperature and post-implantation annealing at 1 GPa of N_2 and a temperature range of 1200–1400 °C. The implantation profile is shown in Fig. 15.7. The high-pressure thermodynamically stabilized the GaN surface without the need for a capping layer. Diffusion budget, related to the diffusion length, was introduced as a convenient engineering parameter for comparing samples annealed at different temperatures and for different times [15]. The diffusion budget, *DB(T,t)*, is defined as:

$$DB = Ae^{-\frac{T_0}{T}}t,$$

where $T_0 = E_g/2k$, E_g is the material bandgap, k is the Boltzmann constant, and A is a scaling parameter chosen to be 2 s^{-1} [15].

The p-type conductivity of the implanted region is a function of Mg activation and compensation, i.e., the concentration of substitutional Mg_{Ga} (N_A) and total compensation, N_D. Accordingly, to understand the influence of the short-range diffusion associated with damage recovery and bulk diffusion on the formation of Mg_{Ga} and compensation, and hence p-type conductivity, temperature-dependent AC Hall measurements were performed. Interestingly, at 1200 °C and 10 minutes, where damage recovery was observed via XRD, but no significant bulk diffusion of Mg was observed via SIMS, the samples were highly resistive, indicating either low Mg activation (unlikely) or high compensation in the implanted region. In contrast, all samples annealed at higher temperatures or for longer times showed p-type conductivity with high Mg activation, as shown in Fig. 15.8. All these samples also showed different degrees of bulk diffusion of Mg, as shown in Fig. 15.7 and a correlation between the conductivity and diffusion budget may be made. At low diffusion budgets (DB $<$ 0.1 corresponding to 1200 °C and 10 minutes), while damage recovery as measured by XRD was achieved, these samples did not show p-type conductivity. Hence simply crystal damage recovery wasn't sufficient to activate Mg. The primary cause for lack of p-type conductivity was shown to be due to heavy compensation by the implantation-induced point defects [15]. In contrast, samples annealed at intermediate diffusion budgets, ($0.1 < DB < 0.4$), a low Mg ionization energy (~115 meV) was observed with almost complete Mg activation. For higher diffusion budgets (DB $>$ 0.4), significant loss of Mg to the surface was observed with a corresponding reduction in p-type conductivity.

Nevertheless, by estimating the effective Mg doping thickness to be ~1 μm (determined by SIMS), the bulk hole concentrations in the ~mid × 10^{17} cm^{-3} were demonstrated at room temperature, indicating that technologically important hole concentrations can be achieved in GaN by Mg implantation and UHPA.

High compensation at low diffusion budgets and loss of Mg at high diffusion budgets present a unique challenge for shallow implants. A direct control of the formation of compensating defects arising from the implantation damage may be necessary to achieve both hole conductivity and low Mg diffusion. In comparison, annealing of Mg-implanted GaAs without significant bulk diffusion resulted in Mg activation and p-type conductivity, indicating that the fast, short-range rearrangement of atoms associated with damage recovery was sufficient to produce Mg_{Ga} acceptor configuration without significant compensation [42, 43]. Further, increasing the annealing times in GaAs also resulted in bulk diffusion, which was accompanied by increased resistivity, attributed to Mg diffusion to the surface and compensation [42]. This is contrary to the observations in Mg:GaN, which shows strong compensation upon initial damage recovery and p-type conductivity only upon long-range diffusion of Mg atoms. At the moment, it is not clear if there is a significant difference in the p-type conductivity of the implanted and diffused

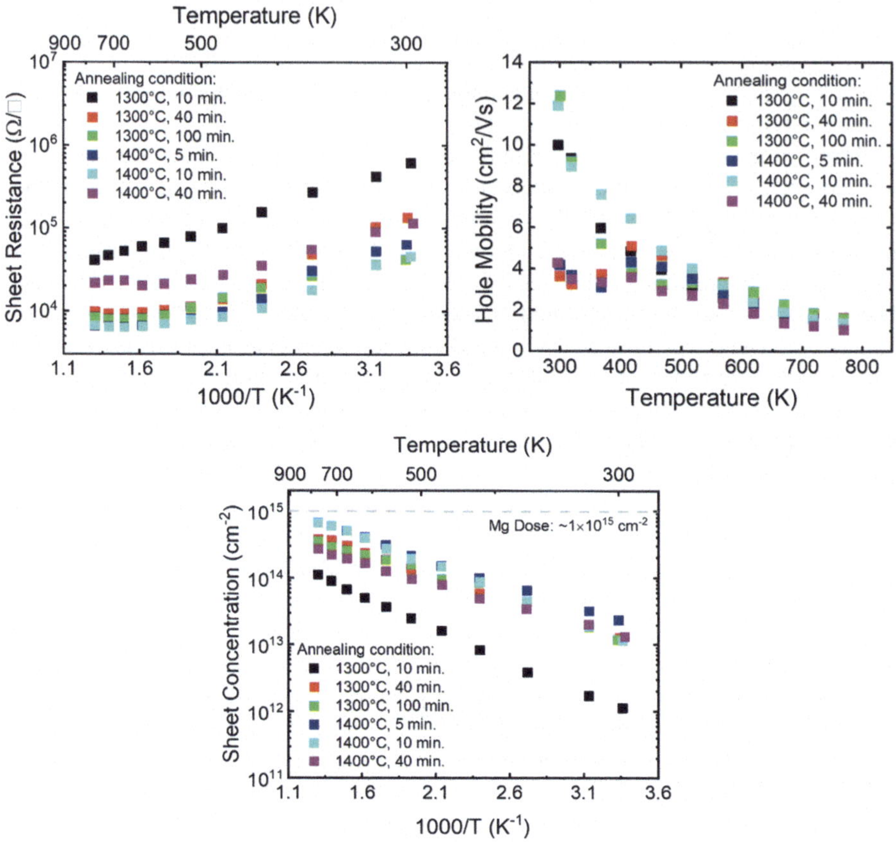

Fig. 15.8 Sheet resistance (top left), hole mobility (top right), and sheet concentration (bottom) of Mg-implanted homoepitaxial GaN as a function of temperature

regions upon high-temperature annealing. Within a single-layer model, we can define an effective Mg activation ratio as the ratio between the maximum observed carrier concentration, that is, $(N_A–N_D)$ measured at high temperature, where all activated but not compensated acceptors produce free holes, and the Mg-implanted dose, as a figure of merit. Although this is an established practice in characterization of semiconductors, one needs to be aware that this method lumps together two effects: (1) the actual Mg activation or positioning of the dopant on the desired lattice site and (2) compensation by a donor point defect, which can be extrinsic or intrinsic.

Figure 15.9 (left) summarizes the dependence of maximum sheet carrier concentration (at 800 K) from Fig. 15.8 as a function of annealing conditions represented by *DB(T,t)*. A general trend is observed with a minimum DB budget needed to achieve p-type conductivity. However, the hole concentration exhibits a "knee" behavior with a maximum value at a certain intermediate DB while decreasing at lower and higher DB.

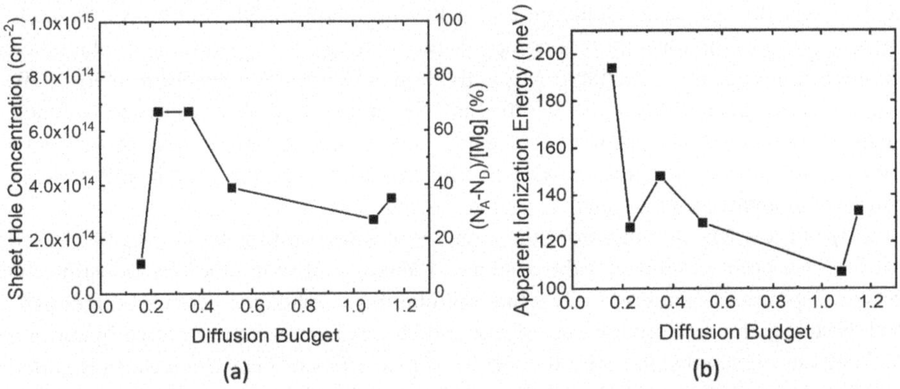

Fig. 15.9 The dependence of hole concentration and *(N_A-N_D)/[Mg]* (left) and apparent activation energy (right) on the diffusion budget

As discussed earlier, the sharp drop in the effective Mg activation at very low diffusion budgets is due to compensation within the implanted region. The source of compensation is implantation-damage-related defects such as V_N. At higher diffusion budgets, the gradual decrease in effective Mg activation is likely due to Mg segregating at the surface, as also observed in GaAs, i.e., a net reduction of the Mg concentration in the bulk GaN, as evident from SIMS shown in Fig. 15.3 [15, 42].

These hypotheses were verified based on the dependence of apparent carrier ionization energy as a function of DB. We expect a higher apparent carrier ionization energy for samples with higher compensation. Thus, based on the dependence of the apparent carrier ionization energy, E_{app}, as a function of diffusion budget, as shown in Fig. 15.9 (right), it is clear that compensation is the primary cause of apparent low effective Mg activation ratios in samples annealed with low diffusion budget. Accordingly, E_{app} decreased from 200 meV for samples annealed at 1300 °C for 10 minutes to 110 meV for samples annealed at 1400 °C for 40 minutes. Hence, we conclude that, while Mg may be activated at lower thermal budgets (1200 °C for 10 minutes), that is, incorporated on the Ga site, the compensation is very high, likely due to point defects associated with the implantation damage. Reducing the compensating point defects is necessary and requires an increase in the diffusion budget. Interestingly, increasing the diffusion budget beyond an optimal value also results in an increase in E_{app}, as shown in Fig. 15.9 (right). This may be explained by the reduction in N_A due to surface accumulation of Mg, resulting in an increased compensation ratio N_D/N_A and, hence, an increase in E_{app}.

5.1 Ion Co-implantation

While one can conclude from the above that larger diffusion budgets are necessary for reducing compensation and increasing effective Mg activation, one needs to realize that this presents challenges in achieving shallow Mg implants. As such, direct control of the formation of compensating defects arising from the implantation damage is necessary to achieve Mg activation without compensation at low DB, where surface accumulation is insignificant. Several point defect control techniques, such as defect quasi-Fermi level control and co-implantation, may need to be invoked to produce shallow active Mg implants [44–47]. Here, we briefly discuss ion co-implantation.

Beyond just Mg implantation, the strategy of ion co-implantation has been used to diminish point defects and the associated leakage currents, thereby boosting the p-type doping efficiency in GaN while maintaining a manageable PIA budget [48–54]. Notably, the co-implantation of Mg and H ions has proven effective in curbing leakage currents and stabilizing the electrical properties of GaN devices, as reported by Narita et al. [48] This technique also helps in reducing the clustering of vacancy-type defects, as confirmed by positron annihilation spectroscopy (PAS) [55]. Additionally, the co-implantation of N or F ions into Mg-doped GaN has been shown to reduce green luminescence, which is attributed to the compensating V_N defects [52, 56]. Recently, Kano et al. highlighted that sequential N implantation in Mg-implanted GaN significantly enhanced the concentration of Mg and suppressed the coarsening of extended defects after annealing. Specifically, the N ion implantation increased the solubility of Mg in GaN and resulted in the formation of nanoscale interstitial-type defects, which did not grow or annihilate post-annealing. Additionally, it reduced the concentration of vacancy-type extended defects, suggesting an overall improvement in the distribution and stability of Mg within the GaN matrix [57].

6 Device Demonstrations Based on Ion Implanted p-GaN Activated Via UHPA

Mg implantation into GaN represents a critical technological step required to realize p-type selective area doping for various devices such as GaN Junction Barrier Schottky (JBS) diodes, Junction Field Effect Transistors (JFETs), and Super-junction diodes and transistors with high voltage blocking capability. Specifically, we describe implementation in junction barrier Schottky diodes (JBS). Ideally, the JBS diode should combine the low turn-on voltage and low ON-resistance (R_{ON}) properties of a Schottky barrier diode, and low leakage and high breakdown voltage of the p-n junction diode. A key barrier to implementation of vertical GaN JBS on GaN substrates was the lack of effective selective area p-type doping. The technological challenge was overcome by utilizing the UHPA method, where Khachariya et al. achieved near-ideal performance in GaN Junction Barrier Schottky (JBS) diodes. The breakdown voltage of implantation-formed p-n diodes surpassed 900 V with minimal leakage current [58]. To achieve 1 kV JBS diodes, a 5 μm-thick n^-GaN epilayer was grown using a vertical, cold-wall, radio frequency

(RF) heated, low-pressure metalorganic chemical vapor deposition (MOCVD) reactor on an ammonothermal n^+-GaN substrate. The net carrier concentration (N_D-N_A) of the n^- GaN drift layer was extracted to be ~1.3×10^{16} cm^{-3} using capacitance-voltage (C-V) measurements at 1 MHz. To achieve a shallow box profile (~100 nm) of implanted Mg at a concentration of ~2×10^{19} cm^{-3}, a two-step implantation process with ion energies of 75 keV and 25 keV and corresponding doses of 2×10^{14} cm^{-2} and 4.4×10^{13} cm^{-2}, respectively, were used with a tilt angle of 7°. After the implantation, activation annealing was performed at 1300 °C for 30 minutes at a high pressure of 400 MPa in an N_2 ambient to prevent decomposition. The activation anneal condition chosen corresponded to a diffusion budget of around 0.36, resulting in high Mg activation. An ideality factor of 1.03 indicated near-ideal Schottky behavior. The differential R_{ON} for the JBS diode was 0.6 mΩ·cm^2 and is the lowest value among all GaN-based JBS reported in the literature [58]. These results indicate that high-quality Schottky contacts can be realized even after UHPT, and thus the surface characteristics are preserved during the high-temperature activation anneal with the help of N_2 overpressure.

Finally, to achieve 2 kV class GaN JBS diodes, the n-type drift region thickness was increased to 13 μm with free carrier concentration level of ~1×10^{16} cm^{-3}. Again, the selective area p-type doping was realized using the Mg implantation using previously mentioned conditions. Following the implantation, activation anneal was done at 1300 °C for 30 min at a pressure of 400 MPa in N_2 ambient. To reduce the electric field crowding at the device edge and to achieve the breakdown voltage close to the parallel plate breakdown voltage, junction edge termination (JTE) and floating field rings (FFRs) were implemented using the same Mg implantation conditions. Figure 15.10a shows the schematic cross-section of the fabricated JBS diode. In addition to JBS diodes, Schottky and p-n diodes were also realized on the same wafer for comparison. Figure 15.10b shows the forward bias I-V characteristics for both diodes on a linear scale. The V_{ON} for the Schottky diode and JBS diode was 0.7 V. As shown in Fig. 15.10c, the current rectification ratios (I_{ON}/I_{OFF}) at ±3 V for both diodes are ~10^{12}. The ideality factor (n) for both diodes was 1.01, indicating near-ideal Schottky behavior. The specific R_{ON} for the Schottky diode is 1 mΩ·cm^{-2}, whereas for the JBS diode is 1.9 mΩ·cm^{-2}. Figure 15.10d shows the reverse bias I-V comparison between the Schottky diodes with and without edge termination, JBS, and p-n diodes. As seen in the graph, the Schottky diode without any ET breaks down at only 200 V. In comparison, the Schottky diode with Mg-implanted p-type ET shows a breakdown of 1100 V. The breakdown voltage of the JBS diode is 1900 V, which is among the highest BV values reported for GaN JBS. In comparison, the p-n diode with implanted p-GaN also shows a breakdown of 2000 V, which is also the highest reported breakdown voltage of a GaN p-n diode realized using Mg implantation. Accordingly, the performance of vertical GaN JBS diodes achieved with implantation and UHPA annealing is benchmarked against other state-of-the-art GaN JBS and Schottky diodes reported in the literature using R_{ON} versus breakdown voltage graph as shown in Fig. 15.11. As seen in the graph, the JBS diodes presented here perform favorably compared to the other devices.

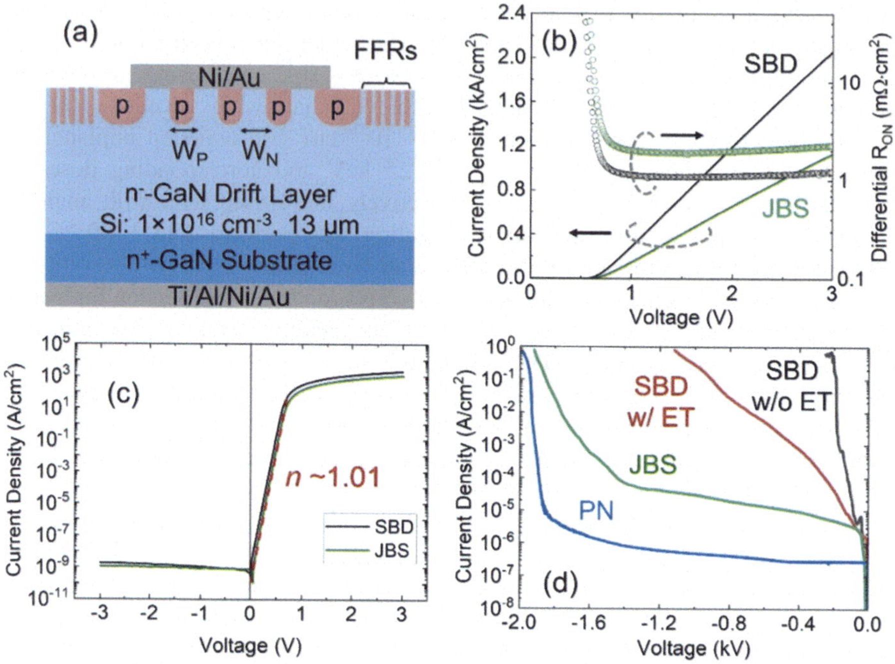

Fig. 15.10 (**a**) A schematic cross-section of the JBS diode. For the JBS diode. The width of p-GaN (W_P) is 3 μm, and n⁻GaN (W_N) is 4 μm. Comparison of room temperature forward bias I-V characteristics of the Schottky barrier and JBS diode in (**b**) linear and (**c**) semi-log scale. (**d**) Reverse bias I-V characteristics comparison between Schottky without and with edge termination (ET), JBS, and p-n diodes

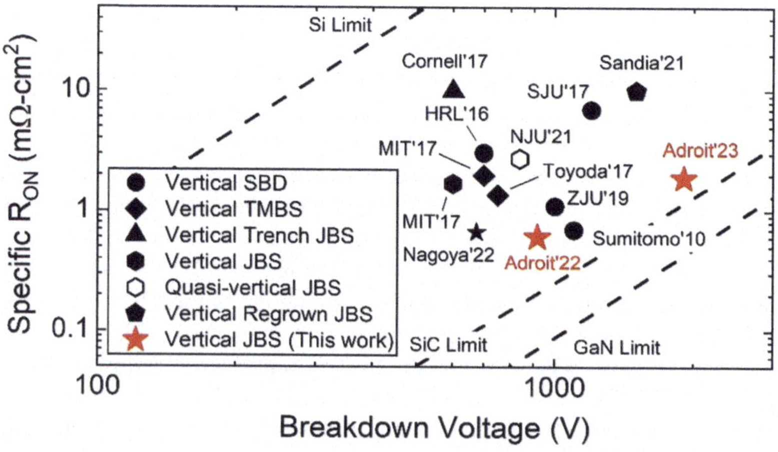

Fig. 15.11 The specific R_{ON} vs. breakdown voltage performance comparison of experimentally reported GaN Schottky and JBS diodes. Closed symbols represent vertical GaN-on-GaN diodes, and open symbols represent quasi-vertical diodes

7 Concluding Remarks

GaN-based high-power switches can be realized by p-type selective area doping and Mg implantation with UHPA based damage recovery and dopant activation. This provides the most promising outlook as demonstrated by kV class JBS diodes with near-unity ideality factors and record low on-resistance. For a wider range of applications, future research may be directed toward overcoming several remaining challenges including the necessary high diffusion budgets required for achieving low Mg ionization energies <150 meV and high Mg activation >70%. Furthermore, efforts should be directed in making the process compatible with standard semiconductor industrial practices including the capability of handling wafers of multiple sizes. Demonstration of successful dopant activation and the realization of practical devices with UHPA at pressures of only 400 MPa, typical pressures used during the HIP process, supports the feasibility of UHPA as a viable industrial process.

References

1. H. Amano, Y. Baines, E. Beam, M. Borga, T. Bouchet, P.R. Chalker, M. Charles, K.J. Chen, N. Chowdhury, R. Chu, C.D. Santi, M.M.D. Souza, S. Decoutere, L.D. Cioccio, B. Eckardt, T. Egawa, P. Fay, J.J. Freedsman, L. Guido, O. Häberlen, G. Haynes, T. Heckel, D. Hemakumara, P. Houston, J. Hu, M. Hua, Q. Huang, A. Huang, S. Jiang, H. Kawai, D. Kinzer, M. Kuball, A. Kumar, K.B. Lee, X. Li, D. Marcon, M. März, R. McCarthy, G. Meneghesso, M. Meneghini, E. Morvan, A. Nakajima, E.M.S. Narayanan, S. Oliver, T. Palacios, D. Piedra, M. Plissonnier, R. Reddy, M. Sun, I. Thayne, A. Torres, N. Trivellin, V. Unni, M.J. Uren, M.V. Hove, D.J. Wallis, J. Wang, J. Xie, S. Yagi, S. Yang, C. Youtsey, R. Yu, E. Zanoni, S. Zeltner, Y. Zhang, J. Phys. D. Appl. Phys. **51**, 163001 (2018)
2. J.Y. Tsao, S. Chowdhury, M.A. Hollis, D. Jena, N.M. Johnson, K.A. Jones, R.J. Kaplar, S. Rajan, C.G. Van de Walle, E. Bellotti, C.L. Chua, R. Collazo, M.E. Coltrin, J.A. Cooper, K.R. Evans, S. Graham, T.A. Grotjohn, E.R. Heller, M. Higashiwaki, M.S. Islam, P.W. Juodawlkis, M.A. Khan, A.D. Koehler, J.H. Leach, U.K. Mishra, R.J. Nemanich, R.C.N. Pilawa-Podgurski, J.B. Shealy, Z. Sitar, M.J. Tadjer, A.F. Witulski, M. Wraback, J.A. Simmons, Adv. Electron. Mat. **4**, 1600501 (2018)
3. S. Chowdhury, M.H. Wong, B.L. Swenson, U.K. Mishra, IEEE Electron Device Lett. **33**, 41 (2012)
4. D. Ji, Y. Yue, J. Gao, S. Chowdhury, IEEE Trans. Electron Dev. **63**, 4011 (2016)
5. D. Ji, S. Chowdhury, IEEE Trans. Electron Dev. **62**, 2571 (2015)
6. C.J. Pan, G.C. Chi, Solid State Electron. **43**, 621 (1999)
7. G.C. Chi, C.H. Kuo, J.K. Sheu, C.J. Pan, Mater. Sci. Eng. B **75**, 210 (2000)
8. Y.-J. Yang, J.-L. Yen, F.-S. Yang, C.-Y. Lin, Jpn. J. Appl. Phys. **39**, L390 (2000)
9. G.W. Pickrell, A.M. Armstrong, A.A. Allerman, M.H. Crawford, K.C. Cross, C.E. Glaser, V.M. Abate, J. Electron. Mater. **48**, 3311 (2019)
10. "Etched-and-Regrown GaN *pn*-Diodes With 1600 V Blocking Voltage," can be found under. https://ieeexplore.ieee.org/abstract/document/9359663 (n.d.)
11. K. Fu, H. Fu, H. Liu, S.R. Alugubelli, T.-H. Yang, X. Huang, H. Chen, I. Baranowski, J. Montes, F.A. Ponce, Y. Zhao, Appl. Phys. Lett. **113**, 233502 (2018)
12. L. Rubin, J. Poate, Ind. Phys. **9**, 12 (2003)
13. V. Heera, D. Panknin, W. Skorupa, Appl. Surf. Sci. **184**, 307 (2001)

14. J.K. Sheu, C.J. Tun, M.S. Tsai, C.C. Lee, G.C. Chi, S.J. Chang, Y.K. Su, J. Appl. Phys. **2002**, 91 (1845)
15. M.H. Breckenridge, J. Tweedie, P. Reddy, Y. Guan, P. Bagheri, D. Szymanski, S. Mita, K. Sierakowski, M. Boćkowski, R. Collazo, Z. Sitar, Appl. Phys. Lett. **118**, 022101 (2021)
16. Y. Wang, K. Huynh, M.E. Liao, H.-M. Yu, T. Bai, J. Tweedie, M.H. Breckenridge, R. Collazo, Z. Sitar, M. Bockowski, Y. Liu, M.S. Goorsky, Phys. Status Solidi B **257**, 1900705 (2020)
17. J.C. Zolper, J. Cryst. Growth **178**, 157 (1997)
18. J. Karpiński, J. Jun, S. Porowski, J. Cryst. Growth **66**, 1 (1984)
19. E.P. Carlson, D.W. Cunningham, I.C. Kizilyalli, ECS Trans. **86**, 3 (2018)
20. H.W. Choi, M.A. Rana, S.J. Chua, T. Osipowicz, J.S. Pan, Semicond. Sci. Technol. **17**, 1223 (2002)
21. W. Khalfaoui, T. Oheix, G. El-Zammar, R. Benoit, F. Cayrel, E. Faulques, F. Massuyeau, A. Yvon, E. Collard, D. Alquier, Phys. Status Solidi A **214**, 1600438 (2017)
22. T. Niwa, T. Fujii, T. Oka, Appl. Phys. Express **10**, 091002 (2017)
23. J.D. Greenlee, B.N. Feigelson, T.J. Anderson, M.J. Tadjer, J.K. Hite, M.A. Mastro, C.R. Eddy, K.D. Hobart, F.J. Kub, J. Appl. Phys. **116**, 063502 (2014)
24. M.J. Tadjer, B.N. Feigelson, J.D. Greenlee, J.A. Freitas, T.J. Anderson, J.K. Hite, L. Ruppalt, C.R. Eddy, K.D. Hobart, F.J. Kub, ECS J. Solid State Sci. Technol. **5**, P124 (2015)
25. B.N. Feigelson, T.J. Anderson, M. Abraham, J.A. Freitas, J.K. Hite, C.R. Eddy, F.J. Kub, J. Cryst. Growth **350**, 21 (2012)
26. S.R. Aid, T. Uneme, N. Wakabayashi, K. Yamazaki, A. Uedono, S. Matsumoto, Phys. Status Solidi A **214**, 1700225 (2017)
27. S.G. Sundaresan, M. Murthy, M.V. Rao, J.A. Schreifels, M.A. Mastro, C.R. Eddy, R.T. Holm, R.L. Henry, J.A. Freitas, E. Gomar-Nadal, R.D. Vispute, Y.-L. Tian, Semicond. Sci. Technol. **22**, 1151 (2007)
28. G.S. Aluri, M. Gowda, N.A. Mahadik, S.G. Sundaresan, M.V. Rao, J.A. Schreifels, J.A. Freitas Jr., S.B. Qadri, Y.-L. Tian, J. Appl. Phys. **108**, 083103 (2010)
29. S. Porowski, I. Grzegory, D. Kolesnikov, W. Lojkowski, V. Jager, W. Jager, V. Bogdanov, T. Suski, S. Krukowski, J. Phys. Condens. Matter **14**, 11097 (2002)
30. A.G. Jacobs, B.N. Feigelson, J.K. Hite, C.A. Gorsak, L.E. Luna, T.J. Anderson, F.J. Kub, Phys. Status Solidi A **217**, 1900789 (2020)
31. M.H. Breckenridge, *Ion Implantation into the III-Nitrides* (North Carolina State University, 2021)
32. K. Sierakowski, R. Jakiela, B. Lucznik, P. Kwiatkowski, M. Iwinska, M. Turek, H. Sakurai, T. Kachi, M. Bockowski, Electronics **9**, 1380 (2020)
33. K.T. Jacob, G. Rajitha, J. Cryst. Growth **311**, 3806 (2009)
34. K. Sumida, K. Hirukawa, H. Sakurai, K. Sierakowski, M. Horita, M. Bockowski, T. Kachi, J. Suda, Appl. Phys. Express **14**, 121004 (2021)
35. H. Sakurai, T. Narita, K. Hirukawa, S. Yamada, A. Koura, K. Kataoka, M. Horita, N. Ikarashi, M. Bockowski, J. Suda, T. Kachi, *2020 32nd International Symposium on Power Semiconductor Devices and ICs (ISPSD)* (2020), pp. 321–324
36. H. Sakurai, M. Omori, S. Yamada, Y. Furukawa, H. Suzuki, T. Narita, K. Kataoka, M. Horita, M. Bockowski, J. Suda, T. Kachi, Appl. Phys. Lett. **115**, 142104 (2019)
37. A. Kumar, W. Yi, J. Uzuhashi, T. Ohkubo, J. Chen, T. Sekiguchi, R. Tanaka, S. Takashima, M. Edo, K. Hono, J. Appl. Phys. **128**, 065701 (2020)
38. T. Nakashima, E. Kano, K. Kataoka, S. Arai, H. Sakurai, T. Narita, K. Sierakowski, M. Bockowski, M. Nagao, J. Suda, T. Kachi, N. Ikarashi, Appl. Phys. Express **14**, 011005 (2020)
39. K. Iwata, H. Sakurai, S. Arai, T. Nakashima, T. Narita, K. Kataoka, M. Bockowski, M. Nagao, J. Suda, T. Kachi, N. Ikarashi, J. Appl. Phys. **127**, 105106 (2020)
40. K. Huynh, Y. Wang, M.E. Liao, J. Tweedie, P. Reddy, M.H. Breckenridge, R. Collazo, Z. Sitar, K. Sierakowski, M. Bockowski, X. Huang, M. Wojcik, M.S. Goorsky, J. Appl. Phys. **135**, 025104 (2024)

41. H. Sakurai, T. Narita, M. Omori, S. Yamada, A. Koura, M. Iwinska, K. Kataoka, M. Horita, N. Ikarashi, M. Bockowski, J. Suda, T. Kachi, Appl. Phys. Express **13**, 086501 (2020)
42. H. Tews, R. Neumann, A. Hoepfner, S. Gisdakis, J. Appl. Phys. **67**, 2857 (1990)
43. Y.K. Yeo, Y.S. Park, P.W. Yu, J. Appl. Phys. **50**, 3274 (1979)
44. P. Reddy, M.P. Hoffmann, F. Kaess, Z. Bryan, I. Bryan, M. Bobea, A. Klump, J. Tweedie, R. Kirste, S. Mita, M. Gerhold, R. Collazo, Z. Sitar, J. Appl. Phys. **120**, 185704 (2016)
45. P. Reddy, F. Kaess, J. Tweedie, R. Kirste, S. Mita, R. Collazo, Z. Sitar, Appl. Phys. Lett. **111**, 152101 (2017)
46. K. Morizuka, T. Nozu, K. Tsuda, M. Azuma, Electron. Lett. **22**, 315 (1986)
47. S. Yamahata, S. Adachi, Appl. Phys. Lett. **52**, 1493 (1988)
48. T. Narita, T. Kachi, K. Kataoka, T. Uesugi, Appl. Phys. Express **10**, 016501 (2016)
49. T. Oikawa, Y. Saijo, S. Kato, T. Mishima, T. Nakamura, Nucl. Instrum. Methods Phys. Res., Sect. B **365**, 168 (2015)
50. S. Takashima, K. Ueno, R. Tanaka, H. Matsuyama, M. Edo, K. Nakagawa, Normally-off MOSFET Properties Fabricated on Mg Implanted GaN Layers. In Proceedings of the Extended Abstracts of the 2017 International Conference on Solid State Devices and Materials, Sendai, Japan, 19–22 September; pp. 1075–1076 (2017)
51. R. Tanaka, S. Takashima, K. Ueno, H. Matsuyama, M. Edo, K. Nakagawa, Appl. Phys. Express **12**, 054001 (2019)
52. M. Takahashi, A. Tanaka, Y. Ando, H. Watanabe, M. Deki, M. Kushimoto, S. Nitta, Y. Honda, K. Shima, K. Kojima, S.F. Chichibu, K.J. Chen, H. Amano, Phys. Status Solidi B **257**, 1900554 (2020)
53. A.Y. Polyakov, M. Shin, M. Skowronski, R.G. Wilson, D.W. Greve, S.J. Pearton, Solid State Electron. **41**, 703 (1997)
54. A.Y. Polyakov, N.B. Smirnov, A.V. Govorkov, S.J. Pearton, R.M. Frazier, G.T. Thaler, C.R. Abernathy, J.M. Zavada, Semicond. Sci. Technol. **19**, 1169 (2004)
55. A. Uedono, H. Iguchi, T. Narita, K. Kataoka, W. Egger, T. Koschine, C. Hugenschmidt, M. Dickmann, K. Shima, K. Kojima, S.F. Chichibu, S. Ishibashi, Phys. Status Solidi B **256**, 1900104 (2019)
56. H. Sakurai, T. Narita, K. Kataoka, K. Hirukawa, K. Sumida, S. Yamada, K. Sierakowski, M. Horita, N. Ikarashi, M. Bockowski, J. Suda, T. Kachi, Appl. Phys. Express **14**, 111001 (2021)
57. E. Kano, J. Uzuhashi, K. Kobayashi, K. Ishikawa, K. Sawabe, T. Narita, K. Sierakowski, M. Bockowski, T. Ohkubo, T. Kachi, N. Ikarashi, Phys. Status Solidi Rapid Res. Lett., 2400074 (2024)
58. D. Khachariya, S. Stein, W. Mecouch, M.H. Breckenridge, S. Rathkanthiwar, S. Mita, B. Moody, P. Reddy, J. Tweedie, R. Kirste, K. Sierakowski, G. Kamler, M. Bockowski, E. Kohn, S. Pavlidis, R. Collazo, Z. Sitar, Appl. Phys. Express **15**, 101004 (2022)

Chapter 16
High-Energy Ion Implantation Using Electrostatic Accelerators

Dannie Steski, Thomas Kubley, Charles Carlson, Florian Krippendorf, Michael Rub, Robert Koch, and Reza Ghandi

1 Introduction

The properties of silicon carbide (SiC), such as wide band gap and high voltage breakdown, give SiC a significant advantage over silicon for the next generation of medium- and high-frequency applications. In addition, the high thermal conductivity of SiC allows it to operate in high-temperature environments. However, ultrahigh blocking (>3.3 kV) SiC unipolar switches and diodes suffer from high conduction losses at elevated temperatures [1], and SiC bipolar devices such as IGBTs exhibit a prohibitively high forward voltage drop of 3 V [2]. Therefore, in medium voltage (MV) power conversion systems, the advantage of SiC technology is diminished compared to the incumbent Si IGBT technology. SiC superjunctions (SJs) can break the unipolar conduction limit and offer an improved trade-off between specific on-resistance and blocking voltage for MV-class applications. To date, 1.2–3.3 kV multi-epitaxial SiC SJ devices have been demonstrated [3, 4]. Fabricating such devices requires several iterations of epi regrowth due to shallow projectile depth

D. Steski (✉) · T. Kubley · C. Carlson
Brookhaven National Laboratory, Upton, NY, USA
e-mail: steski@bnl.gov; tkubley@bnl.gov; ccarlson@bnl.gov

F. Krippendorf
mi2-factory GmbH, Jena, Germany

M. Rub
mi2-factory GmbH, Jena, Germany

University of Applied Sciences Jena, Jena, Germany

R. Koch
University of Applied Sciences Jena, Jena, Germany

R. Ghandi
GE Aerospace Research, Niskayuna, NY, USA
e-mail: ghandi@ge.com

I. C. Kizilyalli et al. (eds.), *Gallium Nitride and Related Materials*, The Materials Research Society Series, https://doi.org/10.1007/978-3-031-83056-3_16

Fig. 16.1 BNL's MP-class Tandem Van de Graaff. MP7 is in the foreground and MP6 in the background

of implanted atoms in SiC using conventional process tools. Additionally, 6.5 kV trench-refill SiC SJ devices have been shown to suffer from excessive leakage at high blocking voltages due to crystallographic defects from a complex refill process [5]. For silicon devices, the ions are implanted at low energies and then thermally diffused to form deep junctions. Unfortunately, unlike silicon, silicon carbide has low dopant diffusion rates even at high temperatures. The energy of conventional ion implanters (<1 MeV) limits the implantation depth in silicon carbide to approximately 1 μm. To form deep junctions in SiC, much higher implantation energies are needed. Implanting nitrogen and aluminum ions to a depth of 15 μm requires beam energies of 28 and 60 MeV, respectively [6]. One solution is to take advantage of the well-defined beam energies and small energy spread of electrostatic accelerators. This allows for a localized deposition of the ions with a narrow width. By changing the beam energy in discrete steps, a series of overlapping implantations can build a uniform implantation profile from the maximum implantation depth to the surface of the wafer.

A new class of vertical doped-pillar SJ devices is being developed at GE Aerospace and BNL that utilizes MeV-level implantations to form deep p-type and n-type superjunction pillars. This multi-epitaxial SJ approach is implemented in high-voltage SiC devices with a smaller number of regrowth steps compared to previously reported SJ devices with pillars formed using lower implant energy processes.

At Brookhaven National Laboratory (BNL), a high-energy implantation [7] system has been developed that provides uniform concentration of implanted ions from the surface to depths of greater than 15 μm. This system uses BNL's 15 MV MP-class Tandem Van de Graaff [8] (Fig. 16.1) to accelerate the ions for implantation. The system is described below along with results of an implementation and future improvements.

2 BNL Tandem Van de Graaff Facility

The BNL Tandem Van de Graaff Facility [9] (Fig. 16.2) consists of two MP-class Tandem Van de Graaff, each capable of operating independently at a terminal voltage above 14 MV. Originally built to study low-energy nuclear physics, the Tandems are now used for applied research as well as an injector into the Booster synchrotron at BNL. The Tandem Van de Graaff facility has two target rooms where applied research is conducted. Target Room 4 is used primarily for single event effects (SEE) testing and detector calibration. Target Room 2 has a specially designed beam line for micropore filter material production. This beam line has been modified to perform high-energy implantation in silicon carbide.

The terminal voltage of the Tandems is continuously variable from approximately 14.4 MV to <1 MV. Below a terminal voltage of a couple of MV, the transmission of the ion beam through the accelerator drops dramatically, impacting the available ion current. To achieve a uniform concentration of ions from the maximum depth to the surface of the wafer would require varying the ion beam energy from 10s of MeV to a few keV. Although this is possible in principle, it is challenging for any single accelerator to deliver ion beams over such a wide range.

To overcome this problem, a stationary aluminum energy-absorbing foil is installed immediately upstream of the SiC wafers. The energy-absorbing foil reduces the ion energy from 10s of MeV (where the transmission of the ions through the machine is high) to the very low energies needed to implant ions near the surface of the wafer. This provides high ion beam intensities even at low ion beam energies.

For implantation in silicon carbide, the ions of interest are boron, nitrogen, aluminum, and phosphorus. For a variety of reasons, BNL decided to concentrate on nitrogen and aluminum. The negative ions needed for injection into the Tandems are created in a cesium sputter ion source. These ion sources can provide a wide variety of ions from across the periodic table. The intensity of the ion beam is dependent on the ease of forming a negative ion.

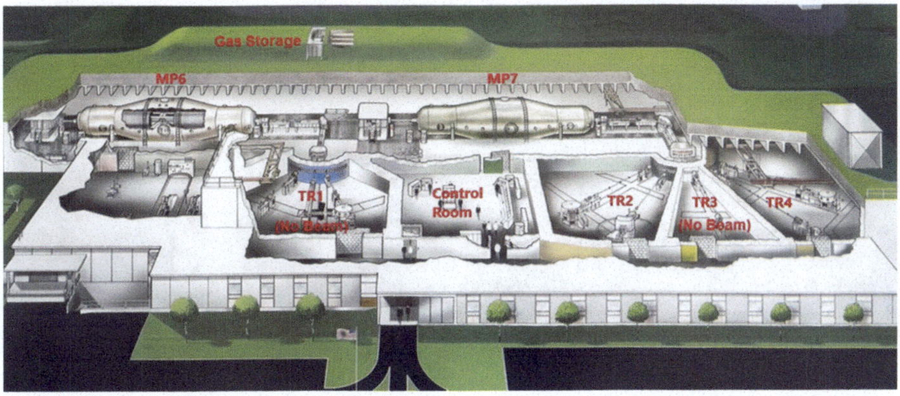

Fig. 16.2 BNL Tandem Van de Graaff facility

Nitrogen does not form a negative ion, but a large current of CN- can be produced by using a cathode pressed with a mixture of graphite and boron nitride [10]. Aluminum does produce a negative ion, and to avoid contamination from silicon, which is a very prolific negative ion, a cathode that is 99.999% pure aluminum is used. To get the maximum aluminum ion beam current, the Al_2- dimer is accelerated rather than the Al- beam, which has a lower intensity. Both the CN- and Al_2- molecules are broken into their constituent positive ions when passed through the stripping gas in the center terminal of the Tandem. The positive ions are then accelerated back to ground potential in the second stage of the Tandem Van de Graaff.

3 BNL High Energy Implanter

The BNL High Energy Implanter is located in Target Room 2, using an existing beam line (Fig. 16.3) originally built for the production of micropore filter material. The unique feature of this beam line is that the diameter of the beam pipe gradually increases from 100 to 710 mm just before the end vacuum chamber. This feature allows the beam to be swept over a wide area.

The high-energy implanter (Fig. 16.4) consists of a pair of horizontal deflector plates, Faraday cups, a stationary aluminum foil energy absorber, and a vertically moving SiC wafer support plate attached to a linear stage. The horizontal deflector plates will sweep the beam horizontally while the linear stage moves the SiC wafers through the swept ion beam. The Faraday cups monitor the beam intensity during the implantation, and the aluminum foil energy absorber degrades the beam energy to allow implantation of the ions to the surface of the SiC wafers. Each part of the high-energy implanter will be discussed separately.

The horizontal deflection electrodes are a pair of metal plates located approximately 14 m upstream of the end station. To ensure that the ion beam is swept linearly, a triangle wave generator [11] is used as the input to a high voltage amplifier [12]. The frequency of the input signal is 400 Hz to have multiple sweeps across the wafer, ensuring uniformity of the implantation.

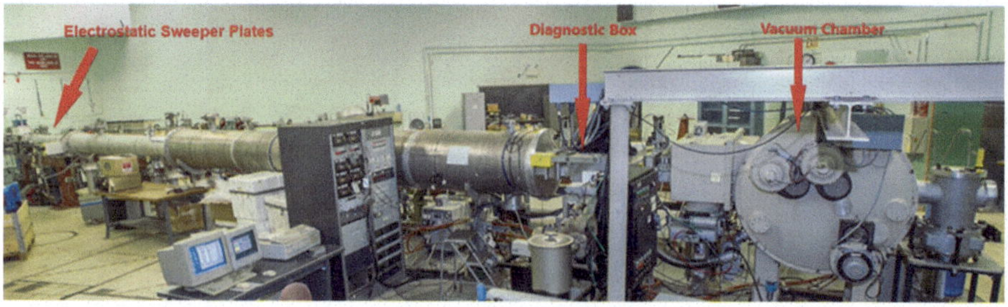

Fig. 16.3 High Energy Implanter beam line in Target Room 2

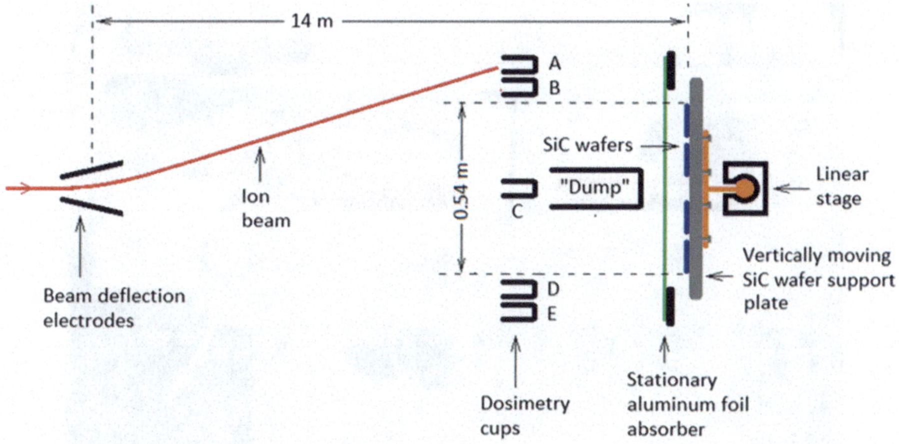

Fig. 16.4 Overhead schematic of high-energy implanter. Not to scale

The Faraday cups, used to monitor the beam intensity, are located in the diagnostic box immediately upstream of the end station. The "Dump" cup is an insertable Faraday cup, 95 mm in diameter and 300 mm long. This cup is magnetically suppressed and gives a precise measurement of the beam current before the start of each implantation with the voltage of the deflection plates turned off. The center cup (labelled C) and the pair of right cups (labelled A and B) and left cups (labelled D and E) are much narrower (10 mm) and are used to monitor the beam during the implantation. The right and left cups make sure that the ion beam is swept over the desired width and centered on the wafers. This is done by increasing the voltage on the deflection plates, and thus the sweep width, until the outer left cup (E) and outer right cup (A) read approximately half of the right inner (B) and left inner (D) cup. The voltage on the deflection plates is adjusted for each ion beam energy. The center cup measures the beam current during the implantation. It is connected to a digital electrometer [13] that provides the accumulated dose to the control system. Once the desired dose is reached, a Faraday cup is automatically inserted upstream to prevent further implantation.

The stationary aluminum foil energy absorber (Fig. 16.5) is located 13 mm upstream from the SiC wafers. The foil is 565 × 57 mm. For aluminum implantations, a foil with an effective thickness of 5.75 μm is used. For nitrogen implantation, the foil is changed and has an effective thickness of 16.9 μm. The main purpose of the aluminum foil is to degrade the beam energy to allow implantation of the ions to the surface of the wafer. As discussed above, the transmission of the ion beam falls dramatically when the terminal voltage of the Tandem is below approximately 2 MV. By using the energy absorber, a higher energy ion beam can be efficiently transported to the foil where its energy is degraded to a few keV, making implantation to the surface of the SiC wafer practical. A secondary benefit of the energy absorber is to increase the energy spread of the ion beam, decreasing the number of

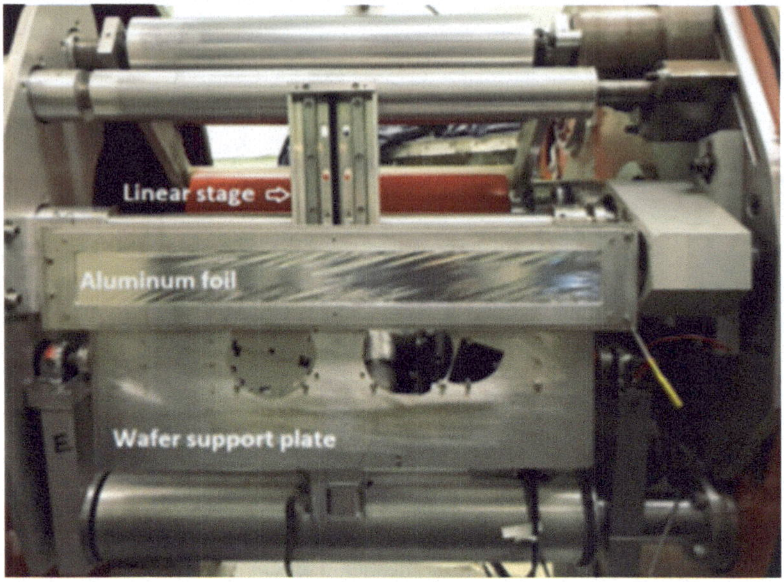

Fig. 16.5 Aluminum foil energy absorber in front of the wafer support plate and the linear stage

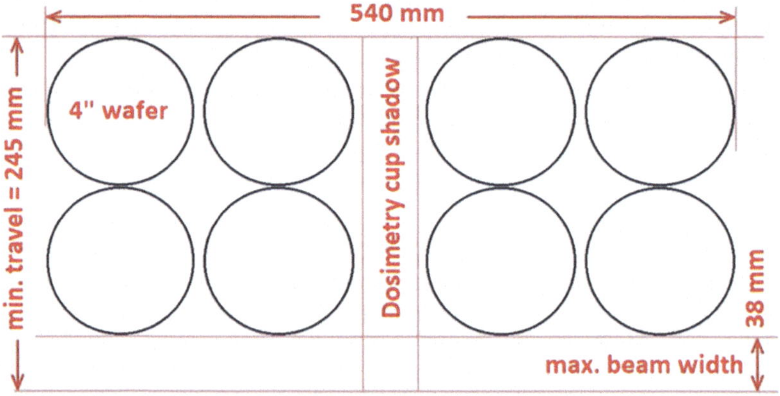

Fig. 16.6 Layout of eight 4″ wafers on the wafer support plate

discrete energy steps required to have a uniform concentration on the surface of the wafer.

The wafers are mounted on a support plate that is connected to a linear stage [14] that drives the wafers vertically through the ion beam. The support plate can hold a maximum of eight 4-inch wafers (Fig. 16.6). The maximum travel of the linear stage is 300 mm at a speed of approximately 10 mm/sec. The relatively slow speed of the linear stage, coupled with the 400 Hz of the ion beam sweep, makes sure that the implantation is very uniform. To further ensure uniform implantation, the wafers make several passes through the beam at each energy. During operation the center cup provides feedback to the control system to adjust the speed of the linear stage.

4 Ion Implantation Results

Each implantation energy must be carefully chosen so that the overlap of the individual implantations results in a uniform concentration of implanted ions from the surface to the maximum depth. To minimize the time required for the implantation, the ion charge state is chosen to maximize the beam intensity available while keeping the terminal voltage of the Tandem within the normal operating range. Keeping these factors in mind, for aluminum, a total of 22 energy steps (Table 16.1) are used. For nitrogen implantation, only 14 energy steps (Table 16.2) are needed.

Mounted on the support plate with the wafers is a small coupon for secondary-ion mass spectrometry (SIMS) analysis to validate the implantation. Figure 16.7 shows the simulation of individual aluminum implantations and the resulting accumulated dose. The SIMS [15] analysis of the implantation is also shown. For this case, the

Table 16.1 Aluminum energies needed for a uniform implantation

Beam energy (MeV)	Terminal voltage (MV)	Charge state Q	Implantation depth (μm)
13.80	3.923	3	0.37
15.80	4.494	3	1.00
18.40	5237	3	1.59
21.00	5.980	3	2.19
23.60	6.723	3	2.78
26.20	7.466	3	3.38
28.80	8.209	3	3.98
31.40	8.951	3	4.58
34.00	7.54	4	5.18
36.60	8.118	4	5.79
39.10	8.673	4	6.37
41.60	9.229	4	6.96
44.10	9.784	4	7.55
46.60	10.340	4	8.13
49.10	10.896	4	8.72
51.50	11.429	4	9.29
53.90	11.962	4	9.86
56.30	10.224	5	10.43
58.70	10.660	5	11.01
61.10	11.096	5	11.58
63.40	11.515	5	12.13
65.70	11.933	5	12.52

Table 16.2 Nitrogen energies needed for a uniform implantation

Beam energy (MeV)	Terminal voltage (MV)	Charge state Q	Implantation depth (μm)
25.86	7.287	3	0.86
27.11	7.640	3	1.50
28.43	8.013	3	2.30
29.68	8.366	3	3.12
30.95	8.725	3	3.91
32.26	9.095	3	4.75
33.59	9.471	3	5.64
34.92	9.847	3	6.55
36.25	10.223	3	7.48
37.58	10.599	3	8.44
38.91	10.975	3	9.42
40.24	11.350	3	10.44
41.57	11.726	3	11.47
42.90	12.102	3	12.53

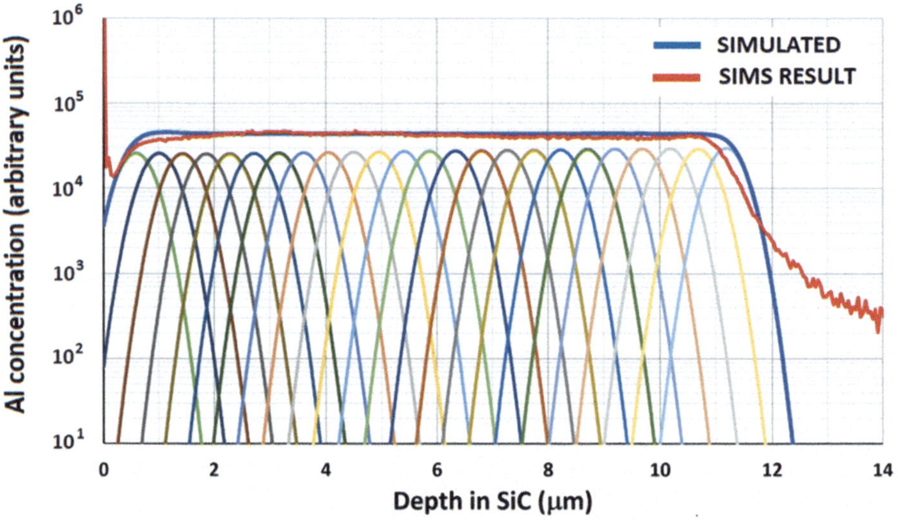

Fig. 16.7 The simulation and SIMS analysis of an aluminum implantation with agreement within 0.87%

simulation and the SIMS analysis agree to within 0.87%. Although other implantations have larger discrepancies, the disagreement was still within the acceptable range. The dosimetry should be accurate to 2%.

A GE team evaluated the feasibility of BNL high-energy implantation by the fabrication of single-layer 2 kV SiC SJ PiN diodes [16]. Figure 16.8 shows a SEM cross section of the SJ structure employing deep-implanted Al-doped and N-doped

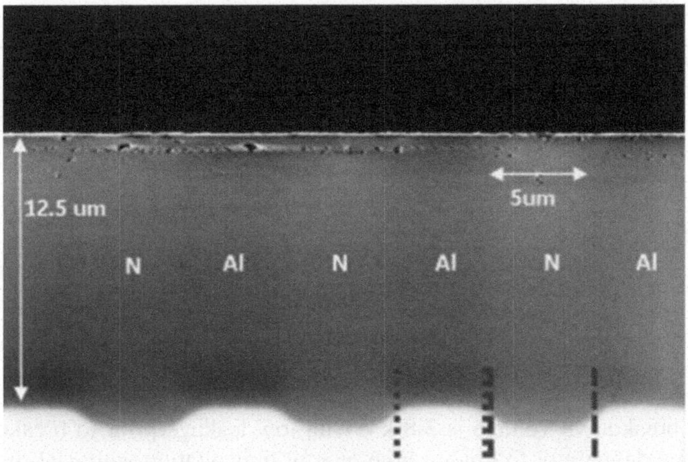

Fig. 16.8 SEM photograph of deep-implanted P/N pillars formed on a SiC test wafer

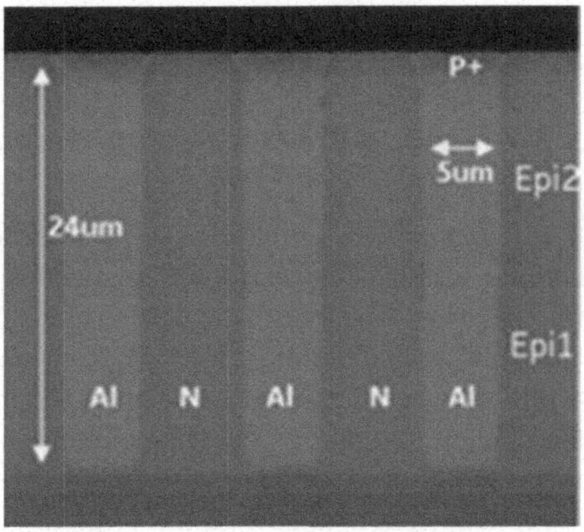

Fig. 16.9 SEM photograph of 3.8 kV deep-implanted SiC SJJBS diode with two rounds of epitaxial overgrowth and high-energy implantations

pillars in SiC formed by high-energy implantation, representing a successful masking and alignment process. This device showed a breakdown voltage of >2.1 kV, which is 500 V higher than the simulated breakdown voltage of SiC non-SJ diodes. GE also reported a successful demonstration of 3.8 kV SiC deep-implanted SJ JBS diodes with two rounds of epitaxial overgrowth (12 μm each) with a total drift layer thickness of 24 μm [17]. P-doped and N-doped pillars were formed using two rounds of high-energy implantations with a maximum junction depth of 12 μm (Fig. 16.9). Top JBS diode is realized through P+ and P-JTE implantation followed by front and backside metallization with Ni as the Schottky contact. Figure 16.10 shows the forward and reverse I-V characteristics of the SJ-JBS diode. The device turns on at 1.4 V and shows $R_{on,sp}$ of 4.5 mΩ.cm^2 at room temperature and 9.6 mΩ.cm^2 at 150 °C, which is ~45% below the SiC unipolar

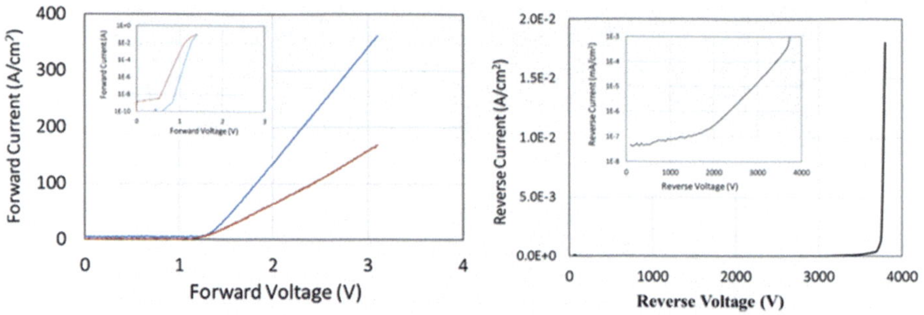

Fig. 16.10 Forward and reverse characteristics of 3.5 kV SiC deep-implanted SJ JBS diodes

limit. The breakdown voltage is 3.8 kV with low leakage prior to breakdown that suggests low defectivity following high-energy implantation, epitaxial overgrowth, and activation anneal at 2000 °C.

5 Future Improvements

5.1 Heating Wafers During Implantation

The implantation of high-energy ions introduces damage to the lattice of the silicon carbide wafers. One way to mitigate this damage is to anneal the wafers during implantation. To investigate this further, the BNL implanter is being upgraded to allow the implantation of ions while the SiC wafers are heated. This upgrade uses 4 wafer heaters mounted on a plate attached to a linear stage. The original room temperature mounting plate was mounted in front of the rollers used to produce the micropore filter material. Unfortunately, the new heated wafer system required more space than was available in the original chamber. Instead of building a new beam line, which would have been prohibitively expensive, it was decided to modify the existing system. The need for micropore filter production to continue at BNL limited the modifications that could be done to the existing system. Fortunately, all of the micropore filter rollers are mounted on the vacuum chamber door, which is mounted on an "I" beam.

The solution was to roll the micropore filter door far enough from the vacuum chamber and attach a new chamber door with the heated wafer equipment mounted on it. The new heated wafer door can be installed on the "I" beam with the overhead crane, allowing the beam line to be used for both applications.

The wafer heaters [18] (Fig. 16.11) are mounted on a plate that is attached to a 600 mm linear stage [14]. As before, the linear stage will drive the wafers vertically through the ion beam that is being swept horizontally by the electrostatic deflection plates. The heaters can accommodate either 4-inch or 6-inch wafers. It is hoped that

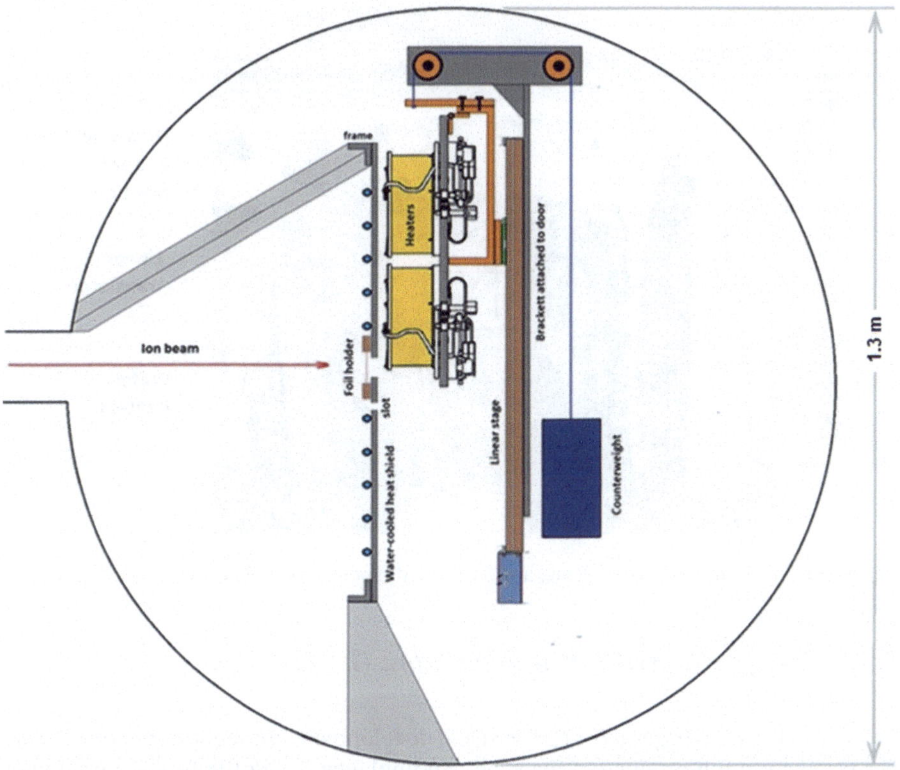

Fig. 16.11 Side schematic of the Heated Wafer system

ultimately the heaters will be able to reach surface temperatures of 1000 °C but are still undergoing testing.

In front of the heaters is a water-cooled (Fig. 16.12) heat shield with the aluminum energy absorber mounted on it. There is also a slot in the heat shield that allows an infrared camera to look directly at the surface of the wafers to measure the surface temperature and the uniformity across the wafer. In addition, there are thermocouples in contact with the surface of the wafer and a second thermocouple inserted into each of the heaters.

Each heater can be controlled independently with the inserted thermocouple used to control the temperature.

The system is undergoing commissioning. A test implantation at temperature [19] was performed but without the linear stage operational. The full system should be operational soon, although it may take additional time to reach the maximum temperature.

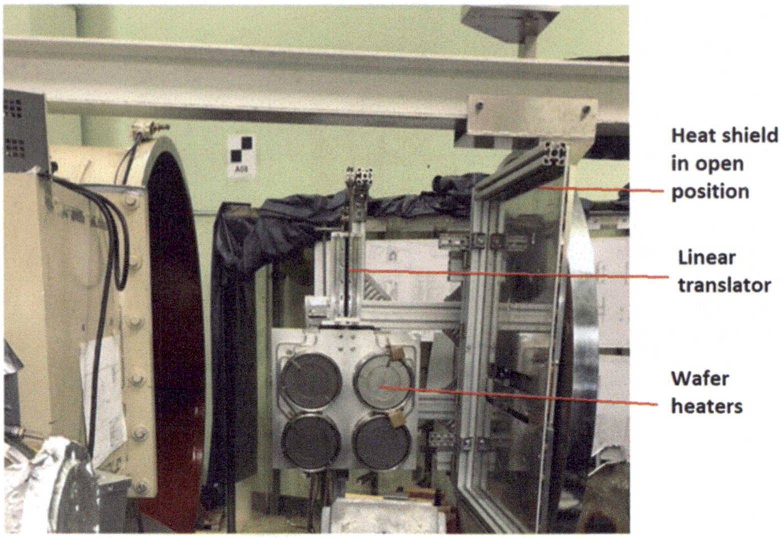

Fig. 16.12 Wafer heaters, linear stage, and heat shield mounted on the new chamber door

5.2 Energy Filtered Ion Implantation (EFII)

The current system is good for research and development but requires too much time per wafer for a full production machine. An implantation profile from 12 μm to the surface requires 14 discrete energies for nitrogen and 22 for aluminum. To first order, the time to change energies and retune the ion beam optics is approximately equal to the implantation time. The total time required for the implantation of a batch of eight wafers is 6–8 h, or approximately 45–60 min per wafer. Although acceptable for the proof of principle stage, this is too long for large-scale efficient production.

The Energy Filtered Ion Implantation (EFII) technology [20] is a novel approach to implanting semiconductor materials over a depth range of several μm with only one high-energy implantation energy. The energy filter itself is a thin, micro-patterned silicon membrane that is inserted into the ion beam, forcing energy loss in the filter material. This filter can take a monoenergetic ion beam and generate a beam with a uniform energy spread (Fig. 16.13). The resulting ion implantation would be uniform from the surface to the maximum depth.

By replacing the uniform energy filter with an energy filter with a different micro-pattern, other implantation profiles such as a graded, stacked, and step can be accomplished. Using these filters eliminates the need for energy changes of the beam and the needed change in the accelerating voltage and beam optics. This has the potential to reduce the implantation time by approximately 50%.

Initial tests have proven to be promising. Using a flat profile EFII, a 90 MeV Al beam was implanted into silicon (Fig. 16.14). This produced a flat profile from the surface to 26 μm. It also reduces the dose rate for specific depths by approximately

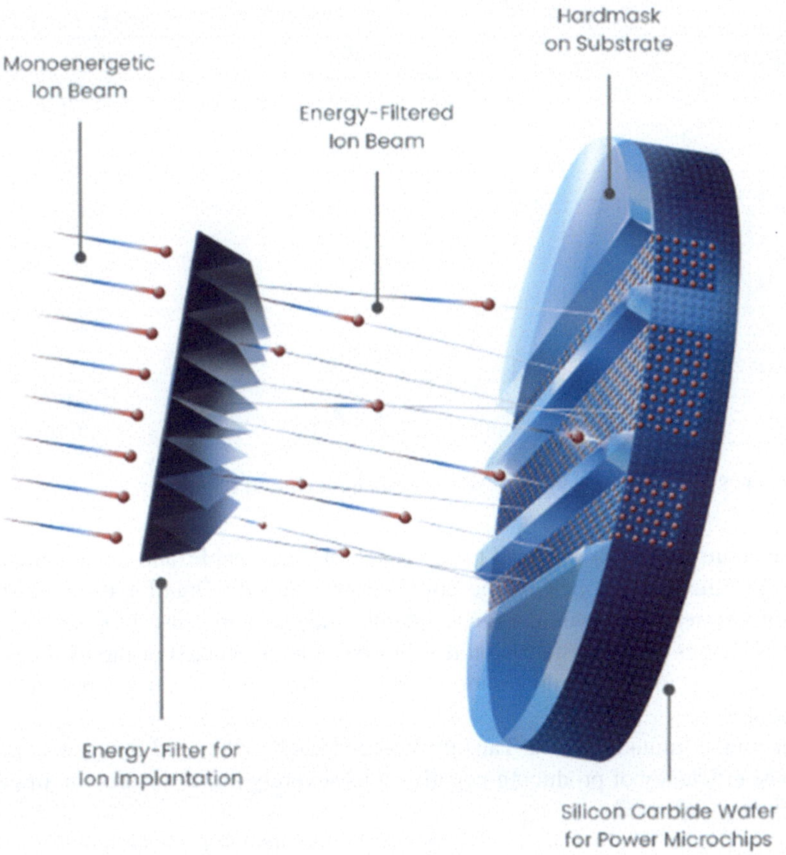

Fig. 16.13 The energy-filter for ion implantation uses a monoenergetic beam to produce a beam with uniform energy spread

20. This has the potential to reduce the lattice damage during the implantation process but will need to be investigated further.

6 High-Throughput Ion Implanter

Although the BNL High Energy Implanter has proven to be a useful machine for research and development of the implantation process, a high-throughput production implanter would need to be specifically designed to increase the wafer throughput.

To minimize downtime, the system would use commercially available parts for the accelerator, ion source, and implantation end station. Each part would need to be computer controlled to eliminate the need to manually retune the ion beam.

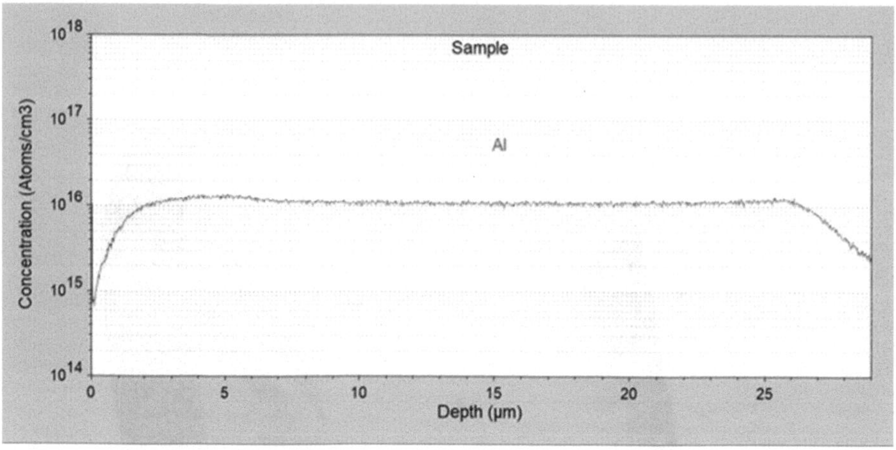

Fig. 16.14 SIMS analysis of 90 MeV Al ion beam through EFII filter

The entire system would need to be very stable, repeatable, and low maintenance. Replacing the moving belt of the conventional Van de Graaff with a solid-state charging system [21] would reduce maintenance and increase repeatability. The energy changes in this system would either be fully automated or the EFII could be employed for uniform implantation with a single ion beam energy from the accelerator.

One of the limitations for a Tandem Van de Graaff is the need to inject a negative ion. The efficiency of producing negative ions severely limits the amount of current that can be generated.

To overcome this limitation. A single-ended positive Van de Graaff (Fig. 16.15) with a positive ion source in the terminal could replace the Tandem Van de Graaff presently used. To keep the terminal voltage below 8 MV while still generating the high ion beam energies needed would require an ion source capable of producing large currents of highly charged ions. One possibility is to use a commercially available Electron Cyclotron Resonance (ECR) ion source [22].

The maximum voltage required of the Van de Graaff is dependent on the highest charge state the ECR ion source can produce with reasonable intensity. To improve the overall reliability and cost of the system, the lowest maximum terminal voltage is desirable. If the maximum beam energy of the nitrogen ions is 42.90 MeV (as in the BNL system), a fully stripped nitrogen ion ($Q = +7$) would require a terminal voltage of a little over 6 MV. This would also satisfy the maximum aluminum energy of 65.70 MeV with a $Q = +11$.

Having the ion source in the terminal of the accelerator would complicate maintenance as the insulating gas inside the pressure vessel would need to be removed before any maintenance could be performed. The reliability and lifetime of the ion source would therefore be critical.

For maximum efficiency, two accelerators would be used. One accelerator would be dedicated to each implanted ion species.

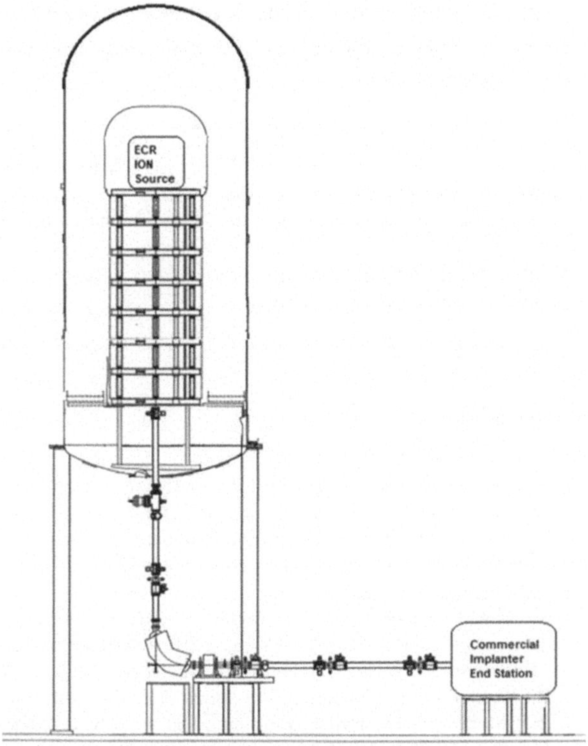

Fig. 16.15 Single-ended Van de Graaff with ECR Ion Source

7 Conclusion

High-energy electrostatic accelerators can overcome the limitations of conventional ion implanters with regard to forming deep junctions in silicon carbide wafers. The BNL High Energy Implanter has demonstrated uniform implantation in SiC from the surface to a depth of >12 μm by combining a high-voltage Tandem Van de Graaff with a stationary aluminum energy absorber. An upgrade to the present system will allow implantations while the wafers are being heated.

The use of energy filter ion implantation (EFII) technology looks promising for the elimination of beam energy changes. This will cut the implantation time of a wafer by approximately 50%. EFII may have an additional benefit by reducing the damage to the SiC lattice during the implantation process.

To lower the cost of SiC devices significantly would require a high-energy implanter with a high wafer throughput. One possible option using commercially available technology would be to combine a single-ended Van de Graaff with a high-charge state ECR ion source. All of the components of the system are mature technologies with high reliability.

Although the work to date has been with SiC, the same technique could be applied to other materials such as gallium nitride (GaN) by changing the species of the implanted ion.

Acknowledgments The information, data, or work presented herein was funded in part by the Advanced Research Projects Agency-Energy (ARPA-E), U.S. Department of Energy, under Award Number DE-AR0000674 and DE-AR00001028. The views and opinions of authors expressed herein do not necessarily state or reflect those of the United States Government or any agency thereof.

Work supported by Brookhaven Science Associates, LLC, under Contract No. DE-SC0012704 with the U.S. Department of Energy.

References

1. A. Bolotnikov et al., IEEE Appl. Power Electron. Conf. Expo., 2445–2452 (2015)
2. L. Han, L. Liang, Y. Kang, Y. Qiu, IEEE Trans. Power Electron. **36**(2), 2080–2093 (2021)
3. T. Tanaka et al., First demonstration of dynamic characteristics for SiC superjunction MOSFET realized using multi-epitaxial growth method, in *2018 IEEE Int. Electron Devices Meet.*, (2019), pp. 8.2.1–8.2.4
4. M. Baba, T. Tawara, T. Morimoto, S. Harada, M. Takei, H. Kimura, Ultra-low specific on-resistance achieved in 3.3 kV-class SiC superjunction MOSFET, in *2021 33rd International Symposium on Power Semiconductor Devices and ICs (ISPSD)*, (2021), pp. 83–86
5. R. Kosugi et al., Breaking the theoretical limit of 6.5 kV-class 4H-SiC super-junction (SJ) MOSFETs by trench-filling epitaxial growth, in *2019 31st International Symposium on Power Semiconductor Devices and ICs (ISPSD)*, (2019), pp. 39–42
6. Transport of ions in Matter (TRIM) in James Ziegler's SRIM-2013 program: http://www.scrim.org
7. P. Thieberger et al., Nucl. Instrum. Meth. Phys. Res. B **442**, 36–40 (2019)
8. P. Thieberger, Nucl. Instrum. Meth. Phys. Res. **220**, 45–53 (1984)
9. The Tandem Van de Graaff Facility at BNL. https://www.bnl.gov/tandem/
10. R. Middleton, A Negative-Ion Cookbook, Dept. of Physics, University of Pennsylvania, Philadelphia, PA
11. Keysight 33500B Waveform Generator
12. TREK model P0622 High Voltage Amplifier
13. ORTEC model 439 Digital Current Integrator
14. ZABER Linear Stage Model X-LRQ300BL-SV1
15. Evans Analytical Group LLC (EAG), Sunnyvale, CA, 94086, USA
16. R. Ghandi, C. Hitchcock, S. Kennerly, Demonstration of 2kV SiC deep-implanted super-junction PiN diodes. Mater. Sci. Forum **1062**, 477–481 (2022)
17. R. Ghandi, et.al. 32nd ISPSD, 2020, pp. 126–129
18. HeatWave Labs Inc, Watsonville, CA, 95076, USA
19. Zeyu Chen et al 2022 Meet. Abstr. MA2022-02 1353
20. mi2-factory GmbH, Moritz-van-Rohr-Strabe 1a, 07745, Jena, Germany https://mi2-factory.com
21. High Voltage Engineering Europa B.V. https://www.highvolteng.com/
22. Pantechnick, https://www.pantechnik.com/

Chapter 17
Solid State Diffusion in GaN

Noah Allen, Kyoung Kweon, Clint Frye, Joel Varley, and Lars Voss

1 Introduction

Selective area doping is commonly applied to Si and SiC to make a multitude of device designs and geometries. However, this ability largely does not exist for doping GaN, particularly for the selective-area formation of p-type regions, which are required for vertical GaN power electronic devices. Significant p-type doping of GaN with Mg during growth was first achieved in 1989 by Amano et al. [1] and in 1991 by Nakamura et al. [2]. Since then, virtually all p-type GaN has been formed by in-situ Mg doping during epitaxial growth. While epitaxial growth is suitable for planar-junction devices, others require selective area doping of lateral junctions or 3D geometries [3–7]. The challenge in creating even simple lateral-junction structures severely limits the application space for GaN devices. Etch and regrowth is possible but presents significant challenges due to plasma damage and maintaining a clean surface after exposure to atmosphere [8, 9]. To this end, ex situ doping in the form of implantation or diffusion is desired as an alternative for selected area doping and doping 3D structures.

For power electronics, p-type doping is necessary to realize many vertical GaN architectures. Further, to make use of the high critical breakdown strength of GaN, the electric field near the device edge needs to be managed through the implementation of p-doped JTEs with precise spatial and dopant concentration control. While many researchers have explored using Mg ion implantation, dopant activation and healing of induced vacancies necessitate annealing temperatures above the decomposition point of GaN at ambient pressures. P-type doping of GaN by Mg ion implantation has been tested over the past two decades, but activation of the Mg dopant and annealing of implant damage requires extreme temperatures, typically

N. Allen · K. Kweon · C. Frye · J. Varley · L. Voss (✉)
Lawrence Livermore National Laboratory, Livermore, CA, USA
e-mail: voss5@llnl.gov

I. C. Kizilyalli et al. (eds.), *Gallium Nitride and Related Materials*, The Materials Research Society Series, https://doi.org/10.1007/978-3-031-83056-3_17

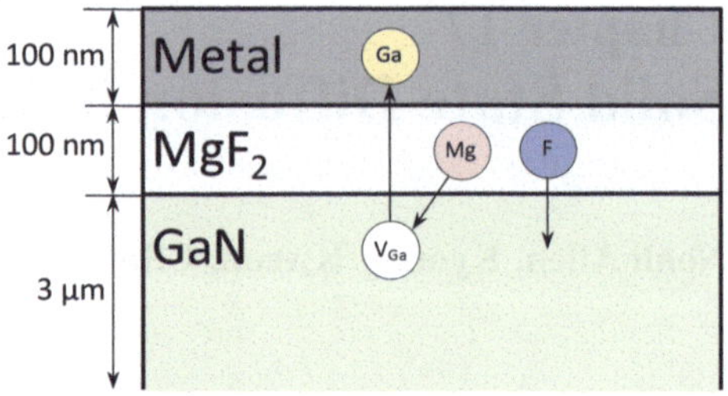

Fig. 17.1 Schematic cross-section of the diffusion bilayer and GAID process

>1200 °C. GaN decomposes severely at this temperature even for very short durations in N_2 overpressure [10–16]. GaN is thermodynamically unstable, and nitrogen vacancies begin to form at temperatures >800 °C [17, 18]. Capping layers to prevent decomposition, used routinely for other semiconductors, are insufficient due to a need for high N_2 vapor pressures [19, 20]. Feigelson et al. have developed a multi-cycle rapid thermal annealing (MRTA) process that demonstrates high doping levels ($>1 \times 10^{17}$ cm^{-3}) but requires highly specialized equipment [15, 16]. Few studies have explored diffusion doping of GaN, due to the extremely low diffusivities of impurities even at temperatures >1000 °C [21–26]. Therefore, a post-growth method of Mg-doping of GaN below 1000 °C is necessary to avoid nitrogen vacancy formation. Additionally, such a process could be more easily integrated into various fabrication processes due to thermal budget constraints on etched and metallized surfaces, allowing for the realization of new device structures and architectures. Diffusion doping could also enable doping of vertical sidewalls, *something very difficult to do using ion implantation.*

P-type GaN formation requires that the magnesium dopant occupies a gallium site (Mg_{Ga}) within the lattice. However, previous attempts of acceptor diffusion from a solid magnesium source did not provide a means to induce gallium vacancies (V_{Ga}) needed for the Mg_{Ga} formation unless performed at temperatures where nitrogen vacancies were also formed. By utilizing a layer stack consisting of a gallide-forming metal and Mg-containing insulator (as the diffusion source), the formation of gallium vacancies at the GaN surface can be thermodynamically favored, permitting Mg_{Ga} formation to occur. A schematic cross-section of the diffusion bilayer and GAID process is depicted in Fig. 17.1. Similar approaches have been demonstrated in both SiC, termed vacancy assisted impurity diffusion (VAID) [27, 28], and diamond [29] at moderate temperatures using an additional element either in the diffusion source or as a separate layer. Additionally, from ab initio simulations, Mg_{Ga} formation can be enhanced further when a low work function metal is used as the gallide-forming metal. Here, when the difference between the metal work function and GaN electron affinity is negative, the conduction band at the surface of the gallium nitride film is bent toward mid-gap, thereby lowering the formation energy of Mg_{Ga}.

In this chapter, we discuss the Gallidation Assisted Impurity Diffusion (GAID) process to form Mg-doped GaN layers by annealing a bilayer stack consisting of a gallidation-forming metal and magnesium fluoride (MgF_2) at 800 °C. Here, we take advantage of a low-temperature gallidation reaction using gallide-forming metals, including platinum [30], gold, palladium, and yttrium. MgF_2 was chosen as the Mg source due to its high thermal stability ($T_m = 1263$ °C) and ease of deposition. We further show the use of in-situ bias can also impact diffusivity during this process.

2 First Principles Analysis of mg Diffusion

For the GAID process to work, both an abundance of Ga vacancy defects and low Mg_{Ga} diffusion barriers are required. We perform hybrid functional theory calculations to better understand the interaction of V_{Ga} and Mg during the GAID process. Figure 17.2a shows the formation energies of V_{Ga} and two possible Mg configurations, incorporated substitutionally on a Ga site (Mg_{Ga}) and Mg interstitial (Mg_i). We include the results for the N-rich limit in Fig. 17.2, which is the most favorable for V_{Ga} and V_N suppression. Our results agree well with previous hybrid calculations of these defects in GaN [31]. As expected for charged defects, the calculated formation energies of V_{Ga}, Mg_{Ga}, and Mg_i and their preferred charge states are strongly dependent on the Fermi level. For V_{Ga}, which can trap one hole or up to three electrons as a ${V_{Ga}}^{-3}$, its formation energy rapidly decreases for Fermi levels that approach the CBM owing to its preferred acceptor state. As shown in Fig. 17.2a, Mg_{Ga} serves as a shallow acceptor with a (0/−) charge transition level at 0.09 eV above the VBM, while Mg_i exists only in a 2+ charge state within the bandgap. The relative stability of Mg_{Ga} and Mg_i varies with respect to the Fermi level. When the Fermi level is located near the VBM, Mg in GaN prefers to stay as Mg_i, whereas Mg_{Ga} becomes dominant as the Fermi level shifts to the CBM. Recent experiments show that both Mg_{Ga} and Mg_i coexist in GaN, but the fraction of Mg_i is enhanced (suppressed) in p-type (n-type) samples, respectively [32, 33], which is consistent with our calculation in Fig. 17.2. Also, as more N-rich conditions lower the chemical potential of Ga as well as Mg (computed from the GaN and Mg_3N_2 at fixed μ_N) [34], the formation energy of Mg_{Ga} (Mg_i) decreases (increases), respectively, so Mg_{Ga} incorporation is preferred over Mg_i even for low Fermi levels in these conditions.

Next, we examine possible migration pathways for Mg incorporation into GaN from the Mg source layer. Figure 17.2 also illustrates the migration barriers associated with ${Mg_i}^{2+}$ and V_{Ga}-assisted Mg diffusion pathways along the [0001] direction in Fig. 17.2b, in which Mg_{Ga} moves into the adjacent V_{Ga} site while leaving a V_{Ga} behind on the site that Mg has just left. As the charge state of V_{Ga} is largely dependent on the Fermi level, the formation energy of the (Mg_{Ga}–V_{Ga}) complex also shows a strong dependence on the Fermi level, as shown in Fig. 17.2a, with various charges of the complex ranging in charge from 0 to −4. It should be noted that when V_{Ga} as well as Mg_{Ga} is negatively charged, the binding interaction between these two becomes unfavorable with a positive $E_b(V_{Ga}/Mg_{Ga})$ as shown in Fig. 17.2a, where E_b of the (V_{Ga}/Mg_{Ga}) complex is computed relative to the isolated V_{Ga} and Mg_{Ga}, i.e., $E_b\,(V_{Ga}/Mg_{Ga}) = E_f\,(V_{Ga}/Mg_{Ga}) - E_f\,(V_{Ga}) - E_f\,(Mg_{Ga})$. However, the binding energy becomes less positive with F substitution on one of the

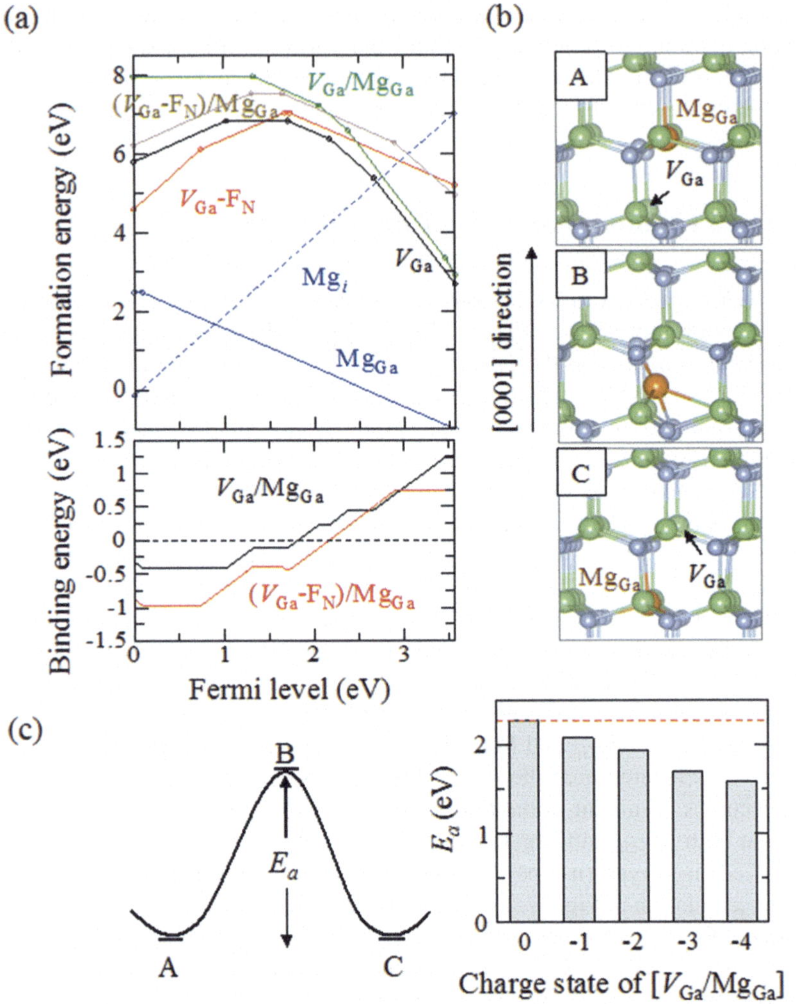

Fig. 17.2 (**a**) Formation and binding energies of V_{Ga} and Mg-related defects with atomic configurations of substitutional Mg at the Ga site and interstitial Mg. (**b**) initial (A), transition (B), and final (C) configuration for V_{Ga} assisted Mg_{Ga} diffusion pathway. (**c**) computed activation energy for the V_{Ga}-assisted Mg_{Ga} diffusion at various charge states. Activation energy for Mg_i diffusion is also shown with a red dashed line

N (F_N) next to V_{Ga}, implying that the donor-type F_N could lower the repulsion between V_{Ga} and Mg_{Ga} and increase the chance to meet them together. More details on fluorine in GaN will be discussed in the following section. The activation energy (E_a) of V_{Ga}-assisted Mg_{Ga} diffusion is computed to be 2.08, 1.94, 1.70, and 1.59 eV, with the charge state of −1, −2, −3, and −4, respectively, as summarized in Fig. 17.2c. In all cases, these migration energies are significantly lower than 2.3 eV for Mg_i^{2+} diffusion, suggesting V_{Ga} may greatly accelerate the diffusion of

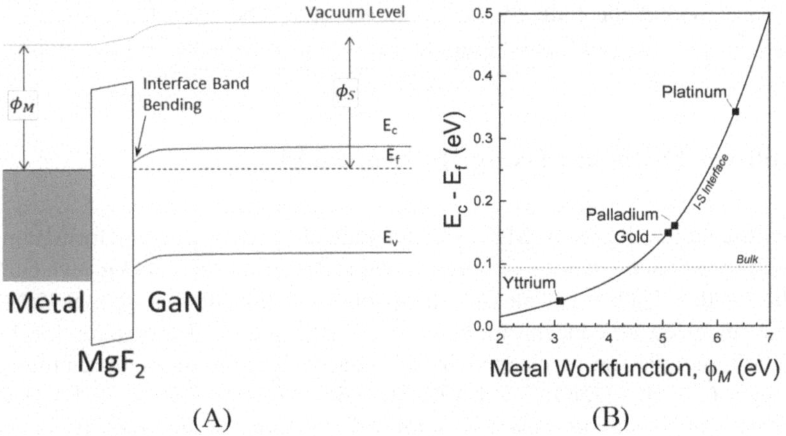

Fig. 17.3 (**a**) band lineup of an M-I-S interface where $\phi_M < \phi_s$ and (**b**) the calculated band bending at the insulator-GaN interface with varying metal work functions

Mg. Indeed, recent experimental studies report that the E_a for migration of Mg_i in Mg-implanted GaN is 1.3–2.0 eV along the *c*-axis [4]. We speculate that V_{Ga}-mediated Mg_{Ga} diffusion could occur as well as Mg_i diffusion in their experiment in the presence of high implantation-induced V_{Ga}, which may lower overall diffusion energies than that solely by the interstitial diffusion mechanism. This tendency clearly suggests that as the Fermi level is increased towards the CBM, Mg_{Ga} diffusion could be enhanced with reduced activation energies. Based on our calculation, we speculate that the driving force for the GAID process and the resulting high Mg concentration is associated with the high Fermi levels that lower the formation energies of V_{Ga} and Mg_{Ga}, which decreases the activation energies of Mg_{Ga} incorporation and diffusion.

Considering these findings, gallide-forming metals were selected such that prior to acceptor formation, the conduction band at the surface of the GaN was shifted toward the fermi level. In the case of an n-type semiconductor, when a metal with a work function (ϕ_M) smaller than that of the semiconductor work function (ϕ_s) interacts with electrons at the semiconductor's surface, the conduction band is shifted downward toward the Fermi-level, while metals with work functions larger than that of the semiconductor will shift the conduction band upward away from the Fermi level. The degree of band bending within the semiconductor is dependent on the work function difference, doping concentration, insulator thickness, and insulator dielectric constant. To understand these effects, a simulation was performed that mimics that of our experiments. Figure 17.3 shows the degree of band bending at the insulator-semiconductor (I-S) interface (solid line) as a function of the metal work function. Here a 100 nm thickness and dielectric constant of 5.45 are used for the MgF_2 layer, and a free carrier concentration of 6×10^{16} cm^{-3} is used for the GaN. The dashed line indicates the energy difference between the conduction band and the

Fermi-level within the bulk of the semiconductor. The expected degree of band bending for yttrium, gold, palladium, and platinum is indicated on the graph.

3 Bilayer Diffusion Using Built-in Field

To investigate the degree of Mg diffusion and subsequent acceptor formation after the GAID process, $6 \times 6\ mm^2$ samples were diced from an MOCVD-grown GaN-on-Sapphire wafer. Hall effect measurements on the 5 μm thick as-grown GaN layer indicate an electron concentration of $\sim 6 \times 10^{16}\ cm^{-3}$. To prepare for bilayer deposition, samples were cleaned in subsequent 10 min soaks of acetone, IPA, aqua regia (3:1 HCl:HNO_3), and 49% HF with thorough rinsing in DI H_2O and drying with an N_2 gun after each step. Immediately after cleaning, each sample was loaded into an electron beam evaporator, and a metal/MgF_2 (100/100 nm) bilayer was deposited over a broad area at $<2 \times 10^{-6}$ Torr. Five samples were fabricated with varying bilayer structures including with no metal, yttrium, gold, palladium, and platinum. Samples were individually loaded into a rapid thermal annealer where the diffusion process was executed. Each sample was annealed at 800 °C for 5 min under a continuous argon flow.

In preparation for SIMS measurements, chemical etching was required to remove the reacted bilayer. This was accomplished with a sequence of boiling aqua regia (3: 1 HCl:HNO_3, 72 °C) for 1 h to remove the reacted metals, 44% KOH at 80 °C for 1 h to oxidize the reacted MgF_2, and 49% HF for 1 h with thorough DI H_2O and N_2 drying after each step. Prior to SIMS measurements, each sample was examined under an optical microscope, and no change in surface roughness was apparent.

Secondary ion mass spectroscopy (SIMS) was performed to measure the concentration profiles of Mg, F and the metallic species to a depth of 200 nm. The depth profiles for each prepared sample are shown in Fig. 17.4 below. Here, a Mg concentration above the SIMS background limit ($2 \times 10^{16}\ cm^{-3}$) is observed in each of the GaN samples, with a peak concentration in the Y/MgF_2-prepared sample reaching values of $10^{20}\ cm^{-3}$ near the surface and $>1 \times 10^{17}\ cm^{-3}$ at 200 nm in both the Y/MgF_2 and Au/MgF_2 samples. This reveals that substantial magnesium diffusion at temperatures below the growth temperature is possible with careful consideration given to the diffusion stack. Next, considering the sample prepared without a metal cap, the magnesium concentration rapidly decreases below the background limit within ~33 nm into the surface, whereas samples prepared with a metal cap (excluding platinum) display more significant concentration profiles. As discussed previously, the enhanced Mg diffusivity is a result of both gallium gettering of the capping metal and surface band bending, resulting in higher density and deeper dopant diffusion. Additionally, bearing in mind the peak Mg concentration in samples annealed with a metal cap, the work function trend presented in Fig. 17.4 agrees with what is observed where higher work function metals should enhance Mg diffusivity. Here, the band bending from the metal-semiconductor work function mismatch is expected to affect diffusivity in the order shown, with yttrium being the

Fig. 17.4 SIMS Measurement of various samples

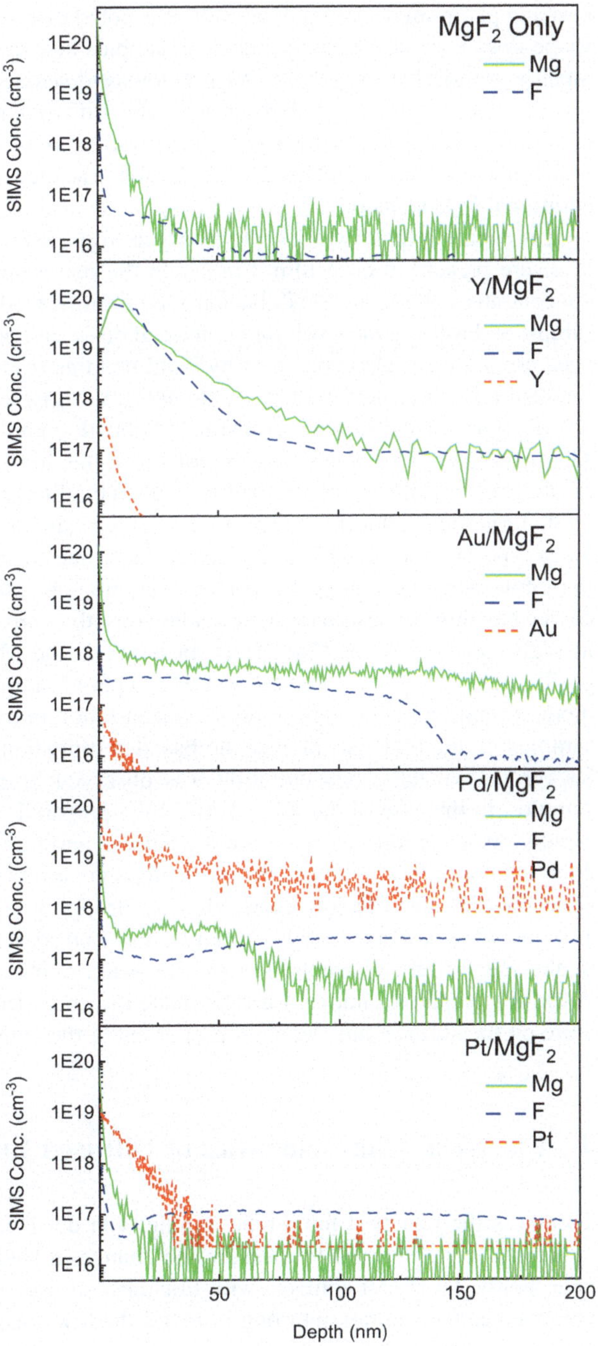

best and platinum the worst. However, it is noted that the concentration depth profile shape does vary significantly between samples and may be caused by the capping metal's gallidation properties. As previously discussed, as gallium vacancies are formed, the metal cap will react with the diffused gallium species, inducing a diffusion gradient across the magnesium fluoride. Without this driving force, gallium will "pile up" at the interface and inhibit vacancy formation, thereby limiting additional Mg diffusion.

It is apparent that both fluorine from the MgF_2 layer and metal from the capping layer are present in each film. Similar to the magnesium concentration, a fluorine concentration above the SIMS background limit (2×10^{15} cm^{-3}) is observed in each sample and follows the work function trend discussed previously. This indicates that F incorporation is also enhanced by band bending at the surface prior to diffusion. Interestingly, the concentration profile in the Y/MgF_2 follows the Mg profile closely but diverges in the no-cap, Au/MgF_2, Pd/MgF_2, and Pt/MgF_2-prepared samples. This suggests that for some bilayer structures, the fluorine behaves independently of the magnesium and can be controlled by the metallic capping layer. In the case of PN diode formation, a fluorine concentration larger than the magnesium concentration in the p-type layer would have negative effects due to its likely incorporation as an immobile negative charge [35]. However, implementation on the periphery of a device has shown use as a termination layer for the purpose of tailoring electric fields in high-voltage devices [36]. Thus, in terms of the Mg/F ratio, the concentration profiles revealed in the Y/ and Au/MgF_2-prepared samples may be beneficial for PN diode formation, while high fluorine concentration in the Pd/ and Pt/MgF_2-prepared samples could find use in fluorine-based termination layers. Finally, an above-background metallic concentration was observed in each of the bilayer-prepared samples. In the case of the Y/ and Au/MgF_2-prepared samples, the metallic species rapidly decays within 25 nm from the surface and may be an indication of surface contamination. The Pd/ and Pt/MgF_2-prepared samples reveal a metallic species concentration that decays more slowly, dropping below background levels at 100 and 75 nm, respectively. Currently, this behavior is not well understood. Due to the magnesium concentration profile peak, depth, and Mg/F ratio, additional Au/MgF_2 prepared samples were fabricated to study the effect of time and temperature on the surface and electrical properties of the GaN films.

4 Electrical Characteristics of Diffused Layers

To assess the potential for p-type doping, van der Pauw configuration Hall effect structures were prepared to measure the change in Hall resistance at varying magnetic fields. Pd/Ni/Au contacts with thickness of 50/75/125 nm were deposited by electron beam evaporation through a metal shadow mask and subsequently annealed at 500 °C in argon for 5 min. The contacts were inset from the edge by ~100 μm to prevent shorting the contacts to the underlying doped layer. Samples were placed at room temperature in a Quantum Design Versalab Physical Property Measurement

Table 17.1 Summary of result from Hall effects measurements of samples diffused at varying times and temperatures

Diffusion Source	Sheet Charge Conc. ($\times 10^9$ cm^{-2})					
	As-grown	800 °C 5 min	800 °C 25 min	800 °C 53 min	858 °C 5 min	915 °C 5 min
Au/MgF2 (100/100 nm)	−14,000	−2.2	−4.9	−0.8	−0.6	+43

Table 17.2 Summary of results from Hall effect measurements performed under varying optical excitation

Diffusion Source	Diffusion Condition	Sheet Charge Conc. ($\times 10^9$ cm^{-2})				
		Dark	940 nm (1.32 eV)	611 nm (2.03 eV)	415 nm (2.99 eV)	365 nm (3.39 eV)
Au/MgF_2 (100/100 nm)	858 °C 5 min	−0.6	−1.5	+1.4	Comp.	−8900

system with a 3 T magnet, and the DC Hall resistance was measured with a Keithley 4200 semiconductor parameter analyzer equipped with a low voltage amplifier. The as-grown sample exhibited n-type conduction at room temperature apparent by the negative slope of the Hall resistance vs. magnetic field data. The electron sheet concentration was found from the slope of the fitted line and calculated to be ~1×10^{13} cm^{-2}. Repeating the measurement on the Au/MgF2 bilayer-annealed samples resulted in an increased slope indicative of lower overall carrier sheet concentration and a large offset in the y-axis from 0 Ω, which is a symptom of non-ohmic or non-uniform contacts. Most samples exhibited an approximate three orders of magnitude reduction in conductivity, implying the Au/MgF_2 bilayer annealing scheme can compensate for the background free carrier concentration. Additionally, annealing the sample at 915 °C for 5 min resulted in a positively sloped Hall resistance vs. magnetic field graph implying hole conduction is dominant. From the slope, a hole sheet concentration of ~3.4×10^{10} cm^{-2} was calculated. Sheet conductivity measurements were attempted for each of the annealed samples, but the data (not shown) was inconclusive (Table 17.1).

From the SIMS of the Au/MgF_2-prepared sample, it is apparent that a high concentration of both magnesium and fluorine exists near the surface, with the F likely compensating the Mg. Thus, Hall measurements were repeated under optical excitation from an LED mounted normal to the sample surface. LEDs with characteristic wavelengths of 940, 611, 415, and 365 nm were selected. Excitation from the 940 nm LED resulted in no change compared to the dark measurements, while excitation from the 415 and 365 nm LEDs resulted in films that exhibited less electron compensation and thus became more n-type. It was found that excitation from a 611 nm (2.03 eV) LED resulted in the most interesting outcome, summarized below in Table 17.2. For samples annealed at 800 °C for both 5 and 53 min, excitation from the 611 nm source resulted in a larger apparent n-type sheet concentration compared to dark measurements. However, for samples annealed at

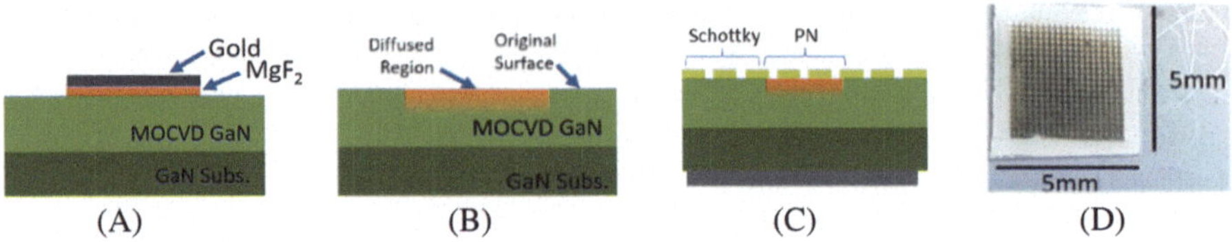

Fig. 17.5 Fabrication process for diffused and non-diffused diode mapping as well as prepared sample

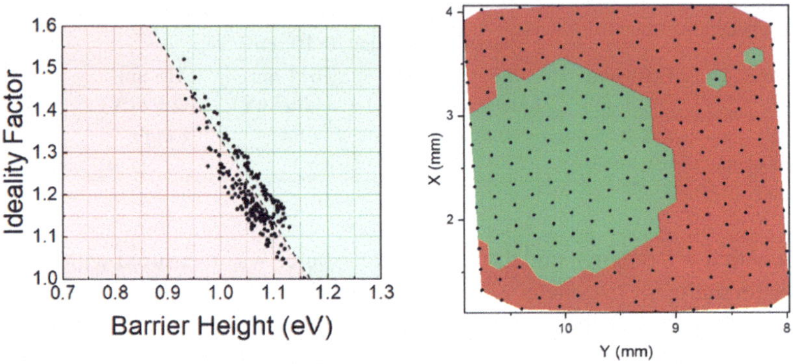

Fig. 17.6 Ideality factor vs. barrier height for diffused (green) and surrounding (red) areas

800 °C for 15 min, 858 °C for 5 min, and 915 °C for 5 min, excitation from the LED resulted in a net increase in the apparent hole sheet concentration, where the former two changed from electron- to hole-dominated conduction compared to dark measurements.

This apparent increase in hole sheet concentration after optical excitation may be indicative of the presence of deep-level traps. Once annealed, the incorporated Mg and F induce a large concentration of hole traps within the forbidden gap. Results imply that a deep-level hole trap exists in Samples C, E, and F with a characteristic activation energy between 1.32 and 2.03 eV from the valence band indicated by no change in the Hall measurements (not shown) with 940 nm optical excitation but a net hole concentration increase upon 611 nm excitation. To understand how magnesium and fluorine could contribute to this phenomenon, first principles simulations were executed.

The device properties of prepared films were examined by fabricating a diode matrix on a selectively diffused GaN sample. The process is outlined below in Fig. 17.5. First, the Au/MgF2 bilayer is deposited by e-beam through a metal mask and then diffused at 800 °C for 5 mins. After diffusion, a backside ohmic contact was defined with Kapton tape, and a topside 150 μm ohmic/Schottky contact was defined on each mesa with a photoresist liftoff step.

Examining the I-V results, the diffusion has an effect on the electrical characteristics; however, the magnitude of this change is not apparent. Figure 17.6 shows the voltage at which the diode reaches 10 μA and plotting the contour maps, it becomes clear that the patterned bilayer does indeed alter the surface of the semiconductor such that the I-V characteristics are shifted on the x-axis. Additionally, a shift in the voltage required to reach 10 μA while still maintaining the thermionic emission-limited region (linear region on a semi-log plot) suggests that the energy barrier at the M-S interface is varying.

As shown in the SIMS profile in Fig. 17.4, we observe high F concentrations incorporated into the GaN during the GAID process that strongly track the Mg profiles. To elucidate the effect of F on the electrical activation of Mg diffused in GaN, we compute the formation energy of F impurities in GaN and assess their expected kinetics. As shown in Fig. 17.7, F interstitial (F_i) exhibits negative-U behavior by changing its preferential charge state from (1+) to (1−) with a large structural change as the Fermi level is increased above 0.87 eV. For the (1+) charge state, F_i is stable in the bridge center site between Ga and N, while for the (1-) charge state, F_i occupies an octahedral interstitial site at the center of a hexagonal channel along the [0001] axis and is bonded with three adjacent Ga atoms. On the other hand, substitutional F at the N site (F_N) serves as a donor with the (2+/0) charge transition state at 0.6 eV below the CBM. Our hybrid DFT calculation suggests that F would prefer to be F_N for low Fermi levels, whereas F_i becomes more favorable when the Fermi level moves towards the CBM. Thus both F_i and F_N donors could act as sources of compensation to any incorporated Mg_{Ga} acceptors for Fermi levels below 0.87 eV. As our interest lies in high Fermi levels, we further examine the diffusion of the negatively charged F_i into GaN along the [0001] direction during the GAID process. In Fig. 17.3b, the F_i in the octahedral site goes to the next equivalent site by passing a basal (0001) plane, and the corresponding E_a is 2.24 eV (see Fig. 17.2). However, we found that the F_i diffusion could be facilitated when it is mediated by V_{Ga}, as demonstrated in Fig. 17.7b and c. With various charge states for F_i and V_{Ga}, $(F_i - V_{Ga})$ also exists in six different charge states ranging from 1+ to 4−; when the Fermi level is higher than 1.99 eV from the VBM, F_i in $(F_i - V_{Ga})$ takes the octahedral interstitial configuration [(A) in Fig. 17.2b], whereas it passivates the V_{Ga} by bonding with one of three undercoordinated N atoms and forms F_{Ga} when the Fermi level is lower than 1.99 eV from the VBM. So, the diffusion pathway in Fig. 17.7b would be applicable to explain the V_{Ga}-mediated F_i diffusion for higher Fermi levels. As shown in Fig. 17.7c, the energy of (A) is lower than that of (C) due to a binding energy between F_i and V_{Ga}, while the energy difference varies depending on $[F_i - V_{Ga}]$ charge states. The activation energies for the V_{Ga}-medated F_i diffusion for charge states ranging from −2 to −4 are included in Fig. 17.3c and are found to strongly decrease by more than 1 eV relative to the isolated F_i. This enhanced F diffusion has been suggested by previous positron annihilation spectroscopy (PAS) and TOF-SIMS experiments in F-implanted GaN, in which enhanced F diffusion as well as formation of $(F_i - V_{Ga})$ complex was observed when sufficient V_{Ga} were available [35, 37]. The reduced activation energies with V_{Ga} suggest enhanced F diffusion in the presence of higher V_{Ga} concentrations similar to that expected for Mg, in which V_{Ga} serves as a diffusion channel for both Mg and F. This can explain why the depth profile for F similarly follows that for

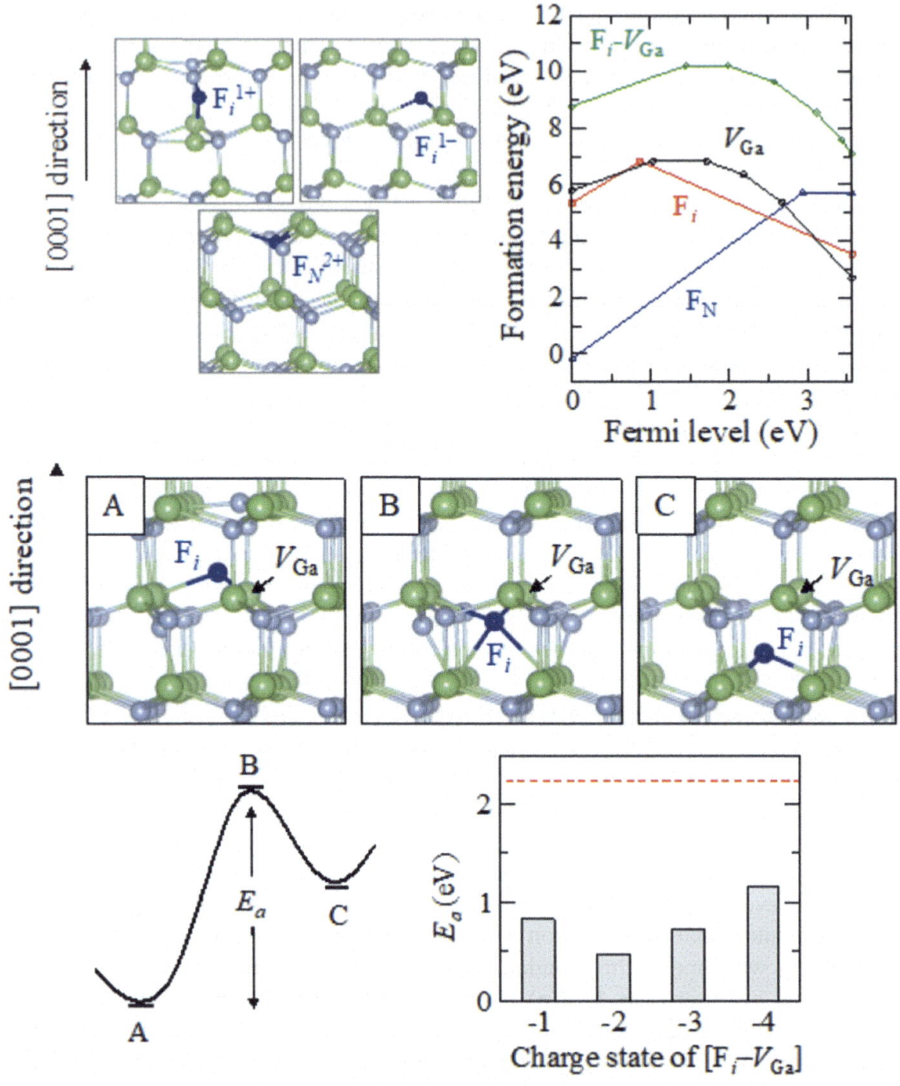

Fig. 17.7 Calculations of F diffusion mechanisms. Top shows possible positions of F within the lattice as well as formation energies associated with these positions. Middle shows interstitial F diffusing through the lattice. Bottom shows the diffusion barrier to diffusion based on the charge state of the F_i-V_{Ga} complex

Mg in SIMS. We note that the diffusion barriers for V_{Ga}-mediated F_i diffusion increase with higher Fermi levels, in contrast with Mg, implying that V_{Ga}-mediated F diffusion and Mg diffusion exhibit distinct behavior that may influence their spatial correlation and time dependence of how these dopants are incorporated. Although during the GAID process, the negatively charged F_i is energetically

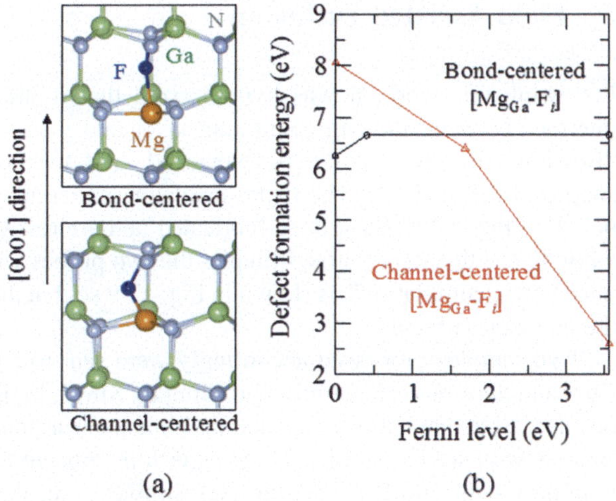

Fig. 17.8 (**a**) atomic configuration of bond-centered and channel-centered ($Mg_{Ga} - F_i$) complex. (**b**) Formation energy of the ($Mg_{Ga} - F_i$) complex with respect to the Fermi level

favorable with high Fermi levels, it can easily convert to the donor-type F_N when the Fermi level shifts down during the Mg activation process, suggesting that the diffused F could become compensating defects in the GaN.

As both Mg and F prefer to diffuse to V_{Ga}-rich regions, these two impurities likely meet and may bind together, leading to the formation of the ($Mg_{Ga} - F_i$) complex. As shown in Fig. 17.8, two stable configurations, bridge position configuration in (a) and channel-centered configuration in (b), are identified for the ($Mg_{Ga} - F_i$) complex due to the amphoteric nature of F_i. From the formation energy calculation, we found that the bond-center configuration exhibits donor-like characteristics with a (1+/0) charge transition level at 0.28 eV, while the channel-centered configuration serves as an acceptor with a deep (1−/2−) charge transition level at 1.67 eV above the valence band. Significantly lower formation energy for the channel-centered configuration for higher Fermi levels than that for the bond-centered configuration suggests that during the GAID process, the channel-centered ($Mg_{Ga} - F_i$) complex is likely formed. However, this acceptor-type ($Mg_{Ga} - F_i$) complex can convert to the donor-type compensating complex during the photoexcitation process. As suggested in $(Mg_{Ga} - F_i)^{1-} + hv \rightarrow (Mg_{Ga} - F_i)^0 + e_{CB}$, an electron in the defect state can be excited to the CBM, resulting in a neutral channel-centered ($Mg_{Ga} - F_i$) complex. As the neutral channel-centered ($Mg_{Ga} - F_i$) complex is not stable, it spontaneously converts to the bond-centered $(Mg_{Ga} - F_i)^0$ complex. The computed excitation energy for this transition is 3.3 eV, and by which, the acceptor-type ($Mg_{Ga} - F_i$) becomes the donor-type complex.

5 Field Assisted Diffusion

Following this work, it was hypothesized that if the Fermi-level position at the interface between the Mg-containing layer and the GaN can be controlled during diffusion, diffusivity could be enhanced, and the compensating defects can be suppressed (Fig. 17.9). The metal-insulator-semiconductor (MIS) structures shown below in Fig. 17.10 have been fabricated and are ready for preliminary experimentation to test this hypothesis. Initially, the two probes will be used to bias the sample into "soft accumulation" as shown in Fig. 17.9 so that the Fermi level lies at or above the conduction band.

Two samples processed accordingly were sent to EAG Laboratories for magnesium and fluorine concentration profiling by SIMS. Figure 17.11 reveals two distinct profiles. Note that EAG performed multiple measurements in different spots on each sample, with similar results. The magnesium concentration in the sample measured with no in-situ bias has a profile very similar to results from previous experiments

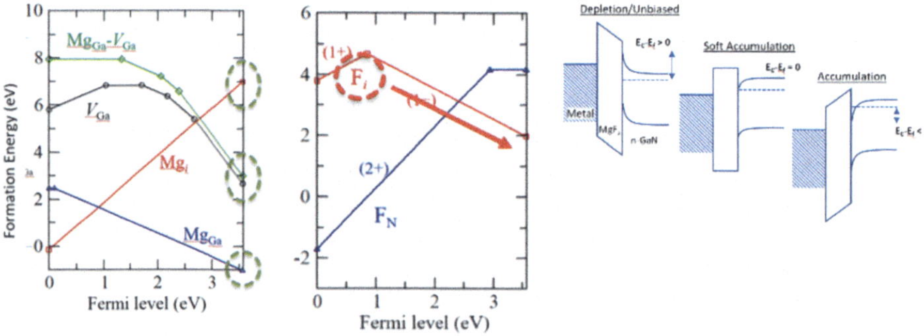

Fig. 17.9 A metal-insulator-semiconductor band profile under varying bias conditions

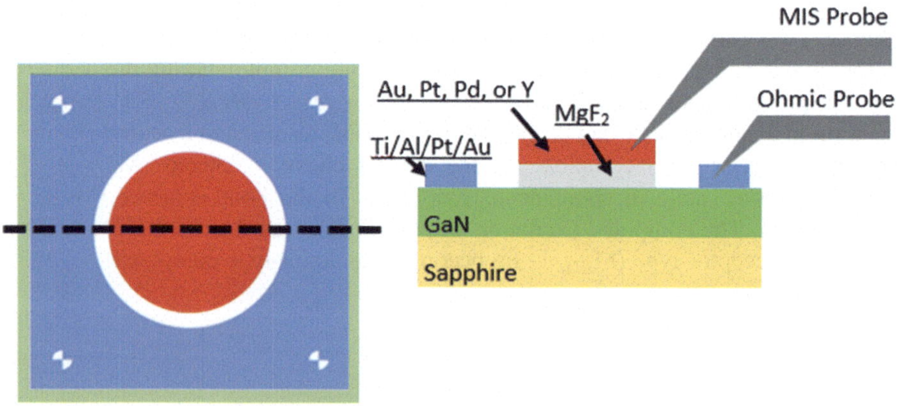

Fig. 17.10 (Left) Metal-insulator-semiconductor (MIS) device structure used for experimenting with bias control of the interface fermi-level according to the (Right) theoretical band profile

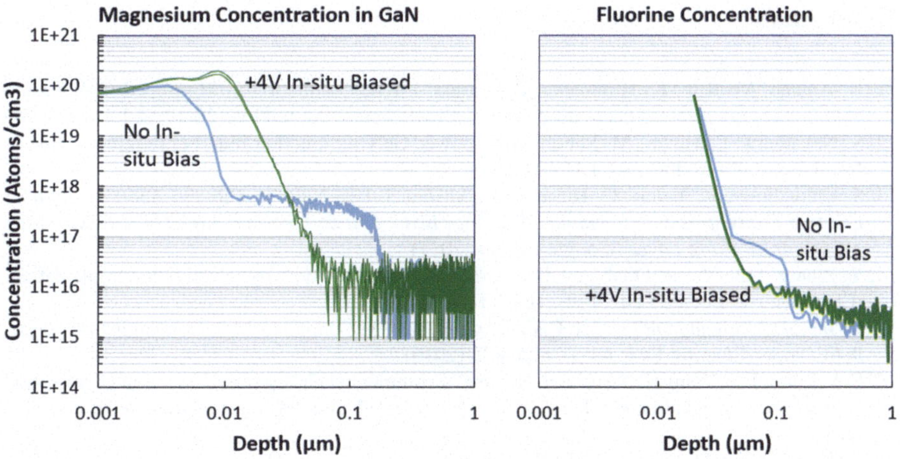

Fig. 17.11 SIMS measurement results for the depth-dependent (Left) Mg and (Right) F concentrations

where the GaN layer was grown on sapphire, showing a long tail into the GaN. Applying a +4 V in-situ bias during the annealing process appears to increase the surface concentration while also extending the >1e18 cm^{-3} crossing point from 10 to 35 nm. Additionally, the +4 V in-situ bias suppresses the fluorine concentration when compared to the sample processed with no in-situ bias. Note that due to the use of a different sputtering ion to measure F, depth resolution is decreased; thus, near-surface concentrations are difficult to resolve. These results are exciting and indicate that the core premise of surface Fermi-level tailoring is correct.

6 Conclusion

We have demonstrated the ability to diffuse Mg into GaN at temperatures as low as 800 °C, with room temperature p-type conductivity measured by Hall apparent at 915 °C annealing temperatures. Further work is required to fully exploit the potential of this technique and to understand the limitations. This has significant implications for GaN- and III-N-based electronics and optoelectronics, as selective area doping at temperatures consistent with other process steps is now possible. We expect that further improvement is possible using better diffusion sources such as Mg_3N_2, optimizing Mg and gallidation layer thicknesses, and additional experimentation of an in-situ bias to improve lower Mg_{Ga} formation energy.

Acknowledgments This work was supported in part by the U.S. Department of Energy Advanced Research Projects Agency (ARPA-E) PNDIODES Program. This work was performed under the auspices of the U.S. Department of Energy by Lawrence Livermore National Laboratory under Contract DE-AC52-07NA27344, LLNL-BOOK-865817.

References

1. H. Amano, M. Kito, K. Hiramatsu, I. Akasaki, P-type conduction in Mg-doped GaN treated with low-energy electron beam irradiation (LEEBI). Jpn. J. Appl. Phys. **28**(12A), L2112 (1989)
2. S. Nakamura, T. Mukai, M. Senoh, High-power GaN pn junction blue-light-emitting diodes. Jpn. J. Appl. Phys. **30**(12A), L1998 (1991)
3. S. Chowdhury, R. Yeluri, C. Hurni, U.K. Mishra, I. Ben-Yaacov, *Current Aperture Vertical Electron Transistors with Ammonia Molecular Beam Epitaxy Grown P-Type Gallium Nitride as a Current Blocking Layer* (Google Patents, 2017)
4. Y. Zhang et al., Vertical GaN junction barrier Schottky rectifiers by selective ion implantation. IEEE Electron Device Lett. **38**(8), 1097–1100 (2017)
5. D. Ji et al., Normally OFF trench CAVET with active Mg-doped GaN as current blocking layer. IEEE Trans. Electron Dev. **64**(3), 805–808 (2016)
6. E.A. Jones, F.F. Wang, D. Costinett, Review of commercial GaN power devices and GaN-based converter design challenges. IEEE J. Emerg. Selected Top. Power Electron. **4**(3), 707–719 (2016)
7. J. Hu, Y. Zhang, M. Sun, D. Piedra, N. Chowdhury, T. Palacios, Materials and processing issues in vertical GaN power electronics. Mater. Sci. Semicond. Process. **78**, 75–84 (2018)
8. D. Storm et al., Critical issues for homoepitaxial GaN growth by molecular beam epitaxy on hydride vapor-phase epitaxy-grown GaN substrates. J. Cryst. Growth **456**, 121–132 (2016)
9. D.F. Storm et al., Surface preparation of freestanding GaN substrates for homoepitaxial GaN growth by rf-plasma MBE. J. Vac. Sci. Technol. B:Nanotechnol. Microelectron. **35**(2), 02B109 (2017)
10. S. Pearton, C. Vartuli, J. Zolper, C. Yuan, R. Stall, Ion implantation doping and isolation of GaN. Appl. Phys. Lett. **67**(10), 1435–1437 (1995)
11. J. Chan et al., Thermal annealing characteristics of Si and Mg-implanted GaN thin films. Appl. Phys. Lett. **68**(19), 2702–2704 (1996)
12. N. Parikh et al., Ion implantation of epitaxial GaN films: Damage, doping and activation. Nucl. Instrum. Methods Phys. Res., Sect. B **127**, 463–466 (1997)
13. B. Pong et al., Structural defects and microstrain in GaN induced by mg ion implantation. J. Appl. Phys. **83**(11), 5992–5996 (1998)
14. C. Ronning, E. Carlson, R.F. Davis, Ion implantation into gallium nitride. Phys. Rep. **351**(5), 349–385 (2001)
15. B. Feigelson et al., Multicycle rapid thermal annealing technique and its application for the electrical activation of Mg implanted in GaN. J. Cryst. Growth **350**(1), 21–26 (2012)
16. T. Anderson et al., Activation of Mg implanted in GaN by multicycle rapid thermal annealing. Electron. Lett. **50**(3), 197–198 (2014)
17. D.-H. Youn et al., Investigation on the p-type activation mechanism in Mg-doped GaN films grown by metalorganic chemical vapor deposition. Jpn. J. Appl. Phys. **38**(2R), 631 (1999)
18. C.A. Hurni, J.R. Lang, P.G. Burke, J.S. Speck, Effects of growth temperature on Mg-doped GaN grown by ammonia molecular beam epitaxy. Appl. Phys. Lett. **101**(10), 102106 (2012)
19. S. Porowski et al., Annealing of GaN under high pressure of nitrogen. J. Phys. Condens. Matter **14**(44), 11097 (2002)
20. J. Unland, B. Onderka, A. Davydov, R. Schmid-Fetzer, Thermodynamics and phase stability in the Ga–N system. J. Cryst. Growth **256**(1–2), 33–51 (2003)
21. C. Pan, G. Chi, The doping of GaN with Mg diffusion. Solid State Electron. **43**(3), 621–623 (1999)
22. T. To, A. Djurisic, M. Xie, W. Fong, C. Surya, Doping of GaN by Mg diffusion, in *2002 Conference on Optoelectronic and Microelectronic Materials and Devices. COMMAD 2002. Proceedings (Cat. No. 02EX601)*, (IEEE, 2002), pp. 75–78
23. Y.-J. Yang, J.-L. Yen, F.-S. Yang, C.-Y. Lin, P-type GaN formation by Mg diffusion. Jpn. J. Appl. Phys. **39**(5A), L390 (2000)

24. S.J. Chung, B. Karunagaran, C.H. Hong, H.J. Lee, E.K. Suh, Photocurrent spectroscopy investigations of Mg-related defect levels in p-type GaN. J. Korean Phys. Soc. **48**(3), 460–463 (2006) [Online]. Available: <Go to ISI>://WOS:000236038100022
25. J. Zolper, Ion implantation in group III-nitride semiconductors: A tool for doping and defect studies. J. Cryst. Growth **178**(1–2), 157–167 (1997)
26. R. Jakiela, A. Barcz, E. Dumiszewska, A. Jagoda, Si diffusion in epitaxial GaN. Phys. Status Solidi C **3**(6), 1416–1419 (2006)
27. C.-C. Tin et al., Oxide film assisted dopant diffusion in silicon carbide. Thin Solid Films **518**(24), e118–e120 (2010)
28. C.-C. Tin, S. Mendis, M.T. Tin, T. Isaacs-Smith, J.R. Williams, A new approach in impurity doping of 4H-SiC using silicidation. J. Appl. Phys. **114**(24), 244502 (2013)
29. J.-H. Seo et al., Thermal diffusion boron doping of single-crystal natural diamond. J. Appl. Phys. **119**(20), 205703 (2016)
30. S. Gasser, E. Kolawa, M.-A. Nicolet, Thermal reaction of Pt film with< 110> GaN epilayer. J. Vac. Sci. Technol. A **17**(5), 2642–2646 (1999)
31. J.L. Lyons, A. Alkauskas, A. Janotti, C.G. Van de Walle, First-principles theory of acceptors in nitride semiconductors. Phys. Status Solidi B **252**(5), 900–908 (2015). https://doi.org/10.1002/pssb.201552062
32. U. Wahl et al., Lattice location of Mg in GaN: A fresh look at doping limitations. Phys. Rev. Lett. **118**(9) (2017). https://doi.org/10.1103/PhysRevLett.118.095501
33. U. Wahl et al., Lattice location studies of the amphoteric nature of implanted Mg in GaN. Adv. Electron. Mater. **7**(9) (2021). https://doi.org/10.1002/aelm.202100345
34. G. Miceli, A. Pasquarello, Self-compensation due to point defects in Mg-doped GaN. Phys. Rev. B **93**(16), 165207 (2016). https://doi.org/10.1103/PhysRevB.93.165207
35. M. Wang, L. Yuan, C. Cheng, C. Beling, K. Chen, Defect formation and annealing behaviors of fluorine-implanted GaN layers revealed by positron annihilation spectroscopy. Appl. Phys. Lett. **94**(6), 061910 (2009)
36. S. Han, S. Yang, K. Sheng, Fluorine-implanted termination for vertical GaN Schottky rectifier with high blocking voltage and low forward voltage drop. IEEE Electron Device Lett. **40**(7), 1040–1043 (2019)
37. M. Wang, L. Yuan, K. Chen, F. Xu, B. Shen, Diffusion mechanism and the thermal stability of fluorine ions in GaN after ion implantation. J. Appl. Phys. **105**(8), 083519 (2009)

Chapter 18
Transmutation Doping of GaN

Jae W. Kwon

1 Introduction

The neutron transmutation doping (NTD) of semiconductors was unveiled in 1961 and has been widely used in many nuclear reactor facilities to produce precisely controlled and homogeneously distributed dopants throughout a bulk semiconductor crystal. The dopants are produced through the radioactive decay of an isotope of the host element into an isotope of a different element, the dopant. High-quality n-type doped silicon crystals with an incredibly homogeneous doping profile can be produced through the conversion of the host Si to the dopant phosphorus. This process can also be used in other semiconductor materials such as GaN. Figure 18.1 illustrates how Ge is created from Ga by neutron reactions. In particular, we found that effective neutron transmutation doping of GaN happens when thermal or epithermal neutrons interact with ^{69}Ga and ^{71}Ga to form unstable ^{70}Ga and ^{72}Ga, respectively. The half-lives of the ^{70}Ga and ^{72}Ga are 21.1 min and 14.1 h, respectively, and they undergo beta decay to ^{70}Ge and ^{72}Ge.

Neutrons during the NTD process in GaN cause three main reactions to occur: ^{69}Ga(n,γ) ^{70}Ge, ^{71}Ga(n,γ) ^{72}Ge, and ^{14}N(n,p)^{14}C, as shown in the reaction equations below. Through the (n,γ) process, gallium (Ga) first creates unstable Ga, which then decays to yield stable germanium (Ge) isotopes. Carbon-14 (^{14}C) is created when nitrogen (N) reacts with other elements (n, p). The following formulas can be used to represent these three reactions: The symbol (X*) indicates that element X is unstable.

J. W. Kwon (✉)
Infinity Power LLC, San Diego, CA, USA
e-mail: enerjkwon@infinitypower.energy

I. C. Kizilyalli et al. (eds.), *Gallium Nitride and Related Materials*, The Materials Research Society Series, https://doi.org/10.1007/978-3-031-83056-3_18

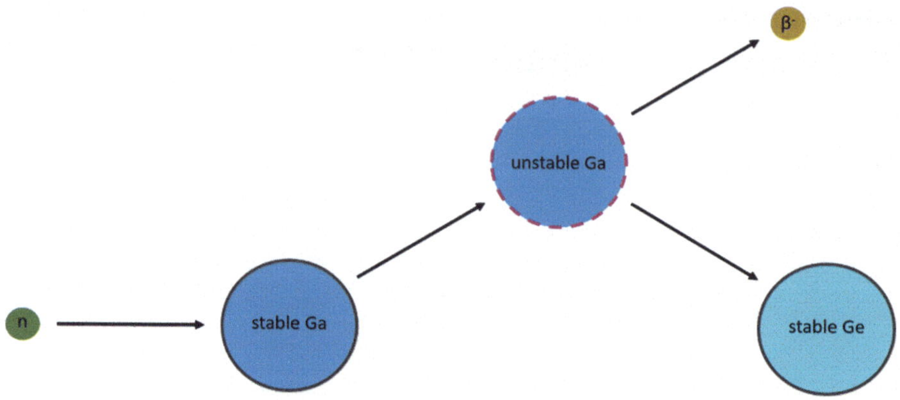

Fig. 18.1 Gallium is converted to germanium by neutron capture and subsequent beta decay

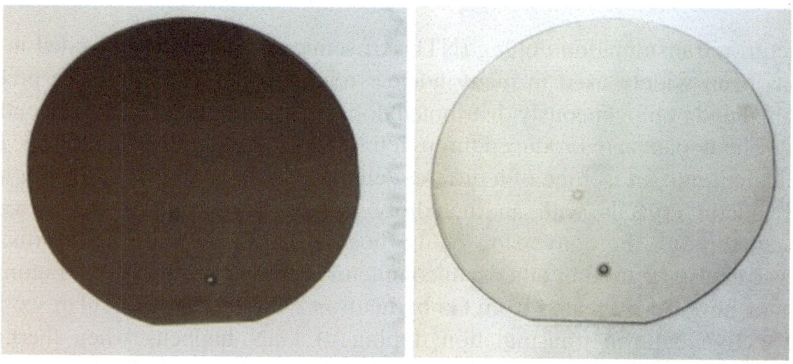

Fig. 18.2 Photos of irradiated GaN wafers. (**a**) 2″ GaN wafer after irradiated for 60 h and (**b**) after the annealing process [1]

$$
\begin{aligned}
{}^{69}_{31}Ga + {}^{1}_{0}n &\rightarrow {}^{70}_{31}Ga^{*} \rightarrow {}^{70}_{32}Ge + {}^{0}_{-1}\beta;\\
{}^{71}_{31}Ga + {}^{1}_{0}n &\rightarrow {}^{72}_{31}Ga^{*} \rightarrow {}^{72}_{32}Ge + {}^{0}_{-1}\beta;\\
{}^{14}_{7}N + {}^{1}_{0}n &\rightarrow {}^{14}_{6}C^{*} + {}^{1}_{1}p;
\end{aligned}
$$

As seen in Fig. 18.2a, the wafers become discolored after being exposed to radiation. As demonstrated in Fig. 18.2b, an annealing procedure helps restore the optical property of the GaN wafers, which are clear prior to irradiation. The annealing procedure uses a pressurized chamber filled with ultra-pure nitrogen at 331 kPa in which the wafers were annealed at 950 °C to prevent thermal dissociation.

2 Superior Doping Uniformity by Neutrons

NTD can provide spatially excellent uniformity throughout the whole volume of the substrate. Secondary-ion mass spectroscopy (SIMS) and gamma-ray spectroscopy were used to measure the amounts of Ge isotopes (^{72}Ge and ^{70}Ge) and ^{14}C produced, and the results show good agreement between the two methods. SIMS is a strong analytical method that offers comprehensive details about the molecular and elemental makeup of materials. Secondary ions are extracted by sputtering with a primary ion beam and are then measured with a mass spectrometer to determine the elemental composition. All SIMS measurements were carried out at the EAG Laboratory for accurate mass analysis of the principal dopants.

Because ^{14}C has extremely low radioactivity, the SIMS technique is the sole method available for ^{14}C measurements. On the other hand, there is detectable radioactivity in Ga radioisotopes that will convert to Ge isotopes. We have devised a precise technique for measuring the doping concentration in our NTD samples. Our innovative way of counting gamma-ray spectroscopy allows us to accurately determine the doping content of both ^{70}Ge and ^{72}Ge. Since the SIMS approach can only measure the overall doping concentration, including the initial doping concentration prior to irradiation, and naturally has a greater error range as the atomic number grows, measuring the dopant synthesis rate by irradiation has proven to be highly helpful.

SIMS measurements on opposite sides of an NTD-processed 350 μm GaN wafer with a concentration of 10^{18} Ge atoms/cm^3 are presented in Fig. 18.3. The results of the analysis demonstrated the superior uniformity of the germanium doping concentration across the entire wafer in the thickness direction and verified the approximations utilized in the gamma-ray spectroscopy. It is not surprising that there is a very little neutron flux drop-off through the 350 μm wafer thickness. The GaN wafer thickness of 350 μm has a flux reduction of 7.7×10^{-5}%, as calculated using the

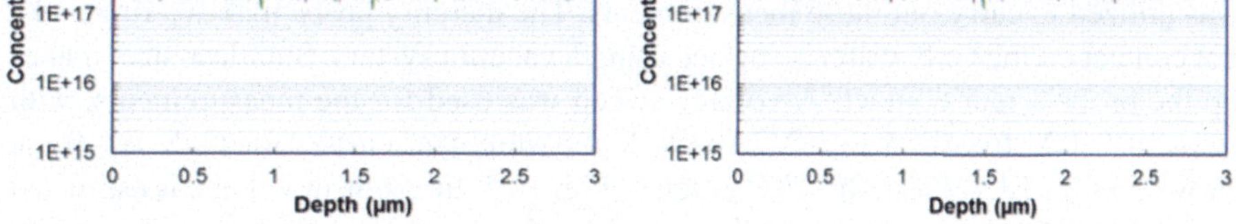

Fig. 18.3 SIMS data showing consistent doping concentration of both ^{70}Ge and ^{72}Ge in a 10^{18} doped GaN wafer [1]

thermal cross sections for ^{69}Ga and ^{71}Ga, which are 4.61 barns and 1.68 barns, respectively [1].

In comparison to SIMS, which has an error range of ±10%, our new gamma-ray spectroscopy counting approach for Ge doping concentration measurements has a reduced error range of ±3.5%. The high accuracy of the gamma-ray spectroscopy counting approach can be seen by the nearly identical values from SIMS when taking into account the amounts of Ge doping. The doping concentrations of Ge isotopes can be computed using the concentration equations [2, 3] in accordance with the conditions of radioactive decay. We have derived a new formula for the transmutation-derived atom density (doping concentration) of ^{72}Ge (CT[Ge-72]):

$$\mathbf{C}_{\mathbf{T[Ge-72]}} = \frac{C_{\text{Ga-72}}\lambda_{\text{Ga-72}}\left[t_i + \frac{1}{\lambda_{\text{Ga-72}}}e^{\left(-\lambda_{\text{Ga-72}}t_i\right)}\right]}{E_e\varepsilon_e\left[1 - e^{\left(-\lambda_{\text{Ga-72}}t_i\right)}\right]e^{\left(-\lambda_{\text{Ga-72}}t_{\text{decay}}\right)}\left[1 - e^{\left(-\lambda_{\text{Ga-72}}t_l\right)}\right]}$$

where t_i is irradiation time [s], $\lambda_{Ga\text{-}72}$ is a decay constant of ^{72}Ga from the end of irradiation to infinity [$1.37 \times 10^{-5}\ s^{-1}$], $C_{Ga\text{-}72}$ is the measured count rate for ^{72}Ga in the 833 KeV peak [cps], E_e is a branching ratio of ^{72}Ga (0.9563), ε_e is the energy-dependent detector efficiency at the 833 KeV peak, the chosen time period to count the sample [s], t_{decay} is decay time from the end of irradiation to gamma-ray spectroscopy measurement [s], and t_l is live time, or counting time.

Since the ^{70}Ga activity has a short half-life ($T_{1/2} = 21.14$ minutes), it is generally difficult to determine the ^{70}Ge transmutation doping concentration by gamma-ray spectroscopy counts. Nevertheless, our new equation may be used to compute the atom density of ^{70}Ge by transmutation (CT[Ge-70]) using the measured ^{72}Ga data:

$$\mathbf{C}_{\mathbf{T[Ge-70]}} = PR_{\text{Ga-70}}A_{\text{Ga-69}}\,\rho_{\text{GaN}}\left[t_i + \frac{1}{\lambda_{\text{Ga-70}}}\left\{e^{\left(-\lambda_{\text{Ga-70}}t_i\right)} - 1\right\}\right]R_{\text{Ga-72}}$$

Where $A_{Ga\text{-}69}$ is ^{69}Ga atoms per GaN mass [4.32×10^{21} atoms/g], ρ_{GaN} is the density of GaN [6.15 g/cm^3], $\lambda_{Ga\text{-}70}$ is a decay constant of ^{70}Ga from the end of irradiation to infinity [$5.46 \times 10^{-4}\ s^{-1}$], $PR_{Ga\text{-}70}$ is the production rate of ^{70}Ga calculated by the product of the neutron flux and neutron capture cross-section of ^{69}Ga, t_i is the irradiation time [s], and $R_{Ga\text{-}72}$ is the ratio of measured ^{72}Ga activity to calculated ^{72}Ga activity through the production rate of ^{72}Ga and takes into account neutron self-shielding corrections.

C-V measurements were also used to characterize the GaN wafers. A Hg probe was utilized to carry out these measurements. The mercury probe makes a ring and a dot contact on the GaN wafer's surface using a vacuum system. Stainless steel makes up the probe's rear contact. A voltage sweep was used for the measurements, with steps of 0.1 V from −5 to 0 V. The C-V measurement curves for GaN B1–9 are shown in Fig. 18.4. From the $1/C^2$ graph in Fig. 18.5, the built-in voltage is estimated to be 1.15 V. Moreover, the following formula is used to determine the carrier concentration based on the C-V measurement findings.

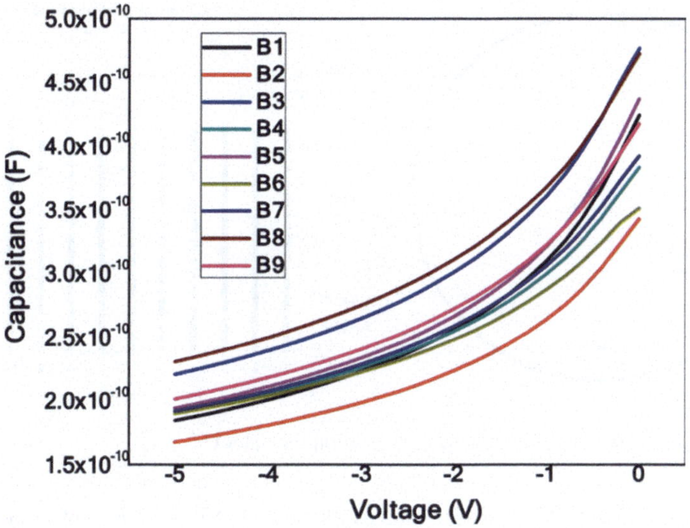

Fig. 18.4 The C-V measurements of GaN B1–9 samples using a mercury probe

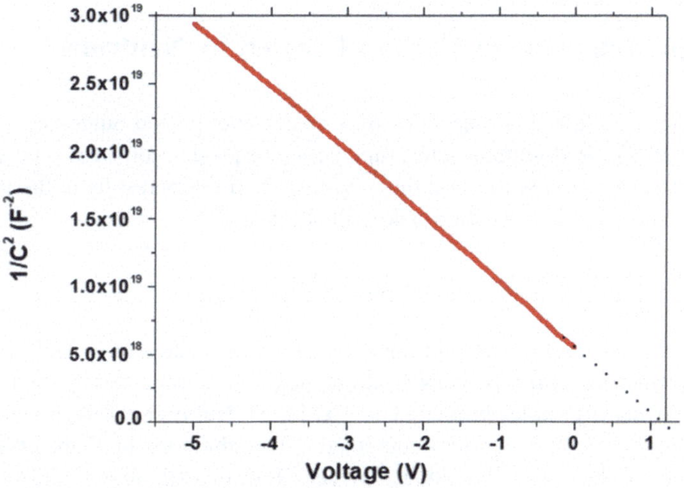

Fig. 18.5 The $1/C^2$ vs V plot of the GaN-B1 wafer

$$\rho = \frac{2}{A^2 q \varepsilon_0 \varepsilon m}$$

where m is the slope of the $1/C^2$ curve, ε_0 is the permittivity of space (8.854×10^{-12} F/m), ε is the permittivity of GaN (8.9), and ρ is the carrier concentration. The area of the probe contact is represented by A (0.00398×10^{-4} m^2). The 13 spots on wafer

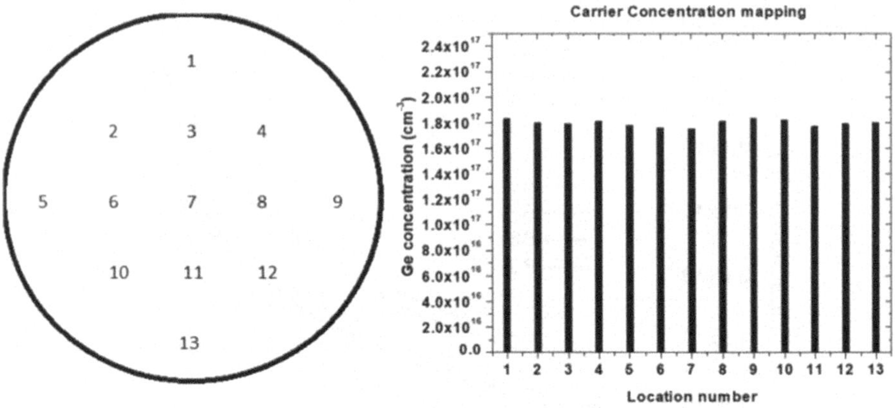

Fig. 18.6 The carrier concentration of each location on wafer C [1]

C were selected for C-V measurements as illustrated in Fig. 18.6, and a very uniform carrier concentration was measured over the 2″ GaN wafer.

3 P-type Behaviors of Carbon Created by Neutrons

The presence of additional impurities may create unexpected outcomes, even if the generation of Ge is dominant and crucial for the production of n-type GaN. The transmutation of nitrogen through the $^{14}N(n,p)^{14}C$ process results in the production of carbon, as the nuclear reaction below illustrates:

$$^{14}_{7}\mathrm{N} + ^{1}_{0}\mathrm{n} \rightarrow ^{\mathbf{14}}_{\mathbf{6}}\mathbf{C}^{*} + ^{1}_{1}\mathrm{p}$$

Carbon-14 is hypothetically thought to be a p-type dopant in GaN [4, 5]. The electrical properties of n-type GaN could be negatively impacted by the recombination of electrons with holes produced by ^{14}C [4, 5]. But regrettably, it has proven to be exceedingly challenging to understand ^{14}C behaviors in GaN crystals. The behavior of ^{14}C has been the subject of very few investigations, and research on the effects of neutron irradiation on GaN-based devices has just recently begun [6, 7]. One of the most challenging aspects of studying the behavior of carbon in GaN is finding out when ^{14}C in GaN results in holes created by the ^{14}C.

Depending on an incident particle's energy, each atomic nucleus has a chance of experiencing a particular response related to the species of incident particle. Figure 18.7 illustrates the neutron capture cross section for both Ga and N at various neutron energies. From the figure, the Ge generation would be larger because the cross-sections of Ga are higher than those of ^{14}N over nearly the whole energy range. That being said, by varying neutron energy, the production of Ge and ^{14}C can be impacted.

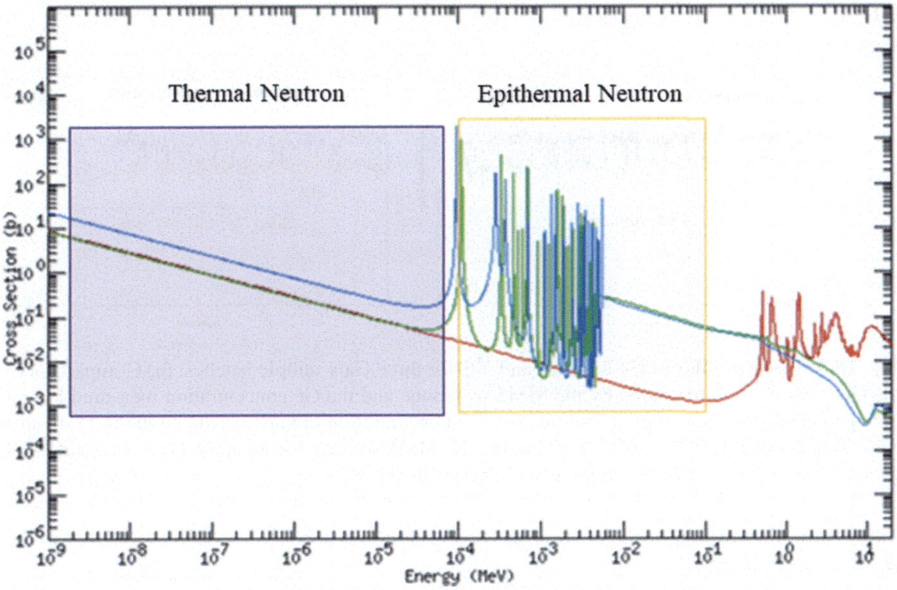

Fig. 18.7 Cross-sections graph[8] of primary neutron transmutation reactions on GaN: $^{69}Ga(n,\gamma)^{70}Ga^* \rightarrow\ ^{70}Ge$ (green), $^{71}Ga(n,\gamma)^{72}Ga^* \rightarrow\ ^{72}Ge$ (blue), $^{14}N(n,p)^{14}C^*$ (red). Except for the reactions above, other transmutation reactions on GaN are negligible since the reactions have relatively much lower cross-sections than the primary neutron transmutation reactions on GaN [9]

We created and compared Schottky-barrier diodes utilizing GaN irradiated in two distinct reactor areas—the dominant thermal neutron (DTN) region, which is the irradiation position with a high thermal neutron flux, and the dominant epithermal neutron (DEN) area, which is the irradiation position with a high epithermal neutron flux—in order to investigate the carbon-14 production behavior. Because ^{69}Ga and ^{71}Ga have far greater absorption cross-sections than ^{14}N in the epithermal neutron energy range of 0.55 eV to 100 KeV, the transmutation rate of Ga to Ge increases significantly in the DEN area and lowers the formation of ^{14}C. There are numerous peaks in the ^{69}Ga and ^{71}Ga epithermal excitation functions that are two to three orders of magnitude greater than the thermal excitation functions.

In order to distinctly verify the production of transmuted dopants, three batches of samples were processed in different regions of the reactor. Two batches were exposed to radiation in the DTN region for a duration of 600 hours, and another batch was exposed in the DEN area for 20 hours. SIMS measurements of the three samples are shown in Fig. 18.8. Checking the ^{14}C concentration is the most reliable way to verify the transmutation production because ^{14}C is a radioactive isotope that is hardly found in nature.

The Ge doping concentrations of GaN irradiated in the DTN region and DEN area were computed using the neutron flux data in Table 18.1. Ratios of ^{14}C to Ge concentration have been computed using the measurement data. Both the first and second batches of GaN samples had a similar ratio of ^{14}C to Ge concentration

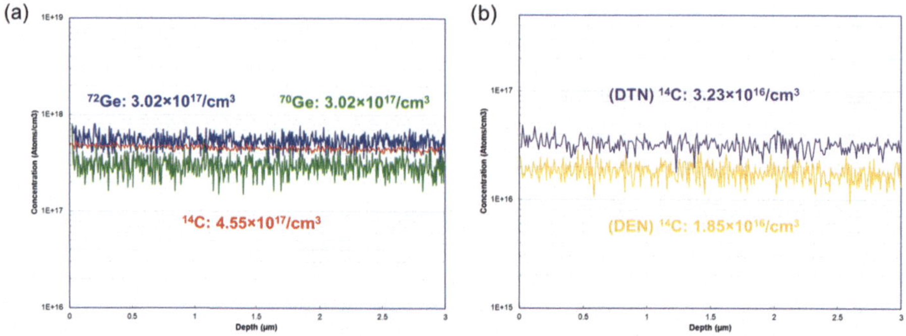

Fig. 18.8 SIMS profiles of Ge isotopes and ^{14}C for three GaN sample batches. (**a**) Comparison of the Ge concentration measured by the SIMS technique and the Ge concentration measured by our new gamma-ray spectroscopy technique. (b) ^{14}C concentration of GaN irradiated in the DTN area for 600 hours and the DEN area for 20 hours. The SIMS measurements in the DTN area and DEN area are the second and third sample batches respectively [8]

Table 18.1 Information of the neutron fluxes and resulting dopants for samples exposed in the DTN and DEN areas [8]

GaN sample batch	1st (Comparison for SIMS)	2nd (DTN area)	3rd (DEN area)
Thermal flux (/cm^2/s)	3.76×10^{12}	3.01×10^{12}	1.19×10^{6}
Epithermal flux (/cm^2/s)	1.25×10^{11}	1.00×10^{11}	2.60×10^{12}
Fast flux (/cm^2/s)	1.52×10^{11}	1.22×10^{11}	1.67×10^{13}
^{14}C Concentration (/cm^3)	4.55×10^{17}	3.23×10^{16}	1.85×10^{16}
Ge Concentration (/cm^3)	8.29×10^{17}	6.62×10^{16}	2.79×10^{17}

(~50%) and were irradiated in the same position. When taking into account the neutron flux ratio in Table 18.1, the ratio corresponds with the absorption cross-section ratio of Ga isotopes (^{71}Ga and ^{69}Ga) and ^{14}N displayed in Fig. 18.7. The Ge concentrations in the first batch of GaN samples, as determined by gamma-ray spectroscopy counting technique and SIMS methods, were 8.29×10^{17}/cm^3 and 8.43×10^{17}/cm^3, respectively.

The third batch of GaN samples was exposed to radiation in a completely different location in the nuclear reactor (the DEN region). The measurement in the DEN area revealed a concentration of 2.79×10^{17}/cm^3 for Ge doping and 1.85×10^{16}/cm^3 for ^{14}C impurities. The production of Ge doubled, and the generation of carbon nearly halved in comparison to the doping concentrations of ^{14}C and Ge in the DTN area. The ^{14}C to Ge ratio drastically decreased (6.6%). The reduction in ^{14}C generation is evident when the irradiation period is taken into consideration. The results indicate that the sample treated in the DEN area had a considerably lower NTD production ratio of ^{14}C over ^{70}Ge and ^{72}Ge than the sample treated in the DTN area. Stated otherwise, the same amount of Ge isotopes produced in the DEN area results in eight times less ^{14}C contamination. This is consistent with the capture cross-section data displayed in Fig. 18.7.

3.1 Optical Characteristics

Photoluminescence (PL) measurements were used to examine the behavior of ^{14}C as an acceptor in GaN and explore the prospect of ^{14}C functioning as a p-type dopant. A 325 nm He-Cd laser with a 60 mW beam output was used as the excitation light for the room-temperature PL studies. The PL spectrum may be impacted by the annealing process; thus, untreated samples were annealed in the same conditions as the irradiated materials.

Broad PL spectra due to defects are seen in Fig. 18.9. The yellow luminescence (YL) band region is the area that has changed the greatest. The sample's YL band has a maximum intensity of 2.25 eV prior to irradiation. Gallium vacancy-containing defect complexes are typically associated with the location of the YL band in GaN [10, 11]. It is interesting to note that after irradiation, the YL bands' peak location dropped to 2.17 eV while their intensity rose. Isolated nitrogen site (C_N) deficiencies and ^{14}C generation were the primary causes of the YL band's redshift [12–16]. The intensity difference supports the reasonable assumption that the isolated C_N defects are replaced by ^{14}C generated by neutron irradiation. Compared to the DEN area, the DTN area produces a significantly higher amount of ^{14}C. While the intensity of PL spectra does not allow us to compute the impurity concentration accurately, we can reasonably estimate the variation of irradiation byproducts, and we conclude that ^{14}C is the primary factor responsible for moving the YL band. P-type conductivity has recently been used to confirm the acceptor nature of C_N defects [12–16]. It is thought that ^{14}C has an activation energy between 0.9 and 1.1 eV as a p-type dopant based on

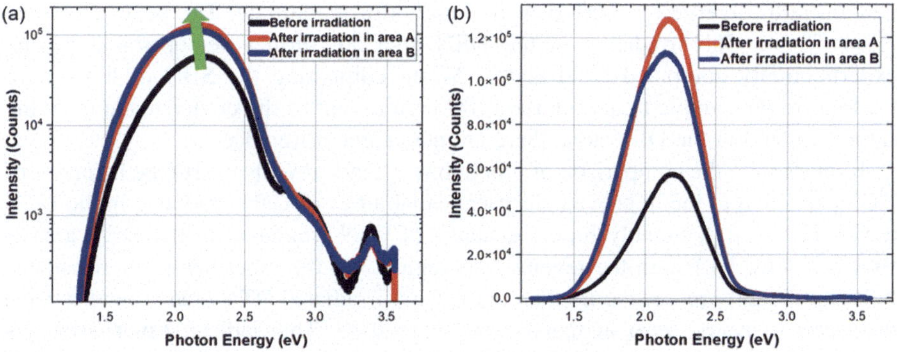

Fig. 18.9 PL spectra of GaN before and after irradiation in area A or area B: (**a**) log scale PL spectra and (**b**) linear scale PL spectra. The green arrow indicates that the peak position around the yellow luminescence was redshifted after irradiation due to carbon production [8]

the redshifted PL spectra with intensity differences and the aforementioned references. Later, ^{14}C's p-type behavior is further examined using electrical properties.

3.2 Electrical Characteristics

We have been particularly interested in the effects of ^{14}C generation on GaN's n-type properties. Sputter-deposited metal layers were used to create Schottky barrier diodes (SBDs) on samples processed in different areas of the reactor. Pt/Au (80 nm/40 nm) was deposited as a Schottky contact on the polished side of the GaN samples using a shadow mask with holes of 600 nm diameter. Ohmic connections were fabricated using Ti/Au (30 nm/150 nm) deposited on the unpolished side of the GaN samples. The effects of the carbon reduction were evident in the SBDs fabricated from samples with various ^{14}C/Ge concentrations, as demonstrated in Fig. 18.10. Using junction resistance (R_j) and the thermionic emission theory, the current (I) across an SBD under a forward bias ($V \geq 3kT/q$) is expressed as [17].

$$I = I_0 e^{\frac{qV}{\eta kT}}$$

$$I_0 = AA^{**} T^2 e^{\frac{q}{kT}\Phi_b}$$

$$R_j = \frac{\partial V}{\partial I}$$

where q is the charge of an electron (1.6×10^{-19} C), A** is the effective Richardson constant (35 A/($cm^2 \eta K^2$) [18], Φ_b is the Schottky barrier height (SBH), I_0 is the saturation current, V is the biased voltage, and A is the diode area (5.03×10^{-3} cm^2). Figure 18.10a and b, respectively, display the I-V characteristics of SBDs made from GaN samples that were exposed to radiation for 60 hours in the DTN area and 20 hours in the DEN area. The samples treated in the DTN and DEN areas clearly differ from one another, as the I-V curves demonstrate. Leakage current and rectification ratio diverge by more than five orders of magnitude. The better electrical properties of GaN-irradiated in the DEN area are further demonstrated by the experimentally computed SBH and η. When comparing the SBH of the device made from the sample treated in the DTN area to that of the device made from the sample treated in the DEN area, there are noticeable differences.

Remarkably, the comparison of junction resistance reveals the p-type behavior of ^{14}C in addition to the improved electrical capabilities of GaN irradiated in the DEN region [19–22]. A notably high resistance of GaN irradiated in the DEN area is shown in Fig. 18.10c in the reverse bias region. On the other hand, the resistance falls significantly more sharply than in the GaN-irradiated DTN area. The resistance decreases to nearly zero in the forward bias zone. This indicates that SBH has excellent control over electrons as the majority carrier. Stated differently, because there are comparatively fewer holes from ^{14}C acting as acceptors than there are electrons, holes acting as minority carriers have no effect on the forward current.

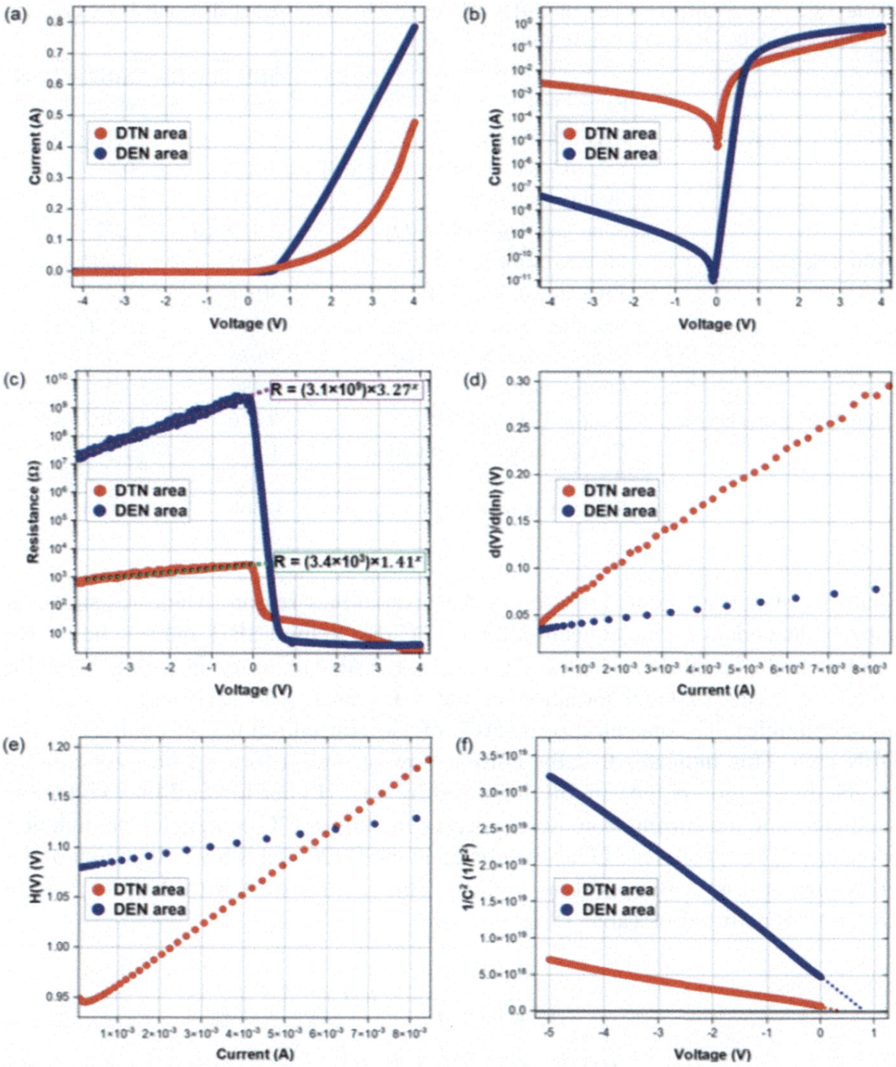

Fig. 18.10 Electrical characteristics of GaN irradiated in the DTN area for 60 hours (red) and in the DEN area for 20 hours (blue). (**a**) Linear scale I-V curves. (**b**) Log scale I-V curves. (**c**) P-type behavior of ^{14}C on junction resistance curves. At a reverse bias region, it is assumed that the junction resistance can be expressed as exponential functions: $(3.1 \times 10^3) \times 1.41^x$ in the DTN area and $(3.1 \times 10^9) \times 3.27^x$ in the DTN area, respectively. (**d, e**) Curves obtained from Cheung's method: (**d**) a curve of d(V)/d(lnI) versus I and (**e**) a curve of H(I) versus I. (f) $1/C^2$-V curves. By using these measurements above, electrical parameters of GaN irradiated in the DTN area and in the DEN area are tabulated in Table 18.2

Table 18.2 Electrical parameters of SBDs fabricated on GaN irradiated in the DTN area for 60 hours and in the DEN area for 20 hours [26]

Method		Parameters	SBD in DTN area	SBD in DEN area
From I-V characteristics		Rectification ratio	183	2.46×10^7
		Schottky barrier height (Φ_b) [eV]	0.452	0.800
		Ideality factor (η)	1.711	1.200
		Shunt resistance (R_{Sh}) [Ω]	2.91×10^3	2.77×10^9
From Cheung's functions	$\frac{dV}{d(\ln I)}$	Ideality factor (η)	1.789	1.307
		Series resistance (R_S) [Ω]	30.93	7.41
	H (V)-I	Schottky barrier height (Φ_b) [eV]	0.521	0.826
		Series resistance (R_S) [Ω]	30.73	7.46
		Ideality factor (η)	1.789	1.307
From C-V characteristics		Built-in Potential [eV]	0.286	0.763
		Charge carrier concentration (CCC) [10^{17}/cm^3]	1.32	2.93
		Schottky barrier height (Φ_b) [eV]	0.363	0.845

Sharp resistance gradients are seen in the reverse bias region, where resistance is entirely dependent on the concentration of ^{14}C due to high SBH and few holes. By comparing the junction resistances, it was also possible to verify the p-type behavior of ^{14}C in terms of GaN radiation in the DTN area. The DTN area exhibits a non-negligible ^{14}C concentration (~50% of Ge concentration), in contrast to the DEN area. This indicates that the current flow at both a forward bias zone and a reverse bias region is measurably impacted by ^{14}C. Higher forward bias voltage is needed to stop the current from holes because the higher ^{14}C content lowers junction resistance. The resistance of GaN irradiated in the DTN area is lower than that of the DEN area at a high forward bias voltage. This is consistent with the ^{14}C and Ge concentrations listed in Table 18.1.

$$\frac{dV}{d(lnI)} = \frac{kT}{q}\eta + R_S I$$

$$H(V) = V - \frac{\eta kT}{q} ln\left(\frac{I}{AA^{**}T^2}\right)$$

$$H(V) = R_S I + \eta \Phi_b$$

The series resistance (RS), η, and SBH in the non-linear forward bias area were calculated using Cheung's functions using the formulas above [23]. The SBDs exhibit linear behavior in Fig. 18.10d and e, where they are plotted dV/(d(lnI)) versus I and H(V) versus I, respectively. GaN irradiation in the DEN area exhibits lower η and higher SBH than the DTN area, similar to I-V properties. More than four times less GaN is irradiated in the DEN area than in the DTN area. The graphs of dV/(d(lnI)) against I and H(V) versus I nearly match in terms of RS values, suggesting that the electric parameters ascertained by Cheung's functions are legitimate and consistent. Because of the increased trap-assisted recombination and the hole accumulation that results in an elevated space charge close to the junction, the ^{14}C is thought to behave as acceptors. As the ratio of Ge/^{14}C grows, the behavior results in an increase in SBH and a drop in η and RS. There is a notable improvement in RS due to the high SBH and improved η in the DEN area. The capacitance-voltage properties of the GaN SBDs were measured at a frequency of 1 MHz, and the 1/C^2-V characteristics are plotted in Fig. 18.10f. The built-in potential, charge carrier concentration, and SBHs are derived from these plots using the following equations [24].

$$\frac{1}{C^2} = \frac{2\left(V_{bi} - \frac{kT}{q} - V\right)}{A^2 \times q \times N \times \varepsilon_S}$$

$$\Phi_b = V_{bi} + \frac{kT}{q} \, ln \, \left(\frac{N_C}{N}\right)$$

where ε_S is the dielectric constant of GaN (7.88 × 10^{-11} F/m), N_C is the effective density of states at the conduction band in GaN (4.3 × T1.5 × 10^{14}/cm^3), V_{bi} is the built-in potential, and N is the charge carrier concentration (CCC) of the GaN SBDs. The measured V_{bi} and CCC are 0.286 eV and 1.32 × 10^{17}/cm^3 for the DTN region, whereas for the DEN area, they are 0.763 eV and 2.93 × 10^{17}/cm^3, respectively. The linear scale I-V curve in Fig. 18.10a is matched by the values of V_{bi} in each location, and the CCCs agree with the total doping concentrations in Table 18.1. The SBHs of GaN SBDs irradiated in the DEN and DTN areas are 0.845 eV and 0.363 eV, respectively, based on the observed CCC and V_{bi}. There is a slight difference between the SBH values determined by I-V characteristics and Cheung's functions and those determined by C-V characteristics. Neutron irradiation can cause some structural, line, or point defects to increase or decrease, which adds to the lateral inhomogeneity in the SBHs at the metal/semiconductor interface [25]. Table 18.2 summarizes and tabulates all measured electric parameters of SBDs that were exposed to radiation for 60 hours in the DTN area and 20 hours in the DEN area.

4 Selective Area Doping by Neutrons

According to their energy, neutrons in reactor physics can be broadly classified into three groups: thermal neutrons, which have an energy of less than eV; epithermal neutrons, which have an energy of between one eV and 100 keV; and fast neutrons, which have an energy of more than 100 keV. The three reactions discussed above have high cross-sections in the energy range from thermal to epithermal neutrons, as Fig. 18.11a illustrates. On the other hand, as shown in Fig. 18.11b, gadolinium (Gd) isotopes have extremely high neutron absorption cross-sections and are

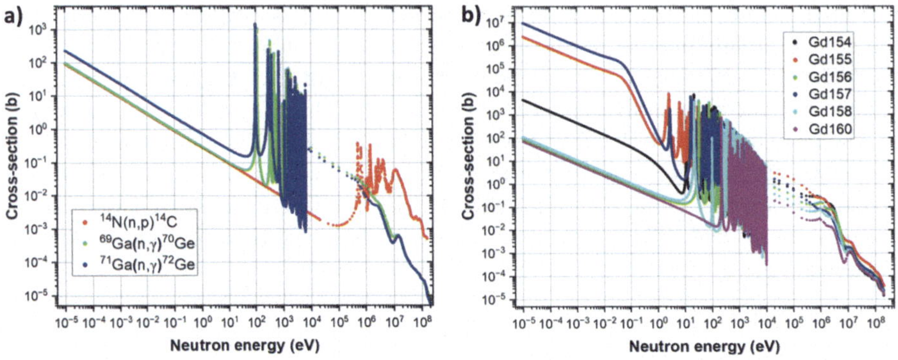

Fig. 18.11 (a) Cross-sections for major reactions in GaN during the NTD: $^{69}Ga(n,\gamma)^{70}Ge$, $^{71}Ga(n,\gamma)^{72}Ge$, and $^{14}N(n,p)^{14}C$. (b) neutron absorption cross-sections of stable Gd isotopes: ^{154}Gd, ^{155}Gd, ^{156}Gd, ^{157}Gd, ^{158}Gd, and ^{160}Gd [26]

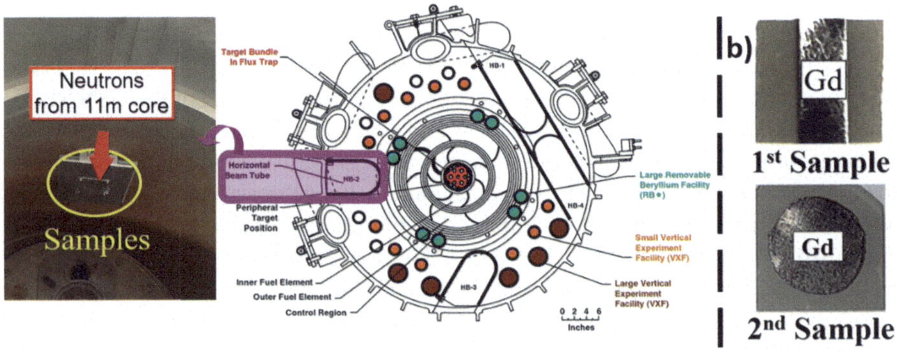

Fig. 18.12 (a) Appearance of the HB-2D port where the sample is mounted, 11 m from the core in HFIR of ORNL [26]

frequently used as neutron absorbers [27]. Ge generated by the NTD process can be selectively doped into GaN by using neutron shielding with Gd masks with patterns.

The following formula is used to determine the attenuation of the neutron dosage caused by a Gadolinium (Gd) shadow mask:

$$\frac{I(t_x)}{I_0} = e^{-N \cdot t_x \cdot \sigma}$$

where $I(t_x)$ is the neutron flux after it passes through the Gd mask and I_0 is the original neutron flux without any attenuation. The atomic number density (N) of natural Gd is 3.025×10^{22} atoms/cm^3, the thickness of the Gd shadow mask is t_x, and the thermal neutron reaction cross section (σ) is 49,300 barn for natural Gd. The neutron dosage transmission rate, which is dependent on the thickness of a Gd mask, is expressed as $I(t_x)/I_0$.

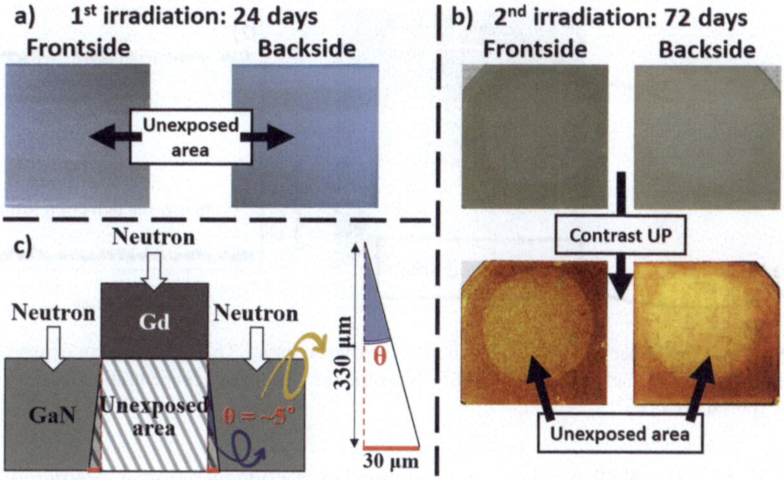

Fig. 18.13 (**a**) optical images of the first irradiation sample, (**b**) optical images of the second irradiation sample, and (**c**) measurement for neutron scattering angle in GaN [26]

The ORNL High Flux Isotope Reactor (HFIR), which is capable of performing directed neutron irradiation, was used to irradiate GaN samples. The samples were put on an HB-2D tube in the HFIR, as Fig. 18.12a illustrates. For a good shielding effect, Gd masks with a thickness of 250 μm were employed. Two Gd masks with distinct patterns were present; one had a rectangular design while the other had a circular one.

Thermal neutrons with 5.518×10^9 neutrons/cm^2/s of flux were employed throughout the irradiations, and the distance between the core and the sample container is 11 m. Two GaN samples with the rectangular pattern were exposed to radiation for a total of 82 days during the first irradiation and 60 days during the second.

The damage caused by neutron irradiations made the border line more distinct as the irradiation time rose. The portions covered by Gd masks in Fig. 18.13 that are not exposed are brighter than those that are, and they even resemble GaN prior to irradiation. It was exceedingly hard to cut off Gd with great precision. Based on a rough assessment from the photos, the neutron scattering angle is estimated to be approximately 5 degrees.

As seen in Fig. 18.14a, the change in carbon concentration was examined using EDS to verify whether the NTD process in GaN truly took place and created byproducts. Some carbon from the wafer growth process is already present in the untreated GaN samples. The ^{14}C doping concentration was determined using SIMS to ensure an accurate comparison. The ^{14}C concentration in the unexposed area was below the detection limit, but a significant amount of ^{14}C was detected in the exposed region, demonstrating that the NTP process was carried out selectively and the Gd mask was seldom penetrated by neutrons. The GaN samples were initially heavily doped with Ge (~10^{18}/cm^3), so it was not possible to measure the

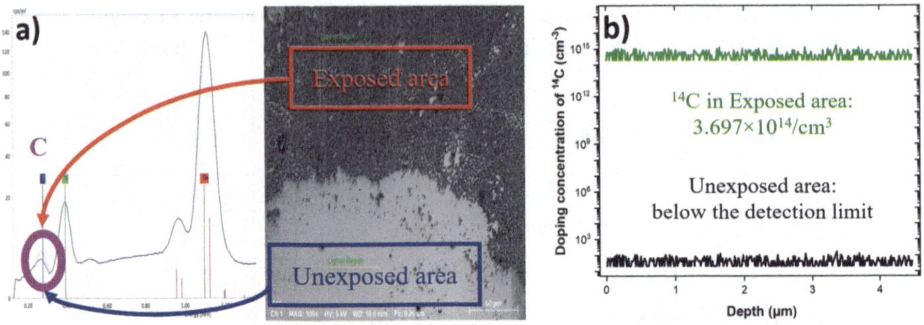

Fig. 18.14 (**a**) EDS spectra in the exposed and unexposed areas. The red box is an exposed area, and the blue box is an unexposed area. (**b**) SIMS profile for ^{14}C doping concentration in the exposed and unexposed area [26]

Ge created by the NTP process. The generated Ge concentration is predicted to be approximately $10^{15} \sim 10^{16}/cm^3$ based on the ^{14}C doping concentration.

5 Defect Curing Effects by Neutrons

The impact of neutron irradiation on defects in GaN samples with varying fluence times was investigated. Secondary-ion mass spectroscopy (SIMS) was used to quantify the transmutation concentration of Ge isotopes and ^{14}C for the various fluence times. Hall-effect measurements, current-voltage (I-V), and capacitance-voltage (C-V) on n-type GaN SBDs were performed to measure the electrical properties. Photoluminescence (PL) spectroscopy and optical imaging were used to study the optical properties of the GaN samples. Synchrotron X-ray topography (SXRT) was employed to examine the neutron-induced defect fluctuations and compared to the device performances [28, 29].

Different neutron irradiation periods (5, 25, 75, and 100 hours) were conducted in the area GR; these were referred to as GR5, GR25, GR75, and GR100, respectively. Based on neutron energy, three categories were identified from the flux spectra of neutrons. Fast neutrons have energies greater than 100 keV, epithermal neutrons have energies between 0.625 eV and 100 keV, and thermal neutrons have energies less than 0.5 eV, with a most likely energy of 0.025 eV [30]. Three types of neutron flux were measured and summarized in a fluence form (neutron flux × irradiation period) in Table 18.3 using neutron flux wires.

Untreated and irradiated GaN samples (GR5, GR25, GR75, and GR100) had their I-V characteristics assessed on a linear scale (Fig. 18.15a) and a log scale (Fig. 18.15b), respectively. While the I-V curves in the linear scale plot are quite similar, in the log scale plot, each I-V curve has a distinctly different rectification ratio. The rectification ratio and leakage current decrease with increasing irradiation time. Reverse leakage current at −1 V is used to measure the ϕ_b and n. Shunt

Table 18.3 Neutron fluence information from the neutron flux monitoring and doping concentration of Ge isotopes (^{70}Ge and ^{72}Ge) and ^{14}C measured by SIMS[31]

Parameters	Untreated	GR5	GR25	GR75	GR100
Thermal neutron fluence [Neutrons/cm^2]	×	1.5×10^{18}	5.74×10^{18}	1.90×10^{19}	2.68×10^{19}
Epithermal neutron fluence [Neutrons/cm^2]	×	5.41×10^{16}	1.68×10^{17}	2.81×10^{17}	6.95×10^{17}
Fast neutron fluence [Neutrons/cm^2]	×	1.20×10^{17}	5.21×10^{17}	1.47×10^{18}	1.97×10^{18}
Doping concentration of Ge isotopes (^{70}Ge & ^{72}Ge) [cm^{-3}]	~ 8×10^{17}	1.02×10^{18}	1.68×10^{18}	3.53×10^{18}	4.85×10^{18}
Doping concentration of ^{14}C [cm^{-3}]	0	9.66×10^{16}	4.97×10^{17}	1.45×10^{18}	2.11×10^{18}

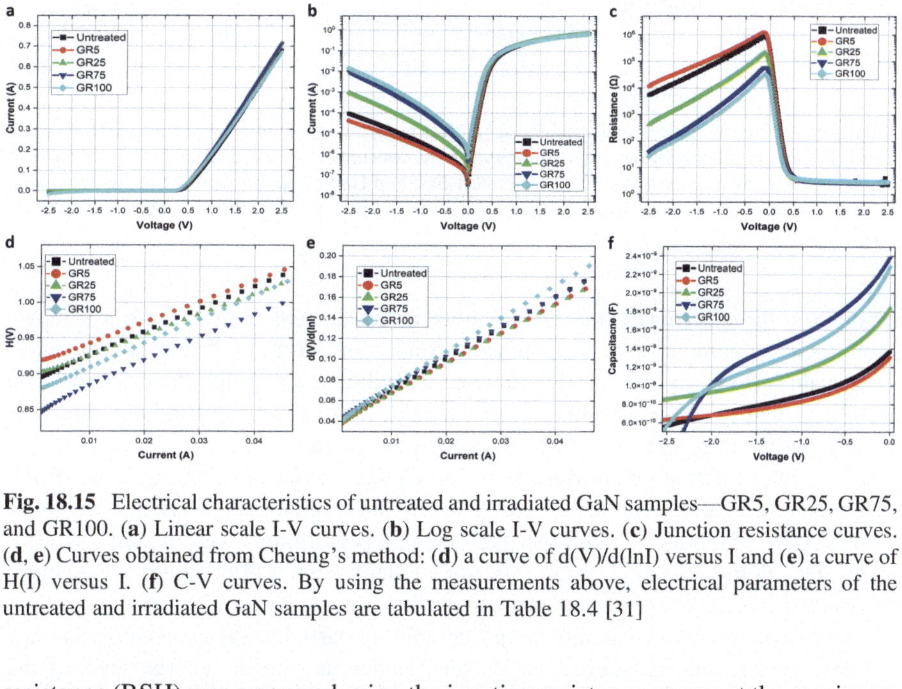

Fig. 18.15 Electrical characteristics of untreated and irradiated GaN samples—GR5, GR25, GR75, and GR100. (**a**) Linear scale I-V curves. (**b**) Log scale I-V curves. (**c**) Junction resistance curves. (**d, e**) Curves obtained from Cheung's method: (**d**) a curve of d(V)/d(lnI) versus I and (**e**) a curve of H(I) versus I. (**f**) C-V curves. By using the measurements above, electrical parameters of the untreated and irradiated GaN samples are tabulated in Table 18.4 [31]

resistance (RSH) was measured using the junction resistance curves at the maximum resistance point, which is located around the turn-on voltage, as shown in Fig. 18.15c. Table 18.4 tabulates the electrical parameters.

GR5 exhibits superior features compared to other models in all electrical metrics, which is rather intriguing. It appears that certain faults, such as dislocation, point, line, or structural defects, are eliminated while the other, perhaps insignificant consequences of ^{14}C generation, are concealed. While the electrical characteristics associated with the forward bias current (μ_e and σ) were somewhat increased, the reduced defect density effectively suppressed the leakage current. Aside from GR5

Table 18.4 Electrical parameters of Schottky barrier diodes (SBDs) fabricated on untreated and irradiated GaN samples—untreated, GR5, GR25, GR75, and GR100

	Parameters	Untreated	GR5	GR25	GR75	GR100
From I-V Characteristics	Rectification ratio	14,937	31,441	1781	204	117
	Schottky barrier height ϕ_b [eV]	0.584	0.614	0.527	0.488	0.474
	Ideality factor n	1.496	1.466	1.744	2.066	2.162
	Shunt resistance R_{SH}	1.59 MΩ	3.88 MΩ	219 kΩ	65.4 kΩ	37.6 kΩ
From Cheung's functions	Schottky barrier height ϕ_b [eV]	0.646	0.669	0.612	0.584	0.573
	Ideality factor n	1.389	1.367	1.400	1.433	1.524
From C-V Characteristics	Schottky barrier height ϕ_b [eV]	0.740	0.756	0.715	0.669	0.595
From Hall-effect measurements	Electron mobility μ_e [cm^2/V·s]	253	256	243	168	126
	Conductivity σ [1/Ω·m]	120	125	101	43	30
	Charge carrier concentration [$10^{18}/cm^3$]	−2.80	−3.05	− 2.59	− 1.60	−1.48

Rectification ratio, Schottky barrier height (ϕb), ideality factor (n), shunt resistance (RSH), electron mobility (μe), conductivity (σ), and charge carrier concentration were obtained from the I-V characteristics, Cheung's method, C-V characteristics, and Hall-effect measurements. In the charge carrier concentration, a minus sign means that electrons are the majority charge carriers in the SBDs [31]

irradiation, the electrical properties of both untreated and irradiated GaN samples show a clear degradation with increasing irradiation duration. Table 18.2 displays lower values for every electrical parameter, including the rectification ratio, electron mobility, ideality factor, Schottky barrier height, shunt resistance, and conductivity.

The higher trap and hole buildup, which raises the space charge close to the junction, may facilitate recombination. The charge carrier concentration, ϕ_b, RSH, μ_e, σ, and rectification ratio are all decreased by the behavior of ^{14}C as an acceptor or a hole. After 75 hours of radiation, the device's performance significantly deteriorated. The defect curative impact is not evident from GR75 since the neutron transmutation effects are more dominant.

Since neutrons are mass- and energy-containing particles, they can cause damage to GaN crystals and introduce new defects. Depending on the energy level of the neutron, the damage varies significantly. When thermal neutrons are used to irradiate a GaN wafer, the wafer becomes discolored and turns brown [1]. The discoloration was visibly restored upon annealing. On the other hand, GaN crystals exposed to fast neutrons that have a large energy (up to 2 ~ 3 MeV) caused the GaN samples to turn black. The electrical and optical properties of the black GaN samples are absent. It is possible to recover the electrical and optical properties after annealing. As the neutron fluence rises, the strength of recovery decreases. As seen in Fig. 18.16a, the color depth of the irradiated GaN samples indicates the degree of recovery.

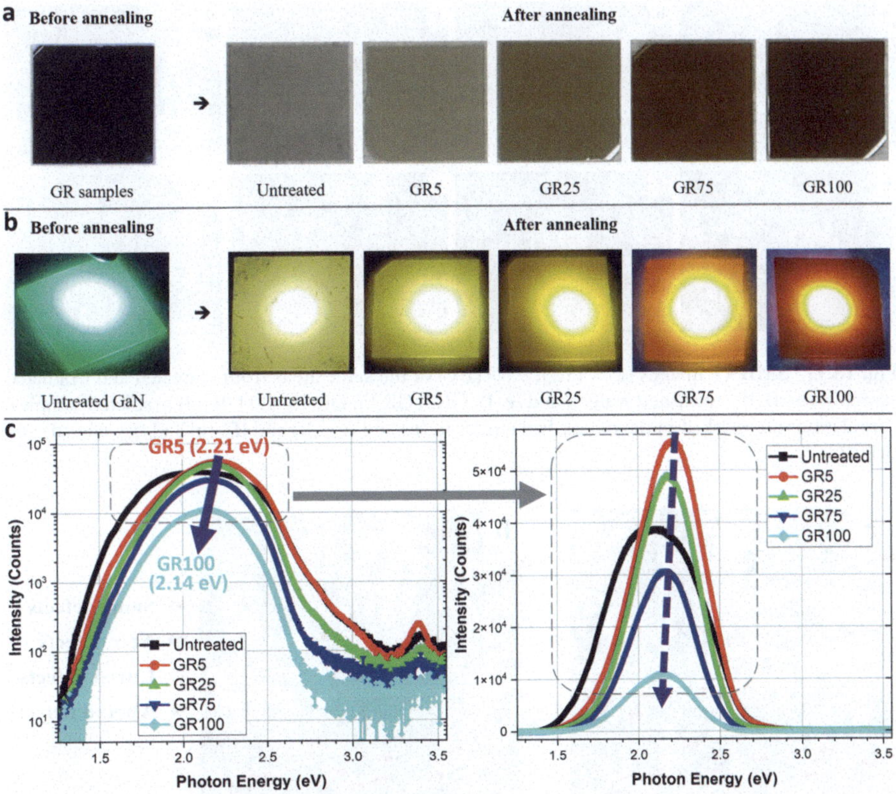

Fig. 18.16 Optical characteristics of untreated and irradiated GaN samples—untreated, GR5, GR25, GR75, and GR100. (**a**) Pictures of untreated and irradiated GaN samples obtained from an optical microscope. There is no significant change in untreated GaN before and after annealing, while some annealing defects were created after annealing. For irradiated (GR) samples, there are no optical properties before annealing regardless of neutron fluences. After annealing, some optical properties were restored. (**b**) Optical photographs of luminescence from untreated and irradiated GaN samples excited by a 325 nm He-Cd laser. (**c**) Photoluminescence (PL) spectra of untreated and irradiated GaN on a log scale (left) and a linear scale (right). The peak of irradiated GaN samples redshifts from 2.21 to 2.14 eV as the neutron fluence increases. The intensity of each PL spectrum decreases [31]

The fluorescence from untreated and irradiated GaN samples stimulated by a 325 nm He-Cd laser source is shown optically in Fig. 18.16b. Green luminescence (GL) band region, associated with gallium vacancy (V_{Ga}) containing defects, particularly the V_{Ga}-$3H_i$ complex, is where the untreated GaN emits green light prior to annealing [32]. Following annealing, a yellow luminescence (YL) band region exhibits freshly formed flaws, while the defects on the GL region are diminished. It is reasonable to anticipate that the annealing process will eliminate a hydrogen (H) atom and change V_{Ga}-$3H_i$ into V_{Ga}-$2H_i$, V_{Ga}-2ON-H_i, and V_{Ga}-ON-$2H_i$ [33, 34]. The light emitted by GR samples is more red than yellow when the neutron

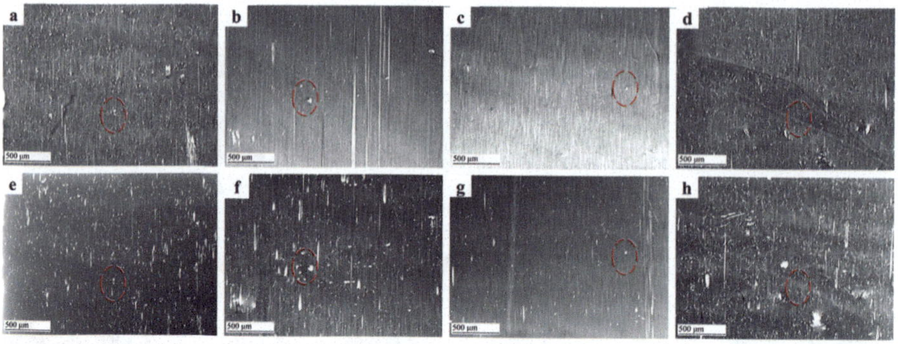

Fig. 18.17 SMBXT images (g = $11\bar{2}4$; 9.37 keV) of the same areas from untreated and irradiated samples. (**a**–**d**): the untreated wafers; and (**e**–**f**): GR5, GR25, GR 75 and GR100 irradiated samples. Correlations are marked in the red dashed circles in (**a**) and (**e**), (**b**) and (**f**), (**c**) and (**g**), and (**d**) and (**h**) [31]

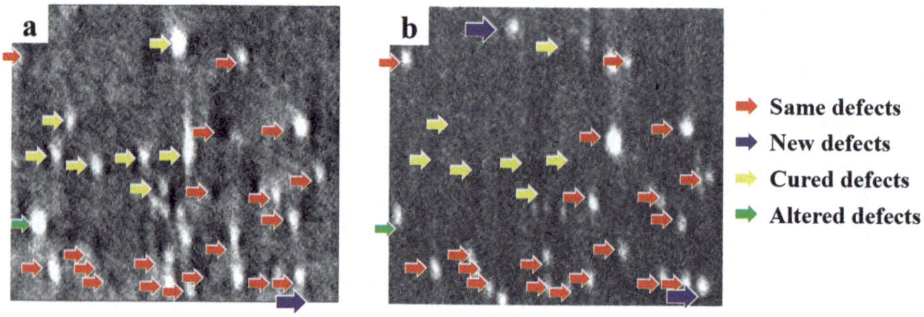

Fig. 18.18 Detailed analysis in a small area of the GR5 sample. (**a**) is a selected region in the untreated sample, and (**b**) is the same area after the sample is treated in the condition of GR5 [31]

fluence increases for irradiated GaN materials, as shown in Fig. 18.16c. The redshift of the peak of the PL spectra observed by the irradiated GaN samples is 2.14 eV (GR100) instead of 2.21 eV (GR5). The production of ^{14}C and the elimination of isolated nitrogen (N) site defects are two hypotheses that could account for the redshift [11, 12, 16]. It is possible that the isolated N site defects are replaced by ^{14}C transmuted from N during neutron irradiations. It is evident that when the neutron fluence increases, the intensity of the PL spectra also decreases in addition to the changed peak position. Since GR5 produced ^{14}C with a negligible quantity, the intensity of the PL spectrum was higher than that of untreated GaN. However, the complex mixes of new defects created by the neutron irradiations—point defects, stacking faults, and dislocation loops—are sufficient to deteriorate the optical characteristics [35].

Figure 18.17 displays Synchrotron monochromatic beam X-ray topography (SMBXT) pictures of samples that were both irradiated and untreated. The threading dislocations exhibit a nearly uniform distribution, with the untreated wafers

Table 18.5 Effect of neutron irradiation flux on the changes in dislocation distribution [31]

Type	GR5	GR25	GR75	GR100
Same [%]	65.39	77.47	44.58	44.64
Cured [%]	29.66	0.59	25.30	23.33
New [%]	2.00	21.21	27.30	27.16
Altered [%]	2.96	0.74	2.81	4.87

exhibiting an average threading dislocation density of 1.97×10^4 cm^{-2}. On the enlarged topography, a one-to-one correlation is carried out to examine the impact of radiation on the dislocation distribution. A further extended region of the association is shown in Fig. 18.18. It is seen that following irradiation, the topography's dislocation contrast distribution slightly changes. After irradiation, contrast from certain dislocations is not evident, but contrast from a few new dislocations can be detected. It appears that the damage from the neutron irradiation procedure causes a change in the dislocation distribution. Atoms are displaced, and vacancies and interstitial atoms are created as part of the damage. These point defects may interact with the preexisting dislocations due to heat damage from neutron irradiation, which could cause the dislocations to climb. As a result, dislocation configurations may fluctuate, which may explain why these dislocations have changed in location.

Table 18.5 provides a summary of the dislocations in Fig. 18.17 that were meticulously counted in order to investigate the impact of neutron irradiation. Research shows that samples with higher neutron fluence (GR25, GR75, and GR100) have more created dislocations than cured ones, while samples with lower neutron fluence (GR5) had more cured dislocations than freshly generated ones. This research shows that when neutron fluence is above the requirements for GR5, additional defects are introduced by radiation damage, which outweighs the benefit of neutron defect healing.

6 Photonuclear Transmutation Doping

We found that photonuclear transmutation doping (PNTD) can be identified as one of the most promising methods for high-grade zinc (Zn) doping of GaN. We observed p-type properties from the Zn-doped GaN, which was produced in conjunction with Argonne National Laboratory (ANL) by modulating the beam energy for the PNTD method. GaN with a substantial concentration of Zn was obtained, and using secondary-ion mass spectrometry (SIMS) measurements, the uniform distribution of Zn in the crystal was verified. Furthermore, gamma-ray spectroscopy, a Monte Carlo simulation program, and a nuclear data library were utilized to model the transmutation doping concentration of Zn. The concentration of Zinc generated by the PNTD was found to be $2.25 \times 10^{17}/cm^3$ [36]. Diodes made from Zn-doped GaN have I-V characteristics that support the p-type, and Hall-effect measurement revealed that the hole carrier concentration was $8.802 \times 10^{16}/cm^3$ [36].

7 Proton Transmutation Doping

Protons can be used as another nuclear transmutation doping method for Zn-doped p-type GaN. Using a simulation of the proton transmutation doping (PTD) process, we discovered that Zn production is most dominant at the proton energies of approximately 46 MeV [36]. Nevertheless, because of ^{68}Ge's long half-lifetime ($T1/2 = 270.95$ days), the PTD method will still raise some residual radioactivity issues [36].

References

1. R. Barber, Q. Nguyen, J. Brockman, J. Gahl, J. Kwon, Thermal neutron transmutation doping of Gan semiconductors. Sci. Rep. **10**, 16295 (2020)
2. MARLAP Manual Volume II: Chapter 16, Data Acquisition, Reduction, and reporting for Nuclear-Counting instrumentation (2019)
3. R.R. Hart et al., Measurements of 3 1P concentrations produced by neutron transmutation doping of silicon, in *Neutron Transmutation Doping in Semiconductors*, ed. by J.M. Meese, (Springer US, Boston, MA, 1979), pp. 345–354. https://doi.org/10.1007/978-1-4684-8249-2_28
4. M. Matsubara, E. Bellotti, A first-principles study of carbon-related energy levels in GaN. I. Complexes formed by substitutional/interstitial carbons and gallium/nitrogen vacancies. J. Appl. Phys. **121**, 195701 (2017)
5. J.L. Lyons, A. Janotti, C.G. Van de Walle, Effects of carbon on the electrical and optical properties of InN, GaN, and AlN. Phys. Rev. B **89**, 35204 (2014)
6. H. Sakurai et al., Effects of the sequential implantation of Mg and N ions into GaN for p-type doping. Appl. Phys. Express **14**, 111001 (2021)
7. M.J. Tadjer et al., Selective p-type doping of GaN:Si by Mg ion implantation and multicycle rapid thermal annealing. ECS J. Solid State Sci. Technol. **5**, P124 (2016)
8. J. Kim, J.W. Kwon, p-Type behavior of carbon on n-type GaN Schottky-Barrier diodes doped by neutron transmutation effects. IEEE Trans. Electron Devices **71**, 1474–1480 (2024)
9. A.J. Koning et al., TENDL: Complete nuclear data library for innovative nuclear science and technology. Nuclear Data Sheets **155**, 1–55 (2019)
10. M.A. Reshchikov et al., Defect-related photoluminescence from ammono GaN. J. Appl. Phys. **129**, 095703 (2021)
11. E. Gaubas et al., Study of spectral and recombination characteristics of HVPE GaN grown on ammono substrates. Mater. Sci. Semicond. Process. **91**, 341–355 (2019)
12. F. Zimmermann et al., Current status of carbon-related defect luminescence in GaN. Phys. Status Solidi A **218**, 2100235 (2021)
13. M.A. Reshchikov, Measurement and analysis of photoluminescence in GaN. J. Appl. Phys. **129**, 121101 (2021)
14. F. Liang et al., Role of Si and C impurities in yellow and blue luminescence of unintentionally and Si-doped GaN. Nanomaterials **8**, 1026 (2018)

15. M.A. Reshchikov et al., Two charge states of the C_N acceptor in GaN: Evidence from photoluminescence. Phys. Rev. B **98**, 125207 (2018)
16. R. Piotrzkowski, M. Zajac, E. Litwin-Staszewska, M. Bockowski, Self-compensation of carbon in HVPE-GaN:C. Appl. Phys. Lett. **117**, 012106 (2020)
17. E.H. Rhoderick, A. Rothwarf, Metal–semiconductor contacts. Phys. Today **32**, 66–71 (1979)
18. Y. Zhou et al., Temperature-dependent electrical characteristics of bulk GaN Schottky rectifier. J. Appl. Phys. **101**, 024506 (2007)
19. V. Gopal, Origin of the shunt currents and their influence on the ideality factor (η) of a p-n junction. J. Electron. Electr. Eng. **10**, 180–183 (2022)
20. L.D. Rao, V.R. Reddy, Electrical parameters and series resistance analysis of Au/Y/p-InP/Pt Schottky barrier diode at room temperature. AIP Conf. Proc. **1731**, 120020 (2016)
21. V. Gopal, N. Gautam, E. Plis, S. Krishna, Modelling of current-voltage characteristics of infrared photo-detectors based on type—II InAs/GaSb super-lattice diodes with unipolar blocking layers. AIP Adv. **5**, 097132 (2015)
22. P. Chattopadhyay, The effect of shunt resistance on the electrical characteristics of Schottky barrier diodes. J. Phys. D. Appl. Phys. **29**, 823 (1996)
23. S.K. Cheung, N.W. Cheung, Extraction of Schottky diode parameters from forward current-voltage characteristics. Appl. Phys. Lett. **49**, 85–87 (1986)
24. S.M. Sze, K.N. Kwok, *Physics of Semiconductor Devices* (2006), pp. 134–196. https://doi.org/10.1002/9780470068328.ch3
25. R. Padma, G. Nagaraju, V. Rajagopal Reddy, C.-J. Choi, Effect of annealing temperature on the electrical and structural properties of V/p-GaN Schottky structures. Thin Solid Films **598**, 236–242 (2016)
26. J. Kim, M. Frontzek, L. Crow, J.W. Kwon, Selective thermal neutron transmutation doping with Gd masks in GaN semiconductors. Solid State Electron. **199**, 108510 (2023)
27. T. Piotrowski, Neutron shielding evaluation of concretes and mortars: A review. Constr. Build. Mater. **277**, 122238 (2021)
28. Y. Liu et al., Synchrotron X-ray topography characterization of high quality ammonothermal-grown gallium nitride substrates. J. Cryst. Growth **551**, 125903 (2020)
29. B. Raghothamachar, M. Dudley, G. Dhanaraj, X-Ray Topography Techniques for Defect Characterization of Crystals (2010)
30. Z.A. Kulage, C.H. Castano, S. Usman, G. Mueller, Characterization of the neutron flux energy spectrum at the Missouri University of Science and Technology Research Reactor (MSTR). Nucl. Eng. Des. **261**, 174–180 (2013)
31. J. Kim, Y. Liu, B. Raghothamachar, M. Dudley, J.W. Kwon, Defect-curing effects of fast neutrons on n-type GaN. Mater. Chem. Phys. **315**, 128934 (2024)
32. P. Arifin et al., Room-temperature photoluminescence of Mg-doped GaN thin films grown by plasma-assisted MOCVD. AIP Adv. **10**, 045123 (2020)
33. M.A. Reshchikov et al., The effect of annealing on photoluminescence from defects in ammonothermal GaN. J. Appl. Phys. **131**, 035704 (2022)
34. M.A. Reshchikov, Photoluminescence from vacancy-containing defects in GaN. Phys. Status Solidi A **220**, 2200402 (2023)
35. M. Grabowski et al., The impact of point defects in n-type GaN layers on thermal decomposition of InGaN/GaN QWs. Sci. Rep. **11**, 2458 (2021)
36. J. Kim, A. Patapenka, S. Chemerisov, J. W. Kwon, Zn-Doped P-Type Gallium Nitride Through Nuclear Transmutation Doping Processes. https://www.mrs.org/meetings-events/presentation/2022_mrs_fall_meeting/2022_mrs_fall_meeting-3804502

Part V
Materials Characterization Techniques

Chapter 19
Deep Level Defect Spectroscopy and Electron Beam-Induced Current Characterization of GaN Junctions

Andrew M. Armstrong, Andrew A. Allerman, Andrew Aragon, Daniel Feezell, François Léonard, and A. Alec Talin

1 Introduction

GaN offers increased power density for power electronics compared to Si [1–8] because the former can sustain much larger electric fields owing to its much larger band gap energy (E_g) [9]. For GaN, the critical electric is ~3.9 MV/cm [7] compared to 0.3 MV/cm for Si, and this advantage enables greater than 100x higher breakdown voltage at a given doping level than Si can achieve. However, GaN power electronics development has been limited because selective area doping has been difficult to achieve. Developing selective area doping methods for GaN is necessary because advanced vertical power devices including metal-oxide semiconductor field effect transistors and junction barrier Schottky diodes require selective area doping. Mg ion implantation can form p-type GaN regions selectively in an n-GaN field [10], but kV operation with low leakage has yet to be demonstrated using this method. Selective area epitaxy can also form p-regions within n-GaN by using plasma-based etching followed by p-GaN regrowth, but plasma etch-induced defects cause severe junction leakage [11].

This chapter focuses on understanding the electronic properties of defects in etched-and-regrown GaN p-n diodes caused by inductively coupled plasma (ICP)

A. M. Armstrong (✉) · A. A. Allerman
Sandia National Laboratories, Albuquerque, NM, USA
e-mail: aarmstr@sandia.gov

A. Aragon
Intel, Rio Rancho, NM, USA

D. Feezell
University of New Mexico, Albuquerque, NM, USA

F. Léonard · A. A. Talin
Sandia National Laboratories, Livermore, CA, USA

I. C. Kizilyalli et al. (eds.), *Gallium Nitride and Related Materials*, The Materials Research Society Series, https://doi.org/10.1007/978-3-031-83056-3_19

etching, the impact of these defects on p-n diode electrical performance, and methods to mitigate their impact. Photocapacitance can measure the deep level defect properties and concentration as a function of etch method and post-etch surface treatment, and these data correlate well with etched-and-regrown p-n diode reverse bias leakage and breakdown voltage. Similar trends for etch-induced defect introduction and methods for their mitigation are found for *c*-plane and *m*-plane p-n diodes, suggesting that defect mitigation strategies should have comparable efficacy for basal plane and sidewall surfaces formed by ICP. ICP etching produces significant surface and sub-surface damage, and elevated defect profiles propagate tens of nanometers into subsequently regrown layers. Nonetheless, post-ICP treatments prior to regrowth can greatly mitigate the impact of etch-induced defects on p-n diode electrical performance. Indeed, 1.6 kV etched-and-regrown GaN p-n diodes have been demonstrated [12]; however, their performance still lags that of continuously grown GaN p-n diodes that have achieved 7.8 kV [13]. In addition to photocapacitance, electron beam-induced current (EBIC) can be used to assess the impact of etched-induced defects. Depth-dependent EBIC can determine the change in minority carrier diffusion length for both the p- and n-type regions of p-n diodes as a function of etch method and post-etch treatment. Depth-dependent EBIC is non-destructive by modeling the dependence of the beam-accelerating voltage. This methodology was validated through normal-collector EBIC measurements, which require cleaving the device but provide direct evaluation of diffusion lengths. Methods for high-resolution EBIC evaluation of p-n diodes are also discussed.

2 Etch-Induced Deep Level Defects and Mitigation of Their Impact on GaN p-n Diode Electrical Performance

This section describes using deep level defect spectroscopy to correlate quantitatively the unintentional introduction of crystal defects through ICP etching and excess reverse leakage current in etched-and-regrown GaN p-n diodes. Basal plane-related defects were studied using planar etch and subsequent regrowth on *c*-plane diodes. Similar studies of *m*-plane GaN p-n diodes served as a proxy for sidewall-related defects that would exist in three-dimensional wells etched into the *c*-plane, such as those used in junction Schottky barrier (JBS) diodes. Realization of kV-class etched-and-regrown p-n diodes is then described using post-ICP etch treatments followed by p-type regrowth and fabrication of a p-n diode, including a junction termination extension (JTE).

2.1 Etch Damage and Related Deep Levels in c-plane GaN p-n Diodes

Figure 19.1 shows three different test device structures used to examine the impact of ICP etch damage and its mitigation in p-n diodes [14]. A continuously grown diode without dry etching, labeled as CT, was used for a reference. An etched-and-regrown diode (labeled as ICP) with an n-type layer that was ICP dry-etched to remove 0.4 μm of material prior to the regrowth of the p-type GaN without any post-ICP treatment, was used to identify etch-induced defects. A third sample, labeled ICP + AZ400K, was nominally identical to the ICP sample except that, prior to regrowth, it was treated with AZ400K (2% KOH and 5–15% potassium borates in water by weight) prior to regrowth.

The GaN devices were grown (and regrown) by metal organic chemical vapor phase deposition (MOCVD) on commercially available, hydride vapor phase epitaxy (HVPE) n-type GaN substrates ($n > 10^{18}$ cm^{-3}). They consisted of a 10-μm-thick, n-type, Si-doped ($3–5 \times 10^{16}$ cm^{-3}) drift layer and a 0.4-μm-thick, p-type GaN layer ([Mg] ~ $2–4 \times 10^{19}$ cm^{-3}). For the ICP and ICP + AZ400K devices, the ICP etch process was a low-damage, $Cl_2/BCl_3/Ar$ ICP etch recipe with powers of 125 W for the ICP power and 10 W for the radio frequency (RF) power. AZ400K treatment used a temperature of 80 °C for 10 min followed by a deionized water rinse and a nitrogen dry prior to regrowth. No measurable etch rate was observed with the AZ400K treatment, so its effects were likely limited to the surface or immediate sub-surface regions. Devices were defined by mesa isolation, and electron beam deposition of Pd/Au p-type and Ti/Al n-type ohmics were used to complete simple p-n diodes. These test devices did not have a JTE.

Figure 19.2 compares the reverse current-voltage (I-V) characteristics of the devices with various etch conditions (including an ICP + RIE etched-and-regrown

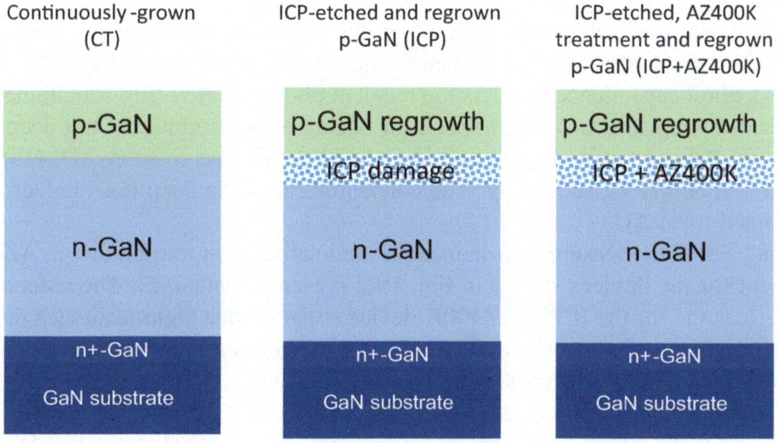

Fig. 19.1 Epitaxial structure of etched-and-regrown diodes for I-V analysis

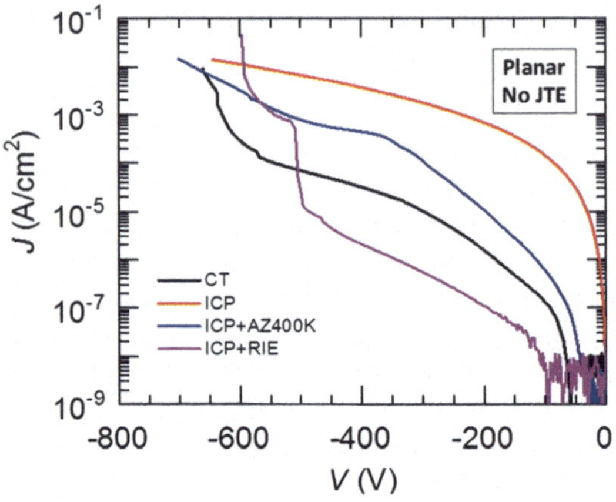

Fig. 19.2 Reverse *I-V* data showing increased reverse leakage for the ICP device relative to the CT device. The ICP + AZ400K device shows much lower leakage compared to ICP p-n diode, but the leakage remains higher than the CT device. The ICP + RIE diode achieves leakage comparable to the CT diode, suggesting its post-CIP etch treatment mitigated significant ICP-enhanced defectivity

p-n diode that is discussed later in this section). Reverse leakage currents are 2–3 orders of magnitude higher for the ICP diode compared to the CT device for $V < 500$ V. The excess leakage current is attributed to ICP etch-induced defects inducing additional recombination current. The discrepancy in reverse *I-V* behaviors between the CT and ICP devices at relatively low reverse bias confirms that the impact of IC etch-induced damage on diode electrical quality can be studied without need for elaborate field management to achieve kV operation. Compared to the ICP device, the ICP + AZ400K diode has reverse leakage that is several orders of magnitude lower, especially as low reverse bias. This observation suggests that the AZ400K treatment removes or otherwise mitigates ICP etch-induced defectivity near the surface. However, the ICP + AZ400K reverse leakage was still approximately 10x greater than the CT diode, indicating residual etch damage.

Steady-state photocapacitance (SSPC) deep level defect spectroscopy was used to understand better the nature of etch-induced defects and their dependence on AZ400K treatment. SSPC uses monochromatic light to photoionized (or photoneutralized) deep level defects in the depletion region of the p-n diodes. The excess space-charge in the depletion region caused by defect photoabsorption creates a photocapacitance, ΔC. As the incident photon energy ($h\nu$) is incrementally increased, a step in the $\Delta C(h\nu)$ spectrum occurs at approximately the deep level energy (E_o). An associated technique, deep level optical spectroscopy (DLOS) [15], is used to quantify E_o accurately. The concentration of the deep level defect, N_t, is determined from ΔC.

Using SSPC to measure etch-induced N_t and subsequent mitigation via AZ400K treatment for the devices shown in Fig. 19.2 presents challenges. The reduction in reverse leakage for the ICP + AZ400K device suggests that significant etch damage occurs near the surface of the n-GaN. This etched surface becomes the metallurgic junction upon p-GaN regrowth. However, SSPC averages its sensitivity over the entire depletion region. In the case of an asymmetric p^+-n junction, as was the case for all diodes discussed here, the depletion region is restricted to the n-side. However, due to the low n-GaN doping used in high voltage p-n diodes, this depletion

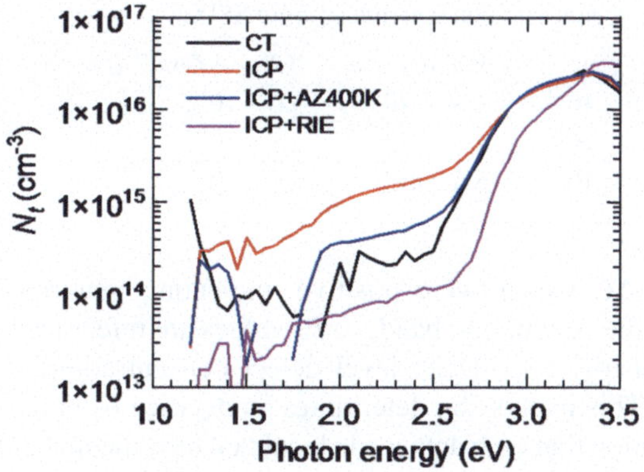

Fig. 19.3 SSPC data showing increased concentration of the $E_c - 1.90$ eV deep level for the ICP sample compared to the CT sample. The ICP + AZ400K sample has much lower concentration of the $E_c - 1.90$ eV deep level compared to the ICP sample, but it remains elevated compared to the CT sample. The ICP + RIE sample shows similar concentration for the $E_c - 1.90$ eV deep level as the CT sample, which matches the trend in reverse leakage for the low-doped p-n diodes

region is often ~1 μm in depth below the p-n junction at 0 V. Thus, SSPC is not well suited to examine differences in etch-induced defects for the devices in Fig. 19.2 because the effects of AZ400K treatment are largely restricted to the metallurgic interface, while SSPC samples a bulk-like volume.

To increase SSPC sensitivity to etch-induced defects near the metallurgic junction of etched-and-regrown diodes, a second set of diodes were grown like those shown in Fig. 19.1 except that the n-GaN doping was increased to 10^{18} cm^{-3}. The magnitude of the photocapacitance induced by deep level defects near the regrowth interface ($\Delta C_{,\text{int}}$) is [16]

$$\Delta C_{\text{int}} = \frac{N_{t,\text{int}}}{2} \frac{C_0}{N_d} \frac{x_{\text{int}}^2}{x_d^2} \propto \frac{1}{N_d x_d^3} \propto \sqrt{N_d}, \tag{19.1}$$

where $N_{t,\text{int}}$ is the average volumetric defect concentration near the regrowth interface, N_d is the net doping concentration, $x_{,\text{int}}$ is the spatial extent of the interfacial defects near the regrown junction, C_0 is the junction capacitance in the dark, and x_d is the depletion depth. Increasing the Si doping by 100x thus increases SSPC sensitivity to the regrown interface by 10x relative to SSPC measurements of typical, low-doped drift layers.

The SSPC data in Fig. 19.3 compare N_t data for the heavily doped sample set, which are summarized in Table 19.1 (the additional sample ICP + RIE is discussed below in this section). The same labeling scheme is used for the low-doped diodes used for *I-V* analysis in Fig. 19.2 and the heavily doped diodes used for SSPC in Fig. 19.3 to identify their etch and post-etch conditions. All samples in Fig. 19.3 show a change in slope at ~1.7, 2.5, and 3.2 eV, and DLOS analyses [14] (not discussed here) determined the associated E_o values of 1.90, 2.90, and 3.2 eV,

Table 19.1 Deep level concentration determined from SSPC

	CT (cm^{-3})	ICP (cm^{-3})	ICP + AZ400K (cm^{-3})	ICP + RIE (cm^{-3})
E_c − 1.90 eV	2.0×10^{14}	1.8×10^{15}	5.0×10^{14}	1.1×10^{14}
E_c − 2.90 eV	2.0×10^{16}	1.7×10^{16}	2.1×10^{16}	1.5×10^{16}
E_c − 3.2 eV	6.0×10^{15}	5.3×10^{15}	5.0×10^{15}	1.7×10^{16}

respectively. All ΔC values were positive, indicating majority carrier (electron) photoemission to the conduction band, so E_o values are referenced to the conduction band minimum energy (E_c). Deep level defect concentrations were calculated as $N_t = 2N_d\Delta C/C_0$. This expression determines an average N_t in the depletion region. Given the expectation that etch damage is localized near the metallurgic junction, the N_t data presented in Fig. 19.3 and Table 19.1 likely underestimate etch-induced defectivity, but the lack of knowledge regarding x_{int} vitiates a more accurate determination. Nonetheless, meaningful correlations were found between SSPC and reverse *I-V* characteristics.

Concentrations of the E_c − 2.90 and E_c − 3.2 eV deep levels are very similar for the CT, ICP, and ICP + AZ400K devices, which suggests that these deep levels exist in as-grown n-GaN and are not related to ICP etch damage. Conversely, N_t for the E_c − 1.90 eV deep level is much larger for the ICP sample compared to the CT sample. Moreover, the ICP + AZ400K sample shows a marked reduction in the E_c − 1.90 eV deep level compared to the ICP sample but still has an elevated N_t compared to the CT sample. This trend in N_t for the E_c − 1.90 eV defect state follows that found for reverse leakage current at relatively low V in Fig. 19.2 for the low-doped devices. Thus, the concentration of the E_c − 1.90 eV deep level is enhanced by ICP etching and contributes to excess reverse leakage but can be largely, though not fully, mitigated by AZ400K treatment of the etched surface prior to regrowth.

As discussed above, the strong response of the E_c − 1.90 eV concentration to AZ400K surface treatment suggests that the defect is spatially concentrated near the etched-and-regrown junction and that accurate determination of N_t is difficult using SSPC. Equation (19.1) shows that if, for example, the majority of E_c − 1.90 eV deep level defects lie within 10 nm of the etched surface for these diodes with $x_d = 150$ nm (measured by capacitance-voltage), then N_t averaged over the entire depletion region underestimates the etch-enhanced N_t concentration at the regrown interface by >200x.

Given the uncertainties in the depth dependence of ICP etch-related defects before and after AZ400K, additional studies included a different device design to enable depth profiling of defects. Figure 19.4 shows the epitaxial structure of the new device structures, including a continuously grown n-GaN region, an ICP etched-and-regrown n-GaN region, and an ICP etched-and-regrown n-GaN region that received a post-etch AZ400K treatment prior to regrowth. Like the preceding studies, these samples are labeled CT, ICP, and ICP + AZ400K, respectively. N-type Schottky devices were fabricated by defining diodes using mesa isolation and deposition of an Ni Schottky contact on the epitaxial surface and a Ti/Al ohmic contact on the backside of the n-GaN substrate. For the ICP and the ICP + RIE

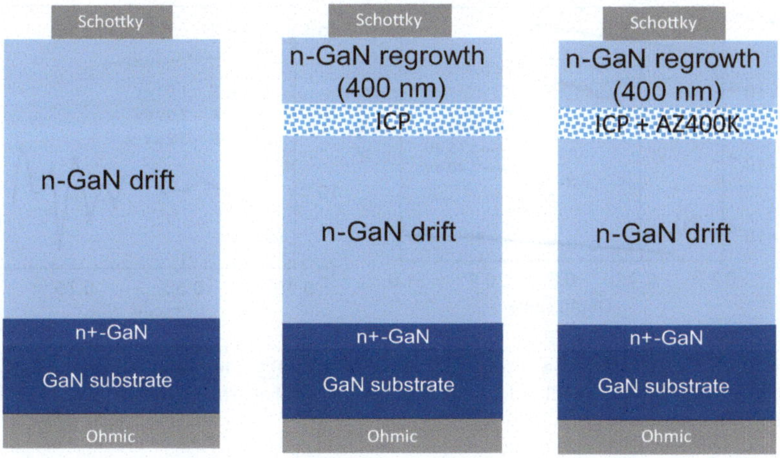

Fig. 19.4 Epitaxial structure of etched-and-regrown Schottky diodes enabling depth profiling of defect states near the regrowth interface

devices, the etched region is contained near the middle of the Schottky diode depletion region, which enabled defect depth profiling using the lighted capacitance-voltage (LCV) method [17].

LCV operates under a similar principal as SSPC, but specific values of $h\nu$ are selected to photoionize deep levels individually, and a capacitance-voltage measurement is taken to measure the change in effective doping after deep level photoabsorption. For example, Fig. 19.3 shows that $h\nu = 2.40$ eV photoionizes just the $E_c - 1.90$ eV deep levels while leaving the $E_c - 2.90$ and $E_c - 3.2$ eV levels unperturbed. The change in effective doping measured by capacitance-voltage (*C-V*) in the dark and under 2.40 eV illumination gives the depth dependence of the $E_c -$ 1.90 eV deep level concentration. Similarly, the change in effective doping under 2.40 eV and 3.15 eV illumination gives the depth profile of the $E_c - 2.90$ eV deep level, and the change in effective doping between 3.15 eV and 3.30 eV illumination gives the depth profile of the $E_c - 3.2$ eV deep level.

Figure 19.5 shows the effective doping levels under illumination for the CT sample. The change in effective doping is too small to resolve in Fig. 19.5a for the $E_c - 1.90$ eV deep level, but the corresponding changes are much larger for the $E_c -$ 2.90 eV and $E_c - 3.2$ eV levels. Figure 19.5b shows the differences of these curves, yielding the depth profile of the deep levels. As expected for the CT sample without etching, the as-grown deep levels have an approximately uniform concentration.

Figure 19.6 shows the dark *C-V* and LCV data for the ICP sample. There are several key differences compared to the CT sample. The dark *C-V* of the ICP sample shows a spike in the doping profile 0.4 μm below the Schottky metal. This doping spike is due to unintentional Si incorporation at the surface of the n-GaN surface prior to regrowth [18, 19]. Such Si spikes typically occur for GaN regrowth regardless of whether the surface has been etched. While this Si spike is deleterious for the electrical properties of regrown p-n diodes [6], it is fortuitous for defect depth

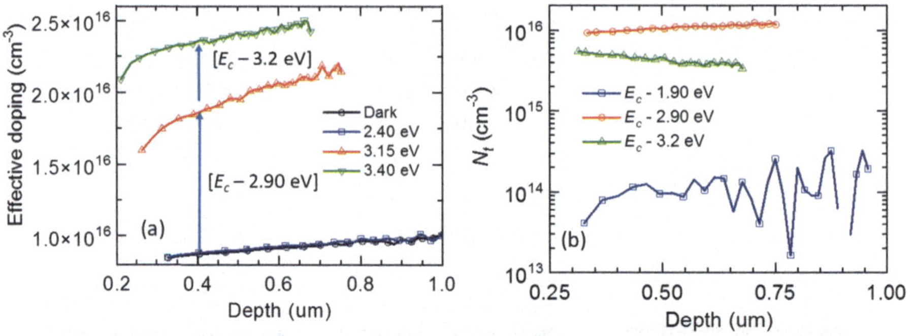

Fig. 19.5 (**a**) Effective doping profile for the CT Schottky diode measured via *C-V* under illumination and (**b**) deep level defect depth profile calculated from differences in the effective doping profiles shown in (a)

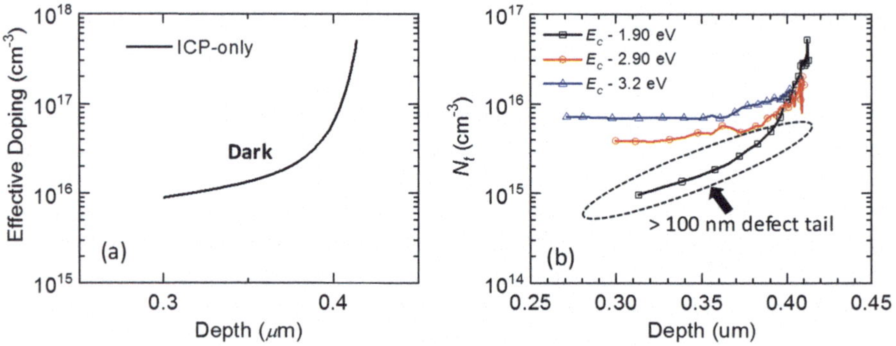

Fig. 19.6 (**a**) Dark and (**b**) lighted *C-V* data for the ICP Schottky diode. The doping spike near 0.4 μm marks the etched-and-regrown interface. Elevated deep level concentrations are observed above the regrowth interface for the E_c − 1.90 eV etch-enhanced deep level

profile studies because it provides an unambiguous marker of the regrowth interface. The LCV data in Fig. 19.6b show that the concentrations of the E_c − 2.90 and E_c − 3.2 eV deep levels are approximately constant above the etched-and-regrown interface, whereas the E_c − 1.90 eV exhibits an elevated deep level concentration extending >100 nm above the regrowth interface. This observation indicates that, without post-etch mitigation, ICP etch-enhanced defectivity propagates into regrown material and that attempts to bury the etch-induced defects below a p-n junction should use several hundred nanometers of regrown n-GaN prior to growth of the p-GaN region.

Figure 19.7 shows LCV data on defect profiles in the ICP + AZ400K Schottky sample. Unlike the ICP Schottky sample, the ICP + AZ40K device was able to sustain large enough reverse bias during LCV measurements to deplete through the unintentional Si spike at the regrowth interface to probe deep level defects below the etched-and-regrown surface. Despite the AZ400K post-etch treatment, elevated

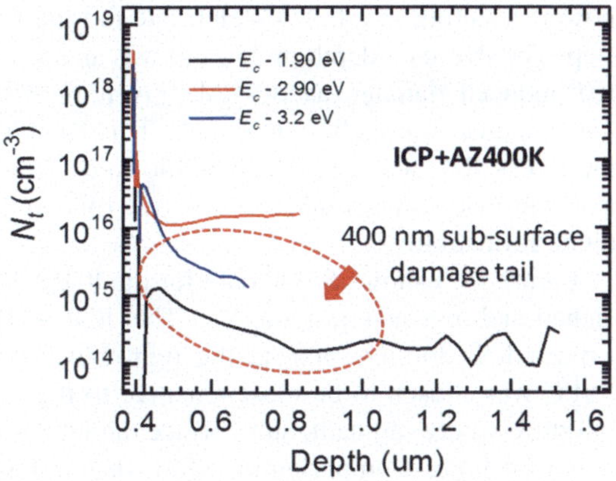

Fig. 19.7 Deep level depth profile data for the ICP + AZ400K Schottky diode. The doping spike near 0.4 *mm* marks the etched-and-regrown interface. Elevated deep level concentrations are observed below the regrowth interface for the $E_c - 1.90$ eV etch-enhanced deep level and the $E_c - 3.2$ eV deep level

concentration of the $E_c - 1.90$ eV deep level was observed ~400 nm below the regrown interface. This finding suggests that the ICP etch causes substantial sub-surface damage that is not ameliorated by the AZ400K treatment, which is consistent with the higher reverse leakage observed in the ICP + AZ400K data in Fig. 19.2 compared to the CT device. It is noteworthy that the concentration of the $E_c - 1.90$ eV deep level near the etched-and-regrown interface in the ICP + AZ400K Schottky diode in Fig. 19.7 is significantly lower than what was observed in the ICP Schottky device in Fig. 19.6, which reinforces the trend observed by SSPC for ICP etch and AZ400K post-etch treatment. Figure 19.7 also indicates higher concentration of the $E_c - 3.2$ eV deep level ~ 300 nm below the etched-and-regrown interface compared to the underlying crystal bulk. This deep level has been attributed to carbon impurities incorporating on the nitrogen sub-lattice [20], so it is possible that ICP damage also creates excess nitrogen vacancies that then enable preferential carbon occupancy on the nitrogen site in the tail of the etch damage.

While the AZ400K post-ICP surface treatment proves capable of significantly reducing etch-related deep level defect concentration and reverse leakage, it is not able to restore either characteristic to the levels observed for continuously grown p-n diodes. Remnant, sub-surface damage observed in Fig. 19.7 suggests that complete mitigation of ICP etch defectivity requires the removal of the damaged n-GaN material. A slow reactive ion etch (RIE) process was employed to remove ICP-damaged material. RIE is a physical etch process that is often expected to damage the crystal; however, previous work used the same RIE conditions to form low-leakage, recessed gates for AlGaN high-electron-mobility transistors [21]. Slow (~ 3 Å/s) ICP etching of n-GaN also enables kV-class performance for etched-and-regrown p-n diodes [22].

Another etched-and-regrown Schottky diode was fabricated to test the efficacy of RIE to remove ICP-damaged material without introducing additional damage. The epitaxial structure and fabrication steps were similar to the samples shown in Fig. 19.4 except that following ICP etching, the wafer was then etched to a depth of 270 nm by RIE using 45 sccm CF_4 + 5 sccm O_2 gas chemistry with 25 W RF power, resulting in a 0.3 Å/s etch rate to remove defective material left by the ICP step. The RIE etch depth of 270 nm was as deep as was practical to remove as much ICP-induced damage as possible given time and tool constraints (etch time of 150 minutes) for such a slow etch. This sample is labeled ICP + RIE. After RIE, the ICP + RIE was also received the same AZ400K treatment as previous etched-and-regrown samples discussed above followed by n-GaN regrowth and Schottky diode formation.

Figure 19.3 shows SSPC data for the ICP + RIE Schottky diode compared to the etched-and-regrown p-n diodes. The ICP + RIE Schottky diode exhibited the expected Si doping spike at the regrowth interface, and the applied bias during SSPC was chosen to be most sensitive to the region ~400 nm below the regrowth interface where remnant, sub-surface damage was observed for the ICP + AZ400K Schottky diode. Comparing the SSPC data in Fig. 19.3 shows that the concentration of the E_c − 1.90 eV deep level in the ICP + RIE Schottky diode is similar to that observed in the as-grown CT p-n diode, and the N_t data for the ICP + RIE are included in Table 19.1 for comparison. This observation verifies that the RIE treatment removes substantial ICP etch-induced defectivity and suggests that etched-and-regrown p-n diodes incorporating a similar post-ICP treatment should achieve superior reverse voltage characteristics compared to the ICP + AZ400K treatment.

Etched-and-regrown p-n diodes with a thick (10 μm), low-doped (3×10^{16} cm^{-3}) n-GaN drift layer using the same ICP + RIE treatment as described above were grown, fabricated, and characterized [12]. These p-n diodes employed a multi-step JTE [23] to exceed performance of simple, planar diodes and are labeled ICP + RIE (JTE). Anodes had a circular geometry with a 150 μm diameter. JTE step etch depths of 200, 260, and 300 nm in the *p*-GaN were patterned using multiple photolithography steps and ICP etch conditions described above. The etch depth increased with increasing distance from the anode. The step widths were 50 μm starting from the edge of the anode. A final ICP etch through the junction was performed to provide electrical isolation.

Forward and reverse voltage performance for ICP + RIE (JTE) devices are shown in Fig. 19.8. The diodes exhibited low forward leakage with a turn-on voltage of 2.9 V, current density of 0.8 kA/cm^2 at 4 V, and a differential, specific on-resistance ($R_{on,sp}$) = 1.1 mΩ.cm^2. The reverse *I-V* data in Fig. 19.8b show low leakage less than 1 nA up to 1250 V and maximum reverse voltage of 1610 V prior to breakdown. Comparison of $R_{on,sp}$ and maximum reverse voltage for various regrown, etched-and-regrown, and continuously grown GaN p-n diodes are shown in Fig. 19.9. The ICP + RIE(JTE) shows excellent performance compared to other regrown and etched-and-regrown p-n diodes. However, despite the mitigation of etch damage using RIE following ICP etching, etched-and-regrown GaN diodes require further improvements to reach equivalence to continuously grown diodes. The discrepancy could be due to (1) accelerated breakdown due to remnant etch-related defects, (2) excess resistance due to the drift region being thicker than necessary for the realized V_{max}, (3) non-optimized JTE, or a combination thereof.

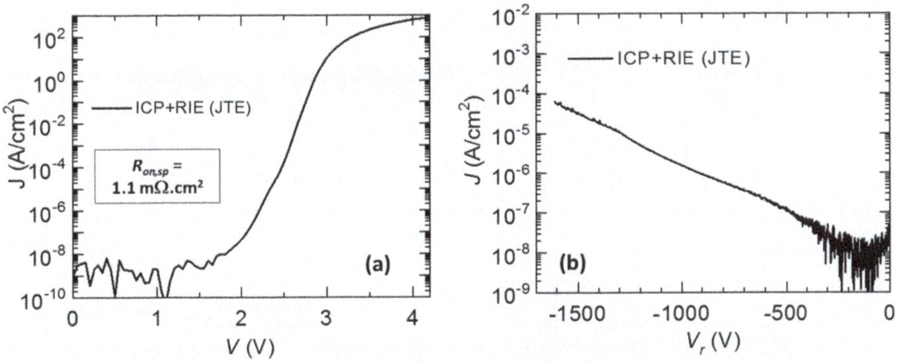

Fig. 19.8 (**a**) Forward and (**b**) reverse characteristics of an etched-and-regrown p-n diode using the ICP + RIE etch and post-etch conditions prior to p-GaN regrowth. P-n diodes included a five-step JTE to control the lateral electric field spatial profile to increase the breakdown voltage

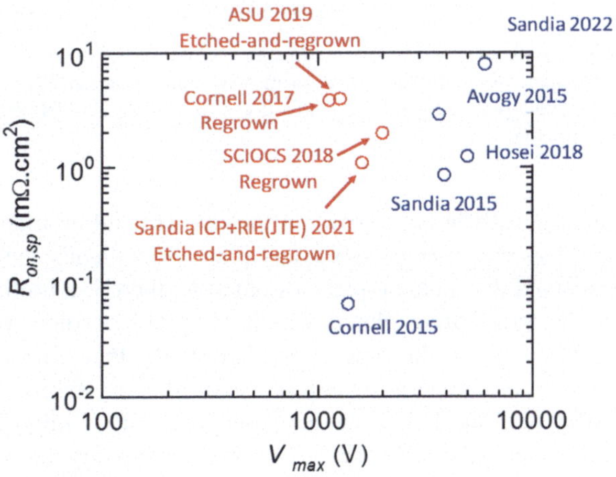

Fig. 19.9 $R_{on,sp}$ vs. maximum voltage for various regrown or etched-and-regrown p-n diodes (red symbols) and continuously grown p-n diodes (blue symbols). The ICP + RIE (JTE) p-n diodes show excellent performance compared to other etched-and-regrown diodes but still lag continuously grown devices

3 Interfacial Impurity and Etch-Induced Deep Level Defects and Mitigation of Their Impact on m-Plane GaN Diode Electrical Performance

Two important crystalline planes within Wurtzite GaN crystal structures are schematically represented in Fig. 19.10a. In Wurtzite GaN the III-V species are Ga and N, respectively. These planes are identified in Fig. 19.10a as the polar *c*-plane with miller indices (0001) and the nonpolar *m*-plane with miller indices $(10\bar{1}0)$. Each plane is identified as polar or nonpolar due to discontinuities in spontaneous and piezoelectric polarization from the configuration of the III-V species within the Wurtzite crystalline lattice [24]. A cross-sectional crystalline lattice structure of GaN for basal *c*-plane and *m*-plane is represented schematically in Fig. 19.10b. An

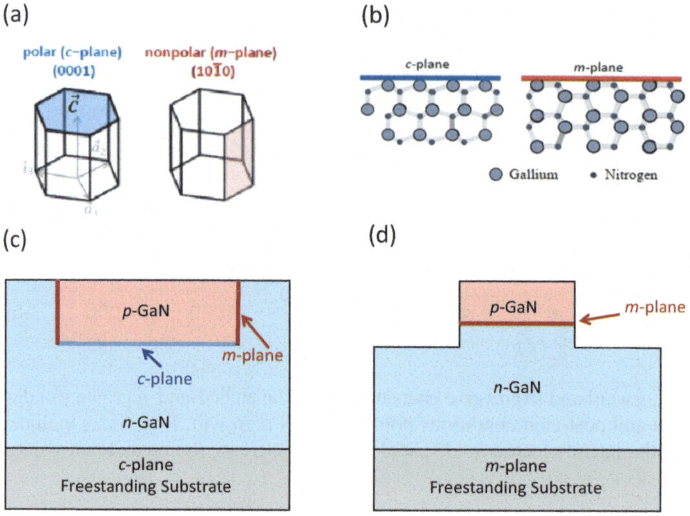

Fig. 19.10 (**a**) Crystal orientation representation of m-plane and c-plane in Wurtzite GaN (**b**) Schematic of crystalline lattice for m-plane and c-plane in Wurtzite GaN (**c**) Schematic of relevant crystalline planes for selective area-doped p-n diode, and (**d**) planar vertical p-n diode on m-plane orientation

important difference between the two crystalline planes is the arrangement of the III-V species (Ga and N) at any plane, specifically the terminating plane. It may be observed that in the *c*-plane orientation, there is a uniform amount of either Ga or N at the terminating planes which may be classified as either Ga-polar or N-polar termination. In the case of *m*-plane GaN, there is an equal amount of Ga and N species at the termination plane of the crystalline structure. This fundamental difference has led to an increased amount of impurity incorporation in *m*-plane GaN due to the presence of N atoms at the nonpolar surface [25].

The complex vertical designs of devices such as junction field effect transistors (JFETs), and current aperture vertical electron transistors (CAVETs) typically require selectively doped regions in their architecture [26–28]. Therefore, to investigate and develop selectively doped regions we will consider the most basic selectively doped device which is the vertical selectively doped *p-n* diode device. By considering this device we can more easily analyze and understand the challenges that arise through implementing selectively doped regions on more complex devices. In these complex device structures, there are more than one crystalline planes of GaN which constitute the planes of the regrown at the junctions.

A schematic representation of a vertical selectively doped *p-n* diode device architecture with the *m*-plane orientation highlighted is shown in Fig. 19.10c. As can be observed, *m*-plane junctions are present on the sidewalls of the trenches within selectively doped regions. However, studying the sidewall junctions directly is difficult to accomplish. Thus, investigation on planar diodes allows us to investigate *m*-plane junctions more easily. The availability of free-standing *m*-plane substrates further enables this investigation. Therefore, the junctions at the sidewall of the trenches within vertical selective area-doped *p-n* diodes will be studied more easily by investigating *m*-plane planar diodes Fig. 19.10d.

The effects of interfacial impurities on *m*-plane GaN *p–n* diodes are investigated in order to observe the impurity incorporation at regrown interfaces and observe the effects they have on the device electrical characteristics. The *p-n* diode epitaxial layers were grown using MOCVD on free-standing *m*-plane GaN substrates. A ~ 0.95° miscut toward the minus *c*-direction is used on the substrate from Mitsubishi Chemical Corporation (MCC). This miscut, along with growing in N_2 carrier gas, was shown to effectively mitigate the formation of pyramidal hillocks [29]. The growth conditions consisted first of a 2-μm-thick *n*-type template layer ($[Si] = 6 \times 10^{17}$ cm^{-3}) grown as the baseline material to ensure proper morphology for the subsequent layers. Subsequently, a 250 nm-thick n++ (($[Si] = 5 \times 10^{18}$ cm^{-3}) layer which may constitute a top-side *n*-contact layer for purposes of lateral devices. Next, a 5 μm-thick *n*-type drift region is grown using the minimum controllable *n*-type doping for the MOCVD system ($n_0 = 6 \times 10^{16}$ cm^{-3}). Now, the *n*-GaN part of the p-n diode structure is complete. A total of five subsequent processes/treatments will intentionally introduce impurities at the top-side interface of the aforementioned epitaxial layer. The five samples were grown: (1) continuously grown p–n diode; (2) regrown *p*-GaN grown after 1 week outside the chamber in N2 box; (3) regrown p-GaN grown after an in-chamber interruption for 10 min; (4) regrown p-GaN grown after 1 week outside the chamber in N_2 box with N_2 blow; and (5) regrown p-GaN grown after the sample was exposed to acetone, isopropyl alcohol, and deionized (DI) water followed by hydrofluoric acid (HF) and DI rinse. Sample 1 is used as the baseline for comparison between the subsequent treatments. The treatments in samples 2–4 were performed to replicate impurity incorporation at the regrown surface due to different exposure conditions. Sample 2 was intended to incorporate impurities from keeping the sample outside the chamber within a contained N_2 purged box. Sample 3 was intended to incorporate impurities from within the chamber during a temperature cycle within the MOCVD. Sample 4 was intended to replicate incorporation due to keeping the sample outside the chamber within a contained N2 purged box along with a N2 blow that will bring impurities to the surface. The surface treatment for sample 5 was performed to observe if interfacial impurities associated with ACE/IPA/DI could be removed using HF. Subsequent p-GaN layers are grown after the 5 previously mentioned interruptions/treatments. The p-type dopant CP_2Mg is flowed along with H_2 in order to constitute the acceptor-type doping. The layer constituting the *p*-GaN part of the *p-n* junction is grown with an intentional concentration of $[Mg] = 1 \times 10^{19}$ cm^{-3} with a thickness of 400 nm. A highly doped *p*++-type layer follows with an intentional concentration of $[Mg] = 1 \times 10^{20}$ cm^{-3} which constitutes the *p*-contact layer of the diode. The sample is then annealed at 700 ° C for 15 min in order to activate the Mg atoms within the *p*-type layers. A schematic of the fabricated diodes with the placement of the regrowth interface and the 5 subsequent regrowth/treatments are shown in Fig. 19.11.

The effects of the interfacial impurities on the electrical performance of the 5 samples described in the previous sections will be investigated in this section.

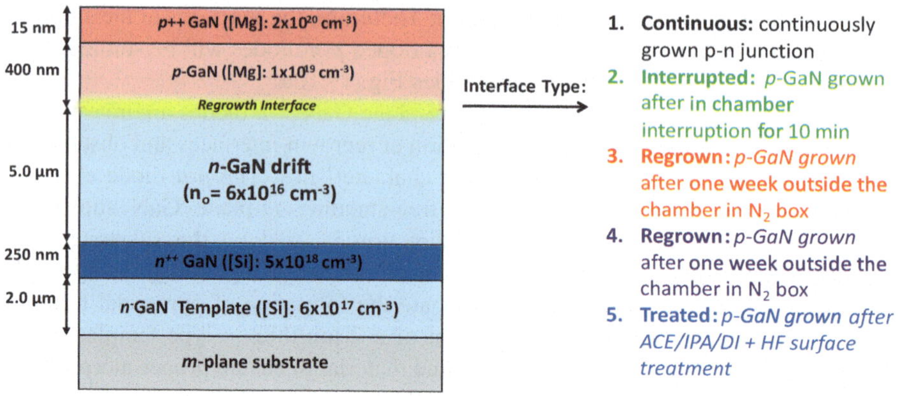

Fig. 19.11 Epitaxial stack for regrown/treated vertical *m*-plane GaN *p-n* diodes showing the five different surface treatments done after *n*-GaN layers are grown

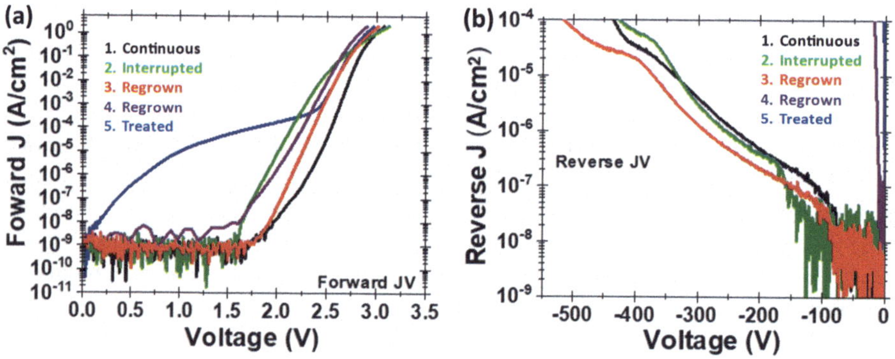

Fig. 19.12 Current density-voltage curves of characteristics of selected interrupted/treated *m*-plane GaN *p-n* diodes in (**a**) forward regime and (**b**) reverse regime

Moderate levels of all species (Si, O, C) will be found to have a minimum effect on the diode performance in the forward and reverse. However, the performance of the diodes becomes compromised at higher interfacial impurity levels, and individual impurities are concluded to have a different effect on the device performance in the forward and/or reverse regimes. The current density–voltage (*J–V*) characteristics of a single representative diode, out of approximately ten measured diodes on each sample, are shown in Fig. 19.12. In forward bias, all the samples showed very low off-state forward leakage (~1 nA/cm^2) below 1.5 V, except for sample 5. At higher voltages (>1.5 V) but still below turn-on, however, the off-state leakage currents differ for all the samples. All diodes showed similar turn-on voltages (~2.9 V). The extracted ideality factors (n) as a function of applied voltage for all diodes show nonideal diode behavior ($n > 1$) that often indicates Shockley–Read–Hall (SRH)-dominant recombination current within the depletion region of the diodes [30].

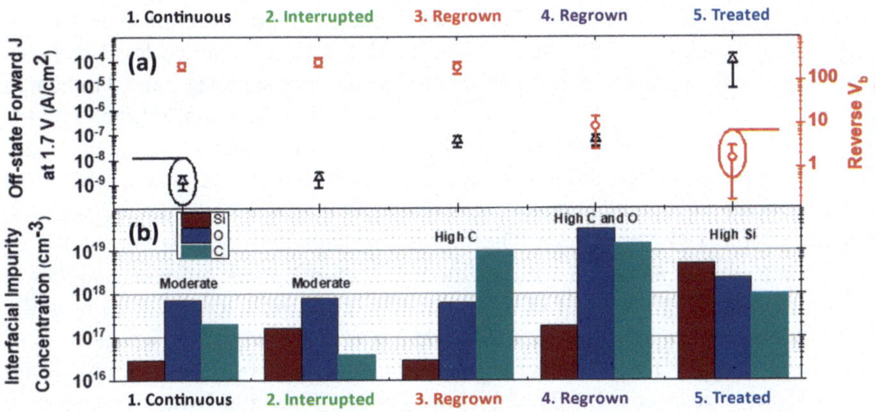

Fig. 19.13 (**a**) V_b and off-state forward current density at 1.7 V, (**b**) interfacial impurity concentrations (cm^{-3}) at the metallurgical junction

The *J*–*V* results are correlated with impurity levels by using impurity profiles obtained from SIMS analysis to understand the impacts of interfacial impurities (types and concentrations) on the diode performance. The impurities between *n*-GaN and *p*-GaN may be resolved along with sharp spikes in impurities at the interface. The correlations between forward off-state leakage current (defined at forward current at 1.7 V), reverse blocking voltage (V_b) (defined as the reverse voltage at 1 $\mu A/cm^2$), and impurity concentration at the interface are shown in Fig. 19.13. Statistical (~ ten diodes per sample) forward off-state leakage current and reverse blocking voltage (V_b) are shown in Fig. 19.13a with error bars. The interfacial impurity concentration shown in Fig. 19.13b is defined as the peak of the impurity spikes observed at the regrowth interface in the SIMS data. The SIMS data were also separately analyzed by integrating over the spikes to obtain the total impurity concentration which produced similar trends.

The analysis reference samples 1 (continuously grown p–n diode) and 2 (regrown p-GaN grown after 1 week in an N_2 box) showed that moderate levels of impurities ($\sim 5 \times 10^{17}$ cm^{-3}) at the junction yielded no significant degradation of the reverse ($V_b > 200$ V) and forward (<2 nA/cm^2 off-state leakage at 1.7 V) characteristics. It is observed that the O levels within the drift region may differ by an order of 2.5–4 times. This difference was attributed to background impurity levels within the MOCVD at the time of growth. However, performance of the diodes was very similar which indicated that these differences in O within the *n*-GaN drift region did not degrade the device performance at these O levels. Comparison of sample 3 (regrown *p*-GaN grown after an in-chamber interruption for 10 minutes and reference sample 1 (or sample 2) showed that high levels of carbon ($\sim 1 \times 10^{19}$ cm^{-3}) with moderate levels of silicon and oxygen ($<1 \times 10^{18}$ cm^{-3}) did not adversely affect the reverse blocking voltage ($V_b > 200$ V). However, the forward off-state leakage current was increased (from <2 nA/cm^2 to ~60 nA/cm^2 at 1.7 V) compared with the reference sample 1. Sample 4 (regrown p-GaN grown after 1 week outside in N_2 box + N_2 blow) contained high levels of O and C, (~ 3×10^{19} cm^{-3} and ~ 1×10^{19} cm^{-3}, respectively) with a moderate level of Si (~ 2×10^{17} cm^{-3}). Comparison of sample 4 with the reference samples 1 and 3 indicated that the high level of oxygen (~ 3×10^{19} cm^{-3}) or the combination of high O and C was responsible for the degradation in the reverse performance, given that high levels of C impurities alone did not significantly alter the reverse characteristics. The reverse blocking voltage of sample 4 was reduced (V_b ~ 8.5 V) compared with the reference sample 1 ($V_b > 200$ V). The forward leakage ($J = 60$ nA/ cm^2 at 1.7 V) was similar to that of sample 3 with a similar C impurity level, but with a higher O level. Therefore, a high level of O was primarily responsible for a reduction in the reverse blocking voltage in sample 4. Sample 5 was characterized by a high level of Si (~ 5×10^{18} cm^{-3}) with moderate levels of O and C (~ 2×10^{18} cm^{-3} and ~ 1×10^{18} cm^{-3}, respectively). The blocking voltage of sample 5 was tremendously reduced (from $V_b > 200$ V to ~1.6 V) and the forward leakage was considerably increased (from <2 nA/cm^2 to ~100 μA/cm^2 at 1.7 V) compared with the reference sample 1. Thus, we observed that large Si spikes strongly affect both forward and reverse characteristics of the diodes. Also, high carbon spikes at the junction affects primarily the forward characteristics with high off-state leakage current in forward bias (~100x higher forward leakage current compared with a reference diode), whereas the reverse characteristics are not severely affected ($V_b > 200$ V at 1 μA/ cm2). High O spikes at the junction strongly affect the reverse blocking voltage (V_b ~ 8.5 V) compared with the reference sample 1 ($V_b > 200$ V) with nominal effect of the off-state forward current density. It is also shown that regrown diodes with impurity (Si, O, and C) levels below (~ 5×10^{17} cm^{-3}) show comparable forward and reverse results with the reference continuously grown diodes. Thus, the effect interfacial impurities at the regrown junction of vertical m-plane GaN *p-n* diodes has been investigated with some key takeaways on impurity species incorporation and concentration.

An investigation of regrown and etched-and-regrown *m*-plane GaN vertical Schottky diodes by steady-state photocapacitance (SSPC) by comparing changes in the defect concentration relative to a continuously grown film, will be presented in the following sections. It will be shown that elevated defect concentrations are localized at regrowth interfaces even without a plasma etch. Also, regrowth on dry-etched surfaces produced even higher peak defect concentrations near the regrowth interface, and elevated defectivity propagated into the regrown region. Post-etch surface treatment using AZ400K, prior to regrowth, reduced the defect concentration at the regrown interface. We also introduced a photoelectrochemical (PEC) wet etch treatment to remove dry etch-enhanced damage at the regrowth interface of the etched-and-regrown diodes. Since active removal of macro-amounts of material is occurring using an etch with significant physical component, it may be expected that the process yields some sort of damage to the resulting crystalline lattice. As will be presented subsequently, damage is present at the surface of the crystalline lattice and, below the surface, which leads to nonideal surface for regrowth and propagation into the subsequently regrown material. Furthermore, crystalline defects can manifest into deep level traps within the band gap of GaN.

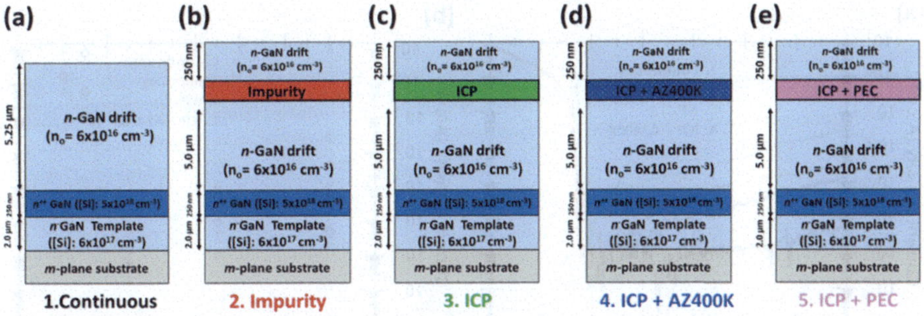

Fig. 19.14 Schematic of vertical m-plane GaN Schottky didoes with the following surface treatment: (**a**) continuously grown, (**b**) impurity interface, (**c**) ICP interface, (**d**) ICP + AZ400K interface, and (**e**) ICP + PEC interface

Spectroscopy will be used to identify these levels and their concentration, and their effects on the electrical performance will be presented. Thus far, we have investigated vertical m-plane GaN p-n diode structures for interfacial impurities at regrown junctions. However, the implementation of Schottky diodes in this section enables a more precise control over the depletion depth as a function of applied voltage. This will allow for precise depth resolution when conducting spatially varying optical spectroscopy, which will be elaborated in the following sections. Additionally, benefits of the Schottky diode arise due to the device unipolar device operation, which allows for detailed analysis of the majority carriers only.

GaN Schottky diode structures were grown through MOCVD on free-standing *m*-plane GaN substrate with a $-0.95°$ miscut toward the minus *c*-direction from MCC with *n*-type Si doping of 1×10^{17} cm^{-3}. The epitaxial growth consists of a 2-μm-thick *n*-type GaN buffer layer ([Si] $\sim 6 \times 10^{17}$ cm^{-3}) followed by a 250-nm-thick n + − GaN ([Si] $\sim 5 \times 10^{18}$ cm^{-3}) and a subsequent 5-μm-thick lightly doped *n*-type GaN with net doping $\sim 6 \times 10^{16}$ cm^{-3} to serve as a drift region. Schematics of the Schottky diode structures are shown in Fig. 19.14.

A total of five samples with various treatments at the regrowth interface were studied. Sample 1 (reference sample) was continuously grown without any interruption, etching, or surface treatment to serve as a reference baseline as-grown material for comparison. Sample 2 contained a regrowth interface without etching but with a solvent cleaning acetone/isopropanol/deionized water (Ace/IPA/DI) to intentionally introduce impurities at the regrowth interface.

Sample 3 was regrown after 400 nm of dry etching using a PlasmaTherm Apex SLR inductively coupled plasma (ICP) etcher. Sample 4 was ICP etched using the same conditions as those for sample 3 but with a subsequent AZ400K treatment at 80 °C for 10 min prior to the regrowth. Sample 5 was also ICP-etched (using the same conditions as those for samples 3 and 4) followed by a subsequent photoelectrochemical (PEC) etch (with an etch depth of 200 nm) in a 0.01 M KOH in DI at room temperature with a mercury arc lamp positioned to have a surface flux of 100 mW/cm^2.

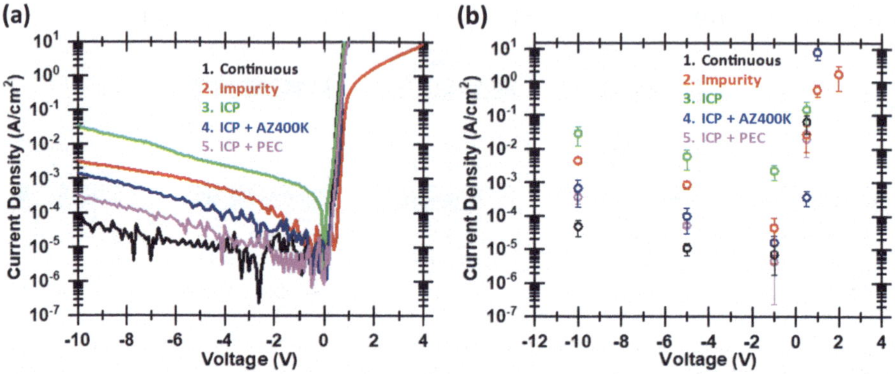

Fig. 19.15 (**a**) *J-V* characteristics of selected representative Schottky diodes from each sample. (**b**) *J– V* mean and standard deviation of statistical data

The current density–voltage (*J–V*) characteristics of a single representative diode out of ~10 measured vertical m-plane GaN Schottky diodes on each wafer are shown in Fig. 19.15a. Statistical mean and standard deviation of ~10 measured diodes on each sample are shown in Fig. 19.15b at specific reverse and forward voltages of interest. According to the *J–V* results, the continuous reference sample (sample 1) shows the lowest reverse leakage current of ~ 4×10^{-5} A/cm^2 at −10 V among all the samples tested and a turn-on voltage of 0.7 V. A comparison of *J–V* data between the continuously grown (sample 1) and the regrown (sample 2) diodes indicates the influence of a regrowth process (without an etch) on current leakage. Sample 2 shows an ~100× larger reverse *J* compared to that of the reference (sample 1) (~3×10^{-3} A/cm^2 compared to ~4×10^{-5} A/cm^2 at −10 V). A comparison of sample 3 shows the highest reverse leakage of ~3×10^{-2} A/cm^2 at −10 V of all samples with ~1000× increase in the reverse leakage compared to the reference (sample 1). The forward turn-on voltage of sample 3, however, is similar to that of the reference (sample 1) (0.7 V), indicating that dry etch-enhanced damage has a more prominent effect on the reverse characteristics than on the forward characteristics of the Schottky diodes. An additional testing of AZ400K wet-treated ICP etched-and-regrown diodes (sample 4) shows a reduction in the reverse *J* (~1×10^{-3} A/cm^2) compared to the ICP etched-and-regrown diodes (sample 3).

Thus, the removal of ICP-etched damage regions using AZ400K shows a slight improvement in the reverse blocking voltage. Further comparison of an ICP-etched (sample 3) and subsequent PEC-etched (sample 5) diode shows further reduction of the reverse *J* (~3×10^{-4} A/cm^2). The decrease in the reverse leakage current of sample 5 is attributed to the removal of the dry etch-enhanced damage using PEC and shows that PEC was more effective at reducing the reverse leakage current than by AZ400K. The forward turn-on voltage is similar (0.7–0.8 V) for all samples. However, sample 2 exhibits a higher R_{on} than the rest of the samples. Complete ionization has not been confirmed, but the higher R_{on} may be attributed to impurity compensation which may affect R_{on}. The elevation in deep levels may also indicate

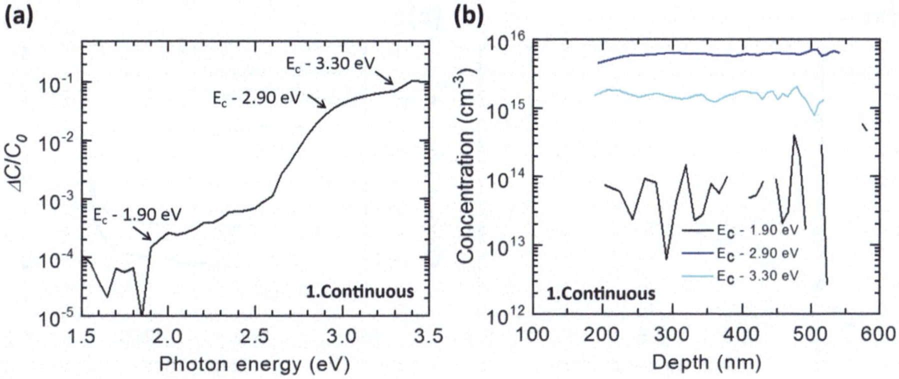

Fig. 19.16 (**a**) SSPC spectrum and (**b**) depth profile of the continuously grown Schottky diode (sample 1)

other phenomenon such as increased recombination or trapping at the regrown junction which may also lead to higher R_{on}.

SSPC was used to examine the type of deep level defects and LCV was used to measure the depth profile of defect concentrations in the regrown diodes. For the case of uniform doping, SSPC can quantify N_t. However, it was found in this study that the regrowth interface often was accompanied by a large spike in the doping. This rapidly varying doping concentration made conventional SSPC near the regrowth interfaces difficult from the standpoint of establishing Nt. Therefore, SSPC data are shown in as $\Delta C/C_0$ to focus only on defect energy thresholds. Defect concentration was measured using LCV which measures the depth profile of a defect's concentration by measuring the increase in the apparent doping concentration when the deep level defect is photoionized by continuous wave monochromatic illumination. The extent of the measured depth profiles of capacitance–voltage and LCV was determined largely by the existence of (or lack of) a UID doping spike at the regrown interface. For the regrown samples without PEC etch, the large UID doping spike at the regrowth interface restricted the depletion region and, therefore, C-V and LCV sensitivity to near the regrowth interface. For the regrown sample with the PEC etch, the UID doping spike was not evident, and this allowed the depletion region to move much deeper into the sample. For the case of multiple defect states, the measurement is repeated with successively larger photon energies chosen to optically excite defects one at a time.

The SSPC spectrum of the continuously grown diode (sample 1) is presented in Fig. 19.16a and is very similar to that of *c*-plane *n*-GaN grown on free-standing GaN [31], and similar E_o values are found in this study. Values of $E_o = 1.9$ eV, 2.9 eV, and 3.3 eV were determined for the deep level signature between 1.20 and 2.60 eV, 2.60 and 3.20 eV, and 3.30 and 3.45 eV, respectively. The origin of the $E_c - 1.90$ eV deep level is likely related to a native defect caused by atomic displacement during growth [32]. Defect levels near $E_c - 2.9$ eV in n-GaN are associated with either C or Ga vacancy, and it is likely that both defect sources are present in the film

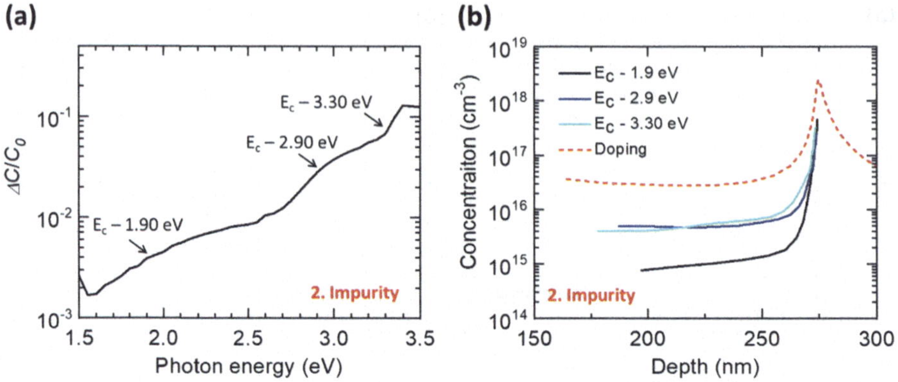

Fig. 19.17 (**a**) SSPC spectrum and (**b**) depth profile of the regrown impurity Schottky diode (sample 2)

[33, 34]. The E_c − 3.3 eV defect state is likely the same as a level at E_c − 3.28 eV that is often ascribed to C impurities [35]. Previous reports on *m*-plane n-type GaN show a level at E_c − 3.31 eV, which was also attributed to C [36].

The depth profiling of the deep levels for sample 1 using the LCV method is presented in Fig. 19.16b. Spatially uniform defect concentrations were observed for the as-grown *n*-type *m*-plane GaN, which is expected for a continuously grown film. The defect concentrations of the deep levels are similar to continuously grown *c*-plane-oriented *n*-GaN on free-standing GaN substrates, indicating that the concentration of point defects, and especially impurity-related defects, can be well controlled for epitaxial growth of *m*-plane GaN [31]. The low defect density observed in the continuously grown sample corresponds well with the observed lowest reverse leakage characteristic in the *J*–*V* results ($\sim 4 \times 10^{-5}$ A/cm^2 at −10 V) among all the samples tested in this study (Fig. 19.15a).

Spectroscopy of a regrown Schottky diode (sample 2) was also studied to investigate the role of interfacial impurities and regrowth without any dry etching on defect levels in GaN. The SSPC spectrum is shown in Fig. 19.17a with the net doping profile measured by *C*–*V* and the defect concentration profile measured by LCV in Fig. 19.17b The SSPC spectrum in Fig. 19.17a was measured with a reverse voltage, $V_r = 0$ V, corresponding to a depletion region from 0 to 200 nm below the surface, i.e., 75 nm above the regrowth interface. The fill pulse (+0.5 V) was sufficient to reset the diode capacitance to near its dark value. The SSPC spectrum of the regrown diode is qualitatively similar to that of the continuously grown diode, implying that the same type of deep level defect exists in both samples. However, the profile of these defect states in the regrown sample shows a strong depth dependence that is closely correlated to the regrowth interface. The doping profile shows a large spike at the depth of ~275 nm from the sample surface (Fig. 19.17b), corresponding to the regrowth interface. The spike in the doping profile of the regrown sample was previously correlated with Si and/or O impurity spikes buried at the regrowth interface [37]. The peak N_t for all observed deep levels exceeds 10^{17} cm^{-3} just

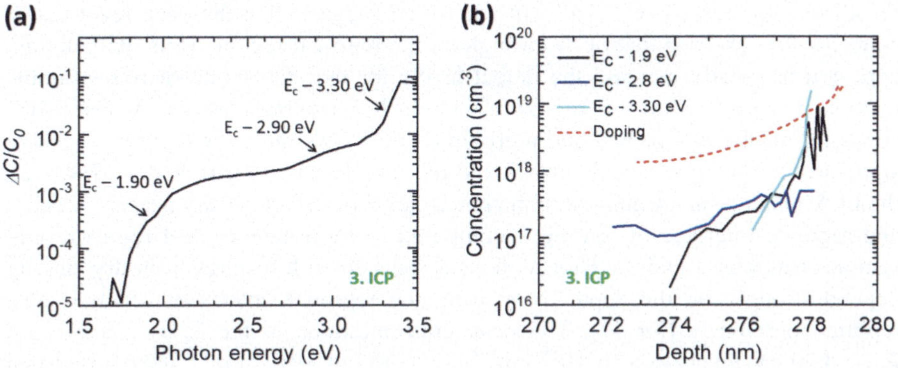

Fig. 19.18 (**a**) SPPC spectrum and (**b**) depth profile of the ICP etched-and-regrown Schottky diode (sample 3)

above the regrowth interface. However, at only ~25 nm above the interface, N_t for the E_c – 2.9 and E_c – 3.30 eV defect levels reach those of the as-grown continuously grown diode (Fig. 19.17b).

Considering prior impurity investigation of regrown m-plane GaN diodes indicating sharply peaked C concentration profiles at the regrowth interface, the observed trend here is also consistent with C-related defects being the origin of the E_c – 2.9 eV and E_c – 3.3 eV levels [37]. The large density of the Ec – 1.9 eV level (relative to the continuously grown sample) persists even over 75 nm above the regrowth interface, suggesting that elevated concentration of this defect level propagates to the upper layers during the regrowth process. A possible scenario is that the thermal desorption during the substrate ramp-up (prior to initiating the epitaxial regrowth) generates point defects at the regrowth interface, which may then diffuse to the upper layers. A comparison of *J*–*V* data (Fig. 19.15a) between the continuously grown (sample 1) and the regrown (sample 2) diodes shows an ~100× larger reverse *J* compared to that of the reference (sample 1), which is likely due to the increased concentration of the E_c – 1.9 eV deep levels. The E_c – 1.9 eV state is energetically located near mid-gap, which may give rise to efficient band-to-band tunneling and increase both forward and reverse leakages [37, 38]. Furthermore, the gradually "saturated" forward current from sample 2 after 0.5 V may also be attributed to the increased interfacial impurities. One possible mechanism for such a behavior is an increased recombination at the regrowth junction. As the impurity spikes are highly localized to the regrowth interface, it may adversely affect the performance of regrown *p*–*n* diodes, especially if the regrowth interface is located right at the metallurgic junction, where the electric field is the highest [38].

The SSPC and LCV results of the baseline etched-and-regrown diode (sample 3) are presented in Fig. 19.18. Similar to the regrown diode (sample 2), the etched-and-regrown diode (sample 3) also showed a spike in the doping at the regrowth interface. The depletion width was pinned at the regrowth interface, indicating an increase in net doping near the regrown interface. The Si doping level was the same for all the structures (~6×10^{16} cm^{-3}), which suggests that the deep level defect concentration exceeds that of Si throughout the regrown region. Thus, ICP etching appeared to greatly increase the defect levels for the etched-and-regrown sample (sample 3) compared to just regrowth without ICP etching (sample 2). The SSPC spectrum for the ICP etched-and-regrown diode shows the same deep level defect signatures as the continuously grown and regrown (without etch) diodes. However, the LCV data reveal a significantly increased level of defect density near the etched-and-regrown interface of sample 3 compared to the reference and regrown-only samples (samples 1 and 2). Thus, defects associated with features from dry etching depend strongly on the depth away from the regrowth interface, which is also reported previously [39–41]. The peak concentrations of the E_c − 1.9 eV and E_c − 3.30 eV levels exceed 10^{18} cm^{-3} compared to ~10^{17} cm^{-3} for the regrown diode.

The defect concentration near the regrown surface could not be probed by LCV due to the fully depleted regrown drift region. However, the results from the regrown diode imply that the high concentration of the E_c − 1.9 eV relative to the nominal Si doping level persists well into the region above the regrowth interface. Dry etch sub-surface damage has been previously shown to extend ~500 Å below the surface, which is attributed to the creation of nitrogen vacancy-related shallow donors [42–44]. However, due to the increased charge interface at the regrown interface, the depletion region was not able to extend into the sub-surface region below the regrown interface for analysis. Electrical testing of sample 3 shows the highest reverse leakage of all samples. The interpretation of high defect density propagating to the surface is accompanied with ~1000× increase in reverse leakage of sample 3 compared to the reference (sample 1) (Fig. 19.15a).

The effects of dry etch and regrowth on the electrical performance of diodes have also been reported on c-plane GaN [22, 30, 45]. Increased leakage current in p-n diodes grown on the *c*-plane in both the forward and reverse biases has been correlated with dry etch-enhanced defects (E_c − 1.9 eV) [14]. The origin of the E_c − 1.9 eV level is likely to be an intrinsic point defect, considering that it exists at a low concentration in a continuously grown material, has elevated concentration at a non-plasma-etched regrowth interface (where thermal desorption may occur), and an even larger concentration at a plasma-etched regrowth interface, given that atomic displacement from ion bombardment is likely to occur during any plasma-based dry etching.

The gallium interstitial (Ga_i) has been suggested as the source for the E_c − 1.9 eV deep level [31], as its density increase coincides with the emergence of point defect clustering during proton irradiation of *c*-plane *n*-GaN. Also, due to the relatively high mobility of Ga_i at room temperature [46], the point defect clustering may occur with a high probability considering Ga_i as the point defect.

The effect of a 10 min AZ400K at 80 °C treatment of the dry-etched GaN material (sample 4) was also investigated to see whether the elevated defect levels could be mitigated near the regrowth interface. AZ400K is a KOH-based solution that etches GaN very slowly [47], but it could be sufficient to remove some surface defects. The results of SSPC (Fig. 19.19a) and LCV (Fig. 19.19b) of the ICP + AZ400K are

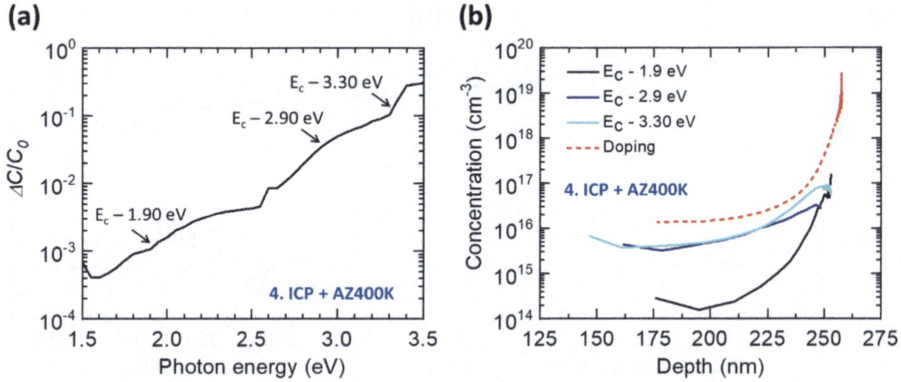

Fig. 19.19 (**a**) SSPC spectrum and (**b**) depth profile of the ICP + AZ400K etched-and-regrown (sample 4) Schottky diode

shown. The same three defect levels as those for the other samples were observed in the SSPC spectrum. The peak N_t for all deep levels was drastically reduced for the ICP + AZ400K diode (sample 4) compared to the ICP baseline etched-and-regrown diode (sample 3) and was comparable to those for the regrown diode without ICP. The tail of the elevated defects from the regrowth interface into the bulk (Fig. 19.19b) was much longer for the ICP + AZ400K compared to the regrown without an etch sample. However, N_t for all deep levels reached the values observed in the reference continuously grown diode (sample 1). This is especially notable for the E_c – 1.9 eV deep level with a high N_t for the regrown sample (sample 2) compared to the reference (sample 1). This reduction in N_t for the E_c – 1.9 eV level explains the reduced reverse leakage for the ICP + AZ400K diode (sample 4) compared to the regrown diode without an etch (sample 2) (Fig. 19.17a).

As we have observed, ICP etching is likely to generate sub-surface damage via ion bombardment which may not be readily removed with etchants with slow etch rates, such as AZ400K. While our results show that AZ400K is marginally effective in reducing defect levels associated with regrowth, additional methods for removing dry etch-enhanced damage should be investigated to reduce the defects further. PEC etching has previously been used to etch significant depths into GaN [48–51]. Therefore, it may be a suitable candidate for etching away damaged material caused by a dry etching process. For this purpose, the PEC etch rate was calibrated to approximately 2 nm/min of n-GaN (6×10^{16} cm^{-3}) using 0.01 M KOH in deionized (DI) water at room temperature with a surface flux of 100 mW/cm^2. The SSPC spectrum and depth profile of the ICP-etched (200 nm deep) plus a subsequent PEC etching (400 nm deep) diode (sample 5) are presented in Fig. 19.20. The same group of deep levels was observed in the ICP + PEC sample as in the other samples in this study (Fig. 19.20a). No doping spike at the regrowth interface (~500 nm depth) was observed (Fig. 19.20b) in the ICP + PEC sample (sample 5) using *C-V*. However, the presence of Si (~5×10^{17} cm^{-3}) and O (~6×10^{17} cm^{-3}) "spikes" with a full-width

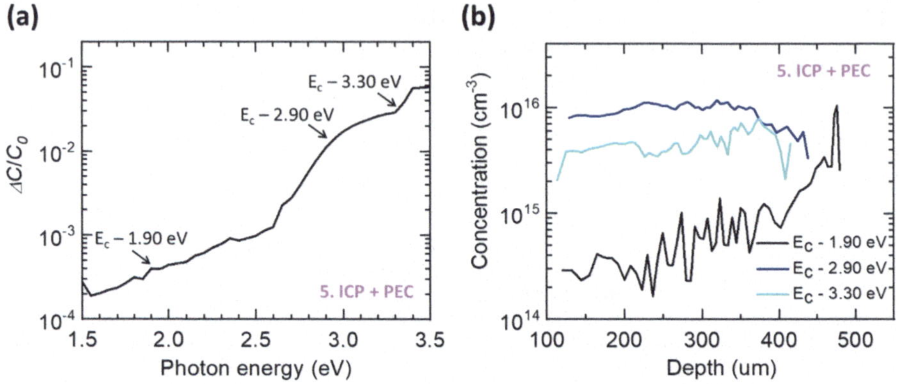

Fig. 19.20 (**a**) SSPC spectrum and (**b**) depth profile of the ICP + PEC etched-and-regrown Schottky diode (sample 5). The depth profile in (b) is taken near the regrowth interface

Fig. 19.21 SIMS depth profile showing impurities Si, O, and C of the ICP + PEC etched-and-regrown Schottky diode (sample 5)

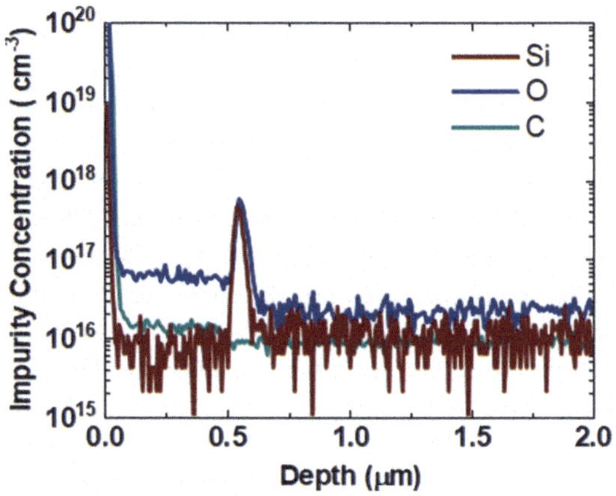

half max (FWHM) of ~33 nm and ~37 nm, respectively, is observed by secondary ion mass spectrometry (SIMS) in Fig. 19.21.

The detection limits (sensitivity) of the analysis are Si = 3×10^{15} cm^{-3}, O = 1×10^{16} cm^{-3} with a depth resolution of 12.5 Å. It is unclear whether the lack of a doping spike in *C-V* at the regrowth interface is a result of the PEC treatment or simply due to the sporadic nature of surface contamination by ambient exposure [37, 38]. Interfacial impurities are likely to increase with ambient exposure and surface treatments. However, a lack of a doping spike (i.e., charge) does not necessarily indicate a lack of interfacial impurities as seen from the comparison of SIMS and *C-V*. Therefore, the interfacial impurities are not completely ionized. A defect profile comparison of the ICP + AZ400K (Fig. 19.19b) and ICP + PEC (Fig. 19.20b) shows that the latter has a slightly lower N_t for the $E_c - 1.9$ eV level in the bulk but much lower (~10^{16} cm$^{-3)}$ N_t near the regrowth interface.

Table 19.2 Summarized deep level defect density results for all the samples investigated

Surface treatment	Above regrown interface			Near regrown interface		
	E_c- 1.9 eV (x10^{14} cm^{-3})	E_c- 2.9 eV (x10^{14} cm^{-3})	E_c- 3.3 eV (x10^{14} cm^{-3})	E_c- 1.9 eV (x10^{14} cm^{-3})	E_c- 2.9 eV (x10^{14} cm^{-3})	E_c- 3.3 eV (x10^{14} cm^{-3})
N/A	1	30	70	N/A	N/A	N/A
Ace/IPA/DI	2	20	30	2000	1600	3300
ICP (400 nm)	10	200	1000	85,600	5030	81,600
ICP (400 nm) + AZ400K	1	30	30	300	300	800
ICP (400 nm) + PEC (200 nm)	4	90	30	30	50	70

The decrease in the reverse leakage current of sample 5 (3×10^{-4} A/cm^2) as compared to the ICP-etched only (sample 3) (3×10^{-2} A/cm^2) is attributed to the removal of the dry etch-enhanced damage based on the reduction of all deep levels in sample 6 compared to sample 3. Also, the $E_c - 2.9$ eV deep level is the highest Nt in the ICP + PEC diodes. Thus, the effect of this level is unlikely to contribute to the reverse leakage, as the ICP + PEC diodes coincide with the lowest reverse J compared to all other regrown samples (samples 2–4). This result is consistent with our previous report of a high-C interfacial layer which corresponds to the $E_c - 2.9$ eV deep level showing little effect on reverse leakage.[37]

A summary of the deep level defect densities for all the samples investigated is presented in this section. Table 19.2 summarizes the deep level defect densities for all the samples in this study near the regrown interface (close to the peak defect densities in the depth profile) and away from the interface in the bulk region, where the defect density spectrum is flat away from the tail regions of the interface. The regrown sample (sample 2) did not show significant enhancement in any of the three deep levels (namely, $E_c - 1.9$ eV, $E_c - 2.9$ eV, and $E_c - 3.3$ eV) in the bulk region, while a localized enhancement in the deep levels near the regrowth interface was observed. The ICP etched-and-regrown sample, on the other hand, shows a significant enhancement of all the three deep levels in the bulk and near the regrowth interface. AZ400K (sample 4) and PEC (sample 5) treatments both significantly reduce the deep level defect densities away from the regrown interface down to the levels close to those for the reference continuously grown sample (sample 1). The main difference between samples 4 and 5 is that localized defects near the regrowth interface were largely eliminated in the PEC-treated sample (sample 5).

In addition, the ICP + PEC (sample 5) and the regrown without an etch (sample 2) samples have similar N_t for the $E_c - 1.9$ eV level in the regrown bulk, while the latter shows >10× higher leakage. Hence, the large spike in $E_c - 1.9$ eV defect N_t ($>10^{17}$ cm^{-3}) for the regrown diode (and missing for the ICP + PEC sample), is likely to be the main contributor to leakage rather than background levels $N_t \sim 10^{15}$ cm^{-3} in the regrown bulk. The ICP + PEC sample also shows a lower N_t for the $E_c - 1.9$ eV level near the regrown interface even when compared to the regrown sample without a plasma etch, which may imply that the PEC not only removes dry etch-enhanced damage, but also may provide an epitaxial surface that is less susceptible to the formation of defects during the regrowth process. PEC may also be considered to replace the dry etch process to define the selective area-regrown region as it introduces significantly reduced levels as compared to ICP etching. However, PEC etching is not able to fully eliminate dry etch-enhanced defect levels. Also, the concentration of the E_c − 2.9 eV level in the bulk is somewhat higher, while the E_c − 3.3 eV concentration is almost the same in sample 5 compared to other samples. The wet treatment methods suggested in this work (both AZ400K and PEC) enable reduced damage, while maintaining smooth epitaxial-ready surfaces.

In summary, SSPC was performed on etched-and-regrown *m*-plane *n*-type Schottky diodes. As-grown levels were identified at E_c − 1.9 eV, E_c − 2.9 eV, and E_c − 3.3 eV with concentrations below $N_t \sim 10^{16}$ cm^{-3}. The origin of the E_c − 1.9 eV level is likely to be an intrinsic point defect introduced during growth where thermal desorption may occur. Defect levels near E_c − 2.9 eV are associated with either C or Ga vacancies, and the E_c − 3.3 eV defect state is likely ascribed to carbon impurities. In addition, localized levels associated with impurities and regrowth were observed for regrowth without an etch. However, by introducing an ICP etch process before regrowth a ~100, 000× increase in the E_c − 1.9 eV level with an emergent E_c − 3.3 eV level as compared to as-grown materials is observed near the regrowth surface. Thus, atomic displacement from ion bombardment is likely to occur during any plasma-based dry etching resulting in an even larger defect concentration at a plasma-etched regrowth interface. Depth profiling shows that the effect of these plasma-enhanced defects is not localized and may propagate several hundred nanometers away from the etched interface. This is indicated by an observed tail for all defect levels. Further, *J*-*V* testing confirms the effect of these elevated levels as leakage current was ~1000× higher as compared to the as-grown material. Therefore, wet treatments were investigated to mitigate the elevated defect levels enhanced by dry etching. An AZ400K dip was introduced after ICP etching due to its ability to slowly etch GaN. The deep levels were significantly reduced near the regrown interface along with an ~100× reduction in reverse leakage current compared to the ICP-etched only sample. An observed tail of the elevated defects from the regrowth interface into the bulk was again observed indicating only a partial mitigation of the dry etch-enhanced damage. Thus, a PEC wet etch treatment after ICP-etching was investigated due to its ability to etch GaN at a higher rate than AZ400K. It was observed that the subsequent PEC treatment reduced the E_c − 1.9 eV level. Reverse leakage was also reduced by ~1000× compared to the ICP-etched only sample. No observable tail in the depth profiling indicated that the deep levels enhanced by ICP etching did not propagate away from the regrowth interface in the ICP + PEC sample. Therefore, the wet treatment methods suggested in this work (both AZ400K and PEC) enable mitigation of ICP dry etch-enhanced damage on n-type m-plane GaN, while maintaining smooth epitaxial-ready surfaces. Similar wet etching methods have been implemented on the *c*-plane and are part of ongoing investigation on etched-and-regrown *c*-plane diodes. By providing solutions to dry-etched-enhanced damage using wet treatments, the next generation of selective-doped high-power vertical electronics devices on GaN may be realized.

4 Spatially Resolved Device Characterization Using Electron Beam-Induced Current

In the previous sections, we discussed how etch-induced defects impact the performance of GaN p-n diodes. However, an open question remains: what is the mechanism by which defects impact performance? Answering this question requires approaches that can link atomistic defects to macroscopic device properties, for example, by understanding mesoscopic properties. One such property is the carrier diffusion length: it is impacted by defects and determines macroscopic device performance. While carrier diffusion length can be measured in bulk materials using various techniques, the complex topology of GaN devices and the spatially non-uniform defect distributions require a technique that can measure the carrier diffusion length with high spatial resolution, such as electron beam-induced current (EBIC).

As illustrated in Fig. 19.22, electron-hole pairs are locally generated with a focused electron beam from a scanning electron microscope, while the current across the device is measured. Local electric fields in the device separate the electron-hole pairs to generate a current; those electron-hole pairs are either generated directly in the field region or diffuse to the field region. Rastering the electron beam allows identification of these mechanisms and quantitative extraction of the diffusion length. EBIC can be performed in a top-down fashion where spatial information is obtained by varying the e-beam penetration depth by changing the e-beam energy, or in cross-sectional mode where the device is cleaved.

We performed top-down and cross-sectional EBIC on four different devices [52]. GaN epilayers were grown by MOCVD at 200 torr using trimethylgallium, ammonia, hydrogen, and nitrogen on commercially available, HVPE n-type GaN substrates with carrier concentrations $>10^{18}$ cm^{-3}. Device layers consisted of a 10-μm-thick, *n*-type Si-doped drift layer ([Si] ~ 3×10^{16} cm^{-3}) and a ~0.4-μm-thick, p-type GaN layer ([Mg] ~ 3×10^{19} cm^{-3}). Silane diluted in nitrogen was used for the n-type dopant and Bis(cyclopentadienyl)magnesium in solution was used as the p-type dopant. N-type (p-type) layers were grown at a pyrometer temperature of 1050 °C (980 °C).

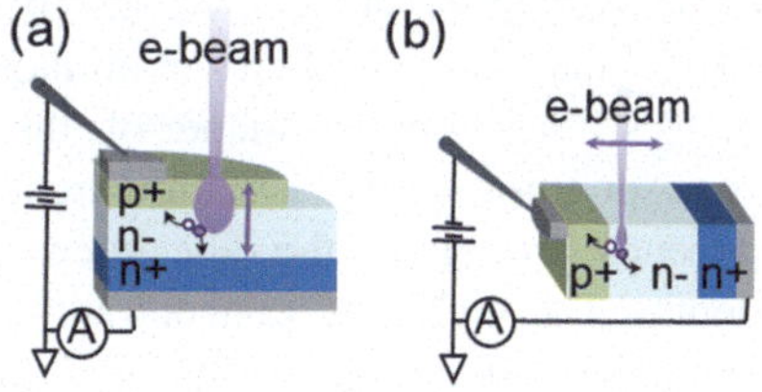

Fig. 19.22 Illustration of (**a**) top-down and (**b**) cross-sectional EBIC measurements

Three different device types were used. A continuously grown diode without dry etching and a 358 nm p-layer was used for a reference, and two types of etched-and-regrown samples were studied. The etch process consisted of $Cl_2/BCl_3/Ar$ ICP with 125 W for the ICP power and 10 W for the radio frequency (RF) power. The drift layer of one device (Regrown 1) was dry-etched to remove 0.4 μm of material prior to the regrowth of 474 nm of p-type GaN. Two additional devices (Regrown 2 and 3) were etched-and-regrown (328 nm of p-type GaN) using the same process, except that prior to regrowth a wet chemical treatment was used. The etched drift layer was treated with AZ400K (2% KOH and 5–15% potassium borates in water by weight) at a temperature of 80 °C for 10 min (AZ400K) followed by a deionized water rinse and a nitrogen dry prior to regrowth. Pd/Au (20 nm/30 nm) p-type contacts were formed using e-beam deposition followed by a rapid thermal anneal. A trench was etched around each device using ICP to electrically isolate adjacent devices, and a blanket, backside, n-type contact metal of Ti/Al (5 nm/200 nm) was deposited.

The cross-sectional EBIC measurements were performed with a 5 keV accelerating voltage with the diodes mounted on a 70° tilt stage. For depth-dependent EBIC measurements, diodes were mounted with top contact planes perpendicular to the e-beam, and currents generated in the mesa region outside of the contact area and away from the cleave were evaluated as a function of e-beam-accelerating voltage. The specimen current at each accelerating voltage was measured using a Faraday cup.

Figure 19.23a shows cross-sectional SEM and EBIC images from the four devices. The SEM contrast allows us to delineate the various device regions, including the p-type layer, the n-type layer, and the top contact. This detailed identification of the interfaces provides correlation with the EBIC images, which show a bright horizontal line at the p-n junction. Thus, the EBIC magnitude is largest where an internal field is present; however, as Fig. 19.23b shows, substantial EBIC is observed up to a micron away from the junction on the n-type side, and up to 250 nm on the p-type side. All four devices behave similarly on the p-type side, but significant differences exist on the n-type side, with the regrown devices showing a narrower EBIC profile compared to the continuously grown diode. At a position x far from the junction, the EBIC is proportional to exp.($-x/L_D$) with L_D the carrier diffusion length. Fits to the data in Fig. 19.23b give electron diffusion lengths around 50 nm in the p-type region for the four devices. On the n-type side, the hole diffusion length is on the order of 340 nm for the continuously grown diode; for the regrown diodes a value of 350 nm is found for diode 1, with diode 2 showing a reduction to about 230 nm, while diode 3 has a much lower value of 120 nm.

To further explore this trend in the diffusion lengths and reduce uncertainties due to cleaving of the devices, we also performed top-down EBIC. Figure 19.23b shows EBIC images acquired on a continuously grown diode and a regrown diode. In these images, the circular shape of the diodes is visible, including an outer ring corresponding to the isolation trench. The EBIC from the center of the devices depends strongly on the e-beam energy as a consequence of the varying penetration depths. For GaN, the concentration of electron-hole pairs generated at distance z from the junction is given by $h(z) = (3.207/R) \exp(-A(z/R - 0.11)^2)$, with

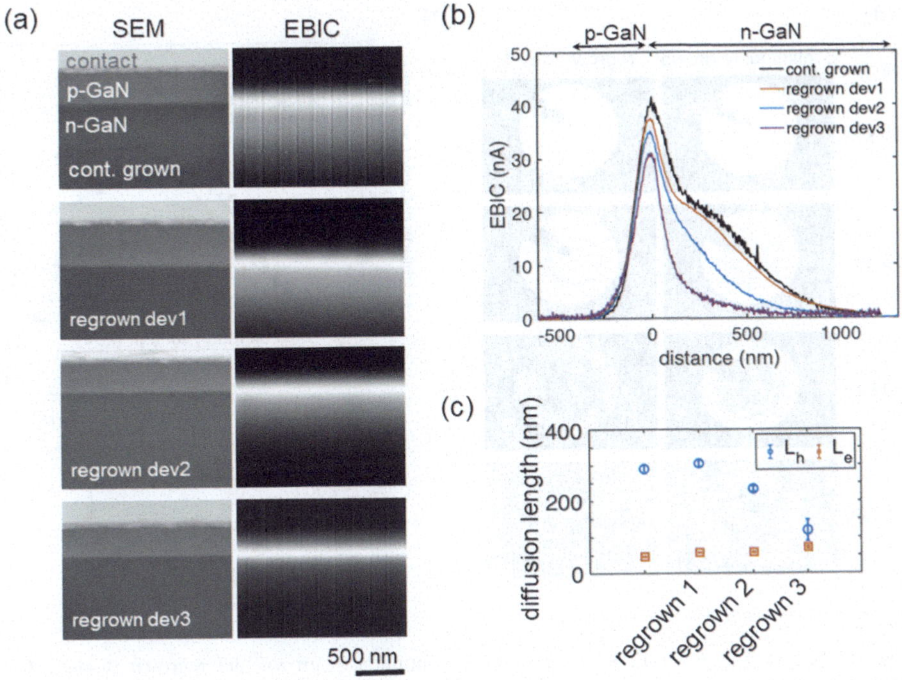

Fig. 19.23 (**a**) Cross-sectional SEM and EBIC images for different p-n diodes. (**b**) Vertical linecuts through the EBIC images for each device

$R(\mu m) = 0.0132 E_b(\text{keV})^{1.75}$, $A = 42.8$ for $z < 0.11R$, and $A = 16.5$ for $z \geq 0.11$ [53]. Thus, as the e-beam energy E_b is increased, the centroid of the e-h plume located at R is deeper in the material.

We define the charge collection efficiencies [54] as $\eta = I_{\text{EBIC}}/(I_s E_b f/E_i)$, where I_{EBIC} is the measured EBIC current, I_S is the current of e-beam electrons impinging on the device, f is the fraction of the e-beam energy that gets absorbed by the material, and E_i is the average energy required for the generation of one e-h pair. For the specific case of GaN, $f/E_i = 0.08\ \text{eV}^{-1}$ [53]. The measured charge collection efficiency $\eta = \eta_p + \eta_{\text{dep}} + \eta_n$ is the sum of the charge collection efficiencies within the depletion region, η_{dep}, and in the p-type and n-type layers outside of the depletion region, η_p and η_n.

In the depletion region where an electric field is present, charge carriers are assumed to be collected with 100% efficiency; charge carriers outside the depletion region must diffuse to the depletion region to be collected. With the p-n junction located at $z = 0$ and no surface recombination the charge collection efficiencies are $\eta_p = \int_{w_p}^{H} h(z) \exp\left(\frac{z - w_p}{L_e}\right) dz$, $\eta_{dep} = \int_{-w_n}^{w_p} h(z) dz$, and $\eta_n = \int_{-w_n} h(z) \exp\left(-\frac{z + w_n}{L_h}\right) dz$ [54]. Here H is the distance between the p-n junction and the p-type contact, w_n

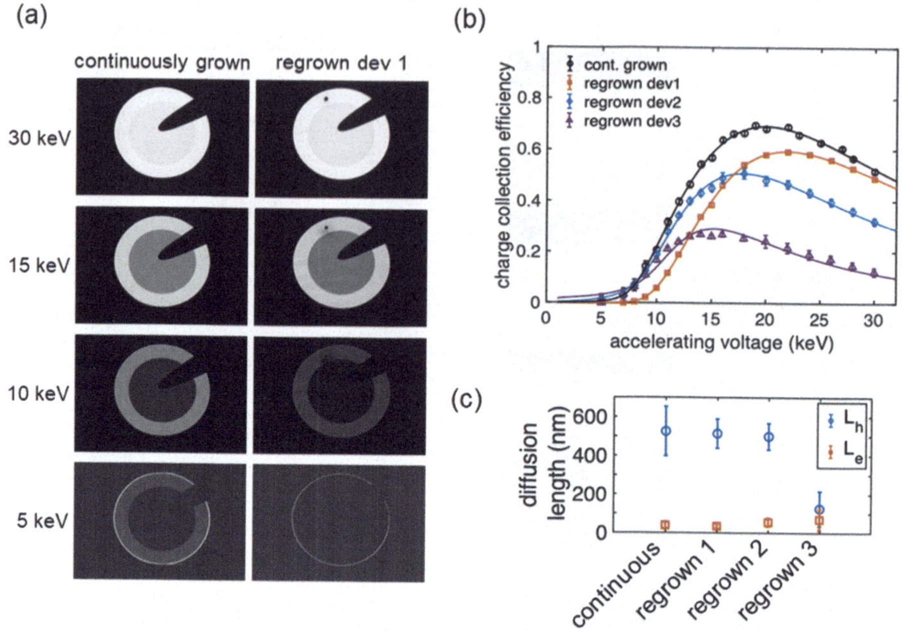

Fig. 19.24 (**a**) Top-down EBIC images for a continuously grown and regrown diodes. (**b**) Dependence of the EBIC on the e-beam-accelerating voltage. (**c**) Extracted diffusion lengths for holes (h) and electrons (e)

and w_p are the depletion widths on the n- and p- sides, L_n is the hole diffusion length in the n-type material, and L_e is the electron diffusion length in the p-type material.

With these expressions, we can now analyze the data of Fig. 19.24a by looking at the charge collection efficiency as a function of the e-beam energy when the e-beam is located at the center of the circular devices. Figure 19.24b shows that all four devices exhibit the same general behavior where the charge collection efficiency increases as the e-beam energy increases from 5 keV to 15–20 keV, and then decreases as E_b is further increased. This trend can be qualitatively understood from the penetration depth of the e-beam in GaN: at 5 keV, $R = 220$ nm, and most of the e-h pairs are generated in the p-type region, while at 8 keV, $R = 0.5$ μm and the e-h pairs are generated near the p-n junction. Further increase of the e-beam energy locates the plume in the n-type region. (Note that the top metal contact reduces these penetration depths somewhat.)

To extract quantitative values for the diffusion lengths, we fit the experimental data to the above expressions for the charge collection efficiency. The solid lines in Fig. 19.24b show excellent agreement with the experimental data, allowing us to extract values for L_h in the range 40–80 nm across devices, while L_p varied from a high of 500 nm for the continuously grown diode down to 100 nm for regrown device 3. The values for the diffusion in the p-type material are in good agreement between the cross-sectional and top-down EBIC approaches. While the values for diffusion in the n-type side vary by a factor of two between the two techniques, the general trend among the four devices is the same.

The above experiments and analysis suggest that two different EBIC approaches for assessing carrier transport in continuously grown and etched-and-regrown GaN p-n diodes can provide similar insight into material properties. The top-down approach is non-destructive and easier to implement and may thus provide a more routine methodology to assess device properties. In terms of the devices themselves, the electron diffusion length in the p-layer was found to be similar in continuous and regrown diodes. However, the hole diffusion length in the n-layer near the interface was degraded in the regrown samples regardless of whether a wet etch treatment was performed after ICP etching, indicating that wet etching may correct surface damage but not impact sub-surface etch damage. Remarkably, we observed an etched-and-regrown diode with properties similar to a continuously grown diode, indicating that achieving high-quality etched-and-regrown materials and interfaces is possible but that it requires careful control and/or preservation of material quality even in the n-layer.

5 Concluding Remarks

This chapter describes the characterization of deep levels enhanced by ICP etching, their impact on electrical properties, and performance of p-n diodes and their mitigation by various post-etch treatments. ICP etching of n-GaN drift regions within regrown p-n diodes oriented on the c-plane induced a significant increase in reverse leakage. Photocapacitance studies correlated this increased leakage with increased concentration of a deep level defect at $E_c - 1.9$ eV. The $E_c - 1.9$ eV deep level was observed in p-n diodes without ICP etching. Taken together, these results suggest that the $E_c - 1.90$ eV deep level is an intrinsic defect, possible Gai, whose concentration increases with ICP etching and whose deep level in the band gap provides a mechanism for reverse leakage. It was observed by depth-resolved deep level defect profiling that defectivity was highest at the etched surface, which subsequently becomes the metallurgic junction in regrown p-n diodes, and elevated defectivity extends both above and below the metallurgic junction. Similar trends were observed for m-plane-oriented p-n diodes, which suggests that ICP etching causes similar defects on basal and non-basal planes with similar effect on device performance. Removal of ICP-damaged material in the n-GaN layer prior to regrowth greatly reduced reverse leakage, but etched-and-regrown diodes still did not perform to the same level as continuously grown p-n diodes. A slow RIE etch was found to remove defective material after ICP etching of c-plane diodes, and 1.6 kV etched-and-regrown p-n diodes were realized. For m-plane devices, PEC was also observed to remove damage in ICP-etched n-GaN and greatly reduce leakage in subsequently regrown diodes. EBIC characterization of etched-and-regrown c-plane diodes was performed to understand the impact of ICP damage on minority carriers' diffusion length on both sides of the p-n junction. Top-down EBIC, which non-destructive but only indirectly measures diffusion length, and cross-sectional EBIC, which is destructive but directly measures diffusion length, were described. It was found that electron diffusion length on the p-side was not impacted by ICP etching due to the high concentration of Mg dopants. Conversely, the hole diffusion length was reduced for etched-and-regrown p-n diodes, likely due to the etch-enhanced defects observed by photocapacitance. Interestingly, one etched-and-regrown p-n diode exhibited nearly the same hole diffusion length in the ICP-etched n-GaN region. This was not a typically repeatable result; however, it does demonstrate that recovery of the electrical properties of the ICP-etched n-GaN region is possible but requires conditioning and treatments that are potentially subtle and challenging to standardize.

Acknowledgments This work was supported in part by the U.S. Department of Energy Advanced Research Projects Agency (ARPA-E) PNDIODES Program.

Sandia National Laboratories is a multi-mission laboratory managed and operated by National Technology & Engineering Solutions of Sandia, LLC (NTESS), a wholly owned subsidiary of Honeywell International Inc., for the U.S. Department of Energy's National Nuclear Security Administration (DOE/NNSA) under contract DE-NA0003525. This written work is authored by an employee of NTESS. The employee, not NTESS, owns the right, title, and interest in and to the written work and is responsible for its contents. Any subjective views or opinions that might be expressed in the written work do not necessarily represent the views of the U.S. Government. The publisher acknowledges that the U.S. Government retains a non-exclusive, paid-up, irrevocable, worldwide license to publish or reproduce the published form of this written work or allow others to do so, for U.S. Government purposes. The DOE will provide public access to results of federally sponsored research in accordance with the DOE Public Access Plan.

References

1. Z. Hu, K. Nomoto, B. Song, M. Zhu, M. Qi, M. Pan, X. Gao, V. Protasenko, D. Jena, H.G. Xing, Near unity ideality factor and Shockley-Read-Hall lifetime in GaN-on-GaN p-n diodes with avalanche breakdown. Appl. Phys. Lett. **107**(24), 243501 (2015)
2. I.C. Kizilyalli, A.P. Edwards, O. Aktas, T. Prunty, D. Bour, Vertical power p-n diodes based on bulk GaN. IEEE Trans. Electron Devices **62**(2), 414–422 (2015)
3. I.C. Kizilyalli, A.P. Edwards, H. Nie, D. Bour, T. Prunty, D. Disney, 3.7 kV Vertical GaN PN Diodes. IEEE Electron Device Lett. **35**(2), 247–249 (2014)
4. H. Ohta, N. Kaneda, F. Horikiri, Y. Narita, T. Yoshida, T. Mishima, T. Nakamura, Vertical GaN p-n junction diodes with high breakdown voltages over 4 kV. IEEE Electron Device Lett. **36**(11), 1180–1182 (2015)
5. H. Ohta, K. Hayashi, F. Horikiri, M. Yoshino, T. Nakamura, T. Mishima, 5.0 kV breakdown-voltage vertical GaN p–n junction diodes. Jpn. J. Appl. Phys. **57**(4S), 04FG09 (2018)
6. Z. Hu, K. Nomoto, M. Qi, W. Li, M. Zhu, X. Gao, D. Jena, H.G. Xing, 1.1-kV vertical GaN p-n diodes with p-GaN regrown by molecular beam epitaxy. IEEE Electron Device Lett. **38**(8), 1071–1074 (2017)
7. A.M. Armstrong, A.J. Fischer, J.R. Dickerson, M.W. Moseley, M.H. Crawford, M.P. King, A.A. Allerman, R.J. Kaplar, M.S. van Heukelom, J.J. Wierer, High voltage and high current density vertical GaN power diodes. Electron. Lett. **52**(13), 1170–1171 (2016)
8. L. Yates, B.P. Gunning, M.H. Crawford, J. Steinfeldt, M.L. Smith, V.M. Abate, J.R. Dickerson, A.M. Armstrong, A. Binder, A.A. Allerman, R.J. Kaplar, Demonstration of >6.0-kV

breakdown voltage in large area vertical GaN p-n diodes with step-etched junction termination extensions. IEEE Trans. Electron Devices **69**(4), 1931–1937 (2022)

9. J.L. Hudgins, G.S. Simin, E. Santi, M.A. Khan, An assessment of wide bandgap semiconductors for power devices. IEEE Trans. Power Electron. **18**(3), 907–914 (2003)
10. M.J. Tadjer, B.N. Feigelson, J.D. Greenlee, J.A. Freitas, T.J. Anderson, J.K. Hite, L. Ruppalt, C.R. Eddy, K.D. Hobart, F.J. Kub, Selective p-type doping of GaN: Si by Mg ion implantation and multicycle rapid thermal annealing. ECS J. Solid State Sci. Technol. **5**(2), P124–P127 (2016)
11. S. Kotzea, A. Debald, M. Heuken, H. Kalisch, A. Vescan, Demonstration of a GaN-based vertical-channel JFET fabricated by selective-area regrowth. IEEE Trans. Electron Devices **65**(12), 5329–5336 (2018)
12. A.M. Armstrong, A.A. Allerman, G.W. Pickrell, M.H. Crawford, C.E. Glaser, T. Smith, Etched-and-regrown GaN pn-diodes with 1600 V blocking voltage. IEEE J. Electron Devices Soc. **9**, 318–323 (2021)
13. Y. Xu, V.G.T. Vangipuram, V. Talesara, J. Cheng, Y. Zhang, T. Hashimoto, E. Letts, D. Key, H. Zhao, W. Lu, 7.86 kV GaN-on-GaN PN power diode with BaTiO3 for electrical field management. Appl. Phys. Lett. **123**(14), 142105 (2023)
14. G.W. Pickrell, A.M. Armstrong, A.A. Allerman, M.H. Crawford, C.E. Glaser, J. Kempisty, V.M. Abate, Investigation of dry-etch-induced defects in >600 V regrown, vertical, GaN, p-n diodes using deep-level optical spectroscopy. J. Appl. Phys. **126**(14), 145703 (2019)
15. A. Chantre, G. Vincent, D. Bois, Deep-level optical spectroscopy in GaAs. Phys. Rev. B **23**(10), 5335–5359 (1981)
16. P. Blood, J.W. Orton, The electrical characterization of semiconductors: Majority carriers and electron states, in *The Electrical Characterization of Semiconductors: Majority Carriers and Electron States*, (Academic Press, San Diego, CA, 1992), p. 375
17. A. Armstrong, A.R. Arehart, S.A. Ringel, A method to determine deep level profiles in highly compensated, wide band gap semiconductors. J. Appl. Phys. **97**(8), 083529 (2005)
18. G.W. Pickrell, A.M. Armstrong, A.A. Allerman, M.H. Crawford, K.C. Cross, C.E. Glaser, V.M. Abate, Regrown vertical GaN p–n diodes with low reverse leakage current. J. Electron. Mater. **48**(5), 3311–3316 (2019)
19. H. Fujikura, K. Hayashi, F. Horikiri, Y. Narita, T. Konno, T. Yoshida, H. Ohta, T. Mishima, Elimination of macrostep-induced current flow nonuniformity in vertical GaN PN diode using carbon-free drift layer grown by hydride vapor phase epitaxy. Appl. Phys. Express **11**(4), 045502 (2018)
20. A. Armstrong, A.R. Arehart, D. Green, U.K. Mishra, J.S. Speck, S.A. Ringel, Impact of deep levels on the electrical conductivity and luminescence of gallium nitride codoped with carbon and silicon. J. Appl. Phys. **98**(5), 053704 (2005)
21. B.A. Klein, E.A. Douglas, A.M. Armstrong, A.A. Allerman, V.M. Abate, T.R. Fortune, A.G. Baca, Enhancement-mode Al0.85Ga0.15N/Al0.7Ga0.3N high electron mobility transistor with fluorine treatment. Appl. Phys. Lett. **114**(11), 112104 (2019)
22. K. Fu, H. Fu, X. Huang, H. Chen, T.-H. Yang, J. Montes, C. Yang, J. Zhou, Y. Zhao, Demonstration of 1.27 kV etch-then-regrow GaN p-n junctions with low leakage for GaN power electronics. IEEE Electron Device Lett. **40**(11), 1728–1731 (2019)
23. L. Lin, J.H. Zhao, Simulation and experimental study of 3-step junction termination extension for high-voltage 4H-SiC gate turn-off thyristors. Solid State Electron. **86**, 36–40 (2013)
24. F. Bernardini, V. Fiorentini, D. Vanderbilt, Spontaneous polarization and piezoelectric constants of III-V nitrides. Phys. Rev. B **56**(16), R10024–R10027 (1997)
25. S.C. Cruz, S. Keller, T.E. Mates, U.K. Mishra, S.P. DenBaars, Crystallographic orientation dependence of dopant and impurity incorporation in GaN films grown by metalorganic chemical vapor deposition. J. Cryst. Growth **311**(15), 3817–3823 (2009)
26. J.C. Zolper, R.J. Shul, A.G. Baca, R.G. Wilson, S.J. Pearton, R.A. Stall, Ion-implanted GaN junction field effect transistor. Appl. Phys. Lett. **68**(16), 2273–2275 (1996)

27. S.C. Binari, W. Kruppa, H.B. Dietrich, G. Kelner, A.E. Wickenden, J.A. Freitas Jr., Fabrication and characterization of GaN FETs. Solid State Electron. **41**(10), 1549–1554 (1997)
28. S. Chowdhury, B.L. Swenson, U.K. Mishra, Enhancement and depletion mode AlGaN/GaN CAVET with Mg-ion-implanted GaN as current blocking layer. IEEE Electron Device Lett. **29**(6), 543–545 (2008)
29. R.M. Farrell, D.A. Haeger, X. Chen, C.S. Gallinat, R.W. Davis, M. Cornish, K. Fujito, S. Keller, S.P. DenBaars, S. Nakamura, J.S. Speck, Origin of pyramidal hillocks on GaN thin films grown on free-standing m-plane GaN substrates. Appl. Phys. Lett. **96**(23), 231907 (2010)
30. K. Fu, H. Fu, H. Liu, S.R. Alugubelli, T.-H. Yang, X. Huang, H. Chen, I. Baranowski, J. Montes, F.A. Ponce, Y. Zhao, Investigation of GaN-on-GaN vertical p - n diode with regrown p -GaN by metalorganic chemical vapor deposition. Appl. Phys. Lett. **113**(23), 233502 (2018)
31. K.C. Collins, A.M. Armstrong, A.A. Allerman, G. Vizkelethy, S.B.V. Deusen, F. Léonard, A.A. Talin, Proton irradiation effects on minority carrier diffusion length and defect introduction in homoepitaxial and heteroepitaxial n-GaN. J. Appl. Phys. **122**(23), 235705 (2017)
32. S. Limpijumnong, C.G. Van de Walle, Diffusivity of native defects in GaN. Phys. Rev. B **69**(3), 035207 (2004)
33. T. Ogino, M. Aoki, Mechanism of yellow luminescence in GaN. Jpn. J. Appl. Phys. **19**(12), 2395 (1980)
34. A. Hierro, S.A. Ringel, M. Hansen, J.S. Speck, U.K. Mishra, S.P. DenBaars, Hydrogen passivation of deep levels in n–GaN. Appl. Phys. Lett. **77**(10), 1499–1501 (2000)
35. A. Armstrong, C. Poblenz, D.S. Green, U.K. Mishra, J.S. Speck, S.A. Ringel, Impact of substrate temperature on the incorporation of carbon-related defects and mechanism for semi-insulating behavior in GaN grown by molecular beam epitaxy. Appl. Phys. Lett. **88**(8), 082114 (2006)
36. T.A. Henry, A. Armstrong, K.M. Kelchner, S. Nakamura, S.P. DenBaars, J.S. Speck, Assessment of deep level defects in m-plane GaN grown by metalorganic chemical vapor deposition. Appl. Phys. Lett. **100**(8), 082103 (2012)
37. A. Aragon, M. Monavarian, I. Stricklin, G. Pickrell, M.H. Crawford, A.A. Allerman, A.M. Armstrong, D. Feezell, Interfacial impurities and their electronic signatures in high-voltage regrown nonpolar m-plane GaN vertical p–n diodes. Phys. Status Solidi A **217**(7), 1900757 (2019)
38. M. Monavarian, G. Pickrell, A.A. Aragon, I. Stricklin, M.H. Crawford, A.A. Allerman, K.C. Celio, F. Léonard, A.A. Talin, A.M. Armstrong, D. Feezell, High-voltage regrown nonpolar m-plane vertical p-n diodes: A step toward future selective-area-doped power switches. IEEE Electron Device Lett. **40**, 387–390 (2019)
39. J.G. Nedy, N.G. Young, K.M. Kelchner, Y. Hu, R.M. Farrell, S. Nakamura, S.P. DenBaars, C. Weisbuch, J.S. Speck, Low damage dry etch for III-nitride light emitters. Semicond. Sci. Technol. **30**(8), 085019 (2015)
40. C.-Y. Fang, W.-J. Huang, E.Y. Chang, C.-F. Lin, M.-S. Feng, Etching damages on AlGaN, GaN and InGaN caused by hybrid inductively coupled plasma etch and photoenhanced chemical wet etch by Schottky contact characterizations. Jpn. J. Appl. Phys. **42**(7R), 4207 (2003)
41. X.A. Cao, S.J. Pearton, G.T. Dang, A.P. Zhang, F. Ren, J.M.V. Hove, GaN n- and p-type Schottky diodes: Effect of dry etch damage. IEEE Trans. Electron Devices **47**(7), 1320–1324 (2000)
42. X.A. Cao, H. Cho, S.J. Pearton, G.T. Dang, A.P. Zhang, F. Ren, R.J. Shul, L. Zhang, R. Hickman, J.M. Van Hove, Depth and thermal stability of dry etch damage in GaN Schottky diodes. Appl. Phys. Lett. **75**(2), 232–234 (1999)
43. K.J. Choi, H.W. Jang, J.-L. Lee, Observation of inductively coupled-plasma-induced damage on n-type GaN using deep-level transient spectroscopy. Appl. Phys. Lett. **82**(8), 1233–1235 (2003)

44. X.A. Cao, A.P. Zhang, G.T. Dang, F. Ren, S.J. Pearton, R.J. Shul, L. Zhang, Schottky diode measurements of dry etch damage in n- and p-type GaN. J. Vac. Sci. Technol. A **18**(4), 1144–1148 (2000)
45. K. Fu, J. Zhou, X. Deng, X. Qi, D.J. Smith, S.M. Goodnick, Y. Zhao, H. Fu, X. Huang, T.H. Yang, C.Y. Cheng, P.R. Peri, H. Chen, J. Montes, C. Yang, Reverse leakage analysis for as-grown and regrown vertical GaN-on-GaN Schottky barrier diodes. IEEE J. Electron Devices Soc. **8**(8949530), 74–83 (2020)
46. K.H. Chow, G.D. Watkins, A. Usui, M. Mizuta, Detection of interstitial Ga in GaN. Phys. Rev. Lett. **85**(13), 2761–2764 (2000)
47. D. Zhuang, J.H. Edgar, Wet etching of GaN, AlN, and SiC: A review. Mater. Sci. Eng. R. Rep. **48**(1), 1–46 (2005)
48. V. Meyers, E. Rocco, K. Hogan, S. Tozier, B. McEwen, I. Mahaboob, F. Shahedipour-Sandvik, Removal of dry-etch-induced surface layer damage from p-GaN by photoelectrochemical etching. J. Electron. Mater. **49**(6), 3481–3489 (2020)
49. M.S. Minsky, M. White, E.L. Hu, Room-temperature photoenhanced wet etching of GaN. Appl. Phys. Lett. **68**(11), 1531–1533 (1996)
50. C.O. Holder, J.T. Leonard, R.M. Farrell, D.A. Cohen, B. Yonkee, J.S. Speck, S.P. DenBaars, S. Nakamura, D.F. Feezell, Nonpolar III-nitride vertical-cavity surface emitting lasers with a polarization ratio of 100% fabricated using photoelectrochemical etching. Appl. Phys. Lett. **105**(3), 031111 (2014)
51. Y. Gao, M.D. Craven, J.S. Speck, S.P. Den Baars, E.L. Hu, Dislocation- and crystallographic-dependent photoelectrochemical wet etching of gallium nitride. Appl. Phys. Lett. **84**(17), 3322–3324 (2004)
52. K.C. Celio, A.M. Armstrong, A.A. Talin, A.A. Allerman, M.H. Crawford, G.W. Pickrell, F. Léonard, Carrier diffusion lengths in continuously grown and etched-and-regrown GaN pin diodes. IEEE Electron Device Lett. **42**(7), 1041–1044 (2021)
53. E.B. Yakimov, What is the real value of diffusion length in GaN? J. Alloys Compd. **627**, 344–351 (2015)
54. M. Kittler, K.-W. Schröder, Determination of semiconductor parameters and of the vertical structure of devices by numerical analysis of energy-dependent EBIC measurements. Phys. Status Solidi A **77**(1), 139–151 (1983)

Chapter 20
Ion Beam Analysis of GaN Surfaces and Interfaces

Joshua J. P. Cooper, Sam Frisone, Jiaheng He, Guanjie Cheng, Zhirong Zhang, and Rachel S. Goldman

1 Introduction

Wide-bandgap semiconductors such as GaN are promising for high-power electronics due to higher breakdown voltages and lower on-resistances compared to conventional Si-based devices [1, 2]. Vertical GaN p-i-n devices are expected to provide improved thermal management and reduced leakage current in comparison with their lateral counterparts [3, 4]. One of the main challenges in realizing vertical GaN p-i-n devices is the development of selective-area p-type doping techniques for lateral junctions. Despite its relatively high ionization energy of 145–245 meV [5, 6], Mg is the most effective p-type dopant for GaN to date [7, 8]. Strategies for selective-area Mg doping have included masked ion implantation and/or dopant diffusion; however, these approaches are limited to shallow junction formation and require high temperature annealing. A promising alternative involves a combination of selective-area etching followed by selective-area regrowth. For all selective p-type doping approaches, a detailed understanding of the influence of processing on the structure and properties of GaN surfaces and interfaces intended for use in vertical GaN p-i-n devices is needed.

To date, ion beam analysis techniques, including channeling Rutherford backscattering spectrometry (RBS/c) and elastic recoil detection analysis (ERDA) have been used to quantitatively determine composition, uniformity, impurity, and elemental depth profiles of major, minor, and trace elements in group III-V nitrides [9]. For example, RBS/c studies of heteroepitaxial GaN on sapphire [10–16], bulk GaN [17, 18], and implanted GaN [17, 19] have revealed χ_{min} values ranging from

J. J. P. Cooper · S. Frisone · J. He · G. Cheng · R. S. Goldman (✉)
University of Michigan, Ann Arbor, MI, USA
e-mail: rsgold@umich.edu

Z. Zhang
University of California Berkeley, Berkeley, CA, USA

I. C. Kizilyalli et al. (eds.), *Gallium Nitride and Related Materials*, The Materials Research Society Series, https://doi.org/10.1007/978-3-031-83056-3_20

1.2% to 4%; however, RBS/c studies of homoepitaxial GaN have not been reported. ERDA studies to date are limited to Mg-implanted GaN, revealing a Mg-induced increase in sub-surface [H] [17].

Therefore, in this chapter, we present new understanding of relationships between structure, chemistry, and electronic states at GaN surfaces and interfaces intended for use in vertical GaN p-i-n devices. In particular, we present investigations of depth-dependent crystallinity, atomic displacements, impurity incorporation, and electronic states using ion beam analysis techniques in conjunction with chemical and electronic characterization methods. First, we describe the ion beam analysis and chemical and electronic characterization methods, including Rutherford backscattering spectrometry (RBS), ERDA, and cathodoluminescence (CL) spectroscopy. Next, we describe investigations of surface treatments for selective-area regrowth of GaN substrates and regrown epitaxial GaN layers. The subsequent sections of the chapter focus on p-doped GaN layers prepared by Mg implantation and Mg/N co-implantation plus gyrotron annealing and electric field-assisted Mg diffusion. In the final section of the chapter, we compare and contrast the GaN surfaces and interfaces prepared by epitaxy/etching/regrowth, Mg ion implantation plus gyrotron annealing, and electric field-assisted Mg diffusion, suggesting promising avenues for future investigations.

2 Methods

In this section, we describe the methods used to examine the depth-dependent crystallinity, atomic displacements, impurity incorporation, and electronic states. First, we describe several RBS-based approaches including 2D planar channeling maps, as well as profiles of angular yields, channeling yields, and random yields, which are used together to examine depth-dependent crystallinity and atomic displacements. In addition, we describe ERDA, which is used to determine the depth-profile of [H]. Both RBS/c and ERDA were conducted using a 1.7 MV National Electrostatic Corporation tandem ion accelerator equipped with a fully-automated five-axis goniometer, with key details provided in Table 20.1, as described by Cooper et al. [20]. We also describe CL spectroscopy used to probe the depth-dependent electronic states associated with both Mg dopants and unintentional impurities at GaN surfaces and regrown interfaces. The CL spectroscopy was performed in a TESCAN RISE scanning electron microscope (SEM) equipped with a Gatan MonoCL4 detector [21, 22].

2.1 Channeling Rutherford Backscattering Spectrometry

Ion channeling is the focusing effect that occurs when a well-collimated beam of charged particles impinges upon a crystal in a direction of high atomic symmetry.

Table 20.1 Tabulated parameters for Rutherford Backscattering Spectrometry (RBS) and Elastic Recoil Detection Analysis (ERDA) measurements including incident particle, incident ion beam energy, incident ion beam angle (relative to surface normal), detector angle (relative to incident beam direction), and detected species

Experiment	Incident ion	Incident ion beam energy (MeV)	Incident beam angle (°)	Detector angle (°)	Detected species
RBS	H^+	1.5	0	160	Ga, N
RBS	He^{2+}	2.5	70	170	Ga, N
ERDA	He^{2+}	2.5	70	30	H

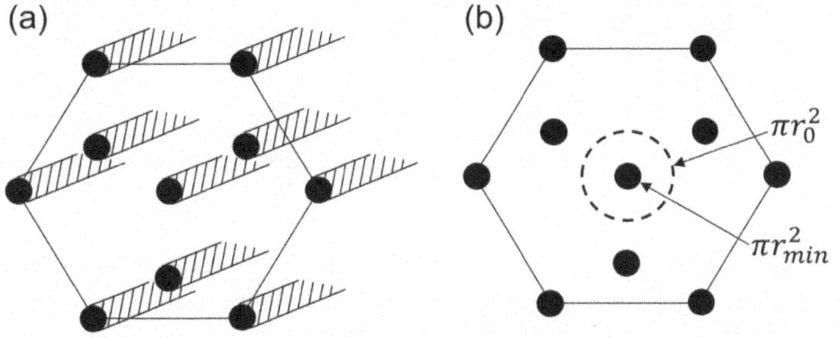

Fig. 20.1 (**a**) Schematic illustration of primitive cell for hexagonal close-packed crystal, with black circles representing atoms, hashed regions representing a column of atoms with radius r_{min}, and the remaining white spaces representing ion channels with radius r_0. (**b**) Within πr^2_0 of each column of atoms, particles channel only for $r > r_{\text{min}}$. Thus, the fraction channeled is $\frac{r_0^2 - r_{\text{min}}^2}{r_0^2}$

As illustrated in Fig. 20.1a, when an ion beam is directly aligned along a crystallographic direction, the ions are steered by small-angle scattering collisions between rows and planes of atoms. As shown in Fig. 20.1b, the fraction of ions that are channeled is estimated as

$$f_{\text{channel}} = \frac{\pi r_0^2 - \pi r_{\text{min}}^2}{\pi r_0^2} \tag{20.1}$$

where r_{min} is the atomic radius and r_0 is the radius of the ion channel surrounding the atom.

In the vicinity of the atomic row, the incident particles experience fewer close-encounters with atoms outside a small region surrounding the atomic rows, resulting in a reduced backscattered yield. In contrast, in the vicinity of atomic displacements, a fraction of the incident particles is de-channeled, leading to an increased probability for close-encounter collisions, resulting in an increased backscattered yield.

The fraction of ions that are not channeled is equivalent to the so-called minimum yield

Table 20.2 Structural and electronic properties of as-grown and etched GaN surfaces: ratios of channeling to random Rutherford backscattering spectrometry (RBS) yields summed over the top 100 nm, the so-called minimum yield, χ_{min}; fraction of displaced Ga, N_d/N; minimum yield from angular yield scans, Y_{min}; half-angle at half-minimum of angular yield scans, $\psi_{1/2}$ (°); lateral displacement of Ga atoms, r_x (nm); near band edge cathodoluminescence (NBE CL) emission intensity normalized to the emission of the as-grown GaN; Schottky barrier heights (SBH, eV); and ideality factors. For this analysis, χ_0 and ψ_0 are defined as the χ_{min} and the $\psi_{1/2}$ of the as-grown GaN

Surface	χ_{min}	N_d/N	Ymin	$\psi_{1/2}$ (°)	r_x(nm)	NBE CL	SBH (eV)	Ideality factor, n
as-grown	0.010	–	0.02	0.51	0.007	I_0	1.11	1.07
TBCl	0.011	0.002	0.02	0.51	0.007	$0.80I_0$	1.11	1.12
ICP	0.012	0.003	0.03	0.48	0.007	$0.25I_0$	0.66	1.79
ICP + TBCl	0.010	0.000	0.02	0.49	0.007	$0.54I_0$	0.96	1.18

$$\chi_{min} = 1 - f_{channel} = \frac{r_{min}^2}{r_0^2} \tag{20.2}$$

The minimum yield, χ_{min}, is often defined as the ratio of the "channeling" yield to the "random" yield. The "channeling" yield is provided by the RBS spectra collected along the channel axis, and the "random" yield is obtained by rotating the goniometer in θ and ϕ about the channel axis. The fraction of displaced Ga atoms over the depths of interest, N_d/N, is estimated via a comparison of the minimum yields of the processed GaN layers with that of a high-quality GaN substrate, as follows:

$$\frac{N_d}{N} = \frac{\chi_{min} - \chi_0}{1 - \chi_0} \tag{20.3}$$

where N is the Ga atomic density of the crystal and χ_0 is the minimum yield for a high-quality GaN substrate standard [23]. In this work, we use the as-grown GaN listed in Table 20.2 as the standard.

Another indication of the crystalline quality is the average displacement of the atoms, r_x, given by:

$$\psi_{1/2}/\psi_0 = \left(\ln\left[(Ca/r_x)^2 + 1\right]\right) / \left(\ln\left[(Ca/\rho)^2 + 1\right]\right) \tag{20.4}$$

where $C^2 = 3$, the Thomas-Fermi Screening radius is $a = 0.099$ Å, the thermal vibration amplitude is $\rho = 0.0620$ Å [24–27], and $\psi_{1/2}$ and ψ_0 are the half-angle at half-depths from angular yield scans for the processed GaN films and that of a high-quality GaN substrate standard, respectively. The as-grown GaN listed in Table 20.2 is also utilized as the standard for ψ_0.

To obtain $\psi_{1/2}$ and ψ_0, the angular-dependence of RBS yields (so-called "angular yield scans") were measured at ±3° about the $[11\bar{2}0]$ and $[1\bar{1}00]$ directions. The half-angle at half depth is determined from the width of the "well" at the half-depth

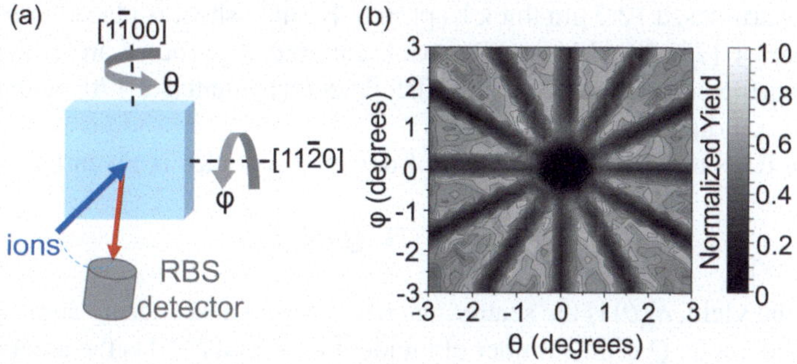

Fig. 20.2 (**a**) Illustration of Rutherford backscattering spectrometry (RBS): He or H ions (blue) are initially incident upon the GaN sample, and backscattered ions (red) are collected by a silicon surface-barrier detector. The automated ϕ-tilts (about the $[11\bar{2}0]$ axis) and θ-rotations (about the $[1\bar{1}00]$ axis), controlled by a five-axis goniometer, are also shown. (**b**) Normalized (0001) planar RBS channeling map, with yields integrated over the top 100 nm

of the normalized angular scans. In general, smaller values of $\psi_{1/2}$ correspond to larger values of the lateral atomic displacements, r_x.

For these experiments, the samples are mounted with copper tape onto the stage of a five-axis goniometer with computer control of polar (θ) and azimuthal (ϕ) rotations, as illustrated in Fig. 20.2a. Preliminary alignment to the sample surface normal is achieved using a laser that is coaxial with the incident ion beam.

Following sample surface alignment, a 2D planar scan is performed using the θ- and ϕ-ranges of ±4° with a step size of 0.4°. The θ and ϕ values are adjusted so that the minimum yield was set at $\theta = 0$ and $\phi = 0$. Next a high resolution 2D planar scan is performed with θ- and ϕ-ranges of ±3° with a step size of 0.2°, as shown in Fig. 20.2b for the [0001] channel of GaN. For the "random" spectra, the goniometer is set to 5° in ϕ away from the channel axis, and the backscattered signal is collected while the goniometer is rotated in θ (i.e., "rocked") from −5 to +5° about the channel axis, in steps of 0.2°. Both the channeling and random yield profiles are collected using a silicon surface barrier detector equipped with an amplification system and multichannel analyzer which bins the signal into "channels" that are linearly proportional to the particle energy.

2.2 Elastic Recoil Detection Analysis

ERDA is a complementary ion beam analysis technique that enables quantitative analysis of light elements—in this case, it is used to probe the depth dependence of [H]. For ERDA, 2.5 MeV α particles are incident on the GaN at 70° to the surface normal, and recoiled H atoms are detected by a Si surface-barrier detector (the ERDA detector) located at 30° with respect to the incident beam. To filter out scattered α particles, a 12.7 μm thick Kapton polyimide sheet is placed in front of the ERDA detector [28, 29]. Meanwhile, backscattered α particles are collected with another Si surface-barrier detector (the RBS detector) situated at 170° with respect to the incident ion direction.

For both RBS and ERDA, the yield of detected particles is given by:

$$Y = \sigma_i(\theta) \cdot \Omega \cdot Q \cdot N_s[i] \tag{20.5}$$

where Y is the yield, $\sigma_i(\theta)$ is the scattering cross-section at scattering angle θ, Ω is the detector solid angle, Q is the number of incident ions, and $N_s[i]$ is the areal density of target atoms i. The scattering cross-sections, $\sigma_i(\theta)$, for RBS of Ga and N were calculated using simulation of nuclear reaction analysis (SIMNRA) [30], and those for ERDA of H were taken from Ref. [31].

Due to the shallow angles of ion incidence and detection, accurate determination of the solid angles for the RBS detector (Ω_{RBS}) and ERDA detector (Ω_{ERDA}) is challenging. Instead, the ratio, $\Omega_{\mathrm{RBS}}/\Omega_{\mathrm{ERDA}}$, is determined using a target with known dimensions and [H] concentration, namely, a 1.5 cm × 1.5 cm sized Kapton sheet. Given the N_s value for Kapton, and the fact that $Q_{\mathrm{RBS}} = Q_{\mathrm{ERDA}}$ during simultaneous RBS and ERDA measurements, the measured Y_{RBS} and Y_{ERDA} for Kapton were used to determine $\Omega_{\mathrm{RBS}}/\Omega_{\mathrm{ERDA}} = 0.36$ for these experiments.

For the processed GaN films, simultaneous RBS and ERDA measurements were used to ensure $Q_{\mathrm{RBS}} = Q_{\mathrm{ERDA}}$. Using the measured value of Y_{RBS}, the known N_s for pure GaN, and the $\Omega_{\mathrm{RBS}}/\Omega_{\mathrm{ERDA}}$ ratio discussed above, $Q\Omega_{\mathrm{RBS}}$ and $Q\Omega_{\mathrm{ERDA}}$ were determined. Then, using the measured value of Y_{ERDA}, along with the reported $\sigma(\theta)$ for ERDA of H, N_s was computed, and [H] was determined for each measured GaN film.

2.3 *Channel-to-Energy-to-Depth Conversion*

For these ion beam analysis experiments, signals or "yields," consisting of counts vs channel, are obtained. Thus, to quantify depth profiles, both channel-to-energy and energy-to-depth conversions are needed. For example, for RBS, the channel-to-energy conversion is achieved by comparing the positions of Ga-edges for measured and simulated random yields from GaN. For ERDA, the positions of the H-edges for measured and simulated recoiled H yields from Kapton are compared. In both cases, the measured yield vs channel is fit to the simulated yield vs channel obtained using the SIMNRA code [30], with fitting parameters including the number of particles × Ω, detector resolution, and energy per channel, as described in Fig. 20.3a.

For the energy-to-depth conversion, the energy lost to collisions within the solid is considered, as illustrated in the inset in Fig. 20.3b. At a depth x, the energy of emitted ions is estimated as:

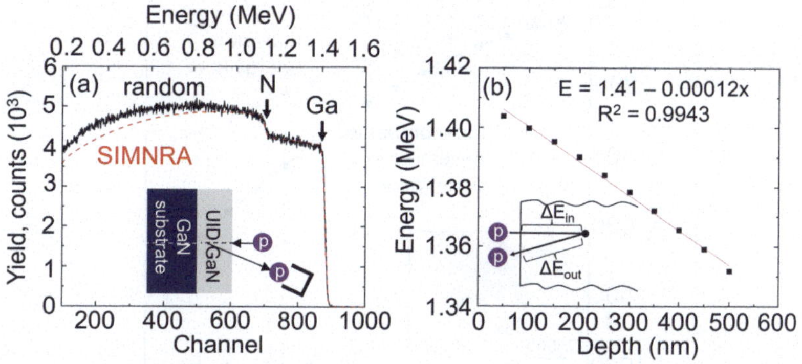

Fig. 20.3 (**a**) Plot of measured (black) and SIMNRA computed (red) "random" Rutherford backscattering spectrometry (RBS) yield vs channel, for example, UID-GaN. The best-fit to the data was determined by adjusting the number of particles (1.2×10^{11} protons·sr), detector resolution (23 keV), and energy per channel (1.58 keV/channel) within the SIMNRA simulation. (**b**) Backscattered energy vs depth, *E(x)*, from SIMNRA simulations of 1.5 MeV protons incident on GaN. An illustration of energy loss due to ion stopping power in a solid is shown as an inset. The depth-dependent energy loss allows for depth-sensitive measurements using a fixed ion energy

$$E(x) = E_0 - \int_0^x \frac{dE}{dx'}(E(x'), x')dx' \tag{20.6}$$

where the integral is the depth and energy-dependent ion stopping of the target sample. Differentiating this equation with respect to x gives:

$$\frac{dE}{dx} = -\epsilon(E) \tag{20.7}$$

where $\epsilon(E)$ is the energy-dependent stopping cross-section. Using SIMNRA [30], which incorporates stopping cross-section data from Ziegler [32], the emitted ion energy vs depth, *E(x)*, is computed, assuming equivalent channeling and random stopping powers [33]. In Fig. 20.3b, the backscattered energy vs depth, *E(x)*, is plotted for 1.5 MeV protons incident on GaN. Using a linear-least squares fit for *E (x)*, the RBS yield vs energy spectra is converted to yield vs depth. Similar approaches were used for the energy-to-depth conversions for the recoiled H yields in ERDA.

2.4 Cathodoluminescence Spectroscopy

CL spectra were collected in a Tescan RISE SEM equipped with a Gatan MonoCL4 Plus system. Within the SEM chamber, a parabolic mirror and light guide were used to collect emitted photons and couple them into an ex situ monochromator and

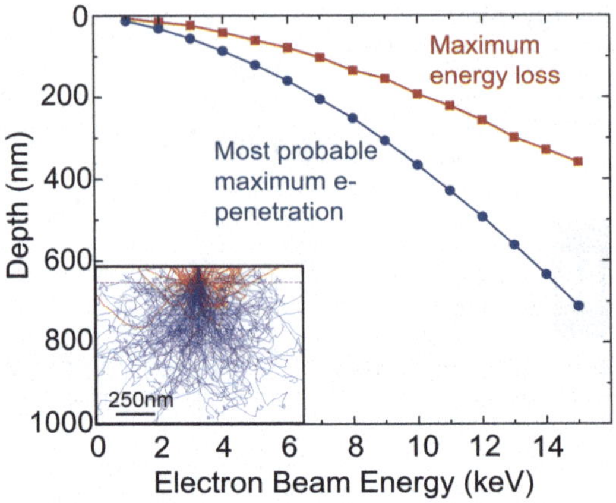

Fig. 20.4 CASINO simulations of electron trajectories in GaN: depths of maximum energy loss (red) and most probable electron penetration depth (blue) as a function of beam energy. For each beam energy, the simulated number of electrons vs maximum penetration depth was box-car smoothed and fit with a Gumbel distribution [35]. For the inset, red and blue electron trajectories indicate electrons which escape and do not escape the sample, respectively. Image processing of the corresponding electron trajectories was used to locate the highest density regions. (Reprinted from Ref. [21], with permission from Elsevier)

photomultiplier tube. The excitation depths for various electron-beam energies were estimated from CASINO Monte Carlo simulations of electron trajectories in GaN [34], as shown in Fig. 20.4. The average of the depths of maximum energy loss and most probable electron penetration was used to select appropriate beam energies (typically 5–10 keV) to preferentially probe the surfaces and buried interfaces. The electron beam spot size and current were selected to maintain power density at <0.34 W/cm^2. During electron beam scanning over 40×40 μm^2 regions, room temperature CL spectra were collected in the range 300–750 nm (4.12–1.65 eV) using a 50 nm blaze diffraction grating.

2.5 *Schottky Barrier Height Measurements*

In preparation for measurements of Schottky barrier heights, the GaN surfaces were etched with 1:1 HCl:HF for 60 s. Diodes with 80 μm diameter were prepared by e-beam evaporation of 20 nm/50 nm Ni/Au through a shadow mask onto the etched surfaces, followed by deposition of 20 nm/200 nm/20 nm/50 nm Ti/Al/Ni/Au layers on the backside of the GaN substrates. I–V measurements were performed at room temperature using a Keithley Model 4200 Semiconductor Characterization System, and the Schottky barrier heights and ideality factors were extracted from the forward-bias data.

3 Results and Discussion

3.1 Epitaxy, Etching, and Regrowth

In this section, we present investigations of GaN structures prepared by epitaxy, etching, and regrowth. In particular, the influence of ambient exposure and/or chlorine-based etching approaches on the structure and properties of GaN surfaces and regrown layers are described. Following a review of recent etching/regrowth approaches toward vertical GaN devices, we describe the sequence of metalorganic-vapor-phase epitaxy (MOVPE), chlorine-based chemical etching, and regrowth used to prepare the structures. We then present data from RBS/c, ERDA, CL, and secondary ion mass spectrometry (SIMS).

Selective-area etching and regrowth have emerged as a promising approach for vertical GaN devices on various substrates. For GaN substrates, both vertical p-n and n-p GaN devices have been demonstrated [36–39]. To date, the limited performance of vertical GaN devices, including low breakdown voltages and high leakage currents, has been attributed to imperfections at the regrowth interface [40, 41]. It has been suggested that the regrowth interface disrupts the GaN stoichiometry, resulting in interfacial N vacancies [42]. In the vicinity of the regrowth interfaces, C, O, and Si concentrations have been quantified [42, 43], and strategies to reduce their concentrations have been explored. For example, annealing in NH_3 has been shown to reduce the surface/interface concentrations of C and O [44, 45]. To reduce interfacial Si concentrations, Cl-based plasma and dry etching [46], wet chemical etching, and UV-ozone cleaning [43, 45] have been utilized. However, the influence of these surface/interface treatments on the crystallinity and electronic states is just beginning to be explored [21, 47].

For these investigations, UID-GaN layers were grown on n-type GaN surfaces using MOVPE; they were subsequently processed with two chlorine-based etchants, tertiary-butyl chloride (TBCl), performed within the MOVPE system, and/or Cl_2/BCl_3 inductively coupled plasma (ICP), performed ex situ, as illustrated in

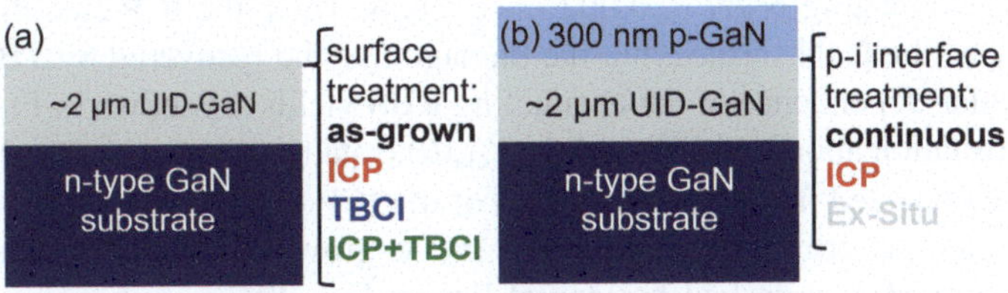

Fig. 20.5 GaN structures prepared by epitaxy, etching, and regrowth: (**a**) surface-treated UID-GaN layers grown on n-type GaN substrates, and (**b**) p-GaN layers (3×10^{19} cm^{-3}) grown on surface-treated UID-GaN layer structures on n-type GaN substrates. For (**a**), the surface treatments include ex situ inductively coupled plasma (ICP), in situ tertiary-butyl chloride (TBCl), and ex situ ICP etching followed by in situ TBCl exposure (ICP + TBCl), in comparison with as-grown UID-GaN (as-grown). For (**b**), the interface treatments include air exposure (ex situ) and ICP etching (ICP), in comparison with continuously grown layers (continuous)

Fig. 20.5a and b. For select samples, in situ and/or ex situ etchings were followed by MOVPE growth of a p-GaN layer, as illustrated in Fig. 20.5b. Additional discussion of TBCl etching is given in Chap. 12 of this book.

Our ion beam analysis studies reveal that the all in situ prepared GaN surfaces have the lowest fraction of displaced Ga atoms, the highest near band edge (NBE) CL emissions, and nearly ideal diode characteristics. Meanwhile, for the GaN p-i-n structures, the fraction of displaced Ga atoms, summed from the surface to the regrown p-i interface, is about >2× higher than those of the etched surfaces, presumably due to the displacement of Ga by Mg incorporation in the p-layers. In addition, the all in situ prepared GaN p-i-n structures have the highest donor-acceptor pair (DAP) CL emission. Interestingly, observation of yellow luminescence (YL) at the ex situ prepared regrown p-i interfaces is consistent with the recent hypothesis that YL is due to a surface state associated with air exposure [48].

3.1.1 Surface-Treated UID-GaN Layers

The [0001] channeling RBS spectra from an as-grown (black) GaN film as well as those from ICP (red), ICP + TBCl (green), and TBCl (blue) etched GaN films, are shown in Fig. 20.6a. In addition, the random RBS spectrum from the as-grown GaN is plotted (black dashed). In Fig. 20.6a, a high backscattered yield, with well-defined N- and Ga-edges, is apparent for the random spectrum, with significantly reduced yields for the channeling spectra, as expected for single crystals. The close-up view of the channeling spectra in Fig. 20.6b allows us to quantify changes in crystallinity, at specific depths, following etching.

To quantify the depth-dependent crystallinity, we consider the fractions of backscattered ions using the minimum yield, χ_{min}, defined as the ratio of the channeling yield to the random yield, summed over the depth of interest, in this case, the top 100 nm. As shown in Table 20.2, χ_{min} values of 0.01 are observed for both the as-grown and TBCl-etched GaN, suggesting a fraction of displaced Ga, $N_d/N < 0.002$. For the ICP-etched GaN, χ_{min} increases to 0.012, resulting in $N_d/N \approx 0.003$. Interestingly, a subsequent TBCl etch (i.e., "ICP + TBCl") reduces χ_{min} to 0.01, indicating $N_d/N < 0.002$ once again. Thus, the ICP etch disrupts the crystallinity of the GaN surface, but the damage may be recovered by TBCl etch.

The angular-dependence of the channeling RBS yields are shown in Fig. 20.7 for (a) $Y_{Ga}(\phi)$ (rotation about $[11\bar{2}0]$) and (b) $Y_{Ga}(\theta)$ (rotation about $[1\bar{1}00]$). $Y_{Ga}(\theta)$ and $Y_{Ga}(\phi)$ were collected from the top 100 nm of the ICP + TBCl etched, TBCl etched, ICP etched, and as-grown GaN films. For the as-grown GaN, the minimum yield, $Y_{min,Ga}$, illustrated as a dashed horizontal line in Fig. 20.7a, is apparent at $\theta = 0$, where the incident beam is aligned along the [0001] channel. For the as-grown GaN, the $Y_{min,Ga}$ values are essentially identical for $Y_{Ga}(\theta)$ and $Y_{Ga}(\phi)$. In addition, well-developed shoulders are apparent near the channel edges for $Y_{Ga}(\theta)$, but they are absent for $Y_{Ga}(\phi)$, presumably due to differences in flux focusing at the $(11\bar{2}0)$ and $(1\bar{1}00)$ surfaces [18, 49]. The presence of well-defined shoulders suggests a highly crystalline structure, as evidenced by the low values of χ_{min} and N_d/N in Table 20.2.

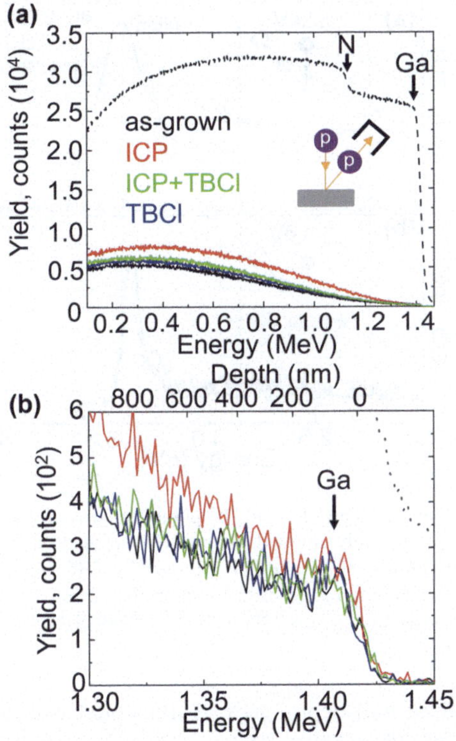

Fig. 20.6 (**a**) Random and channeling Rutherford backscattering spectrometry (RBS) yield vs backscattered particle energy for GaN following MOVPE growth (black), and post-growth surface treatments including inductively coupled plasma (ICP) etching (red), tertiary-butyl chloride (TBCl) etching (blue), and ICP + TBCl etching (green), with close-up view of RBS spectra in the vicinity of the surface in (**b**)

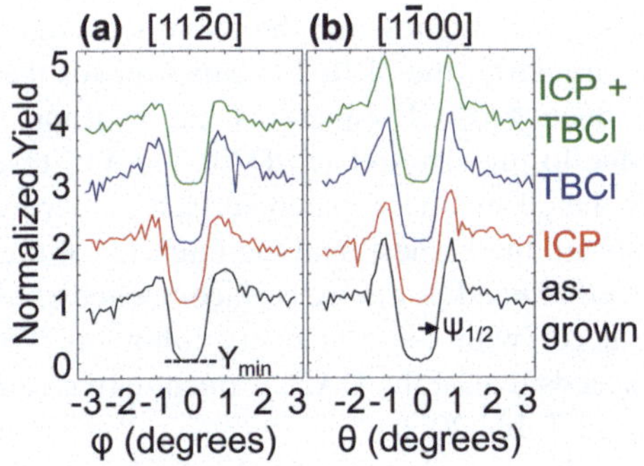

Fig. 20.7 Angular-dependence of the [0001] channeling Rutherford backscattering spectrometry (RBS/c) yields, rotated about the (**a**) $[11\bar{2}0]$ and (**b**) $[1\bar{1}00]$ axes. These angular yield scans were collected from the top 100 nm of the GaN following MOVPE growth (black), ICP etching (red), TBCl etching (blue), and ICP + TBCl etching (green). Example minimum yields, $Y_{min,Ga}$, and half-width at half-depth, $\psi_{1/2}$, are illustrated by the dashed horizontal line in (a) and the horizontal arrow in (b), respectively

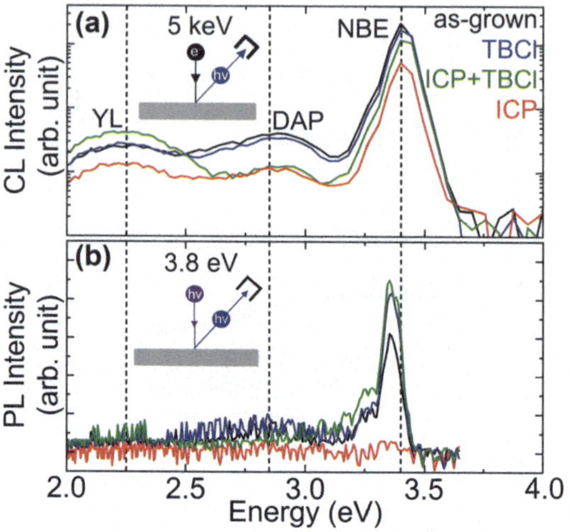

Fig. 20.8 (**a**) Cathodoluminescence (CL) and (**b**) photoluminescence (PL) spectra of MOVPE-grown UID-GaN (black), inductively coupled plasma (ICP) etched UID-GaN (red), tertiary-butyl chloride (TBCl) etched UID-GaN (blue), and ICP followed by TBCl etched UID-GaN (green). For the CL in (**a**), regions in the vicinity of the surfaces are probed using a beam energy of 5 keV. For the PL in (**b**), the incident laser energy was 3.8 eV. (b) is reproduced from Ref. 46, with permission from Elsevier

Finally, the half-width at half-depth, $\psi_{1/2}$, is illustrated with a horizontal arrow in Fig. 20.7b. As summarized in Table 20.2, $\psi_{1/2}$ values of $0.49 \pm 0.02°$ are apparent for all of the GaN surfaces, consistent with earlier measurements of GaN [12, 15, 18]. The corresponding displacements of Ga atoms from the [0001] row, r_x, are 0.007 nm, similar to the thermal vibration amplitude.

We now consider the luminescence from the surface-treated UID-GaN layers, via the CL and photoluminescence (PL) spectra in Fig. 20.8a and b. Both the CL and PL spectra reveal NBE emissions at $\simeq$3.4 eV. For the CL shown in Fig. 20.8, the intensity of NBE emission is highest for the as-grown UID-GaN and lowest for the ICP-etched GaN. Similarly, the NBE PL emission for the ICP-etched GaN is lowest, namely, it is quenched. CL emissions are also apparent at 2.85 eV and 2.2 eV, presumably due to impurity-related DAP and YL, respectively. It is interesting to note that for the all in situ prepared surfaces, as-grown and TBCl-etched, the NBE CL emissions are the highest, and the intensity of the DAP is higher than that of the YL. On the other hand, for those surfaces etched ex situ via ICP, the NBE CL emissions are reduced (with some recovery following TBCl etching) and the intensity of the YL exceeds that of the DAP. A quantitative comparison of the NBE CL intensities is listed in Table 20.2.

Finally, we compare the performance of Schottky barrier devices prepared with various GaN surface treatments, as listed in Table 20.2. For all cases, Ni/Au contacts were deposited on the GaN surfaces and current–voltage (I–V) characteristics were measured at room temperature. The Schottky barrier heights and ideality factors for each surface treatment are summarized in Table 20.2. For the as-grown and TBCl etched GaN surfaces, similar Schottky barrier heights (1.11 eV) and ideality factors

(~1.1) are apparent. In contrast, the ICP etch leads to the lowest Schottky barrier height (0.66 eV) and highest ideality factor (1.79), indicating non-ideal diode behavior, likely associated with the interfacial atomic displacements, i.e., increased N_d/N, discussed above. For the ICP + TBCl etch, the Schottky barrier height is increased to 0.96, with an ideality factor of 1.18. The nearly-ideal diode behavior for the ICP + TBCl-etched GaN diode correlates well with the decrease in atomic displacements, i.e., reduced N_d/N, discussed above.

3.1.2 GaN p-i-n Structures

Figure 20.9 presents [0001] channeling RBS spectra from the continuous (black), ex situ (gray), and ICP etched (red) GaN p-i-n structures, in comparison with a random RBS spectrum from a UID GaN layer prior to p-GaN growth. In Fig. 20.9a, a high backscattered yield, with well-defined N- and Ga-edges, is apparent for the random spectrum, with significantly reduced yields for the channeling spectra. To quantify

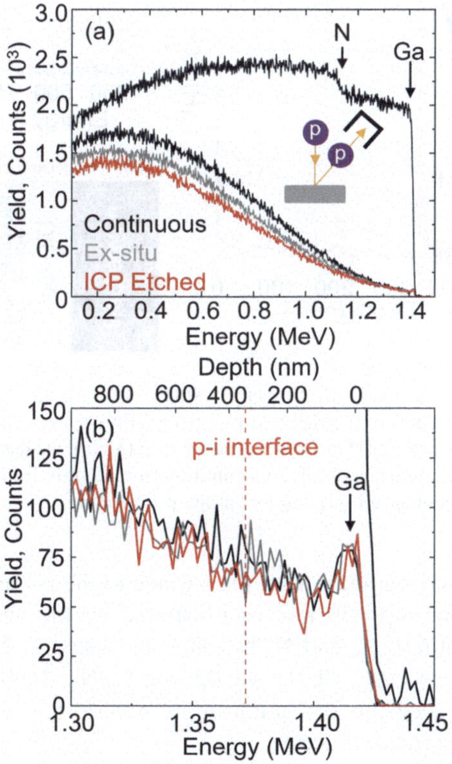

Fig. 20.9 (**a**) Random and channeling Rutherford backscattering spectrometry (RBS) yield vs backscattered particle energy for MOVPE-grown p-GaN layers grown on UID GaN layers, with close-up view of RBS spectra in the vicinity of the p-i interface in (**b**). The black, gray, and red spectra are data from the p-GaN grown on UID GaN without interruption (continuous), following air exposure and re-insertion into the MOVPE (ex situ), and following air exposure, ICP-etching, and re-insertion into the MOVPE (ICP etched). The position of the p-i interface is indicated by a vertical dashed line in (b)

Table 20.3 Structural and electronic properties of continuous, ex situ, and ICP-etched p-i-n GaN structures: ratios of channeling to random Rutherford backscattering spectrometry (RBS) yields summed over top 350 nm, the so-called minimum yield, χ_{min}; the fraction of displaced Ga, N_d/N; interfacial [Si] and [H]; interfacial cathodoluminescence (CL) emission intensities from Fig. 20.11b: yellow luminescence (YL), donor-acceptor pair (DAP), and near band edge (NBE) normalized to the CL emission of the continuous p-i-n GaN structure. For this analysis, χ_{min} and $\psi_{1/2}$ for the as-grown GaN from Table 20.2 are defined as χ_0 and $\psi_{1/2}$, respectively

p-i-n structure	χ_{min}	N_d/N	[Si] (10^{18} cm^{-3})	[H] (10^{20} cm^{-3})	YL	DAP	NBE
Continuous	0.016	0.007	0.1	2.2	I_0	I_0	I_0
Ex situ	0.017	0.008	1.1	1.8	$1.80I_0$	$0.86I_0$	$0.44I_0$
ICP	0.014	0.005	70	10	$2.98I_0$	$0.47I_0$	$1.11I_0$

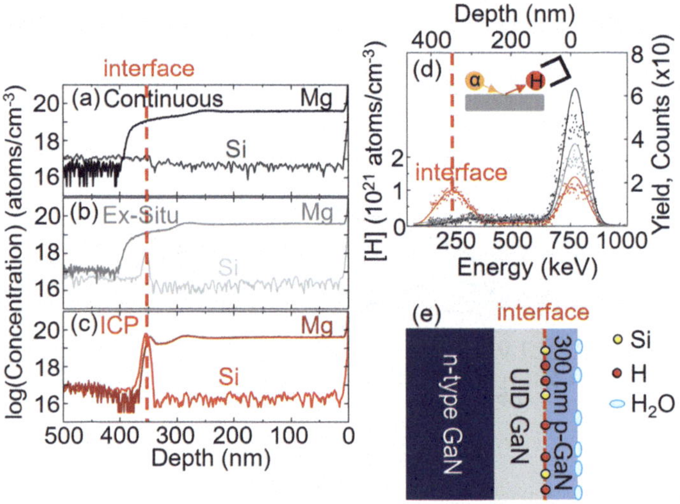

Fig. 20.10 Secondary ion mass spectrometry (SIMS) and elastic recoil detection analysis (ERDA) data: depth-dependence of [Mg] and [Si] from SIMS for (**a**) continuous (black), (**b**) ex situ (gray), and (**c**) ICP etched (red) GaN p-i-n structures. In (a)–(c), the dashed red line indicates the p-i interface. (**d**) Depth-dependence of [H] concentration from ERDA. (**e**) Illustration of ICP-etched GaN p-i-n structure with enhanced [H] and [Si] at the p-i interface

the depth-dependent crystallinity from the surface to the p-i interface, we compute χ_{min} summed over the top 350 nm, as illustrated by the vertical dashed line in Fig. 20.9b. The resulting χ_{min} and N_d/N values are listed in Table 20.3. For all of the GaN p-i-n structures, $\chi_{min} = 0.015 \pm 0.02$ with $N_d/N = 0.007 \pm 0.002$, indicating that any atomic displacements of Ga during air exposure and/or ICP etching were restored during the regrowth process.

The depth-dependence of the impurity concentrations were determined using a combination of SIMS and ERDA, as shown in Fig. 20.10, with the surface to bulk spanning from the right to the left. In Fig. 20.10, the SIMS data of (a) continuous,

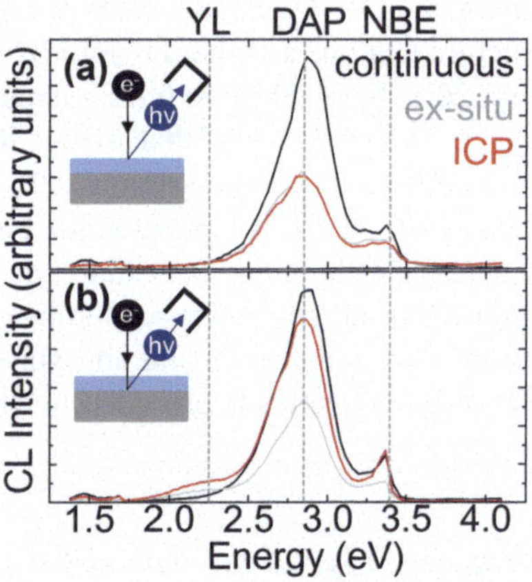

Fig. 20.11 Cathodoluminescence (CL) spectra from continuous (black), ex situ (gray), and ICP-etched (red) GaN p-i-n structures. To probe the surface and p-i interface, the CL spectra are generated with electron beam energies of (**a**) 5 keV and (**b**) 10 keV, respectively

(b) ex situ, and (c) ICP-etched GaN p-i-n structures, reveal similar depth-dependences of [Mg], with the highest [Mg] ($\simeq 5 \times 10^{19}/\text{cm}^3$) in the p-GaN layers, and the lowest [Mg] ($\leq 10^{17}/\text{cm}^3$) in the UID-GaN layers. The SIMS data also reveal similar depth-dependences of [Si], with the lowest [Si] ($\lesssim 10^{17}/\text{cm}^3$) in the p-GaN layers, and slightly higher [Si] ($\simeq 10^{17}/\text{cm}^3$) in the UID-GaN layers. In the vicinity of the p-i interface, indicated by the vertical red dashed line at the depth of 350 nm in Fig. 20.10a–c, the [Mg] monotonically decreases across the interface (from the surface to bulk) for (a) continuous and (b) ex situ p-i interfaces. However, for the ICP-etched p-i interface in Fig. 20.10c, an enhancement of [Mg] to $\simeq 4 \times 10^{20}/\text{cm}^3$ is apparent. It is interesting to note that enhancements of [Si] are also apparent at the ex situ and ICP-etched p-i interfaces in Fig. 20.10b and c, with peak concentrations of $10^{18}/\text{cm}^3$ and $\simeq 7 \times 10^{19}/\text{cm}^3$, respectively. Meanwhile, the depth-dependence of the [H] in Fig. 20.10d reveals $\simeq$3–10 nm of surface H_2O for the continuous, ex situ, and ICP-etched GaN p-i-n structures. For all samples, [H] drops to $\simeq 10^{19}/\text{cm}^3$ within 280 nm of the surface, and a slight increase in [H] to $\simeq 10^{20}/\text{cm}^3$ at $\simeq$300 nm depth is apparent. In addition, for the ICP-etched GaN p-i-n structure, an enhancement of [H] to $\geq 10^{21}/\text{cm}^3$ is apparent at the p-i interface. Thus, while air exposure leads to elevated [Si] at the p-i regrowth interface, ICP etching results in a further enhancement of [Si], as well as elevated [H] and [Mg], as illustrated in Fig. 20.10e.

We now consider the electronic properties of the surface and p-i interface for the continuous, ex situ, and ICP-etched GaN p-i-n structures. As shown in Fig. 20.11, CL spectra were generated with electron beam energies of (a) 5 keV and (b) 10 keV, which probe the surface and p-i interface, respectively. For the near-surface regions in Fig. 20.11a, the emissions observed at 2.85 eV and 3.4 eV are attributed to DAP and NBE emission, respectively. For the ex situ and ICP-etched GaN p-i-n structures, the relative intensities of DAP and NBE are similar. Meanwhile, the continuous GaN p-i-n structure exhibits a higher DAP to NBE intensity ratio. For the p-i interface regions in Fig. 20.11b, the DAP to NBE intensity ratios for continuous and ex situ are essentially unchanged, while that of the ICP-etched structure has increased substantially. In addition, a new emission at 2.25 eV—the YL—is apparent. Interestingly, the intensity of the YL is highest at the buried p-i interface within the ICP-etched p-i-n structure, with an intermediate intensity at the buried p-i interface within the ex situ p-i-n structure. These findings are consistent with the attribution of YL to a surface state induced during air exposure [48]. A quantitative comparison of the CL emission intensities is provided in Table 20.3.

3.2 Ion Implantation and Gyrotron Annealing

In this section, we present investigations of GaN structures prepared by Mg implantation and Mg/N co-implantation, followed by gyrotron annealing. Following a brief review of gyrotron annealing and its potential impact on vertical GaN device technology, we describe the sequences of box implantation/co-implantation and gyrotron annealing used to prepare the structures. We then present RBS/c data, emphasizing the impact of the implantation and annealing processes on GaN crystallinity. More details of gyrotron annealing on Mg implanted GaN are reviewed in Chap. 14 of this book.

Due to the need for post-growth annealing to drive off passivating H from precursors used in GaN MOVPE [50, 51], ion implantation continues to be a viable alternative for vertical p-n devices [52, 53]. However, p-n device fabrication using ion implantation is limited to shallow junction formation, and requires both masking for lateral selectivity and high temperature annealing to resolve lattice damage. Recently, microwave annealing was proposed as a promising alternative due to its promise for lateral precision in shallow junction formation, without the need for ultra-high temperature annealing [54]. Specifically, a gyrotron microwave generator [55] enables radiative heating with an ultra-high temperature ramp rate (10,000 °C/s). Although PL studies suggest that gyrotron annealing of Mg implanted GaN has enabled substitutional incorporation of Mg, Mg_{Ga}, the limited doping efficiency was attributed to high concentrations of V_N generated during implantation [54, 56]. Aiming to mitigate the formation of V_N [57, 58], thereby increasing the maximum hole concentration, Mg/N co-implantation followed by gyrotron annealing was also explored [59, 60].

For these investigations, MOVPE-grown UID GaN layers were implanted with a three-step box-implant targeting depths/concentrations of 500 nm GaN with $[Mg] = 1 \times 10^{19}$ cm^{-3} and 250 nm GaN with $[Mg] = [N] = 1 \times 10^{19}$ cm^{-3}, as illustrated in Fig. 20.12 and Table 20.4. All ion implantations were performed at 7° tilt with 23° twist relative to the c-plane to prevent ion channeling. For a subset of the implanted GaN layers, gyrotron annealing was performed using 15 pulses at 1000 °C at a pressure of 4 MPa in N_2 gas [54, 59].

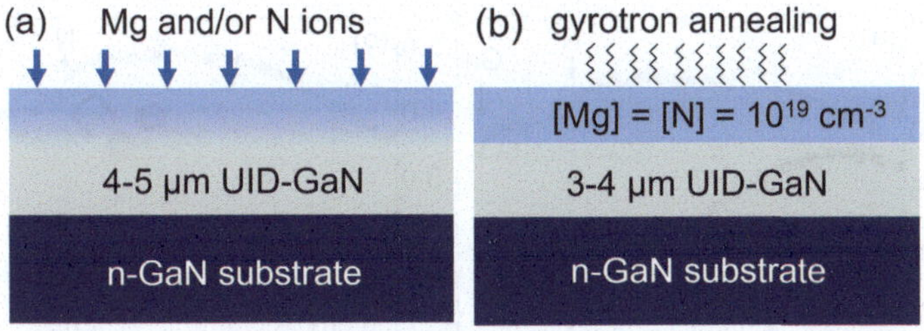

Fig. 20.12 Epitaxial UID GaN layers prepared with box implants, before and after gyrotron annealing: (**a**) Mg implantation or Mg/N co-implantation was achieved using 3-step box implants described in Table 20.4 and (**b**) following gyrotron annealing, the targeted structures consist of 500 nm GaN with [Mg] = 1×10^{19} cm^{-3} or 250 nm GaN with [Mg] = [N] = 1×10^{19} cm^{-3}

Table 20.4 Approach for box implantation of GaN with Mg or Mg + N, including targeted implanted species, depths, and concentrations. The ion energies and ion doses for each step are also provided

Species	Depth (nm)	[Mg] (10^{19} cm^{-3})	[N] (10^{19} cm^{-3})	Mg Ion Energy (keV)	Mg Ion Dose (10^{14} cm^{-2})	N Ion Energy (keV)	N Ion Dose (10^{14} cm^{-2})
Mg	500	1.0	N/A	–	–	–	–
				100	0.9	N/A	N/A
				200	1.3	N/A	N/A
				400	3.3	N/A	N/A
Mg, N	250	1.0	1.0	–	–	–	–
				100	0.9	90	0.7
				180	0.9	150	1.0
				270	3.0	240	3.0

The [0001] channeling RBS spectra for as-grown GaN (black), GaN following Mg implantation (cyan, dashed), and GaN following Mg implantation plus gyrotron annealing (cyan, solid) are shown in Fig. 20.13a and c. Similarly, the [0001] channeling RBS spectra for as-grown GaN, GaN following Mg/N co- implantation (magenta, dashed), and GaN following Mg/N co-implantation plus gyrotron annealing (magenta, solid) are shown in Fig. 20.13b and d. For both cases, RBS/c for the as-grown GaN is plotted in black, with dashed (solid) lines corresponding to random (channeling) spectra in Fig. 20.13a–d. In Fig. 20.13a and b, high backscattered yields, with well-defined N- and Ga-edges, are apparent for the random spectra, with significantly reduced yields for the channeling spectra, as expected for single crystals. The close-up views of the channeling spectra in Fig. 20.13c and d allow us to quantify changes in crystallinity, at specific depths, following implantation and implantation plus gyrotron annealing.

To quantify the depth-dependent crystallinity, we consider the fractions of backscattered ions using the minimum yield, χ_{min}, summed over the top 100 nm.

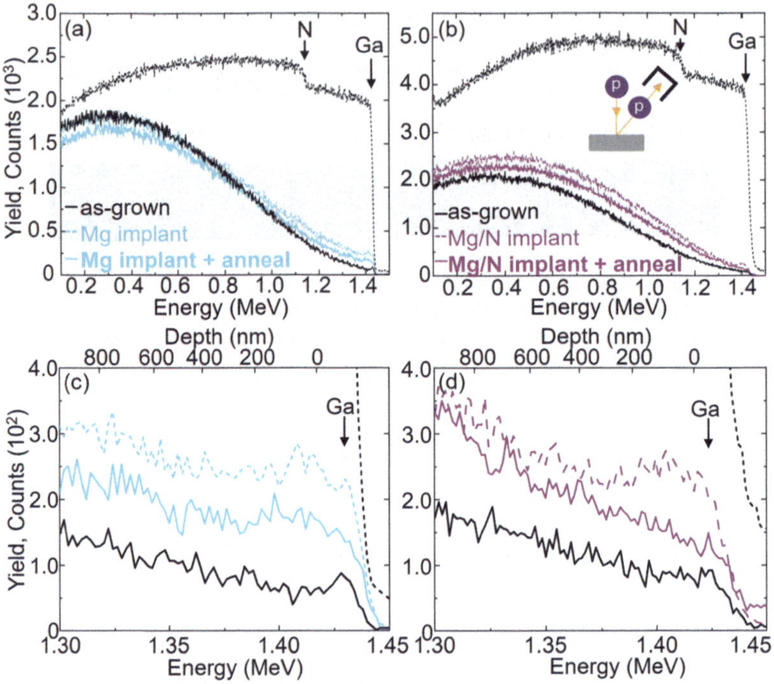

Fig. 20.13 (**a**) and (**b**) Random and channeling Rutherford backscattering spectrometry (RBS) spectra from p-GaN prepared by ion implantation plus gyrotron annealing, with close-up views of the RBS spectra in (**c**) and (**d**). The dashed and solid black spectra in (a)–(d) are random and channeling data from MOVPE-grown UID GaN prior to ion implantation (as-grown). The dashed and solid cyan spectra are channeling data following Mg implantation and Mg implantation + gyrotron annealing, respectively. Finally, the dashed and solid magenta spectra are channeling data following Mg/N co-implantation and Mg/N co-implantation plus gyrotron annealing. For both Mg implantation and Mg/N co-implantation, implantation increases the channeling backscattered yield, with a partial decrease following gyrotron annealing

Table 20.5 Crystallinity of the Mg-implanted and Mg/N co-implanted GaN, before and after gyrotron annealing: ratios of the channeling to random yields, summed over the top 100 nm, the so-called minimum yield, χ_{min}; the fraction of displaced Ga, N_d/N; minimum yield from angular yield scans, Y_{min}; half-angle at half-minimum of angular yield scans, $\psi_{1/2}$ (°); and lateral displacement of Ga atoms, r_x (nm). For this analysis, χ_{min} and $\psi_{1/2}$ for the as-grown GaN from Table 20.2 are defined as χ_0 and $\psi_{1/2}$, respectively

Implanted GaN	χ_{min}	N_d/N	Y_{min}	$\psi_{1/2}$ (°)	r_x (nm)
As-grown	0.018	0.009	N/A	N/A	N/A
Mg	0.058	0.049	N/A	N/A	N/A
Mg + gyrotron	0.041	0.032	N/A	N/A	N/A
As-grown	0.022	0.013	0.03	0.55	0.006
Mg/N	0.063	0.054	0.10	0.55	0.006
Mg/N + gyrotron	0.036	0.027	0.05	0.57	0.006

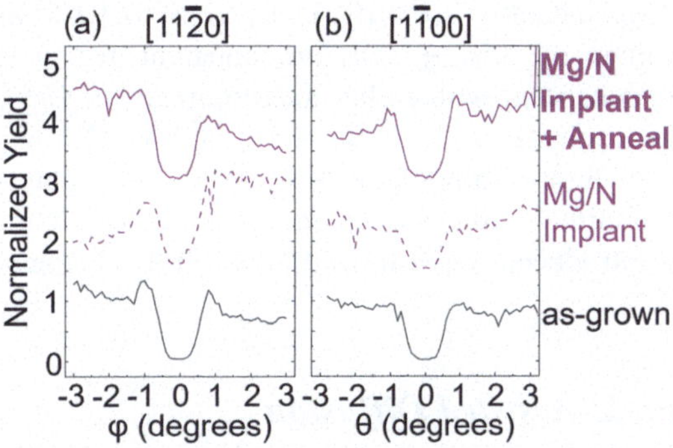

Fig. 20.14 Angular-dependence of the [0001] channeling Rutherford backscattering spectrometry (RBS/c) yields, rotated about the $[11\bar{2}0]$ and $[1\bar{1}00]$ axes, collected from the top 100 nm, for MOVPE-grown GaN prior to ion implantation (as-grown, black), following Mg/N co-implantation (magenta, dashed), and following Mg/N co-implantation plus gyrotron annealing (magenta, solid)

As shown in Table 20.5, for the as-grown UID GaN "reference" samples, $\chi_{min} = 0.020 \pm 0.002$, indicating a fraction of displaced Ga, $N_d/N = 0.011 \pm 0.002$. We note that the values of χ_{min} and N_d/N for these GaN reference samples are systematically higher than those of the as-grown GaN in Table 20.2, likely due to differences in the starting substrates and MOVPE-growth of the GaN layers.

For the Mg implanted GaN layers, the minimum yield increases to $\chi_{min} = 0.058$, indicating an increase in the fraction of displaced Ga to $N_d/N = 0.049$. Following gyrotron annealing, the crystallinity is partially recovered, with the channeling yield decreasing to $\chi_{min} = 0.041$, and the fraction of displaced Ga decreasing to $N_d/N = 0.032$.

When Mg and N are simultaneously implanted, the minimum yield increases further to $\chi_{min} = 0.063$, indicating an increase in the fraction of displaced Ga to $N_d/N = 0.054$. Interestingly, following gyrotron annealing, the crystallinity recovery is substantial, with the channeling yield decreasing to $\chi_{min} = 0.036$, and the fraction of displaced Ga decreasing to $N_d/N = 0.027$. These results support the hypothesis that the excess N provided by Mg/N co-implantation enhances recovery of crystallinity during gyrotron annealing.

The angular-dependence of the RBS/c yields are shown in Fig. 20.14 for (a) $Y_{Ga}(\phi)$ (rotating about the $[11\bar{2}0]$) and (b) $Y_{Ga}(\theta)$ (rotating about the $[1\bar{1}00]$). $Y_{Ga}(\theta)$ and $Y_{Ga}(\phi)$ were collected from the top 100 nm of the Mg/N co-implanted, Mg/N co-implanted plus gyrotron annealed, and as-grown GaN films. For the as-grown GaN, the $Y_{min,Ga}$ values are essentially identical for $Y_{Ga}(\theta)$ and $Y_{Ga}(\phi)$. For these data, shoulders are apparent near the channel edges for the as-grown Y_ϕ, presumably due to flux-focusing, as described earlier. The presence of well-defined shoulders suggests a highly crystalline structure, as evidenced by the low values of χ_{min} and N_d/N in Table 20.5.

As summarized in Table 20.5, $Y_{min} = 0.03$ for the as-grown GaN, similar to the values for MOVPE-grown GaN reported in Table 20.2. Following Mg/N co-implantation, Y_{min} increases to 0.10, recovering to 0.05 following gyrotron annealing. $\psi_{1/2}$ values of 0.56 ± 0.01°are apparent for as-grown and Mg/N co-implanted GaN, consistent with earlier measurements of GaN [12, 15, 18]. The corresponding displacements of Ga atoms from the [0001] row, r_x, are 0.006 nm, similar to the thermal vibration amplitude. Since Mg/N co-implantation and gyrotron annealing have a minimal influence on the $\psi_{1/2}$ values, it is likely that those processes have not introduced impurities into the [0001] channels.

3.3 Electric Field-Assisted Diffusion

In this section, we present investigations of GaN structures prepared by an electric field-assisted diffusion process. Following a brief review of electric field-assisted diffusion and its potential impact on vertical GaN device technology, we describe the sequences of deposition and annealing used to prepare the structures. We then present RBS/c data, emphasizing the impact of the annealing and Mg diffusion processes on GaN crystallinity.

Selective solid-state diffusion of Mg into GaN [61, 62] has been suggested as an alternative approach to achieve vertical p-n devices. However, high concentrations of V_{Ga} are needed to facilitate Mg_{Ga} formation and diffusion. Typically, the high temperatures needed to produce V_{Ga} also lead to surface N loss and the formation of V_N. Recently, an alternative approach, based upon vacancy-assisted impurity diffusion (VAID) [63, 64], has been proposed for selective p-type doping of GaN with Mg [65]. In brief, the approach consists of deposition of a magnesium-containing insulator and a gallate-forming metal atop GaN, as shown in Fig. 20.15, for the combination of MgF_2 and gold. It is hypothesized that the band-bending introduced at the MgF_2/GaN interface lowers the GaN conduction band edge toward mid-gap, thereby lowering the formation energy of V_{Ga}. Furthermore, it is hypothesized that the built-in electric field provided by the metal (Au)-insulator (MgF_2)-semiconductor (GaN) structure enhances Mg diffusion into GaN.

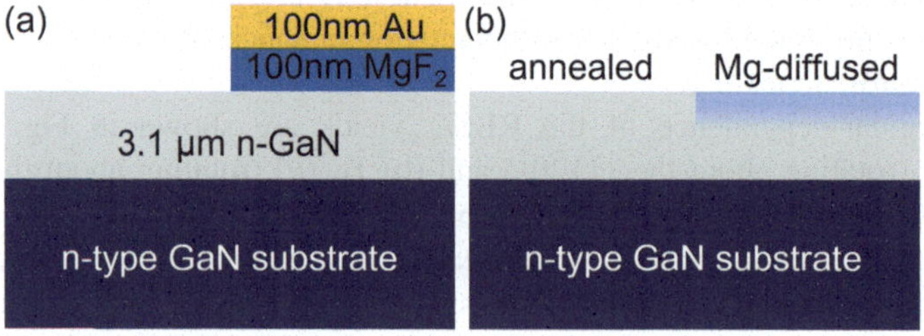

Fig. 20.15 Epitaxial n-GaN layers for studies of electric field-assisted Mg diffusion in GaN illustrated before and after heat-treatment: (**a**) a metallic bilayer consisting of 100 nm Au deposited on 100 nm MgF_2 was deposited through a shadow mask blocking half of a GaN layer and (**b**) following annealing for 50 min at 700, 800, or 900 °C, the structures consist of both "annealed" reference and "Mg-diffused" regions

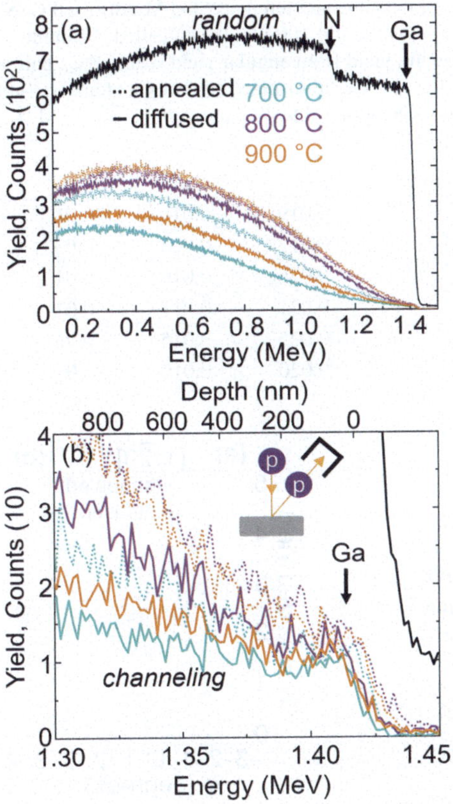

Fig. 20.16 (**a**) Random and channeling Rutherford backscattering spectrometry (RBS) spectra from GaN prepared by annealing to 700 °C (teal), 800 °C (purple), and 900 °C (orange), both with (solid) and without (dashed) Mg diffusion; close-up views of the RBS spectra are shown in (**b**). The black spectra in (a) and (b) correspond to representative random data, in this case for the 700 °C annealed reference GaN

For these investigations, a metallic bilayer consisting of 100 nm gold deposited onto 100 nm MgF_2 was deposited through a shadow mask blocking half of several Si-doped GaN layers, as illustrated in Fig. 20.15. Each structure was annealed in Ar_2 gas for 50 min at 700, 800, and 900 °C, forming both "annealed" reference and "Mg diffused" regions.

The [0001] channeling RBS spectra for annealed reference GaN (dashed) and Mg-diffused GaN (solid) are shown in Fig. 20.16a and b, with 700 °C (teal), 800 °C (purple), and 900 °C (orange). For comparison, a representative random spectrum, in this case from the 700 °C annealed reference GaN, is plotted in black. In Fig. 20.16a, a high backscattered yield, with well-defined N- and Ga-edges, is apparent for the random spectra, with significantly reduced yield for the channeling spectra, as expected for single crystals. The close-up view of the channeling spectra in Fig. 20.16b allow us to quantify changes in crystallinity, at specific depths, following annealing and Mg-diffusion.

Table 20.6 Crystallinity of the annealed reference and Mg-diffused GaN: ratios of the channeling to random yields, summed over the top 100 nm, the so-called minimum yield, χ_{min}; the fraction of displaced Ga, N_d/N; minimum yield from angular yield scans, Y_{min}; half-angle at half-minimum of angular yield scans, $\psi_{1/2}$ (°); and lateral displacement of Ga atoms, r_x (nm). For this analysis, χ_{min} and $\psi_{1/2}$ for the as-grown GaN from Table 20.2 are defined as χ_0 and $\psi_{1/2}$, respectively

Processing T (°C)	Mg Diffused?	χ_{min}	N_d/N	Y_{min}	$\psi_{1/2}$ (°)	r_x (nm)
700	N	0.019	0.010	0.01	0.58	0.006
700	Y	0.017	0.008	0.03	0.52	0.006
800	N	0.026	0.016	0.01	0.52	0.007
800	Y	0.021	0.011	0.01	0.53	0.007
900	N	0.025	0.015	0.02	0.54	0.007
900	Y	0.020	0.010	0.01	0.75	0.004

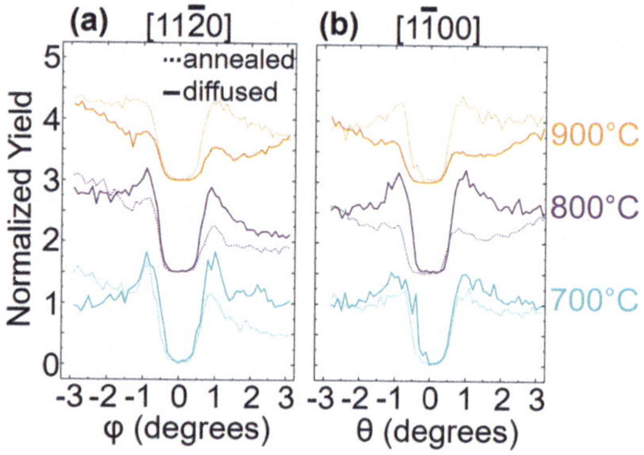

Fig. 20.17 Angular-dependence of the [0001] channeling Rutherford backscattering spectrometry (RBS/c) yields, rotated about the $[11\bar{2}0]$ and $[1\bar{1}00]$ axes, collected from the top 100 nm, for GaN annealed to 700 °C (teal), 800 °C (purple), and 900 °C (orange), both with (solid) and without (dashed) Mg diffusion

To quantify the depth-dependent crystallinity, we consider the fractions of backscattered ions using the minimum yield, χ_{min}, summed over the top 100 nm. As shown in Table 20.6, for the annealed reference layers, χ_{min} ranges from 0.019 to 0.026, indicating a fraction of displaced Ga, N_d/N = 0.010–0.016. We note that the values of χ_{min} and N_d/N for these annealed GaN reference layers are higher than those of the as-grown GaN in Table 20.2, presumably due to Ga and/or N loss at the surface during annealing. However, the χ_{min} and N_d/N are similar to those of the as-grown GaN in Table 20.5.

For the Mg-diffused regions, as shown in Table 20.6, χ_{min} ranges from 0.017 to 0.021, indicating a fraction of displaced Ga, N_d/N = 0.008–0.011. This decrease in χ_{min} and N_d/N for the Mg-diffused regions supports the hypothesis that MgF_2 has lowered the formation energy for V_{Ga}, enabling the formation and diffusion of Mg_{Ga}. It is worth noting that the 700 °C Mg-diffused GaN layer exhibits the lowest values of χ_{min} and N_d/N, with crystallinity similar to that of the continuous p-i-n GaN structure from Table 20.3.

The angular-dependence of the RBS/c yields are shown in Fig. 20.17a $Y_{Ga}(\phi)$ (rotating about the $[11\bar{2}0]$) and (b) $Y_{Ga}(\theta)$ (rotating about the $[1\bar{1}00]$). Spectra were collected from the top 100 nm of the GaN layers prepared by annealing to 700 °C, 800 °C, and 900 °C, both with (solid) and without (dashed) Mg diffusion. For the 700 °C annealed reference and Mg-diffused GaN layers, the $Y_{min,Ga}$ values are essentially identical for $Y_{Ga}(\theta)$ and $Y_{Ga}(\phi)$. Furthermore, for the 700 °C Mg-diffused layers, well-defined shoulders are apparent near the channel edges for Y_{ϕ}, presumably due to flux-focusing, as described earlier. The presence of well-defined shoulders suggests a highly crystalline structure, as evidenced by the low values of χ_{min} and N_d/N in Table 20.6.

As summarized in Table 20.6, $Y_{min} \leq 0.03$ for all of the annealed references and Mg-diffused GaN regions is similar to the values reported for the as-grown and etched GaN structures discussed in Table 20.2. For processing temperatures of 700 and 800 °C, $\psi_{1/2}$ values of 0.55 ± 0.03° are apparent—consistent with earlier measurements of GaN [12, 15, 18]. The corresponding displacements of Ga atoms from the [0001] channel, r_x, are 0.006–0.007 nm, similar to the thermal vibration amplitude. Since Mg-diffusion at 700 and 800 °C has a minimal influence on the $\psi_{1/2}$ values, it is likely that those processes have not introduced impurities into the [0001] channels. However, for the 900 °C Mg-diffused GaN regions, a substantial increase in $\psi_{1/2}$ and corresponding decrease in r_x are apparent. While the value of r_x is apparently lower than the thermal vibration amplitude for GaN, the increase in $\psi_{1/2}$ suggests that this process has effectively removed impurities from the [0001] channel, yielding Mg incorporated primarily as Mg_{Ga}. This finding is consistent with the reported achievement of hole conduction for the highest processing temperature (915 °C) for Au-assisted Mg diffusion [65].

4 Concluding Remarks

In this chapter, we described ion beam analysis techniques and their role in developing new understanding of relationships between structure, chemistry, and electronic states at GaN surfaces and interfaces intended for use in vertical GaN p-i-n devices. Following a description of the ion beam analysis and chemical and electronic characterization methods, we presented investigations of surface treatments for selective-area regrowth of GaN substrates and regrown epitaxial layers, as well as p-doped GaN layers prepared by Mg ion implantation plus gyrotron annealing and electric field-assisted Mg diffusion. In this section, we summarize the main findings and make suggestions for future opportunities.

For the surface-treated UID-GaN structures, the lowest fraction of near-surface atomic displacements and correspondingly most ideal diode characteristics are observed for the all in situ prepared GaN structures. Meanwhile, the highest fraction of near-surface atomic displacements—with correspondingly far from ideal diode characteristics—are apparent for the ICP-etched structures. Fortunately, the near-surface crystallinity and ideal diode characteristics are restored by subsequent TBCl etching.

For the GaN p-i-n structures, the fraction of displaced Ga atoms is nearly doubled, presumably due to the displacement of Ga by Mg incorporation in the p-type layers. Air exposure prior to p-type layer growth leads to elevated [Si] at the p-i regrowth interface, and ICP etching further enhances [Si], [H], and [Mg] at that interface. The resulting DAP emission intensity is reduced by the interfacial impurities, and YL emission is observed due to a surface state associated with air exposure.

For GaN structures prepared by Mg implantation and Mg/N co-implantation, subsequent gyrotron annealing provides partial recovery of crystallinity. Interestingly, the excess N provided by co-implantation enhances the crystallinity recovery during gyrotron annealing, with minimal introduction of impurities into the [0001] channels, presumably due to primarily substitutional Mg incorporation.

Finally, for GaN structures prepared by electric field-assisted Mg diffusion, we consider both the annealed reference and the Mg-diffused regions. Although the near-surface crystallinity is likely degraded by Ga and/or N loss at the surface during annealing, the Mg-diffused regions exhibit a substantial recovery of crystallinity, with near-surface atomic displacements comparable to those of the continuous GaN p-i-n structures prepared by MOVPE. For all annealing temperatures, the Mg-diffusion process has minimized the introduction of impurities into the [0001] channels, with 900 °C annealing apparently leading to the most effective substitutional Mg incorporation.

Looking to the future, these findings suggest that successful implementation of vertical p-n GaN devices may require hybrid approaches, such as localized Mg ion implantation and/or localized electric field-assisted Mg diffusion in combination with epitaxy [66]. For example, a stable focused ion beam (FIB) source for Mg was recently demonstrated [22], and plans are underway to integrate a FIB-based Mg source with molecular beam epitaxy in order to achieve both lateral and vertical selectivity in p-type doping of GaN.

For further insight into Mg dopant incorporation mechanisms and the interplay between structure, chemistry, and electronic states at GaN surfaces and interfaces, a combined experimental-computational approach is needed. In particular, we suggest using density functional theory (DFT) to generate relaxed structures of GaN crystals containing substitutional and/or interstitial impurities. These structures would then be used as input in combined Monte Carlo-molecular dynamics (MC-MD) ion channeling simulations of angular yield scans for comparison to experiment, as described in Refs [20, 67]. This combined experimental-computational approach will enable new understanding of dopant incorporation mechanisms and the influence of point defects on the properties of GaN surfaces and interfaces intended for use in vertical GaN p-i-n devices.

References

1. H. Nie, Q. Diduck, B. Alvarez, A.P. Edwards, B.M. Kayes, M. Zhang, G. Ye, T. Prunty, D. Bour, I.C. Kizilyalli, 1.5-kV and 2.2-mΩ -cm^2 Vertical GaN Tran-sistors on Bulk-GaN Substrates. IEEE Electron Device Lett. **35**(9), 939–941 (2014)
2. X. Liu, F. Lin, J. Li, Y. Lin, J. Wu, H. Wang, X. Li, S. Huang, Q. Wang, H.-C. Chiu, H.-C. Kuo, 1.7-kV vertical GaN-on-GaN Schottky barrier diodes with helium-implanted edge termination. IEEE Trans. Electron Devices **69**(4), 1938–1944 (2022)
3. Y. Zhang, M. Sun, Z. Liu, D. Piedra, H.-S. Lee, F. Gao, T. Fujishima, T. Palacios, Electrothermal simulation and thermal performance study of GaN vertical and lateral power transistors. IEEE Trans. Electron Devices **60**, 2224 (2013)
4. T. Pu, U. Younis, H.-C. Chiu, K. Xu, H.-C. Kuo, X. Liu, Review of recent progress on vertical GaN-based PN diodes. Nanoscale Res. Lett. **16**, 101 (2021)
5. G. Kipshidze, V. Kuryatkov, B. Borisov, Y. Kudryavtsev, R. Asomoza, S. Nikishin, H. Temkin, Mg and O codoping in p-type GaN and AlxGa1-xN (0<x <0.08). Appl. Phys. Lett. **80**, 2910 (2002)
6. S. Brochen, J. Brault, S. Chenot, A. Dussaigne, M. Leroux, B. Damilano, Dependence of the Mg-related acceptor ionization energy with the acceptor con- centration in p-type GaN layers grown by molecular beam epitaxy. Appl. Phys. Lett. **103**, 032102 (2013)
7. J.K. Sheu, G.C. Chi, The doping process and dopant characteristics of GaN. J. Phys. Condens. Matter **14**, R657 (2002)
8. B. McEwen, M.A. Reshchikov, E. Rocco, V. Meyers, K. Hogan, O. Adrieiev, M. Vorobiov, D.O. Demchenko, F. Shahedipour-Sandvik, MOCVD growth and characterization of Be-Doped GaN. ACS Appl. Electron. Mater. **4**, 3780 (2022)
9. J. Kennedy, A. Markwitz, H.J. Trodahl, B.J. Ruck, S.M. Durbin, W. Gao, Ion Beam analysis of amorphous and nanocrystalline group III-V nitride and ZnO thin films. J. Electron. Mater. **36**, 472 (2007)
10. K.L. Enisherlova, V.S. Kulikauskas, V.V. Zatekin, T.F. Rusak, N.B. Gladysheva, I.I. Razgulyaev, AlGaN/GaN heterostructure study using Rutherford backscattering spectrometry. J. Surf. Investig. X-Ray Synchrotron Neutron Tech. **5**, 626 (2011)
11. B. Sundaravel, E.Z. Luo, J.B. Xu, I.H. Wilson, W.K. Fong, L.S. Wang, C. Surya, Ion channeling studies on mixed phases formed in metalorganic chemical vapor deposition grown Mg-doped GaN on Al2O3(0001). J. Appl. Phys. **87**, 955 (2000)
12. M. Lind'en, R. Hellborg, Crystalline order and disorder in epitaxial GaN. Nucl. Instrum. Methods Phys. Res. B **68**, 170 (1992)
13. M. Usman, A. Hall'en, A. Nazir, Ion implantation induced nitrogen defects in GaN. J. Phys. D **48**, 455107 (2015)
14. B. Hollander, S. Mantl, M. Mayer, C. Kirchner, A. Pelzmann, M. Kamp, S. Christiansen, M. Albrecht, H.P. Strunk, Ion channeling studies of GaN layers on c-oriented sapphire. Nucl. Instrum. Methods Phys. Res. B **136**, 1248 (1998)
15. A. Turos, R. Ratajczak, K. Pagowska, A. Stonert, L. Nowicki, P. Caban, Nitrogen sublattice analysis in GaN by non-Rutherford He-ion scattering. Nucl. Instrum. Methods Phys. Res. B **266**, 1897 (2008)
16. S. Kurai, T. Saimei, M. Konishi, S. Kubo, T. Taguchi, *Identification of Homogenous and Inhomogenous Structures of NH_3 GS-MBE-Grown GaN Epilayers by Ion-Channeling Studies, in Compound Semiconductors 2001* (CRC Press, 2002)
17. N. Nishikata, K. Kushida, T. Nishimura, T. Mishima, K. Kuriyama, T. Nakamura, Evaluation of lattice displacement in Mg-implanted GaN by Rutherford backscattering spectroscopy. Nucl. Instrum. Methods Phys. Res. B **409**, 302 (2017)
18. A. Turos, L.J. Nowicki, A. Stonert, M. Leszczyn'ski, I. Grzegory, S. Porowski, Ion channeling study of GaN single crystals. Mater. Sci. Forum **248–249**, 419 (1997)
19. A. Redondo-Cubero, K. Lorenz, R. Gago, N. Franco, S. Fern´andez-Garrido, P.J.M. Smulders, E. Mun˜oz, E. Calleja, I.M. Watson, E. Alves, Breakdown of anomalous channeling with ion

energy for accurate strain determination in GaN-based heterostructures. Appl. Phys. Lett. **95**, 051921 (2009)

20. J.J.P. Cooper, T. Jen, A. Novak, Z. Xi, L. Qi, F.U. Naab, Y.Q. Wang, R.S. Goldman, Onset of tetrahedral interstitial formation in GaAsN alloys. Appl. Phys. Lett. **124**, 161108 (2024)
21. A. Chang, B. Li, S. Wang, S. Frisone, R. Goldman, J. Han, L. Lauhon, Unveiling the influence of selective-area-regrowth interfaces on local electronic properties of GaN p-n junctions for efficient power devices. Nano Energy **102**, 107689 (2022)
22. M. Titze, A. Katzenmeyer, S. Frisone, J.A. Ohlhausen, A. Flores, D. Campbell, B. Li, Y. Wang, J. Han, E.S. Bielejec, R.S. Goldman, Comparison of Mg-based liquid metal ion sources for scalable focused-ion-implantation doping of GaN. AIP Adv. **14**, 045326 (2024)
23. W.-K. Chu, J.W. Mayer, M.-A. Nicolet, Chapter 8 – Use of channeling techniques, in *Backscattering Spectrometry*, ed. by W.-K. Chu, J.W. Mayer, M.-A. Nicolet, (Academic Press, 1978), pp. 223–275
24. L.C. Feldman, J.W. Mayer, S.T. Picraux, Chapter 3 – Particle distributions within the channel, in *Materials Analysis by Ion Channeling*, ed. by L.C. Feldman, J.W. Mayer, S.T. Picraux, (Academic Press, San Diego, 1982), pp. 61–87
25. Y. Yamamura, W. Takeuchi, T. Kawamura, *The Screening Length of Interatomic Potential in Atomic Collisions*, vol No. 45 (National Institute for Fusion Science, 1998)
26. A. Yoshiasa, K. Koto, H. Maeda, T. Ishii, The mean-square relative displacement and displacement correlation functions in tetrahedrally and octahedrally coordinated A8B8-N crystals. Jpn. J. Appl. Phys. **36**, 781 (1997)
27. K. Kubota, T. Nishimura, K. Kuriyama, T. Nakamura, Evaluation of lattice displacement and electrical property of Zn-ion implanted GaN by Rutherford backscattering spectrometry. Nucl. Instrum. Methods Phys. Res. B **451**, 70 (2019)
28. K. Kunimune, Y. Soeda, S. Itami, K. Kikuta, Adhesive Polyimide Film, US5300627A (5 April 1994)
29. S. Damache, S. Djaroum, S. Ouichaoui, L. Amari, D. Moussa, Stopping powers and energy loss straggling for (0.9–3.4) MeV protons in a Kapton polyimide thin film. Nucl. Instrum. Methods Phys. Res. B **383**, 164 (2016)
30. M. Mayer, *SIMNRA User's Guide* (2020)
31. V. Quillet, F. Abel, M. Schott, Absolute cross section measurements for H and D elastic recoil using 1 to 2.5 MeV 4He ions, and for the 12C(d,p)13C and 16O(d,p1)17O nuclear reactions. Nucl. Instrum. Methods Phys. Res. B **83**, 47 (1993)
32. J.F. Ziegler, J.P. Biersack, The stopping and range of ions in matter, in *Treatise on Heavy-Ion Science: Volume 6: Astrophysics, Chemistry, and Condensed Matter*, ed. by D.A. Bromley, (Springer US, Boston, MA, 1985), pp. 93–129
33. M.L. Swanson, L. Shao, Ion Channeling, in *Handbook of Modern Ion Beam Materials Analysis, Second Edition*, (Materials Research Society, Warrendale, PA, 2009), pp. 247–284
34. D. Drouin, P. Hovington, R. Gauvin, CASINO: A new Monte Carlo code in C language for electron beam interactions. Scanning **19**, 20–28 (1997)
35. E.J. Gumbel, Le valeurs extr^emes des distributions statistiques. Annales de l'institut Henri Poincar'e **5**, 115 (1935)
36. W. Li, K. Nomoto, K. Lee, S.M. Islam, Z. Hu, M. Zhu, X. Gao, J. Xie, M. Pilla, D. Jena, H.G. Xing, Activation of buried P-GaN in MOCVD-regrown vertical structures. Appl. Phys. Lett. **113**, 062105 (2018)
37. J. Wang, L. Cao, J. Xie, E. Beam, R. McCarthy, C. Youtsey, P. Fay, High voltage, high current GaN-on-GaN p-n diodes with partially compensated edge termination. Appl. Phys. Lett. **113**, 023502 (2018)
38. K. Fu, H. Fu, X. Huang, H. Chen, T.-H. Yang, J. Montes, C. Yang, J. Zhou, Y. Zhao, Demonstration of 1.27 kV etch-then-regrow GaN p-n junctions with low leakage for GaN power electronics. IEEE Electron Device Lett. **40**, 1728 (2019)

39. C. Yang, H. Fu, V.N. Kumar, K. Fu, H. Liu, X. Huang, T.-H. Yang, H. Chen, J. Zhou, X. Deng, J. Montes, F.A. Ponce, D. Vasileska, Y. Zhao, GaN Vertical- Channel junction field-effect transistors with regrown p-GaN by MOCVD. IEEE Trans. Electron. Devices **67**, 3972 (2020)
40. J. Hu, Y. Zhang, M. Sun, D. Piedra, N. Chowdhury, T. Palacios, Materials and processing issues in vertical GaN power electronics. Mater. Sci. Semicond. Process. **78**, 75 (2018)
41. H. Fu, K. Fu, C. Yang, H. Liu, K.A. Hatch, P. Peri, D. Herath Mudiyanselage, B. Li, T.-H. Kim, S.R. Alugubelli, P.-Y. Su, D.C. Messina, X. Deng, C.-Y. Cheng, R. Vatan Meidanshahi, X. Huang, H. Chen, T.-H. Yang, J. Zhou, A.M. Armstrong, A.A. Allerman, E.T. Yu, J. Han, S.M. Goodnick, D.J. Smith, R.J. Nemanich, F.A. Ponce, Y. Zhao, Selective area regrowth and doping for vertical gallium nitride power devices: Materials challenges and recent progress. Mater. Today **49**, 296 (2021)
42. R.A. Ferreyra, B. Li, S. Wang, J. Han, Selective area doping of GaN toward high-power applications. J. Phys. D **56**, 373001 (2023)
43. K. Fu, H. Fu, X. Deng, P.-Y. Su, H. Liu, K. Hatch, C.-Y. Cheng, D. Messina, R.V. Meidanshahi, P. Peri, C. Yang, T.-H. Yang, J. Montes, J. Zhou, X. Qi, S.M. Goodnick, F.A. Ponce, D.J. Smith, R. Nemanich, Y. Zhao, The impact of interfacial Si contamination on GaN-on-GaN regrowth for high power vertical devices. Appl. Phys. Lett. **118**, 222104 (2021)
44. J. Hite, M.A. Mastro, Understanding interfaces for homoepitaxial GaN growth, in *Wide Bandgap Semiconductor-Based Electronics*, (IOP Publishing, 2020)
45. M. Noshin, R. Soman, X. Xu, S. Chowdhury, A systematic study of the regrown interface impurities in unintentionally doped Ga-polar c-plane GaN and methods to reduce the same. Semicond. Sci. Technol. **37**, 075018 (2022)
46. B. Li, S. Wang, M. Nami, J. Han, A study of damage-free in-situ etching of GaN in metalorganic chemical vapor deposition (MOCVD) by tertiarybutylchloride (TBCl). J. Cryst. Growth **534**, 125492 (2020)
47. P.-Y. Su, H. Liu, C. Yang, K. Fu, H. Fu, Y. Zhao, F.A. Ponce, Lateral and vertical growth of Mg-doped GaN on trench-patterned GaN films. Appl. Phys. Lett. **117**, 102110 (2020)
48. Y. Turkulets, N. Shauloff, O.H. Chaulker, Y. Shapira, R. Jelinek, I. Shalish, The GaN(0001) yellow-luminescence-related surface state and its interaction with air. Surf. Interfaces **38**, 102834 (2023)
49. D.N. Wijesundera, K.B. Ma, X.M. Wang, B.P. Tilakaratne, L. Shao, W.-K. Chu, The role of flux-focusing in the origin of shoulders in ion channeling angular scans. Phys. Lett. A **376**, 1763 (2012)
50. B. Paszkiewicz, R. Paszkiewicz, A. Szyszka, M. Wosko, W. Macherzynski, M. Tlaczala, R. Kudrawiec, M. Syperek, J. Misiewicz, E. Dumiszewska, W. Strupinski, Study of the activation process of Mg Dopant in GaN:Mg layers. Phys. Status Solidi C **3**, 579 (2006)
51. S. Huang, M. Ikeda, F. Zhang, M. Zhang, J. Zhu, S. Zhang, J. Liu, Achievable hole concentration at room temperature as a function of Mg concentration for MOCVD-grown p-GaN after sufficient annealing. J. Semicond. **45** (2024)
52. J.D. Greenlee, B.N. Feigelson, T.J. Anderson, J.K. Hite, K.D. Hobart, F.J. Kub, Symmetric multicycle rapid thermal annealing: Enhanced activation of implanted dopants in GaN. ECS J. Solid State Sci. Technol. **4**, 382 (2015)
53. T.J. Anderson, J.D. Greenlee, B.N. Feigelson, J.K. Hite, K.D. Hobart, F.J. Kub, Improvements in the annealing of mg ion implanted GaN and related devices. IEEE Trans. Semicond. Manuf. **29**, 343 (2016)
54. K. Hogan, S. Tozier, E. Rocco, I. Mahaboob, V. Meyers, B. McEwen, F. Shahedipour-Sandvik, R. Tompkins, M. Derenge, K. Jones, M. Shevelev, V. Sklyar, A. Lang, J. Hart, M. Taheri, M. Reshchikov, Novel Gyrotron Beam Annealing Method for Mg-Implanted Bulk GaN, in *2019 IEEE International Reliability Physics Symposium (IRPS)*, (2019), pp. 1–6
55. U. Singh, N. Kumar, A.K. Sinha, Gyrotron and its electron beam source: A review. J. Fusion Energ. **31**, 489 (2012)
56. V. Meyers, E. Rocco, T.J. Anderson, J.C. Gallagher, M.A. Ebrish, K. Jones, M. Derenge, M. Shevelev, V. Sklyar, K. Hogan, B. McEwen, F. Shahedipour- Sandvik, p-type conductivity

and damage recovery in implanted GaN annealed by rapid gyrotron microwave annealing. J. Appl. Phys. **128**, 085701 (2020)
57. K.T. Liu, Y.K. Su, S.J. Chang, Y. Horikoshi, Magnesium/nitrogen and beryllium/nitrogen coimplantation into GaN. J. Appl. Phys. **98**, 073702 (2005)
58. A. Uedono, R. Tanaka, S. Takashima, K. Ueno, M. Edo, K. Shima, K. Kojima, F. Chichibu, S. Ishibashi, Dopant activation process in Mg-implanted GaN studied by monoenergetic positron beam. Sci. Rep. **11**, 20660 (2021)
59. V. Meyers, E. Rocco, K. Hogan, B. McEwen, M. Shevelev, V. Sklyar, K. Jones, M. Derenge, F. Shahedipour-Sandvik, P-type conductivity and suppression of green luminescence in Mg/N co-implanted GaN by gyrotron microwave annealing. J. Appl. Phys. **130**, 085704 (2021)
60. V. Meyers, E. Rocco, B. McEwen, M. Shevelev, V. Sklyar, F. Shahedipour-Sandvik, Defect-mediated diffusion of implanted Mg in GaN: Suppressing dopant redistribution by sequential thermal and microwave annealing. J. Appl. Phys. **133**, 155701 (2023)
61. T. To, A.B. Djurisic, M.H. Xie, W.K. Fong, C. Surya, Doping of GaN by Mg Diffusion, in *2002 Conference on Optoelectronic and Microelectronic Materials and Devices. COMMAD 2002. Proceedings (Cat. No.02EX601)*, (2002), pp. 75–78
62. Y. Itoh, S. Lu, H. Watanabe, M. Deki, S. Nitta, Y. Honda, A. Tanaka, H. Amano, Substitutional diffusion of Mg into GaN from GaN/Mg mixture. Appl. Phys. Exp. **15**, 116505 (2022)
63. C.C. Tin, S. Mendis, M.T. Tin, T. Isaacs-Smith, J.R. Williams, A new approach in impurity doping of 4H-SiC using silicidation. J. Appl. Phys. **114**, 244502 (2013)
64. J.-H. Seo, H. Wu, S. Mikael, H. Mi, J.P. Blanchard, G. Venkataramanan, W. Zhou, S. Gong, D. Morgan, Z. Ma, Thermal diffusion boron doping of single-crystal natural diamond. J. Appl. Phys. **119**, 205703 (2016)
65. J.B. Varley, N.P. Allen, C. Frye, K.E. Kweon, V. Lordi, L. Voss, Field Assisted Interfacial Diffusion Doping through Heterostructure Design, US20210257463A1 (19 August 2021)
66. M.A. Briere, In-situ dopant implantation and growth of a III-nitride semiconductor body, US Patent 8.093,597 B2 (10 January 2012)
67. T. Jen, G. Vardar, Y.Q. Wang, R.S. Goldman, Identifying the dominant interstitial complex in dilute GaAsN alloys. Appl. Phys. Lett. **107**, 221904 (2015)

Chapter 21
Structural and Chemical Defect Characterization for Selectively Doped GaN

Yekan Wang, Kenny Huynh, Michael Liao, and Mark Goorsky

1 Introduction

Gallium nitride is a compound semiconductor with a large bandgap (3.4 eV). Owing to certain favorable material properties, high-voltage electrical power switches based on GaN offer advantages over those currently produced using Si [1]. Power electronics represents an increasingly important technology—as noted by the US Department of Energy, electricity accounted for ~40% of primary energy consumption in the United States in 2015. Power electronics switching devices typically convert electricity into forms that are more useful for consumer electronics, the electricity grid, power supplies, solar and wind power, automotive, ship propulsion, and aerospace. These devices are projected to play a significant and growing role in delivering this electricity. It has been estimated that as much as 80% of electricity could pass through power electronics by 2030. Achieving high-power conversion efficiency in these systems, however, requires low-loss power semiconductor switches. Figure 21.1 shows the beneficial properties of GaN compared to silicon.

Y. Wang
Intel Corp, Chandler, AZ, USA
e-mail: zjjxwyk@ucla.edu

K. Huynh
Information Sciences Institute, University of Southern California, Marina del Rey, CA, USA
e-mail: kennyh@isi.edu

M. Liao
Apex Microdevices, West Chester, OH, USA
e-mail: michael@apexmds.com

M. Goorsky (✉)
Materials Science and Engineering, University of California Los Angeles, Los Angeles, CA, USA
e-mail: goorsky@seas.ucla.edu

I. C. Kizilyalli et al. (eds.), *Gallium Nitride and Related Materials*, The Materials Research Society Series, https://doi.org/10.1007/978-3-031-83056-3_21

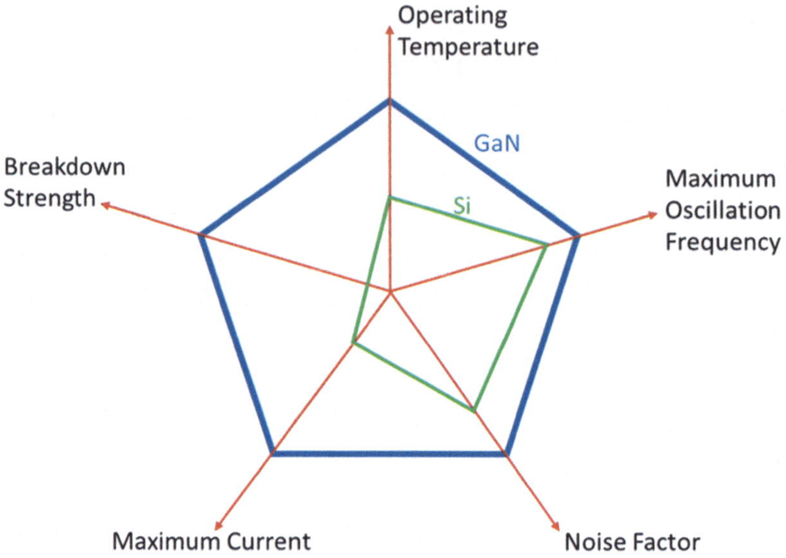

Fig. 21.1 The beneficial materials properties of GaN compared to silicon for high-voltage switching operation

Besides a superior breakdown field strength, the properties of GaN support high temperatures (300–400 °C), high currents, and high switching speeds.

In spite of these benefits, there remain technological challenges associated with GaN that have, until recently, limited the use of gallium nitride for high-voltage switching applications. The importance of substrate crystalline perfection is demonstrated for p-i-n diode structures. One specific challenge for GaN for high-power electronic devices is to develop advances in the process of selective area doping p-type regions, in this case by using ion implantation to selectively make specific regions p-type using magnesium as the dopant implant species. High-quality substrates are also important for understanding the implantation of Mg in GaN. Here, we demonstrate that harmful defects can form during certain annealing steps, but through careful study of the transformation and eradication of these defects, materials suitable for high performance can be achieved.

2 Advances in GaN Substrates

The initial studies of doping GaN via ion implantation were performed on substrates or templates with high defect densities. This was initially due to the fact that most GaN consisted of layers epitaxially deposited on silicon or sapphire substrates. Over time, through the development of different buffer layers and growth optimization, better quality material could be grown on silicon or sapphire (and is used in GaN-based white light emitting diodes), but the dislocation density remained high

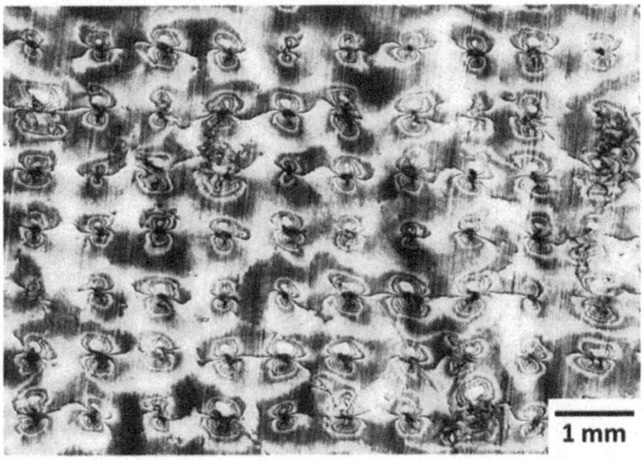

Fig. 21.2 (11–24) double crystal X-ray topography set of superimposed images of a section of a 50 mm diameter A-DEEP wafer

(~10^8–10^9 cm^{-2}). The development of vertical GaN power devices depends heavily on the availability of large-size bulk GaN substrates with low defect density. Due to the high melting point (>2500 °C) of GaN and high equilibrium nitrogen pressure necessary for congruent melting of GaN, it is not feasible to grow GaN substrate from a melt-growth approach, such as the Czochralski method. Therefore, GaN substrates are grown by other techniques [2].

One major effort for state of the art from approximately 2000–2015 was free-standing native GaN wafers produced by hydride vapor phase epitaxy. Two main variants of this growth technique have been developed to minimize the defect density during the HVPE bulk growth. The first method involves growing freestanding GaN substrates with periodically positioned inversion domains known as "cores." The core position is fixed by using a patterned, foreign substrate with a feature spacing of approximately 1 mm. Examples include Dislocation Elimination by the Epitaxial growth with inverse pyramidal Pits (DEEP) and Advance-DEEP (A-DEEP) technologies [3, 4]. During HVPE growth, threading dislocations propagate at the cores and concentrate at the centers of cores, while dislocations in regions between the cores are significantly reduced. As a result, the defects in the A-DEEP GaN substrates vary strongly by location. The dislocation density away from the core centers can be as low as 10^4 cm^{-2} but close to the core centers is reported to be up to 10^8 cm^{-2} [5]. Figure 21.2 is a (11–24) double crystal X-ray topography image from an A-DEEP substrate. This is a "rocked" image, meaning that X-ray sensitive film was exposed, the sample was rotated on omega—rocking curve axis—and another image was exposed. This step repeats and a super-position of these images results [6, 7]. Such measurements are useful when the sample diffracts over a large angular range and features are spread over a wide angular range. In Fig. 21.2, the cores are located at the center of each looped feature. The loops or "wings" extending from each core are iso-deformation contours. One can clearly see the regions between the cores are of very high crystalline quality (uniform intensity).

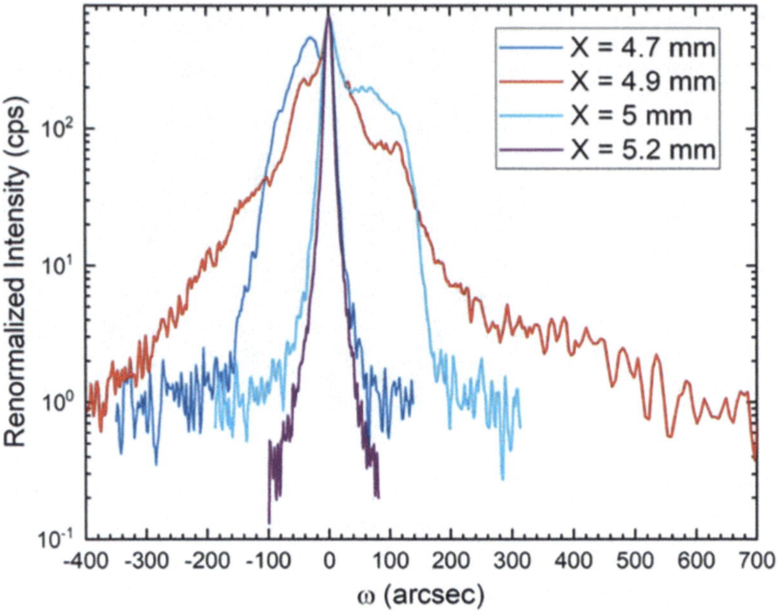

Fig. 21.3 Triple-axis X-ray diffraction rocking curves from p-i-n diodes fabricated on an A-DEEP substrate. The distances in the legend correspond to distances from the wafer flat and are further explained in Fig. 21.4. (From [10], used with permission of IOP Publishing conveyed through Copyright Clearance Center, Inc.)

The second technology is represented by a technique referred to as the Void-Assisted Separation (VAS) method [8, 9]. The VAS method involves the formation of voids between the template (MOCVD GaN on sapphire) and the subsequently grown HVPE-GaN. A thin metallic layer (Ti) is deposited on the surface of the template and annealed in an ammonia atmosphere to form a TiN nanonet. After the cooling process of the HVPE growth, the thick GaN layer was separated from the template by the many voids generated under the TiN nanonet. During HVPE growth, high-density threading dislocations in the GaN template layer were effectively blocked by the TiN nanonet. The dislocation density in the VAS GaN substrate is reduced to an average of $\sim 10^6$ cm^{-2} and the distribution is more uniform compared to A-DEEP GaN substrate [8].

A recent study compared the impact of the defect distributions in GaN substrates on the leakage current in GaN vertical Schottky and PIN diodes for both the A-DEEP and VAS substrates [10]. Identical epitaxial layer deposition and device fabrication were performed for both sets of structures. Noting that extended defects that exist in the substrate will be replicated in the epitaxial layer, Fig. 21.3 shows triple-axis diffraction rocking curves with p-i-n diodes processed on an A-DEEP wafer at different positions across the processed wafer surface. It is noted that the full width at half maximum (FWHM) is typically used as a metric for crystal quality. In this case, the FWHM for all of the scans is ~25 arcsec; which suggests relatively

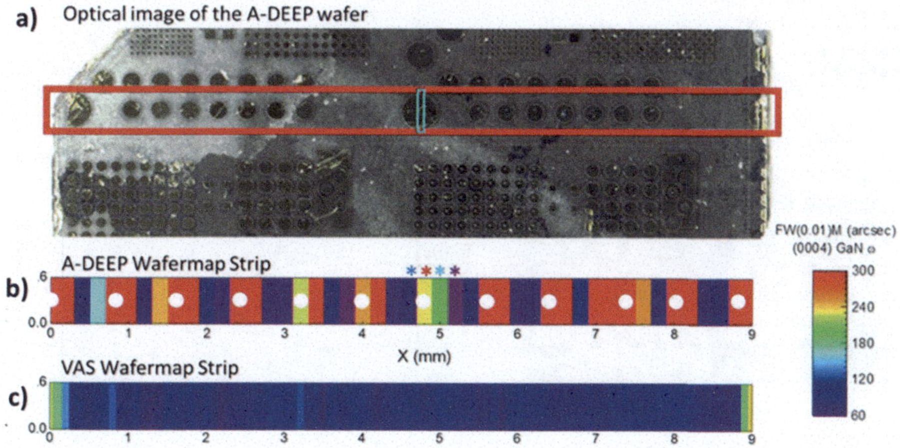

Fig. 21.4 (**a**) is an optical image of a processed p-i-n diode wafer with the red rectangle showing the region that was mapped using TAD rocking curves and the blue rectangle representing the 0.2 mm × 0.6 mm beam height. (**b**) shows the FW(0.01)M from the A-DEEP wafer as a function of position and the white dots correspond to the core regions. The color-coded asterisk features at ~5 mm position correspond to the rocking curves in Fig. 21.3. (**c**) shows a similar map from p-i-n diodes on a VAS substrate. (From [10], used with permission of IOP Publishing conveyed through Copyright Clearance Center, Inc.)

high quality, but it is in the wings of the rocking curve that the major differences can be observed. Therefore, a measure of the full width at one-hundredth (FW(0.01)M) or one-thousandth (FW0.001M) of the maximum provides a distinctive measurable difference among the rocking curves. Plotting the rocking curves on a log intensity scale shows the differences clearly.

Figure 21.4a shows an optical image of the wafer section from which the measurements are taken. The X-ray beam size is reduced to 0.2 mm × 0.6 mm to better distinguish between the core features and the regions between the cores. The rocking curves show dramatic differences as the measurement is taken over different wafer positions. There is not a strong correlation between the FWHM and any processing step, but there is a strong relationship between the diffraction peak width at the full width at one-hundredth of the maximum (FW(0.01)M) and the relative distance to the "core" regions in the A-DEEP substrates. Figure 21.4b shows that the greatest crystalline distortions in terms of the rocking curve width occur at the core regions (represented by the white dots), but between the core regions, the FW(0.01)M gets as low as 70 arcsec, which is typical of higher quality GaN substrate material. The VAS wafer map strip, Fig. 21.4c, shows the rocking curve widths that are much more uniform due to the absence of the core defect structures in the VAS material.

The connection between the defective regions and device performance is described in Fig. 21.5. Figure 21.5a shows a "rocked" double crystal X-ray topography image for the p–i–n diodes fabricated on the A-DEEP wafer. Here, the periodic nondiffracting regions (white features) were observed around the core

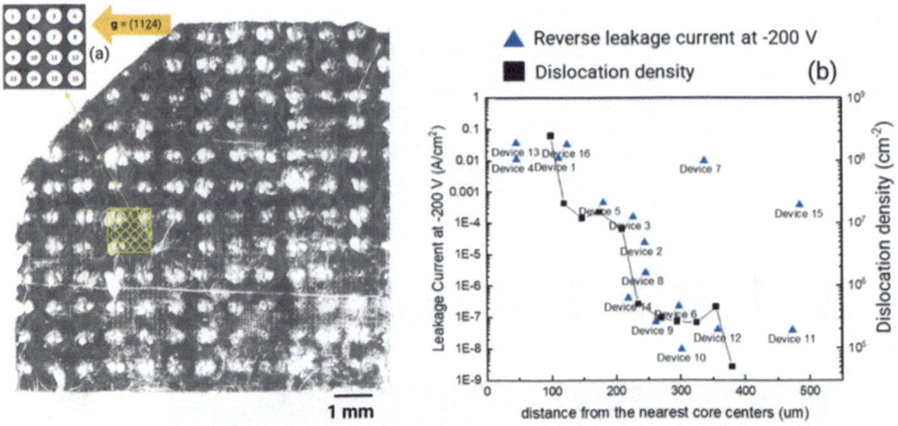

Fig. 21.5 (**a**) (11–24) double crystal topography rocked image with diode structure highlighted in yellow. Other diode patterns are also imaged, demonstrating the usefulness of this topography technique to directly relate device location with the underlying defect structure. (**b**) The reverse leakage current at −200 V for the 16 devices (diameter = 200 μm) and their relative positions were measured with respect to the nearest core center. A literature report [5] of the dislocation density at various distances from the core centers is overlaid on the leakage current plot showing good correlation to the measured leakage. (From [10], used with permission of IOP Publishing conveyed through Copyright Clearance Center, Inc.)

centers. I–V measurements were performed for 16 devices and their positions are highlighted with a yellow circle pattern. In this arrangement, the four devices on each of the corners (Devices 1, 4, 13, and 16) are closest to the core centers. The reverse leakage current at −200 V from the 16 devices was plotted with their distance to the nearest core centers in Fig. 21.5b. Those four devices (1, 4, 13, and 16) closest to core centers showed the highest leakage current, above 0.01 A cm^{-2} at −200 V demonstrating that the devices with the highest leakage current occur at the highest substrate threading dislocation density regions. Overall, as the device gets further away from the core centers, the leakage current decreases. There are a couple of exceptions like device 7 and device 15, where an abnormally high leakage current was observed. We speculate the higher leakage current for those devices to be associated with other factors such as metal contact issues. However, there were no devices located on or adjacent to the core regions that possessed low leakage current. Overall, the lowest leakage current at −200 V is just below 10^{-8} A cm^{-2}, which is over six orders of magnitude lower than the leakage associated with diodes over the high dislocation density core regions.

The VAS-based structures show a different behavior. The threading dislocation density is on the order of 10^6 cm^{-2} over the entire wafer. The comparison of leakage currents for devices on both wafers is shown in Fig. 21.6. The leakage current for the VAS-based structures is rather uniform over the distances associated with the core periodicity of the A-DEEP structures. Of the eight devices measured, all but one had a leakage current in the 10^{-6} A·cm^{-2} range. The uniformity of the devices reflects the uniformity of the defect concentration. It is interesting to note, however, that the

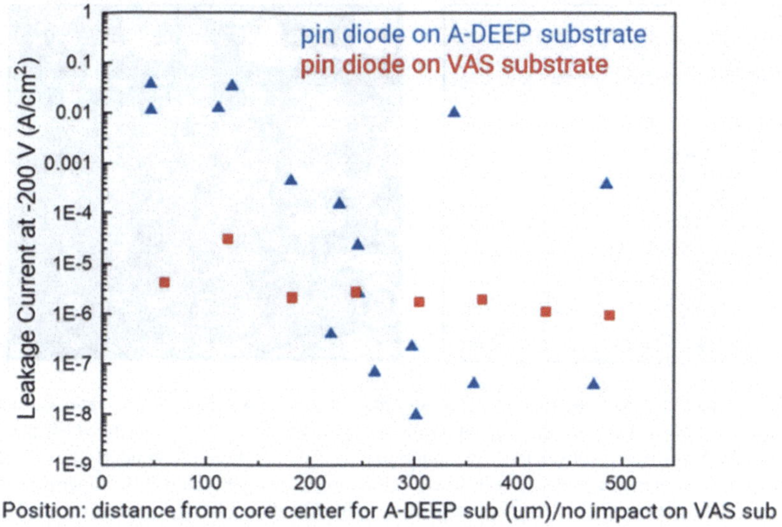

Fig. 21.6 Leakage current at −200 V from the eight devices from the VAS wafer plotted with the leakage current from the 16 p-i-n diodes from the A-DEEP wafer [10]

lowest leakage devices are from the A-DEEP substrate where the best devices (~10^{-8} A·cm^{-2} leakage current) correspond to the regions of the wafer that have a ~ 10^4 cm^{-2} dislocation density. Thus, the best and worst devices are produced on the A-DEEP substrates, and more uniform but intermediate-performance devices are from the VAS substrate-based structure.

Substrate defects also have an impact on the morphology of epitaxial layers. Thick homoepitaxial GaN layers were grown using MOCVD to study the origins of macrosteps and macroterraces that give rise to optically hazy surfaces [11]. To fully utilize GaN in the high-power device technology space, thick drift layers on the order of tens of micrometers on homoepitaxial substrates are needed to fabricate devices operating in the ~kV regime [1, 10, 12–17]. Maintaining smooth surfaces during the growth of thick GaN layers will be essential for optimizing vertical GaN devices because it has been demonstrated that surfaces free of surface morphology (e.g., hillocks and macrosteps/macroterraces) are required to achieve consistent and well-performing devices [1]. Work by Fujikura and Kanno [18] varied growth parameters and substrate miscut angle to study the surface features that gave rise to optically hazy GaN surfaces. The hazy morphology corresponded to wavy surface features of meandering bunched steps consisting of macrosteps. Dot-core GaN substrates were used as defect-engineering samples and showed that localized lattice tilt distortions serve as nucleation sites for these macrosteps and macroterraces to form. Growth over a flat, defect-free material resulted in a featureless material, i.e., surfaces suitable for devices. However, growth over a nondot-core GaN substrate with a threading dislocation density of >10^6 cm^{-2} also resulted in macroterrace

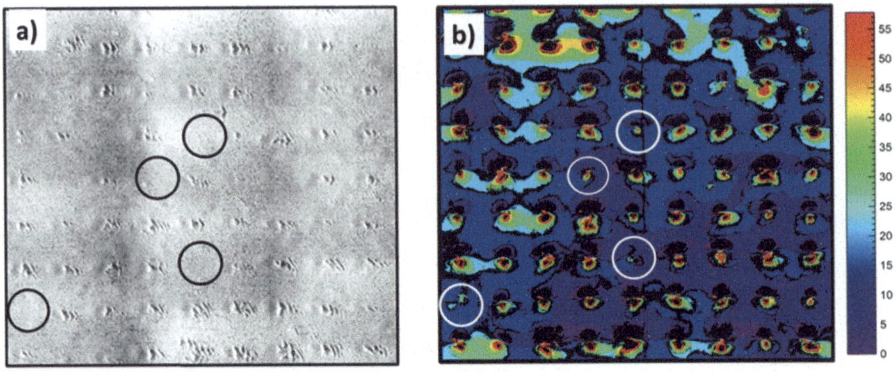

Fig. 21.7 Optical microscope images of (**a**) an optically specular region of the wafer surface. The macrosteps and macroterraces are aligned along the [10], which is also along the substrate miscut direction. Black circled regions in (**a**) correspond to places where there is an underlying core, but no macrostep or macroterrace feature is observed. Lattice tilt maps generated from superimposed and colorized single exposure X-ray topography images for (**b**) the same region as for (**b**). The total range for the lattice distortion map is ~58″ along the rocking curve axis, and the step size is ~7″ between each subsequent exposure. Black corresponds to the regions of the material that diffract outside the total ~58″ scanning range. White circles in (**b**) correspond to the regions where no macrosteps or macroterraces are observed optically on the surface as shown in (**a**) [11], reprinted with permission of AIP Publishing

features made up of hillocks (screw dislocations or nanopipes), demonstrating the benefits of low dislocation density substrates for GaN growth and defect studies.

Figure 21.7a shows the surface of an 11 μm thick GaN epitaxial layer on a dot-core substrate. The observation of the formation of macroterraces forming initially at the dot-core features, but not in the intercore region, is presented. Note the open circles which are regions where cores are expected but the macroterrace features are not present. Figure 21.7b is a lattice distortion map, which is similar to previously described rocked double crystal X-ray topography images. Here, however, the individual images taken at different rocking angles are color-coded to provide angular information that is otherwise lost with standard multiple-image topographs. The region of the wafer is the same for both the optical image and the topograph. These lattice distortion maps show that areas with the same color correspond to where the lattice is aligned and diffracting simultaneously at a given angle along the rocking curve axis. Areas with different colors correspond to regions that are tilted or strained from each other and consequently diffract at a different angle. Having a high color contrast over a relatively short spatial distance, e.g., on the cores where the colors range from purple to red over a ~0.5 mm distance, corresponds to high lattice distortion–curvature in this case because the distortion is due to lattice tilt. While all cores exhibit localized lattice tilt distortion, it can be seen from Fig. 21.7 that not all the core distortions are identical. The distortion field typically spans radially over a ~0.5 mm distance and >12,000 arcsec (3.33°). Interestingly, the distortions over the cores (~1500–~6000) circled in Fig. 21.7b are smaller (~0.20 mm distorted region), and on the surface, there are no macrosteps nor macroterraces over these same cores as shown in Fig. 21.7a. Having the lattice tilt distortions confined to a smaller area, thus resulting in a larger fraction of less defective and flatter material around the core, clearly suppresses the formation of these macrostep and macroterrace features.

Thick homoepitaxial GaN layers were grown using MOCVD to study the origins of macrosteps and macroterraces that give rise to optically hazy surfaces. Dot-core GaN substrates were used as defect-engineering samples and showed that localized lattice tilt distortions serve as nucleation sites for these macrosteps and macroterraces to form. Growth over a flat, defect-free material resulted in a featureless material, i.e., surfaces suitable for devices. However, growth over a nondot-core GaN substrate with a threading dislocation density of $>10^6$ cm^{-2} also resulted in macroterrace features made up of hillocks (screw dislocations or nanopipes), demonstrating the benefits of low dislocation density substrates for GaN growth and defect studies.

To date, the highest quality GaN substrate is produced by using the ammonothermal method [19]. The ammonothermal growth approach is an analog of the hydrothermal method, used in the industrial production of quartz. Briefly, during growth, GaN source crystals are dissolved in supercritical ammonia in a reactor (hundreds of MPa). The solution is then transported to another chamber, where the solution is supersaturated, and crystallization of GaN occurs on top of GaN seed crystals. To facilitate ammonothermal growth, there needs to be an appropriate temperature gradient between the dissolution and crystallization zones. The typical growth temperature is in the range of 400–750 °C [20]. Some mineralizers are added to ammonia in the dissolution zone to enhance the solubility of GaN. Depending on the mineralizers, the growth can be proceeded in a different environment: basic or acidic. The growth rate of ammonothermal GaN is on the order of tens of μm/day, which is relatively slow. However, it produces GaN substrate with the lowest defect density (on the order of 10^4 cm^{-4}) from any bulk GaN substrate [20]. Ammonothermal-grown substrates became more widely available recently and these wafers with known impurity content and high crystalline perfection led to a much better understanding of defects that are introduced due to the implantation and annealing process (in the device active regions) as opposed to being present in the preimplanted substrate).

For comparison of the crystalline quality of good A-DEEP material and the GaN produced by the ammonothermal technique, Fig. 21.8 shows (11–24) double crystal X-ray topographs from each. The topograph from the ammonothermal wafer is from a single exposure, demonstrating that the lattice planes are flat to within a few arcsec over this range. The A-DEEP image is similar to those above and includes six exposures with an interexposure step size of 10 arcsec.

Overall, the availability of high-quality, 50 mm diameter GaN has made it feasible to conduct a wide variety of studies on GaN, in particular ion implantation. The lower perfection of other GaN substrates and the rather large X-ray diffraction peak widths and high defect densities associated with GaN on silicon, for example, will mask defects introduced and formed during the ion implantation process.

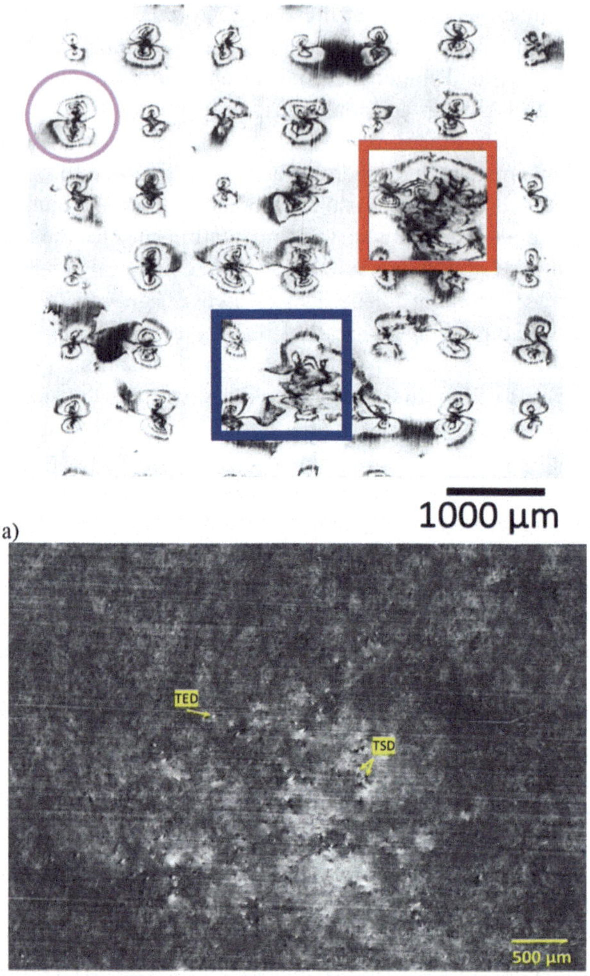

Fig. 21.8 (11–24) double crystal X-ray topographs: (**a**) Rocked image covering 60 arcsec range showing distorted core regions (heavily distorted regions in red and blue box highlights; typical distorted region in a pink circle) and (**b**) single topography image from ammonothermal GaN substrate showing an undistorted crystal with the entire image diffracting over just a few arcsec

3 Ion Implantation Background

Ion implantation is a widely used technique in semiconductor device fabrication to introduce dopant species into the near-surface region. Implantation is attractive because it provides a means to introduce dopants into specific regions of the surface. Implantation also allows for controlled dopant depth profiles to be produced. The general implant process starts with an ionized source. The dopant ion species are accelerated in an evacuated chamber and only species with a specific mass-to-charge ratio are guided toward the wafer target. The bias applied between the source region and the wafer target is such that the ions possess an energy of keV for a bias of kV. As the dopant species hit the target, they decelerate due to collisions with the lattice atoms and with electrons in the target. The depth of penetration of the implanted species can be modeled with a modified Gaussian distribution with the key parameters being the depth at which the maximum concentration of ions stops (projected range Rp) and the distribution of ions is defined by the straggle, ΔRp, which is the distance from the projected range where the concentration is 1/exp(0.5) (~0.606) of the projected range (maximum) concentration.

The interaction of the high-energy implant species and the host atoms results in varying degrees of lattice distortions. First, the implanted species can simply become embedded in the lattice framework and cause local elastic strains. Next, the implant species can collide strongly with the host atoms, and transfer some of its energy to the host atoms. If the energy is great enough, the host atom will be knocked off its lattice site producing a vacancy, and can also embed in the crystal as an interstitial species, but not at a lattice site and thus also locally strain the surrounding lattice. Third, if the energy transferred from the implant species to the host atom is great enough, the host atom can be energetic enough to knock other host atoms off their lattice sites, producing what is known as a damage cascade. In order for the implanted species to become electrically active dopants, the structure can be annealed to allow the species to diffuse to lower energy configurations, preferably to return to the original lattice structure with the implanted species occupying crystallographic lattice sites; in this case, the Mg on Ga sites. A general rule of thumb is that an effective implant anneal temperature should be on the order of two-thirds of the melting point of the host material [21]. However, depending on the annealing temperature, the implant-related defects may form secondary defects which prevent the return to a pristine crystallographic structure and also reduce the fraction of implanted species that occupy lattice sites (the fraction of implanted species that become electrically active dopants is the activation efficiency). These secondary defects can be extended defects such as stacking faults, crystal structure changes from, for example cubic to hexagonal, dislocation loops, or precipitates.

Ion implantation is commonly used for silicon-based devices. Among the many benefits of using implantation are (i) the ability to controllably introduce dopant species into a planar structure through controlling the dose during the implant and the implanted species energy, (ii) the ability to introduce dopant in specific regions through using a mask with openings, and (iii) the ability to controllably introduce not only specific species but even to control the isotope type of the implanted species. However, ion implantation is not as widely employed for compound semiconductors. There are several reasons for this. First, epitaxial processes, particularly heteroepitaxy, are common for compound semiconductors, and dopant incorporation is included during the epitaxial growth. Second, the high temperatures (2/3 T_{melt}) typically required for dopant activation are problematic for many III-V compound semiconductors because the group V vapor pressure over these materials is rather high; for GaN, the nitrogen vapor pressure is one atm at ~840 °C [22]. Therefore, one must either utilize a stable capping layer—one that neither reacts nor physically dissociates during the dopant activated anneal, combined with a novel high-temperature spike rapid thermal anneal processes [23–26]—or to anneal under ultra-high pressure (~1 GPa at 1400 °C) [27]. Fourth, the crystalline quality of the implanted layer can strongly impact dopant activation and defect formation, and can mask the ability to separate defects introduced by implantation from those either present in the preimplanted wafer or defect configurations that are caused by the presence of the preimplant defects.

4 Prior Advancements in Dopant Ion Implantation in GaN

Several excellent review articles tracing the trajectory of dopant implantation in GaN are available. These reviews [21, 28, 29] demonstrated that both n-type and p-type dopant activation were achievable, but that n-type doping was somewhat more successful than p-type doping. For p-type doping, Mg became the dopant of choice (for epitaxy as well as implantation), due, among other reasons to its lower dopant ionization energy than other Group II elements. Initially, typical implantation annealing temperatures for p-type doping were in the range of 1000–1100 °C, [30], increasing to 1200 °C [31], and then up to 1400 °C [32], thus getting nearer the 2/3 T_{melt} "rule." The work by Anderson and Greenlee and by Breckenridge highlighted two distinct processes for ion implantation annealing. Anderson, Greenlee, and colleagues [23–26] demonstrated the concept of Symmetrical Multicycle Rapid Thermal Annealing (SMRTA) in which the first step is a 30 min anneal at ~950–1000 °C, followed by several (10–20) metastable high-temperature pulses (1400–1500 °C) separated by 60 s, and finally an additional 30 min anneal (950–1050 °C). The structure includes a capping layer (SiN, AlN) which provides stability at the lower temperature annealing steps; the high-temperature rapid thermal annealing pulses are of short enough duration that the cap does not degrade. Effective Mg ion activation and recovery of the implant-induced strain were achieved using the SMARTA approach for GaN on sapphire structures [33]. Alternately, ultra-high pressure annealing (UHPA) involves annealing in a nitrogen overpressure of one GPa [34]. The high pressure maintains the surface stoichiometry and prevents nitrogen dissociation from the surface without the need for a capping layer. With UHPA, higher temperatures can be used as many mask materials for GaN breakdown at temperatures above 1400 °C. While the SMARTA approach includes rapid temperature increases (after a single, lower temperature first step), the UHPA approach typically occurs with very slow heating rates (~0.5 °C/s) and thus significant annealing, diffusion, or reactions may take place during the ramp up to the final temperature and the ramp down to room temperature. Note that these examples are representative of a particular time frame and that there are many other references that could be included here that show this same overall trend. A more detailed assessment of the recent, high-temperature implant activation annealing research is provided below.

5 Recent Advances in p-Type Implantation and Annealing of GaN

Improvements in substrate quality, epitaxial growth, and advances in ion implantation annealing provided the foundation for achieving high p-type activation efficiency in GaN. This is an important prerequisite for the manufacture of high-power, high-performance transistors. An initial study in this sequence [35] noted that Mg-implanted GaN exhibited Mg-containing clusters when annealed at 1100–1300 °C (5 min, rapid thermal annealing 1 atm N_2, epitaxial layer on an HVPE substrate). No evidence for electrical activation was provided, however. On the other hand, Breckenridge et al. [32] determined that full dopant activation was achieved at 1400 °C (using UHPA and an epitaxial layer on an ammonothermal substrate) with significant diffusion; and also observed lower activation at 1300 °C. Iwata et al. [36] observed inversion domains that contained Mg upon annealing Mg-implanted GaN at 1300 °C (0.5 min, 1 GPa N_2, epitaxial layer, substrate not reported) but not at 1480 °C (0.5 min). Likewise, Uzuhashi et al. [37] compared Mg-implanted GaN (epitaxial layer and HVPE substrate with 10^6 cm^{-2}—this is likely the VAS or similar-type substrate) annealed at 1300 °C with an AlN mask (5 min. 0.1 MPa N_2) and observed the significant formation of Mg-containing defects. Mg-implanted GaN annealed by UHPA (5 min, 1400 °C, 1 GPa N_2), while still possessing Mg-containing defects, showed a much lower concentration of these Mg-containing defects and higher (~70%) activation efficiency. With those studies, the understanding of Mg ion implantation and annealing improved significantly. However, the prior studies do not provide information about the thermal stability of the Mg-trapping defects or a significant comparison of results from UHPA-treated samples at various annealing temperatures.

Here, we build upon the previous work of Mg-implanted GaN by providing a fundamental understanding that directly relates to defects and electrical performance. Based on the defect characterization method with X-ray scattering and electron microscopy techniques that we developed previously [10, 11, 32], this study aimed to assess: (i) the key residual defects after annealing that have a significant impact on the dopant activation efficiency; and (ii) the kinetics of the defect formation and dissolution at different temperatures. In this work, the stability of the Mg-enriched defects formed at 1300 °C was investigated by subsequent high-temperature (>1400 °C) annealing. These results are anticipated to guide the key processing steps and requirements to achieve high-activation efficiency p-type doping for vertical GaN devices.

In the next section, ion implantation of Mg and the impact of annealing up to 1300 °C are addressed [33]. In the following section, higher annealing temperatures are studied. The structures are 1.5 μm GaN epitaxial layers on (0001) ammonothermal GaN substrates. The magnesium ions (2×10^{14} cm^2 dose) were implanted at room temperature at an accelerating voltage of 100 keV. The peak concentration was calculated to be 1.6×10^{19} cm^{-3} at a depth of ~100 nm.

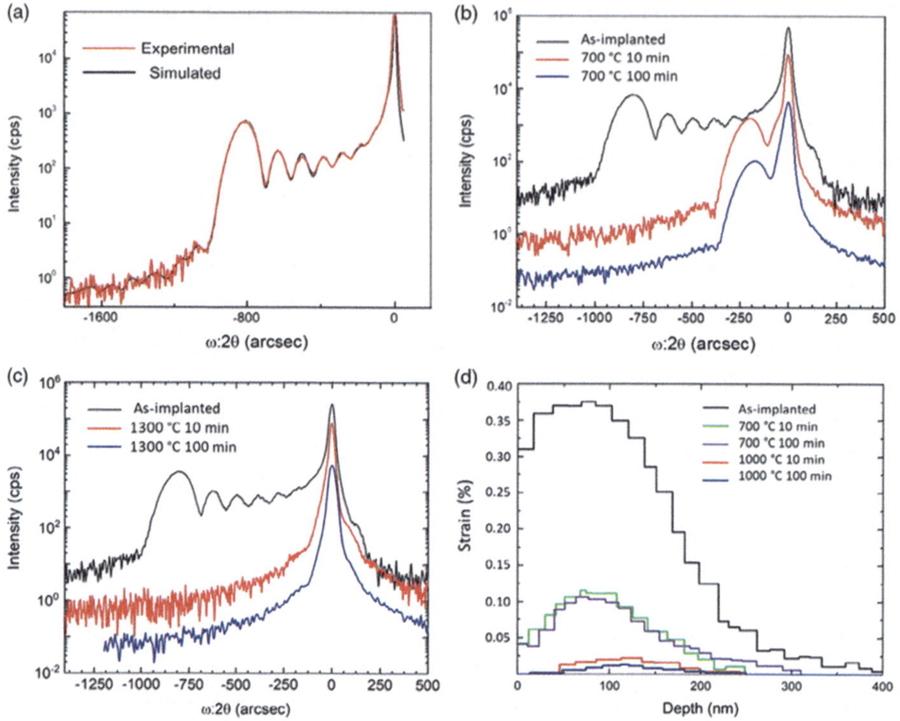

Fig. 21.9 (**a**) Triple-axis XRD ω:2θ scan of the as-implanted sample and a simulated scan using following the profile in (d). (**b**) Triple-axis XRD ω:2θ scans after annealing at 700 °C for 10 and 100 min. (**c**) Triple-axis XRD ω:2θ scans after annealing at 1300 °C for 10 and 100 min. (**d**) Simulated strain profiles before and after annealing. No residual strain remains after annealing at 1300 °C for 10 min [33], reprinted with permission of John Wiley and Sons

Postimplantation annealing was conducted at three different temperatures (700, 1000, and 1300 °C) for 10 and 100 min under UHPA conditions (1 GPa N_2).

Figure 21.9 shows X-ray diffraction (0004) ω:2θ scans from the samples implanted with a 100 keV, 2×10^{14} cm^{-2} dose. This corresponds to a vertical scan (dotted line) in Fig. 21.13a below. Figure 21.9a is the scan from the as-implanted sample. A series of fringes to the left of the main (0004) substrate peak (referenced to zero arcsec) correspond to the effect of the implant. The implant species embed within the lattice and knock a small fraction of the target atoms off their lattice sites and these species also embed in the lattice. These have the effect of dilating the lattice. However, the implant depth is very small compared to the thickness of the wafer. Therefore, the unstrained bulk of the substrate constrains the lattice parameter in-plane, allows for further increase in strain perpendicular to the surface, and produces a pseudomorphic implanted layer. The out-of-plane lattice strain follows a distorted Gaussian profile. A simulated scan with the strain profile as shown in Fig. 21.9d is included in Fig. 21.9a. Figure 21.9b shows how the defects reconfigure after annealing at 700 °C and reduce the strain. Interestingly, the first few

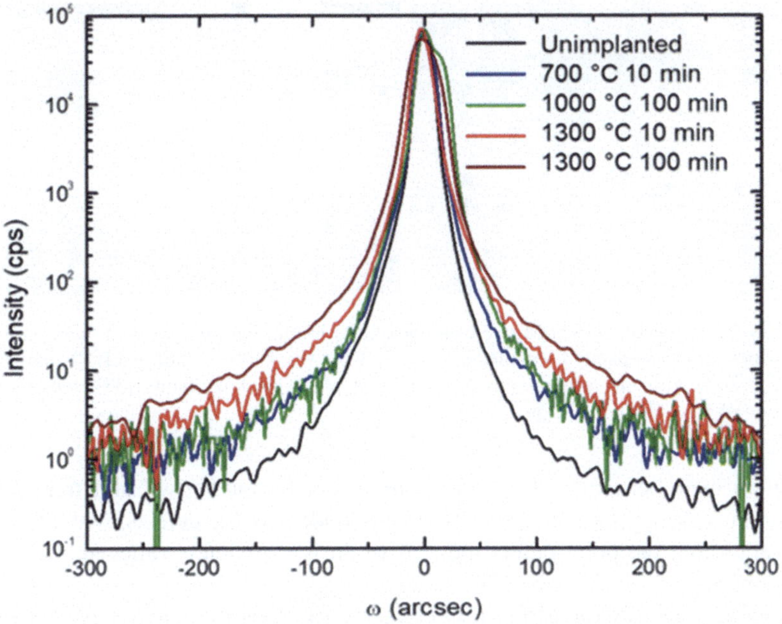

Fig. 21.10 Triple-axis (0004) rocking curve measurement showing diffuse scattering in implanted and annealed samples compared to an unimplanted wafer [33], reprinted with permission of John Wiley and Sons

minutes at this annealing step seem to have the most significant impact with only a small additional change over the next 90 min. Figure 21.9c shows a similar change after annealing at 1300 °C, with the further observation that the point defect-induced strain is completely relieved after only a few minutes. Figure 21.9d shows the strain profiles from simulated scans after each annealing step.

The removal of the strain has two interesting aspects. First, the removal of the strain does not necessarily mean that the implanted species and the knocked-on host atoms return to lattice sites. Secondary defects may form and these defects might not distort the host lattice in a manner which includes significant strain. However, the presence of such extended defects can impact the host crystal in other ways. Figure 21.10 shows the triple-axis diffraction rocking curve widths from the (0004) reflection. Compared to an unimplanted wafer (and note that as-implanted wafers exhibit very similar rocking curves to the unimplanted case), the diffuse scattering from the annealed structures indeed exhibits more diffuse scattering. Generally, all of the annealed samples show extended diffuse scattering but the higher temperatures and longer times produce more diffuse scattering. Indeed, the sample annealed at 1300 °C shows the greatest amount of scattering. Just as in the results of the ADEEP substrates, the differences are noted primarily in the tails of the diffraction peaks. Typically, diffuse scattering of this sort is related to extended defects. This indicates that there are extended point defect clusters or dislocations

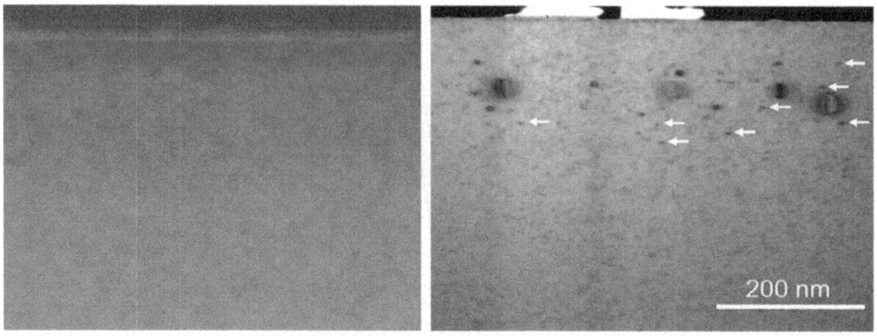

Fig. 21.11 Cross-sectional STEM two beam condition bright-field images with different diffraction vectors: (**a**) as-implanted sample, g = <11–20>; (**b**) 1300 °C annealed 10 min sample, g = <11–20> with the arrows showing the basal plane stacking faults [33], reprinted with permission of John Wiley and Sons

present in these structures [38], but their origin is not known and transmission electron microscopy measurements best help to identify the defects, with the further observation that the point-defect-induced strain is completely relieved after only a few minutes.

Scanning transmission electron microscopy (STEM) measurements confirm the interpretation from the X-ray rocking curve measurements that extended defects are present in the annealed structures. Figure 21.11a below shows the STEM image under two beam (11–24) imaging conditions of an as-implanted structure. No clear discernable features are present, which is consistent with the rocking curve widths observed for the as-implanted sample—the absence of diffuse scattering indicated that high levels of extended defects were not present. However, for the sample annealed at 1300 °C for 10 min (Fig. 21.11b), several types of defects are present. Two beam condition scanning transmission electron microscope (STEM) bright-field images with the diffraction vector g = <11–20> are shown for the as-implanted and for the sample annealed at 1300 °C, 10 min sample. Extended defects were observed to a depth of ~200 nm (which corresponds to the straggle added to the projected range) and consist of three different morphologies: lines, dots, and loops. The small line-shape defects oriented parallel to the sample surface (highlighted with small arrows in Fig. 21.11b are basal plane stacking faults (BSFs). BSFs have been previously observed in implanted GaN as well as in other wurtzite III–V semiconductors [39–41]. The loops are confirmed to be vacancy dislocation loops with these measurements. Small defect cluster dots are also visible.

The high-resolution TEM image of the BSFs in 1300 °C, 10 min sample is shown in Fig. 21.12 in which nanoscale regions of cubic GaN are observed within the hexagonal phase. Higher magnification images in these highlighted regions show the lattice fitted with ball-and-stick models and confirms a change in the stacking sequence in planes from ABAB (wurtzite) to ABCABC (zincblende) and the zincblende [111] is parallel to wurtzite [0001]. Such transformations had been observed in 4H SiC after implantation, which included cubic regions [42]. This transformation is generally related to the stacking fault energy of the material. The

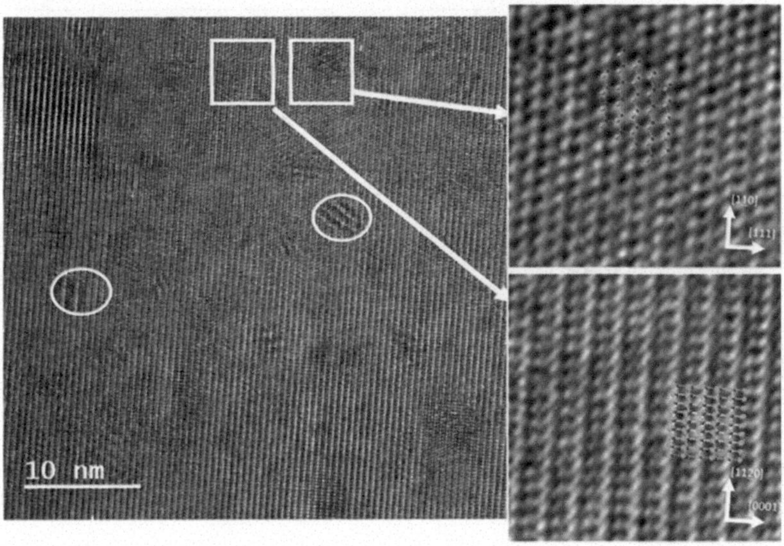

Fig. 21.12 HRTEM image of the implanted region in 1300 °C 10 min sample. The ball-and-stick model confirms a change in the stacking sequence in planes from ABAB (wurtzite) to ABCABC (zincblende). The two circles highlight regions with Moiré fringes [33], reprinted with permission of John Wiley and Sons

stacking fault energy of GaN is higher than SiC, but the formation is stacking faults in GaN is well known, so a similar transition here is not unexpected. Small nanoscale regions also show Moire fringes, indicating that within the thickness of the TEM sample (~80 nm), there is rotation of the lattice planes and can also be ascribed to dislocations.

Another aspect of the strain relief process is that the strain removal appears to be a transient process—very rapidly at first and then almost imperceptibly further as was noted above by the small changes when comparing an anneal for 10 min versus 100 min at either 700, 1000, or 1300 °C. One consideration that is lost or ignored in many of the studies of GaN annealing using UHPA is that the ramp-up times can be substantial compared to the annealing time at the peak temperature. The typical ramp rate published for UHPA is on the order of 0.5 °C/s. Therefore, the time spent between 1200 °C and 1300 °C (>3 min) is longer than a 0.5-min anneal at 1300 °C and is in the same order as 5-min anneals which are more common under UHPA. This suggests that the time at the peak temperature may not provide sufficient detail about the thermal environment of the sample, especially when trying to assess the initial fast change in defect concentration—that may be associated with the role of annealing during the ramp-up step.

One other observation that stems from the X-ray scattering measurements is that the strain induced by the implantation is pseudomorphic. In other words, the strained implanted region - orders of magnitude thinner than the unstrained remainder of the epitaxial layer and the substrate - experiences a compressive in-plane stress, so the

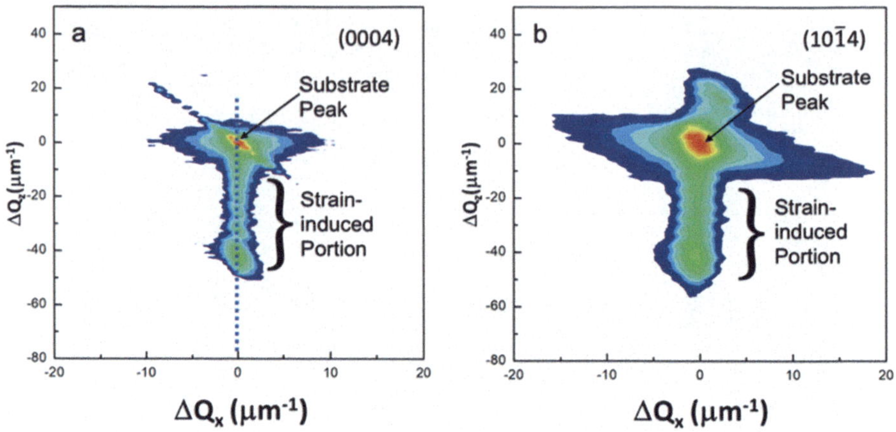

Fig. 21.13 X-ray reciprocal space map for the as-implanted sample showing (**a**) the GaN (0004) RELP and (**b**) the (10–14) RELP. The dashed vertical line in (**a**) corresponds to the ω:2θ scan in Fig. 21.9

layer can only expand perpendicular to the surface. This is demonstrated in Fig. 21.13 below which includes the (0004) and the (10–14) regions of reciprocal space. The origin of each graph is the substrate reciprocal lattice point. The vertical alignment of the strain-induced portion with the substrate peak in the (10–14) graph confirms that the implanted "layer" is commensurate with the substrate and hence is pseudomorphic.

As noted above, annealing of Mg ion-implanted Ga using UHPA at temperatures of 1300 °C [32] indicated that partial activation efficiency was attained and nearly full activation was achieved after annealing at 1400 °C. Mg segregation in inversion domains had been observed 20 years ago in highly Mg-doped epitaxial layers with Mg concentration above the $10^{19}\,cm^{-3}$ level [43, 44] and was determined that these inversion domains play a role in the free carrier reduction in highly Mg-doped GaN. However, the question remained how these inversion domains affect the dopant activation efficiency in ion-implanted material and if they remain stable after annealing at higher temperatures and/or for longer times. In our earlier study, postimplantation annealing at 1300 °C for 10 and 100 min resulted in changes in both defect concentration and defect structures. Therefore, when the annealing temperature approaches ~two-thirds of the GaN melting point, it is expected that there would be further lattice rearrangement and reconfiguration of the defect structures. Additionally, none of the earlier studies assessed the stability of the Mg-enriched defects that formed at 1300 °C after subsequent higher temperature annealing.

We leveraged the current review of Mg-implanted GaN to provide a fundamental understanding that directly relates to defects and electrical performance. Based on the defect characterization method with X-ray scattering and electron microscopy techniques that we developed previously, this study aimed to assess (i) the key

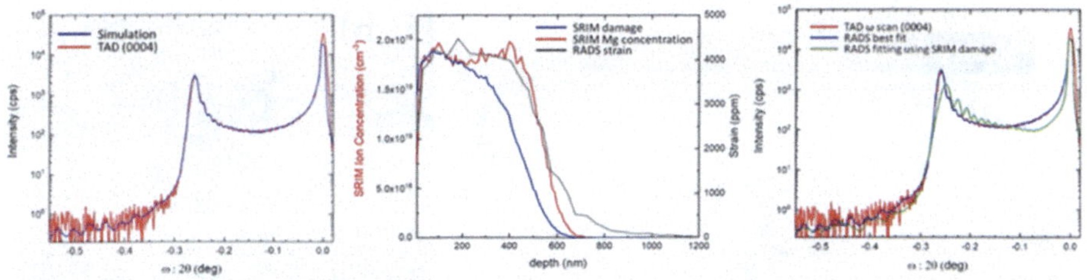

Fig. 21.14 The (0004) triple-axis diffraction ω:2θ scan for the as-implanted sample and the simulation that matches the experimental profile are superimposed in (**a**). (**b**) shows the initial magnesium concentration associated with the multiple implants as well as the strain profile that produced the simulated diffraction pattern in (**a**) as well as the displaced atoms (damage) profile due to the implant. (**c**) includes a simulated pattern that would be expected from the displaced atoms alone

residual defects after annealing that have a significant impact on the dopant activation efficiency and (ii) the kinetics of the defect formation and dissolution at different temperatures. In this work, the stability of the Mg-enriched defects formed at 1300 °C is investigated by subsequent high-temperature (>1400 °C) annealing. These results are anticipated to guide the key processing steps and requirements to achieve high-activation efficiency p-type doping for vertical GaN devices.

In the subsequent study, GaN epitaxy films with a thickness of 2.5 μm were on ammonothermal GaN (0001) substrates. The Mg was implanted at room temperature using six different accelerating voltages (350, 200, 150, 100, 60, and 25 keV) at a total dose level of 1×10^{15} cm^{-2} to produce a uniform concentration thickness layer. The maximum concentration was calculated to be 1.8×10^{19} cm^{-3} from the surface to a depth of ~500 nm. Postimplantation annealing was performed on the implanted wafers at temperatures of 1000, 1300, 1400, and 1500 °C under UHPA conditions (with a ramp rate of 0.65 °C/min) for a duration of either 10 min or 100 min. Multiple annealing sequences were also performed. For example, a wafer was annealed at 1000 °C, analyzed, and then further annealed at, for example, 1300 °C (or annealed at 1300 °C and then at 1450 °C).

The triple-axis X-ray ω:2θ line scans near the GaN (0004) peak for the as-implanted sample is shown in Fig. 21.14a. The presence of the peak intensity and a plateau of scattered intensity between that peak and the substrate peak is typical for such a multistep implant. Figure 21.14b shows the strain model (RADS strain) that fits the experimental curve and also includes the Mg implant concentration profile as well as the damage profile expected from the implant conditions. The strain profile appears to track the Mg implant concentration, while most of the displaced species are expected to be closer to the surface. However, a simulated rocking curve from just the displaced host atoms is shown in Fig. 21.14c for comparison. Therefore, based on this observation, and other implanted structures, the overall strain profile is likely to be associated with a combination of the implanted species and the distortions introduced by the displaced target atoms. As

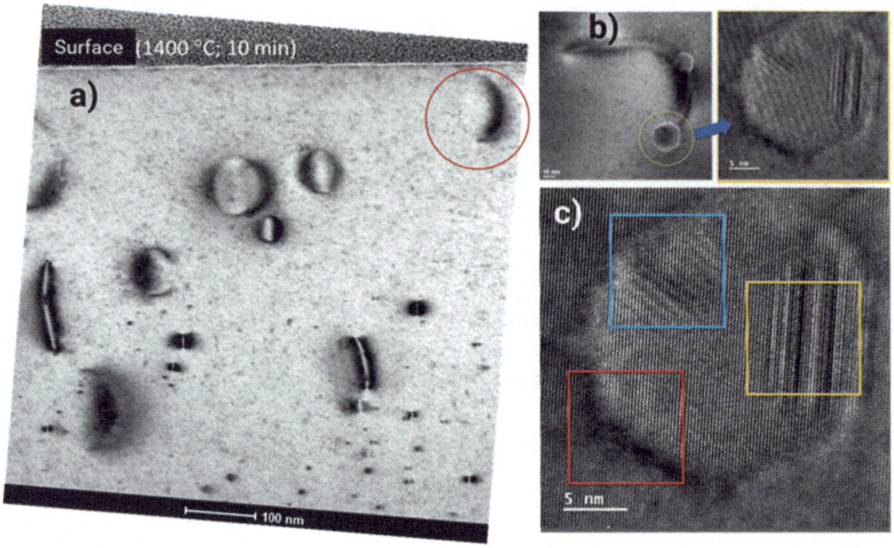

Fig. 21.15 (**a**) (11–20) Two beam STEM images from sample annealed at 1400 °C; 10 min. (**b**) HRTEM image from the defect circled in part (**a**). (**c**) Hexagonal feature details near the edge of the loop structure are observed in (**b**)

with the single implant samples, the implant-induced strain was completely removed after each annealing sequence.

The samples annealed at 1400 and 1500 °C showed very similar defect structures and represented by the annealing at 1400 °C; 10 min. Figure 21.15a shows that the prominent features after annealing at and above 1400 °C are prismatic dislocation loops. Because those loops lie on crystallographic planes that are not perfectly parallel to the sample orientation, i.e., (10–10), only a portion of the loop may be visible. On many of these loops are small patches, as observed under HRTEM conditions in Fig. 21.15b, and sections of which are highlighted in Fig. 21.15c. Within the patch, there are two interesting observations. First is the presence of stacking faults boxed in red and evident in the FFT in Fig. 21.15c. Next along the perimeter of the patch, there are additional diffraction spots that exhibit the same pattern that was observed along the perimeter of the dislocation loop seen in Fig. 21.15c. This corresponds well with the speculation from Iwata et al. [36] that the small patches along these loops are nanoscale versions of the dislocation loops themselves. On some of the dislocation loops, there are some patches with darker contrast (examples are given with yellow arrows). They were observed in all three samples annealed at and above 1400 °C. These small patches include regions with high stacking fault densities (blue and gold boxed regions) and cubic (red boxed region) nanophase. Note that magnesium agglomeration is absent; there is no clustering around the extended defects. The absence of any clustering or trapping of Mg is consistent with related studies which show that the implant activation efficiency is approximately unity under these annealing conditions.

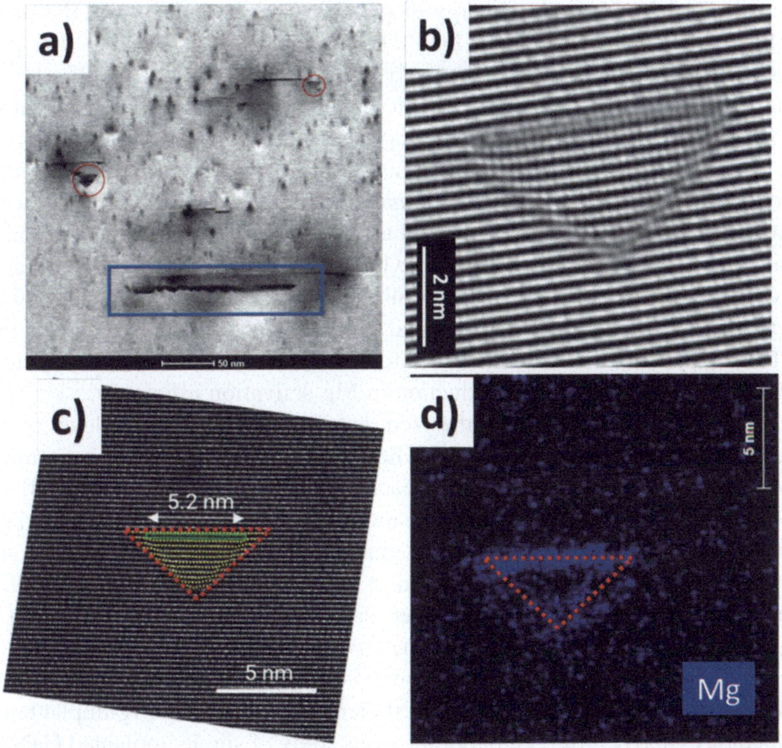

Fig. 21.16 (**a**) STEM image of the 1300 °C 10 min sample showing examples of pyramidal inversion domains (circled in red) and a trapezoidal inversion domain (boxed in blue); (**b**) HRTEM image of a pyramidal inversion, (**c**) domain showing the {11–23} facets (highlighted with dotted white lines) and the extra layer of atoms near the base (boxed in green); (**c**) BF image of a pyramidal inversion domain; (**d**) EDX map showing the Mg segregation in the pyramidal inversion domain shown in (**c**) [47], (**a**) reprinted with permission of AIP Publishing

The 1300 °C 10 min sample exhibits a different defect structure than the samples annealed at higher temperatures. As shown in Fig. 21.16, triangular and trapezoidal shape defects were observed. An unmarked high-resolution TEM measurement shows one of the triangular-shaped defects (Fig. 21.16b). The triangular defects showed structural characteristics which indicate that they are pyramidal inversion domains (PIDs). The base of the PIDs is along the (0001) plane and the sidewalls (highlighted with dotted lines) are along the (11–23) planes, inclined at ~47°, as shown by the high-resolution TEM image in Fig. 21.16c. In postprocessing, the images were aligned and the series of images were compared to atomic models reported in the literature of PIDs that describe that extra plane of atoms along the basal plane [44]. The lattice is distorted inside the PID when compared to the surrounding GaN matrix. An additional atomic layer (highlighted with a green box) was observed near the base of the pyramid (basal plane of GaN), causing lattice bending of the following ten layers. For an individual PID [the STEM BF image is shown in Fig. 21.16c], the energy-dispersive X-ray analysis (EDX) map in Fig. 21.16d shows the Mg signal at the PID. From the EDX map and the line profile, an increased Mg signal is observed at the base of the PID (~9 at. % or ~8×10^{21} cm^{-3}), where the extra layer of atoms is located, with a decrease in the Ga signal at the same position. This indicates that a compound including Mg and N forms in the PIDs. The PID defects are in the initial stages of development and will eventually evolve into the trapezoidal inversion domains (TIDs). In fact, the center part of the largest TID in Fig. 21.16a (boxed in blue) consists of a series of small, aligned pyramids that form a sawtooth pattern, which is the basis of our claim. Plan view TEM of these defects observed in epitaxially grown GaN also supports the claim that these triangular defects have a pyramidal base [45] and are not triangular prisms [46]. Thus, it appears that both the PID and TID defects contain electrically inactive Mg atoms and contribute to the reduction in Mg activation efficiency.

Vennégues et al. [43, 44] hypothesized that introducing a higher level of Mg in GaN results in the formation of Mg_3N_2 where inactive Mg atoms are accommodated into other sites or phases; a conclusion reached by Hansen et al., as well [48]. However, our results do not support the formation of Mg_3N_2, which is characterized by two monolayers of Mg at the inversion domain boundary. The PIDs and TIDs are not observed in any of the samples annealed at and above 1400 °C in this work. This suggests that the PIDs and TIDs either do not form or are dissolved at high temperatures (≥1400 °C) and are formed only at temperatures <1300 °C after implantation and annealing and after epitaxial growth at 1150 °C. This provides insight for improving dopant activation efficiency by annealing Mg-implanted GaN at temperatures above 1400 °C. In our previous study of single implanted GaN, PIDs were not observed when annealed at 1300 °C for 10 min. While the peak concentration of Mg from the single implant is $1.6 \times 10^{19}\,cm^{-3}$ in the as-implanted case, after annealing, diffusion causes the peak concentration to be reduced. Simulations of the Mg concentration profile after annealing at 1300 °C for 10 min (diffusivity of $2.9 \times 10^{-12}\,cm^2/s$) reveal a reduction to ~$2 \times 10^{18}\,cm^{-3}$. This agrees well with studies of PIDs reported in the literature for both epitaxial grown and implanted GaN where the Mg concentration threshold to form PIDs is above $1 \times 10^{19}\,cm^{-3}$. As noted earlier, Breckenridge et al. [32] reported that annealing at 1400 °C leads to higher dopant activation (~10% at room temperature, ~100% at 500 °C, because the Mg level is not that shallow—at ~Ev + 0.2 eV), compared to annealing at 1300 °C (<1% at room temperature, ~40% at 500 °C). Based on the model from Remmele et al., Mg bound only at the inverted base in the PIDs and TIDs is estimated at $1.6 \times 10^{18}\,cm^{-3}$, which is ~10% of the total Mg content [49]. This represents the lower bound of the Mg in PIDs and TIDs because the EDX clearly shows higher Mg concentrations within the entire defect [for example, Fig. 21.16d] and not only at the base. Our study complements the earlier study [32] by providing the structural analysis of the residual defects and showing that the Mg segregation in PIDs and TIDs is a key mechanism for the reduction of dopant activation efficiency and can be avoided if the annealing temperature is ≥1400 °C.

The stability of the PIDs and some insight into the kinetics of their formation and dissolution were assessed through a combination of annealing steps [47]. First, the

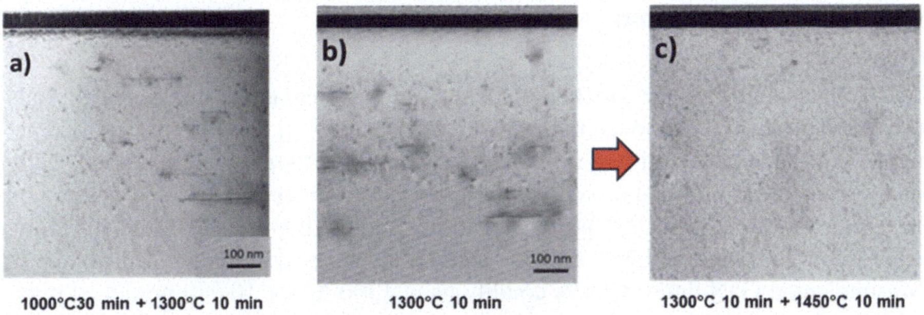

Fig. 21.17 (**a**) STEM image of a sample annealed at 1000 °C, 30 min and 1300 °C, 10 min with a defects structure indistinguishable from the sample annealed at 1300 °C alone (Fig. 21.16a or Fig. 21.17b). (**b**) STEM of the sample annealed at 1300 °C, 10 min and (**c**) the sample from (**b**) was reannealed at 1450 °C, 10 min

role of lower temperature annealing steps on the formation of the PID defects was investigated. As noted in our earlier study, [33] annealing at 1000 °C for 10 min relieves the implant-induced strain. However, the impact of such a lower temperature anneal (or equivalently, a slow ramp rate) on the formation of the Mg-containing PIDs had not been previously addressed. Figure 21.17a shows the TEM images of a sample annealed at 1000 °C for 30 min followed in the chamber by annealing at 1300 °C for 10 min. The PID and related defects are present after the two-step anneal with a density comparable to a 1300 °C step alone (Fig. 21.16). This indicates that the atomic rearrangements after the lower temperatures do not have a significant impact on the PID formation, which, in turn, suggests that the ramp rate (~0.5 °C/s here) does not influence the defect formation at 1300 °C, although the formation of such defects after rapid thermal annealing has not been widely studied.

On the other hand, Fig. 21.17b and c shows the TEM images from a sample that had been annealed at 1300 °C for 10 min (and thus possesses the PID and TID structures) and brought back to room temperature and then reannealed at 1450 °C for 10 min. The PIDs are absent after the higher temperature annealing step, confirming that defects that may form during the ramp-up and anneal at 1300 °C can be dissolved by subsequent high-temperature annealing, supporting our earlier assessment that PID defects formed at lower temperatures can be removed during higher temperature annealing, and can also explain why others [37] observed an Mg-containing defect after a shorter (5 min) anneal at 1400 °C as well as a slower ramp rate to 1400 °C (to include more time in the temperature range of 1300 °C to form PIDs and related defects during the ramp-up and a shorter time to dissolve them at a higher temperature).

6 Concluding Remarks

The understanding of the kinetics of defects in Mg-implanted and annealed GaN provide a pathway to produce high-power vertical GaN diodes. The defect characteristics of substrates and Mg-implanted homoepitaxial GaN were assessed. High quality substrates eliminate many of the growth and processing-related issues that have plagued GaN grown on foreign substrates and the importance of the annealing conditions to achieve high dopant activation were described. Annealing of Mg-implanted GaN under UHPA conditions at 1300 °C for 10 min produced pyramidal and trapezoidal inversion domains, as well as prismatic dislocation loops. The inversion domains show clear Mg segregation and produce electrically inactive Mg that limits dopant activation efficiency. Increasing the annealing temperature ≥1400 °C under the same condition results in a decrease in residual loop defect density with no sign of Mg segregation at these defects. Pyramidal and trapezoidal inversion domains also do not exist with annealing temperatures ≥1400 °C. This study provides the structural analysis of the residual defects, showing Mg segregation in PIDs and TIDs is one of the mechanisms for inactive Mg and can be avoided if the postimplant annealing temperature is in the range of 1400 °C, even if there are lower temperature annealing steps prior to the ≥1400 °C step. Results from this work are expected to help achieve higher activation efficiency of p-type doping for devices including vertical GaN device structures.

Acknowledgments This research was supported through the ARPA-E PNDIODES program under Contract No. DE-AR0001116 at UCLA, and the thought-provoking discussions with Dr. Isik Kizilyalli and Dr. Eric Carlson are gratefully acknowledged. This research used resources of the Advanced Photon Source, a US Department of Energy (DOE) Office of Science User Facility operated for the DOE Office of Science by Argonne National Laboratory under Contract No. DE-AC02-06CH11357 with assistance from Dr. X. Huang and Dr. M. Wojcik at the 1-BM beamline of the Advanced Photon Source, Argonne National Laboratory. TEM sample preparation and measurements were performed at the Center for Nanoscale Materials, Argonne National Laboratory. A special thanks to our collaborators at Adroit Materials and North Carolina State (J. Tweedie, R. Collazo, and Z. Sitar) and colleagues at the Institute of High-Pressure Physics, Polish Academy of Sciences (K. Sierakowski and M. Bockowski supported by the Polish National Science Centre through Project No. 2018/29/B/ST5/00338 as well as by the Polish National Centre for Research and Development through Project No. TECHMATSTRATEG-III/0003/2019-00).

References

1. I.C. Kizilyalli et al., Vertical power p-n diodes based on bulk GaN. IEEE Trans. Electron Devices **6**(2), 414 (2015)
2. M. Boćkowski et al., Recent progress in crystal growth of bulk GaN. Acta Phys. Pol. A **141**(3), 167 (2022)
3. K. Motoki et al., Growth and characterization of freestanding GaN substrates. J. Cryst. Growth **237**, 912 (2002)

4. K. Motoki et al., Dislocation reduction in GaN crystal by advanced-DEEP. J. Cryst. Growth **305**(2), 377 (2007)
5. K. Motoki, Development of gallium nitride substrates. SEI Tech. Rev. **70**, 28 (2010)
6. M. Jackson et al., Determination of stress distribution in III–V single crystal layers for heterogeneous integration applications. Phys. Status Solidi A **204**(8), 2675 (2007)
7. S. Milita et al., X-ray diffraction imaging investigation of silicon carbide on insulator structures. Appl. Phys. A Mater. Sci. Process. **75**, 621 (2002)
8. Y. Oshima et al., Preparation of freestanding GaN wafers by hydride vapor phase epitaxy with void-assisted separation. Jpn. J. Appl. Phys. **42**, L1 (2003)
9. A. Usui et al., Role of TiN film in the fabrication of freestanding GaN wafers using hydride vapor phase epitaxy with void-assisted separation. Phys. Status Solidi A **194**, 572 (2002)
10. Y. Wang et al., Investigation of the reverse leakage behavior and substrate defects in vertical GaN Schottky and PIN diodes. ECS J. Solid State Sci. Technol. **11**(6), 065006 (2022)
11. M.E. Liao et al., Origins of epitaxial macro-terraces and macro-steps on GaN substrates. J. Appl. Phys. **133**(18), 185701 (2023)
12. A.M. Armstrong et al., High voltage and high current density vertical GaN power diodes. Electron. Lett. **52**(13), 1170 (2016)
13. J.R. Dickerson et al., Vertical GaN power diodes with a bilayer edge termination. IEEE Trans. Electron Devices **63**(1), 419 (2015)
14. K. Fu et al., Demonstration of 1.27 kV etch-then-regrow GaN p - n junctions with low leakage for GaN power electronics. IEEE Electron Device Lett. **40**(11), 1728 (2019)
15. W.-C. Ho et al., The study of high breakdown voltage vertical GaN-on-GaN p-i-n diode with modified Mesa structure. Crystals **10**(8), 712 (2020)
16. K.C. Celio et al., Carrier diffusion lengths in continuously grown and etched-and-regrown GaN pin diodes. IEEE Electron Device Lett. **42**(7), 1041 (2021)
17. A.M. Armstrong, Etched-and-regrown GaN pn-diodes with 1600 V blocking voltage. IEEE J. Electron Devices Soc. **9**, 318 (2021)
18. H. Fujikura et al., Roughening of GaN homoepitaxial surfaces due to step meandering and bunching instabilities and their suppression in hydride vapor phase epitaxy. Appl. Phys. Lett. **113**, 152101 (2018)
19. R. Dwiliński et al., Bulk ammonothermal GaN. J. Cryst. Growth **311**(10), 3015 (2009)
20. R. Kucharski et al., Growth of bulk GaN crystals. J. Appl. Phys. **128**, 050902 (2020)
21. C. Ronning, Ion implantation into gallium nitride. Phys. Rep. **351**(5), 349 (2001)
22. J. Karpiński et al., High pressure thermodynamics of GaN. J. Cryst. Growth **66**(1), 11 (1984)
23. T.J. Anderson et al., Activation of Mg implanted in GaN by multicycle rapid thermal annealing. Electron. Lett. **50**(3), 197 (2014)
24. J.D. Greenlee et al., Multicycle rapid thermal annealing optimization of Mg-implanted GaN: Evolution of surface, optical, and structural properties. J. Appl. Phys. **116**, 063502 (2014)
25. T.J. Anderson et al., Improved vertical GaN diodes with mg ion implanted junction termination extension. ECS Trans. **75**(12), 93 (2016)
26. J.D. Greenlee et al., Characterization of an Mg-implanted GaN p–i–n diode. Phys. Status Solidi A. **212**, 12 (2015)
27. M.H. Kuball et al., High-pressure high-temperature annealing of ion-implanted GaN films monitored by visible and ultraviolet micro-Raman scattering. J. Appl. Phys. **87**(6), 2736 (2000)
28. J.C. Zolper, Ion implantation in group III-nitride semiconductors: A tool for doping and defect studies. J. Cryst. Growth **178**, 157 (1997)
29. F. Roccaforte et al., Ion implantation doping in silicon carbide and gallium nitride electronic devices. Micro **2**(1), 23 (2022)
30. S.J. Pearton et al., Ion implantation doping and isolation of GaN. Appl. Phys. Lett. **67**(10), 1435 (1995)
31. T. Narita et al., P-type doping of GaN $(000\bar{1})$ by magnesium ion implantation. Appl. Phys. Express **10**(1), 16501 (2017)

32. M.H. Breckenridge et al., High Mg activation in implanted GaN by high temperature and ultrahigh pressure annealing. Appl. Phys. Lett. **118**(2), 022101 (2021)
33. Y. Wang et al., Strain recovery and defect characterization in Mg-implanted homoepitaxial GaN on high-quality GaN substrates. Phys. Status Solidi B **257**(4), 1900705 (2020)
34. K. Sierakowski et al., High pressure processing of ion implanted GaN. Electronics **9**(9), 1380 (2020)
35. A. Kumar et al., Atomic-scale quantitative analysis of implanted mg in annealed GaN layers on free-standing GaN substrates. J. Appl. Phys. **126**(23), 235704 (2019)
36. K. Iwata et al., Defect evolution in Mg ions implanted GaN upon high temperature and ultrahigh N2 partial pressure annealing: Transmission electron microscopy analysis. J. Appl. Phys. **127**(10), 105106 (2020)
37. J. Uzuhashi et al., Atomic-scale investigation of implanted Mg in GaN through ultra-high-pressure annealing. J. Appl. Phys. **131**(18), 185701 (2022)
38. U. Beck et al., Diffuse x-ray rods and scattering from point defect clusters in ion implanted silicon. Appl. Phys. Lett. **76**, 2698 (2000)
39. S. Leclerc et al., Mechanisms of damage formation in Eu-implanted AlN. J. Appl. Phys. **112**, 073525 (2012)
40. F. Gloux et al., The structure of crystallographic damage in GaN formed during rare earth ion implantation with and without an ultrathin AlN capping layer. Superlattice. Microst. **40**, 300 (2006)
41. A. Turos, On the mechanism of damage buildup in gallium nitride. Radiat. Effects Defects Solids **168**, 431 (2013)
42. J. Pezoldt et al., Polytype transitions in ion implanted silicon carbide. Nucl. Instrum. Methods Phys. Res. B **80**, 943 (1993)
43. P. Vennégués et al., Pyramidal defects in metalorganic vapor phase epitaxial Mg-doped GaN. Appl. Phys. Lett. **77**, 880 (2000)
44. P. Vennégués et al., Atomic structure of pyramidal defects in Mg-doped GaN. Phys. Rev. B **68**, 235214 (2003)
45. Z. Liliental-Weber et al., Defects in p-doped bulk GaN crystals grown with Ga polarity. J. Cryst. Growth **281**(1), 125 (2005)
46. E. Kano et al., Atomic resolution analysis of extended defects and Mg agglomeration in Mg-ion-implanted GaN and their impacts on acceptor formation. J. Appl. Phys. **132**, 065703 (2022)
47. K. Huynh et al., Dissolution of Mg-enriched defects in implanted GaN and increased p-type dopant activation. J. Appl. Phys. **135**(2) (2024)
48. Hansen et al., Phys. Stat. Sol. (b) **228**(2), 353–356 (2001)
49. T. Remmele et al., Pyramidal inversion domain boundaries revisited. Appl. Phys. Lett. **99**, 141913 (2011)

Chapter 22
Secondary Ion Mass Spectroscopy (SIMS) Analysis of GaN Epitaxial Films

Temel Buyuklimanli

1 Introduction

With the increasing demand for GaN high-power transistors, much effort has been put into the research and development for optimization of high-quality epitaxial film growth with reduced manufacturing costs. The impressive performance of the wide bandgap high electron mobility transistors (HEMTs) stems from the specific material properties associated with GaN. Compared to GaAs-based materials, which have also been used for generating power in the microwave region, GaN has a larger peak electron velocity; a higher thermal stability; and a larger band gap. All of these traits make GaN a very appropriate material for the channel of a HEMT. In such structures, it is often paired with AlGaN to form a two-dimensional electron gas (2DEG), which lies at the very heart of the device, dictating its electrical characteristics. Selecting the best substrate for the manufacture of GaN HEMTs was one of the early challenges for device developers. From a performance perspective, a GaN substrate is ideal, because it ensures a perfect lattice match between the epilayers and the foundation. If such a substrate was cost-effective, it would allow the use of a relatively thin buffer layer, because its role would only be to isolate the 2DEG from the substrate—thick buffer layers would not be needed to prevent high levels of defects in the active region of the device. But today GaN substrates cannot be grown in sufficiently large sizes to make their price commercially viable.

Consequently, GaN HEMTs are often grown on foreign substrates. One common choice is single-crystal SiC, which combines good electrical and thermal conductivity with a lattice constant that is close to that of GaN—the difference is just 3%. However, while not as expensive as GaN, SiC can still be expensive, particularly relative to silicon.

T. Buyuklimanli (✉)
Eurofins | EAG laboratories, Sunnyvale, CA, USA
e-mail: Temel.Buyuklimanli@me.eurofinsus.com

I. C. Kizilyalli et al. (eds.), *Gallium Nitride and Related Materials*, The Materials Research Society Series, https://doi.org/10.1007/978-3-031-83056-3_22

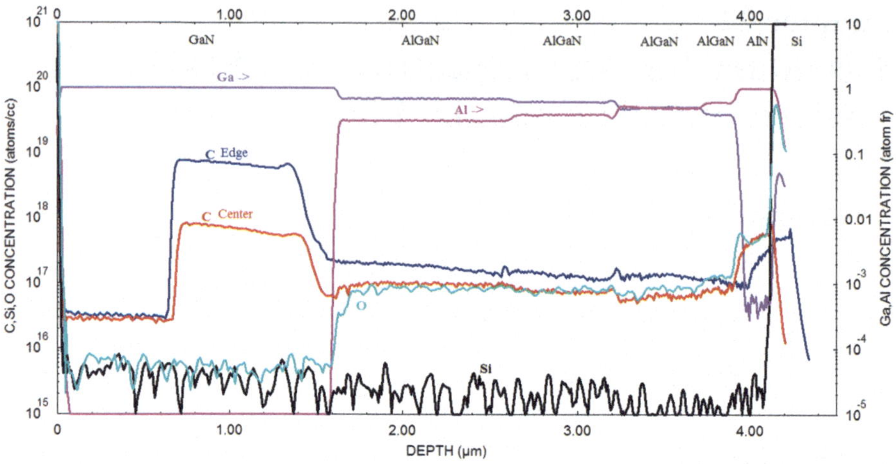

Fig. 22.1 The graph illustrates the concentration profiles of various elements as a function of depth in micrometers, with labeled regions: GaN, AlGaN, AlN and Si. The y-axis on the left shows C (red and blue lines), Si (black line), and O (cyan line) concentrations in atoms/cm^3, while the right y-axis shows Ga (purple line) and Al (pink line) concentrations in group III atom fraction. The blue line represents the C measured near the edge of a wafer, while the red line shows the C measured near the center of the same wafer. All the rest of lines were acquired at the center of the wafer. The graph highlights differences in C concentrations across a 200 mm wafer

A cheaper alternative is sapphire, but it has poor thermal conductivity—a disadvantage for high-power devices—and its lattice mismatch with GaN is 13%. Therefore, a more popular low-cost option is silicon, which has long been the choice of substrate for high-volume semiconductor production. Its great strength is that it has a large base of established manufacturing tools and processes. However, it also has its weaknesses, including a very large lattice mismatch with GaN that gives rise to a high density of defects in the grown epilayers. To prevent defect density in the active layers from being so high that device performance is unacceptable, thick AlGaN buffer layers are inserted between the substrate and the 2DEG-forming layers. An example of SIMS data from this type of epi film is shown in Fig. 22.1. The STEM cross-section image of this epi film in Fig. 22.2 shows the defects decreasing in density toward the surface. Some dislocation defects may extend all the way to the surface and create "pit" defects, which can cause complications in the characterization of the top thin layers.

Because of the challenging growth conditions, proper characterization of these epi films becomes very important. One frequently used metrology for epitaxial films is Secondary Ion Mass Spectroscopy (SIMS), because of its ability to detect alloy, dopant, and impurity elemental concentrations, as well as the layer stack information. Figure 22.1 shows an overlay of SIMS data acquired from the center and edge of a wafer to study differences in C (carbon) dopant incorporation at different stages of epi growth. In this example, large differences in C concentrations across the wafer were observed between the center and edge, suggesting poor growth temperature

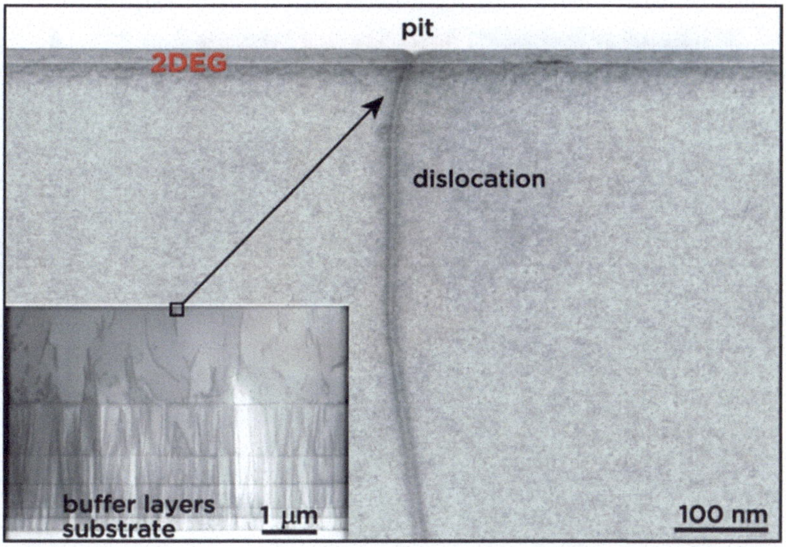

Fig. 22.2 Cross-sectional STEM image of defects in the buffer layers caused by Si lattice mismatch with AlGaN/GaN epi layers. Sometimes these defects lead to imperfections near the 2DEG and surface regions

uniformity. However, for reliable evaluations of the epitaxy film quality, appropriate calibration methods must be used.

2 Secondary Ion Mass Spectrometry (SIMS)

Major analytical techniques for materials characterization are shown in Fig. 22.3. Many of these techniques are applicable to compound semiconductors such as III-nitrides. Dynamic, or profiling, SIMS covers the widest range of concentrations from 10 ppt (1×10^{12} atoms/cm^3) to 100 atomic % (~5×10^{22} atoms/cm^3). Furthermore, all elements in the periodic table, including H, can be detected using SIMS. This unique capability makes SIMS almost essential in semiconductor materials R&D, failure analysis, and production monitoring.

A basic schematic of the SIMS technique is shown in Fig. 22.4. A rastered focused ion beam (FIB) accelerated to the surface of a sample causes ejection/sputtering of particles (some charged and some neutral) from the surface. The ionized particles, i.e., secondary ions, with specific charge are then collected with the help of an electric field applied toward the sample surface. The collected ions are directed into the mass spectrometer for filtering one specific mass at a time. The ions which made it through the mass spectrometer are then counted and amplified by a detector to generate secondary ion signals. Depending on how the data are collected,

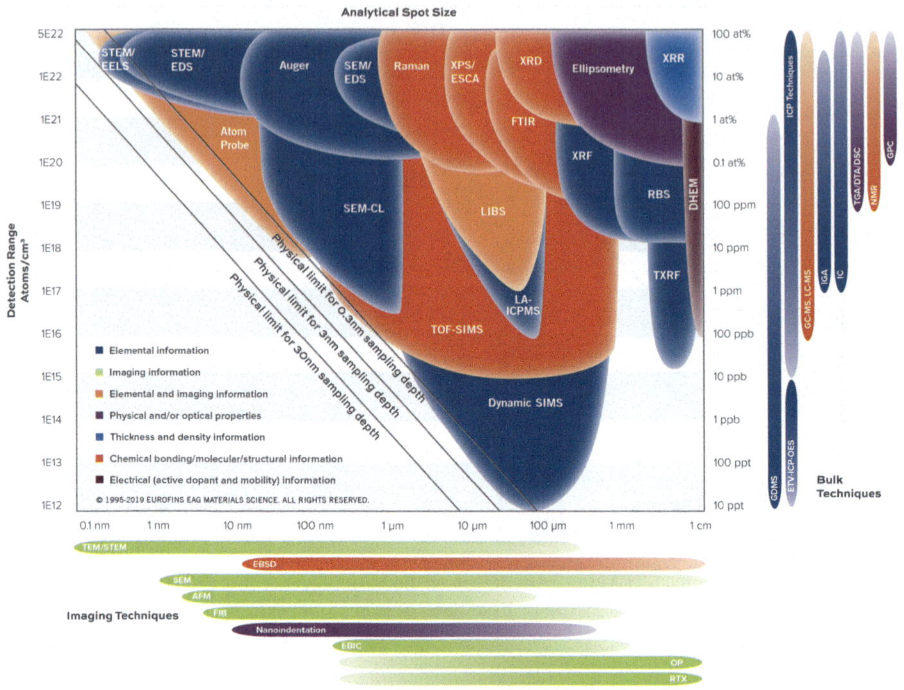

Fig. 22.3 This chart depicts various analytical techniques based on their analytical spot size, detection range and information they provide. The x-axis represents the analytical spot size, ranging from 0.1 nm to 1 cm, while the y-axes show the detection range from 1×10^{12} to 5×10^{22} atoms/cm^3 on the left and 10 ppt to 100 atom % on the right. The techniques are categorized by the type of information they provide, such as chemical and physical properties, separated in different colors. (Source https://www.eag.com/techniques/)

one can generate a mass spectrum, depth profiles of elements, or two-dimensional maps of specific masses. In this chapter, we will discuss depth profiling.

3 Quantitative SIMS Analysis of GaN/AlGaN Epi Films

For useful analysis, quantification of the SIMS intensities to convert them to concentrations is crucial. This is usually accomplished by determining a Relative Sensitivity Factor (RSF) using a reference material which has a stoichiometry closely matching that of the layer where the dopant or impurity element is to be quantified. However, doing so in multilayered compound semiconductor structures is not trivial, requiring many reference materials, or standards, for complete calibration of all layers in the stack of interest. An example of how the secondary ion signals change with the stoichiometry of the material from where the signal is generated is shown in Fig. 22.5. The samples on this graph (AlGaN films with %

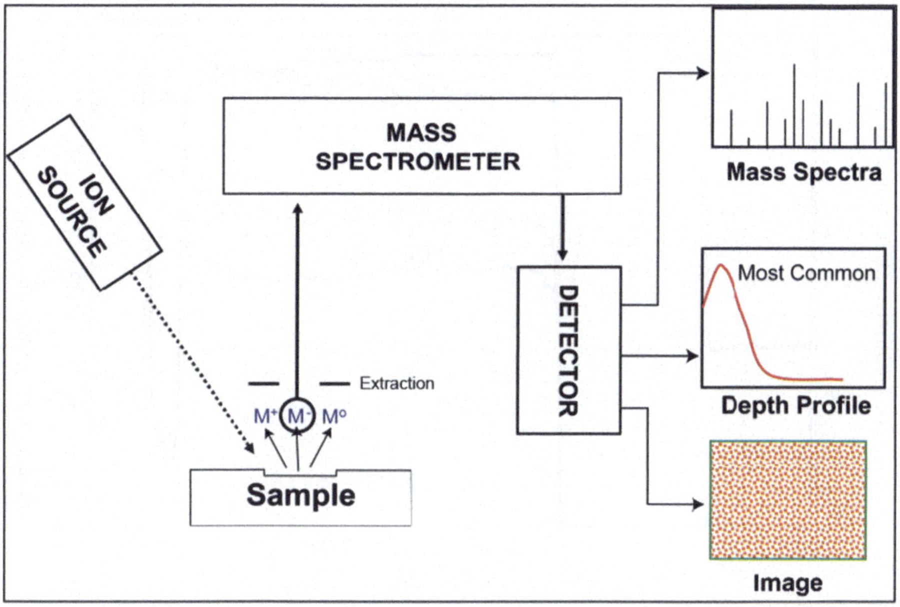

Fig. 22.4 Basic schematic of a SIMS instrument. The mass spectrometer analyzes the secondary ions ejected as a result of an incident primary ion beam energy transfer. Outputs include a mass spectrum, a depth profile, or a map of specific elements in the analytical area

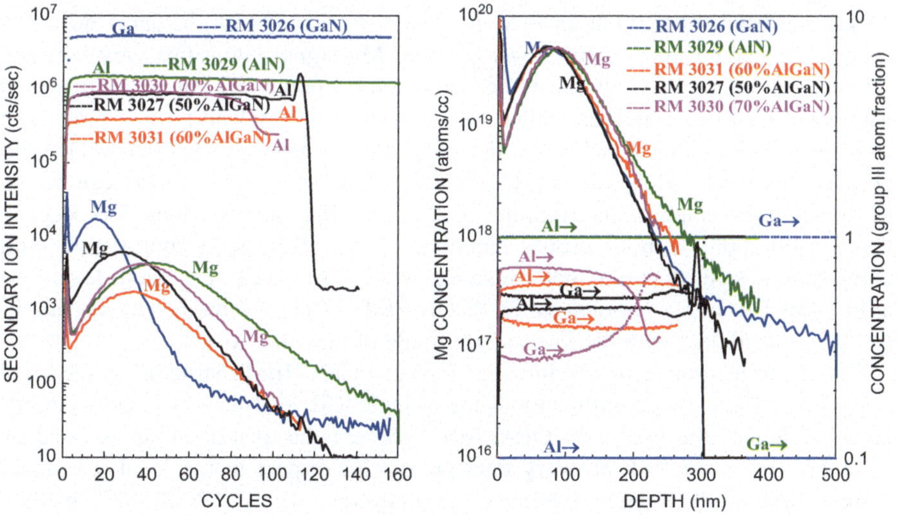

Fig. 22.5 An overlay of SIMS data acquired from five different AlGaN films with Al = 0–100 atom% which were Mg ions implanted with the same dose. The left graph shows intensities of Mg, Al, and Ga, as detected. The right graph shows the same data after quantification using the known ion implant dose. These were intended to be used for Mg quantification in different nitride materials

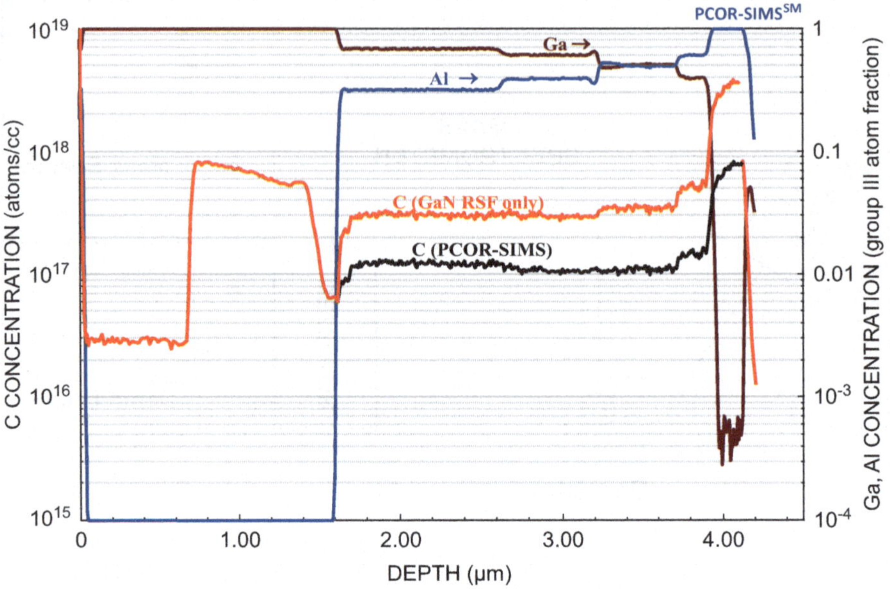

Fig. 22.6 Carbon dopant SIMS profile in GaN HEMT with AlGaN buffer layers. The red curve is quantified using only GaN reference material, whereas the black curve uses the aluminum composition at every data point (PCOR-SIMS^{SM}) to quantify the Carbon SIMS signals with minimum "matrix effect" artifacts, and therefore is more accurate than the red curve

Al varying from 0 to 100 atom%) were all ion implanted with the same dose and energy. The graph on the left shows how the Mg signal intensities vary between samples, although they all have the same total amount of Mg in them. Such variations in SIMS, which are called "matrix effects," may lead to inaccurate, and sometimes misleading, conclusions if signals are not properly calibrated [1]. With proper calibrations, all the Mg curves align to yield the same total Mg concentration, as shown in the graph on the right in Fig. 22.5. Similar complications also occur in depth calibration: without proper handling of the changes in sputter rate with composition, layer thickness determination will be incorrect. A method developed at Eurofins EAG Laboratories, called PCOR-SIMS^{SM} [1], overcomes these complications to yield data with the lowest magnitude of known errors.

The quantification error varying with %Al in a GaN HEMT structure is shown in Fig. 22.6. The carbon quantification using a single RSF leads to ~2x error in AlGaN layers. Whereas the use of PCOR-SIMS^{SM}, where the quantification is done by applying a relevant RSF at every data point with respect to the local aluminum composition, results in more accurate C concentrations. The PCOR-SIMS^{SM} method also provides accurate layer thickness information, as discussed above. Therefore, the individual AlGaN and C-doped GaN layer thicknesses are measured with <5% error.

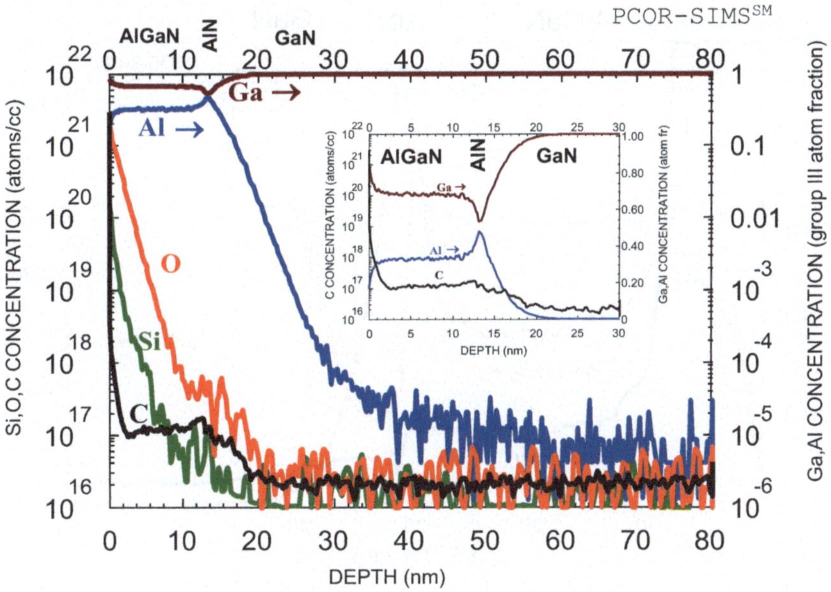

Fig. 22.7 The graph illustrates high-depth resolution SIMS analysis of barrier, channel, and 2DEG regions of a GaN HEMT epitaxial film. The y-axis on the left shows the concentrations of Si, O, and C in atoms/cm3, while the y-axis on the right shows the Ga and Al concentration as a group III atom fraction. The x-axis represents the depth in nanometers. The inset graph provides a more detailed view of the Al and Ga composition changes plotted on a linear scale. The labels "AlGaN," "AlN," and "GaN" indicate layers detected.

Another important aspect of the SIMS analysis is "depth resolution" or the ability to analyze very thin layers and abrupt interfaces. Depending on the area of specific interest, SIMS analyses can be tailored to provide the desired information. For example, the analysis in Fig. 22.6 was designed to obtain accurate data for low C concentrations in a 4 μm thick epi film. However, in this mode, the data in the top 50 nm (at the sample surface) are almost indiscernible. On the other hand, with low enough impact energy to minimize atomic mixing, the top few nm thick layers can be analyzed. Furthermore, using proper sample preparation, C levels down to the $\sim 1 \times 10^{17}$ atoms/cm^3 level can be measured. An example of this is the 2DEG region of a GaN HEMT epitaxial film as shown in Fig. 22.7. This figure shows SIMS is capable of resolving a few nm thick layers, such as the surface ~2 nm GaN cap, ~12 nm AlGaN barrier, and ~2 nm AlN spike layers. The $\sim 1 \times 10^{17}$ atoms/cm^3 C impurity level is clearly measurable although with a ubiquitous high level of C on the surface. For detecting such low C levels, surface hydrocarbons and other environmental adsorbates need to be removed before the analysis. The O and Si in the AlGaN extending from the surface are artifacts due to the mixing of surface contamination. This high-depth resolution capability allows investigation of differences across a wafer, or series of samples, for epi quality.

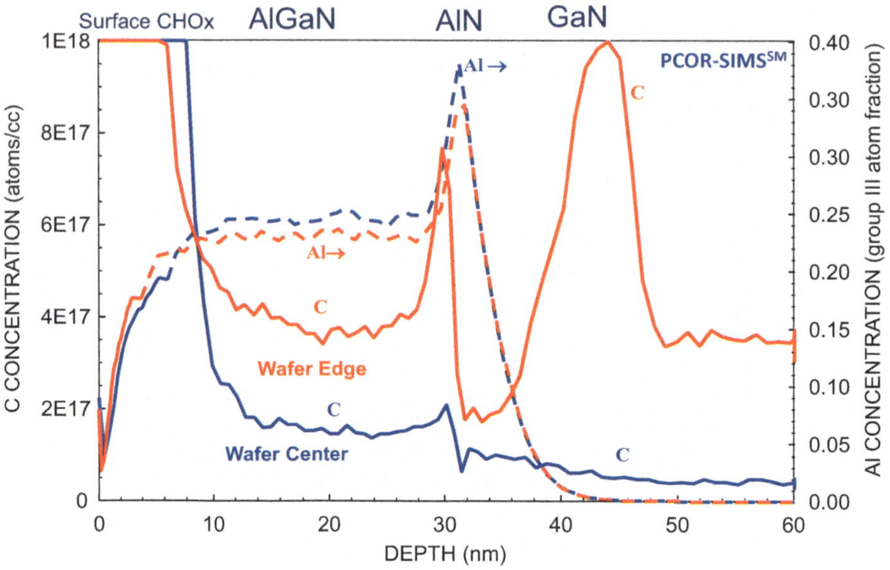

Fig. 22.8 High-depth resolution SIMS analysis of a GaN HEMT epi wafer acquired from the center (blue) and the edge (red) of the same wafer. The nm resolution and good detection limit for C allow detailed investigation of epi growth across a wafer

An example of this type of investigation is shown in Fig. 22.8, where an overlay of data collected from the same epi region, but at different wafer positions shows substantial differences in carbon concentration. The Al signal collected simultaneously with C pinpoints the exact location of the C peaks detected. On the same wafer, the data near the edge show significant C contamination spikes at the AlGaN/AlN interface, possibly due to growth interruption, while this was not detected near the center of the wafer. This difference probably indicates a growth temperature gradient laterally across the wafer leading to C contamination.

Physical defects, which are often present in III-nitride epi films, cause artifacts in the SIMS data leading to potential incorrect evaluation of the data if one is not careful. A few examples of these artifacts are shown in Figs. 22.9 and 22.10. The screw dislocation gaps/pits either get filled in with the material on top or expose the material below. This leads to spurious signals, giving a false impression of interdiffusion between layers. Therefore, in such cases, additional morphological checks using techniques like Scanning Transmission Electron Microscopy (STEM) or Atomic Force Microscopy (AFM) are necessary to make an accurate assessment of the film quality.

The use of complementary techniques is also necessary for a comprehensive examination of GaN materials. One such case (Fig. 22.11) is the characterization of Mg ion-implanted and annealed GaN to make a low-resistance p-GaN contact. In this example, multiple implants with varying doses were initially profiled with SIMS

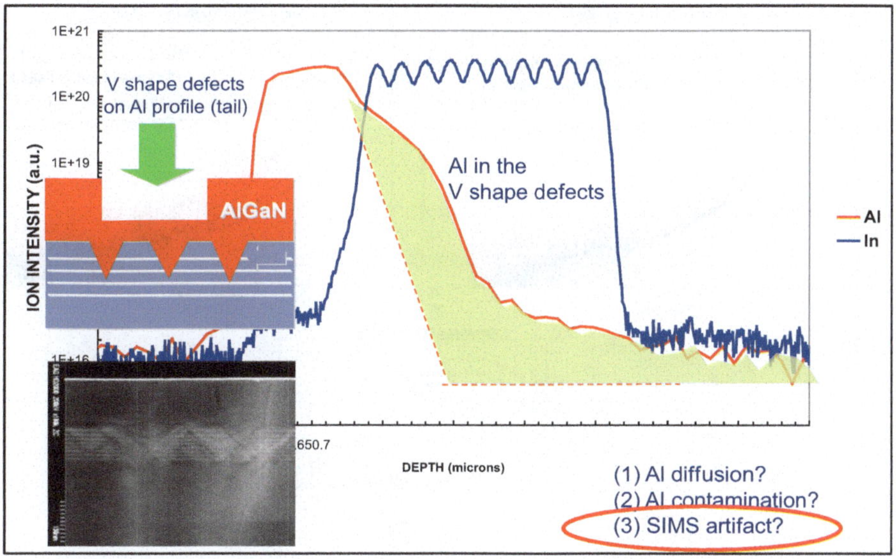

Fig. 22.9 The V-shape defects in the InGaN superlattice get filled in with AlGaN during the deposition of the AlGaN layer on top. This leads to the Al signal (red) being observed well into the InGaN layers (shown in blue) which could be interpreted as Al diffusion into InGaN or Al contamination during the InGaN growth. However, in reality, the tail is an artifact due to the morphology of the epitaxial layers. The expected Al profile (in the absence of defects) is the dotted red line

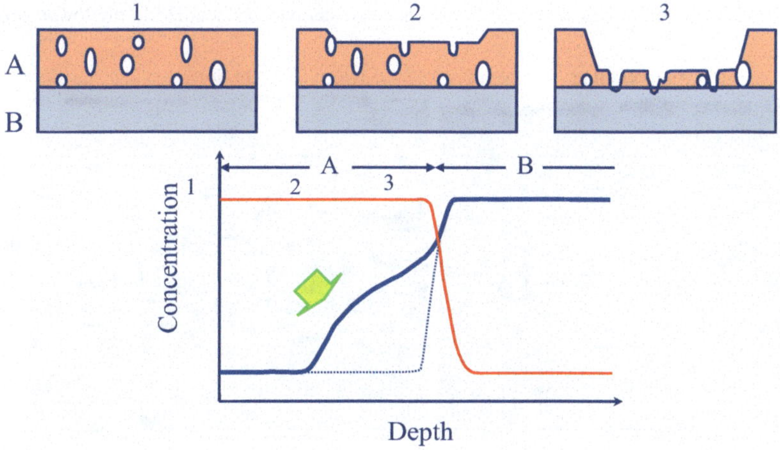

Fig. 22.10 The pinholes or dislocations in layer "A" lead to roughness at the bottom of the SIMS crater while layer "A" (red) is being sputtered away. As the crater bottom reaches the A/B interface, the roughness caused by the defects exposes layer "B" (blue) before layer "A" is completely sputtered away. This could be interpreted as "B" layer out-diffusion, although in reality, it is an artifact due to the pinhole or dislocation defects in layer "A"

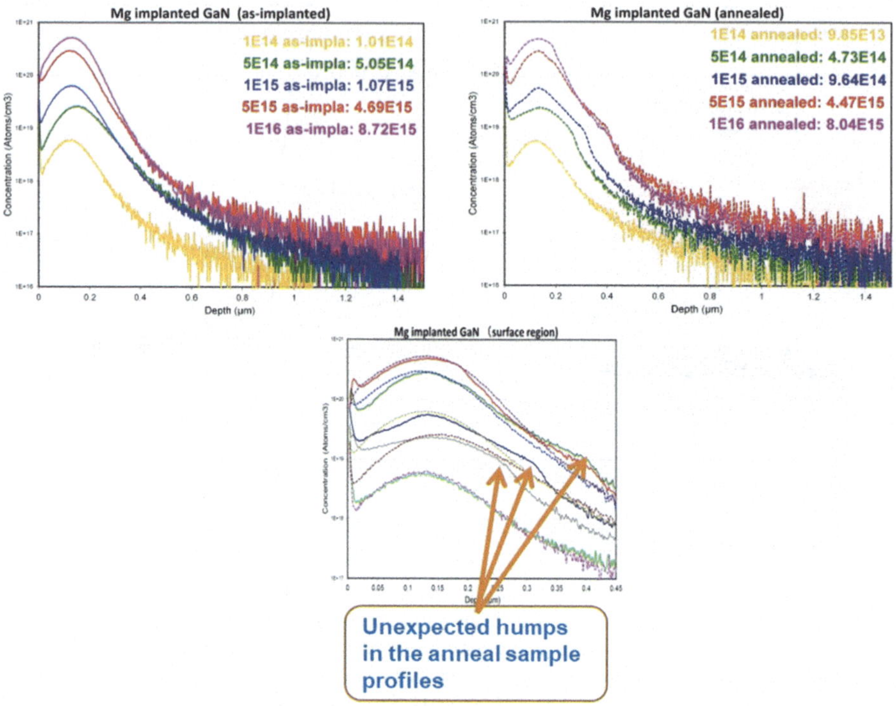

Fig. 22.11 The upper graphs show SIMS profiles of various Mg ion implants into GaN before and after annealing, as part of the p-contact development process. The high-dose implants have unexpected inflection points or humps in the annealed samples, as detailed in the lower graph

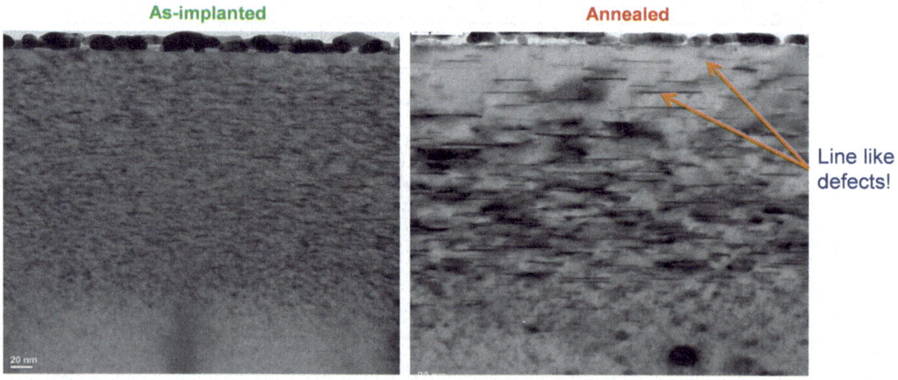

Fig. 22.12 STEM images of the high-dose Mg ion-implanted GaN before (left image) and after (right image) annealing. The point-like crystal damage in the as-implanted sample has been mostly recovered after annealing; however, some new line-like defects have formed

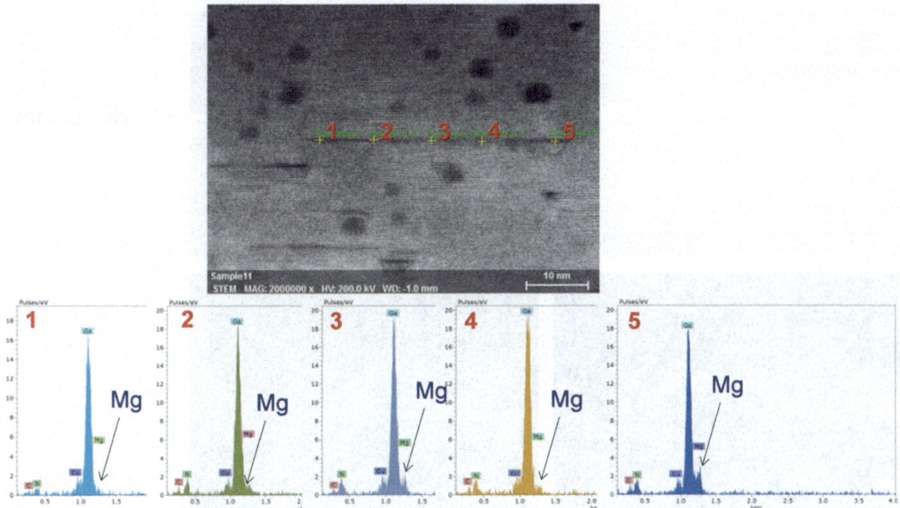

Fig. 22.13 The high-resolution STEM-EDS analysis of various locations of the line defects showed Mg enrichment, suggesting Mg clustering

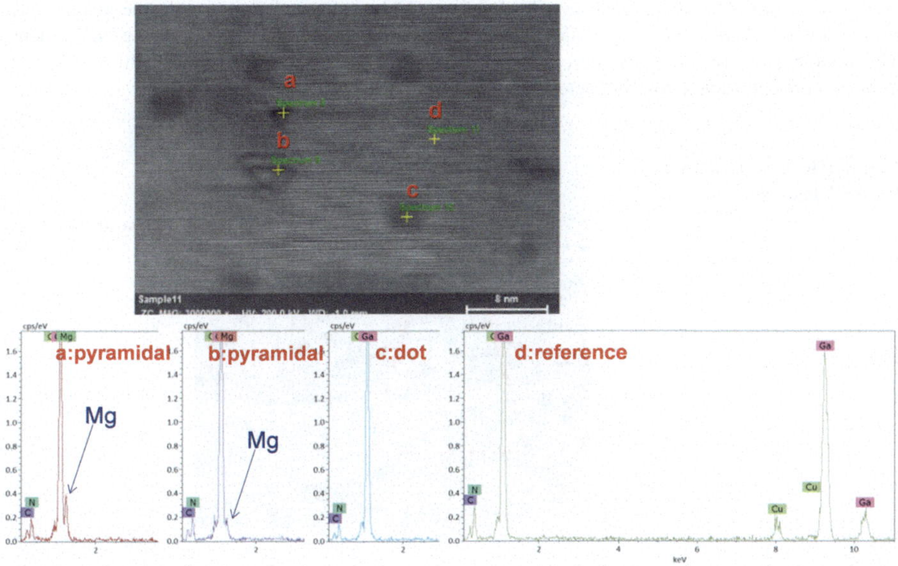

Fig. 22.14 The EDS spectra of defects in the annealed GaN sample after 1E16 atoms/cm^2 Mg ion implantation. A reference spectrum was also collected from a defect-free location (area "d")

and showed interesting inflection points in the annealed samples. To understand what might have been the cause of these features, STEM (Scanning Transmission Electron Microscopy) analyses of these samples were performed.

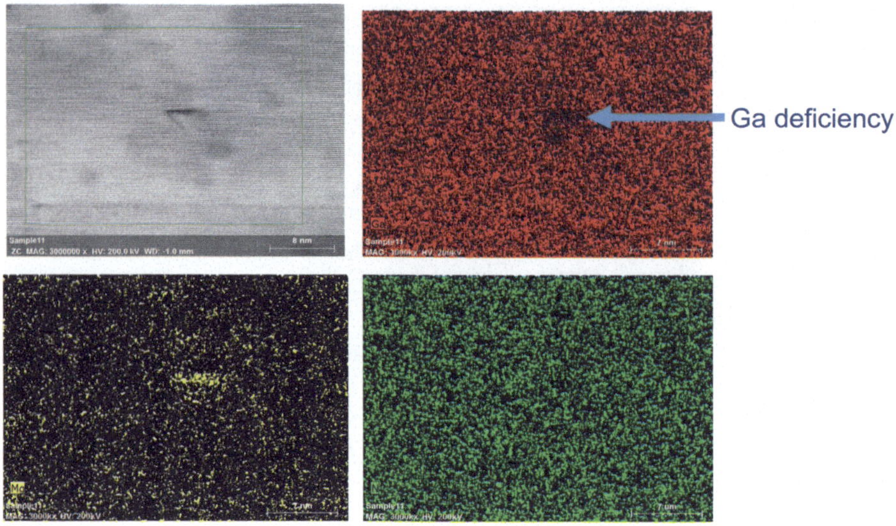

Fig. 22.15 The figure shows STEM and EDS images and maps of regions where pyramidal defects were detected. The top left image is the electron micrograph of the sample showing the location of the line defect. The top right image is the gallium (Ga) distribution map shown in red. The arrow points to a darker region indicating Ga deficiency. The bottom left image is the magnesium (Mg) distribution map shown in yellow. The brighter spot at the center suggests Mg enrichment. The bottom right image is the nitrogen (N) distribution map shown in green, which indicates uniform concentration across the detected area

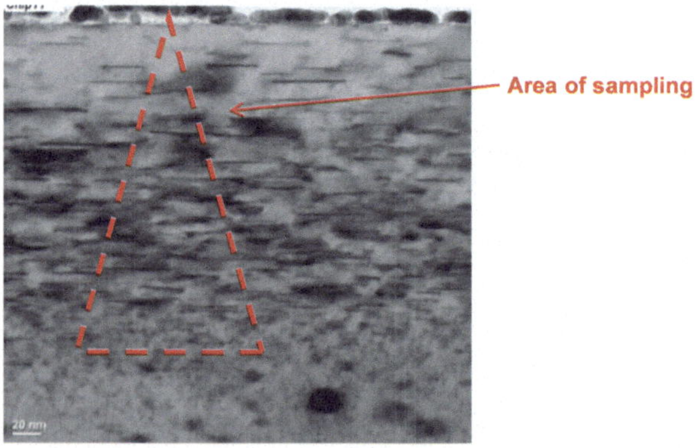

Fig. 22.16 The area used for APT analysis

The cross-section STEM images (Fig. 22.12) show that the ion implantation damage is mostly recovered, but line-like defects have been formed in between recrystallized GaN matrix material. The density of these defects seemed to be higher at the depth where bumps were observed in the SIMS profiles. For chemical identification of these features, Energy-Dispersive X-ray Spectroscopy (EDS) was performed (Fig. 22.13), where it was observed that the line defects were rich in Mg.

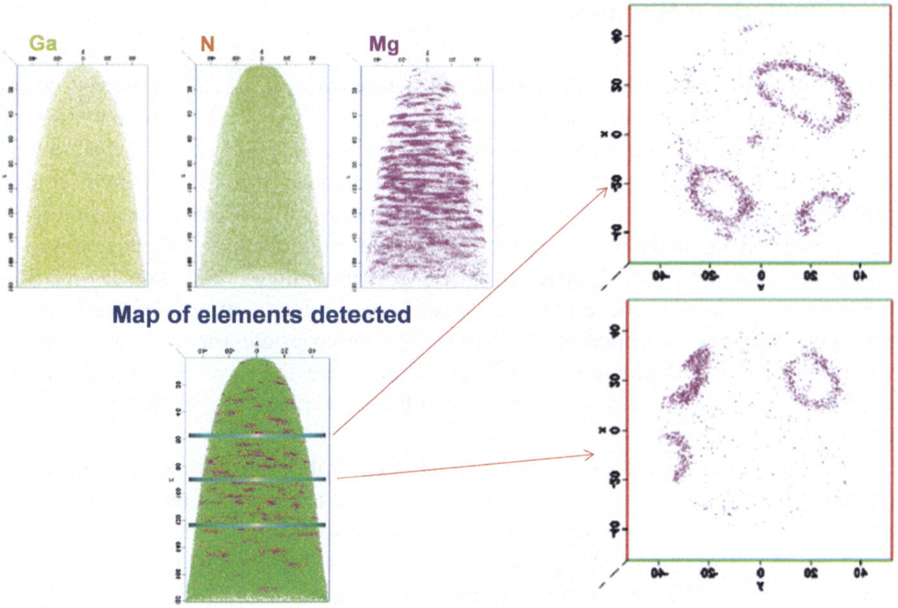

Fig. 22.17 The images and maps of elements detected by APT. The slices observed from the top-down on the right clearly show the clustering of Mg in a ring pattern

In addition to line defects, pyramidal and dot defects also were detected (Fig. 22.14). However, both the pyramidal and dot defects were much smaller than the line defects. Just like the line defects, these defects also showed Mg enrichment, but no Mg was detected in the crystalline "reference" region (area "d").

Figure 22.15 shows a more detailed higher resolution EDS map of a pyramidal defect and its surroundings. The data suggest Ga depletion and Mg enrichment in the defect, while the N distribution was uniform across the same region. This again points to Mg clustering.

For further understanding of the Mg-rich defects, Atom Probe Tomography (APT) was performed to image the atoms in 3D. A triangular-shaped sample (Fig. 22.16) was cut out from the area of interest using focused ion beam sample preparation, to be imaged using APT.

The signals of Ga, N, and Mg acquired from the region shown in Fig. 22.16 imaged in 3D showed evenly distributed Ga and N (Fig. 22.17). However, Mg exhibited slices of clusters. These clusters observed from the top appear to be rings.

It appears that, after annealing high-dose Mg ion implantation led to ring-like clustering in some places, instead of occupying substitutional sites in the GaN crystal lattice, which would cause poor activation of these dopant atoms. These results suggest that the inflection in the SIMS profiles of annealed samples may indicate clustering thus poor activation rate.

4 Concluding Remarks

For useful analysis, quantification of the SIMS intensities is crucial. However, doing so in compound semiconductors is not trivial, requiring many reference materials, or standards, for calibration. The PCOR-SIMS^{SM} method [1] addresses all the calibration issues in III-V materials to minimize quantification errors. This method also provides more accurate layer thickness measurements in complex layer structures.

To achieve the analytical objectives and proper understanding of epi quality for failure analysis and R&D, different SIMS instrumentation and data acquisition methods may need to be employed. Therefore, understanding SIMS analytical protocol tradeoffs is important to choose the appropriate method. In complex compound semiconductor epitaxial films, often more than one SIMS analysis is necessary, each done under optimized conditions for the needed information of interest.

Recognizing artifacts in the SIMS data due to morphological imperfections in the crystal structure is as important. These artifacts may lead to incorrect interpretation of the SIMS data and an inaccurate assessment of the material quality. Additional analyses by TEM or STEM are often necessary to determine that the features detected in the SIMS data are not artifacts.

Acknowledgments Charles W. Magee, Wei Ou, Ozgur Celik, Kazuteru Takahashi, and Kazue Shingu of Eurofins EAG Laboratories.

Reference

1. C.W. Magee, T.H. Buyuklimanli, J. Vac. Sci. Technol. B **41**, 030803 (2023)

Part VI
Gallium Nitride Materials Properties for Power Electronics Devices

Chapter 23
Intrinsic and Extrinsic Thermal Conductivities of Gallium Nitride and Silicon Carbide

Yang Lu, David G. Cahill, and Qiye Zheng

1 Introduction

Over the past two decades, power electronics has undergone a transformative revolution, driven by the adoption of wide bandgap semiconductor (WBGS) devices, particularly those based on silicon carbide (SiC) and gallium nitride (GaN) [1–3]. The superior material properties of WBGSs, such as high electron mobility, exceptional thermal and chemical stability, and large breakdown electric field, have made them indispensable in the power semiconductor industry [1, 4] for applications in electric grids, data centers, electric vehicles, and renewable energy integration [1, 4]. For example, compared to their Si counterparts, WBGS devices exhibit a significantly lower specific ON state resistance for a given breakdown voltage [1, 2, 4]. This enables reduced capacitances, lower switching losses, and higher operating frequencies, which facilitate the miniaturization of overall power electronics systems with enhanced power density and improved efficiency. Among the plethora of WBGSs, GaN and SiC are currently the most extensively studied and commercially established in power electronics (and other applications such as light-emitting diodes [5] and photovoltaic inverters [6]) due to their superior heat-conducting capability, successful growth, and ease of fabrication.

Despite the remarkable electrical properties of WBGSs, the high-power density leads to elevated operating temperatures, which adversely affect both device lifetime

---

Yang Lu and Qiye Zheng contributed equally with all other contributors.

Y. Lu · Q. Zheng (✉)
Department of Mechanical and Aerospace Engineering, The Hong Kong University of Science and Technology, Hong Kong, China
e-mail: qiyezheng@ust.hk

D. G. Cahill
Department of Materials Science and Engineering, Materials Research Laboratory, University of Illinois, Urbana, IL, USA

I. C. Kizilyalli et al. (eds.), *Gallium Nitride and Related Materials*, The Materials Research Society Series, https://doi.org/10.1007/978-3-031-83056-3_23

and performance [1, 2, 7]. Consequently, the thermal conductivity (κ) of the WBGSs and the thermal boundary conductance (*G*) between the layers that govern the local hotspots are of utmost importance. In nondegenerate semiconductors, thermal transport near room temperature is dominated by phonons, i.e., the quantized units of lattice vibration, and their mutual scattering due to the intrinsic anharmonicity. Under ideal conditions, the strong bonds in GaN and SiC that contribute to their wide electronic bandgap also result in a κ greatly exceeding that of Si and Ge [8–10]. However, various extrinsic defects in real-world materials can strongly scatter phonons, impeding heat transport through a variety of different mechanisms. For example, dopants used to modify the carrier concentration introduce impurity scattering of phonons; heteroepitaxy, which enables the scalable growth of devices, also introduces dislocations that hinder phonon transport in these materials [8, 11–13]. The micro- or nano-scale thickness of the WBGS layers and the grain boundaries can induce boundary scattering of acoustic phonons [9, 13]. Therefore, a thorough understanding of the thermal transport properties of WBGSs, especially GaN and SiC, encompassing both intrinsic and extrinsic factors is a fundamental step for addressing the thermal management challenges and optimizing the performance of WBGS for microelectronics [14, 15].

To this end, it is important to develop experimental techniques that offer high precision, resolution, and speed for the characterization of thermal properties. Thermoreflectance-based methods enable the measurement of κ and mapping of the device temperature profile [2]. These methods, in combination with the characterization of composition and structure, can facilitate the construction of the processing-structure-property-performance relationships that are critical for material and device optimization in the industry.

Recent reviews have discussed device thermal management [1, 7, 16] and thermal metrologies [2] for WBGS materials. In this chapter, we focus on fundamental issues and characterization of the intrinsic and extrinsic factors that influence thermal transport in GaN and SiC. The chapter is organized as follows: Sect. 2 explores the thermal transport mechanisms in WBGSs, examining the influence of both intrinsic and extrinsic factors, with a special emphasis on the latter. Section 2 also discusses the temperature dependence and anisotropy of κ and their impact on device reliability and performance. Section 3 provides an overview of thermoreflectance-based techniques for characterizing the thermal transport properties of WBGSs, particularly GaN and SiC, including their operating principles, implementations, and applications. Section 4 reviews the types, formation mechanisms, engineering methods, and characterization of defects, such as point, line, and planar defects, and their effects on thermal transport in GaN and SiC. Finally, Sect. 5 highlights challenges and opportunities for thermal management of GaN- and SiC-based devices. The objective of this chapter is to stimulate further research and development of GaN and SiC applications by providing an overview of the fundamental understanding of their thermal transport properties and the extrinsic factors influencing them.

2 Thermal Transport Mechanisms in WBGS and the Influence of Intrinsic and Extrinsic Factors

2.1 Bandgaps and Thermal Conductivities of Common Semiconductors

We first overview the general bandgaps and thermal conductivity κ properties of GaN and SiC in comparison with representative semiconductors which are typically categorized into narrow bandgap semiconductors (NBGSs) (<2.3 eV), WBGSs, and ultrawide bandgap semiconductors (UBGSs) (>4.0 eV) [2, 17]. Figure 23.1 presents a compilation of bandgaps and bulk room-temperature (RT) thermal conductivities for some common semiconductors in the three categories.

Among NBGSs, Si (1.12 eV) and Ge (0.63 eV), sometimes referred to as first-generation semiconductors [18], exhibit RT κ of ≈140 and ≈60 W m^{-1} K^{-1}, respectively [19–21]. GaAs (1.4 eV) and InP (1.35 eV), known as second-generation semiconductors, have relatively low RT κ below 70 W m^{-1} K^{-1} [22–24]. Cubic BAs (1.81 eV), which have drawn wide attention in the past five years, stand out with an

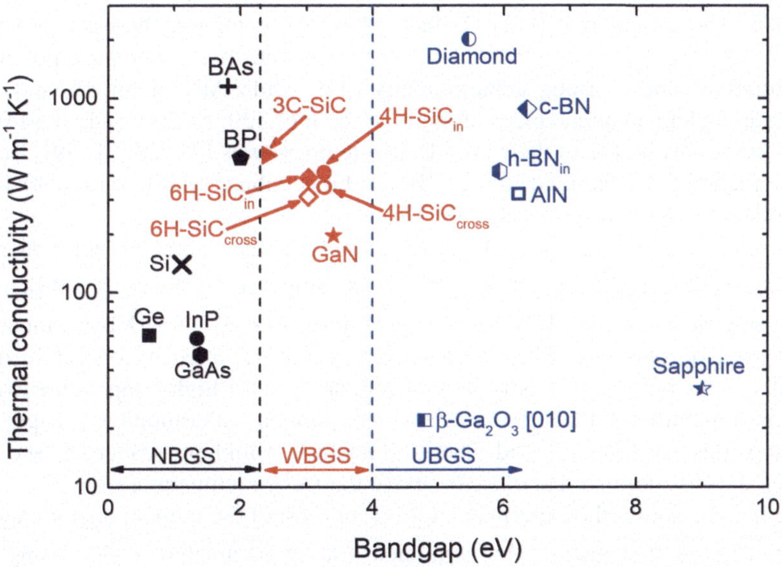

Fig. 23.1 Comparison of bandgaps and bulk RT κ of semiconductors with natural isotope abundance. Data presented are from experiments in the literature, including Si [28, 41], Ge [28, 41], GaAs [22, 23], InP [22, 24], BAs [23, 25, 26, 28], BP [27], wurtzite GaN [8, 28], 4H-SiC_{in} [8, 28], 4H-SiC_{cross} [8, 28], 6H-SiC_{in} [29, 30], 6H-SiC_{cross} [29, 30], 3C-SiC [29, 30], diamond [23, 28, 37], c-BN [28, 38], h-BN_{in} [22, 42–44], AlN [28, 39], β-Ga_2O_3 [010] [22, 36], and sapphire [22, 45]. The symbols are colored based on the material categories: black for NBGSs, red for WBGSs, and blue for UBGSs. The subscripts "in" and "cross" represent the in- and cross-plane components of the κ tensor along the direction perpendicular or parallel to the *c*-axis, respectively

exceptional RT κ measured to be as high as 1000–1300 W m^{-1} K^{-1} due to the suppression of 3-phonon scatterings [25, 26]. Cubic BP also has a relatively large bandgap and RT κ of 2.0 eV and ≈500 W m^{-1} K^{-1} [27]. Third-generation semiconductors, represented by GaN (3.4 eV) and SiC (3.26, 3.02, and 2.4 eV for 4H-, 6H,- and 3C-SiC, respectively), are classified as WBGS and possess relatively high RT κ which are 200–220 W m^{-1} K^{-1} for wurtzite GaN, 320–470 W m^{-1} K^{-1} for anisotropic hexagonal SiC (see Fig. 23.1 for details), and ≈500 W m^{-1} K^{-1} for isotropic 3C-SiC [8, 28–31]. Such values are significantly higher than that of traditional NBGS Si and Ge.

Note that SiC has more than 250 polytypes [32]. We focus on the most industrially relevant 4H, 6H, and 3C polytypes in this chapter. GaN also has zincblende and wurtzite polytypes [33], although the experimental κ data for the zincblende phase is lacking, it is predicted to be similar to the wurtzite one according to first-principles calculation [34]. Thus, we only discuss the wurtzite GaN and refer to it as GaN below for brevity. By isotope enrichment (see more discussion in Sect. 4), the κ of isotopically enriched ^{71}GaN parallel and normal to the c-axis is predicted to reach 385 and 359 W m^{-1} K^{-1} respectively, thanks to the coupling of heat-carrying acoustic phonons with optic phonons [10, 35].

In the UBGS category, β-Ga_2O_3 (4.8 eV), diamond (5.47 eV), BN (5.94 and 6.35 eV for h- and c-BN, respectively), AlN (6.23 eV), and sapphire (9 eV) are included. The monoclinic β-Ga_2O_3 has a low RT κ ranging between ≈9.5 and ≈22 W m^{-1} K^{-1} for the [100, 010] directions, due to the complex polyhedral configurations and a strong anharmonicity [36], while AlN, diamond, and c-BN with natural element abundance exhibit high RT κ of 320, ≈2200, and ≈880 W m^{-1} K^{-1}, respectively due to their high Debye temperatures [23, 28, 37–39]. Isotopically enriched ^{12}C diamond and c-^{11}BN exhibit enhanced RT κ of ≈3000 and ≈1650 W m^{-1} K^{-1}, respectively [38, 40].

The intrinsic RT κ for many dielectric crystals can be estimated by the modified Leibfried-Schlömann equation, $\kappa = A\frac{\overline{M_a}\theta_a^3\delta n^{1/3}}{\gamma^2 T}$, proposed by Slack [46–49], where A is an empirical constant, $\overline{M}$ is the average atomic mass, θ_a is the Debye temperature of the acoustic phonons, δ^3 is the average volume occupied by one atom of the crystal, n is the number of atoms per unit cell, and γ is the high-temperature average Grüneisen parameter that characterizes the phonon anharmonicity. Rigorously speaking, this equation is valid when the acoustic Umklapp scattering (see Sect. 2.2, Eq. 23.4) is dominant at close or above the Debye temperature.

Figure 23.2 shows the experimental RT κ for most of the semiconductor materials in Fig. 23.1 plotted against the combined scaling parameter $\frac{\overline{M}\theta_a^3\delta n^{1/3}}{\gamma^2}$ using their corresponding values from the literature, following the early work by Slack [46–48]. Except for a few outliers, a clear correlation can be seen between the two quantities. For crystals with light atoms and simple crystal structures and hence a high θ_a, such as diamond ($\theta_a \approx 2240$ K) and 3C-SiC ($\theta_a \approx 1200$ K), a high κ can be expected. Materials with low Debye temperature and high anharmonicity (large γ) such as GaAs ($\theta_a \approx 360$ K) and InP ($\theta_a \approx 425$ K) tend to have lower κ. A large deviation from the Slack model can be seen in BAs due to the complex phonon

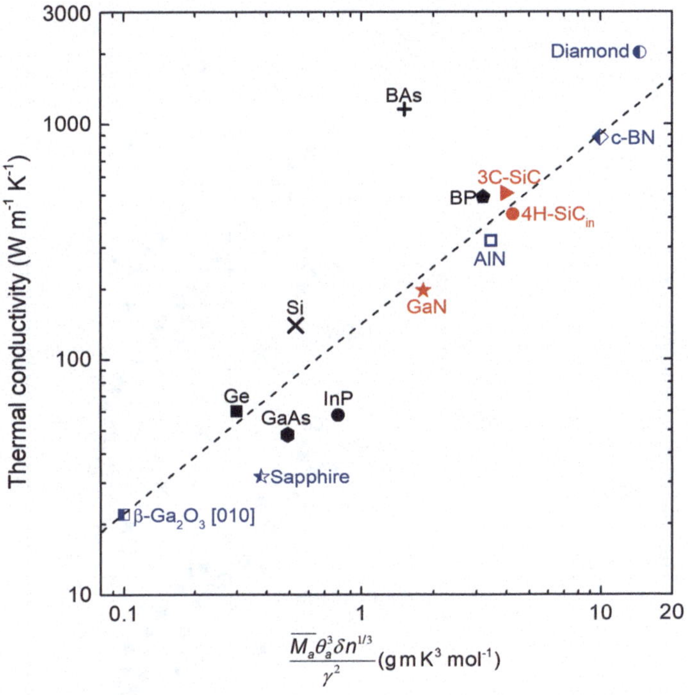

Fig. 23.2 Measured room temperature thermal conductivity of some semiconductors in Fig. 23.1 as a function of the Slack model scaling parameter $\overline{M_a}\theta_a^3\delta n^{1/3}/\gamma^2$ with corresponding values taken from the literature [9, 46, 50–55]. The dashed line is a guide to the eyes

scattering selection rule that cannot be captured by this simple phenomenological model [25].

Figure 23.3 compares the temperature-dependent bulk κ of GaN and SiC with that of NBGSs and UBGSs. Different crystal structures of SiC and BN, as well as the in-plane and cross-plane κ of 4H- and 6H-SiC, are included in Fig. 23.3. Above 300 K, GaN and SiC exhibit higher κ than most NBGSs, except for BAs. When compared to UBGSs, the κ of GaN and SiC are significantly higher than that of β-Ga_2O_3 and sapphire, comparable to the κ of h-BN_{in} and AlN, but lower than that of diamond and c-BN. Such relatively high κ values at elevated temperatures are important for the application of GaN and SiC in power electronics.

With increasing temperature, the enhanced electron-phonon scattering reduces the carrier mobility near the device junctions, which attenuates the I-V characteristic of the electronic device and results in greater self-heating [56, 57]. Such localized hot spots can give rise to nonuniform stress fields and consequently decrease device lifetime. The relatively high κ in conjunction with the high mobility in GaN and SiC at elevated temperatures can effectively mitigate this undesirable self-heating to improve device reliability. To realize the full potential of these materials, it is crucial to minimize extrinsic defects that adversely influence phonon transport.

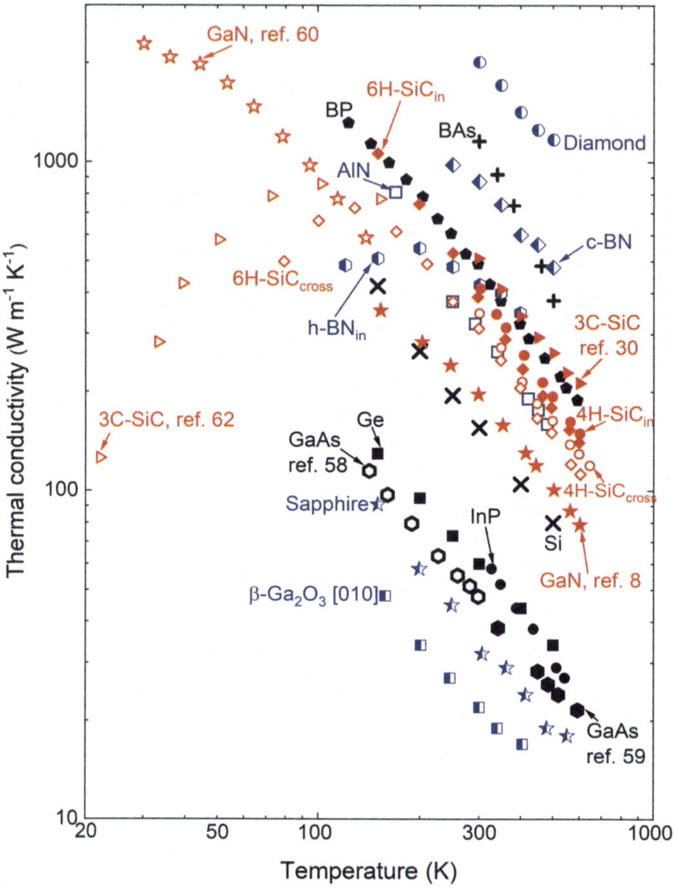

Fig. 23.3 Temperature-dependent bulk κ of the three categories of semiconductors including Si [41], Ge [41], GaAs [58, 59], InP [24], BP [27], BAs [23], wurtzite GaN [8, 60], 4H-SiC$_{in}$ [8], 4H-SiC$_{cross}$ [8], 6H-SiC$_{in}$ [30, 61], 6H-SiC$_{cross}$ [12, 30], 3C-SiC [30, 62], diamond [23], c-BN [38], h-BN$_{in}$ [42], AlN [39], β-Ga_2O_3(010) [36], and sapphire [45]. All κ data presented are from experiments in the literature with color coding the same as Fig. 23.1

2.2 *Basics of Phonon Thermal Transport*

In the context of kinetic theory, the phonon thermal conductivity κ tensor is given by [63, 64].

$$\kappa^{\alpha\beta} = \sum_{\lambda} C_{\lambda} v_{\lambda}^{\alpha} v_{\lambda}^{\beta} \tau_{\lambda} \qquad (23.1)$$

based on the Boltzmann transport equation (BTE) under single mode relaxation time approximation (SMRTA) [65], where the sum is performed over the first Brillouin zone which accounts for the phonon density of states. C_λ represents the volumetric heat capacity for a phonon mode λ, ν_λ^α and ν_λ^β are the group velocity along the orthogonal α and β directions, and τ_λ is the phonon lifetime of mode λ, respectively.

The phonon lifetime is determined by intrinsic phonon-phonon (p-p) and extrinsic phonon-defects scattering. P-p scattering can be categorized into normal (*N*) phonon and Umklapp (*U*) scattering. *U* scattering, in which the phonon momentum is not conserved, is a resistive process which leads to the 1/*T* decay of κ at high temperatures. In contrast, *N* scattering, where the momentum of interacting phonons is conserved, does not directly contribute to κ but redistributes the phonon population, influencing resistive processes. Within SMRTA, assuming that different scattering mechanisms are independent of each other, the total relaxation time τ_C (phonon lifetime) in Eq. (23.1) for a given phonon mode can be estimated using Matthiessen's rule by summing the scattering rates (the inverse of lifetime) from all mechanisms as

$$\tau_C^{-1} = \tau_N^{-1} + \tau_U^{-1} + \tau_E^{-1}, \tag{23.2}$$

where the first two terms on the right side represent the intrinsic *N* phonon scattering rate and *U* phonon scattering rate, respectively. The last term denotes the total phonon-extrinsic defects scattering rate. Deviation from Matthiessen's rule can occur when interdependent scattering occurs, e.g., phonon-assisted impurity scattering of electrons in metals at low temperatures [66].

2.2.1 Intrinsic Phonon-Phonon Scattering

To elucidate the fundamental principles governing phonon transport in WBGSs, an overview of the analytical forms of phonon scatterings within the well-known phenomenological Callaway model is presented [67–69]. Similar to the Holland model [70], it hinges on macroscopic properties (e.g., Debye temperature, sound velocity, and the Grüneisen parameter) of the material and typically requires curve fittings to deduce empirical parameters for extrinsic scattering processes. Despite its limitations, the Callaway model is computationally accessible and physically insightful and hence extensively adopted in the study of group IV and III–V semiconductors [69, 71]. We also briefly discuss the state-of-the-art first-principles-based calculation of κ in GaN and SiC in Sects. 2.2.1 and 2.2.2.

The Callaway model focuses on the transport of the heat-carrying acoustic phonons and formulates the scattering rate based on different branches of phonons. The common three-phonon (3 ph) *N*-scattering rate $\tau_N^{-1}$is given by [22, 69, 72].

$$\tau_N^{-1} = \frac{k_B^b \gamma_0^2 V_0^{(a+b-2)/3}}{\hbar^{b-1} M v^{a+b}} \omega^a T^b, \tag{23.3}$$

where k_B is the Boltzmann constant. ħ is the reduced Planck constant. γ_0 is the Grüneisen parameter. M is the average mass of atoms in the unit cell. V_0 is the crystal's volume per atom. v is the average phonon group velocity for the corresponding phonon branch. ω is the phonon frequency. The coefficients a and b depend on the crystal symmetry, the type of phonon mode, and the temperature range. In the case of a hexagonal crystal-like GaN, *(a,b)* = (2,3) or (2,1) for longitudinal acoustic (LA) phonons and *(a,b)* = (1,4) or (1,1) for transverse acoustic (TA) phonons at low or high temperatures, respectively. This term is prone to the uncertainty from the choice of parameters, e.g., the values of *(a,b)* [71], which can affect the reliability.

The 3 ph U-scattering rate $\tau_U^{-1}$is typically expressed as [22, 72, 73].

$$\tau_U^{-1} = \frac{\hbar \gamma_0^2}{M \Theta v^2} \exp\left(-\frac{\Theta}{3T}\right) \omega^2 T, \tag{23.4}$$

where Θ is the Debye temperature for the corresponding phonon branch. This process leads to the 1/T dependence of κ and the Leibfried-Schlömann (Slack) equation. For GaN, the typical values of the coefficients and the full expression of the κ calculation based on Debye-type phonon dispersions can be found in Refs. [69, 71].

The SMRTA used in the Callaway model and some BTE solutions treats N scattering as purely resistive and cannot capture the restricted phonon scattering when multiple scattering selection rules are activated. Thus, it typically underestimates κ, resulting in large errors for materials with high κ [35, 74, 75].

Today, first-principles-based full BTE calculation beyond the SMRTA has been developed to more accurately predict κ, leveraging density functional theory (DFT) [8, 30, 35, 75]. Since the first study of Si and Ge with a remarkable agreement with the experiment in 2007 [76], such "parameter-free" BTE methods in the context of quantum perturbation theory have been applied to study thermal transport in a variety of materials, e.g., wurtzite and zincblende semiconductors like WGBSs, and have become a reliable standard practice [10, 27, 75, 77–81]. The intrinsic κ from such calculations has demonstrated good agreement with measured data for bulk GaN and SiC [8, 61, 62, 78].

Recently, high-order phonon-phonon scattering, especially four-phonon (4 ph) scattering, has been found to play a crucial role in determining κ of materials such as BAs [23, 25] and AlSb [74], InP [82], BSb [82], and GaN [8, 35, 83], especially at temperatures above 300 K, in addition to the intrinsic 3 ph scattering expressed by Eqs. (23.3) and (23.4) [25, 35, 84, 85]. In these materials, the large acoustic-optic bandgap partly restricts the 3 ph scattering processes subject to the selection rules which do not forbid 4 ph scattering. Only by including both 3 ph and 4 ph scatterings can the theoretical calculation achieve a good agreement with the experimental data.

This indicates that although the higher-order 4 ph scattering is weaker than the 3 ph one in most IV-group materials, it can still be important in many III–V materials with large atomic mass ratios including GaN [8, 35, 82] when the phonon energy gap restricts the scattering phase space of the 3 ph processes. By contrast, phonon transport in SiC with a small acoustic-optic bandgap can be accurately predicted by first-principles-based BTE calculations considering only the 3 ph process [8, 30].

2.2.2 Extrinsic Phonon-Defect Scattering

Extrinsic defects can also scatter phonons, resulting in a shorter phonon lifetime (Eq. (23.2)) and a reduction in κ. Understanding the effect of defects on κ is crucial, as real crystals are rarely pure or ideal. For uniform dense materials, defects that affect the thermal transport properties can be categorized into three groups: (1) 0D point defects (vacancies, impurities, alloying, isotopes, and free electrons), (2) 1D line defects (dislocations), and (3) 2D planar defects (grain, twin, and crystal boundaries, stacking faults, and interfaces) [72]. Typical GaN and SiC in electronic devices rarely contain 3D defects such as precipitates, clusters, or voids, although voids were proposed to be intentionally introduced in GaN to terminate dislocations in special cases [86]. Thus, we focus on the first three types of defects. The typical semi-empirical scattering models for these mechanisms are presented in this section. Further discussions will be provided in Sect. 4.

Phonon-point defects (PD) scattering originates from the lattice disorder caused by vacancies and substitutional impurities (including isotopes), as well as interactions with free carriers introduced by doping. Notably, the phonon-free-carrier scattering usually gives a negligible contribution to the change of κ in undoped and low-doped GaN [71, 83]. Conventional PD scattering rate τ_{PD}^{-1} for acoustic phonon can be quantified by [9, 71, 72].

$$\tau_{PD}^{-1} = \frac{V_0 \Gamma_{LD}}{4\pi v^3}\omega^4 + \frac{E_1^2 m_e^2 k_B T}{2\pi \hbar^4 \rho v} \times \left\{ \frac{\hbar\omega}{k_B T} - \ln\left[\frac{1+\exp\left(\frac{\hbar^2\omega^2}{8 m_e v^2 k_B T} + \frac{m_e v^2}{2k_B T} - \frac{E_F}{k_B T} + \frac{\hbar\omega}{2k_B T}\right)}{1+\exp\left(\frac{\hbar^2\omega^2}{8 m_e v^2 k_B T} + \frac{m_e v^2}{2k_B T} - \frac{E_F}{k_B T} - \frac{\hbar\omega}{2k_B T}\right)} \right] \right\}, \tag{23.5}$$

The first term accounts for the Rayleigh scattering contribution from neutral defects which usually have a much smaller size than phonon wavelengths. Here, Γ_{LD} is the strength of scattering stemming from mass, force constant, and strain perturbations induced by the vacancies, impurities, alloying, and isotopes. The second term is the contribution of phonon-carrier scattering where E_1 is the deformation potential for the carrier-acoustic phonon interaction. m_e is the carrier effective mass. ρ is the mass density. E_F is the Fermi level measured from the bottom of the conduction band. The scattering strength Γ_{LD} can be calculated by

$$\Gamma_{\mathrm{LD}} = \left\{\sum_i f_i\left(1-\frac{M_i}{\overline{M}}\right)^2 + 8f_i\left[\gamma_0(1+Q)\left(\frac{\Delta R_i}{\overline{R}}\right)\right]^2\right\} + \left\{2\sum_i\left(f_i\frac{M_i}{\overline{M}}\right)^2\Gamma_i\right\}, \tag{23.6}$$

where the first brace represents the scattering strength of vacancies, impurities, and alloying, and the second brace denotes the scattering strength of isotopes. In Eq. (23.6), f_i and M_i are the fractional concentration and the atomic mass of the ith atom species, respectively, such that the average atomic mass of the crystal is given by $\overline{M} = \sum_i f_i M_i$. Q is a constant that depends on the anharmonic difference between the nearest-neighbor bonds of the impurity and those in the host crystal. ΔR_i is the average local displacement induced by the ith atom species, and $\overline{R}$ is the average nearest-neighbor atomic radius. Γ_i is the scattering parameter for each isotope which can be expressed as $\Gamma_i = \sum_j c_j^i\left(\frac{m_j^i - M_i}{M_i}\right)^2$. m_j^i is the atomic mass of jth isotope of the ith atom species, and $M_i = \sum_j c_j^i m_j^i$, and $c_j^i$is the fractional atomic natural abundance.

For materials with predominant PD scattering, the early Klemens theory provides a closed-form expression for κ versus defect concentration [87, 88], which is still widely used today as a highly descriptive for III–V semiconductors [89, 90]. This model predicts the ratio of the defective solid's lattice κ to that of a reference pure solid (κ_0) as $\frac{\kappa_{PD}}{\kappa_0} = \frac{\tan^{-1}u}{u}$, where $u = \left[\frac{(6\pi^5 V_0^2)^{\frac{1}{3}}}{2k_B v}\kappa_0 \Gamma_{LD}\right]^{1/2}$ [91].

Phonon-dislocation scattering (D) arises from the interaction with both the dislocation cores and the associate strain fields, which can play a significant role in GaN and SiC due to the wide presence of dislocations in these materials [8, 92–94]. The acoustic phonon D-scattering rate is [72, 95].

$$\tau_D^{-1} = \tau_{\mathrm{core}}^{-1} + \tau_{\mathrm{strain}}^{-1} \tag{23.7}$$

$$\tau_{\mathrm{core}}^{-1} = \eta N_D \frac{V_0^{4/3}}{v^2}\omega^3 \tag{23.8}$$

$$\tau_{\mathrm{strain}}^{-1} = 0.06\eta\gamma_0^2\omega\left(N_S b_S^2 + N_E b_E^2 P_E^2\right) \tag{23.9}$$

$$P_E^2 = \frac{1}{2} + \frac{1}{24}\left(\frac{1-2\sigma}{1-\sigma}\right)^2\left[1+\sqrt{2}\left(\frac{v_L}{v_T}\right)^2\right]^2, \tag{23.10}$$

where η is the weight factor to account for the mutual orientation of the direction of the temperature gradient and the dislocation line. N_D, N_S, and N_E are the total, screw, and edge dislocation densities, respectively. b_S and b_E are the magnitude of Burgers vectors for the screw and edge dislocations. σ is the Poison ratio. v_L and v_T are the longitudinal and transverse velocities.

For phonon-boundaries (B) scattering, the concept of phonon mean free path (PMFP), given by $\nu_\lambda^\beta \tau_\lambda$, is important. PMFP is influenced by all the intrinsic and extrinsic factors mentioned, ranging from 10^0 to 10^4 nm in semiconductors. As the scattering rate increases with temperature and defect concentrations (Eqs. 23.3–23.10), the PMFP decreases accordingly. The broad frequency range of phonons with different modes necessitates the analysis of spectral PMFP [96, 97] and distribution of κ in WBGSs beyond the traditional gray model [61, 78]. In GaN, 60% of the heat is conducted by phonons in a narrow frequency range of 5–7 THz (9% of the frequency span in GaN) which corresponds to PMFP of a few hundred nanometers [78].

When a material's dimension or grain size becomes comparable to the length scales of PMFP, B scattering becomes significant, resulting in a reduction of κ with shrinking dimensions. This is especially the case in submicron thin films, polycrystalline composites, and nanowires of GaN and SiC [7, 98–100]. In the simplified gray model, the B-scattering can be expressed as [72, 98].

$$\tau_B^{-1} = \frac{\nu}{\mathrm{FL}}, \tag{23.11}$$

where L is the dimension of the sample and F is a dimensionless scaling parameter. FL can be regarded as an effective MFP, which is essentially determined by the thickness of the sample for epilayers. More accurate treatment of the boundary scattering can be achieved by using the Fuchs-Sondheimer theory, which takes into account the boundary roughness and spectral phonons [31].

Since defects often break the periodic and symmetric nature of crystals, it is relatively difficult to implement the ab initio BTE method to investigate the effect of defects due to the need for a large enough simulation cell to accommodate enough atoms. Recent advances in DFT on finite-element basis enabled the study of large-scale complex materials involving many thousands of atoms [101–103], which is yet to be widely used for thermal transport studies. For the phonon transport part, BTE solver collections, e.g., almaBTE and GiftBTE have been developed and used to predictively tackle phonon transport in not only perfect bulk crystals but also alloys, thin films, and multiscale structures with size features in the nm-μm range [104, 105]. Leveraging almaBTE, a full ab initio defect perturbation-based Green's function method has been used to study the impact of PD phonon scattering on κ in GaN [106]. Such ab initio Green's-function method has also been used to study the dislocation effect in GaN [107] using finite-sized supercells. MD simulations provide an important alternative simulation technique, particularly in the study of GaN and SiC materials with complex nanostructure, interfaces [108, 109], and defects [110, 111].

2.3 Temperature Effects of κ and Its Impact on Device Performance

As depicted in Fig. 23.3, the κ of WBGS typically decreases with increasing temperature in the range from 200 K to 600 K. According to Eqs. (23.1) and (23.4), κ should scale with 1/*T* at above 1/10 to 1/2 of the Debye temperature when the 3 ph *U* scattering (proportional to *T*, at high temperature) dominates [112]. For the convenience of the discussion, we define the temperature dependence of κ using the slope of lnκ versus ln*T* [48].

$$\varepsilon = -\left(\frac{\partial \ln \kappa}{\partial \ln T}\right)_P. \tag{23.12}$$

Based on the deviation from the high-temperature 1/*T* behavior at a given temperature, the contribution to ε can be phenomenologically divided into four components as [48].

$$\varepsilon = 1 + \eta_{\text{th}} + \eta_{\text{op}} + \eta_{4p} \tag{23.13}$$

where 1 is the contribution to ε when only three-acoustic-phonon scattering processes (so-called *aaa* scattering) are considered, as shown in Eq. (23.4). η_{th} is the contribution to ε from the thermal expansion of crystal under heating [113]. η_{op} is the contribution to ε from the scattering of acoustic phonons by optic phonons [48, 113, 114]. η_{4p} is the contribution of 4 ph scattering to ε, which is significant in some crystals [48], such as BAs and GaN.

The contributions from η_{th} and η_{op} at temperatures below Θ typically lead to a ε greater than 1. For GaN and SiC in 300–850 K, $1.10 < \varepsilon < 1.50$. For η_{4p}, calculations showed that the 4 ph scattering rate typically follows $\tau_4^{-1} \sim T^2$, c.a., the linear dependence for the 3 ph process [25, 84]. This results in a deviation from the 1/*T* scaling in GaN even above its Θ(636 K). In fact, due to the relatively strong contribution of 4 ph scattering which further increases with temperature, ε in GaN increases with temperature, reaching ≈1.5 at 850 K [8]. By contrast, ε in SiC gradually decreases with temperature and approaches 1 at $T > 850$ K since the 3 ph *U* scattering dominates its thermal transport at high temperatures.

Defects also influence ε, as illustrated in Fig. 23.4. Since the phonon-defects scattering is largely temperature-independent, as shown in Eqs. (23.5–23.11), the presence of defects can suppress the effect of the intrinsic *U* scattering, leading to a smaller ε at moderate to high temperatures, which has been shown for GaN and SiC with point defects [8, 115], dislocations [11], and crystal boundaries [7, 98–100]. The temperature corresponding to the maximum value of κ, denoted as $T_{\max}$, usually occurs at approximately 1/10 the traditional atom density-based Θ for bulk crystals with intrinsic κ, which results from a combination of the p-p and phonon boundary scattering. When the temperature is well above the $T_{\max}$, approaching Θ, p-p interaction typically dominates the total scattering process, and κ(*T*) becomes

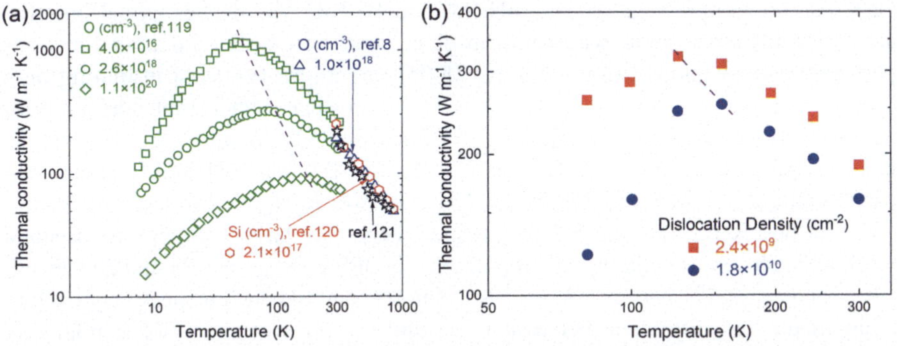

Fig. 23.4 The impact of defects on the temperature dependence of κ_{GaN}. (**a**) The temperature variation of κ in GaN is affected by the O dopant with different concentrations [8, 119] and Si dopant [120]. All materials are bulk except for the material represented by the black star (thickness is about 1.1 μm) [121], and its dopant type and concentration are not given in the literature. (**b**) The effect of dislocation density on the temperature dependence of κ in GaN films with a thickness of ~3.3 μm [11]. The purple dashed line illustrates how the defect concentration or density affects the position of T_{max}—the temperature corresponding to the maximum value of κ_{GaN}. All κ data presented are derived from experiments in the literature

almost defect-independent [60, 116], as shown for various dielectrics [117] and for GaN with varied oxygen impurity concentration and dislocation density in Fig. 23.4.

On the other hand, at temperatures below T_{max}, when the number of excited phonons decreases and U scattering is weak [12, 22], the phonon-defects scatterings are more pronounced [11, 118], leading to a smaller slope of κ as shown in Fig. 23.4. The position of T_{max} is also affected by the defect concentration or density. As they increase, T_{max} usually shifts toward higher temperatures, as indicated by the purple dashed line in Fig. 23.4.

The temperature effects of κ can impact the reliability and analysis of WBGS-based devices. For example, in GaN high-electron-mobility transistors (HEMTs), the nonlinear ε influenced by extrinsic defects can affect the I-V curve and device performance [122]. Also, the channel temperature from simulation for GaN HEMT can be largely underestimated when using a constant κ, deviating from experimental results, compared to using a realistic temperature-dependent κ [123–125]. The discrepancy becomes more significant at higher operating powers, reaching up to 50% [123]. For GaN devices on high-κ substrates such as SiC with smaller ε, the discrepancy is much lower compared to Si and sapphire substrates [123, 125].

2.4 *Anisotropy of κ and Topological Phonons*

The κ of many WBGSs including the common polytypes of GaN and SiC are anisotropic. Exceptions are cubic 3C-SiC [30] and theoretically predicted cubic

GaN. As shown in Figs. 23.1, 23.2, 6H-SiC and 4H-SiC with the hexagonal structure are noticeably anisotropic, where the "in-plane" and "cross-plane" denote directions perpendicular or parallel to the c-axis ([0001] direction). For semi-insulating bulk c-plane 6H and 4H-SiC wafers grown by a physical vapor transport in prior experimental measurements [8], κ_{cross} = 345 and 320 W m^{-1} K^{-1} at 300 K; κ_{cross} = 85 and 78 W m^{-1} K^{-1} at 850 K for 4H- and 6H-SiC, respectively. Meanwhile, κ_{in} = 415 and 390 W m^{-1} K^{-1} at 300 K; κ_{in} = 150 and 140 W m^{-1} K^{-1} at 600 K for 4H- and 6H-SiC, respectively. Similarly, another experimental work on such semi-insulating SiC reports that at 300 K, κ_{cross} = 324 and 273 W m^{-1} K^{-1} and κ_{in} = 471 and 393 W m^{-1} K^{-1} for 4H- and 6H-SiC, respectively [31]. In comparison, the difference between κ_{cross} and κ_{in} for wurtzite GaN is relatively small, in 5–10% at >300 K according to first-principles calculation as confirmed in the measurement of consistent results for c- m- and a-plane-oriented low-dislocation density (<10^7 cm^{-2}) GaN samples [8, 34, 126]. A recent experimental work obtained similar results of κ_{in} = 215 W m^{-1} K^{-1} and κ_{cross} = 225 W m^{-1} K^{-1} low-defect-density and isotopically natural c-plane GaN [127]. As a reference, the RT κ of the UBGS β-Ga_2O_3 in the [010] direction is about 22 W m^{-1} K^{-1}, while it is only ≈9.5 and ≈14 W m^{-1} K^{-1} in the [100, 001] directions, respectively [36, 128]. Thus, for GaN, 6H-SiC, 4H-SiC, and β-Ga_2O_3, the RT thermal anisotropy ratios, $\kappa_{in}/\kappa_{cross}$, are approximately 1.05, 1.2, 1.3, and 2.5, respectively.

According to Eq. (23.1), the anisotropy of κ is induced by the anisotropic velocities v, density of states, and mode-dependent lifetime τ_q (scattering rate) of phonons [54, 61]. The anisotropic v and density of states are usually caused by the asymmetry in crystal structure which influences the phonon dispersion [92, 129]. Additionally, the phonon-focusing effect can also amplify the anisotropic κ at low temperatures in wurtzite GaN, 4H-SiC, and 6H-SiC, especially for samples with small geometric aspect ratios [130–132]. This effect is induced by the noncollinearity of group velocities v with respect to the wave vectors of heat-carrying phonons which generally true even in cubic crystals [130–132]. It mainly appears at sufficiently low temperatures (~10s K), when the PMFP is large, and the B scattering dominates other processes. In this case, phonons, especially transverse modes, are channeled along a few specific directions with large v thereby enhancing κ along those directions in elastically anisotropic samples.

On the other hand, several factors can contribute to anisotropy in the mode-dependent τ_q of phonons along different directions. The first is the intrinsic crystal structure and the corresponding phonon topology. For example, in layered h-BN, the weak interlayer bonding results in a small lifetime for phonons propagating along the c-axis leading to a small κ_{cross}, and an RT-anisotropy ratio as high as 80 [44]. In addition, the crystal structure can also cause phonon dispersion nesting behavior resulting in nearly parallel sections between phonon branches in the phonon dispersion. This mechanism has been found in wurtzite GaN, creating a vast number of 3 ph scattering channels along the in-plane direction than that of the cross-plane direction, reducing κ_{in}, and suppressing the RT-anisotropy ratio [34, 126].

Such anisotropy in κ can affect the performance of WBGS-based devices. For example, when the fin channel is oriented along the [100] direction, the channel temperature rise of a single fin field effect transistor could be reduced by 30% compared to devices with a fin channel aligned to the [010] direction, due to the anisotropy of κ in β-Ga_2O_3 [128]. Similar effects should also be expected in hexagonal SiC and GaN devices since the extrinsic defects induced in the growth process can amplify the anisotropy in κ [92, 129].

Topological surface phonon with novel transport characteristics that are robust and protected by the topological properties of the bulk system is an emerging area in research [133]. A recent work found that although wurtzite GaN is a topologically trivial phonon system, the introduction of Al atoms in GaN induces a topological phonon phase transition, evidenced by the emergence of Weyl phonon states in AlN and AlGaN [134]. The AlGaN/GaN heterostructure may host topological surface states where phonon backscattering is forbidden, which has the potential to improve heat dissipation across AlGaN/GaN interfaces to enhance the device's thermal management.

Besides these intrinsic effects, extrinsic defects, such as highly oriented dislocations, can also induce anisotropic τ and κ [92] by selectively scattering phonons in specific directions, which will be discussed in Sect. 4.

3 Thermoreflectance-Based Thermal Characterization Techniques for WBGSs

An efficient and reliable thermal metrology is crucial for the WBGS material development. In many cases, the capability of spatial thermal property scanning/mapping is useful for resolving the inhomogeneity along the in- and cross-plane direction. Meanwhile, direct temperature mapping of the working microdevices is important for hotspot profiling and thermal management. Before discussing the thermoreflectance-based method, we review a few common thermal characterization tools that have been used in the study of WGBSs.

Both contact and noncontact methods [135] have been used to measure the thermal properties of GaN and SiC. Among the contact methods, steady-state methods, e.g., longitudinal heat flow ("cut bar"), or guarded hot plate [136, 137], have the longest history which can provide highly accurate measurement (uncertainty <5%) when done properly. This kind of method has been used to study the κ in GaN for over a decade [119, 138, 139], which enables the study of their low-temperature behavior that is sometimes not easily attainable with noncontact methods. However, these approaches are usually slow (taking hours), require large sample volumes with regular shapes, and lack spatial mapping capability [2].

The electrothermal 3ω method is a versatile thin films and interfaces contact metrology method that is widely used to study thermal transport in WBGS materials [113, 140, 141]. In a typical application of the 3ω method, the metal heater/sensor layer on the sample surface needs to be patterned and electrically connected, which can be time-consuming [2, 142]. Furthermore, an electrically insulating layer must be added between the metal heater/sensor and GaN [140, 143], which inevitably introduces additional thermal resistance that affects the analysis.

Among noncontact methods, laser flash method has been widely used in materials' thermal diffusivity characterization and has been applied in GaN [120] and SiC [144]. This method lacks spatial resolution and typically requires the application of a sprayed coating to ensure high absorptivity and emissivity for semi-transparent samples like GaN and SiC, a procedure that is usually less controllable than the physical vapor deposition of metal layers. It may also suffer from errors in sample thickness and finite pulse especially when the sample is thin and nonopaque (e.g., GaN and SiC wafers) [145, 146]. Raman thermometry offers the advantage of directly probing the material of interest without the need for a metal transducer layer, despite requiring separate Raman peak calibrations [147], which has been applied in the study of κ and temperature distribution in SiC- and GaN-based HEMTs [148–152]. However, it faces a few limitations [147, 153]. Raman peak shifts and linewidth variations, typically indicative of temperature changes, are also sensitive to stress, disorder, impurities, or contamination, leading to calibration inaccuracies. Furthermore, precise measurement of absorbed laser power is usually required but difficult, impacting the measurement accuracy, although transient or anti-Stokes/Stokes ratio-based Raman methods can mitigate this issue [154]. The first problems can be alleviated by dispersing Raman-sensitive diamond or other oxide microparticles on the device as thermometry which has been used for GaN devices [148, 149]. Dispersed nanodiamond particles can also be used leveraging the nitrogen-vacancy center-based thermometry for temperature mapping in GaN HEMTs [155]. The use of nanoparticles inevitably complicates the sample preparation and may interfere with the device's thermal profile.

3.1 Principles, Implementation, and Application of Thermoreflectance-Based Techniques

Thermoreflectance-based techniques rely on the principle of thermoreflectance, that is, a material's optical reflectance (R) changes with temperature (ΔT). For common metals (e.g., Al, Au, and Cu), $\Delta R/R$ is linearly proportional to ΔT when ΔT is small, i.e., [2].

$$\frac{\Delta R}{R} = \frac{1}{R}\frac{dR}{dT}\Delta T = C_{\mathrm{th}}\Delta T \tag{23.14}$$

where dR/dT or the lumped C_{th} is denoted by the thermoreflectance coefficient. With the use of a metallic transducer layer that acts as both a laser absorber for heating and an optical thermometer, thermoreflectance methods can typically avoid the nonthermal signals and tedious calibration in the Raman method. The main types of thermoreflectance techniques today include time-domain, frequency-domain, steady-state thermoreflectance (TDTR, FDTR, SSTR, respectively), and thermoreflectance thermal imaging (TTI). These techniques have become powerful tools and are extensively used for measuring the thermal properties (e.g., κ, G, and volumetric heat capacity C_v) and probing temperatures in WBGS materials and devices respectively [2].

It is worth mentioning that C_{th} depends on the wavelength, and there are significant differences in the C_{th} spectra of different materials. For example, the C_{th} spectra of Au have a positive peak and a negative valley located at ~470 nm and ~ 530 nm, respectively, while the C_{th} of Al reaches a maximum at ≈800 nm [2]. In general, to avoid the reflectance change of semiconductor materials due to optically excited carriers, WBGSs are coated with metal transducers (e.g., Al or Au) in the κ measurements. Typically, C_{th} itself does not need to be calibrated in TDTR and FDTR as they utilize the thermal phase lag or the ratio of in- and out-of-phase signal rather than the absolute temperature (see below) in the data analysis. By contrast, TTI can be used to directly obtain the device temperature distribution without a transducer layer, which requires additional device calibrations (steady-state environmental heating) [156].

Figure 23.5 shows the schematic of TDTR, FDTR, SSTR, and TTI systems (left), and their corresponding responses (right). In general, these method uses a strong pump beam (TDTR, FDTR, SSTR) or device excitation (TTI) as a heat source to create a temperature rise in the sample, and a probe beam is used to interrogate the sample surface temperature change via the principle of thermoreflectance. On the other hand, there are some differences among these techniques in the characteristics of pump and probe beams (e.g., pulsed or continuous wave), optical path layout, and experimental data processing models. Here, the standard implementation and unique features of each technique are briefly discussed.

TDTR typically uses a pulsed laser with a high-repetition rate (e.g., 80 MHz) to generate sub-picosecond pulse trains. The laser output is split into a pump beam, which is optically modulated (typically in the range of 0.1–20 MHz) to heat the sample, and a probe beam, which is typically modulated at a relatively low frequency delayed with respect to the pump beam up to several nanoseconds through a mechanical delay stage, to monitor the surface temperature evolution of the sample. The reflected probe beam is collected by a photodetector and then demodulated by a lock-in amplifier at the pump modulation or the sum frequency of the pump and probe in the heterodyne detection [157]. The complex thermoreflectance signal is recorded as a function of the delay time, as shown in Fig. 23.5. The thermal properties can be extracted by fitting the data via nonlinear regression using a thermal transport model (e.g., Feldman transfer matrix model [158, 159]).

FDTR differs from TDTR in that it can utilize both pulsed- and continuous-wave (cw) lasers. FDTR monitors the thermal-induced amplitude or phase lag between the pump and probe beam as a function of modulation frequency, as shown in Fig. 23.5. Apart from these differences, other optical configurations, thermal transport model, and sensitivity analysis of FDTR are similar to those of TDTR [2]. Spatially scanned FDTR with a modulation frequency in 100–100 kHz, sometimes called thermal-wave microscopy, has also been developed and applied to semiconductor materials [160, 161].

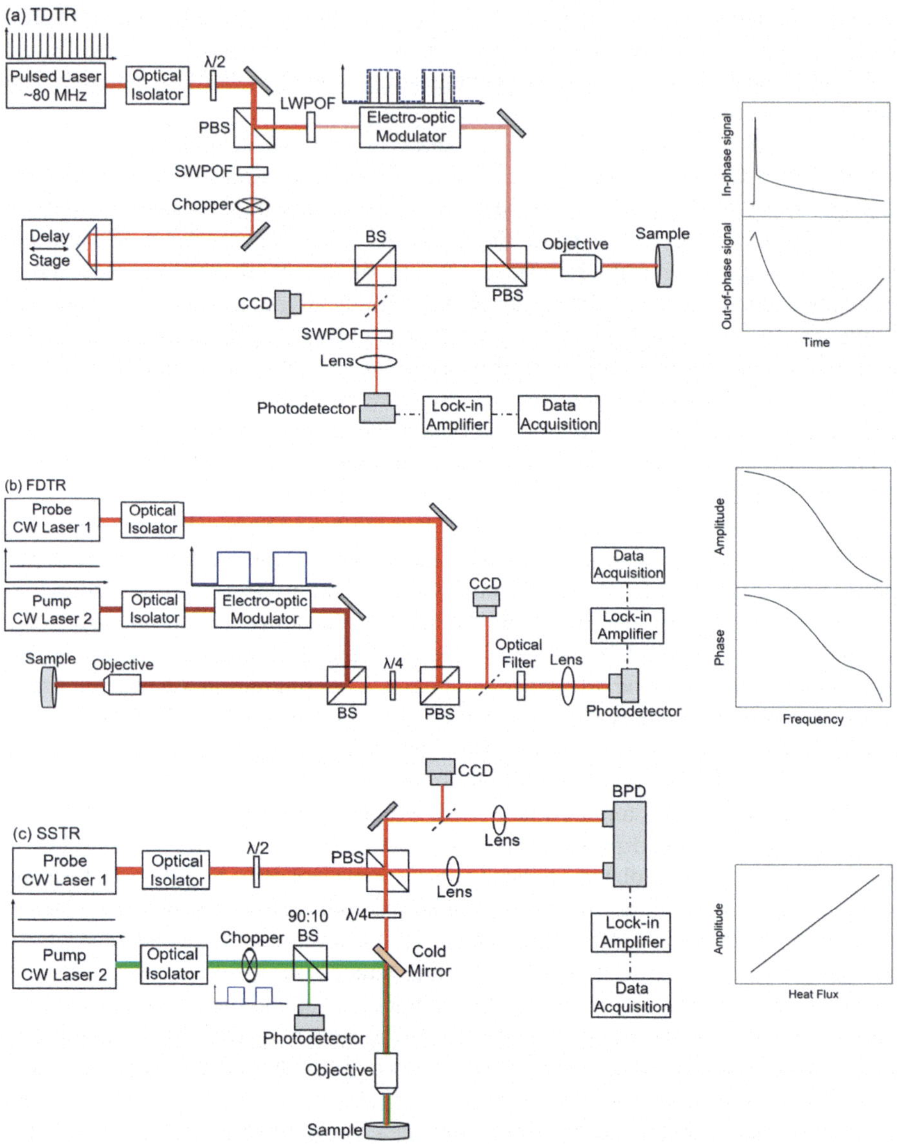

Fig. 23.5 Schematic of (**a**) TDTR, (**b**) FDTR, (**c**) SSTR, and (**d**) TTI systems, and their corresponding responses. The acronyms λ/2, λ/4, PBS, BS, L/SWOPF, and BPD stand for half-wave plate, a quarter-wave plate, polarizing beam splitter, beam splitter, long/short-wave pass optical filter, and balanced photodetector, respectively

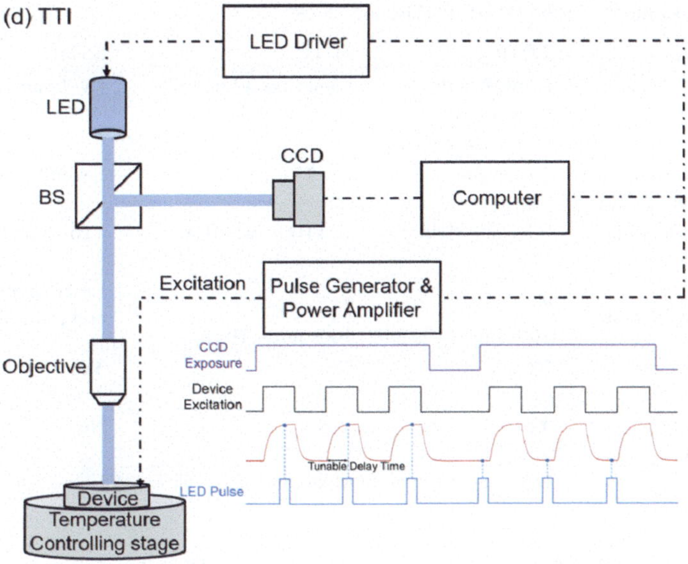

Fig. 23.5 (continued)

SSTR is a recently developed technique that employs cw lasers as pump and probe beams, similar to FDTR. A key difference between them is the modulation frequency of the pump beam. In FDTR, the modulation frequency is typically on the order of kHz to MHz, while in SSTR, it is on the order of 10s of Hz. Due to the low-modulation frequency in SSTR, the sample has sufficient time to reach a steady state, and the thermoreflectance response is dominated by κ, independently of volumetric heat capacity C_V. This means that the C_V does not need to be known in advance when fitting κ, which is needed for the analysis of TDTR and FDTR data [2, 162]. Another key difference is the way signals are collected and processed. SSTR measures the actual power of the pump heat flux at the sample surface and the resultant steady-state temperature rise by the probe beam. A linear relationship between temperature rise and heat flux by varying the pump power is utilized to extract the κ of materials by fitting the slope of the data with a thermal transport model. Since SSTR uses absolute temperature rise during data processing, it is necessary to precalibrate the proportionality constant that depends on C_{th}, electronic gains, and optic factors, which influence the accuracy of SSTR [2]. The comparisons between TDTR, FDTR, and SSTR are shown in Table 23.1.

The lateral resolution in FDTR and TDTR is mainly determined by the maximum value of the $1/e^2$ laser spot size and the in-plane thermal penetration length in the sample defined as the distance at which the temperature decays to $1/e$ of its value at the spot center. For high-frequency pump modulation at f, $d_p^{\text{in}} = \sqrt{\frac{\kappa_{\text{in}}}{\pi f C_V}}$ for bulk samples. When d_p^{in} is small, the resolution is determined by the spot size which is typically 0.8–10 μm for visible and near-infrared lasers in TDTR and FDTR

Table 23.1 Comparisons of TDTR, FDTR, and SSTR

Techniques	TDTR	FDTR	SSTR
Laser	Femtosecond pulsed laser	Femtosecond pulsed laser or cw laser	Cw laser
Wavelength	400 ~ 1064 nm		
Focused spot size w ($1/e^2$ radius)	2 ~ 25 μm	2 ~ 25 μm	Several μm ~ mm
Pump beam modulation frequency f	0.2 ~ 20 MHz	10 kHz ~ 50 MHz	~10–200 Hz
Thermal characteristic length	$d_p = \sqrt{\frac{\kappa}{\pi f C_V}}$: 100s nm ~ 10s μm	$d_p = \sqrt{\frac{\kappa}{\pi f C_V}}$: 100s nm ~ 10s μm	Laser spot size: several μm ~ mm
Is C_{th} calibration needed?	No	No	Yes
Is C_v needed for extracting κ?	Yes	Yes	No
Cost	High	Medium	Low

[162]. The through-plane resolution is often described by the thermal penetration depth, which is approximately $d_p^{\text{cross}} = \sqrt{\frac{\kappa_{\text{cross}}}{\pi f C_V}}$ (for bulk samples) at relatively high frequency. d_p^{cross} is typically on the scale of microns for high-frequency measurements (for GaN, $d_p^{\text{cross}} \approx 1.5$ μm for $f = 10$ MHz at room temperature). By contrast, the in- and cross-plane spatial resolution of SSTR is on the order of the spot size, larger than that of TDTR and FDTR, but smaller than those in laser flash [163]. The accuracy of these methods depends on both the experimental signal-to-noise ratio and the error propagation in the thermal model. The overall error is 5–10% when done properly [159].

Differing from TDTR, FDTR, and SSTR that detect the areal averaged information within the probe beam with a photodetector, TTI uses wide-field LED pulses to illuminate the device under test (DUT), which is typically electrically excited, and records the reflected light and hence the surface temperature evolution from the device by a CCD camera via thermoreflectance. The delay time between LED illumination and device excitation can be varied continuously to obtain a series of thermal images independent of the camera frame rate [164, 165]. Prior to measuring the device, a calibration of $C_{th}(x, y)$ should be implemented. During the measurement, the $\Delta R(x, y)/R(x, y)$ can be monitored to calculate the temperature rise directly, and finite-element models are typically used to help extract the thermal properties of the materials. It is worth mentioning that the precise alignment between $C_{th}(x, y)$ and $\Delta R(x, y)/R(x, y)$determines the accuracy of the TTI technique which occasionally requires per-pixel transient calibration [2, 166]. The spatial resolution of TTI is limited by optical diffraction of the microscope system, which can achieve 150–200 nm if deep-ultraviolet light and a high NA (>0.6) objective lens can be used [166–168].

3.2 *Application of Thermoreflectance-Based Techniques in GaN and SiC*

Due to minimal restrictions on sample shape and size, high throughput, accuracy, and spatial resolution, thermoreflectance-based techniques have been widely used to investigate thermal transport in GaN and SiC. Here we only list a few examples. Beechem et al. [9] measured the κ of n- and p-type doped GaN films with thicknesses of 3 ~ 4 μm using TDTR with a focus on the impurity and carrier concentration effect. Zheng et al. used TDTR to measure the κ of GaN with natural isotope abundance and ^{71}GaN in 150 to 850 K, revealing that the κ of ^{71}GaN is ~15% higher than that of GaN with natural isotope abundance (60% ^{69}GaN and 40% ^{71}GaN) [8]. In low-dislocation density ($< 10^7$ cm^{-2}) and moderate doping ($< 10^{19}$ cm^{-3}) bulk and homoepitaxial GaN, κ is insensitive to crystal orientation (c- m- or a-plane) and doping concentration. This work also investigated the temperature-dependent in- and cross-plane κ of semi-insulating 4H- and 6H-SiC, showing good agreement with first-principles calculation. Tong et al. [169] employed TDTR to measure the RT κ of $In_xGa_{1-x}N$ alloys with x ranging from 0.08 to 0.98, and found that the κ followed a bathtub trend, being strongly suppressed at intermediate compositions ($0.2 < x < 0.9$). Park and Bayram [95] investigated the κ of GaN with different dislocation densities using TDTR. By combining their results with an empirical model, the deflection point where the κ of GaN starts to drop is approximately 10^8 cm^{-2}. Ziade et al. [100] reported FDTR measurements of κ for GaN films with thicknesses in 15 ~ 1000 nm, revealing a significant size effect as the κ increased from ~5 to ~120 W m^{-1} K^{-1}. Qian et al. studied the semi-insulating and n-type (N-doped 1×10^{19} cm^{-3}) 4H-SiC and 6H-SiC using TDTR and found that the doping can reduce the κ of 4H-SiC by ~7% at 300 K [31]. Li et al., used TDTR in combination with BTE to show that the phonon-dislocation scattering is weaker than suggested by previous measurements for GaN films of a few micron thickness, where dislocation density 1.80×10^{10} cm^{-2} reduces κ by ~25% [11]. SSTR has been used to study Al/GaN/GaN, and Al/AlN/sapphire, with advantages in measuring the κ of buried substrates (GaN and sapphire) that are challenging for TDTR [170]. Various publications on temperature characterization of GaN devices via TTI have been covered in a prior review paper [2].

3.3 *Spot Size and Modulation Frequency Dependence of Apparent κ in Thermoreflectance-Based Measurements*

The apparent κ of some semiconductors, such as Si [171], 3C-SiC [30], InGaAs [172], and GaN [11] obtained from TDTR and FDTR can sometimes depend on the characteristic length determined either by the laser spot size [11, 30, 171] or the thermal penetration depth [11, 172], when they become comparable or smaller than the dominant PMFPs of the materials. The mechanism behind this phenomenon is typically explained by the quasi-ballistic phonon transport [12, 173, 174], in which some phonons can experience multiple scattering events and transport diffusively, whereas others travel ballistically across the length scale without scattering. As a result, these phonons fail to establish a local thermal equilibrium and have a strongly suppressed contribution to the measured thermal transport properties and the definition of local temperature becomes difficult [12, 159]. Since the thermal model in analysis is usually based on Fourier's law, the apparent κ fitted from such models becomes a spot-size or frequency dependent due to thermal nonequilibrium [175]. Thus, caution should be exercised when conducting TDTR measurements of the apparent κ of materials like GaN and SiC, where a large portion of long-MFP phonons contribute to κ.

3.4 Challenges in Thermoreflectance-Based Measurement

Thermoreflectance methods typically require highly specular surfaces with rms roughness of less than 15 nm [159]. Although this can be challenging for some composites or ceramics, the smoothness of the bulk and thin film GaN and SiC are usually sufficient. Yet, two challenges are still present in thermoreflectance-based techniques for studying WBGSs. Firstly, for both the κ measurement and device temperature mapping in TTI, the spatial resolution of these far-field optical techniques is limited by diffraction and cannot directly approach the sub-100 nm regime. Nanostructured transducer has to be used to study the short PMFP in TDTR [175–178], while TTI requires image deconvolution to achieve sub-wavelength resolution using the point spread function (PSF) of the imaging system which is often hard to obtain or rely on hypothesis [179]. Further incorporation of techniques from pump-probe super-resolution microscopy [180] may help overcome the limitation in TTI.

Secondly, the measurement of κ at extremely high or low temperatures is not simple. Recently, high-melting-point transducers, such as Pt and HfN, have enabled thermoreflectance-based techniques to measure κ at high temperatures up to approximately 1000 K [159]. To further extend the upper-temperature limit, the development of novel transducers with superior properties, including higher melting points and thermoreflectance coefficients, is necessary. Due to the high melting points of GaN (2500 °C) and SiC (2730 °C), performing TDTR/FDTR without transducers at above 1000 K can circumvent the problem of transducers melting, and enable measurements at higher temperatures [159]. However, this requires careful selection of pump and probe wavelengths to ensure effective heating and a sufficiently high C_{th} in which the photo-carrier effect on reflectance needs to be taken into account [1]. At cryogenic temperatures below 30 K, thermoreflectance-based techniques face challenges due to the low C_v of materials at these temperatures, which can lead to significant temperature rise even with minimal laser power. This can cause changes in C_{th} during the measurement, resulting in a nonlinear relationship between the thermoreflectance signal and the sample surface temperature change, thereby introducing artifacts [159]. To address this issue, a combination of strategies can be employed, such as using a large spot size to reduce low-power density and selecting a transducer or wavelength that suppresses the absorbance [159, 181].

4 Effects of Extrinsic Defects on Thermal Transport in GaN and SiC

4.1 Point Defects

Point defects in semiconductor crystals are atomic-scale entities that disrupt the lattice periodicity [182, 183]. Vacancies occur when an atom is missing from its regular lattice site, while interstitials happen when an atom occupies a position different from a regular lattice site. Substitutional impurities refer to impurity atoms substituting host lattice atoms while an antisite defect forms in the compound when an atom from one sublattice is placed in the another. Defect complexes can also form due to the clustering of mobile point defects at the growth temperature.

Various instruments can characterize these point defects. Positron annihilation spectroscopy (PALS) can determine vacancy type and concentration [184, 185], while secondary ion mass spectroscopy (SIMS) [8, 9, 95] and photoluminescence (PL) [186] can measure doping type and concentration. Hall effect measurements can quantify electrical conductivity and carrier mobility [132, 187], and Rutherford backscattering Spectrometry (RBS) [169] and X-ray diffraction (XRD) [169, 188] can determine alloying fraction.

Point defects can be engineered through methods such as doping, annealing, and growth techniques. Doping by thermal diffusion and ion implantation deliberately introduces "foreign" impurity atoms, or dopants (e.g., Si, O, Mg, and Fe [72]), into the host lattice to manipulate the electronic properties of semiconductors. The dopants can act as interstitial or substitutional impurities [189, 190]. Annealing can place ions substitutionally on lattice sites, making it a special postprocessing step after ion implantation [183, 191–193].

In the growth process, by alloying multiple compounds and isotope engineering [69], WBGS materials are tunable in the bandgap, lattice constant, and thermal transport [194]. For example, the main growth techniques of GaN include hydride vapor phase epitaxy (HVPE), Na flux method, metalorganic chemical vapor deposition (MOCVD), molecular beam epitaxy (MBE), and ammonothermal (AT). For bulk GaN growth in the industry, native GaN seeds from the AT method are frequently used in HVPE growth [195]. Impurities such as H, O, C, Si, Mg, Mn, and Fe can infiltrate GaN from the substrate, reactor walls, and precursors, with their concentrations significantly affected by growth conditions such as temperature, pressure, and Ga/N ratio [196, 197]. Typically, bulk GaN from the AT method tends to have a relatively high impurity compared with those from Na flux and HVPE [195].

As discussed in Sect. 2, point defects can affect thermal transport in GaN and SiC. For example, using TDTR, Beechem et al. [9] found that the κ of n-type GaN films remained constant when the carrier concentration increased from 10^{15} cm^{-3} to 10^{18} cm^{-3}. However, for the p-type GaN films, a ~ 30% decrease in κ was observed when the carrier concentration increased from 10^{17} cm^{-3} to 10^{18} cm^{-3}. By combining their results with a phonon scattering model, they discovered that H impurity concentrations exceeding 10^{18} cm^{-3} affect the κ of GaN films, while the critical concentrations for Si and Mg are about 10^{20} cm^{-3}. For bulk GaN, Zheng et al. [8] showed that the κ of bulk GaN with low-dislocation density (<10^7 cm^{-2}) is not strongly affected by impurity if the doping level is lower than 10^{19} cm^{-3} while the high concentration of impurity (Mn, H, and O dopant at ~2 × 10^{19} cm^{-3}) can reduce RT κ by ~30%. Importantly, although the isotope enrichment was predicted to enhance κ by ~80% [35], the high impurity level (2.6 × 10^{19} cm^{-3} Al, and a total of 1.2 × 10^{19} cm^{-3} of other metal impurity) suppresses the κ enhancement to ~15% in the ^{71}GaN homoepitaxial layer studied by Zheng et al. [8].

In SiC, the impurity effect on κ has also been reported. Based on temperature-dependent TDTR measurements and doping control, Cheng et al. reported a stronger phonon scattering by boron dopants compared to nitrogen dopants in >110 μm SiC homoepitaxial layer when they are on the order of 10^{18} to 10^{19} cm^{-3} [12]. The point defect scattering with such doping level mainly does not seem to affect the RT κ, and plays a noticeable role mainly at <200 K. A recent first-principles calculation investigated the κ in 3C-SiC with 10^{16} to 10^{21} cm^{-3} point defect concentration and found that phonon-electron scattering is significant in reducing κ for N-doped SiC when point defect is lower than 10^{20} cm^{-3}, while PD scattering is dominant in B-doped SiC across all concentrations due to its resonant behavior and large ionization energy. Using laser flash, Wei et al. showed that the nitrogen doping 5 × 10^{18} cm^{-3} can reduce the RT thermal diffusivity by ~10% [144].

Figures 23.6 and 23.7 demonstrate the influence of point defects on the RT κ of GaN from experiments and calculation. Introducing foreign impurity atoms such as Si, O, Mg, Fe, and Al can reduce the κ of GaN when the dopant concentration reaches a certain level. The critical doping concentrations at which κ starts to reduce significantly are approximately 1 × 10^{19} cm^{-3} for Si, 2 × 10^{19} cm^{-3} for Mg and O, and 5 × 10^{19} cm^{-3} for Fe. The main reason for such a PD effect is the increased phonon-PD scattering that decreases phonon lifetime, as indicated by Eqs. (23.5) and (23.6). Dopant incorporation can also change the bond strength between atoms or atomic masses which slightly modifies the crystal's band structure, hence slightly affecting the group velocity and ultimately κ [198, 199].

Figure 23.7a shows the effect of Al atom substitution in $Al_xGa_{1-x}N$ alloy. A typical bathtub trend similar to metallic alloys can be observed: The RT κ changes sharply with x at low ($0 \leq x \leq 0.1$) and high ($0.9 \leq x \leq 1.0$) Al composition, while remaining relatively stable for $0.1<x<0.9$ [188]. A similar phenomenon is seen in Fig. 23.7b for the isotope effect of mixing ^{71}Ga with ^{69}Ga in GaN derived from the first-principles BTE calculation which shows the high sensitivity of κ to the isotope disorder and indicates that the isotopically pure ^{69}GaN and ^{71}GaN can provide a high RT κ ≈ 440 W m^{-1} K^{-1}.

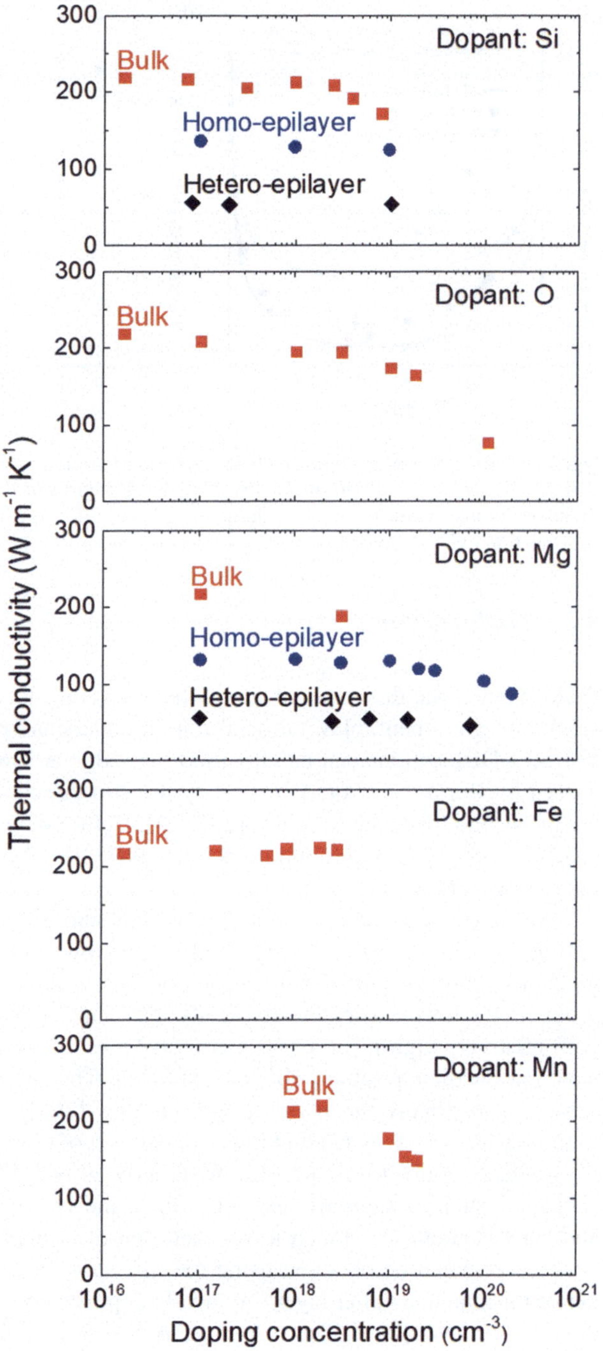

Fig. 23.6 Impurity effects in GaN. RT κ of bulk GaN and epitaxial layers as functions of the doping concentration with different types of dopants [8, 72, 116]

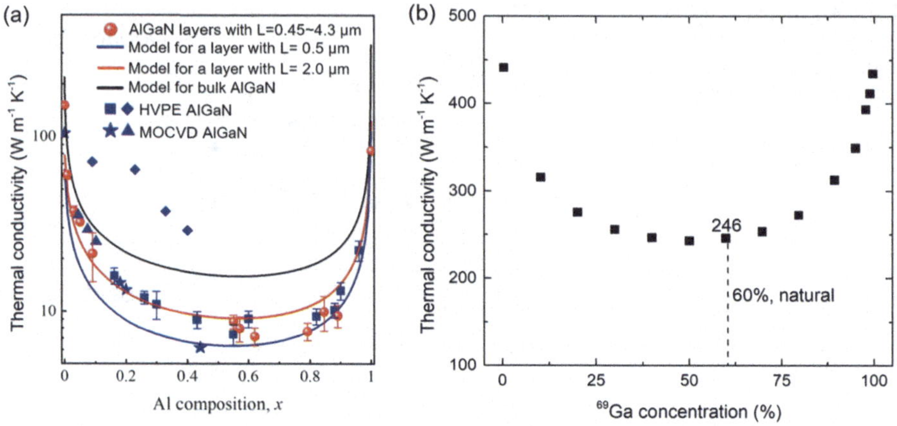

Fig. 23.7 Alloying and isotope effects in GaN. (**a**) Measured (symbols) and calculated (lines). RT κ of $Al_xGa_{1-x}N$ as a function of Al composition adapted from Ref. [188] under the terms of the Creative Commons Attribution 4.0 International license (CC BY 4.0). (**b**) Calculated RT κ of GaN as a function of ^{69}Ga isotope concentration adapted from Ref. [200]

4.2 *Dislocations*

Dislocations, one-dimensional line defects causing local lattice distortions due to atomic misalignment, play a crucial role in understanding thermal transport in GaN and SiC. Dislocations can be classified into edge, screw, and mixed types, characterized by Burgers vectors that specify the slip direction with respect to the orientation of the dislocation core [201, 202]. In GaN and SiC, dislocations are characterized by cathodoluminescence, XRD [8, 9], and transmission electron microscopy (TEM) [11, 92].

Akin to point defects, dislocations in GaN and SiC can be engineered primarily during the material growth stage. Different growth methods can produce crystals with varying dislocation densities (ρ). For GaN epilayers, the ammonothermal (AT) produces the lowest ρ crystals (typically $\leq 10^4$ cm^{-2}), while HVPE results in crystals with relatively high ρ (up to 10^{15} cm^{-2}). MOCVD and MBE-grown GaN crystals have intermediate ρ (about 10^{10} cm^{-2}) [203]. The substrate used for GaN epilayer growth also affects the ρ of the grown film [204]. Although homoepitaxy can minimize lattice mismatch and lower ρ, the cost of GaN substrates for such purpose is still high. GaN films are still frequently grown by heteroepitaxy on foreign substrates such as sapphire and SiC. Since the lattice mismatch between 6H-SiC and GaN is about 3%, much lower than that of c-sapphire and GaN (~15%), GaN with lower ρ can be obtained using 6H-SiC substrates [204, 205]. To mitigate the lattice mismatch, a buffer layer is typically deposited on the substrate before growing the GaN epilayer [206]. The type (e.g., AlN [207] and SiN [208]), thickness [207], and treatment [209] of the buffer layer can significantly affect the quality of GaN films.

The 4H-SiC single crystals can be grown by top-seeded solution growth (TSSG), high-temperature chemical vapor deposition (HTCVD), and physical vapor transport (PVT) methods, among which the PVT technology prevails over other growth technologies because of its high maturity, temperature tunability, and low cost of the solid raw materials [210]. The ρ in 150 mm 4H-SiC substrates from PVT is in 10^3–10^4 cm^{-2} which is already low for phonon transport, although it may still influence the electrical performances [211].

As discussed in Sect. 2, dislocations can influence thermal transport in GaN and SiC by introducing phonon scattering via two mechanisms [94]. The first is the scattering of phonons on the dislocation cores, the immediate vicinity of the dislocation line where interatomic bonds are significantly distorted [202]. This short-range interaction is modeled like phonon-point defects scattering. The second mechanism is the long-range scattering by the inhomogeneous strain field of the dislocation lines, which is found to be dominant according to both semi-empirical models (e.g., Carruthers model [213]) and more recent first-principles-based BTE coupled with the perturbative Green's function calculation [11, 107]. Figure 23.8 shows the dependence of RT κ on ρ for GaN. When the ρ is low, κ is insensitive to dislocation density. The deflection point where κ starts to drop as ρ increases is in the range of 10^7 to 10^9 cm^{-2}, which is also affected by other defects or film thickness [9, 11].

Furthermore, dislocations can be utilized to manipulate the anisotropy of κ in GaN and SiC. Figure 23.9 shows the κ anisotropy of GaN samples with different thicknesses and ρ as a function of temperature. Since dislocations are highly oriented along the cross-plane direction, phonons propagating perpendicular to the dislocations (i.e., in-plane direction) are strongly scattered, leading to a significant reduction in κ_{in}, while κ_{cross} is not strongly suppressed. This results in a large anisotropy [11, 92]. Although Carruthers' model outperforms Klemens' model in predicting κ, it cannot explain the low-temperature κ behavior since it does not consider the changes in the interatomic force constants in the neighborhood of defects [92]. In SiC, there is still a lack of systematic experimental investigation on the ρ effect on thermal transport, and theoretical research on this topic is also rare [214].

4.3 Planar Defects and Size Effects

Planar defects, 2D imperfections within crystal planes, encompass grain, twin, crystal boundaries, stacking faults, and interfaces. In polycrystalline materials, single crystals or twins with varied orientations are separated by grain boundaries. Stacking faults occur when the regular atomic sequence is disrupted by a layer's omission or addition [9, 202, 215]. Interfaces form between dissimilar materials, including both homo- and hetero-structures in WBGSs which we broadly categorize as planar defects. Even at atomically perfect interfaces, the Kapitza thermal resistance ($1/G$) exists due to differences in the vibrational properties in different materials that influence the transmission and the local scattering of phonons [216, 217].

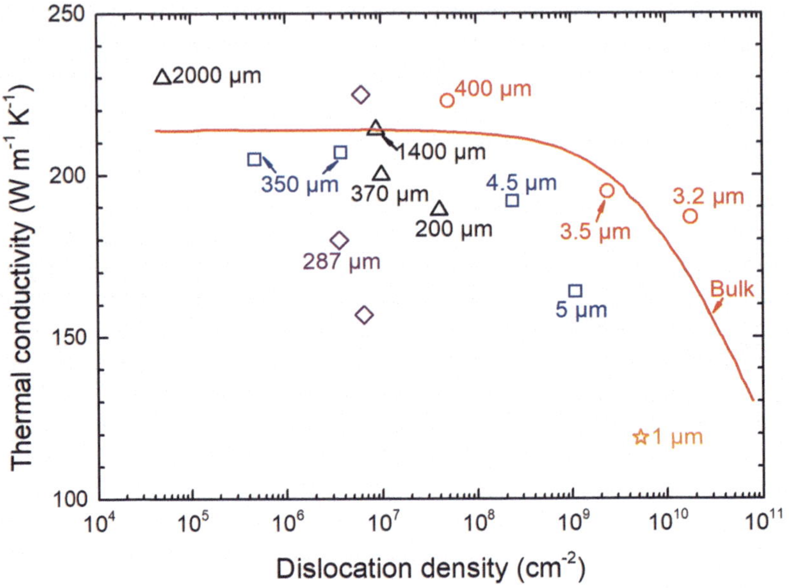

Fig. 23.8 Measured (symbols) and calculated (curve) RT κ of GaN in the cross-plane direction as a function of ρ. The thickness of GaN films is labeled in the figure. Note that the thickness of the samples represented by the other two purple diamonds is not given in the literature. References: red circles and curve [11], blue squares [95], black triangles [113], purple diamonds [212], and orange star [100]

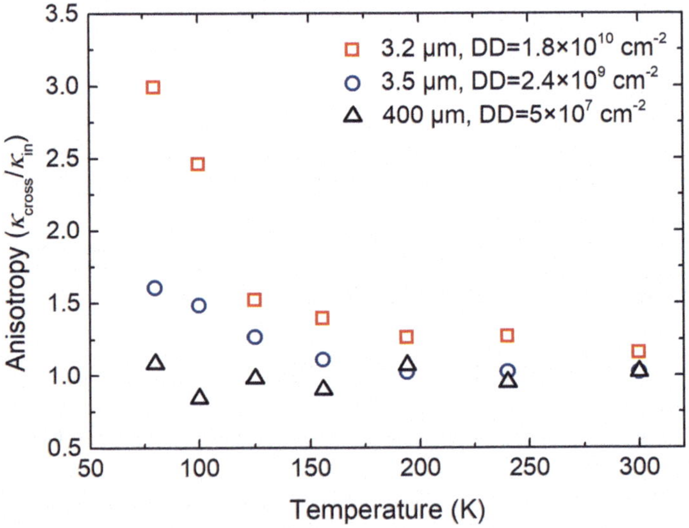

Fig. 23.9 κ anisotropy of GaN with different thickness and ρ as a function of temperature [11]

There are various methods to characterize these planar defects in WBGSs. High-resolution scanning TEM can be used to characterize the grains, twins, stacking faults, and interfaces at the atomic scale [28, 218–222], while electron backscattering diffraction provides information on grain size and orientation with mesoscopic statistics [174, 223]. Film thickness can be measured using TEM, X-ray reflectivity [95], RBS [224], ellipsometry [225], and atomic force microscopy [226]. TDTR can be adapted to perform picosecond acoustic and Brillouin scattering (for transparent materials like GaN and SiC) to measure the film thickness, sound velocity, and the corresponding elastic modulus [227, 228]. The strains and mosaic tilt angle [229] of epilayers can be measured by X-ray reciprocal space mapping [230, 231].

Planar defects can be engineered during the material synthesis stage. For example, one approach to grow GaN epilayers on amorphous substrates is by combining MBE and MOCVD. The grain size of these epilayers can be controlled by the growth time during the MOCVD process [218], and control the thickness of the AlN buffer layer [232] and growth pressure [233]. The grain orientation of III–V films can be tuned by manipulating the substrate and temperatures [223, 234]. The twin density can be reduced by slowing down the SiC film growth rate and increasing the film thickness [220]. Stacking faults in GaN films can be minimized by introducing Al (Ga)N buffer layers between GaN epilayers and substrates [235].

As mentioned in Sect. 2, planar defects can affect thermal transport in GaN and SiC. When the length size of grain, twin, and crystal boundaries becomes comparable with PMFP, phonons with MFPs longer than this length can transport ballistically and experience relatively strong boundary scattering (Eq. 23.11), leading to a reduction in κ [12, 30]. Moreover, the grain and twin boundaries within the crystal introduce thermal boundary resistance, which affects the effective κ of crystals with multigrains and twins [174, 221, 236, 237].

Thin films with varying thicknesses are commonly used in WBGS-based devices. Figure 23.10 illustrates the relationship between the experimental RT κ and film thickness for GaN and SiC, as well as a few other representative WBGSs and UBGSs reported in the literature. The κ of these films monotonically decreases as the thickness decreases, demonstrating a significant size effect. GaN films with a thickness of 1.5 μm and 3C-SiC films with a thickness of 1.75 μm have RT κ that reach ~80% of their bulk values. In general, the thickness effect on the in-plane κ is less pronounced than the cross-plane κ, since the boundary is presented along the direction for the cross-plane heat flux, which can be explained by the Fuchs-Sondheimer theory [238].

The importance of thermal interface conductance *G* is universal in almost all WBGS devices between similar or dissimilar layers grown in different steps [28, 217], which is significantly influenced by factors such as the matching of phonon density of states, the nature of interface bonding, the sound velocity in the materials, and the quality of the interface. Some of these issues can be mitigated by designing heterostructures to include an interlayer between two layers, which helps to bridge the phonon spectra and improve phonon transmission across the interface [28]. For example, a recent work found that the addition of an AlN transition layer leads to a remarkable increase in the *G* of the GaN/SiC interface, particularly at

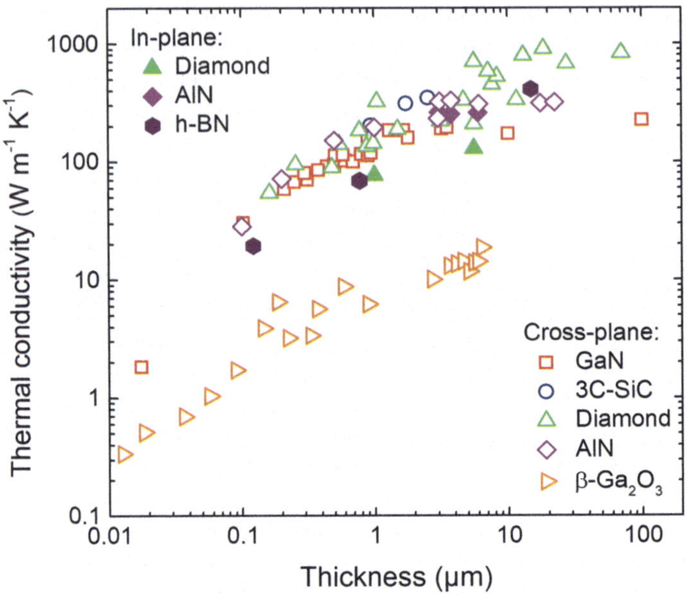

Fig. 23.10 Experimental RT κ of GaN and SiC layers with different thicknesses in comparison with other WBGSs. The symbols in red, blue, green, magenta, purple, and orange represent the κ data of GaN [9, 11, 100, 115, 239, 240], 3C-SiC [30], diamond [2, 118, 241–243], AlN [118, 244, 245], h-BN [118], and β-Ga_2O_3 [222, 246], respectively

elevated temperatures, result from a combination of improved crystalline quality at the interface and the "phonon bridge" effect provided by AlN that enhances the overlap between the phonon spectra of GaN and SiC [109].

GaN and SiC can be used in electronic devices not only as films and bulk materials but also as nanostructures. Quantum dots and nanoparticles are examples of 0D nanostructures, while nanowires, nanotubes, and nanobelts are considered 1D nanostructures. 2D nanostructures including free-standing nanosheets or thin films on substrates are the most common form of these materials. 3D nanostructures can be formed by an ensemble of nanostructures [247]. Among various nanostructures, nanowires have attracted significant research attention in nanoscale electronic devices [247, 248]. Figure 23.11 illustrates the κ of GaN and 3C-SiC nanowires as a function of their diameters or lengths. The κ of these nanowires increases with increasing diameter and length, exhibiting a significant size effect, which is attributed to ballistic thermal transport, as discussed earlier.

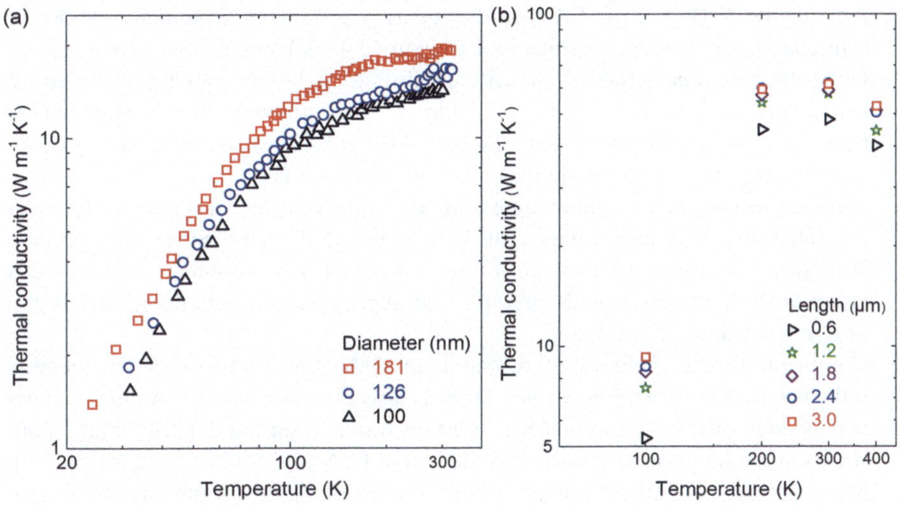

Fig. 23.11 κ of nanowires. (**a**) κ of GaN nanowires with different diameters, where the length is ~5 μm [249]. (**b**) κ of 3C-SiC nanowires with different lengths, where the cross-section sizes are 250 nm in width and 150 nm in thickness [250]. All κ data presented are derived from experiments in the literature

5 Concluding Remarks

This chapter has provided an overview of the intrinsic and extrinsic factors that influence the thermal conductivity (κ) of GaN and SiC, which have gained fast-growing interest and have been widely used in power semiconductor devices due to their superior electrical and thermal properties. We introduced the basics of phonon thermal transport, including both intrinsic (three- and four-phonon scattering, phonon topological behavior, and phonon-focusing effect) and extrinsic factors (phonon-point defects, line defects, and planar defects scattering). We also discussed the temperature dependence and anisotropy of κ and their influence on device reliability and performance. Thermoreflectance-based techniques have been extensively utilized to characterize the thermal transport properties of WBGSs. We have discussed the basic principles and standard implementations of these techniques, compared them with each other, exemplified their use in studying GaN and SiC, and provided a technical outlook. With a focus on extrinsic factors, we have reviewed in detail the types, formation mechanisms, engineering methods, and characterization techniques of defects and how they affect thermal transport in GaN and SiC.

Based on the findings of this review, opportunities for future research are summarized below on the experimental study of thermal transport in GaN and SiC, as well as potential improvements in the thermal management of GaN and SiC-based devices.

1. Among the WBGSs, 3C-SiC stands out as a promising candidate for power semiconductor devices, despite its bandgap (2.4 eV) being lower than that of GaN, 4H-SiC, and 6H-SiC. 3C-SiC exhibits a higher κ than these materials. Moreover, 3C-SiC wafers are available in sizes up to 6 inches and can be heterogeneously integrated with Si and AlN substrates via epitaxial growth, overcoming the current challenges faced by diamond [30].
2. As discussed in Sect. 2.3, the temperature dependence of κ can significantly affect the reliability and performance of WBGS-based devices. Since four-phonon scattering can affect the thermal transport in GaN, it is crucial to consider this effect in III–V materials with large phonon energy gaps at both the materials and device level thermal modeling.
3. Compared to the plethora of research on GaN, the study of the dislocation influence in SiC is extremely rare. There is also insufficient data on the thickness dependence of κ in hexagonal SiC. Further research on the experimental reconstruction of the phonon transport in defective GaN and SiC in comparison with first-principles calculation may benefit the material development and device thermal management.
4. As discussed in Sect. 4, defects can be strategically employed to tune κ and its temperature dependence in GaN and SiC, allowing for the creation of materials with desired κ. For example, interface engineering [118] can be used to enhance interfacial thermal conductance while highly oriented dislocations [11] can be used to manipulate the anisotropy ratio of κ. Further experimental study on the enhancement or manipulation of thermal transport in these materials is needed.

References

1. Y. Qin, B. Albano, J. Spencer, et al., Thermal management and packaging of wide and ultra-wide bandgap power devices: A review and perspective. J. Phys. D. Appl. Phys. **56**, 093001 (2023)
2. C. Yuan, R. Hanus, S. Graham, A review of thermoreflectance techniques for characterizing wide bandgap semiconductors' thermal properties and devices' temperatures. J. Appl. Phys. **132**, 220701 (2022)
3. M. Buffolo, D. Favero, A. Marcuzzi, et al., Review and outlook on GaN and SiC Power devices: Industrial state-of-the-art, applications, and perspectives. IEEE Trans. Electron Devices **71**, 1344–1355 (2024)
4. Y. Zhong, J. Zhang, S. Wu, et al., A review on the GaN-on-Si power electronic devices. Fund. Res. **2**, 462–475 (2022)
5. S. Nakamura, Nobel lecture: Background story of the invention of efficient blue InGaN light emitting diodes. Rev. Mod. Phys. **87**, 1139–1151 (2015)
6. J.Y. Tsao, S. Chowdhury, M.A. Hollis, et al., Ultrawide-bandgap semiconductors: Research opportunities and challenges. Adv. Electron. Mater. **4**, 1600501 (2017)
7. J. Cho, Z. Li, M. Asheghi, K.E. Goodson, Near-junction thermal management: Thermal conduction in gallium nitride composite substrates. Annu. Rev. Heat Transfer **18**, 7–45 (2015)
8. Q. Zheng, C. Li, A. Rai, et al., Thermal conductivity of GaN, ^{71}GaN, and SiC from 150 K to 850 K. Phys. Rev. Mater. **3**, 014601 (2019)
9. T.E. Beechem, A.E. McDonald, E.J. Fuller, et al., Size dictated thermal conductivity of GaN. J. Appl. Phys. **120**, 095104 (2016)

10. L. Lindsay, D.A. Broido, T.L. Reinecke, Thermal conductivity and large isotope effect in GaN from first principles. Phys. Rev. Lett. **109**, 095901 (2012)
11. H. Li, R. Hanus, C.A. Polanco, et al., GaN thermal transport limited by the interplay of dislocations and size effects. Phys. Rev. B **102**, 014313 (2020)
12. Z. Cheng, W. Lu, J. Shi, et al., Quasi-ballistic thermal conduction in 6H–SiC. Mater. Today Phys. **20**, 100462 (2021)
13. C. Mion, J.F. Muth, E.A. Preble, D. Hanser, Thermal conductivity, dislocation density and GaN device design. Superlattice. Microst. **40**, 338–342 (2006)
14. R.W. Keyes, Figure of merit for semiconductors for high-speed switches. Proc. IEEE **60**, 225 (1972)
15. K. Shenai, The figure of merit of a semiconductor power electronics switch. IEEE Trans. Electron Devices **65**, 4216–4224 (2018)
16. D.-S. Tang, B.-Y. Cao, Phonon thermal transport and its tunability in GaN for near-junction thermal management of electronics: A review. Int. J. Heat Mass Transf. **200**, 123497 (2023)
17. S. An, H. Park, M. Kim, Recent advances in single crystal narrow band-gap semiconductor nanomembranes and their flexible optoelectronic device applications: Ge, GeSn, InGaAs, and 2D materials. J. Mater. Chem. C **11**, 2430–2448 (2023)
18. J. Yang, K. Liu, X. Chen, D. Shen, Recent advances in optoelectronic and microelectronic devices based on ultrawide-bandgap semiconductors. Progr. Quant. Electron. **83**, 100397 (2022)
19. G.A. Slack, Thermal conductivity of pure and impure silicon, silicon carbide, and diamond. J. Appl. Phys. **35**, 3460–3466 (1964)
20. W. Fulkerson, J.P. Moore, R.K. Williams, R.S. Graves, D.L. McElroy, Thermal conductivity, electrical resistivity, and Seebeck coefficient of silicon from 100 to 1300°K. Phys. Rev. **167**, 765–782 (1968)
21. J. Carruthers, T. Geballe, H. Rosenberg, J.M. Ziman, The thermal conductivity of germanium and silicon between 2 and 300 K. Proc. R. Soc. London. Ser. A. Math. Phys. Sci. **238**, 502–514 (1957)
22. R.L. Xu, M. Muñoz Rojo, S.M. Islam, et al., Thermal conductivity of crystalline AlN and the influence of atomic-scale defects. J. Appl. Phys. **126**, 185105 (2019)
23. F. Tian, B. Song, X. Chen, et al., Unusual high thermal conductivity in boron arsenide bulk crystals. Science **361**, 582–585 (2018)
24. I. Kudman, E.F. Steigmeier, Thermal conductivity and seebeck coefficient of InP. Phys. Rev. **133**, A1665–A1667 (1964)
25. S. Li, Q. Zheng, Y. Lv, et al., High thermal conductivity in cubic boron arsenide crystals. Science **361**, 579–581 (2018)
26. J.S. Kang, M. Li, H. Wu, H. Nguyen, Y. Hu, Experimental observation of high thermal conductivity in boron arsenide. Science **361**, 575–578 (2018)
27. Q. Zheng, S. Li, C. Li, et al., High thermal conductivity in isotopically enriched cubic boron phosphide. Adv. Funct. Mater. **28**, 1805116 (2018)
28. T. Feng, H. Zhou, Z. Cheng, L.S. Larkin, M.R. Neupane, A critical review of thermal boundary conductance across wide and ultrawide bandgap semiconductor interfaces. ACS Appl. Mater. Interfaces **15**, 29655–29673 (2023)
29. J.B. Casady, R.W. Johnson, Status of silicon carbide (SiC) as a wide-bandgap semiconductor for high-temperature applications: A review. Solid State Electron. **39**, 1409–1422 (1996)
30. Z. Cheng, J. Liang, K. Kawamura, et al., High thermal conductivity in wafer-scale cubic silicon carbide crystals. Nat. Commun. **13**, 7201 (2022)
31. X. Qian, P. Jiang, R. Yang, Anisotropic thermal conductivity of 4H and 6H silicon carbide measured using time-domain thermoreflectance. Mater. Today Phys. **3**, 70–75 (2017)
32. R. Cheung, *Silicon Carbide Microelectromechanical Systems for Harsh Environments* (Imperial College Press, 2006)
33. H. Joo Lee, H. Ryu, C.-R. Lee, K. Kim, Polytypes in GaN films grown by metalorganic chemical vapor deposition on (0001) sapphire substrate. J. Cryst. Growth **191**, 621–626 (1998)

34. M. Raya-Moreno, R. Rurali, X. Cartoixà, Thermal conductivity for III-V and II-VI semiconductor wurtzite and zinc-blende polytypes: The role of anharmonicity and phase space. Phys. Rev. Mater. **3**, 084607 (2019)
35. X. Yang, T. Feng, J. Li, X. Ruan, Stronger role of four-phonon scattering than three-phonon scattering in thermal conductivity of III-V semiconductors at room temperature. Phys. Rev. B **100**, 245203 (2019)
36. P. Jiang, X. Qian, X. Li, R. Yang, Three-dimensional anisotropic thermal conductivity tensor of single crystalline β-Ga_2O_3. Appl. Phys. Lett. **113**, 232105 (2018)
37. J.E. Graebner, in *Diamond: Electronic Properties and Applications*, ed. by L.S. Pan, D.R. Kania, (Springer US, Boston, 1995), pp. 285–318
38. K. Chen, B. Song, N.K. Ravichandran, et al., Ultrahigh thermal conductivity in isotope-enriched cubic boron nitride. Science **367**, 555–559 (2020)
39. Z. Cheng, Y.R. Koh, A. Mamun, et al., Experimental observation of high intrinsic thermal conductivity of AlN. Phys. Rev. Mater. **4**, 044602 (2020)
40. L. Wei, P.K. Kuo, R.L. Thomas, T.R. Anthony, W.F. Banholzer, Thermal conductivity of isotopically modified single crystal diamond. Phys. Rev. Lett. **70**, 3764–3767 (1993)
41. C.J. Glassbrenner, G.A. Slack, Thermal conductivity of silicon and germanium from 3 K to the melting point. Phys. Rev. **134**, A1058–A1069 (1964)
42. L. Duclaux, B. Nysten, J.P. Issi, A.W. Moore, Structure and low-temperature thermal conductivity of pyrolytic boron nitride. Phys. Rev. B **46**, 3362–3367 (1992)
43. C. Yuan, J. Li, L. Lindsay, et al., Modulating the thermal conductivity in hexagonal boron nitride via controlled boron isotope concentration. Commun. Phys. **2**, 43 (2019)
44. P. Jiang, X. Qian, R. Yang, L. Lindsay, Anisotropic thermal transport in bulk hexagonal boron nitride. Phys. Rev. Mater. **2**, 064005 (2018)
45. S. Burghartz, B. Schulz, Thermophysical properties of sapphire, AlN and $MgAl_2O_4$ down to 70 K. J. Nucl. Mater. **212–215**, 1065–1068 (1994)
46. D.T. Morelli, G.A. Slack, *High Thermal Conductivity Materials*. (Springer, 2006), chap. 2, pp. 37–68
47. G.A. Slack, Nonmetallic crystals with high thermal conductivity. J. Phys. Chem. Solids **34**, 321–335 (1973)
48. G.A. Slack, The thermal conductivity of nonmetallic crystals. Solid State Phys. **34**, 1–71 (1979)
49. X. Qian, J. Zhou, G. Chen, Phonon-engineered extreme thermal conductivity materials. Nat. Mater. **20**, 1188–1202 (2021)
50. C. Toher, J.J. Plata, O. Levy, et al., High-throughput computational screening of thermal conductivity, Debye temperature, and Grüneisen parameter using a quasiharmonic Debye model. Phys. Rev. B **90**, 174107 (2014)
51. G.A. Slack, Effect of isotopes on low-temperature thermal conductivity. Phys. Rev. **105**, 829–831 (1957)
52. J.E.N. Swallow, R.G. Palgrave, P.A.E. Murgatroyd, et al., Indium gallium oxide alloys: Electronic structure, optical gap, surface space charge, and chemical trends within common-cation semiconductors. ACS Appl. Mater. Interfaces **13**, 2807–2819 (2021)
53. K.A. Mengle, E. Kioupakis, Vibrational and electron-phonon coupling properties of β-Ga_2O_3 from first-principles calculations: Impact on the mobility and breakdown field. AIP Adv. **9**, 015313 (2019)
54. Z. Guo, A. Verma, X. Wu, et al., Anisotropic thermal conductivity in single crystal β-gallium oxide. Appl. Phys. Lett. **106**, 111909 (2015)
55. M. Islam, A comprehensive investigation on the physical properties of SiC polymorphs for high-temperature applications: A DFT study based on GGA and hybrid HSE06 exchange correlation functionals. Nucl. Mater. Energy **38**, 101631 (2024)
56. X.-D. Wang, W.-D. Hu, X.-S. Chen, W. Lu, The study of self-heating and hot-electron effects for AlGaN/GaN double-channel HEMTs. IEEE Trans. Electron Devices **59**, 1393–1401 (2012)

57. S.K. Mohanty, Y.-Y. Chen, P.-H. Yeh, R.-H. Horng, Thermal management of GaN-on-Si high electron mobility transistor by copper filled micro-trench structure. Sci. Rep. **9**, 19691 (2019)
58. M.G. Holland, Phonon scattering in semiconductors from thermal conductivity studies. Phys. Rev. **134**, A471–A480 (1964)
59. A. Amith, I. Kudman, E.F. Steigmeier, Electron and phonon scattering in GaAs at high temperatures. Phys. Rev. **138**, A1270–A1276 (1965)
60. R. Rounds, B. Sarkar, T. Sochacki, et al., Thermal conductivity of GaN single crystals: Influence of impurities incorporated in different growth processes. J. Appl. Phys. **124**, 105106 (2018)
61. N.H. Protik, A. Katre, L. Lindsay, et al., Phonon thermal transport in 2H, 4H and 6H silicon carbide from first principles. Mater. Today Phys. **1**, 31–38 (2017)
62. T. Wang, Z. Gui, A. Janotti, C. Ni, P. Karandikar, Strong effect of electron-phonon interaction on the lattice thermal conductivity in 3C-SiC. Phys. Rev. Mater. **1**, 034601 (2017)
63. G.P. Srivastava, *The Physics of Phonons* (CRC Press, 2022)
64. J.M. Ziman, *Electrons and Phonons: The Theory of Transport Phenomena in Solids* (Oxford University Press, 2001)
65. A.J. McGaughey, M. Kaviany, Quantitative validation of the Boltzmann transport equation phonon thermal conductivity model under the single-mode relaxation time approximation. Phys. Rev. B **69**, 094303 (2004)
66. M.C. Karamargin, C.A. Reynolds, F.P. Lipschultz, P.G. Klemens, Lattice thermal conductivity and deviations from Matthiessen's rule for dilute alloys of tin with cadmium. Phys. Rev. B **6**, 3624–3633 (1972)
67. P.B. Allen, Improved Callaway model for lattice thermal conductivity. Phys. Rev. B **88**, 144302 (2013)
68. J. Callaway, Model for lattice thermal conductivity at low temperatures. Phys. Rev. **113**, 1046 (1959)
69. D. Morelli, J. Heremans, G. Slack, Estimation of the isotope effect on the lattice thermal conductivity of group IV and group III-V semiconductors. Phys. Rev. B **66**, 195304 (2002)
70. M. Holland, Analysis of lattice thermal conductivity. Phys. Rev. **132**, 2461 (1963)
71. P.P. Paskov, M. Slomski, J.H. Leach, J.F. Muth, T. Paskova, Effect of Si doping on the thermal conductivity of bulk GaN at elevated temperatures—Theory and experiment. AIP Adv. **7**, 095302 (2017)
72. D. Tran, Thermal conductivity of wide and ultra-wide bandgap semiconductors. Linköping University (2023)
73. G.A. Slack, S. Galginaitis, Thermal conductivity and phonon scattering by magnetic impurities in CdTe. Phys. Rev. **133**, A253 (1964)
74. N. Malviya, N.K. Ravichandran, Failure of the Callaway description of heat flow in boron arsenide and boron antimonide driven by phonon scattering selection rules. Phys. Rev. B **108**, 155201 (2023)
75. L. Lindsay, First-principles Peierls-Boltzmann phonon thermal transport: A topical review. Nanosc. Microsc. Thermophys. Eng. **20**, 67–84 (2016)
76. D.A. Broido, M. Malorny, G. Birner, N. Mingo, D.A. Stewart, Intrinsic lattice thermal conductivity of semiconductors from first principles. Appl. Phys. Lett. **91**, 231922 (2007)
77. A.J. McGaughey, A. Jain, H.-Y. Kim, B. Fu, Phonon properties and thermal conductivity from first principles, lattice dynamics, and the Boltzmann transport equation. J. Appl. Phys. **125**, 011101 (2019)
78. J. Garg, T. Luo, G. Chen, Spectral concentration of thermal conductivity in GaN—A first-principles study. Appl. Phys. Lett. **112**, 252101 (2018)
79. A. Katre, J. Carrete, B. Dongre, G.K. Madsen, N. Mingo, Exceptionally strong phonon scattering by B substitution in cubic SiC. Phys. Rev. Lett. **119**, 075902 (2017)
80. X. Tang, J. Dong, Lattice thermal conductivity of MgO at conditions of Earth's interior. Proc. Natl. Acad. Sci. **107**, 4539–4543 (2010)

81. L. Lindsay, D.A. Broido, T.L. Reinecke, Ab initio thermal transport in compound semiconductors. Phys. Rev. B **87**, 165201 (2013)
82. N.K. Ravichandran, D. Broido, Phonon-phonon interactions in strongly bonded solids: Selection rules and higher-order processes. Phys. Rev. X **10**, 021063 (2020)
83. J. Sun, S. Li, Z. Tong et al., Unexpected weak effects of phonon-electron interactions on the lattice thermal conductivity of high-concentration n-type GaN. arXiv preprint arXiv:2401.02133, (2024)
84. T. Feng, L. Lindsay, X. Ruan, Four-phonon scattering significantly reduces intrinsic thermal conductivity of solids. Phys. Rev. B **96**, 161201 (2017)
85. Z. Han, X. Yang, W. Li, T. Feng, X. Ruan, FourPhonon: An extension module to ShengBTE for computing four-phonon scattering rates and thermal conductivity. Comput. Phys. Commun. **270**, 108179 (2022)
86. S.I. Salah, T.M. Hatem, E.E. Khalil, S.M. Bedair, Embedded void approach effects on intrinsic stresses in laterally grown GaN-on-Si substrate. Mater. Sci. Eng. B **242**, 104–110 (2019)
87. P.G. Klemens, The scattering of Low-frequency lattice waves by static imperfections. Proc. Phys. Soc. Sect. A **68**, 1113 (1955)
88. P.G. Klemens, Thermal resistance due to point defects at high temperatures. Phys. Rev. **119**, 507–509 (1960)
89. C.A. Polanco, L. Lindsay, Thermal conductivity of InN with point defects from first principles. Phys. Rev. B **98**, 014306 (2018)
90. M. Schrade, T.G. Finstad, Using the Callaway model to deduce relevant phonon scattering processes: The importance of phonon dispersion. Phys. Status Solidi B **255**, 1800208 (2018)
91. R. Gurunathan, R. Hanus, M. Dylla, A. Katre, G.J. Snyder, Analytical models of phonon--Point-defect scattering. Phys. Rev. Appl. **13**, 034011 (2020)
92. B. Sun, G. Haunschild, C. Polanco, et al., Dislocation-induced thermal transport anisotropy in single-crystal group-III nitride films. Nat. Mater. **18**, 136–140 (2018)
93. Y. Cheng, M. Nomura, S. Volz, S. Xiong, Phonon–dislocation interaction and its impact on thermal conductivity. J. Appl. Phys. **130**, 040902 (2021)
94. J. Zou, D. Kotchetkov, A.A. Balandin, D.I. Florescu, F.H. Pollak, Thermal conductivity of GaN films: Effects of impurities and dislocations. J. Appl. Phys. **92**, 2534–2539 (2002)
95. K. Park, C. Bayram, Impact of dislocations on the thermal conductivity of gallium nitride studied by time-domain thermoreflectance. J. Appl. Phys. **126**, 185103 (2019)
96. D.G. Cahill, W.K. Ford, K.E. Goodson, et al., Nanoscale thermal transport. J. Appl. Phys. **93**, 793–818 (2003)
97. T. Feng, X. Ruan, Prediction of spectral phonon mean free path and thermal conductivity with applications to Thermoelectrics and thermal management: A review. J. Nanomater. **2014**, 206370 (2014)
98. Y. Chung, A.V. Gubarevich, K. Yoshida, Effects of solid solution, grain size and porosity on the thermal conductivity of aluminum- and boron-added porous silicon carbide ceramics with in-situ grain growth. J. Eur. Ceram. Soc. **43**, 792–804 (2023)
99. D. I. Florescu, F. H. Pollak, T. Paskova, E. Valcheva, B. Monemar, in *2000 IEEE International Symposium on Compound Semiconductors. Proceedings of the IEEE Twenty-Seventh International Symposium on Compound Semiconductors (Cat. No.00TH8498).* (2000), pp. 467–472
100. E. Ziade, J. Yang, G. Brummer, et al., Thickness-dependent thermal conductivity of gallium nitride. Appl. Phys. Lett. **110**, 031903 (2017)
101. P. Motamarri, S. Das, S. Rudraraju, et al., DFT-FE—A massively parallel adaptive finite-element code for large-scale density functional theory calculations. Comput. Phys. Commun. **246**, 106853 (2020)
102. K. Ghosh, H. Ma, M. Onizhuk, V. Gavini, G. Galli, Spin–spin interactions in defects in solids from mixed all-electron and pseudopotential first-principles calculations. npj Comput. Mater. **7**, 123 (2021)

103. R. Zhuravel, H. Huang, G. Polycarpou, et al., Backbone charge transport in double-stranded DNA. Nat. Nanotechnol. **15**, 836–840 (2020)
104. J. Carrete, B. Vermeersch, A. Katre, et al., almaBTE : A solver of the space–time dependent Boltzmann transport equation for phonons in structured materials. Comput. Phys. Commun. **220**, 351–362 (2017)
105. Y. Hu, R. Jia, J. Xu, et al., GiftBTE: An efficient deterministic solver for non-gray phonon Boltzmann transport equation. J. Phys. Condens. Matter **36**, 025901 (2023)
106. A. Katre, J. Carrete, T. Wang, G.K.H. Madsen, N. Mingo, Phonon transport unveils the prevalent point defects in GaN. Phys. Rev. Mater. **2**, 050602 (2018)
107. T. Wang, J. Carrete, N. Mingo, G.K.H. Madsen, Phonon scattering by dislocations in GaN. ACS Appl. Mater. Interfaces **11**, 8175–8181 (2019)
108. C. Yang, J. Wang, D. Ma, et al., Phonon transport across GaN-diamond interface: The nontrivial role of pre-interface vacancy-phonon scattering. Int. J. Heat Mass Transf. **214**, 124433 (2023)
109. R. Li, K. Hussain, M.E. Liao, et al., Enhanced thermal boundary conductance across GaN/SiC interfaces with AlN transition layers. ACS Appl. Mater. Interfaces **16**, 8109–8118 (2024)
110. A. Sarikov, A. Marzegalli, L. Barbisan, et al., Molecular dynamics simulations of extended defects and their evolution in 3C–SiC by different potentials. Model. Simul. Mater. Sci. Eng. **28**, 015002 (2020)
111. A. Hosseini, M.N. Nasrabadi, Investigation of vacancy defects and temperature effects on the GaN bombarding with argon atoms: Molecular dynamics simulation. Mater. Chem. Phys. **271**, 124854 (2021)
112. G. Leibfried, E. Schlömann, *Wärmeleitung in elektrisch isolierenden Kristallen* (Vandenhoeck & Ruprecht, 1954)
113. C. Mion, J.F. Muth, E.A. Preble, D. Hanser, Accurate dependence of gallium nitride thermal conductivity on dislocation density. Appl. Phys. Lett. **89**, 092123 (2006)
114. G.A. Slack, L.J. Schowalter, D. Morelli, J.A. Freitas, Some effects of oxygen impurities on AlN and GaN. J. Cryst. Growth **246**, 287–298 (2002)
115. J. Cho, Y. Li, W.E. Hoke, et al., Phonon scattering in strained transition layers for GaN heteroepitaxy. Phys. Rev. B **89**, 115301 (2014)
116. P. Bagheri, P. Reddy, J.H. Kim, et al., Impact of impurity-based phonon resonant scattering on thermal conductivity of single crystalline GaN. Appl. Phys. Lett. **117**, 082101 (2020)
117. R. Berman, J.C.F. Brock, The effect of isotopes on lattice heat conduction. I. Lithium fluoride. Proc. R. Soc. London. Series A, Math. Phys. Sci. **289**, 46–65 (1965)
118. M.S.B. Hoque, Y.R. Koh, J.L. Braun, et al., High in-plane thermal conductivity of aluminum nitride thin films. ACS Nano **15**, 9588–9599 (2021)
119. A. Jeżowski, O. Churiukova, J. Mucha, et al., Thermal conductivity of heavily doped bulk crystals GaN:O. Free carriers contribution. Mater. Res. Express **2**, 085902 (2015)
120. H. Shibata, Y. Waseda, H. Ohta, et al., High thermal conductivity of gallium nitride (GaN) crystals grown by HVPE process. Mater. Trans. **48**, 2782–2786 (2007)
121. L. Mitterhuber, R. Hammer, T. Dengg, J. Spitaler, Thermal characterization and modelling of AlGaN-GaN multilayer structures for HEMT applications. Energies **13**, 2363 (2020)
122. S. García-Sánchez, I. Íñiguez-de-la-Torre, S. Pérez, et al., Non-linear thermal resistance model for the simulation of high power GaN-based devices. Semicond. Sci. Technol. **36**, 055002 (2021)
123. A. Darwish, A.J. Bayba, H.A. Hung, Channel temperature analysis of GaN HEMTs with nonlinear thermal conductivity. IEEE Trans. Electron Devices **62**, 840–846 (2015)
124. V. Sodan, S. Stoffels, H. Oprins, et al., Fast and distributed thermal model for thermal modeling of GaN power devices. IEEE Trans. Compon. Packag. Manuf. Technol. **8**, 1747–1755 (2018)
125. S. Chakraborty, W. Amir, J.W. Shin, et al., Explicit thermal resistance model of self-heating effects of AlGaN/GaN HEMTs with linear and non-linear thermal conductivity. Materials **15**, 8415 (2022)

126. B. Wei, Q. Cai, Q. Sun, et al., Matryoshka phonon twinning in α-GaN. Commun. Phys. **4**, 227 (2021)
127. D.Q. Tran, T. Paskova, V. Darakchieva, P.P. Paskov, On the thermal conductivity anisotropy in wurtzite GaN. AIP Adv. **13**, 095009 (2023)
128. B. Chatterjee, W. Li, K. Nomoto, H.G. Xing, S. Choi, Thermal design of multi-fin Ga_2O_3 vertical transistors. Appl. Phys. Lett. **119**, 103502 (2021)
129. B. Sun, X. Gu, Q. Zeng, et al., Temperature dependence of anisotropic thermal-conductivity tensor of bulk black phosphorus. Adv. Mater. **29**, 1603297 (2017)
130. A.K. McCurdy, H.J. Maris, C. Elbaum, Anisotropic heat conduction in cubic crystals in the boundary scattering regime. Phys. Rev. B **2**, 4077–4083 (1970)
131. D.A. Chernodoubov, A.V. Inyushkin, Phonon focusing effect on thermal conductivity of hexagonal group III-nitrides and silicon carbide crystals. Phys. Lett. A **384**, 126120 (2020)
132. A.V. Inyushkin, A.N. Taldenkov, J.W. Ager, et al., Ultrahigh thermal conductivity of isotopically enriched silicon. J. Appl. Phys. **123**, 095112 (2018)
133. Y. Liu, Y. Xu, S.-C. Zhang, W. Duan, Model for topological phononics and phonon diode. Phys. Rev. B **96**, 064106 (2017)
134. D.-S. Tang, B.-Y. Cao, Topological effects of phonons in GaN and AlGaN: A potential perspective for tuning phonon transport. J. Appl. Phys. **129** (2021)
135. A. Palacios, L. Cong, M.E. Navarro, Y. Ding, C. Barreneche, Thermal conductivity measurement techniques for characterizing thermal energy storage materials—A review. Renew. Sust. Energ. Rev. **108**, 32–52 (2019)
136. I. O. f. Standardization, *Thermal Insulation: Determination of Steady-State Thermal Transmission Properties: Calibrated and Guarded Hot Box* (International Organization for Standardization, 1994)
137. ASTM E1530-06, Standard test method for evaluating the resistance to thermal transmission of materials by the guarded heat flow meter technique. (2006)
138. A.V. Inyushkin, A.N. Taldenkov, D.A. Chernodubov, V.V. Voronenkov, Y.G. Shreter, High thermal conductivity of bulk GaN single crystal: An accurate experimental determination. JETP Lett. **112**, 106–111 (2020)
139. A. Jeżowski, B.A. Danilchenko, M. Boćkowski, et al., Thermal conductivity of GaN crystals in 4.2–300 K range. Solid State Commun. **128**, 69–73 (2003)
140. A. Filatova-Zalewska, Z. Litwicki, T. Suski, A. Jeżowski, Thermal conductivity of thin films of gallium nitride, doped with aluminium, measured with 3ω method. Solid State Sci. **101**, 106105 (2020)
141. O. Churiukova, A. Jeżowski, P. Stachowiak, et al., Thermal conductivity of donor-doped GaN measured with 3ω and stationary methods. Low Temp. Phys. **41**, 563–566 (2015)
142. Y.K. Koh, S.L. Singer, W. Kim, et al., Comparison of the 3ω method and time-domain thermoreflectance for measurements of the cross-plane thermal conductivity of epitaxial semiconductors. J. Appl. Phys. **105**, 054303 (2009)
143. M. Slomski, P.P. Paskov, J.H. Leach, J.F. Muth, T. Paskova, Thermal conductivity of bulk GaN grown by HVPE: Effect of Si doping. Phys. Status Solidi B **254**, 1600713 (2017)
144. R. Wei, S. Song, K. Yang, et al., Thermal conductivity of 4H-SiC single crystals. J. Appl. Phys. **113** (2013)
145. L. Vozár, W. Hohenauer, Uncertainty of thermal diffusivity measurements using the laser flash method. Int. J. Thermophys. **26**, 1899–1915 (2005)
146. G.V. Kuznetsov, M.D. Kats, The errors when determining thermal characteristics by the laser flash method due to the thickness of the sample and the duration of the heating pulse. Meas. Tech. **55**, 454–458 (2012)
147. T. Beechem, L. Yates, S. Graham, Error and uncertainty in Raman thermal conductivity measurements. Rev. Sci. Instrum. **86**, 041101 (2015)
148. M. Power, J.W. Pomeroy, Y. Otoki, et al., Measuring the thermal conductivity of the GaN buffer layer in AlGaN/GaN HEMTs. Phys. Status Solidi A **212**, 1742–1745 (2015)

149. G. Brocero, Y. Guhel, P. Eudeline, et al., Measurement of self-heating temperature in AlGaN/GaN HEMTs by using cerium oxide micro-Raman thermometers. IEEE Trans. Electron Devices **66**, 4156–4163 (2019)
150. M. Elhajhasan, W. Seemann, K. Dudde, et al., Optical and thermal characterization of a group-III nitride semiconductor membrane by microphotoluminescence spectroscopy and Raman thermometry. Phys. Rev. B **108**, 235313 (2023)
151. B. Zhu, R. Wang, S. Harrison, et al., Thermal conductivity of SiC microwires: Effect of temperature and structural domain size uncovered by 0 K limit phonon scattering. Ceram. Int. **44**, 11218–11224 (2018)
152. K.R. Bagnall, O.I. Saadat, S. Joglekar, T. Palacios, E.N. Wang, Experimental characterization of the thermal time constants of GaN HEMTs via micro-Raman thermometry. IEEE Trans. Electron Devices **64**, 2121–2128 (2017)
153. S. Sandell, E. Chávez-Ángel, A. El Sachat, et al., Thermoreflectance techniques and Raman thermometry for thermal property characterization of nanostructures. J. Appl. Phys. **128**, 131101 (2020)
154. S. Xu, A. Fan, H. Wang, X. Zhang, X. Wang, Raman-based nanoscale thermal transport characterization: A critical review. Int. J. Heat Mass Transf. **154**, 119751 (2020)
155. C. Foy, L. Zhang, M.E. Trusheim, et al., Wide-field magnetic field and temperature imaging using nanoscale quantum sensors. ACS Appl. Mater. Interfaces **12**, 26525–26533 (2020)
156. T. Favaloro, J.H. Bahk, A. Shakouri, Characterization of the temperature dependence of the thermoreflectance coefficient for conductive thin films. Rev. Sci. Instrum. **86**, 024903 (2015)
157. X. Ji, Z. Cheng, E.K. Pek, D.G. Cahill, Thermal conductivity mapping of oxidized SiC/SiC composites by time-domain thermoreflectance with heterodyne detection. J. Am. Ceram. Soc. **104**, 4773–4781 (2021)
158. A. Feldman, Algorithm for solutions of the thermal diffusion equation in a stratified medium with a modulated heating source. High Temp.-High Press. **31**, 293–298 (1999)
159. P. Jiang, X. Qian, R. Yang, Tutorial: Time-domain thermoreflectance (TDTR) for thermal property characterization of bulk and thin film materials. J. Appl. Phys. **124**, 161103 (2018)
160. M. Khafizov, D.H. Hurley, Measurement of thermal transport using time-resolved thermal wave microscopy. J. Appl. Phys. **110** (2011)
161. P. Jiang, D. Wang, Z. Xiang, R. Yang, H. Ban, A new spatial-domain thermoreflectance method to measure a broad range of anisotropic in-plane thermal conductivity. Int. J. Heat Mass Transf. **191**, 122849 (2022)
162. D.H. Olson, J.L. Braun, P.E. Hopkins, Spatially resolved thermoreflectance techniques for thermal conductivity measurements from the nanoscale to the mesoscale. J. Appl. Phys. **126**, 150901 (2019)
163. J.L. Braun, D.H. Olson, J.T. Gaskins, P.E. Hopkins, A steady-state thermoreflectance method to measure thermal conductivity. Rev. Sci. Instrum. **90**, 024905 (2019)
164. D. Kendig, G. Pavlidis, S. Graham *et al.*, in *2018 91st ARFTG Microwave Measurement Conference (ARFTG)*. (2018), pp. 1–4
165. K. Maize, D. Kendig, A. Shakouri, V. Vashchenko, in *2011 International Reliability Physics Symposium*. (2011), pp. EL.4.1–EL.4.6
166. G.T. Hohensee, D. Kendig, E. Pek, et al., Calibrated sub-micron temperature mapping of an operating plasmonic HAMR device by thermoreflectance imaging. MRS Adv. **2**, 3613–3618 (2017)
167. A.E. Helou, P. Komarov, M.J. Tadjer, et al., High-resolution Thermoreflectance imaging investigation of self-heating in AlGaN/GaN HEMTs on Si, SiC, and diamond substrates. IEEE Trans. Electron Devices **67**, 5415–5420 (2020)
168. D.C. Shoemaker, A. Karim, D. Kendig, H. Kim, S. Choi, in *2022 21st IEEE Intersociety Conference on Thermal and Thermomechanical Phenomena in Electronic Systems (iTherm)*. (2022), pp. 1–5
169. T. Tong, D. Fu, A.X. Levander, et al., Suppression of thermal conductivity in $In_xGa_{1-x}N$ alloys by nanometer-scale disorder. Appl. Phys. Lett. **102**, 121906 (2013)

170. M.S.B. Hoque, Y.R. Koh, K. Aryana, et al., Thermal conductivity measurements of sub-surface buried substrates by steady-state thermoreflectance. Rev. Sci. Instrum. **92**, 064906 (2021)
171. A.J. Minnich, J.A. Johnson, A.J. Schmidt, et al., Thermal conductivity spectroscopy technique to measure phonon mean free paths. Phys. Rev. Lett. **107**, 095901 (2011)
172. Y.K. Koh, D.G. Cahill, Frequency dependence of the thermal conductivity of semiconductor alloys. Phys. Rev. B **76**, 075207 (2007)
173. X. Chen, C. Hua, H. Zhang, N.K. Ravichandran, A.J. Minnich, Quasiballistic thermal transport from nanoscale heaters and the role of the spatial frequency. Phys. Rev. Appl. **10**, 054068 (2018)
174. A. Sood, R. Cheaito, T. Bai, et al., Direct visualization of thermal conductivity suppression due to enhanced phonon scattering near individual grain boundaries. Nano Lett. **18**, 3466–3472 (2018)
175. R.B. Wilson, D.G. Cahill, Anisotropic failure of Fourier theory in time-domain thermoreflectance experiments. Nat. Commun. **5**, 5075 (2014)
176. G. Chen, Non-Fourier phonon heat conduction at the microscale and nanoscale. Nat. Rev. Phys. **3**, 555–569 (2021)
177. K.M. Hoogeboom-Pot, J.N. Hernandez-Charpak, X. Gu, et al., A new regime of nanoscale thermal transport: Collective diffusion increases dissipation efficiency. Proc. Natl. Acad. Sci. **112**, 4846–4851 (2015)
178. U. Choudhry, T. Kim, M. Adams, et al., Characterizing microscale energy transport in materials with transient grating spectroscopy. J. Appl. Phys. **130**, 231101 (2021)
179. A. Ziabari, M. Parsa, Y. Xuan, et al., Far-field thermal imaging below diffraction limit. Opt. Express **28**, 7036–7050 (2020)
180. E.S. Massaro, A.H. Hill, E.M. Grumstrup, Super-resolution structured pump–probe microscopy. ACS Photon. **3**, 501–506 (2016)
181. Y. Wang, J.Y. Park, Y.K. Koh, D.G. Cahill, Thermoreflectance of metal transducers for time-domain thermoreflectance. J. Appl. Phys. **108**, 043507 (2010)
182. M. Lannoo, J. Bourgoin, *Point Defects in Semiconductors I: Theoretical Aspects* (Springer, New York, 1981)
183. E.G. Seebauer, K.W. Noh, Trends in semiconductor defect engineering at the nanoscale. Mater. Sci. Eng. **70**, 151–168 (2010)
184. J. Čížek, Characterization of lattice defects in metallic materials by positron annihilation spectroscopy: A review. J. Mater. Sci. Technol. **34**, 577–598 (2018)
185. A. Uedono, S. Takashima, M. Edo, et al., Carrier trapping by vacancy-type defects in mg-implanted GaN studied using monoenergetic positron beams. Phys. Status Solidi B **255**, 1700521 (2017)
186. M.A. Reshchikov, A. Usikov, H. Helava, et al., Evaluation of the concentration of point defects in GaN. Sci. Rep. **7**, 9297 (2017)
187. S. Yue, F. Tian, X. Sui, et al., High ambipolar mobility in cubic boron arsenide revealed by transient reflectivity microscopy. Science **377**, 433–436 (2022)
188. D.Q. Tran, R.D. Carrascon, M. Iwaya, et al., Thermal conductivity of $Al_xGa_{1-x}N$ ($0 \leq x \leq 1$) epitaxial layers. Phys. Rev. Mater. **6**, 104602 (2022)
189. Y. Amit, A. Fasut, O. Milo, et al., How to dope a semiconductor nanocrystal. ECS Trans. **58**, 127–133 (2013)
190. D.J. Norris, A.L. Efros, S.C. Erwin, Doped nanocrystals. Science **319**, 1776–1779 (2008)
191. T.W. Pfeifer, J.A. Tomko, E. Hoglund, et al., Measuring sub-surface spatially varying thermal conductivity of silicon implanted with krypton. J. Appl. Phys. **132**, 075112 (2022)
192. E.A. Scott, K. Hattar, C.M. Rost, et al., Phonon scattering effects from point and extended defects on thermal conductivity studied via ion irradiation of crystals with self-impurities. Phys. Rev. Mater. **2**, 095001 (2018)
193. A. Azarov, A. Galeckas, E. Wendler, et al., Normal and reverse defect annealing in ion implanted II-VI oxide semiconductors. J. Appl. Phys. **122**, 115701 (2017)

194. B. Abeles, Lattice thermal conductivity of disordered semiconductor alloys at high temperatures. Phys. Rev. **131**, 1906–1911 (1963)
195. R. Kucharski, T. Sochacki, B. Lucznik, M. Bockowski, Growth of bulk GaN crystals. J. Appl. Phys. **128**, 050902 (2020)
196. J.L. Lyons, D. Wickramaratne, C.G. Van de Walle, A first-principles understanding of point defects and impurities in GaN. J. Appl. Phys. **129**, 111101 (2021)
197. M.A. Reshchikov, *Semiconductors and Semimetals*, vol 91, chap. 9 (Elsevier, 2015), pp. 315–367
198. B.R. Ortiz, W. Peng, L.C. Gomes, et al., Ultralow thermal conductivity in diamond-like semiconductors: Selective ccattering of phonons from antisite defects. Chem. Mater. **30**, 3395–3409 (2018)
199. Z. Chen, X. Zhang, Y. Pei, Manipulation of phonon transport in thermoelectrics. Adv. Mater. **30**, 1705617 (2018)
200. G. Sun, Z. Xiang, J. Ma, X. Luo, D. Xu, Opposite atom dependence of isotope engineering of thermal conductivity in bulk and 2D GaN. Nanotechnology **34**, 485404 (2023)
201. J.H. You, H.T. Johnson, Effect of dislocations on electrical and optical properties in GaAs and GaN. Solid State Phys. **61**, 143–261 (2009)
202. F.K. Yam, L.L. Low, S.A. Oh, Z. Hassan, Optoelectronics-Materials and Techniques (InTech, Shanghai, 2011), chap. 4, pp. 99–136
203. H. Ravanbakhsh, L. Shekari, Transistors based on gallium nitride (GaN), growth techniques, and nanostructures. J. Theor. Appl. Phys. Lett. **16**, 162236 (2022)
204. C.R. Miskys, M.K. Kelly, O. Ambacher, G. Martínez-Criado, M. Stutzmann, GaN homoepitaxy by metalorganic chemical-vapor deposition on free-standing GaN substrates. Appl. Phys. Lett. **77**, 1858–1860 (2000)
205. S.D. Bhattacharyya, P. Mukhopadhyay, P. Das, D. Biswas, A strategic review of reduction of dislocation density at the heterogenious junction of GaN epilayer on foreign substrate. J. Nano Electron. Phys. **3**, 67–84 (2011)
206. T. Sasaki, T. Matsuoka, Analysis of two-step-growth conditions for GaN on an AlN buffer layer. J. Appl. Phys. **77**, 192–200 (1995)
207. S. Tanaka, S. Iwai, Y. Aoyagi, Reduction of the defect density in GaN films using ultra-thin AlN buffer layers on 6H-SiC. J. Cryst. Growth **170**, 329–334 (1997)
208. S. Sakai, T. Wang, Y. Morishima, Y. Naoi, A new method of reducing dislocation density in GaN layer grown on sapphire substrate by MOVPE. J. Cryst. Growth **221**, 334–337 (2000)
209. H. Ashraf, D.V. Sridhara Rao, D. Gogova, et al., Reduction of the dislocation density in HVPE-grown GaN epi-layers by an in situ SiN_x treatment. J. Cryst. Growth **312**, 595–600 (2010)
210. J. Li, G. Yang, X. Liu, et al., Dislocations in 4H silicon carbide. J. Phys. D. Appl. Phys. **55**, 463001 (2022)
211. I. Manning, Y. Matsuda, G. Chung et al., in *Materials Science Forum*. (Trans Tech Publ, 2020), vol. 1004, pp. 37–43
212. W. Liu, A.A. Balandin, C. Lee, H.Y. Lee, Increased thermal conductivity of free-standing low-dislocation-density GaN films. Phys. Status Solidi A **202**, R135–R137 (2005)
213. P. Carruthers, Theory of thermal conductivity of solids at low temperatures. Rev. Mod. Phys. **33**, 92–138 (1961)
214. Y. Ni, S. Xiong, S. Volz, T. Dumitrică, Thermal transport along the dislocation line in silicon carbide. Phys. Rev. Lett. **113**, 124301 (2014)
215. D. Hull, D.J. Bacon, *Introduction to Dislocations*, vol 37 (Elsevier, 2011)
216. S. Mei, C.J. Foss, L.N. Maurer *et al.*, Boundaries, interfaces, point defects, and strain as impediments to thermal transport in nanostructures, in *2017 IEEE International Reliability Physics Symposium (IRPS)*, 6A-1.1–6A-1.10 (2017)
217. P.E. Hopkins, Thermal transport across solid interfaces with nanoscale imperfections: Effects of roughness, disorder, dislocations, and bonding on thermal boundary conductance. ISRN Mech. Eng. **2013**, 1–19 (2013)

218. J.-W. Min, S.-Y. Bae, W.-M. Kang, et al., Evolutionary growth of microscale single crystalline GaN on an amorphous layer by the combination of MBE and MOCVD. CrystEngComm **17**, 5849–5859 (2015)
219. B.M. Morrow, E.K. Cerreta, R.J. McCabe, C.N. Tomé, Toward understanding twin–twin interactions in hcp metals: Utilizing multiscale techniques to characterize deformation mechanisms in magnesium. Mater. Sci. Eng. A **613**, 365–371 (2014)
220. F. La Via, A. Severino, R. Anzalone, et al., From thin film to bulk 3C-SiC growth: Understanding the mechanism of defects reduction. Mater. Sci. Semicond. Process. **78**, 57–68 (2018)
221. J. Li, Y. Xie, C. Zhang, et al., Stacking fault-induced minimized lattice thermal conductivity in the high-performance GeTe-based thermoelectric materials upon Bi_2Te_3 alloying. ACS Appl. Mater. Interfaces **11**, 20064–20072 (2019)
222. C.J. Szwejkowski, N.C. Creange, K. Sun, et al., Size effects in the thermal conductivity of gallium oxide (β-Ga_2O_3) films grown via open-atmosphere annealing of gallium nitride. J. Appl. Phys. **117**, 084308 (2015)
223. T. Liu, D. Raabe, W.-M. Mao, A review of crystallographic textures in chemical vapor-deposited diamond films. Front. Mater. Sci. China **4**, 1–16 (2010)
224. S. Roa, M. Sandoval, S. Suárez, Rutherford backscattering spectroscopy analysis of the growth quality of chemical bath deposited PbSe thin films. Solid State Sci. **113**, 106545 (2021)
225. J. Lu, K. Yuan, F. Sun, et al., Self-assembled monolayers for the polymer/semiconductor interface with improved interfacial thermal management. ACS Appl. Mater. Interfaces **11**, 42708–42714 (2019)
226. H. Jang, C.R. Ryder, J.D. Wood, M.C. Hersam, D.G. Cahill, 3D anisotropic thermal conductivity of exfoliated rhenium disulfide. Adv. Mater. **29**, 1700650 (2017)
227. Z. Cheng, L. Yates, J. Shi, et al., Thermal conductance across β-Ga2O3-diamond van der Waals heterogeneous interfaces. APL Mater. **7**, 031118 (2019)
228. V.E. Gusev, P. Ruello, Advances in applications of time-domain Brillouin scattering for nanoscale imaging. Appl. Phys. Rev. **5**, 031101 (2018)
229. B. De Vries, U. Wahl, S. Ruffenach, O. Briot, A. Vantomme, Influence of crystal mosaicity on axial channeling effects and lattice site determination of impurities. Appl. Phys. Lett. **103**, 172108 (2013)
230. A.K. Bilgili, Ö. Akpinar, G. Kurtulus, et al., Structural properties of InGaN/GaN/Al_2O_3 structure from reciprocal space mapping. J. Mater. Sci. Mater. Electron. **29**, 12373–12380 (2018)
231. S.K. Jana, P. Mukhopadhyay, S. Ghosh, et al., High-resolution X-ray diffraction analysis of $Al_xGa_{1-x}N/In_xGa_{1-x}N$/GaN on sapphire multilayer structures: Theoretical, simulations, and experimental observations. J. Appl. Phys. **115**, 174507 (2014)
232. E. Arslan, M.K. Ozturk, A. Teke, S. Ozcelik, E. Ozbay, Buffer optimization for crack-free GaN epitaxial layers grown on Si(1 1 1) substrate by MOCVD. J. Phys. D. Appl. Phys. **41**, 155317 (2008)
233. A.E. Wickenden, D.D. Koleske, R.L. Henry, M.E. Twigg, M. Fatemi, Resistivity control in unintentionally doped GaN films grown by MOCVD. J. Cryst. Growth **260**, 54–62 (2004)
234. H.-J. Lee, T. Yao, C. Kim, J. Chang, Use of Polytypes to control crystallographic orientation of GaN. Cryst. Growth Des. **10**, 5307–5311 (2010)
235. J. Bläsing, V. Holý, A. Dadgar, et al., Growth and characterization of stacking fault reduced GaN (1013) on sapphire. J. Phys. D. Appl. Phys. **46**, 125308 (2013)
236. H. Dong, B. Wen, R. Melnik, Relative importance of grain boundaries and size effects in thermal conductivity of nanocrystalline materials. Sci. Rep. **4**, 7037 (2014)
237. A. Porter, C. Tran, F. Sansoz, Intrinsic nanotwin effect on thermal boundary conductance in bulk and single-nanowire twinning superlattices. Phys. Rev. B **93**, 195431 (2016)
238. N.K. Ravichandran, A.J. Minnich, Role of thermalizing and nonthermalizing walls in phonon heat conduction along thin films. Phys. Rev. B **93**, 035314 (2016)
239. F. Mu, Z. Cheng, J. Shi, et al., High thermal boundary conductance across bonded heterogeneous GaN-SiC interfaces. ACS Appl. Mater. Interfaces **11**, 33428–33434 (2019)

240. C. Yuan, J.W. Pomeroy, M. Kuball, Above bandgap thermoreflectance for non-invasive thermal characterization of GaN-based wafers. Appl. Phys. Lett. **113**, 102101 (2018)
241. Y. Zhou, R. Ramaneti, J. Anaya, et al., Thermal characterization of polycrystalline diamond thin film heat spreaders grown on GaN HEMTs. Appl. Phys. Lett. **111**, 041901 (2017)
242. J. Anaya, S. Rossi, M. Alomari, et al., Control of the in-plane thermal conductivity of ultra-thin nanocrystalline diamond films through the grain and grain boundary properties. Acta Mater. **103**, 141–152 (2016)
243. A. Sood, J. Cho, K.D. Hobart, et al., Anisotropic and inhomogeneous thermal conduction in suspended thin-film polycrystalline diamond. J. Appl. Phys. **119**, 175103 (2016)
244. G. Alvarez-Escalante, R. Page, R. Hu, et al., High thermal conductivity and ultrahigh thermal boundary conductance of homoepitaxial AlN thin films. APL Mater. **10**, 011115 (2022)
245. Y.R. Koh, Z. Cheng, A. Mamun, et al., Bulk-like intrinsic phonon thermal conductivity of micrometer-thick AlN films. ACS Appl. Mater. Interfaces **12**, 29443–29450 (2020)
246. Y. Song, D. Shoemaker, J.H. Leach, et al., Ga_2O_3-on-SiC composite wafer for thermal management of ultrawide bandgap electronics. ACS Appl. Mater. Interfaces **13**, 40817–40829 (2021)
247. R. Wu, K. Zhou, C.Y. Yue, J. Wei, Y. Pan, Recent progress in synthesis, properties and potential applications of SiC nanomaterials. Prog. Mater. Sci. **72**, 1–60 (2015)
248. A. Motayed, M. Vaudin, A.V. Davydov, et al., Diameter dependent transport properties of gallium nitride nanowire field effect transistors. Appl. Phys. Lett. **90**, 043104 (2007)
249. C. Guthy, C.-Y. Nam, J.E. Fischer, Unusually low thermal conductivity of gallium nitride nanowires. J. Appl. Phys. **103**, 064319 (2008)
250. R. Anufriev, Y. Wu, S. Volz, M. Nomura, Quasi-ballistic thermal transport in silicon carbide nanowires. Appl. Phys. Lett. **124**, 022202 (2024)

Chapter 24
Breakdown Phenomena in GaN

Jun Suda

1 Introduction

Understanding of breakdown phenomena and pre-breakdown characteristics (leakage current) is very important to realize high-performance power devices. There are two fundamental breakdown mechanisms: Zener breakdown and avalanche breakdown [1].

Zener breakdown is due to band-to-band (valence band to conduction band) tunneling in the presence of a high electric field. The definition of breakdown voltage for tunneling-related breakdown is difficult because the tunneling current gradually increases with increasing voltage on a semilogarithmic scale. There is no clear threshold or breakdown voltage for Zener breakdown. In general, we defined the breakdown voltage at which the current reaches a certain level such as 10^{-4} A/cm^2 which is acceptable for practical use.

Avalanche breakdown is attributed to impact ionization processes. One carrier, for example, one electron enters into a depletion region and is subsequently accelerated by high electric field. For low electric fields, electrons lose their kinetic energy by conventional inelastic processes—namely, phonon scattering. On the other hand, for high electric fields and high carrier energy, there is a chance that a small fraction of the electrons realize sufficiently high energy to excite electrons from valence band to conduction band. When it happens, we call it impact ionization. One electron results in two electrons and one hole. If the depletion layer width is sufficiently large thick, then impact ionization of electrons and holes repeated within the depletion layer—this is the condition for avalanche breakdown. Sometimes, we can observe pre-breakdown current increase called avalanche multiplication; however, in general, avalanche breakdown has a clear threshold—the avalanche breakdown voltage.

J. Suda (✉)
Department of Electronics, Nagoya University, Nagoya, Aichi, Japan
e-mail: suda@nuee.nagoya-u.ac.jp

I. C. Kizilyalli et al. (eds.), *Gallium Nitride and Related Materials*, The Materials Research Society Series, https://doi.org/10.1007/978-3-031-83056-3_24

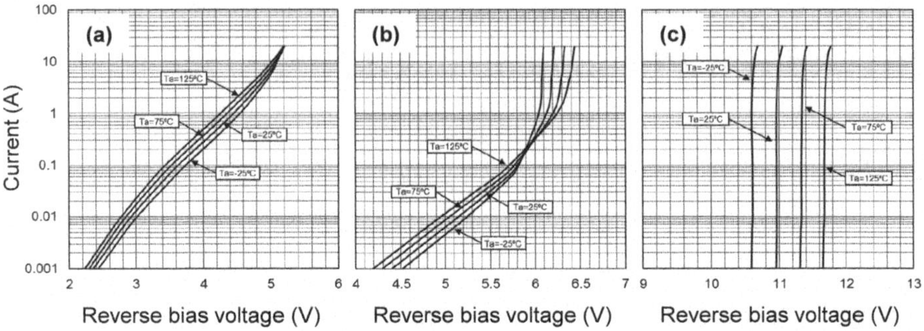

Fig. 24.1 Reverse *I-V* characteristics and its temperature dependence of commercially available Si constant voltage diodes with breakdown voltage (defined at a current level of 10 mA at 25 °C) of (**a**) 5.1 V, (**b**) 6.2 V, and (**c**) 11 V

For Si pn junctions, these breakdown phenomena are used to generate reference voltage in electric circuits. When several times larger voltage is applied to a series connection of a resistor and a pn diode in the reverse direction, the voltage drop at the pn junction is fixed at its breakdown voltage. Since change of the applied voltage does not affect the voltage drop, it can be used as a voltage reference. The pn junctions used for the purpose are called constant voltage diodes, Zener diodes, or avalanche diodes. Reverse current-voltage (*I-V*) characteristics of Si constant voltage diodes are shown in Fig. 24.1. Here, we define breakdown voltage at current level of 10 mA. As shown in Fig. 24.1a, diodes with breakdown voltage less than 5 V are dominated by Zener breakdown. Those with higher breakdown voltage larger than 10 V are dominated by avalanche breakdown as shown in Fig. 24.1c.

For design of power device structures or prediction of leakage current of power devices, material parameters should be known. These are the dependence of the band-to-band tunneling rate and impact ionization coefficients on electric field. In this chapter, recent reports on these material parameters for GaN are reviewed. It should be noted that these parameters are also necessary to design ultra-high-sensitive photodetectors, avalanche photodiodes (APD), or high-frequency negative resistance devices (amplifiers) such as impact-ionization avalanche transit-time (IMPATT) diodes or tunneling-injection transit-time (TUNNETT) diodes [1].

2 Theory

2.1 Zener Breakdown

Zener tunneling or band-to-band tunneling is illustrated in Fig. 24.2. Under a high electric field, electron wave functions of the valence band and conduction band (with the same energy) penetrate the forbidden gap. If the electric field is large enough, there is an overlapping of two wave functions resulting in tunneling with a distance

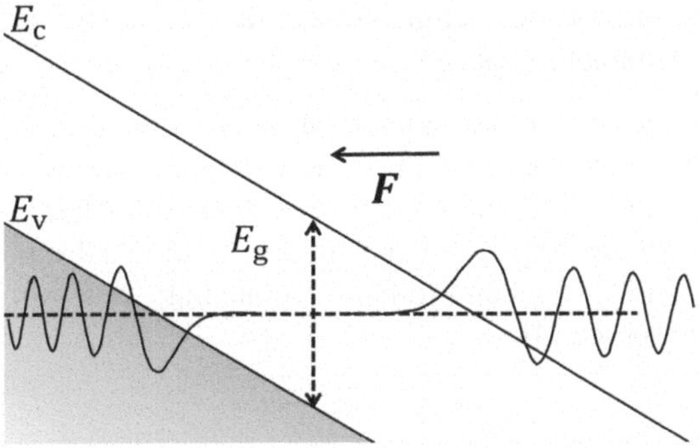

Fig. 24.2 Schematic illustration of Zener tunneling

of E_g/eF, where E_g is energy band gap, and e and F are elementary charge and electric field strength. The tunneling resulted in continuous generation of electron-hole pairs in the depletion layer resulting in off-state reverse current in power devices.

In 1959, Kane analyzed the probability (generation rate) of tunneling in direct-bandgap semiconductors [2]. The obtained result is as follows:

$$G(F) = \frac{e^2 F^2 m_r^{\frac{1}{2}}}{18\pi\hbar^2 E_g^{\frac{1}{2}}} \exp\left\{ -\frac{\pi m_r^{\frac{1}{2}} E_g^{\frac{3}{2}}}{2\hbar e F} \right\} = A F^2 \exp\left(-\frac{B}{F} \right) \quad (24.1)$$

Here, m_r is reduced mass which is calculated from the conduction band and valence band effective masses, m_e and m_h, along the tunneling direction.

$$\frac{1}{m_r} = \frac{1}{m_e} + \frac{1}{m_h} \quad (24.2)$$

The reverse bias current (often referred to as the off-state leakage current) can be estimated by integrating generated carriers for all positions in the depletion layer. If electron-hole recombination in the depletion layer can be neglected, in other words, all carriers are corrected to both sides of the deletion layer, current density can be expressed by the following formula.

$$J = \int e\, G(F(x))\, \mathrm{d}x \quad (24.3)$$

There are some notes for Zener tunneling:

1. It does not have clear threshold, rather it increases with increasing electric field.
2. It does not require a long length to happen. Tunneling occurs at two points with distance of E_g/eF, the order of 10 nm depending on E_g and F.
3. It has a weak temperature dependence since tunneling processes are not thermally activated. However, the bandgap does have a temperature dependence. With

increasing temperature, the bandgap slightly decreases increasing the tunneling rate.

If one has the values of A and B in Eq. (24.1), it is usually sufficient to predict the Zener tunneling current. For example, reverse I-V characteristics of GaAs p-i-n diodes were analyzed by the Kane model [3]. The I-V characteristics were well-explained by the Kane model. It should be noted that there are some advanced models of band-to-band tunneling; however, not so much difference were observed [4].

2.2 Avalanche Breakdown

Avalanche breakdown originates from impact ionization. A schematic illustration of impact ionization is shown in Fig. 24.3a. If the electric field is sufficiently high, the carrier can gain large kinetic energy before scattering. If the kinetic energy is sufficiently large, it can excite electron-hole pairs. Energy and momentum conservation laws should be satisfied at the same time (as a side note, the matrix element for impact ionization is the same as Auger recombination). Because of this, the kinetic energy should be about 1.5 times larger (depending on E-k dispersion of the material) than the energy bandgap to satisfy the momentum conservation law (Fig. 24.3b).

The impact ionization coefficient is defined as the number of electron-hole pairs generated by a carrier per unit distance. Obviously, impact ionization coefficients

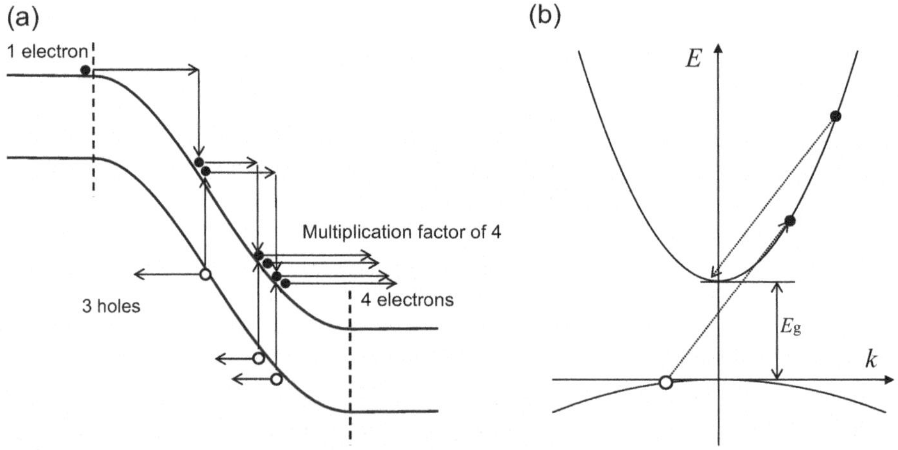

Fig. 24.3 Schematic illustration of impact ionization. (**a**) Electrons gained sufficiently enough energy resulting in one electron-hole pair. (**b**) Both energy and momentum conservation should be satisfied for impact ionization

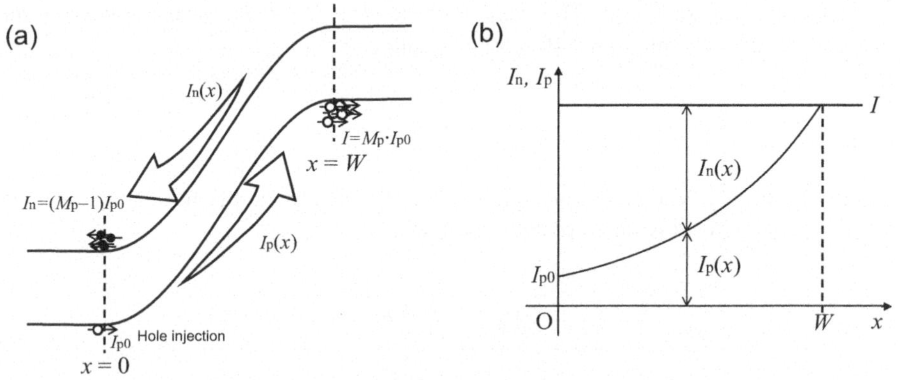

Fig. 24.4 Schematic illustration of impact multiplication in the depletion layer. (**a**) Hole injection I_{po} from the left edge of the depletion layer resulted in total current $I = M_p I_{po}$. (**b**) For steady-state conditions, the total current should be constant everywhere

have an electric field dependence. A physical expression for the ionization coefficients is given by Thornber [5]:

$$\alpha(F) = \frac{eF}{E_{th}} \exp\left(-\frac{F_I}{F\left[1 + \left(\frac{F}{F_P}\right)\right] + F_T} \right)$$

where E_{th} is the high-field effective ionization threshold energy, and F_T, F_P, and F_I are threshold fields for carriers to overcome the deceleration effects of thermal, optical phonon scattering, and ionization scattering, respectively. However, it is often modeled by Chynoweth's empirical formula [6], which is similar to the expression for ionization rate in gas discharge:

$$\alpha(F) = a \exp\left(-\frac{b}{F} \right). \tag{24.4}$$

Here, we consider impact ionization with the depletion layer under a reverse bias condition [1, 7]. Let us assume one minority carrier, a hole, in n-type layer enters into depletion layer. Carrier multiplication will occur due to impact ionization in the depletion layer. Since impact ionization by hole results in both electron and hole generation, impact ionization for electron should be also considered. By considering both carriers, the increase in the hole current can be expressed as:

$$dI_p = I_P\, \alpha_p\, dx + I_n\, \alpha_n\, dx. \tag{24.5}$$

Here, α_p and α_n are the impact ionization coefficients for holes and electrons, respectively. I_p and I_n are the hole current and electron current, respectively. Here, we assume a steady state. The total current ($I = I_p + I_n$) should be constant everywhere. So, we obtain a differential equation,

$$\frac{dI_p}{dx} - (\alpha_p - \alpha_n)I_p = \alpha_n I. \tag{24.6}$$

The solution of the differential equation with boundary conditions of $I_p(W) = I$ and $I_p(0) = I_{p0}$, where W is the depletion layer width [1, 7].

$$I_p(x) = \frac{I\int_0^x \alpha_n \exp\left[-\int_0^x (\alpha_p - \alpha_n)dx'\right]dx + I_{p0}}{\exp\left[-\int_0^x (\alpha_p - \alpha_n)dx'\right]} \tag{24.7}$$

Multiplication factor M_p is defined as I/I_{po}.

$$1 - \frac{1}{M_p} = \int_0^W \alpha_p \exp\left[-\int_0^x (\alpha_p - \alpha_n)dx'\right]dx \tag{24.8}$$

The right-hand side is called the ionization integral (*I.I.*). When it approaches unity, the multiplication factor rapidly increases. It is called avalanche multiplication. When it becomes unity, the multiplication factor becomes infinity, resulting in avalanche breakdown. Avalanche breakdown has a very clear threshold. The breakdown is explained by forming an infinite loop, i.e., the impact ionization of electrons forms a hole, and the hole makes impact ionization to supply electrons within the depletion layer.

The ionization integral can be calculated by electric field distribution within the depletion layer together with electric field dependences of impact ionization coefficients of hole and electron:

$$I.I. = \int_0^W \alpha_p(F(x)) \exp\left[-\int_0^x (\alpha_p(F(x)) - \alpha_n(F(x)))dx'\right]dx \tag{24.9}$$

The critical electric field is widely used to evaluate the figure of merit of power devices (such as Baliga's figure of merit [8]). Strictly speaking, however, the critical electric field (electric field at which avalanche breakdown happens) depends on the depletion layer thickness and electric field distribution within the depletion layer. The critical electric field decreases with increasing depletion layer thickness (which can be realized by lower doping). A fair comparison between various semiconductors can be done by assuming a target breakdown voltage such as 1.2 kV.

There are some notes for avalanche breakdown:

1. It has a clear threshold.
2. It requires a length to happen because infinite loop of multiplication should be established within the depletion layer. Both impact ionization of electron and hole are needed to establish the infinite loop.
3. It has a strong temperature dependence because of the temperature dependence of acoustic and especially optical phonon scattering. With increasing temperature, it becomes more difficult to gain large energy due to frequent scattering, resulting in an increase in the breakdown voltage.

If one has the electric field dependences of impact ionization coefficients $\alpha_n(F)$ and $\alpha_p(F)$, it is good enough to predict avalanche breakdown as well as pre-breakdown avalanche multiplication.

3 Experimental Results

3.1 Design and Fabrication of GaN Vertical PN Diodes

For accurate determination of avalanche and Zener breakdown properties, high-quality devices are required. The devices must satisfy the requirements listed below:

1. The leakage current due to nonideal factors such as threading dislocations, mesa side wall, and impurities should be suppressed sufficiently to assess true GaN properties.
2. Electric field crowding should be eliminated to determine applied electric field in the device precisely.
3. High avalanche capability is needed to ensure breakdown measurements many times without degradation or destruction of the device.

For suppression of leakage current, we utilized bulk GaN substrates with low threading dislocation densities (TDD). Typical TDDs for halide vapor phase epitaxy (HVPE)-grown substrates are around 1×10^6 cm^{-2} [9, 10]. For ammonothermal GaN substrates, the TDD can be less than 10^4 cm^{-2} [11–13]. We also employed a highly sophisticated metal-organic vapor phase epitaxy (MOVPE) method for epitaxial growth of pn junctions on these substrates [14]. Careful attention was paid to reducing metal impurities such as iron (Fe) and incorporation of carbon (C) from source gases. Leakage current at device periphery should be reduced. It will be discussed in the device fabrication section.

3.1.1 Device Design

Electric field crowding can be eliminated by using a proper device structure, i.e., edge termination structure. Field plate, floating field ring (FFR), and junction termination extension (JTE) are commonly used methods [8]. For wide-bandgap

(WBG) semiconductors, field plates are relatively ineffective because a very thick oxide insulating layer is needed to avoid degradation of insulating layer due to high breakdown field of WBG semiconductors. For FFR and JTE, selective doping by ion implantation is needed. However, at that time and still now, selective doping was poorly established for GaN. Thus, we decided to use bevel structures.

It is well known that a reduction of surface electric field can be realized by using a sufficiently small negative bevel angle with a highly graded (diffused) junction in Si power devices [8]. Formation of such a graded junction with high controllability is difficult for GaN growth technology, and graded junctions are unsuitable for analysis of measurement results. So, we consider negative bevel termination for abrupt pn junctions with comparable doping concentrations of p-type and n-type layers. For this case, the depletion region expands for both p-type and n-type layers, resulting in a similar situation of the graded junction [15].

We performed TCAD simulations of bevel mesa structures with various bevel angles and doping concentrations (N_a, N_d) for p and n layers. Electric field distributions with bevel angles of 30° and 10° are shown in Fig. 24.5. The electric field crowding occurs at the edge of the pn junction (Fig. 24.5a). The maximum electric field (E_{max}) is 4.67 MV/cm, 1.5 times higher than the parallel palate electric field (E_{pp}), the maximum electric field for abrupt pn junction with N_a and N_d. As the acceptor concentration N_a decreases, E_{max} decreases and approaches to E_{pp} as seen in Fig. 24.5b and c. It should be noted that the point of maximum electric field moves from the surface to the inside of the devices, and surface electric field is reduced. Further reduction of bevel angle is effective to reduce electric field crowding as shown in Fig. 24.5d.

We confirmed that the results do not depend on the absolute value of doping concentrations but depends on the ratio (N_a/N_d which is equal to W_n/W_p). Since the reduction of electric field crowding is a geometrical effect, the bevel angle and W_n/W_p determine the reduction. Of course, for lower doping structures, large mesa height (very long bevel slope length) is required. TCAD simulation results are summarized in Fig. 24.6. If we accept a slight increase of 5% E_{max}, the hatched area can be used. For example, the bevel angle is 10° and $N_a/N_d < 4$.

3.1.2 Device Fabrication

Formation of bevel structure with small angle is the key process. We developed a device fabrication process by combining thermal reflow of photoresist and dry etching process to transfer resist shape to GaN as shown in Fig. 24.7. We utilize thick photoresist and high-temperature post baking at 200 °C for 3 min to make a lens-like structure. The angle of the lens edge is around 6–10°. A Cl_2-based inductively coupled plasma-reactive ion etching (ICP-RIE) was carried out. Etching rates of photoresist and GaN are almost the same, resulting in successful transferring of the edge angle to GaN. The anode and cathode electrodes were formed by the deposition of Ni/Au on the top p-layer and Ti/Al/Ni on the backside of the GaN

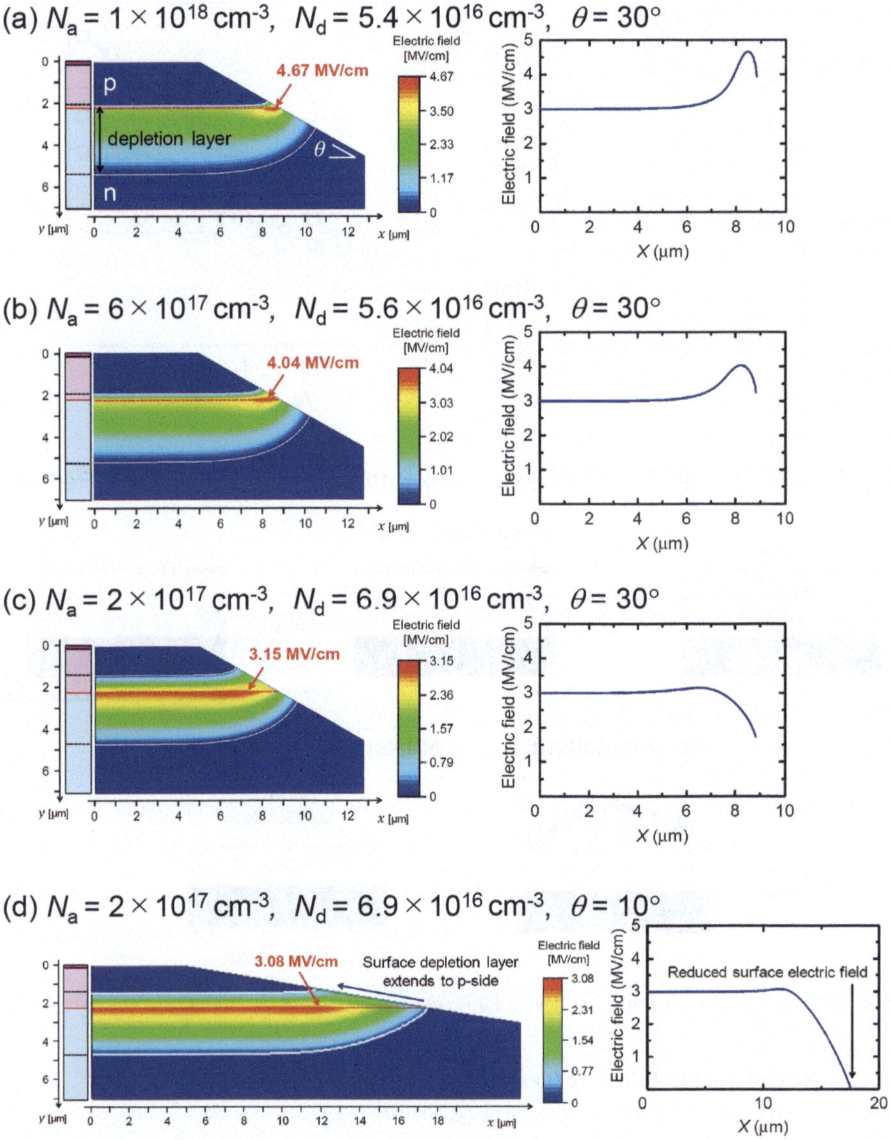

Fig. 24.5 Electric field distribution of negative bevel mesa structures with various doping concentrations and different bevel angles. Electric field strengths along pn junction are shown in the right panels. N_d was slightly modified to keep the total depletion layer width as well as the parallel plate electric field

substrates, respectively. Sintering was carried out at 500–550 °C for several minutes in oxygen ambient [15].

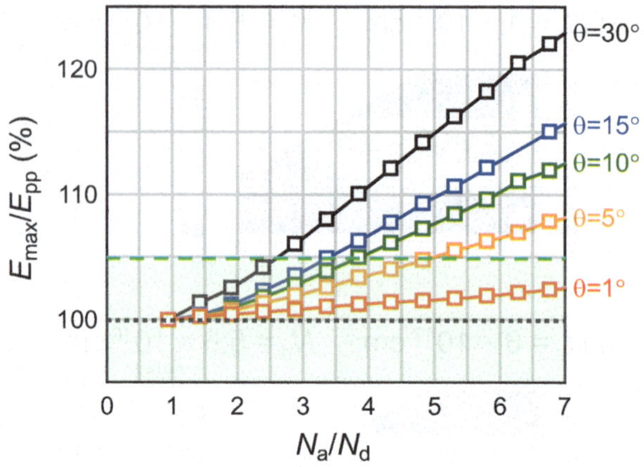

Fig. 24.6 Ratio of E_{max}/E_{pp} as a function of the doping ratio (N_a/N_d) for various bevel angles. E_{max}/E_{pp} of 100% means complete elimination of electric field crowding. If we accept +5% crowding, the shaded area can be used for edge termination

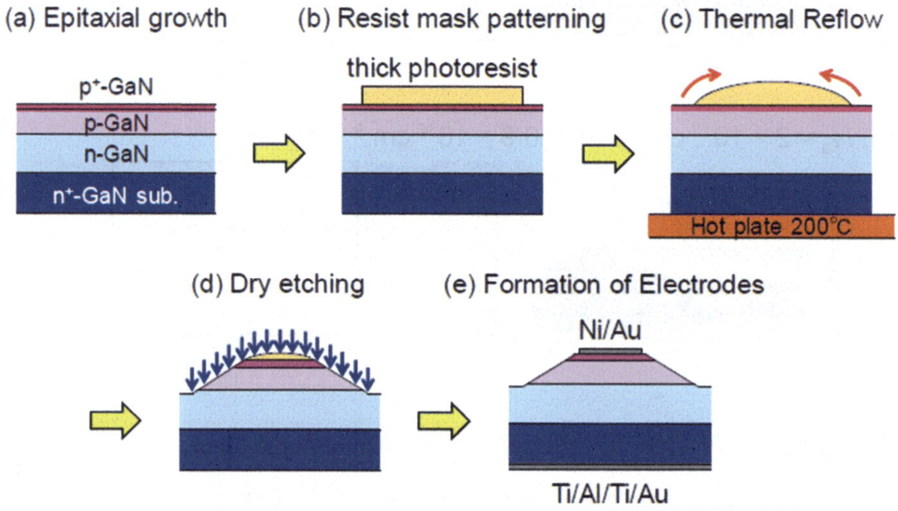

Fig. 24.7 Schematic illustration of device fabrication process

3.2 *Breakdown Characteristics in GaN PN Diodes*

We fabricated many GaN PN diodes. The diodes consist of a top p^+ contact layer with high magnesium (Mg) concentration, a Mg-doped p-type layer, a silicon (Si)-doped n-type layer, and a n^+ bottom layer on the n-type (0001) substrates. We fabricated GaN bevel diodes with various doping concentrations as shown in Table 24.1 [16].

Typical reverse *I-V* characteristics are shown in Fig. 24.8. For lower doped, i.e., high-blocking voltage devices show sharp increases in current at breakdown voltage as shown in Fig. 24.8c, suggesting avalanche breakdown. On the other hand, for higher doped, i.e., low-blocking voltage device shows gradual increases in current as

Table 24.1 Doping concentrations ([Mg] for p-GaN and [Si] for n-GaN) and thicknesses of epitaxial layers of pn diodes. Net doping concentrations obtained by *C-V* measurements are also shown

	PN A	PN B	PN C
p^+-GaN	5×10^{19} cm^{-3} 0.1 μm		
p-GaN	1.6×10^{18} cm^{-3} 2.5 μm	1.3×10^{18} cm^{-3} 2.5 μm	2.7×10^{17} cm^{-3} 2.4 μm
n-GaN	6.3×10^{17} cm^{-3} 2.9 μm	3.3×10^{17} cm^{-3} 2.8 μm	2.3×10^{16} cm^{-3} 6.8 μm
$\frac{N_a N_d}{N_a+N_d}$	3.2×10^{17} cm^{-3}	1.7×10^{17} cm^{-3}	1.5×10^{16} cm^{-3}
V_b (avalanche)	132 V	196 V	1025 V
E_{cr} (avalanche)	3.83 MV/cm	3.38 MV/cm	2.61 MV/cm

shown in Fig. 24.8a, suggesting Zener breakdown. Although the voltage ranges are very different, the observed behavior is very similar to commercially available Si Zener diodes as shown in Fig. 24.1. It should be noted that all devices exhibited good avalanche capability, indicating successful suppression of electric field crowding.

To clarify that the Zener-like behavior of GaN PN diodes comes from the true properties of GaN not from dislocation-related leakage current, we looked for dislocation-free diodes within the same wafer. We found some diodes without any threading dislocations by X-ray topography using synchrotron radiation [17]. We confirmed that no difference was observed in diodes with a few dislocations and without dislocations. It should be noted that PN B diodes shown in Fig. 24.8b are dislocation-free. Therefore, the observed leakage characteristics do not come from dislocations. Even if the Zener current is observed, for example, PN B diodes show an abrupt increase in leakage current at around 200 V, corresponding to avalanche breakdown. Again coexistence of two mechanisms is similar to the case of Si diodes.

We performed capacitance-voltage (*C-V*) measurements to evaluate net doping concentrations as well as depletion layer width. In the calculation, $10.4\epsilon_0$ is used for dielectric constant of GaN along the c-axis [18].

3.3 *Band-to-Band Tunneling Rate of GaN*

We confirmed Zener breakdown for higher doping diodes. We performed a detailed analysis of *I-V* characteristics. Reverse current-voltage characteristics were analyzed by using Eq. (24.1), and we obtained good fitting by using parameters [16]:

$$A = 1.75 \times 10^{21}\ \text{cm}^{-1}\text{s}^{-1}\text{V}^{-2},\quad B = 1.15 \times 10^{8}\ \text{Vcm}^{-1}$$

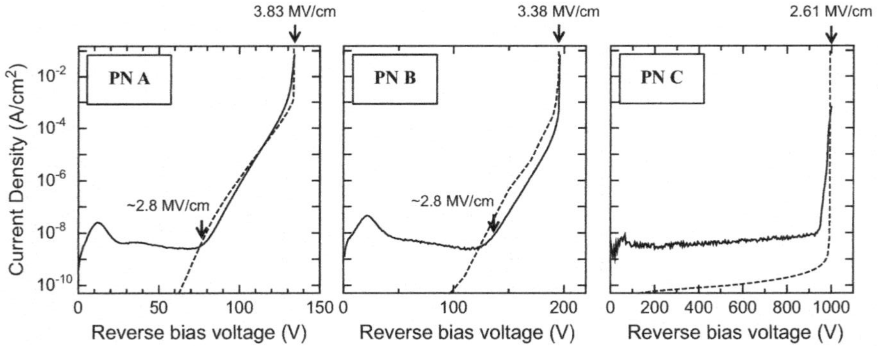

Fig. 24.8 Reverse *I-V* characteristics of GaN pn diodes with different doping concentrations as shown in Table 24.1. Peaks are observed at low-voltage regions. By time constant analysis, we confirmed that they are due to hole emission current from carbon deep level (carbon substituting nitrogen site, C_N) in the p-type layer

These values reproduce reverse *I-V* characteristics of not only one diode but also other low-voltage GaN diodes with different doping concentrations.

A and *B* can be calculated from energy band gap and reduced effective mass. By using E_g of 3.42 eV, electron effective mass of $0.16m_0$ and heavy hole effective mass of $2.03m_0$ along c-axis [19],

$$A_{\text{calc}} = 2.1 \times 10^{20}\ \text{cm}^{-1}\text{s}^{-1}\text{V}^{-2},\quad B_{\text{calc}} = 1.4 \times 10^{8}\ \text{Vcm}^{-1}$$

were obtained. *B* value shows excellent agreement. On the other hand, for *A* value, there is eightfold difference. Other factors such as non-parabolicity of energy band as well as contribution of other hole bands might be considered. However, we can say there is a fair agreement between the experimentally determined and calculated *A* and *B* values.

We observed that the Zener current of GaN PN diodes slightly increases with increasing temperature. The increase can be explained quantitatively by temperature dependence of bandgap at room temperature.

Further investigation will be needed to explain the difference in *A* value. However, our experimentally obtained *A* and *B* values well reproduce device characteristics. The formula (24.1) with obtained values is useful for the design of GaN devices. Although it depends on the depletion layer width, roughly speaking, we should consider Zener tunneling current if the electric field is larger than 2.8–3.0 MV/cm (the tunneling distance is about 11–12 nm).

3.4 Avalanche Breakdown of GaN

Here, we focus on the abrupt increase in reverse current, corresponding to avalanche breakdown. We measured reverse *I-V* characteristics of four other diodes (shown in Table 24.2) at three different temperatures, 223 K, 298 K, and 373 K. The results are shown in Fig. 24.9 [20]. For all devices, the breakdown voltage increases with increasing temperature. All these features clearly indicate avalanche breakdown.

3.5 Impact Ionization Coefficients of GaN

If we can obtain electric field dependence of impact ionization coefficients for electron and hole, we can estimate avalanche breakdown voltage for any device structures. There are many studies on the identification of impact ionization coefficients for Si and GaAs. One of the clearest measurements for this is measurements of photomultiplication of $p^+/i/n^+$ diodes as shown in Fig. 24.10 [21]. By measuring electron-injection and hole-injection photomultiplication, impact ionization coefficients can be calculated easily, thanks to uniform electric field in the middle i-layer.

Table 24.2 Doping concentrations ([Mg] for p-GaN and [Si] for n-GaN) and thicknesses of epitaxial layers of pn diodes. Net doping concentrations obtained by high-voltage C-V measurements (close to breakdown voltage) are also shown. Residual carbon concentrations for all samples are less than 5×10^{15} cm^{-3}

	PN1	PN2	PN3	PN4
p^+-GaN	8×10^{19} cm^{-3} 0.2 μm			
p-GaN	1.2×10^{17} cm^{-3} 3.3 μm	1.4×10^{17} cm^{-3} 2.5 μm	2.7×10^{17} cm^{-3} 2.0 μm	3.9×10^{17} cm^{-3} 1.5 μm
n-GaN	7.6×10^{16} cm^{-3} 3.3 μm	1.2×10^{17} cm^{-3} 2.5 μm	2.1×10^{17} cm^{-3} 2.0 μm	4.1×10^{17} cm^{-3} 1.5 μm
$\frac{N_a N_d}{N_a + N_d}$	4.7×10^{16} cm^{-3}	6.4×10^{16} cm^{-3}	1.2×10^{17} cm^{-3}	1.9×10^{17} cm^{-3}
V_b (avalanche)	480 V	385 V	250 V	180 V
E_{cr} (avalanche)	2.8 MV/cm	2.9 MV/cm	3.2 MV/cm	3.5 MV/cm

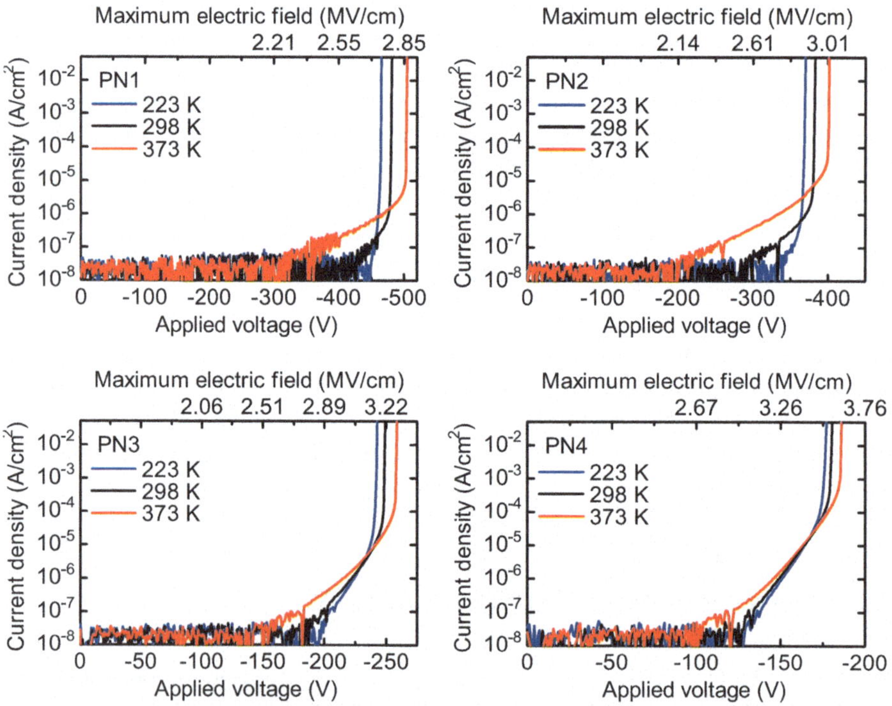

Fig. 24.9 Reverse *I-V* characteristics of four PN diodes showing abrupt breakdown. Measurements were carried out at three different temperatures

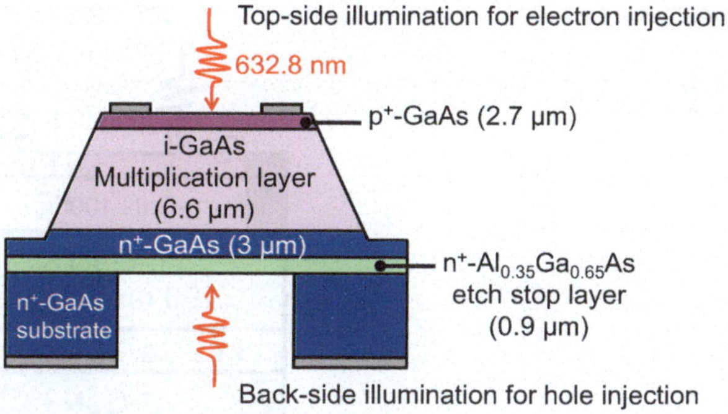

Fig. 24.10 Photomultiplication measurements to determine impact ionization coefficients of hole and electron separately. 632.8 nm laser (penetration depth for GaAs is around 80 nm) was used for excitation

$$\alpha_{\mathrm{n}}(F) = \frac{1}{W} \cdot \frac{M_{\mathrm{n}} - 1}{M_{\mathrm{n}} - M_{\mathrm{p}}} \cdot \ln\left(\frac{M_{\mathrm{n}}}{M_{\mathrm{p}}}\right) \tag{24.10}$$

$$\alpha_{\mathrm{p}}(F) = \frac{1}{W} \cdot \frac{M_{\mathrm{p}} - 1}{M_{\mathrm{p}} - M_{\mathrm{n}}} \cdot \ln\left(\frac{M_{\mathrm{p}}}{M_{\mathrm{n}}}\right) \tag{24.11}$$

By changing the bias voltage, we can obtain electric field dependence of electrons and holes. This method can be expanded for one-sided abrupt pn junction [22].

$$\alpha_{\mathrm{p}}(F) = \frac{1}{M_{\mathrm{n}}(W)} \frac{\mathrm{d}}{\mathrm{d}W} \left[\ln\left(M_{\mathrm{p}}, (W)\right)\right] \tag{24.12}$$

$$\alpha_{\mathrm{n}}(F) = \alpha_{\mathrm{p}}(F) + \frac{\mathrm{d}}{\mathrm{d}W} \left[\ln\left(\frac{M_{\mathrm{n}}(W)}{M_{\mathrm{p}}(W)}\right)\right] \tag{24.13}$$

For GaAs, the $p^{+}/i/n^{+}$ diode as shown in Fig. 24.10 can be fabricated by utilizing $Al_xGa_{1\text{-}x}As$ etching stop layer [21]. Electric field dependence of electron and hole impact ionization coefficients were successfully obtained. However, for GaN, fabrication of such structures is very difficult because of lack of good etch stop layer.

Cao et al. reported utilization of inserting narrower-bandgap $In_xGa_{1\text{-}x}N$ photo absorption layer underneath of GaN pn junction (Fig. 24.11) [23]. In the case of above-bandgap light illumination, the light is absorbed in the top p-GaN layer resulting in electron injection into the depletion layer. On the other hand, in the case of sub-bandgap light illumination ($E_{\mathrm{g\_GaN}} > h\nu > E_{\mathrm{g\_InGaN}}$), the light is selectively absorbed in the InGaN layer, resulting in hole injection into the depletion layer. They successfully measured M_{n} and M_{p}. However, due to insertion of InGaN layer, which has large lattice mismatch to GaN, the crystalline quality of the GaN pn

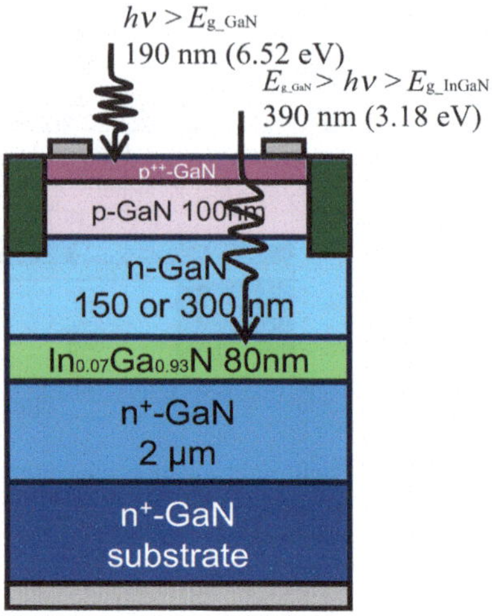

Fig. 24.11 GaN pn diodes with InGaN photoabsorption layer reported by Cao et al. [23]

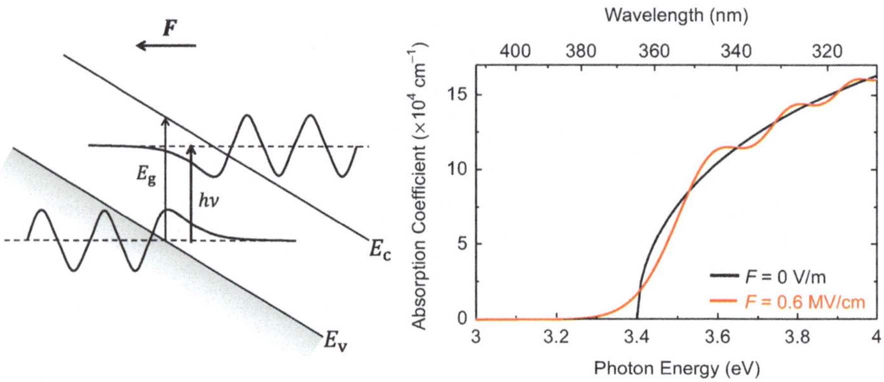

Fig. 24.12 (**a**) Schematic illustration of FK effect (below E_g absorption). (**b**) Calculated absorption coefficient of GaN for various electric field

junction degrades resulting in large leakage current (dark current). Because of this, usable data range was limited.

We developed another method. In our previous study, we observed sub-bandgap optical absorption at GaN with high electric field by the Franz-Keldysh (FK) effect [24, 25]. The theory of FK effect was developed in the 1960s [26, 27]. As shown in Fig. 24.12a, by applying high electric field, electron and hole wave functions penetrated into the forbidden gap just like Zener band-to-band tunneling. Thanks to overlapping of wave function, optical absorption below bandgap can be possible. Another feature of FK effect is the oscillation structure of refractive index and absorption coefficient above the bandgap due to interference of two oscillating shapes of wave functions, which is known as FK oscillation. The calculated absorption coefficient of GaN for various electric fields based on formulation developed by

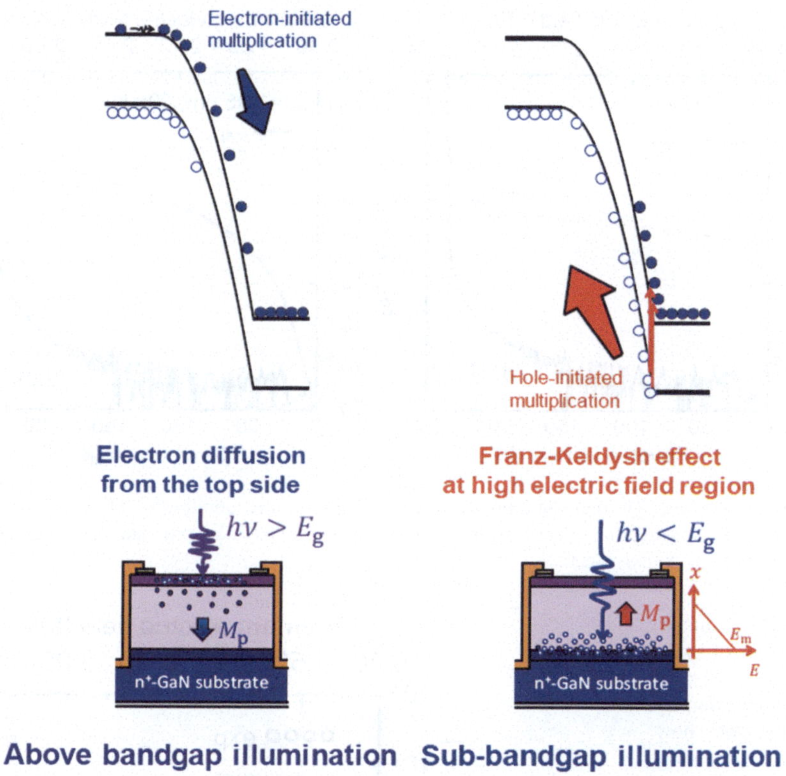

Fig. 24.13 Our approach: inverted pn junction utilizing sub-bandgap absorption due to the FK-effect

Aspnes [28] is shown in Fig. 24.12b [24]. The increase in below bandgap absorption as well as oscillation structure above bandgap are clearly seen.

We focus on the below bandgap absorption. If we fabricate a $p^+/p/n^+$ junction, then we can realize electron injection from top by above-bandgap and hole-injection from bottom by sub-bandgap illumination via the FK effect as shown in Fig. 24.13. Of course, this is not perfect backside hole injection; however, we confirmed that the spatial distribution of absorption area is almost negligible for analysis.

The measured results are shown in Fig. 24.14. The black lines show the dark current. The blue lines are current under illumination of (a) above bandgap (312 nm, 3.96 eV) and (b) below bandgap (405 nm, 3.06 eV). Red dashed lines are calculated photocurrent without considering avalanche multiplications. Details of the calculation are given in references [29]. The measured photocurrent is almost equal to the calculated photocurrent for voltage less than 150 V for above-bandgap illumination and voltage less than 50 V for sub-bandgap illumination, indicating that avalanche multiplication is negligible for these voltage ranges. On the other hand, the measured photocurrent is significantly larger than the calculated photocurrent for higher

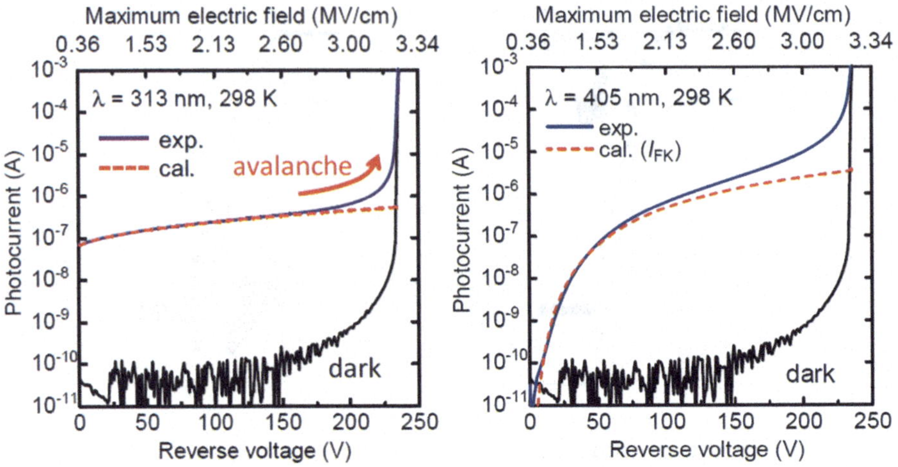

Fig. 24.14 Reverse *I-V* characteristics for dark and illuminated conditions. Dashed lines show calculated photocurrent due to absorption of top p^+-layer and GaN near p/n^+ junction (high electric field region)

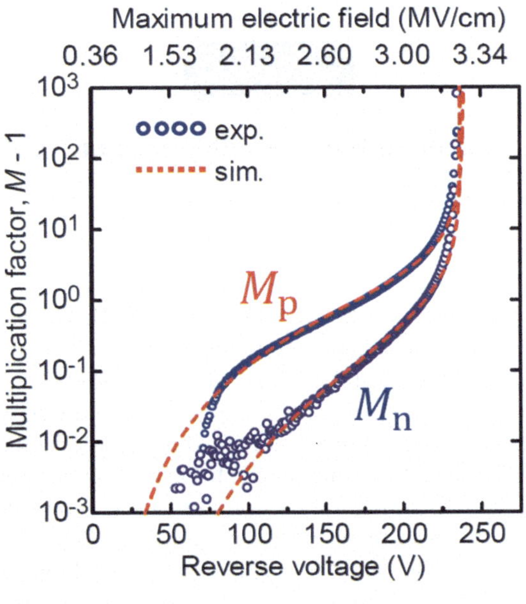

Fig. 24.15 Voltage dependence of avalanche multiplication factors for electron injection and hole injection. Dots show experimental results obtained from Fig. 24.13. Dashed lines are calculated by using obtained empirical expressions

voltage. Multiplication factors can be obtained by dividing measured photocurrent by calculated photocurrent (Fig. 24.15).

The experimentally determined $M_n - 1$ and $M_p - 1$ are shown in Fig. 24.4 by blue dots. We can determine these values for a very wide range from mid-10^{-2} to 10^3. By using formulae (24.12) and (24.13), the electric field dependences of impact ionization coefficients were calculated. The results are shown in Fig. 24.16. In

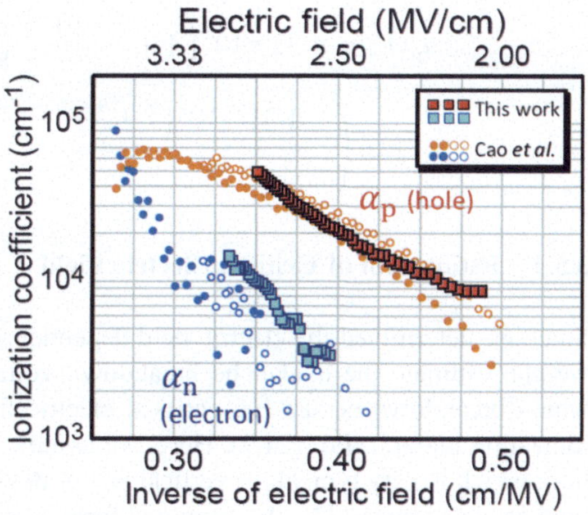

Fig. 24.16 Experimentally obtained impact ionization coefficients of electron and hole. Reported values by Cao et al. are also shown [Cao]

Fig. 24.16, Cao's coefficients are also shown. The impact ionization coefficient of hole is much larger than that of electron. Our result is very close to that reported by Cao. The measurement of $\alpha_n(F)$ is much difficult due to its small value. We successfully obtained the value with relatively small deviation thanks to precise measurements. By fitting these dependencies with Chynoweth's empirical expression, we obtained [30],

$$\alpha_n(F) = 2.69 \times 10^7 \exp\left(-\frac{2.27 \times 10^7 \mathrm{V/cm}}{F}\right) \mathrm{cm}^{-1}$$

$$\alpha_p(F) = 4.32 \times 10^6 \exp\left(-\frac{1.31 \times 10^7 \mathrm{V/cm}}{F}\right) \mathrm{cm}^{-1}$$

M_n and M_p calculated by using obtained expressions with formula (24.8) are shown in Fig. 24.15 by red dashed line. As clearly seen, the calculation well reproduces experimental results. For low multiplication region (M–1 less than 10^{-1}), there are errors. It may be due to measurement error and also insufficient data of impact ionization for electric field <2 MV/cm.

We also measured the temperature dependence of the impact ionization coefficient [30].

$$\alpha_n(F) = 2.69 \times 10^7 \left(1 + \frac{2 \times 10^{-3}(T - 298\ \mathrm{K})}{\mathrm{K}}\right)$$
$$\exp\left(-\frac{2.27 \times 10^7 \left(1 + \frac{5 \times 10^{-4}(T - 298\ \mathrm{K})}{\mathrm{K}}\right) \mathrm{V/cm}}{F}\right) \mathrm{cm}^{-1}$$

$$\alpha_{\mathrm{p}}(F) = 4.32 \times 10^{6}\left(1 + \frac{2 \times 10^{-3}(T - 298\ \mathrm{K})}{\mathrm{K}}\right)$$
$$\exp\left(-\frac{1.31 \times 10^{7}\left(1 + \frac{9 \times 10^{-4}(T - 298\ \mathrm{K})}{\mathrm{K}}\right)\mathrm{V/cm}}{F}\right)\mathrm{cm}^{-1}$$

3.5.1 Estimation of Critical Electric Field

Since we determined the electric field dependence of impact ionization coefficients, we can estimate the avalanche breakdown voltage as well as the electric field at which breakdown occurs (i.e., critical electric field) for any pn junction by using ionization integral, Eq. (24.9). Here, we assume simple p^{+}n one-sided abrupt junction, which is very typical for vertical power devices. The calculated critical electric field can be expressed by the empirical formula widely used in power device design, which was obtained as follows [30]:

$$F_{\mathrm{cr}}(N_{\mathrm{d}}) = \frac{2.40 \times 10^{6}}{1 - \frac{1}{4}\log_{10}\left(\frac{N_{\mathrm{d}}}{10^{16}\mathrm{cm}^{-3}}\right)}\ \mathrm{V/cm}$$

$$V_{\mathrm{b}}(N_{\mathrm{d}}) = \frac{(F_{\mathrm{cr}}(N_{\mathrm{d}}))^{2}}{2eN_{\mathrm{d}}}\epsilon_{\mathrm{GaN}}$$

Reported breakdown voltages and critical electric fields are shown in Fig. 24.17 and Fig. 24.18, respectively. The calculated lines by using obtained expressions are shown in red lines. The avalanche breakdown voltages of our pn diodes are very well reproduced by the expressions, suggesting very successful edge termination by bevel structures. Some reports such as one from Nagoya University and one from Avogy [31–40] are very close to the lines suggesting the excellence of their edge termination. Our impact ionization coefficients were only obtained for >2 MV/cm. To predict much lower doping devices, i.e., devices with breakdown voltage >2 kV, further measurements of impact ionization are necessary by using high-voltage pn diodes with much lower doping.

3.5.2 Unipolar Limit of GaN

Finally, here we mention the unipolar limit (drift layer resistance vs. blocking voltage) of GaN. To calculate the unipolar limit, we need electron mobility along the c-axis. Most of the study of electron mobility is within c-plane. The highest value is 1470 cm^{2}/Vs at room temperature by using ultra-high-purity GaN grown by quartz-free HVPE method [41]. There are some discussions of electron mobility

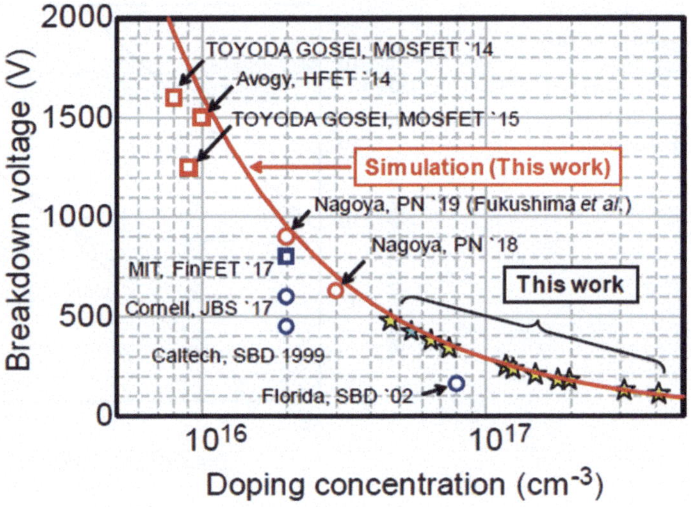

Fig. 24.17 Avalanche breakdown voltages of reported GaN vertical devices [31–40]

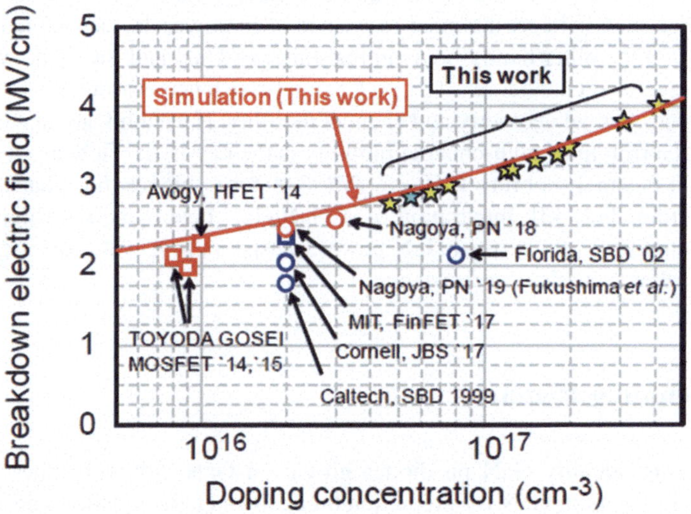

Fig. 24.18 Breakdown of the electric field of reported GaN vertical devices [31–40]

along the c-axis. Some researchers claimed that mobility along the c-axis is higher than within the c-plane because of effective mass anisotropy. However, here, we just use the same values for mobility along the c-axis. The result is shown in Fig. 24.19. The unipolar limit of GaN is much superior to that of Si. For comparison with SiC, since critical electric field is very similar to that of SiC, the on-resistance is not much less than SiC. That of GaN is about 25% smaller than that of SiC. If the mobility along the c-axis is smaller than that with the c-axis, we can expect further reduction.

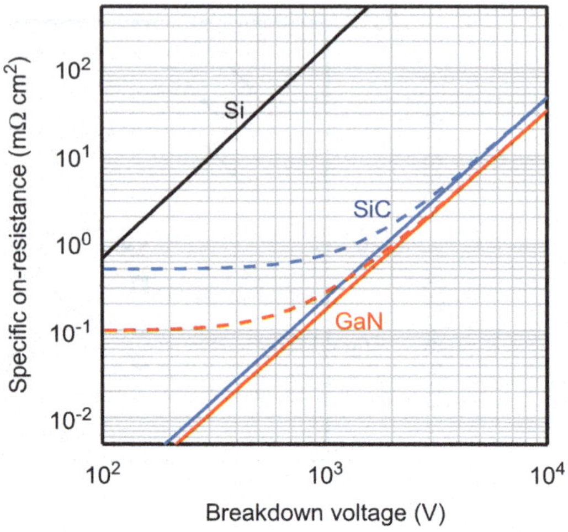

Fig. 24.19 Unipolar limit, drift layer resistance versus avalanche breakdown voltage of Si, SiC, and GaN. On-resistance of GaN is about 25% smaller than that of SiC

Let us think about power transistors such as JFETs and MOSFETs. Blocking voltage around 1–3 kV is the most important (having the largest market in power devices). For such voltage, the main resistance of Si power devices is the drift layer resistance. On the other hand, the main resistance of GaN and SiC is dominated by both drift layer and channel resistances. Channel resistance is around 0.1–1 mΩcm^2 depending on channel geometry and channel mobility. If we think about competition with SiC, realization of high-channel mobility is the key issue. GaN has a chance to realize high-channel mobility (250 cm^2/Vs, 3–5 times higher than that of SiC) in MOS structures [42, 43]. Integration of AlGaN/GaN heterojunction (high-electron mobility transistor) into vertical GaN power transistors [44] will have great advantages of further reduction on-resistance.

4 Concluding Remarks

Thanks to high-quality GaN pn diodes grown on GaN bulk substrates, many of material parameters of GaN for prediction breakdown characteristics were obtained. One of the most important missing parameters is electron drift mobility along the c-axis. Hall-effect measurements of high-quality low-doped nonpolar GaN are needed. For SiC, such an excellent study has already been reported [45]. In addition, for design of edge termination, impact ionization coefficients within the c-plane (along a-axis) are needed. Recently, GaN IMPATT diodes [46–48] and GaN APDs [49, 50] have been reported. Impact ionization coefficients are very important to design those devices.

Now, GaN is just standing at the start line of device development using the avalanche or Zener effect. Further investigation of those phenomena is strongly needed to make a base of the device development.

Acknowledgments The author deeply acknowledges all collaborators for the last 10 years, especially, Prof. Takuya Maeda at the University of Tokyo (former Ph. D. student of the author's group), Dr. Tetsuo Narita at Toyota Central Research and Development Laboratory, Prof. Masahiro Horita, and Prof. Kachi Tetsu at Nagoya University for their great contribution of this work. The author thanks Prof. Tsunenobu Kimoto at Kyoto University and Prof. Hiroshi Amano at Nagoya University for valuable discussions.

This work was supported by several Japanese government programs, the Council for Science, Technology and Innovation (CSTI), the Cross-ministerial Strategic Innovation Promotion Program (SIP), Next-Generation Power Electronics (funding agency: New Energy and Industrial Technology Development Organization (NEDO)), Ministry of Education, Culture, Sports, Science and Technology (MEXT) "Program for Research and Development of Next-Generation Semiconductor to Realize Energy-Saving Society," and MEXT "Innovative Power Electronics Technologies (INNOPEL) Program." Some results were carried out under fundamental research collaboration between Nagoya University and Toyota Motor group.

References

1. S.M. Sze, K.K. Ng, *Physics of Semiconductor Devices*, 3rd edn. (Wiley, New York)
2. E.O. Kane, J. Phys. Chem. Solids **12**, 181 (1959)
3. C. Benz, M. Claassen, J. Appl. Phys. **81**, 3181 (1997)
4. A. Pan, C.-O. Chui, J. Appl. Phys. **116**, 054508 (2014)
5. K.K. Thornber, J. Appl. Phys. **52**, 279 (1981)
6. A.G. Chynoweth, Phys. Rev. **109**, 1527 (1958)
7. B.J. Baliga, *Fundamentals of Power Semiconductor Devices*, 2nd edn. (Springer, Germany)
8. B.J. Baliga, J. Appl. Phys. **53**, 1759 (1982)
9. Y. Oshima, T. Eri, M. Shibata, H. Sunakawa, K. Kobayashi, T. Ichihashi, A. Usui, Jpn. J. Appl. Phys. **42**, L1 (2003)
10. Yoshida, M. Shibata, Jpn. J. Appl. Phys. **59**, 071007 (2020)
11. M. Bockowski, M. Iwinska, M. Amilusik, M. Fijalkowski, B. Lucznik, T. Sochacki, Semicond. Sci. Technol. **31**, 093002 (2016)
12. D. Key, E. Letts, C.-W. Tsou, M.-H. Ji, M. Bakhtiary-Noodeh, T. Detchprohm, S.-C. Shen, R. Dupuis, T. Hashimoto, Materials **12**, 1925 (2019)
13. K. Iso, H. Ikeda, T. Mochizuki, Y. Mikawa, S. Izumisawa, Appl. Phys. Express **13**, 085508 (2020)
14. T. Narita, N. Ikarashi, K. Tomoita, K. Kataoka, T. Kachi, J. Appl. Phys. 124, 165706 (2018)
15. T. Maeda, T. Narita, H. Ueda, M. Kanechika, T. Uesugi, T. Kachi, T. Kimoto, M. Horita, J. Suda, IEEE Elec. Device Lett. **40**, 941 (2019)
16. T. Shoji, T. Narita, Y. Nagasato, M. Kanechika, T. Kondo, T. Uesugi, K. Tomita, S. Ikeda, T. Mori, S. Yamaguchi, Y. Kimoto, J. Kojima, J. Suda, Appl. Phys. Exp. **14**, 114001 (2021)
17. T. Narita, Y. Nagasato, M. Kanechika, T. Kondo, T. Uesugi, K. Tomita, S. Ikeda, S. Yamaguchi, Y. Kimoto, M. Kosaki, T. Oka, J. Kojima, J. Suda, Appl. Phys. Lett. **118**, 253501 (2021)
18. A.S. Barker Jr., M. Ilegems, Phys. Rev. B **7**, 743 (1973)
19. G.D. Chen, M. Smith, J.Y. Lin, H.X. Jiang, S.H. Wei, M. Ashif Khan, C.J. Sun, Appl. Phys. Lett. **68**, 2784 (1996)
20. T. Maeda, T. Narita, H. Ueda, M. Kanechika, T. Uesugi, T. Kachi, T. Kimoto, M. Horita, J. Suda, IEEE Int. Electron Device Meeting **40**, 941 (2018)
21. G.E. Bulman, V.M. Robbins, G.E. Stillman, IEEE Trans. Electron Devices. **32**(11), 2454 (1985)
22. M.H. Woods, W.C. Johnson, M.A. Lampert, Solid State Electron. **13**, 583 (1970)

23. L. Cao, J. Wang, G. Harden, H. Ye, R. Stillwell, A.J. Hoffman, P. Fay, Appl. Phys. Lett. **112**, 262103 (2018)
24. T. Maeda, M. Okada, M. Ueno, Y. Yamamoto, M. Horita, J. Suda, Appl. Phys. Express **9**, 091002 (2016)
25. T. Maeda, T. Narita, M. Kanechika, T. Uesugi, T. Kachi, T. Kimoto, M. Horita, J. Suda, Appl. Phys. Lett. **112**, 252104 (2018)
26. V.W. Franz, Z. Natur forsch. **13(a)**, 484 (1958)
27. L.V. Keldysh, Sov. Phys. JET P. **7**, 788 (1958)
28. D.E. Aspnes, Phys. Rev. **147**, 554 (1966)
29. T. Maeda, T. Narita, H. Ueda, M. Kanechika, T. Uesugi, T. Kachi, T. Kimoto, M. Horita, J. Suda, Appl. Phys. Lett. **115**, 142101 (2019)
30. T. Maeda, T. Narita, S. Yamada, T. Kachi, T. Kimoto, M. Horita, J. Suda, J. Appl. Phys. **129**, 185702 (2021)
31. H. Fukushima, S. Usami, M. Ogura, Y. Ando, A. Tanaka, M. Deki, M. Kushimoto, S. Nitta, Y. Honda, H. Amano, Jpn. J. Appl. Phys. **58**, SCCD25 (2019)
32. Z.Z. Bandi'c, P.M. Bridger, E.C. Piquette, T.C. McGill, Appl. Phys. Lett. **74**, 1266 (1999)
33. J.W. Johnson, A.P. Zhang, W.-B. Luo, F. Ren, S.J. Pearton, S.S. Park, Y.J. Park, J.-I. Chyi, IEEE Trans. Electron Devices **49**(1), 32–36 (2002)
34. T. Maeda, T. Narita, H. Ueda, M. Kanechika, T. Uesugi, T. Kachi, T. Kimoto, M. Horita, J. Suda, IEEE Electron Device Lett. **40**(6), 941–944 (2019)
35. S. Usami, Y. Ando, A. Tanaka, K. Nagamatsu, M. Deki, M. Kushimoto, S. Nitta, Y. Honda, H. Amano, Y. Sugawara, Y.-Z. Yao, Y. Ishikawa, Appl. Phys. Lett. **112**, 182106 (2018)
36. T. Oka, Y. Ueno, T. Ina, K. Hasegawa, Appl. Phys. Express **7**, 021002 (2014)
37. T. Oka, T. Ina, Y. Ueno, J. Nishii, Appl. Phys. Express **8**, 054101 (2015)
38. W. Li, K. Nomoto, M. Pilla, M. Pan, X. Gao, D. Jena, H.G. Xing, IEEE Trans. Electron Devices **64**(4), 869–872 (2017)
39. H. Nie, Q. Diduck, B. Alvarez, A.P. Edwards, B.M. Kayes, M. Zhang, G. Ye, T. Prunty, D. Bour, I.C. Kizilyalli, IEEE Electron Device Lett. **35**(9), 939–941 (2014)
40. Y. Zhang, M. Sun, D. Piedra, J. Hu, Z. Liu, Y. Lin, X. Gao, K. Shepard, T. Palacios, IEDM Tech. Digest. 9.2 (2017)
41. S. Kaneki, T. Konno, T. Kimura, K. Kanegae, J. Suda, H. Fujikura, Appl. Phys. Lett. **124**, 012105 (2024)
42. M. Kanechika, K. Ito, T. Narita, K. Tomota, S. Iwasaki, D. Kikuta, T. Kachi, Phys. Stat. Sol. RRL **2024**, 240010 (2024)
43. K. Ikeyama, H. Tomita, S. Harada, T. Okawa, L. Liu, T. Kawaharamura, H. Miyake, Y. Nagasato, Appl. Phys. Express **17**, 064002 (2024)
44. T. Ueda, Jpn. J. Appl. Phys. **58**, SC0804 (2019)
45. T. Kimoto, Jpn. J. Appl. Phys. **58**, 018002 (2019)
46. D. Ji, B. Ercan, J. Zhuang, L. Gu, J. Riva-Davila, S. Chowdhury, Tech. Digest Device Res. Conf. (2020)
47. S. Kawasaki, Y. Ando, M. Deki, H. Watanabe, A. Tanaka, S. Nitta, Y. Honda, M. Arai, H. Amano, Appl. Phys. Express **14**, 046501 (2021)
48. S. Kawasaki, T. Kumabe, M. Deki, H. Watanabe, A. Takana, Y. Honda, M. Arai, H. Amano, IEEE Trans. Electron Devices **71**, 1408 (2024)
49. J.C. Carrano, D.J.H. Lambert, C.J. Eiting, C.J. Collins, T. Li, S. Wang, B. Yang, A.L. Beck, R.D. Dupuis, J.C. Campbell, Appl. Phys. Lett. **76**, 924 (2000)
50. M.H. Ji, J. Kim, T. Detchprohm, R.D. Dupuis, A.K. Sood, N.K. Dhar, J. Lewis, IEEE Photonics Tech. Lett. **28**, 2015 (2016)

Chapter 25
Theoretical Modeling of Avalanche Currents in GaN

P. D. Yoder

1 Introduction

This chapter begins in Sect. 2 below with an introduction to the theory of avalanche generation in the local approximation. A new and fully analytic model for particle densities and current densities is provided, yielding insight into the avalanche process in layers of finite thickness. The utility of the model is underscored by its dismissal of the notion that breakdown field strength is an intrinsic material property, contrary to conventional wisdom. A discussion of the inference of avalanche coefficients from measurement follows in Sect. 3, beginning with some historical context. The concept of nonlocal transport is introduced via the Boltzmann transport equation in Sect. 3.2, for which original solutions are provided in the low-field limit to study the behavior of near-equilibrium particles injected into a layer of finite thickness supporting both uniform and nonuniform field profiles. A characteristic length scale for the relaxation of the distribution function emerges, exposing limitations to the validity of the local approximation, for which the "dead space" represents a limiting case. Section 3.3. discusses a prominent semi-analytic theory of avalanche current accounting for "dead space" and identifies some shortcomings of the model that undermine its predictive power. The significance of ionization feedback is discussed for relating bulk ionization coefficients to device measurements in Sect. 3.4, and 3.5 introduces a new procedure which is the inference of ionization coefficients from measurements relevant for uniform field p-i-n structures.

P. D. Yoder (✉)
Georgia Institute of Technology, Atlanta, GA, USA
e-mail: doug.yoder@gatech.edu

I. C. Kizilyalli et al. (eds.), *Gallium Nitride and Related Materials*, The Materials Research Society Series, https://doi.org/10.1007/978-3-031-83056-3_25

2 The Local Theory of Impact Ionization

There is much qualitative insight to be gained from fully analytic models of avalanche generation which can be applied to the study of most if not all vertical GaN-based devices. While sacrificing a degree of predictive power, fully analytic models can be used immediately and efficiently, without the heavy time investments required for the programming of more rigorous physical models and the associated burden of computationally intensive simulations. Moreover, analytical models often serve as the theoretical foundation for the inference of avalanche coefficients from measurement. The starting point is the simplifying premise that the average probability for an electron or a hole to initiate an impact ionization event is a function of the local electric field only. *This is known as the local approximation and is strictly valid only for the case of homogeneous material.* The symbols α and β are referred to as impact ionization coefficients or, more simply, avalanche coefficients, In bulk material, α and β may be unambiguously defined as the field-dependent average probabilities per unit distance of net motion along the axis of an electric field that an electron or hole, respectively, will trigger an impact ionization event. Using this definition, α and β are necessarily non-negative. Moreover, they will generally depend on both the direction as well as the magnitude of the electric field, as is the case for wurtzite III-Nitride materials. It follows that the volumetric rates of electron- and hole-initiated impact ionization may be written in one dimension as $\alpha|J_n|$ and $\beta|J_p|$, where the symbols J_n and J_p represent the usual electron and hole current densities. Within a semi-classical framework, the continuity equations for electrons and holes derive from the first moment of the Boltzmann transport equation without approximation and may be written for one spatial dimension in the steady-state limit as

$$\begin{aligned} \frac{dJ_p}{dx} &= \alpha|J_n| + \beta|J_p| + qG \\ \frac{dJ_n}{dx} &= -\alpha|J_n| - \beta|J_p| - qG \end{aligned} \tag{25.1}$$

where the symbol G in this formulation represents the volumetric rate of net electron-hole pair generation due to mechanisms other than impact ionization. In the remainder of this chapter, the electric field will be assumed to point exclusively in the x-direction with $E_x(x)>0$ for all values of its argument, enabling the absolute value operators in Eq. (25.1) to be dropped without further loss of generality.

It is instructive to consider a segment of homogeneous material on the interval $0 < x < W$ with a large and uniform x-directed electric field, into which holes are injected from the left ($J_p(0) = J_{p0}$) and/or electrons are injected from the right ($J_n(W) = J_{n0}$). Limiting cases of this situation resemble the multiplication regions of separate absorption and multiplication avalanche photodiodes (SAM APDs) and the intrinsic regions of p-i-n structures (with the n-side on the left and the p-side on the right) which are ubiquitous in power switching and photodetection applications.

If G is taken to be constant on the interval, e.g., due to uniform optical illumination or perhaps defect-assisted tunneling, the solution to these coupled first-order ordinary differential equations with the aforementioned boundary conditions is:

$$J_p(x) = \frac{\alpha}{\alpha-\beta}\left(1-e^{-(\alpha-\beta)x}\right)\cdot J_{\mathrm{TOT}} + e^{-(\alpha-\beta)x}\cdot J_{p0} + \frac{q}{\alpha-\beta}\left(1-e^{-(\alpha-\beta)x}\right)\cdot G$$
$$J_n(x) = \frac{-\beta}{\alpha-\beta}\left(1-e^{(\alpha-\beta)(W-x)}\right)\cdot J_{\mathrm{TOT}} + e^{(\alpha-\beta)(W-x)}\cdot J_{n0} + \frac{q}{\alpha-\beta}\left(1-e^{(\alpha-\beta)(W-x)}\right)\cdot G \tag{25.2}$$

Elementary principles of electrodynamics dictate that the total current density $J_{\mathrm{TOT}} = J_p(x) + J_n(x)$ is independent of position in one spatial dimension in the steady-state limit, and Eq. (25.2) implies

$$J_{\mathrm{TOT}} = \frac{(\alpha-\beta)e^{(\alpha-\beta)\frac{W}{2}}J_{n0} + (\alpha-\beta)e^{-(\alpha-\beta)\frac{W}{2}}J_{p0} + 2\,q\,\sinh\left(\frac{(\alpha-\beta)W}{2}\right)G}{\alpha e^{-(\alpha-\beta)\frac{W}{2}} - \beta e^{(\alpha-\beta)\frac{W}{2}}} \tag{25.3}$$

One may immediately infer from Eq. (25.3) that the breakdown field strength, i.e., the value of the electric field for which $J_{\mathrm{TOT}} \rightarrow \infty$, is independent of J_{p0}, J_{n0}, and G, meaning that the phenomenon of avalanche breakdown makes no distinction between photogeneration, defect-assisted tunneling, or the details of charge injection. (Static breakdown occurs when the total steady state current density is not merely very large, but is infinite. This is true both in practice (usually preempted by current limiters) and in theory. The local theory of impact ionization predicts static breakdown when the denominator of Eq. (25.3) is zero.) It is also evident from Eq. (25.3) that breakdown can only occur in the presence of ionization feedback with both charge carrier species initiating impact ionization events; avalanche current can grow large but never reach breakdown if only one carrier species is involved. To see this, consider for example that as $\alpha \rightarrow 0$, $J_{TOT} \rightarrow J_{n0} + J_{p0}e^{\beta W} + q(e^{\beta W} - 1)G/\beta$. *llll* Although J_{TOT} can grow quite large, it never becomes mathematically infinite for any finite electric field strength. Breakdown requires ionization feedback. Lastly, we note that in the limit as the electric field strength and hence both α and β tend to zero separately, the expected result $J_{\mathrm{TOT}} \rightarrow J_{n0} + J_{p0} + qGW$ is recovered.

2.1 Electrostatic Self-Consistency

The preceding discussion assumes a constant electric field, independent of both the level of current injection and the amount of avalanche multiplication gain. Indeed, once either or both of these sources of excess charge reach a sufficient threshold,

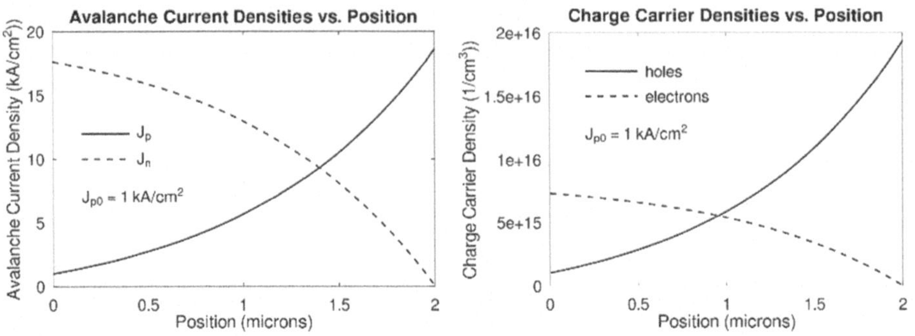

Fig. 25.1 Charge and current densities resulting from hole injection into a 2-micron layer of GaN supporting a 2.2-MV/cm electric field at 300 K. Free carrier densities are already appreciable at a current gain of only 18.6 and can significantly modify the static field closer to breakdown. Avalanche coefficients used for this calculation are taken from Maeda et al. [10]

significant changes to the electric field profile may be observed in practice, and the main assumption under which Eqs. (25.1), (25.2), and (25.3) were derived fails. In the low-field limit when only injection current is the culprit, this phenomenon is observed as the Kirk effect in homojunction and heterojunction bipolar transistors, where the injected charge redistributes the electric field within the collector-base junction, leading eventually to base push-out and a dramatic degradation of modulation bandwidth. Although less well-known, an optical Kirk effect may also be observed in uni-traveling carrier (UTC) photodiodes and separate absorption and multiplication avalanche photodiodes (SAM APD) with quasi-neutral absorbers. A means to establish the relationship between injection current, avalanche gain, and electric field redistribution is therefore necessary for assessing the electrostatic limits of the model presented above.

It is well known that the drift velocities of electrons and holes in bulk semiconductor material tend to saturate and exhibit a weaker dependence on field strength when the electric field is large [1, 2]. Using the common notation v_{nsat} and v_{psat} for electron and hole saturated drift velocities, the carrier densities may be inferred from Eq. (25.2) in a straightforward manner: $p(x) = J_p(x)/(q\ v_{\text{psat}})$, and $n(x) = J_n(x)/(q\ v_{\text{nsat}})$. For present purposes, we take the saturated drift velocity for holes in GaN [3] to be $6 \cdot 10^6$ cm/s, and that of electrons [4] to be $1.5 \cdot 10^7$ cm/s at $T = 300$ K. Figure 25.1 shows the current densities and corresponding carrier densities for the case of hole injection at the level of 1 kA/cm^2 into a high-field region of GaN with a relatively low multiplication factor far from breakdown. In this example, the electron density is dominant toward the cathode, and the larger hole density is dominant toward the anode, which will modify the electric field to exhibit a global maximum at the right boundary, a local maximum at the left boundary, and a global minimum at the point within the layer at which the two carrier densities are equal. If instead one carrier species is everywhere dominant, the minimum and maximum electric field

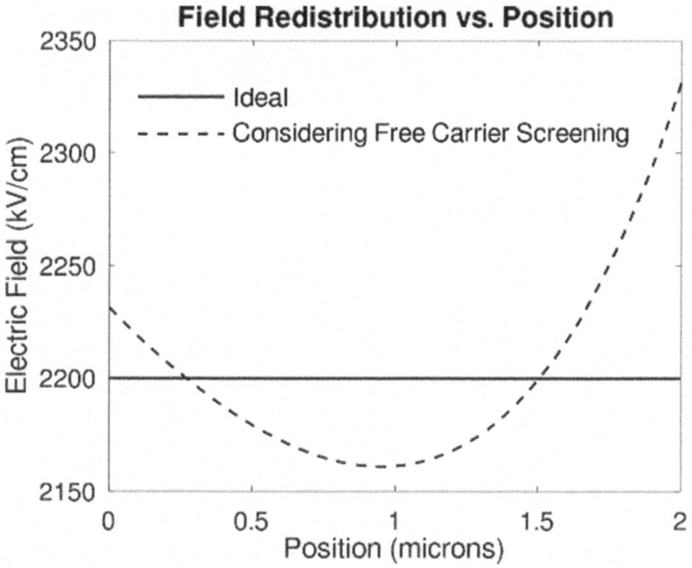

Fig. 25.2 First-order correction to the ideal constant field approximation due to free carrier screening for the case of a 1-kA/cm^2 hole injection current into the left boundary at $x = 0$. This effect is exacerbated by high injection current and large avalanche gain

strengths will appear at the boundaries. It remains to quantify the influence of the carrier densities on the electric field and injection current.

To make further analytical progress and solely for the purpose of quantifying perturbations to the electric field due to the presence of high densities of mobile charge, we assume that the high-field region is bounded on either side by heavily doped quasi-neutral regions with a step-like doping profile. The result which follows in this section is therefore applicable to punchthrough p-i-n structures but would need to be adapted for application to other devices such as SAM APDs. Under constant voltage conditions and without consideration of anode or cathode push-out, the redistribution of the electric field profile due to the presence of mobile charge carriers may be written as follows:

$$\delta E(x) \approx \frac{q}{\epsilon} \int_0^x (p(x') - n(x'))dx' - \frac{q}{\epsilon \cdot W} \int_0^W \int_0^{x'} (p(x'') - n(x''))dx''dx' \quad (25.4)$$

where $\delta E(x) = E(x) - E_0$, with E_0 representing the ideal constant field, and $E(x)$ symbolizing the electric field corrected for free carrier screening. The second term on the right-hand side of Eq. (25.4) affects a rigid shift in the position-independent contribution to the electric field due to constant voltage boundary conditions and will be nonzero whenever the total amounts of positive and negative free charge in the high-field region are unbalanced. Unbalanced charges within the high-field region

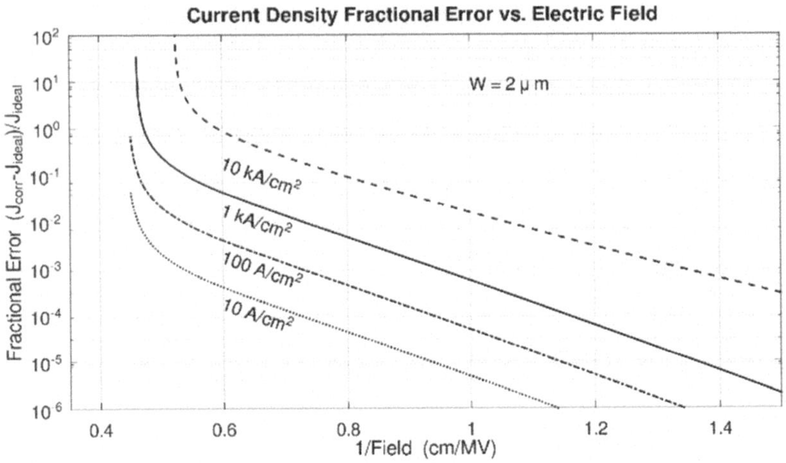

Fig. 25.3 Estimate of fractional error in avalanche current density incurred through neglect of electrostatic self-consistency. A 2-μm thick high-field region is considered here under conditions of various levels of hole current injection. The error becomes increasingly severe near breakdown, and at the highest levels of injection current, rendering non-self-consistent analytic and semi-analytic models unreliable

are screened from the adjacent quasi-neutral regions in p-i-n structures by small shifts in the depletion region boundaries ($\delta W \ll W$), exposing or concealing thin sheets of ionized dopants. It is to be emphasized that while Eq. (25.4) can be solved purely analytically based on Eqs. (25.1), (25.2), and (25.3), it is only a first-order approximation intended for assessing the validity of the constant field approximation upon which the theory presented in this section is based. Application of Eq. (25.4) to the region considered for Fig. 25.1 using free carrier densities calculated under the assumption of a constant field strength of 2.2 MV/cm and a modest injection current of 1 kA/cm^2 reveals appreciable deviation from the ideal flat field profile, as shown in Fig. 25.2. From the foregoing discussion, it is evident that Eqs. (25.1), (25.2), and (25.3) can be rendered invalid should the injection current and/or the avalanche gain become sufficiently large.

A purely analytical and conservative criterion for the validity of the constant field approximation is easily established. Let $E_m = \max \{E_0 + \delta E(x) : 0 < x < W\}$ be the peak field strength within the high field region after correcting for screening by free carriers to first order, and ΔJ be the acceptable error in current density. The electrostatic assumptions of the local theory presented above may then be invalidated under the condition $|\ J_{\text{TOT}}(E_m) - J_{\text{TOT}}(E_0)| > \Delta J$. Figure 25.3 shows the fractional error in avalanche current, $(\ J_{\text{TOT}}(E_m) - J_{\text{TOT}}(E_0))/J_{\text{TOT}}(E_0)$, for the structure considered in Figs. 25.1 and 25.2 at various levels of hole current injection.

It is important to point out that Eq. (25.4) is specific to constant voltage conditions. Although this is relevant for much device-level testing and characterization (including the experimental inference of avalanche coefficients), Eq. (25.4) requires modification when contact and access resistances are significant and for many

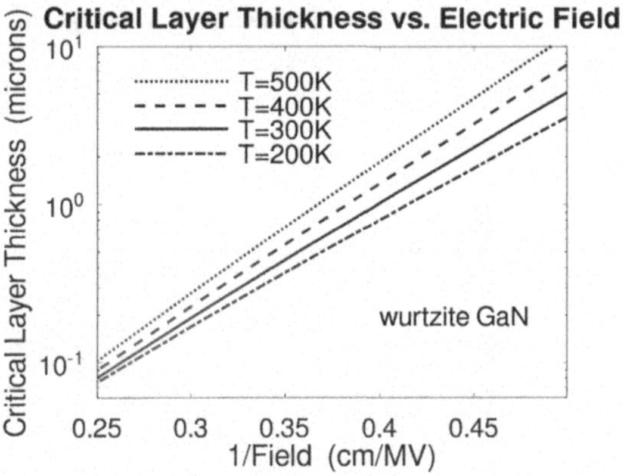

Fig. 25.4 Breakdown field strength is a function of the thickness of the region over which the electric field extends and is not an intrinsic material property. The local theory used here results in an underestimate with respect to predictions which could be made with nonlocal models

circuit-level applications such as those for which the avalanche current drives a series-connected load such as a trans-impedance amplifier, rendering the device bias (and hence the internal field profile) dependent on avalanche current.

2.2 *Breakdown Field Strength*

It has been written that bulk semiconductor materials have a characteristic critical field strength, E_{crit}, above which dielectric breakdown occurs [5–7]. *Rigorously speaking, this is untrue, and E_{crit} is not a fundamental material property.* To see this, we notice from Eq. (25.3) that for *any* fixed field strength, there exists a range of layer thickness $0 < W < W_{\text{crit}}$ for which J_{TOT} is finite, but observe that when $W \rightarrow W_{\text{crit}} = \frac{\ln(\alpha/\beta)}{\alpha - \beta}$, dielectric breakdown occurs. Stated differently, the critical electric field strength at which dielectric breakdown occurs will depend on the thickness of the region over which an electric field extends.

Parameterizations of the bulk ionization coefficients in GaN have recently been inferred from photomultiplication experiments [8–10]. The temperature-dependent avalanche coefficients of Maeda et al. [10] and Eq. (25.3) together imply the relationship between breakdown field strength and critical layer thickness, W_{crit}, depicted in Fig. 25.4. As the critical layer thicknesses are on the same length scale as the high-field intrinsic regions of many contemporary devices, this qualitative insight has direct relevance to device design. In the context of diode structures, the field is understood to include both the applied field and the built-in field of the junction itself, including polarization-induced fields. Finally, we note that the increase in critical layer thickness with temperature is, at a more microscopic level, a direct consequence of the suppression of the high energy tail of the electron and hole distribution functions by carrier-phonon scattering.

Figure 25.5 depicts the relationship between breakdown voltage and layer thickness and reveals what appears to be a nearly linear dependence. Indeed, if

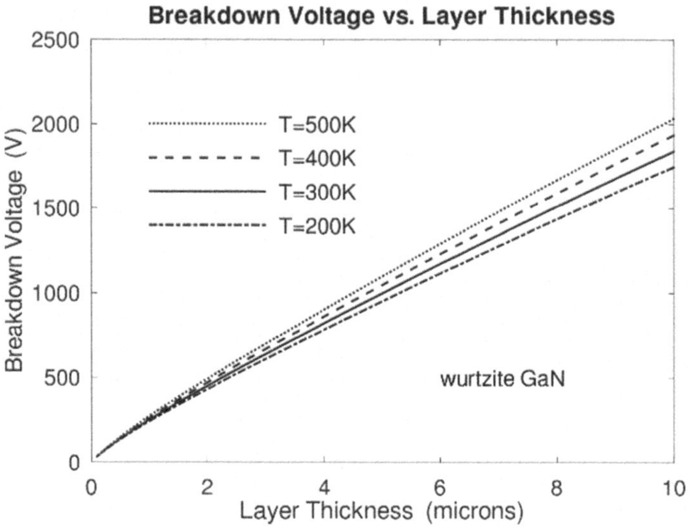

Fig. 25.5 Avalanche-limited breakdown voltage as predicted by the local theory for wurtzite GaN, assuming a built-in junction potential of 3 V. Although the curves appear nearly linear, the breakdown fields implied span the range between 4 MV/cm (for thin layers) and 1.8 MV/cm (at 10 microns). In the limit as layer thickness tends toward infinity, the breakdown field tends toward zero

experimental measurements of breakdown voltage in p-i-n structures are performed over a limited range of layer thicknesses, the empirical inference of a constant breakdown field strength may appear quite reasonable. Yet between 0.1 and 10 microns, the breakdown field implied in Fig. 25.5 varies by over a factor of two.

2.3 *Inhomogeneous Photogeneration*

In many cases of practical interest, the pair generation rate for mechanisms other than impact ionization, G, is nonuniform, and it is instructive to seek solutions to Eq. (25.1) for which $G(x) = G_0\delta(x - x^{'})$, with $0 < x^{'} < W$. Optical illumination of a planar avalanche photodiode, for which the exponential photon absorption profile penetrates into the high-field multiplication region, is one such common example. The symbol G_0 may be interpreted as the electron-hole pair generation rate per unit area in the plane $x = x^{'}$, and we note that the special cases $x^{'} = 0$ and $x^{'} = W$ are equivalent to charge injection from boundaries, involving J_{p0} and J_{n0}, with appropriate scaling. Neglecting for the moment J_{p0} and J_{n0} explicitly, the solutions to Eq. (25.1) with $G(x) = G_0\delta(x - x^{'})$ are as follows:

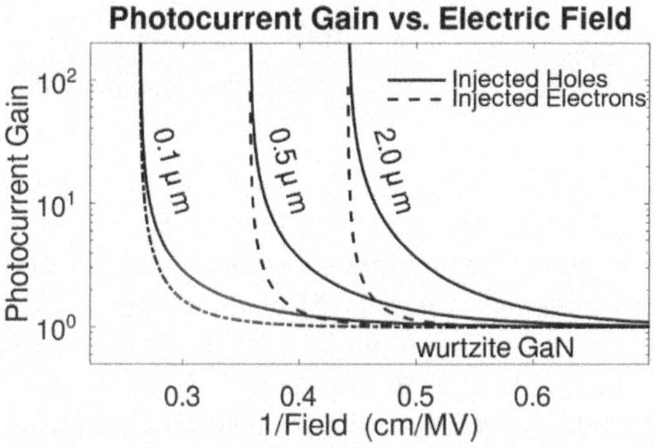

Fig. 25.6 Photocurrent gain as a function of field strength in layers of thickness 0.1, 0.5, and 2.0 μm for the cases of pure hole injection and pure electron injection, calculated according to Eq. (25.8). Although greater gain may be achieved by injecting photogenerated holes rather than electrons into a high field multiplication region in GaN, technological challenges impede the growth of high-quality normal-incidence planar APDs by MOCVD with the p-type quasi-neutral region on the bottom

$$\left.\begin{aligned} J_p(x,x') &= \frac{\alpha}{\alpha-\beta}\left(1-e^{-(\alpha-\beta)x}\right)\cdot J_{T0} \\ J_n(x,x') &= \frac{-\beta}{\alpha-\beta}\left(1-e^{(\alpha-\beta)(W-x)}\right)\cdot J_{T0}+q\cdot e^{(\alpha-\beta)(x'-x)}\cdot G_0 \end{aligned}\right\} ,0<x<x' \quad (25.5)$$

and

$$\left.\begin{aligned} J_p(x,x') &= \frac{\alpha}{\alpha-\beta}\left(1-e^{-(\alpha-\beta)x}\right)\cdot J_{T0}-q\cdot e^{-(\alpha-\beta)(x-x')}\cdot G_0 \\ J_n(x,x') &= \frac{-\beta}{\alpha-\beta}\left(1-e^{(\alpha-\beta)(W-x)}\right)\cdot J_{T0} \end{aligned}\right\} ,x'<x<W \quad (25.6)$$

with

$$\frac{J_{T0}}{q\cdot G_0}=M(x')=\frac{\alpha-\beta}{\alpha\, e^{-(\alpha-\beta)x'}-\beta e^{(\alpha-\beta)(W-x')}} \quad (25.7)$$

where $M(x')$ is convenient to interpret as the multiplication factor, i.e., the expectation of the number of electron-hole pairs created by the absorption of a photon at position $x = x'$, including both the initial photogeneration event and any subsequent avalanche multiplication. For an inhomogeneous non-avalanche electron-hole pair generation profile $G_0(x')$ in the possible presence of current injection into both boundaries, it follows from Eqs. (25.3) and (25.7) that the total current density may be written as follows:

$$J_{\text{TOT}} = \int_0^W q\, M(x')\, G_0(x')\, dx' + \frac{(\alpha-\beta)e^{(\alpha-\beta)\frac{W}{2}}J_{n0} + (\alpha-\beta)e^{-(\alpha-\beta)\frac{W}{2}}J_{p0}}{\alpha e^{-(\alpha-\beta)\frac{W}{2}} - \beta e^{(\alpha-\beta)\frac{W}{2}}} \quad (25.8)$$

Fully analytic expressions for $J_n(x)$ and $J_p(x)$ explicitly considering charge injection at the boundaries also follow likewise from Eqs. (25.2), (25.5), (25.6), and (25.7). Note that this expression for the multiplication gain in Eq. (25.7) is equivalent to the semi-analytic formulation found in Eq. (5) of the pioneering paper on avalanche noise by McIntyre [11] in the limit of uniform field strength. Eq. (25.7) may also be used to assess the benefit of injecting holes vs. electrons into the multiplication region of an APD. Figure 25.6 shows the photocurrent gain of a GaN-based SAM APD for three different multiplication region thicknesses. This figure once again confirms that the breakdown field depends on layer thickness and is independent of the specific carrier species injected into the multiplication region. Most importantly, however, it quantifies the higher gain which can be achieved over a wide range of field strengths by injecting the carrier species whose avalanche coefficient is the largest.

3 Inference of Avalanche Coefficients from Measurement

3.1 Background

The extraction of electron and hole ionization coefficients from diode measurements is neither a new nor trivial exercise. The single greatest complicating factor is that any measurement of avalanche current necessarily includes contributions from both electron- and hole-initiated impact ionization events, and they cannot be decoupled except in the limit of extremely low field strength when avalanche multiplication factors and ionization coefficients are both small, with one coefficient still much smaller than the other. Under these conditions, measurements are prone to the largest errors. This issue is compounded in avalanche studies involving step- or linearly graded p-n junction diodes, where the high-field region involves a (piecewise) linear or parabolic field profile. In such cases, electrons and holes sample a broad range of field strengths throughout the high-field region, rendering the field strength at which the avalanche coefficients are to be inferred ambiguous. Unsurprisingly, many of the earliest attempts to understand the field dependence of impact ionization coefficients from the characterization of p-n diodes [12–16] imposed artificial assumptions that have not stood the test of time.

Even on the same wafer, different junctions exhibit slightly different electrical characteristics. This can be due to factors such as fluctuations in doping, layer thickness, defect density, and contact resistance. Because of this, and the fact that avalanche currents necessarily include contributions from both α and β, early investigations focused on measuring avalanche current twice for each device at each bias level, once with pure hole injection into the high-field region and a second time with pure electron injection. In this way, the measured multiplication factors M_n and M_p could be used to uniquely determine the ionization coefficients. Using the local approximation, integral formulas for the multiplication factors involving the ionization coefficients were first elucidated by Miller [13], accounting for nonuniform field profiles. Assuming a linear field profile, Miller [13] derived approximate expressions for both ionization coefficients in the form of a rapidly converging series expansion, given measurements of M_n and M_p. A similar approach was taken by Lee et al., who instead arrived at approximate closed-form solutions [17]. Exact analytic expressions for the avalanche coefficients in semi-infinite non-punchthrough p^+n and n^+p junctions characterized by linear field profiles were found by Woods et al. [18], with the drawback of requiring differentiation of measured multiplication factors with respect to depletion region width. This was the approach adopted by Maeda et al. [10] in their recent investigation of ionization coefficients in vertical GaN p-i-n structures.

3.2 Nonlocal Charge Transport: Dead Space and Beyond

As previously mentioned, the avalanche current measured using p^+n and n^+p junction diodes will contain contributions from electrons and holes drifting through a broad range of field strengths in the linear field profile. To untangle this situation and associate α and β with unique field strength, Miller, Lee et al., and Woods et al. each used a common differentiation strategy so that the values of α and β extracted at each bias point could be associated unambiguously with the peak field strength at the metallurgical junction. Although mathematically rigorous within the local approximation, this pushes the validity of the local approximation to the limit, as will be shown in Sect. 3.2.2 below. To understand this point clearly, it is necessary to explore the nonlocal theory of charge transport.

The Boltzmann transport equation (BTE) is a seven-dimensional spatio-temporal integro-differential equation of motion for the semiclassical distribution function [1]. To gain insight into nonlocal charge transport in linearly varying electric field profiles such as those found in p^+n and n^+p junction diodes, one may write the time-independent BTE for holes in the relaxation time approximation with a small electric field oriented along the x-direction as follows:

$$\frac{q}{\hbar}E(x)\frac{\partial f\left(x,\vec{k}\right)}{\partial k_x}+\frac{\hbar}{m^*}k_x\frac{\partial f\left(x,\vec{k}\right)}{\partial x}=-\frac{f\left(x,\vec{k}\right)-f_0\left(\vec{k}\right)}{\tau} \quad (25.9)$$

where $f_0\left(\vec{k}\right)$ represents the equilibrium distribution function, and implicit use will be made of the single isotropic parabolic band approximation in this section. Although these assumptions are inappropriate for studying impact ionization, they are ideal for investigating the origin and implications of nonlocal charge transport. For present purposes, $f_0\left(\vec{k}\right)$ will be taken to be of the Maxwell-Boltzmann form, i.e., $f_0\left(\vec{k}\right) = e^{-\left(\frac{\hbar^2}{2m^*}\left|\vec{k}\right|^2 - E_F\right)/k_B T}$. In the limiting case of bulk material with constant field strength $E(x) = E_0$, Eq. (25.9) has the following first-order solution in the low field limit: $f\left(x, \vec{k}\right) \approx \left(1 + \frac{qE_0}{k_B T}\hbar k_x \tau / m^*\right) f_0\left(\vec{k}\right)$. The first moment of this distribution function yields the bulk drift velocity $\hbar \langle k_x \rangle \tau / m^* = (q\tau/m^*)E_0$.

Consider now instead a region of finite thickness, $0 < x < W$, with a linear electric field profile $E(x) = E_0 + \Delta E \frac{x}{W}$. Choosing $\Delta E = -E_0$ corresponds to a linearly decreasing field profile such as within the depletion region of an n$^+$p diode. The first-order solution to Eq. (25.9) has the following suggestive form:

$$
\begin{aligned}
k_x > 0 : f\left(x, \vec{k}\right) &\approx f_0\left(\vec{k}\right) + e^{-\frac{x}{\lambda(k_x)}} f_1\left(x=0, \vec{k}\right) \\
&\quad + \frac{q}{k_B T}\lambda(k_x)E_0\left(1 - e^{-\frac{x}{\lambda(k_x)}}\right) f_0\left(\vec{k}\right) \\
&\quad + \frac{q}{k_B T}\lambda^2(k_x)\frac{\Delta E}{W}\left(\frac{x}{\lambda(k_x)} - \left(1 - e^{-\frac{x}{\lambda(k_x)}}\right)\right) f_0\left(\vec{k}\right) \\
k_x > 0 : f\left(x, \vec{k}\right) &\approx f_0\left(\vec{k}\right) + e^{-\frac{x}{\lambda(k_x)}} f_1\left(x=0, \vec{k}\right) \\
&\quad + \frac{q}{k_B T}\lambda(k_x)E_0\left(1 - e^{-\frac{x}{\lambda(k_x)}}\right) f_0\left(\vec{k}\right) \\
&\quad + \frac{q}{k_B T}\lambda^2(k_x)\frac{\Delta E}{W}\left(\frac{x}{\lambda(k_x)} - \left(1 - e^{-\frac{x}{\lambda(k_x)}}\right)\right) f_0\left(\vec{k}\right)
\end{aligned}
\tag{25.10}
$$

where the boundary conditions $f_1\left(x=0, \vec{k}\right)$ and $f_1\left(x=W, \vec{k}\right)$ are the first-order corrections to the distribution function at the points of forward and backward hole injection, respectively. The function $\lambda(k_x) = \hbar|k_x|\tau/m^*$ may be interpreted as a wavevector-dependent characteristic length scale for relaxation of the distribution function $f\left(x, \vec{k}\right)$ along the axis of the electric field. Indeed, the quantity $\frac{\hbar k_x \tau}{m^*}$ is equivalent to the product of a hole's group velocity and momentum relaxation time, $v_g \tau$. This is not quite the same as the mean free path, which derives from the thermal velocity and reflects an ensemble average.

3.2.1 Hole Injection into a Uniform Field

In a uniform field $E(x) = E_0$, Eq. (25.10) reveals that although $f\left(x, \vec{k}\right)$ varies exponentially with a position close to the plane of forward injection, $f\left(x, \vec{k}\right)$ tends toward the familiar bulk result after a few characteristic relaxation lengths, i.e., $f\left(x > 3\hbar k_x \tau/m^*, \vec{k}\right) \approx f_0\left(\vec{k}\right)\left(1 + \frac{qE_0}{k_B T}\hbar k_x \tau/m^*\right)$. Over the region

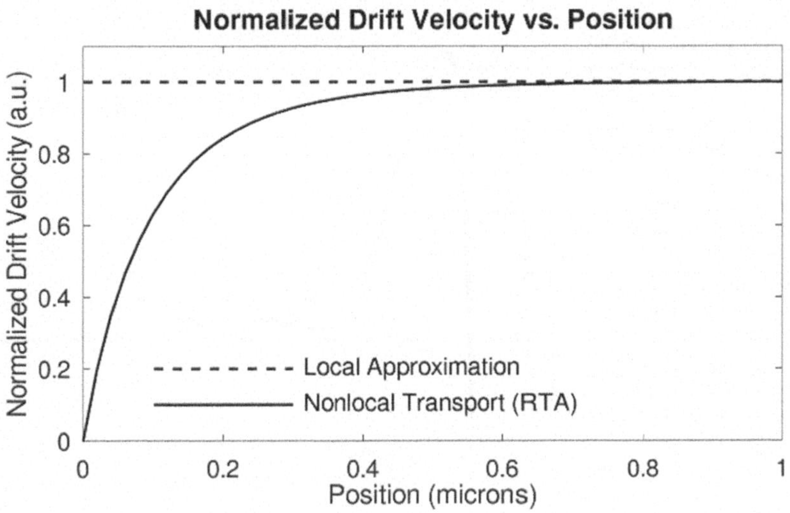

Fig. 25.7 Holes injected from a near-equilibrium distribution into a thick region of constant field exhibit a drift velocity profile that differs substantially near the point of injection from that which is predicted by local theories of charge transport. Drift velocity approaches the bulk value asymptotically, with a characteristic wave vector-dependent length scale $\lambda(k_x) = \hbar |k_x| \tau / m^*$, for which the figure shown reflects an ensemble average. $m^* = 1$ and $\tau = 1$ ps for illustration

$0 < x < 3\,\lambda(k_x), f\left(x, \vec{k}\right)$ transitions from a nearly equilibrium distribution function at $x = 0$ to a nonequilibrium distribution function consistent with a finite field in the bulk material. Stated differently, holes must travel a distance of roughly 3 $\lambda(k_x)$ from the plane of injection at $x = 0$ through a region of constant field before the distribution function $f\left(x, \vec{k}\right)$ is nearly a function of the local electric field alone for holes of wave vector k_x and most of the "memory" of the point of injection has been erased. In this and similar situations, such as the injection of near-equilibrium photogenerated charge into a high-field multiplication region, we say that the charge transport is nonlocal. Evaluating the first moment of Eq. (25.10) for the case of holes injected from an equilibrium distribution into a uniform field profile, the position-dependent average group velocity has been evaluated and depicted in Fig. 25.7 below.

An important observation is that until particles have traveled a few characteristic relaxation lengths from the point of injection, no physical observables (e.g., particle density, velocity, and avalanche coefficients) are well predicted by the local field alone, and the local approximation lacks validity. The so-called "dead space," roughly the minimum distance required for a charged particle to move in a region of high field before it can gain sufficient energy to initiate an impact ionization event, is a special case of this more general "relaxation space." This problem is likely exacerbated by the well-known phenomenon of "velocity overshoot," illustrated for electrons in wurtzite GaN in Fig. 25.8 below. The overshoot is even more

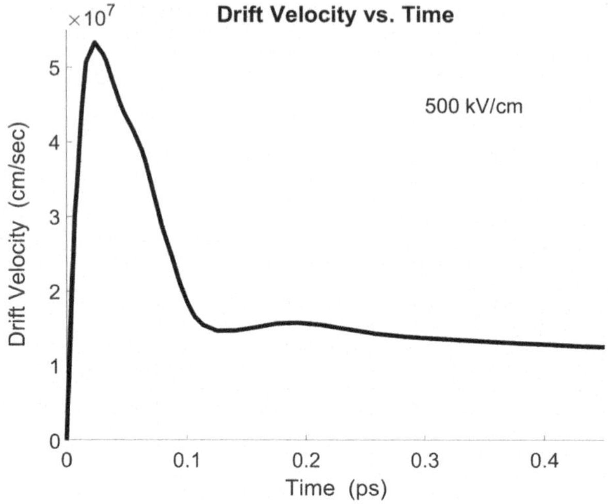

Fig. 25.8 Average velocity of electrons in bulk wurtzite GaN abruptly subjected to a constant electric field along the $\Gamma \rightarrow A$ direction at time $t = 0$ [19]. Larger fields lead to higher "velocity overshoot," and longer transients before a steady state is reached. This situation shown in this figure is equivalent to the injection of equilibrium electrons into a region of a high constant field. For fields above 500 kV/cm, the "relaxation space" can exceed 100 nm for electrons

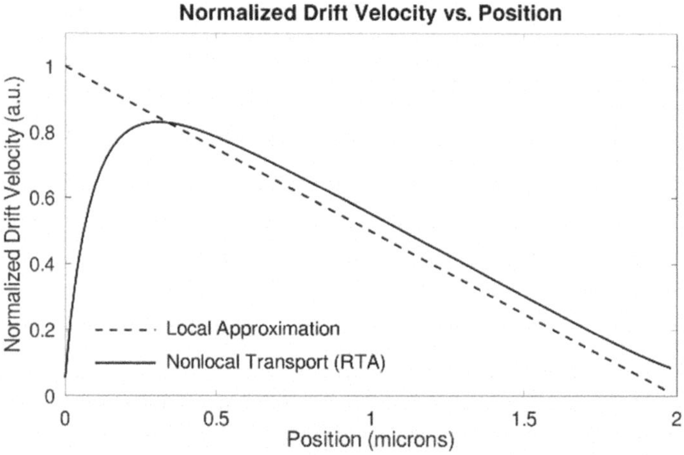

Fig. 25.9 Drift velocity of holes injected from a near-equilibrium distribution into a thick region of a linear field profile, such as found in an n^+p structure. Due to the nonlocal nature of charge transport, holes never quite reach a steady state consistent with the local electric field

pronounced than depicted below at field strengths for which impact ionization is significant. In general, the characteristic relaxation length of high-energy particles is shorter than that of near-equilibrium particles because the relaxation times of high-energy particles are shorter.

3.2.2 Hole Injection into Linear Field Profile

The formula derived by Woods et al. [18] for extracting avalanche coefficients from the differentiation of measured multiplication gains M_n and M_p is exact in the local approximation and has understandably become the method of choice for the extraction of ionization coefficients using n^+p and p^+n non-punchthrough structures. As shown in the previous section, the Boltzmann transport equation implies a characteristic length scale that particles injected into a constant field region must travel before the properties of the ensemble may be approximated by functions of the local field alone, and the local approximation achieves greater validity.

Strictly speaking, however, the local approximation is never valid for the case of nonuniform electric fields, including the linear field profiles within the depletion regions of n^+p and p^+n structures. Figure 25.9 below shows the first moment of Eq. (25.10) evaluated for the case of a linear field profile with the same peak field E_0 as considered in Fig. 25.7 above, but for which $\Delta E = -E_0 \neq 0$, approximating the circumstances of photogenerated holes injected into the depletion region of an n^+p junction from the n^+ quasi-neutral region. Except in the immediate vicinity of the crossover, the local approximation predicts a significantly different drift velocity profile than does the nonlocal formulation of the BTE. The discrepancies increase the steeper the slope of the field, i.e., for higher doping levels in p-n junction diodes, and decrease monotonically as the slope of the field tends toward zero and the result of Fig. 25.7 is recovered. If the relaxation of charge carriers in Fig. 25.7 can be likened to a skater pushed ungracefully onto the ice who eventually manages to recover his balance, the situation depicted in Fig. 25.9 would be like a skater who is never able to recover his balance no matter the available distance.

Figure 25.9 demonstrates that the deficiencies of the local approximation for drift velocity are greatest within the "relaxation space" adjacent to the point of injection. It is likewise true that the deficiencies of the local approximation for the volumetric rates of avalanche generation are greatest in the "dead space" adjacent to the point of injection, which is precisely the location at which ionization coefficients are assigned to specific electric fields in the model of Woods et al. [18]. For this reason, ionization coefficients inferred using this model will always be underestimated, and a simple remedy is not obvious due to the breakdown of the local approximation throughout the entire high-field region.

3.3 *Semi-Analytic Modeling of Nonlocal Impact Ionization*

The question arises as to whether it is possible to formulate a predictive semi-analytic model that accounts for nonlocal high field charge transport without resorting to the fully numerical Monte Carlo solutions of the BTE. An excellent discussion of the role of nonlocal impact ionization may be found in a more recent article by McIntyre [20] based in part on some of the prior work of Hayat et al.

[21]. McIntyre addresses the problem of nonlocal avalanche generation by introducing conditional probability density functions $p_e(x, x')$ and $p_h(x, x')$ for electron- and hole-initiated impact ionization events within a high-field region at points x by electron-hole pairs generated at points x'. Appropriate integrations over x' yield volumetric avalanche generation rates, which in turn may be integrated to calculate diode current. These conditional probability density functions must, however, be calculated based on prior knowledge of nonlocal impact ionization coefficients, which will be different for every conceivable field profile but could in principle be pre-calculated by Monte Carlo simulation for each material and a representative subset of possible field profiles. To the author's knowledge, no such library of nonlocal ionization coefficients exists for any material at the time of this writing. Although McIntyre does provide a semi-analytic method for computing nonlocal avalanche coefficients, it is based on several reasonable yet ad hoc assumptions and introduces additional parameters including *local* field-dependent electron and hole correlation lengths, which themselves must be treated as adjustable parameters if not calculated by Monte Carlo simulation.

Parameterizations of McIntyre's model have been found which correlate well with Monte Carlo calculations of nonlocal avalanche coefficients as well as to measurements of avalanche gain and noise spectral density in APDs [22]. Yet the reliance on adjustable parameters undermines predictive power, and the model itself has yet to be applied to the inference of bulk ionization coefficients from measurement.

Finally, there are two subtle weaknesses of the McIntyre model which deserve mention. First, the model applies the same conditional probability density functions $p_e(x'|x)$ and $p_h(x'|x)$ to all mechanisms of electron-hole pair generation even though the energetic distribution of electrons and holes created through the processes of thermal generation, photogeneration, and especially impact ionization can be quite different. The instant after each impact ionization event, the primary particle and the two secondary particles will share the excess energy that the primary particle had before the collision. For this reason, electron-hole pairs created by impact ionization are more likely to initiate subsequent impact ionization events than electron-hole pairs which are photogenerated or thermally generated with energies much closer to the band edge.

The second and more severe shortcoming of the McIntyre model involves an inconsistency between the definition of pair generation rates and the use made of them. Cold particle generation rates are correctly introduced for electrons and holes as $g_e(x) = \int_x^W [2g_e(x') + g_h(x') + g_0(x')]p_e(x'|x)dx'$ and $g_h(x) = \int_0^x [g_e(x') + 2g_h(x') + g_0(x')]p_e(x'|x)dx'$, respectively, but then used in the continuity equations and subsequent expressions for avalanche gain. This amounts to improperly counting the primary particle in the net avalanche generate rate when in fact only secondary particles should be counted.

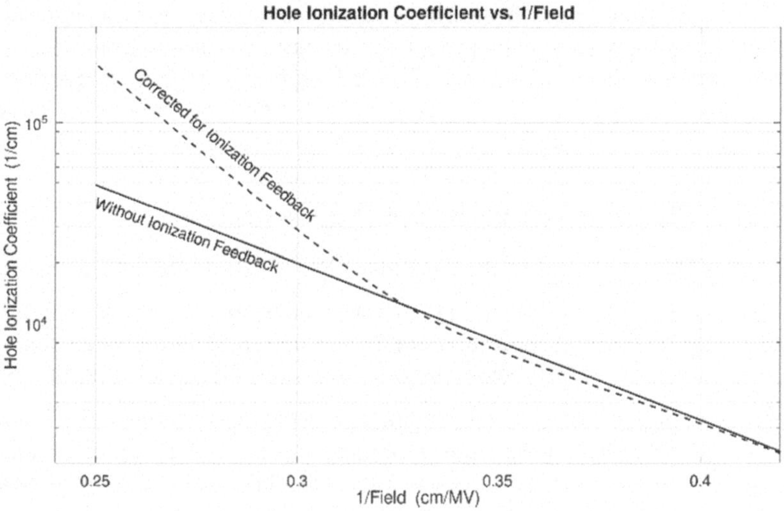

Fig. 25.10 Solid line: hole ionization coefficient extracted from measurement without consideration of ionization feedback [23]. Dashed line: hole ionization coefficient corrected for ionization feedback using α presented in the same study

3.4 The Significance of Impact Ionization Feedback

All attempts to relate bulk ionization coefficients to device measurements will ultimately encounter a common issue. Consider that bulk ionization coefficients make no distinction between particles that were generated at position x and particles that instead drifted to position x from any and all points upstream. Particles *drifting* to position x from points upstream will on average have considerably higher energy; unlike the particles *generated* at x, these particles will experience little or no dead space. This is not a problem in bulk material, where the ionization coefficients for electrons and holes are defined as a weighted average over the position-independent electron and hole distribution functions, respectively, and depend on the electric field alone. In a layer of finite thickness, however, and even when the electric field is uniform, *ionization feedback* ensures that the ratio of "hot carriers" to newly generated carriers of the same species is a highly nontrivial function of position (see Fig. 25.1) which does not scale in a simple way with particle density. Particle distribution functions in regions of finite thickness—even with uniform fields and ignoring the dead space—are therefore intrinsically nonlocal, and increasingly so at higher values of the electric field when ionization feedback plays a larger role.

In a recent study of avalanche coefficients in GaN vertical p-i-n diodes [23], the hole ionization coefficient was determined using the formula $\beta = \ln(M_p)/W$, where M_p is the multiplication factor assuming pure hole injection into a region of thickness W characterized by a uniform field strength. This procedure, as the authors mention, ignores ionization feedback from electron-initiated impact ionization events for purposes of determining β. To determine α, this same study uses a separate non-punchthrough p-i-n structure to extract the electron avalanche coefficient under conditions of pure electron injection into the high-field region, given the field-dependent hole ionization coefficient extracted from the first p-i-n structure. Accounting properly for impact ionization events triggered by both carrier species, it follows from Eqs. (25.3) or (25.8) that the correct expression for the hole ionization coefficient in the uniform field case takes the form $\beta = \frac{\ln(M_p)}{W} + \alpha - \frac{1}{W}\ln\left(1 + \frac{\alpha}{\beta}(M_p - 1)\right)$ rather than simply $\frac{\ln(M_p)}{W}$. It is possible to assess the significance of the neglected terms in this expression by evaluating β with the correct formula using the electron ionization coefficient presented in the same study. The result is shown in Fig. 25.10 below. Ignoring ionization feedback with the formula $\beta = \frac{\ln(M_p)}{W}$ can only result in an overestimation of the dominant carrier's ionization coefficient, and this is exactly what is observed in Fig. 25.10 for field strengths below approximately 3 MV/cm. Yet when β is calculated using the full expression, the corrected value of β exceeds the original β at all field strengths above approximately 3 MV/cm. The published field dependences of β and α become increasingly inconsistent with one another at higher and higher fields, demonstrating the peril of ignoring ionization feedback. For many applications such as photon counting, photon timing, power switching and high-frequency signal amplification, it is precisely at field strengths above static breakdown where the ionization coefficients have the greatest relevance.

3.5 *Inference of Ionization Coefficients from Uniform-Field p-i-n Measurements*

For the extraction of avalanche coefficients in p-i-n structures with uniform field profiles in the intrinsic region, the local theory suggests a relatively simple procedure, requiring only knowledge of M_p and M_n as usual but avoiding the numerically noisy penalty of differentiating measured data. Manipulation of Eq. (25.7) reveals a simple relationship between M_n, M_p, and the avalanche coefficients:

$$\frac{\beta}{\alpha} = \frac{M_p - 1}{M_n - 1} \tag{25.11}$$

It was originally pointed out by Lee et al. [17] that M_p and M_n are related through the expression $\int_0^W (\alpha - \beta)\, dx = \ln\left(\frac{M_n}{M_p}\right)$. In the limiting case of a uniform field profile, this result may be combined with Eq. (25.11) above to yield

$$
\begin{aligned}
\beta &= \frac{1}{W}\frac{M_p - 1}{M_p - M_n}\ln\left(\frac{M_p}{M_n}\right) \\
\alpha &= \frac{1}{W}\frac{M_n - 1}{M_n - M_p}\ln\left(\frac{M_n}{M_p}\right)
\end{aligned}
\tag{25.12}
$$

Due to the nonlocality of charge transport, local ionization coefficients may only be precisely defined in bulk material, where the nonequilibrium distribution functions of electrons and holes are uniform. For this reason and to minimize the influence of the characteristic relaxation length for the nonequilibrium electron and hole distribution functions, the greatest accuracy will be achieved by measuring avalanche current in p-i-n structures with the thickest intrinsic regions and, regrettably, at the lowest avalanche gains.

References

1. K. Hess, *Advanced Theory of Semiconductor Devices* (IEEE Press, New York, 2000)
2. S. Adachi, *Properties of Group-IV, III_V and II-VI Semiconductors* (Wiley, Chichester, 2005)
3. C.G. Rodrigues, R. Vasconcellos, R. Luzzi, J. Appl. Phys. **102**, 0737141–0737147 (2007)
4. P. Siddiqua, W.A. Hadi, M.S. Shur, S.K. O'Leary, J. Mater. Sci. Mater. Electron. **26**, 4475–4512 (2015)
5. S.J. Pearton et al., Mater. Today **5**, 24–31 (2002)
6. B. Gelmont et al., J. Appl. Phys. **74**, 1818–1821 (1993)
7. B.J. Baliga, *Gallium Nitride and Silicon Carbide Power Devices* (World Scientific, New Jersey, 2016)
8. D. Ji, B. Ercan, S. Chowdhury, Appl. Phys. Lett. **115**, 073503 (2019)
9. L. Cao, H. Ye, J. Wang, P. Fay, Appl. Phys. Lett. **112**, 262103 (2018)
10. T. Maeda, T. Narita, S. Yamada, T. Kachi, T. Kimoto, M. Horita, J. Suda, *IEEE International Electron Devices Meeting (IEDM)* (2019), pp. 4.2.1–4.2.4
11. R.J. McIntyre, IEEE Trans. Electron Devices **ED-13**, 164–168 (1966)
12. A.G. Chynoweth, Phys. Rev. **109**, 1537–1540 (1958)
13. S.L. Miller, Phys. Rev. **105**, 1246–1249 (1957)
14. J.L. Moll et al., Solid State Electron. **6**, 147–157 (1963)
15. K.G. McKay, Phys. Rev. **94**, 877–884 (1954)
16. B.M. Wul et al., *Solid-State Physics and Electronics Telecommunications 1* (Academic, New York, 1960)
17. C.A. Lee et al., Phys. Rev. **134**, A761–A773 (1964)
18. M.H. Woods, W.C. Johnson, M.A. Lampert, Solid State Electron. **16**, 381–394 (1973)
19. S. Sridharan et al., IEEE Electron Device Lett. **29**, 1190–1192 (2008)
20. R.J. McIntyre, IEEE Trans. Electron Devices **46**, 1623–1631 (1999)
21. M.M. Hayat et al., IEEE J. Quantum Electron. **28**, 1360–1365 (1992)
22. P. Yuan et al., IEEE Trans. Electron Devices **46**, 1632–1639 (1999)
23. D. Ji et al., Appl. Phys. Lett. **115**, 141602 (2019)

Part VII
Related Wide-Band gap Materials and Devices

Chapter 26
Cubic GaN: Growth, Characterization, and Applications

Jaekwon Lee and Can Bayram

1 Introduction

Although traditional semiconductor devices such as Si-based metal-oxide-semiconductor field effect transistors (MOSFETs) or GaAs-based high-electron-mobility transistors (HEMTs) achieved great success in both chip electronics and radio-frequency (RF) devices, there are emerging needs for even higher power densities (1–10 MW) and high frequencies (MHz–GHz) that require the use of wide-bandgap (WBG) semiconductors for the voltage handling. Applications include but are not limited to electric vehicles [1], high-performance light detection and ranging (Lidar) systems [2], aerospace electronics [3], DC power converters for net-zero grids [4], and energy storage systems [5]. As a result, GaN-based HEMTs utilizing two-dimensional electron gas (2DEG) have achieved great success with commercialization. For instance, it has recently shown that the use of GaN electronics in photovoltaic inverters shows advantages such as two-thirds reduction in system volumes [6] and GaN transistors on 200 mm silicon is also demonstrated [7]. However, there are issues in GaN electronics that are yet to be solved such as normally-on operation in HEMTs [8], high defect density grown on Si (111) wafers [9], current collapse at high-field operation [10], significant nonlinearity at high-frequency operation [11], and the activation of the immersed p-GaN [12].

III-Ns can also form in a zincblende (i.e., cubic) crystal structure, which is the same crystal structure as common III-Vs (e.g., GaAs, InP, or AlSb). There are benefits over hexagonal GaN in terms of power electronics: absence of polarization

*Intel Alumni Endowed Faculty Scholar; Innovative COmpound semiconductoR Laboratory (ICORLAB); Webpage: icorlab.ece.illinois.edu

J. Lee · C. Bayram (✉)
The Grainger College of Engineering, University of Illinois Urbana-Champaign, Urbana, IL, USA
e-mail: jaekwon2@illinois.edu; cbayram@illinois.edu

I. C. Kizilyalli et al. (eds.), *Gallium Nitride and Related Materials*, The Materials Research Society Series, https://doi.org/10.1007/978-3-031-83056-3_26

Table 26.1 Comparison of hexagonal GaN and cubic GaN

Parameter	Hexagonal GaN	Cubic GaN	References
P_{sp} (C/m^2)	−0.034	0	[21]
E_g (300 K) (eV)	3.4	3.2	[21]
$(a_{AlN}\text{-}a_{GaN})/a_{AlN}$	−0.025%	−0.027%	[14]
m_e/m_{hh} (m_0)	0.2/1.92	0.15/0.85	[21]
μ_n/μ_p (cm^{2}V^{-1} s^{-1})	990/150	1150/350	[22]
E_D/E_A (meV)	30/227	30/140	[23–25]
$v_{peak,e}$ (10^7cm/s)	1.9	2.6	[18]
$\Delta E_c^{1\rightarrow 0}(\Gamma)$ (eV)	2	8	[26]

Spontaneous polarization (P_{sp}), bandgap at 300 K (E_g (300 K)), lattice mismatch ($(a_{AlN}\text{-}a_{GaN})/a_{AlN}$), electron/hole effective mass (m_e/m_{hh}), electron/hole mobility (μ_n/μ_p), donor/acceptor activation energies (E_D/E_A), electron peak drift velocity ($v_{peak,e}$), and the energy difference between the first and zeroth order conduction sub-band

in the <100> growth direction due to cubic crystal symmetry [13], smaller electron-heavy hole effective masses [14], smaller Auger loss due to the larger energy difference between the first and zeroth conduction subband [15, 16], lower p-doping activation energy [17], higher hole mobility [17], higher electron peak and saturation drift velocity [18, 19], and CMOS (complementary MOS) compatibility in the case of the growth on Si (100) wafers [20]. Table 26.1 compares the material parameters of hexagonal and cubic GaN in detail. In this regard, promising applications of cubic GaN in power electronics include normally-off HEMTs [27–32], CMOS [33] for logic devices, high-power metal-semiconductor field effect transistors (MESFETs) [34], integrated transistors, superjunction devices, or photoconductive semiconductor switches. As a wide bandgap semiconductor similar to GaAs, cubic GaN-based power electronics is also expected to utilize mature technologies and designs for GaAs electronics.

For any electrical device, the growth of a high-quality bulk template is crucial as the defects limit the electrical performance and cause alloy segregation, especially at the defect site. However, the cubic phase of GaN is a metastable phase in most cases, tending to form in the most stable wurtzite phase. In this regard, the growth of high-quality cubic GaN has always been an issue to be solved in order to translate these material advantages into a functional device. Historically, cubic GaN growth has been widely demonstrated via direct deposition on cubic substrates such as 3C-SiC or GaAs: representative demonstrations include cubic GaN [35–41] and HEMTs [28] on 3C-SiC (100), cubic GaN on GaAs (100) [17, 42–48], and cubic GaN and AlN on MgO (100) [49–55]. However, cubic nitride films grown by such direct deposition often suffer from high defectivity ($\gg 10^5$ cm^{-1}) [40], structural metastability (i.e., hexagonal phase inclusion) [39], and chemical metastability (i.e., the tendency toward spinodal decomposition). Another approach to realizing cubic phase GaN is to stabilize the cubic phase by impurity incorporation of Ga [56] or Mn [57], but the incorporation of impurities reduces the overall quality of the resulting material, which is thus not usable in device applications. Although hexagonal-to-cubic phase-transition (h-to-c-transition) on V-grooved Si (100) [58–

61] partially overcame these issues, they still suffer from roughness and hexagonal shoulder propagation. Recently, phase transition cubic GaN on U-grooved Si (100) utilizing aspect ratio (phase) trapping (ARpT) [20, 62–75] is demonstrated with high bulk IQE of 29–32% [62, 67], high quantum well IQE of 32% [68], the ability to trap or remove hexagonal phase shoulders [65, 67], no random hexagonal GaN inclusion (i.e., 100% phase-pure cubic GaN stripes) [65], scalability [67], and CMOS compatibility [20, 66, 69–71].

In this regard, the growth and characterization of phase-pure, high-quality cubic GaN on the U-grooved Si (100) template is discussed in this chapter. Furthermore, the overgrowth of cubic nitride alloys will be shortly discussed. Finally, the application of cubic GaN in power electronics will be discussed further.

2 Growth and Characterization of Cubic Gallium Nitride

Figure 26.1 (top) shows the cross-sectional sketch, top-view optical photograph, top-view scanning electron microscopy (SEM) image, and cross-sectional SEM image of U-grooved Si (100), prepared through oxide growth, lithography, reactive ion etching (RIE) of oxides, and selective Si wet etching (i.e., exposing {111} planes) via KOH solution. The pattern is generated uniformly all over the 1 cm × 1 cm die, and the pattern dimensions are $p = 677 \pm 3$ nm, $t_d = 94 \pm 3$ nm, $\alpha = 2.9 \pm 0.3°$, and $t_{ox} = 415 \pm 4$ nm.

The U-grooved Si (100) parameters should be carefully designed because the propagation of the hexagonal phase GaN is prevented due to the aspect ratio (phase) trapping (ARpT) [63]. More specifically, by controlling the U-groove parameters, one can achieve a purely cubic top surface. It has been previously shown that

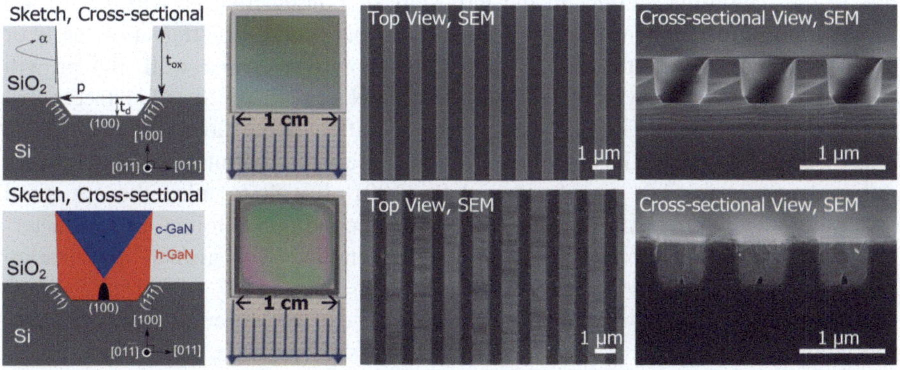

Fig. 26.1 Cubic GaN epitaxy on ~1 cm^2 U-grooved silicon (100) dies is demonstrated. (From left to right) Cross-sectional sketches, top-view photographs, top-view SEM images, and cross-sectional view SEM images of (top row) U-grooved silicon (100) and (bottom row) cubic GaN atop. (Reprinted from J. Lee, Y.C. Chiu, M.A. Johar, and C. Bayram, Appl. Phys. Lett. 121(3), 032101 (2022), with permission from AIP Publishing)

complete cubic GaN coverage without propagating hexagonal GaN shoulders can be achieved when the GaN deposition thickness above Si (100) substrate level (h) equals the critical thickness (h_c) [65]:

$$h_c = \frac{1.06p - 0.75t_d}{1 - \frac{\tan\alpha}{0.71}} \tag{26.1}$$

where the definition of the other parameters is the same as in Fig. 26.1. Figure 26.1 (bottom) shows cubic GaN growth on the U-grooved Si (100) die via metal-organic chemical vapor deposition (MOCVD). First, a low-temperature AlN (LT-AlN) buffer layer deposition and then selective GaN deposition only on silicon {111} planes take place. Those GaN growth lines form an angle of 54.735° with the Si (100) plane. Thus, when two hexagonal phase GaN growth fronts meet, the apex angle (θ) becomes 54.735° × 2 ≈ 109.47°, which is exactly the angle between the two Ga-N bonds in cubic tetrahedral bonding. Furthermore, the hexagonal <0001> and cubic <111> crystal directions are equivalent with the only difference in the stacking sequence (i.e., ABABAB… in hexagonal <0001> while ABCABCABC.. in cubic <111>). These crystallographic principles enable the phase transition from hexagonal GaN to cubic GaN. Such epitaxy enables the growth of pure cubic GaN with a clear phase boundary between the hexagonal and cubic GaN [65], proved by electron backscattering diffraction (EBSD), and enables exceptionally high-quality top surface.

Cubic GaN on U-grooved Si (100) requires a few optimization steps for precise and high-quality growth, as it involves an h-to-c-transition step. The most prominent growth failure can occur if the growth substrate is not prepared well. Any erroneous process in any of the abovementioned steps can result in defective cubic GaN or just hexagonal GaN without phase transition. The erroneous process includes multi (111) facets due to the residue-masking during the KOH etch, poorly defined (111) facets due to the nano-trench effect in the RIE step, or nonselective growth due to too shallow U-grooves. When selective growth does not occur, h-to-c-transition rarely or does not occur. Given that the U-grooved silicon is prepared well, the growth conditions should be favorable for selective GaN growth on Si (111) facets and the formation of cubic GaN. If the growth condition is not favorable, it leads to phase-pure hexagonal GaN growth or cubic GaN discontinuities, which were reported earlier in GaN on V-grooved Si (100) as well to relax the structural instability locally [76]. After the optimum growth condition is found, the characterization is performed to confirm the quality of the template. This growth optimization can be done within a few MOCVD runs.

The characterization method for cubic GaN can be different from ordinary hexagonal GaN film due to different factors such as its substrate, defect type, crystal structure, and dimensions. For instance, the threading dislocations in hexagonal GaN are predominant along <0001> direction, while threading dislocations in cubic GaN are mainly 60°-mixed-type perfect dislocations and stacking faults [77], which are challenging to quantify with the full-width at half-maximum (FWHM) of the X-ray

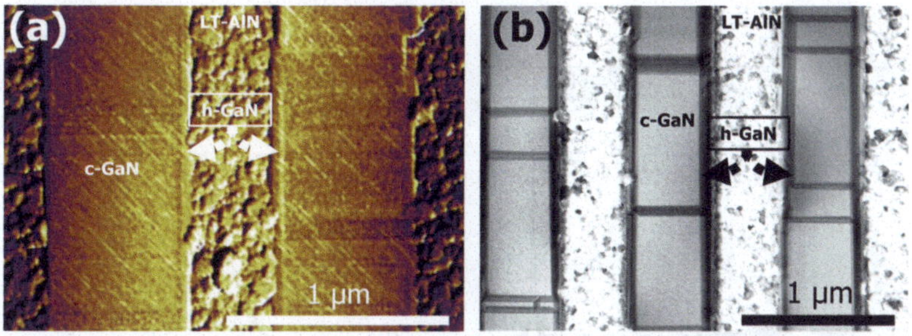

Fig. 26.2 Cubic GaN structural characterization. (**a**) Tapping mode AFM and (**b**) bright field top-view STEM images are shown. The plan-view STEM reveals no threading dislocations or other types of dislocations on the cubic GaN surface. Stacking faults with a density of $3.27 \pm 0.18 \times 10^4\ cm^{-1}$ are observed. Cubic (c-) and hexagonal (h-) GaN and LT-AlN buffers are labeled. (Reprinted from J. Lee, Y.C. Chiu, M.A. Johar, and C. Bayram, Appl. Phys. Lett. 121(3), 032101 (2022), with permission from AIP Publishing)

diffraction (XRD) peaks [78]. Furthermore, the most important characterization goal of cubic GaN films is the hexagonal GaN inclusion, which is usually not the case for normal GaN films. For this goal of characterizing phase purity and defectivity, various characterization methods can be used.

Figure 26.2 shows the structural characterization of the cubic GaN on the U-grooved Si (100) sample, including atomic force microscopy (AFM) and bright-field (BF-) plan-view scanning transmission electron microscopy (STEM). STEM and AFM provide crucial information regarding the defectivity and epitaxial structure. AFM shows the film's root-mean-square roughness, defectivity, and microstructure at the atomic level, while STEM helps mainly in determining the microstructure of materials, defect density and type, and dislocation distribution, in examining crystalline and amorphous structures, and examining the atomic composition. These methods are perfect for inspecting small areas (~ μm^2) at a time. AFM and STEM images confirm the exceptional top surface quality with no threading dislocations on top. Stacking faults with a density of $3.27 \pm 0.18 \times 10^4\ cm^{-1}$ are observed that are mostly perpendicular to the grooves. This density is an order of magnitude lower than what is reported in cubic GaN grown on 3C-SiC (i.e., $10^5\ cm^{-1}$) earlier [40]. From Fig. 26.2b, it can also be observed that narrow hexagonal GaN shoulders are defective, bearing a high density of stacking faults, which coincides with the previous study [62].

Figure 26.3 shows the characterization of cubic GaN on U-grooved Si (100) through high-resolution XRD. XRD is a nondestructive characterization technique that can show film orientation, composition, and the lattice constant of the films. Especially, 2θ-ω scan highlights the orientation of each film, while the rocking curve scan highlights its crystal quality from the FWHM value. Figure 26.3a shows the 2θ-ω scan showing the absence of random hexagonal inclusion, only revealing cubic GaN (200) and silicon substrate peaks. Rocking curve scan in Fig. 26.3b reveals the

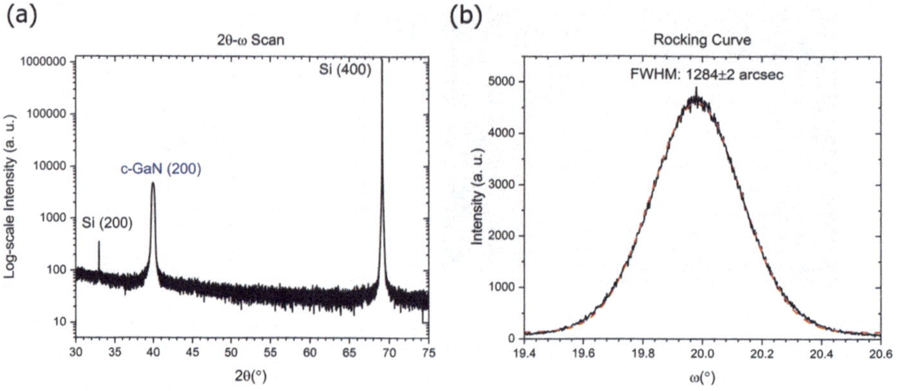

Fig. 26.3 (**a**) 2Theta-omega scan of cubic GaN on U-grooved Si (100) with no hexagonal peaks. (**b**) Rocking curve scan of cubic GaN on U-grooved Si (100) with 1284 ± 2 arcsec of FWHM

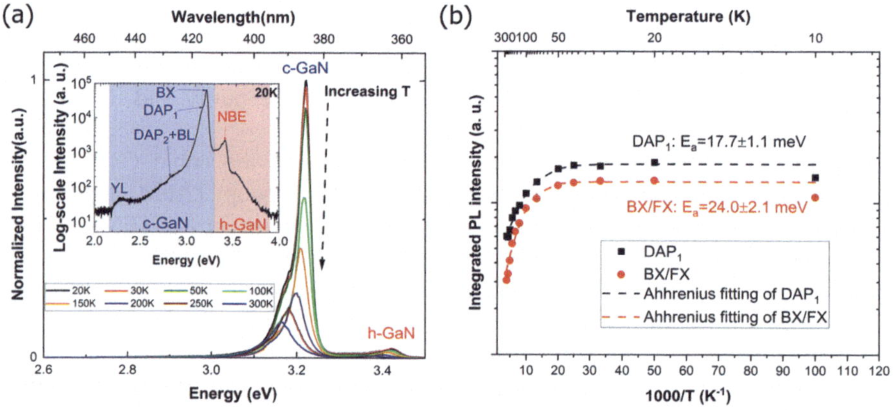

Fig. 26.4 Cubic GaN optical characterization. (**a**) The temperature-dependent PL spectra of the as-grown cubic GaN are shown. The data are normalized to cubic GaN peak intensity at 20 K. The inset shows the low-temperature (20 K) log-scale spectrum and the corresponding peaks. Cubic (c-) and hexagonal (h-) GaN are labeled. (**b**) Arrhenius plots and the calculated activation energies for the donor-acceptor pair (DAP_1) and the bound exciton (BX) peak are shown. 10 K data are excluded from the fitting as suspected outliers. (Adopted from J. Lee, Y.C. Chiu, M.A. Johar, and C. Bayram, Appl. Phys. Lett. 121(3), 032101 (2022), with permission from AIP Publishing)

FWHM of 1284 ± 2 arcsec, which is exceptionally low for 500 nm-thick cubic GaN films [41].

Figure 26.4a shows the photoluminescence (PL) spectra of as-grown cubic GaN at different temperatures ranging from 20 K to 300 K. PL is a useful method to characterize direct bandgap materials such as cubic GaN, revealing its bandgap via the peak intensities and revealing its efficiency from the temperature-dependent change in emission strength. Furthermore, the existence/absence of the peaks reveals the type of defects or impurities existing in the film. Here, a 2 mW continuous-wave diode-pumped solid-state laser ($\lambda = 266$ nm) is used to excite the cubic GaN, and a helium-bath cryostat is used to control the temperature. The spot size of the laser is nearly 400μm × 500μm. The PL spectra reveal two major peaks: the cubic GaN main peak and the hexagonal GaN near band emission (NBE) peak. The multi-peak fitting of the low-temperature cubic GaN main peak results in two Gaussian peaks at 3.19 eV and 3.22 eV, which are the donor-acceptor pair (DAP_1) and the bound exciton/free exciton (BX/FX) peak (Fig. 26.4a inset) [64]. Figure 26.4b shows the Arrhenius plot of the integrated intensity of the DAP_1 and BX/FX cubic GaN peaks. Here, we plot the integrated PL intensity as a function of the inverse of temperature. The integrated intensity (I) obeys [79]:

$$I(T) = \frac{I_0}{1 + C\exp\ (-E_{act}/k_B T)} \tag{26.2}$$

where I_0 and C are fitting coefficients, E_{act} is the activation energy of the luminescence center, k_B is the Boltzmann's constant, and T is the temperature. The fitting reveals the activation energy of 17.7 ± 1.1 meV for DAP_1 and 24.0 ± 2.1 meV for BX/FX. The IQE of the band-to-band transition of cubic GaN (η_{IQE}) is calculated from the following equation:

$$\eta_{IQE} = \frac{I(300K)}{I_{max}} \tag{26.3}$$

where I(300 K) is the integrated intensity of the cubic GaN PL peak (i.e., DAP_1 and BX/FX near 380-390 nm) at 300 K, and I_{max} is the maximum integrated intensity of the cubic GaN peak among all temperatures. This equation assumes that all the non-radiative centers are deactivated at low enough temperatures, such as 10 K or 20 K, thereby assuming 100% IQE at low temperatures. The as-grown cubic GaN sample shows $25.6 \pm 0.9\%$ of IQE. This is slightly lower than our earlier report (i.e., 29% [62]) but is achieved on a much larger areal measurement, measuring hundreds of cubic GaN stripes together.

Figure 26.5 shows the peak energies of DAP_1 and BX/FX peaks as a function of temperature. Here, BX/FX peaks show the transition from BX to FX due to the exciton dissociation by thermal energies. Similarly, DAP_1 changes from donor-acceptor transition (D^0, A^0) to free electron-acceptor (e, A^0) transition. The FX and (e, A^0) can fit into the Varshni's equation [80]:

$$E(T) = E_g(0K) - \frac{\alpha T^2}{T + \beta} + \frac{1}{2} k_B T \tag{26.4}$$

where T is the temperature, $E(T)$ is the peak emission energy at the temperature T, $E_g(0\ K)$ is the bandgap at 0 K, α is the Varshni's coefficient, β is the Debye temperature of GaN assumed to be 600 K, and k_B is the Boltzmann coefficient. Using the activation energy and the peak position of DAP_1 and the bandgap of cubic GaN calculated from the BX peak, the calculated acceptor activation energy is 43.5 ±

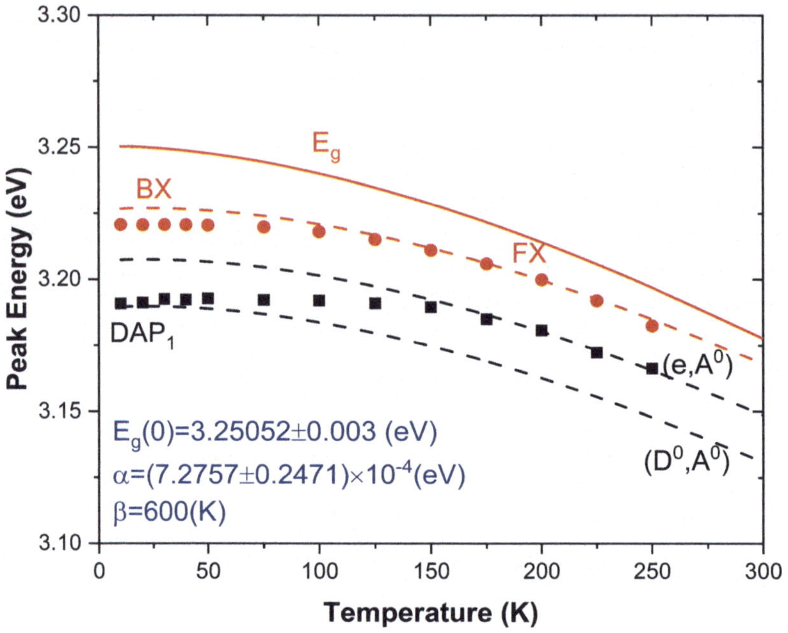

Fig. 26.5 Peak emission energies of DAP_1 and BX/FX peak are shown as a function of temperature. From Varshni's equation and the activation energies of the BX/FX peaks, one can infer the parameters for the temperature-dependent bandgap

4.1 meV. The bandgap of cubic GaN is calculated to be near 3.18 eV, which is lower than what is reported in the literature (3.23 eV [81]) and is attributed to the tensile strain that exists in cubic GaN on U-grooved Si (100) in contrast to the compressive strain on MgO (100) or 3C-SiC (100).

It is shown that one can theoretically achieve a phase-pure cubic GaN top surface without any hexagonal GaN shoulders by carefully engineering the U-grooves per Eq. (26.1). However, for a large (~cm^2) area demonstration as in this work, it is challenging to achieve perfectly designed U-grooves without error in an academic facility. One way to ease these stringent conditions and achieve complete cubic GaN surface coverage is the use of selective etching, removing hexagonal GaN shoulders from imperfect patterning, and achieving different configurations of pure cubic GaN for further fabrication. Three different etching experiments are conducted. In addition to the control sample labeled as (A) as-grown, the three etched samples are labeled as (B) LT-AlN removed, (C) LT-AlN and SiO_2 removed, and (D) LT-AlN, SiO_2, and h-GaN removed. Selective etching involves wet-etch by buffered oxide etchant (BOE) in order to remove oxidized layers and physical etching by prolonged sonication in order to physically break the defective crystal along the crystallographic planes. Sample C and D can be fabricated with the same BOE time, while BOE time must be minimized in sample B. Regarding the sonication time, sample C

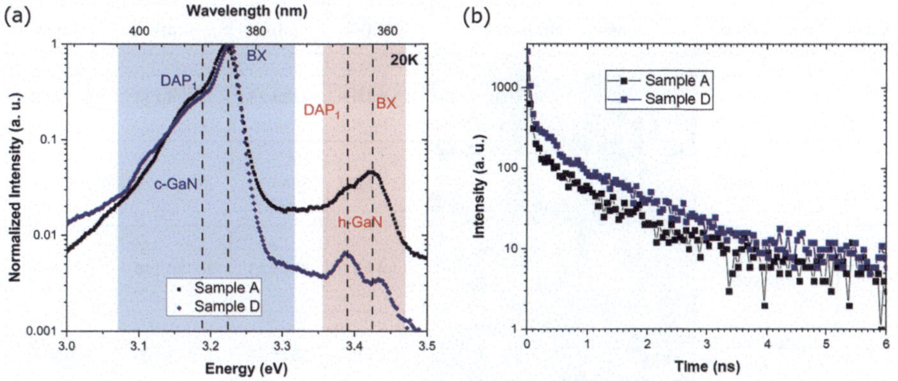

Fig. 26.6 (**a**) Low-temperature PL spectra at 20 K and (**b**) time-resolved PL decay curves of sample A (as-grown) and sample D (LT-AlN, SiO_2, and h-GaN removed)

utilizes the shortest sonication time, and sample B utilizes the longest sonication time, respectively.

In addition to temperature-dependent PL, time-resolved PL (TRPL) measurements are carried out to extract effective carrier lifetimes (τ_{eff}) with a frequency-tripled Ti:Al_2O_3 laser ($\lambda = 266$ nm) under a pulse width of 100 fs and a repetition rate of 80 MHz. The intensity decay curves of all four samples show biexponential decay and fit the following equation:

$$I(t) = C_1 \exp\ \ (-t/\tau_1) + C_2 \exp(-t/\tau_2) + y_0 \tag{26.5}$$

where τ_1 and τ_2 represent the fast and slow decay lifetimes, and C_1 and C_2 are the fast and slow decay amplitudes. Under this condition, τ_{eff} is calculated as [82]:

$$\tau_{eff} = \frac{C_1{\tau_1}^2 + C_2{\tau_2}^2}{C_1\tau_1 + C_2\tau_2} \tag{26.6}$$

From the effective lifetime from TRPL and the IQE values from temperature-dependent PL, radiative (τ_r) and non-radiative (τ_{nr}) recombination lifetimes are calculated per:

$$\eta_{IQE} = \frac{\tau_{eff}}{\tau_r} \tag{26.7}$$

where ${\tau_{eff}}^{-1} = {\tau_r}^{-1} + {\tau_{nr}}^{-1}$. Low-temperature PL and TRPL curves of sample A and sample D are shown in Fig. 26.6.

Figure 26.7 tabulates the results. Recombination physics in sample C is within the error range compared to sample A, whereas the radiative recombination lifetime of sample B increased, pointing to the etch damage due to excess sonication time used in sample B. More importantly, the result reveals that the IQE of sample D increased

Label	Information	Sample SEM Image	IQE (%)	τ_{eff}(ps)	τ_r(ps)	τ_{nr}(ps)
A	As-grown		25.6±0.9	518±28	2025±133	697±59
B	LT-AlN removed		23.9±0.7	564±39	2358±179	741±76
C	LT-AlN & SiO_2 removed	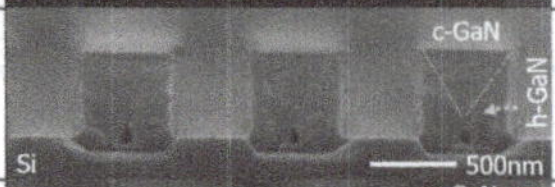	24.4±0.8	508±33	2078±149	672±65
D	LT-AlN & SiO_2 & h-GaN removed	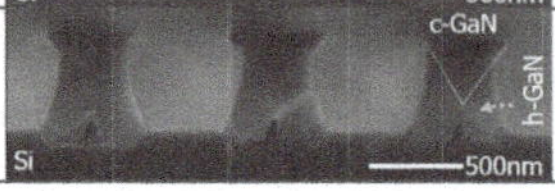	31.6±0.8	551±30	1742±106	806±66

Fig. 26.7 SEM and optical properties of four processed samples are shown. Especially, hexagonal GaN shoulders are removed in sample D, enabling pure cubic GaN on top. (Reprinted from J. Lee, Y.C. Chiu, M.A. Johar, and C. Bayram, Appl. Phys. Lett. 121(3), 032101 (2022), with permission from AIP Publishing)

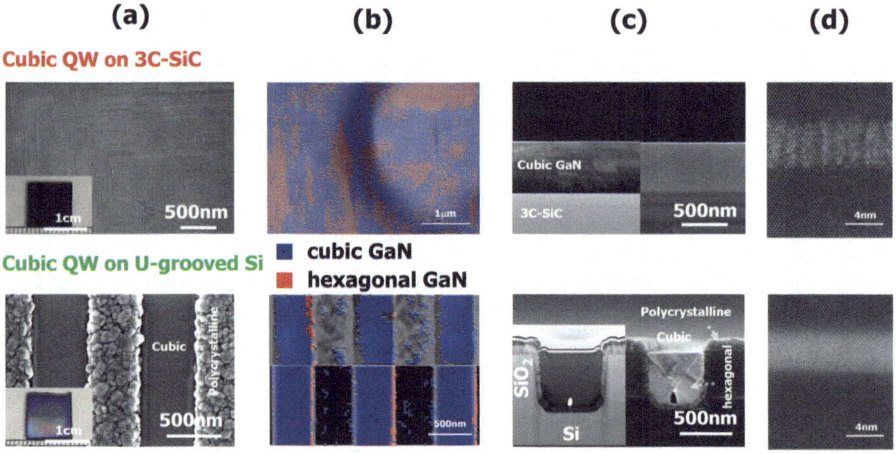

Fig. 26.8 (**a**) Top-view scanning electron microscopy (SEM) images with sample photos as insets, (**b**) electron backscatter diffraction (EBSD) overlay showing phase purities, (**c**) cross-sectional SEM images (inset: bright-field STEM images), and (**d**) closed-up high angle annular dark field STEM images of cubic InGaN/GaN QW on (top) conventional 3C-SiC (100) and (bottom) U-grooved silicon (100). (Adopted from Lee and Bayram [68], with permission from AIP Publishing)

from 25.6 ± 0.9% to 31.6 ± 0.8% compared to sample A, which is a record IQE value higher than our earlier report (i.e., 29% [62]) and is mostly attributed to the decreased radiative recombination lifetime. The removal of hexagonal GaN shoulders accounts for the decrease in the radiative recombination lifetime. The 20 K PL spectrum of sample D in Fig. 26.8a reveals a significant decrease in the hexagonal

GaN peak compared to sample A, indicating that the hexagonal GaN NBE peak in the as-grown PL spectrum mostly comes from the hexagonal GaN shoulders. Additionally, it has been previously shown that stacking faults are created near the cubic GaN/hexagonal GaN interface and the crystal/oxide interface [62], implying that carriers recombine more radiatively due to the removal of defective hexagonal GaN shoulders that are close to both interfaces. Overall, the high IQE of sample D implies that complete cubic GaN surface coverage can be achieved via selective etching without compromising the top surface quality to a large extent.

3 Growth of III-Nitride Alloys and Their Characterization

For power electronics, the growth of cubic AlN or cubic BN and enabling BAlGaN heterojunctions can be useful for various purposes. Table 26.2 tabulates the material parameters of cubic BN, AlN, and GaN. However, it was previously predicted that the introduction of alloys might adversely affect the defectivity [86] or the phase stability [87] in cubic nitrides. It has been recently shown that cubic nitrides grown on purely cubic GaN do not show this issue [68]. In this work, a cubic InGaN/GaN single-quantum well (SQW) is grown on the cubic GaN on a U-grooved Si (100) template using molecular beam epitaxy (MBE). For a comparative study, a similar structure (i.e., 500 nm cubic GaN and the QW) was grown on a commercial 3 μm 3C-SiC (100) film on silicon, namely cubic QW on 3C-SiC.

Figure 26.8 shows optical photographs, top and cross-sectional SEM images, EBSD mapping, and BF and HA-ADF STEM images of the SQW on conventional 3C-SiC (100) and SQW on U-grooved Si (100), whose dimensions are 1 cm × 1 cm. SEM images reveal antiphase domains between striations with different directions and stacking faults in cubic QW on 3C-SiC, while cubic QW on U-grooved Si shows a smooth cubic GaN surface. EBSD overlay images confirm that cubic QW on 3C-SiC shows some hexagonal phase inclusions. In contrast, the SQW grown on

Table 26.2 Comparison of the material parameters in cubic BN, AlN, and GaN

Parameter	Cubic BN	Cubic AlN	Cubic GaN	References
E_g (300 K) (eV)	6.1–6.4	4.9	3.2	[21, 83]
Ec (MV/cm)	2–6	–	2.3	[83, 84]
CB Minima	X	X	Γ	[21, 83]
m_e (m_0)	1.2 (l) 0.26 (t)	0.53 (l) 0.31 (t)	0.15	[21, 83]
$m_{h\ [100]}$ (m_0)	3.16 (hh) 0.64 (lh)	1.02 (hh) 0.35 (lh)	0.85 (hh) 0.24 (lh)	[21, 83]
μ_n/μ_p ($cm^2V^{-1}\ s^{-1}$)	≤200/≤500	200/–	1150/350	[22, 83]
E_D/E_A (meV)	50/190	–/248	30/140	[23–25, 83]
$v_{peak,e}$ (10^7cm/s)	4.3	–	2.6	[18, 85]

Bandgap at 300 K (E_g (300 K)), breakdown field (E_c), conduction band minima (CB minima), electron/hole effective mass (m_e/m_h), electron/hole mobility (μ_n/μ_p), donor/acceptor activation energies (E_D/E_A), and the peak electron drift velocities ($v_{peak,e}$) are shown

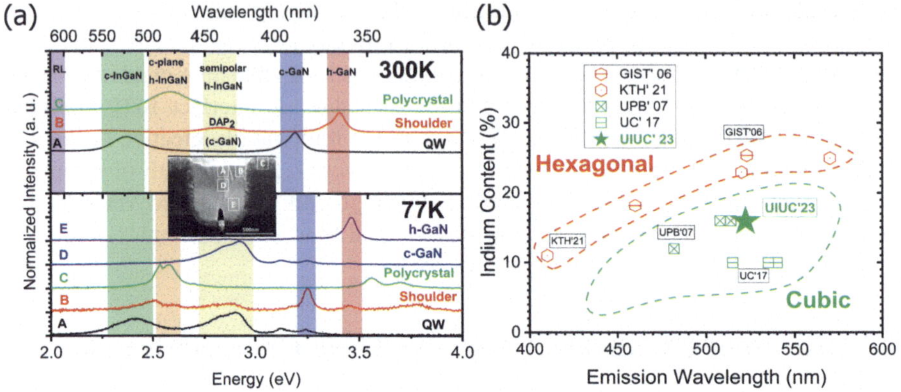

Fig. 26.9 (**a**) (Top) Top view CL at 300 K and (bottom) cross-sectional CL at 77 K. (**b**) Comparison of indium content as a function of emission wavelength in hexagonal and cubic QWs. (Adopted from Lee and Bayram [68], with permission from AIP Publishing)

U-grooved Si (100) has purely cubic middle stripes before (top) and after (bottom) the selective wet-etch. As the cubic phase purity of the top GaN cap layer is 100%, it implies that the phase purity of the underlying $In_XGa_{1-X}N$ active layer is also 100%. In contrast, Figs. 26.8c, d show a striking difference between the two samples. Contrary to conventional cubic QW on 3C-SiC where pipe-like indium clustering is severe and stacking faults are visible, cubic QW on U-grooved Si shows uniform indium distribution free of segregation. Furthermore, the BF-STEM image reveals defects only on phase boundaries and near the oxides, with minimal defects reaching the cubic top surface, coinciding with the template studies. Energy dispersive X-ray spectroscopy (EDS) measurements combined with STEM reveal the structure of GaN film/4.5 nm $In_{0.21}Ga_{0.79}N$/16.5 nm GaN on 3C-SiC substrate and GaN film/ 4.2 nm $In_{0.16}Ga_{0.84}N$/15.3 nm GaN structure on U-grooved Si substrate. It is speculated that the higher average indium content on the 3C-SiC substrate, despite the same growth condition, is the result of the indium segregation at the stacking faults [86].

Cathodoluminescence (CL) can be a useful technique, as cubic AlGaN is a direct bandgap material when Al content is below 70% [26]. Therefore, CL measurements bring light to the position-dependent growth quality as well as the material parameters such as the bowing parameter, when combined with accurate EDS. Fig. 26.9a shows the position-dependent CL spectra of the cubic QW on U-grooved Si (100). Three different areas (i.e., A–C from the inset) from the top at 300 K and five different areas (i.e., A–E) from the cross-section at 77 K are measured with 3 kV acceleration voltage and 2 nA beam current. Region A (QW stripes) shows 525 nm room temperature cubic QW emission and 388 nm cubic GaN emission, region B (hexagonal shoulders) shows 365 nm hexagonal GaN emission and 440 nm semi-polar plane QW emission, and region C (polycrystal) shows 480 nm hexagonal QW emission at room temperature. Cross-sectional CL at 77 K shows a similar trend with the addition of cubic GaN donor-acceptor-pair (DAP_2) peak near 420 nm and AlGaN buffer layer peak near 350 nm, whose intensities increased due to lower temperature. Regions D (bulk cubic GaN) and E (bulk hexagonal GaN) show pure c-GaN and pure h-GaN emission. Room temperature CL measurement over a large area of region A reveals an average peak emission wavelength of 522 nm with FWHM of 152 meV (33 nm). Combined with the STEM and EDS data, Fig. 26.9b compares cubic QW on U-grooved Si with other hexagonal and cubic InGaN QWs in the literature [88–91]. It is shown that cubic QW on U-grooved Si has nearly 7% lower indium content in absolute scale compared to hexagonal QWs, pointing to their smaller bandgap.

4 Application of Cubic Gallium Nitride

One of the most promising applications of cubic GaN power electronics is normally-off transistors [27, 28]. Although normally-off operation is indeed possible in hexagonal AlGaN/GaN HEMTs, they rely on specific techniques such as gate recessing [92], which may increase the leakage current, or p-GaN gate [93], which is not reliable due to the high acceptor activation energy. For cubic AlGaN/GaN HEMTs, one can enable the doping of the AlGaN gate, which can be classified as a modulation-doped structure. This doping enables carriers to move to a low-bandgap material, creating a 2D carrier channel free of ionized impurity scattering from heavy doping. This structure is similar to the early AlGaAs/GaAs heterojunctions [94], where cubic AlGaN/GaN HEMTs have additional advantages over AlGaAs/GaAs HEMTs such as larger bandgap [21], larger conduction band offset [95] leading to higher 2DEG density, or higher electron peak and saturation velocity [96]. The modulation doping can be a δ-doping, created by growing an extremely thin layer of highly doped cubic AlGaN. It is previously shown in the literature that the optimization and the tuning of the threshold voltage are possible in cubic AlGaN/GaN HEMT structure, tuning between normally-on and normally-off behavior [27], as well as positive threshold voltage of 0.6 V is possible [30].

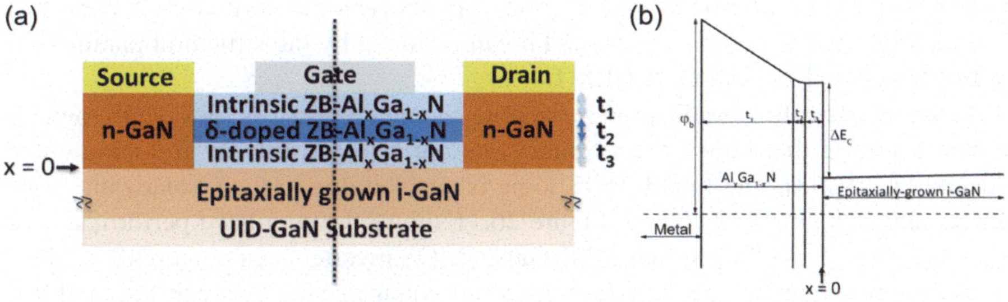

Fig. 26.10 (**a**) Cross-sectional schematic of the δ-doped cubic AlGaN HEMT proposed in the literature and (**b**) the band structure along the cut line in (**a**). (From Grady and Bayram [27]. © IOP Publishing. Reproduced with permission. All rights reserved)

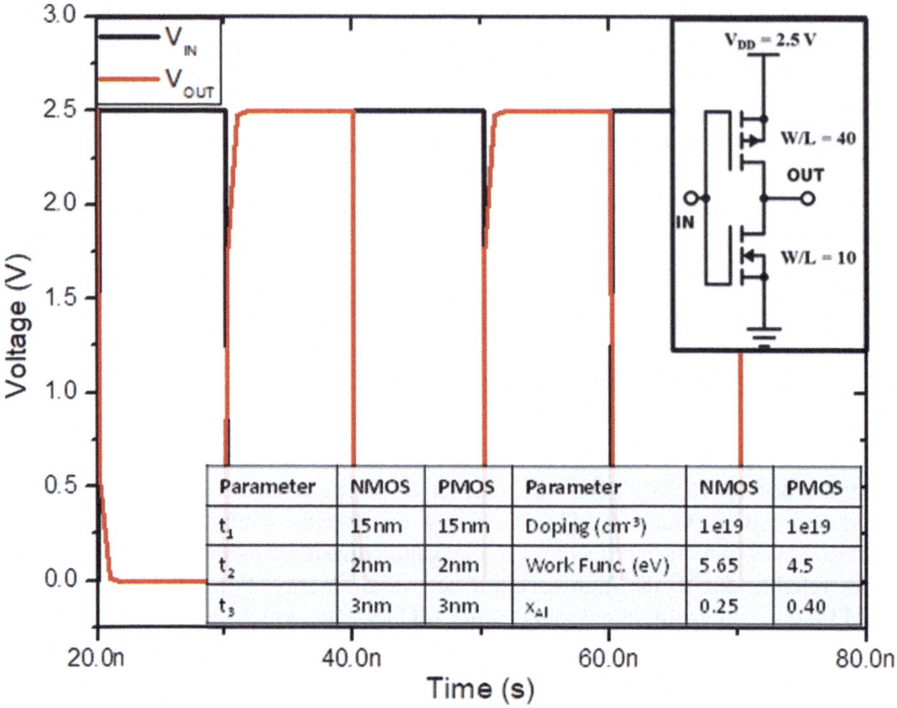

Parameter	NMOS	PMOS	Parameter	NMOS	PMOS
t_1	15nm	15nm	Doping (cm^{-3})	1e19	1e19
t_2	2nm	2nm	Work Func. (eV)	5.65	4.5
t_3	3nm	3nm	x_{Al}	0.25	0.40

Fig. 26.11 Simulated transient response of an insulated gate cubic phase GaN inverter enabled by CMOS

Figure 26.10 shows the suggested HEMT structure reported in the literature [27]. Here, the δ-doping in the AlGaN layer enables high 2DEG concentration on the i-GaN channel. In this work, different effects of each parameter on the threshold voltage and the 2DEG density have been shown. It is shown that the lower the Al content, the thinner the thickness of the δ-doped layer, and the lower the doping density, the threshold voltage increases and normally-off operation is possible. However, this also decreases the 2DEG density, thus contributing to lower output power. Therefore, it is found that the trade-off between the normally-off operation and the high 2DEG density exists, and it can be tuned by the structural parameters, unlike hexagonal AlGaN/GaN HEMTs.

Another interesting application is the hole channel transistors for CMOS logic. It is worth noting that hole channel transistors are also possible by p-doping cubic AlGaN layers [33], and thus CMOS logic is enabled due to their polarization-free nature and δ-doping technology. Figure 26.11 shows the simulated performance of such inverters. It is claimed that cubic GaN CMOS inverters can potentially achieve a maximum frequency of 1.22 GHz with a full voltage swing between V_{DD} and 0 V [33]. The hole channel devices can also be achieved in hexagonal transistors but in a more difficult way such as inverted GaN/AlGaN structure [97], InGaN/GaN

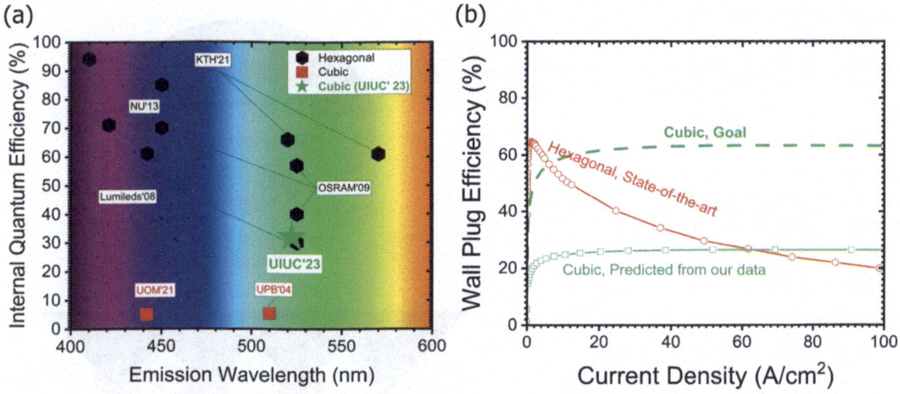

Fig. 26.12 (**a**) Recently reported internal quantum efficiency of cubic InGaN/GaN active layer compared to traditional cubic active layer on 3C-SiC as well as conventional hexagonal active layers. (**b**) Comparison of the state-of-the-art wall-plug efficiency of hexagonal LEDs versus predicted data from the results in (**a**). (Adopted from Lee and Bayram [68], with permission from AIP Publishing)

structure [98], or N-polar epilayers [99]. This makes the co-integration of CMOS challenging, whereas the realization of cubic CMOS on the same platform is easier.

Figure 26.12 shows another promising application of cubic nitrides in photonics: green light-emitting diodes (LEDs) [16, 68, 89, 100–107]. State-of-the-art green LEDs are made with hexagonal InGaN/GaN QWs, but they show extremely low optical power at high current densities due to the material degradation and the fundamental physics rising from piezoelectric polarization [102]. This issue is called the green gap, preventing a more efficient approach to color mixing in solid-state lighting [108]. In this regard, the benefits of cubic nitrides in this area are similar such as absence of polarization in the <100> growth direction due to cubic crystal symmetry [13], smaller bandgap [14], smaller electron-heavy hole effective masses [14], smaller Auger loss [15], shorter radiative recombination lifetime [109], larger optical gain [110], lower p-doping activation energy [17], higher hole mobility [17], and larger conduction band offset between InN/GaN [95]. In this regard, it is expected that cubic nitride LEDs can solve the long-lasting green gap issue and enable high-frequency, droop-free green LEDs [16, 111]. Figure 26.12a compares the recently reported work regarding the internal quantum efficiency (IQE) of the cubic green active layer with the IQE measurements in the literature as a function of emission wavelength for h-LEDs and c-QWs [90, 112–116]. This work shows superior performance over conventional cubic QWs on 3C-SiC (i.e., nearly six times higher IQE with longer peak emission wavelength), where the performance is even comparable to relatively mature hexagonal nitride LEDs. This is thanks to the phase-purity and low defectivity of cubic GaN template, enabled by ARpT. It is worth noting that the measurement here serves as a peak IQE. Therefore, the high current performance might even be favorable for cubic LEDs. Figure 26.12b shows the simulated cubic LED wall-plug efficiency (WPE) curve from the experimental

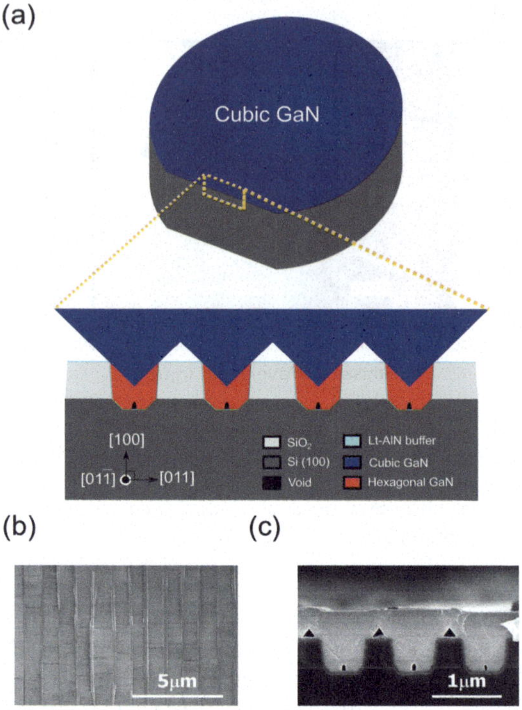

Fig. 26.13 (**a**) Schematic for the cubic phase GaN overgrowth and coalescence growth to achieve 4-inch or larger cubic GaN wafer on CMOS compatible Si (100), (**b**) top view, and (**c**) cross-sectional SEM images of the very first, preliminary result proving the concept of coalescence

data, as well as the WPE curve extracted from CREE Xlamp XP-E2 LED emitting 525 nm light, which shows 20% WPE at 100 A/cm^2. Due to the predicted absence of efficiency droop and weaker voltage penalty in cubic nitride LEDs [16, 117], this result is already expected to exceed state-of-the-art green LED performance at 100 A/cm^2 when packaged completely. The theoretical results suggest even better results: three times higher internal quantum efficiency than hexagonal LED [102], which can be realized from epitaxy optimization.

The applications of cubic GaN also include polarization-free ultraviolet LEDs [118], integrated photonics [20, 66, 69], single-photon emitters [119, 120], room-temperature ferromagnetism [121], and even spintronics [122–124]. For versatile applications, scaling up the wafer size is crucial. Figure 26.13 visualizes the future direction of cubic GaN on U-grooved Si (100). Based on the reported cubic GaN on U-grooved Si (100), the continued growth of cubic GaN via MOCVD results in the expansion of cubic GaN both laterally and vertically under appropriate growth conditions. As mentioned above, the growth of hexagonal shoulders is inhibited via ARpT. Figure 26.13a shows the schematic of such growth that can be scaled up to wafer scale for industry adoption. Similar to lateral epitaxial overgrowth [125], the coalescence of these pure cubic stripes will result in wafer-scale, single-crystal, phase-pure cubic GaN for next-generation power electronics and photonics. Figure 26.13b, c prove the concept of coalescence growth via ARpT, which is the very first, preliminary result on cubic GaN on U-grooved Si (100) platform. Still, extremely narrow hexagonal shoulders remain, which is to be solved via growth condition and pattern optimization.

5 Concluding Remarks

In summary, the growth and characterization of phase-pure cubic GaN on U-grooved Si (100) samples are discussed as well as the growth of cubic nitride alloys and their application. It is shown that cubic GaN on U-grooved Si (100) can enable phase-pure cubic GaN due to the aspect ratio (phase) trapping, with high structural and optical quality as well as scalability that is promising for various applications including power electronics.

It is worth reiterating that all devices fabricated with cubic GaN on U-grooved Si (100) can expand their applications into integrated, CMOS-compatible devices on scaled-up Si (100) substrates. This will accelerate industry adoption, enabling a heterogeneous epitaxy approach for Si-integrated power electronics as well as integrated photonics.

Acknowledgments This work is funded in part by the Advanced Research Projects Agency-Energy (ARPA-E), U.S. Department of Energy, under BREAKERS (Award No. DE-AR0001109) and OPEN 2021 (Award No. DE-AR0001558) program. This work is also funded in part by the Office of Naval Research (ONR) under Award No. N00014-23-1-2423 and in part by the National Science Foundation Faculty Early Career Development (CAREER) Program under award number NSF-ECCS-16-52871. The views and opinions of the authors expressed herein do not necessarily state or reflect those of the United States Government or any agency thereof. J. Lee would also like to acknowledge support from the ECE department at the University of Illinois Urbana-Champaign with the Promise of Excellence Fellowship and the Nick and Katherine Holonyak, Jr. Graduate Student Award. He also acknowledges support from SPIE with Optics and Photonics Education Scholarship.

This work was mainly carried out in the Micro and Nanotechnology Laboratory and Frederick Seitz Materials Research Laboratory Central Research Facilities, University of Illinois Urbana-Champaign, IL, USA. Part of the data was obtained as part of a user project at the Center for Nanophase Materials Sciences (CNMS), which is a US Department of Energy, Office of Science User Facility at Oak Ridge National Laboratory. The authors acknowledge valuable material growth support from SVT Associates, Inc. and valuable characterization support from EUROFINS EAG MATERIALS SCIENCE, LLC.

References

1. C.C. Chan, K.T. Chau, An overview of power electronics in electric vehicles. IEEE Trans. Ind. Electron. **44**, 3 (1997)
2. J. Glaser, How GaN power transistors drive high-performance lidar: Generating ultrafast pulsed power with GaN FETs. IEEE Power Electron. Mag. (2017). https://doi.org/10.1109/MPEL.2016.2643099
3. J. Wang, *Wide bandgap-based power electronics for aerospace applications*, vol 9 (IEEE Power Electron. Mag., 2022), p. 16

4. B. Wang, L. Sun, N. Diao, L. Zhang, X. Guo, J.M. Guerrero, GaN-based step-down power converter and control strategy for hydrogen energy systems. Energy Rep. **9**, 252 (2023)
5. F. Xue, R. Yu, A.Q. Huang, A 98.3% efficient GaN isolated bidirectional DC–DC converter for DC microgrid energy storage system applications. IEEE Trans. Ind. Electron. **64**, 9094 (2017)
6. H. Amano et al., The 2018 GaN power electronics roadmap. J. Phys. D. Appl. Phys. **51**, 163001 (2018)
7. K.-T. Lee et al., GaN devices on a 200 mm Si platform targeting heterogeneous integration. IEEE Electron Device Lett. **38**, 1094 (2017)
8. T. Ishibashi et al., Experimental validation of normally-on GaN HEMT and its gate drive circuit. IEEE Trans. Ind. Appl. **51**, 2415 (2015)
9. J. Meyer, R. Liu, R.D. Schaller, H.-P. Lee, C. Bayram, Systematic study of shockley-read-hall and radiative recombination in GaN on Al_2O_3, freestanding GaN, and GaN on Si. J. Phys. Photonics **2**, 035003 (2020)
10. R. Chu, Y. Zhou, J. Liu, D. Wang, K.J. Chen, K.M. Lau, AlGaN-GaN double-channel HEMTs. IEEE Trans. Electron Devices **52**, 438 (2005)
11. T. Palacios et al., Influence of the dynamic access resistance in the g_m and f_T linearity of AlGaN/GaN HEMTs. IEEE Trans. Electron Devices (2005). https://doi.org/10.1109/TED.2005.856180
12. Y. Ma et al., Activating thick buried p-GaN for device applications. IEEE Trans. Electron Devices **69**, 4224 (2022)
13. T. Kitamura et al., Optical properties of cubic InGaN/GaN multiple quantum Wells on 3C-SiC substrates by radio-frequency plasma-assisted molecular beam epitaxy. Phys. Status Solidi (A) **188**, 705–709 (2001)
14. I. Vurgaftman, J.R. Meyer, Band parameters for nitrogen-containing semiconductors. J. Appl. Phys. **94**, 3675 (2003)
15. K.T. Delaney, P. Rinke, C.G. Van de Walle, Auger recombination rates in nitrides from first principles. Appl. Phys. Lett. **94** (2009). https://doi.org/10.1063/1.3133359
16. Y.-C. Tsai, J.-P. Leburton, C. Bayram, Quenching of the efficiency droop in cubic phase InGaAlN light-emitting diodes. IEEE Trans. Electron Devices **69**, 3240 (2022)
17. C.A. Hernández-Gutiérrez et al., Study of the heavily p-type doping of cubic GaN with Mg. Sci. Rep. **10**, 16858 (2020)
18. R. Brazis, D. Nausewicz, R. Raguotis, Electron drift velocity in high electric fields in zincblende, wurtzite and rocksalt GaN. J. Phys. Conf. Ser. **193**, 012031 (2009)
19. P. Siddiqua, S.K. O'Leary, The sensitivity of the electron transport within bulk zinc-blende gallium nitride to variations in the crystal temperature, the doping concentration, and the non-parabolicity coefficient associated with the lowest energy conduction band valley. J. Appl. Phys. **120**, 095701 (2016)
20. C. Bayram et al., Polarization-free GaN emitters in the ultraviolet and visible spectra via heterointegration on CMOS-compatible Si (100), in *Presented at the SPIE OPTO*, ed. by M. Razeghi, E. Tournié, G.J. Brown, (San Francisco, 2015)
21. I. Vurgaftman, J.R. Meyer, L.R. Ram-Mohan, Band parameters for III–V compound semiconductors and their alloys. J. Appl. Phys. **89**, 5815 (2001)
22. R. Quay, *Gallium nitride electronics*, Springer series in materials science, no. 96 (Springer, Berlin, 2008)
23. Y.-C. Tsai, C. Bayram, Mitigate self-compensation with high crystal symmetry: A first-principles study of formation and activation of impurities in GaN. Comput. Mater. Sci. **190**, 110283 (2021)
24. F. Mireles, S.E. Ulloa, Acceptor binding energies in GaN and AlN. Phys. Rev. B **58**, 3879 (1998)
25. U. Birkle et al., Studies on carbon as alternative P-type dopant for gallium nitride. MRS Internet J. Nitride Semicond. Res. **4**, 526 (1999)

26. Y.-C. Tsai, C. Bayram, Structural and electronic properties of hexagonal and cubic phase AlGaInN alloys investigated using first principles calculations. Sci. Rep. **9**, 6583 (2019)
27. R. Grady, C. Bayram, Simulation of zincblende AlGaN/GaN high electron mobility transistors for normally-off operation. J. Phys. D. Appl. Phys. **50**, 265104 (2017)
28. M. Abe et al., Cubic GaN/AlGaN HEMTs on 3C-SiC substrate for normally-off operation. IEICE Trans. Electron. **E89-C**, 1057 (2006)
29. C. Bayram, R.W. Grady, K. Park, *Normally-off cubic phase GaN (c-GaN) HEMT having a gate electrode dielectrically insulated from a c-AlGaN capping layer* (US10211328B2, 2019)
30. E. Tschumak et al., Nonpolar cubic AlGaN/GaN heterojunction field-effect transistor on Ar+ implanted 3C–SiC (001). Appl. Phys. Lett. **96**, 253501 (2010)
31. R. Granzner et al., Vertical design of cubic GaN-based high electron mobility transistors. J. Appl. Phys. **110**, 114501 (2011)
32. S. Ghosh et al., An unified analytical model for design consideration of doped cubic and undoped hexagonal AlGaN/GaN MIS gate HEMTs. Solid State Electron. **96**, 1 (2014)
33. C. Bayram, R. Grady, K. Park, Novel cubic phase III-nitride complementary metal-oxide-semiconductor transistor technology, in *Quantum Sensing and Nano Electronics and Photonics XV*, ed. by M. Razeghi, G.J. Brown, G. Leo, J.S. Lewis, (SPIE, San Francisco, 2018)
34. M.T. Weber, L. Tirino, K.F. Brennan, et al., IEEE Trans. Electron Devices **50**, 2202 (2003)
35. R.M. Kemper, D.J. As, J.K.N. Lindner, Cubic GaN on nanopatterned 3C-SiC/Si (001) substrates, in *Silicon-Based Nanomaterials*, Springer Series in Materials Science, vol. 187, ed. by H. Li, J. Wu, Z.M. Wang, (Springer New York, New York, 2013)
36. H. Yaguchi et al., Micro Raman and micro photoluminescence study of cubic GaN grown on 3C-SiC(001) substrates by metalorganic vapor phase epitaxy. J. Cryst. Growth **195**, 323 (1998)
37. T.J. Wade et al., MOVPE studies of zincblende GaN on 3C-SiC/Si(001). J. Cryst. Growth **611**, 127182 (2023)
38. L.Y. Lee, M. Frentrup, M.J. Kappers, R.A. Oliver, C.J. Humphreys, D.J. Wallis, Effect of growth temperature and V/III-ratio on the surface morphology of MOVPE-grown cubic zincblende GaN. J. Appl. Phys. **124**, 105302 (2018)
39. A. Gundimeda, M. Frentrup, S.M. Fairclough, M.J. Kappers, D.J. Wallis, R.A. Oliver, Investigation of wurtzite formation in MOVPE-grown zincblende GaN epilayers on $Al_xGa_{1-x}N$ nucleation layers. J. Appl. Phys. **131**, 115703 (2022)
40. L.Y. Lee, M. Frentrup, P. Vacek, M.J. Kappers, D.J. Wallis, R.A. Oliver, Investigation of stacking faults in MOVPE-grown zincblende GaN by XRD and TEM. J. Appl. Phys. **125**, 105303 (2019)
41. M.F. Zscherp et al., AlN buffer enhances the layer quality of MBE-grown cubic GaN on 3C-SiC. Cryst. Growth Des. **22**, 6786 (2022)
42. H. Yang, O. Brandt, K. Ploog, MBE growth of cubic GaN on GaAs substrates. Phys. Status Solidi (B) **194**, 109 (1996)
43. S. Strite, An investigation of the properties of cubic GaN grown on GaAs by plasma-assisted molecular-beam epitaxy. J. Vac. Sci. Technol. B **9**, 1924 (1991)
44. X.H. Zheng et al., Method for measurement of lattice parameter of cubic GaN layers on GaAs (001). J. Cryst. Growth **250**, 345 (2003)
45. C.H. Hong, D. Pavlidis, K. Hong, K. Wang, Phase-controlled metal-organic chemical vapor deposition epitaxial growth of GaN on GaAs(100) using NH3. Mater. Sci. Eng. B **32**, 69 (1995)
46. J. Wu, H. Yaguchi, K. Onabe, R. Ito, Y. Shiraki, Photoluminescence properties of cubic GaN grown on GaAs(100) substrates by metalorganic vapor phase epitaxy. Appl. Phys. Lett. **71**, 2067 (1997)
47. R. Trejo-Hernández, Y.L. Casallas-Moreno, S. Gallardo-Hernández, M. López-López, Crystalline phase purity and twinning of Mg-doped zincblende GaN thin films. Appl. Surf. Sci. **636**, 157667 (2023)

48. Y. Hu et al., Blue luminescence origin and Mg acceptor saturation in highly doped zinc-blende GaN with Mg. J. Alloys Compd. **897**, 163133 (2022)
49. V.D. Compeán-García et al., Effects of Mg incorporation in cubic GaN films grown by PAMBE near Ga rich conditions. Mater. Sci. Semicond. Process. **93**, 196 (2019)
50. M. Pérez Caro, A.G. Rodríguez, E. López-Luna, M.A. Vidal, H. Navarro-Contreras, Critical thickness of β-InN/GaN/MgO structures. J. Appl. Phys. **107**, 083510 (2010)
51. R. Suzuki, A. Kawaharazuka, Y. Horikoshi, MBE growth of GaN on MgO substrate. J. Cryst. Growth **301–302**, 478 (2007)
52. M. Kakuda, K. Makino, T. Ishida, S. Kuboya, K. Onabe, MBE growth of cubic AlN films on MgO substrate via cubic GaN buffer layer. Phys. Status Solidi C **9**, 558 (2012)
53. N. Discharoen, S. Sanorpim, N. Nuntawong, V. Yordsri, S. Kijamnajsuk, K. Onabe, Effect of two-step cubic GaN buffer layer on the radio-frequency plasma-assisted molecular beam epitaxy growth of cubic AlN films grown on MgO (001) substrates. Thin Solid Films **784**, 140065 (2023)
54. M. Kakuda, S. Morikawa, S. Kuboya, R. Katayama, H. Yaguchi, K. Onabe, RF-MBE growth of cubic AlN on MgO (001) substrates via 2-step c-GaN buffer layer. J. Cryst. Growth **378**, 307 (2013)
55. K. Okura, K. Takamiya, S. Yagi, H. Yaguchi, Growth temperature dependence of cubic GaN step structures and cubic InN dot arrays grown on MgO (001) vicinal substrates. Jpn. J. Appl. Phys. **58**, SC1051 (2019)
56. S. Dhara et al., Hexagonal-to-cubic phase transformation in GaN nanowires by Ga+ implantation. Appl. Phys. Lett. **84**, 5473 (2004)
57. Y. Cui et al., Cubic GaN formation in Mn/GaN multilayer films grown on 6H-SiC(0001). Appl. Phys. Lett. **82**, 4666 (2003)
58. S.C. Lee, E.J. Peterson, Y.-B. Jiang, C. Wetzel, S.R.J. Brueck, Elastic variation of quasi-one-dimensional cubic-phase GaN at nanoscale. Cryst. Growth Des. **19**, 5046 (2019)
59. S.C. Lee et al., Growth of GaN on a nanoscale periodic faceted Si substrate by metal organic vapor phase epitaxy, in *2003 International Symposium on Compound Semiconductors: Post-Conference Proceedings (IEEE Cat. No.03TH8767)*, (IEEE, San Diego, 2004)
60. S.C. Lee, Y.-B. Jiang, M.T. Durniak, T. Detchprohm, C. Wetzel, S.R.J. Brueck, Atomic-scale phase transition of epitaxial GaN on nanostructured Si(001): Activation and beyond. Cryst. Growth Des. **16**, 2183 (2016)
61. S.C. Lee, B. Pattada, S.D. Hersee, Y.-B. Jiang, S.R. Brueck, Nanoscale spatial phase modulation of GaN on a V-grooved Si Substrate-cubic phase GaN on Si(001) for monolithic integration. IEEE J. Quantum Electron. **41**, 596 (2005)
62. R. Liu, R. Schaller, C.Q. Chen, C. Bayram, High internal quantum efficiency ultraviolet emission from phase-transition cubic GaN integrated on Nanopatterned Si(100). ACS Photonics **5**(3), 955–963 (2018)
63. C. Bayram et al., Cubic phase GaN on Nano-grooved Si (100) via maskless selective area epitaxy. Adv. Funct. Mater. **24**, 4492 (2014)
64. R. Liu, C. Bayram, Cathodoluminescence study of luminescence centers in hexagonal and cubic phase GaN hetero-integrated on Si(100). J. Appl. Phys. **120**, 025106 (2016)
65. R. Liu, C. Bayram, Maximizing cubic phase gallium nitride surface coverage on nano-patterned silicon (100). Appl. Phys. Lett. **108**, 231106 (2016)
66. C. Bayram, R. Liu, Polarization-free integrated gallium-nitride photonics, in *Presented at the SPIE OPTO*, ed. by M. Razeghi, (San Francisco, 2017)
67. J. Lee, Y.C. Chiu, M.A. Johar, C. Bayram, Structural and optical properties of cubic GaN on U-grooved Si (100). Appl. Phys. Lett. **121**, 032101 (2022)
68. J. Lee, C. Bayram, Green-emitting cubic GaN/In0.16Ga0.84N/GaN quantum well with 32% internal quantum efficiency at room temperature. Appl. Phys. Lett. **124**, 011101 (2024)
69. C. Bayram, R. Liu, Cubic phase light emitters hetero-integrated on silicon, in *2017 IEEE Photonics Conference (IPC)*, (IEEE, Orlando, 2017)

70. C. Bayram et al., Gallium nitride on silicon for consumer and scalable photonics, in *Presented at the SPIE OPTO*, ed. by M. Razeghi, (San Francisco, 2013)
71. C. Bayram et al., Engineering future light emitting diodes and photovoltaics with inexpensive materials: Integrating ZnO and Si into GaN-based devices, in *Presented at the SPIE OPTO*, ed. by F.H. Teherani, D.C. Look, D.J. Rogers, (San Francisco, 2013)
72. C. Bayram, M.A. Johar, *Large Area Synthesis of Cubic Phase Gallium Nitride on Silicon*, US20230235480A1, (2023)
73. C. Bayram, M.A. Johar, *Large Area Synthesis of Cubic Phase Gallium Nitride on Silicon*, US20230238246A1, (2023)
74. C. Bayram, R. Liu, *Maximizing Cubic Phase Group III-nitride on Patterned Silicon*, US10027086B2, (2018)
75. C. Bayram, C.-W. Cheng, D.K. Sadana, K.-T. Shiu, *Dual Phase Gallium Nitride Material Formation on (100) Silicon*, US9048173B2, (2015)
76. S.C. Lee, Y.-B. Jiang, M. Durniak, C. Wetzel, S.R.J. Brueck, Initial stage of cubic GaN for heterophase epitaxial growth induced on nanoscale v-grooved Si(001) in metal-organic vapor-phase epitaxy. Nanotechnology **30**, 025711 (2019)
77. M. Frentrup et al., X-ray diffraction analysis of cubic zincblende III-nitrides. J. Phys. D. Appl. Phys. **50**, 433002 (2017)
78. M.A. Moram, M.E. Vickers, X-ray diffraction of III-nitrides. Rep. Prog. Phys. **72**, 036502 (2009)
79. J. Krustok, H. Collan, K. Hjelt, Does the low-temperature Arrhenius plot of the photoluminescence intensity in CdTe point towards an erroneous activation energy? J. Appl. Phys. **81**, 1442 (1997)
80. Y.P. Varshni, Temperature dependence of the energy gap in semiconductors. Physica **34**, 149 (1967)
81. G. Ramírez-Flores, H. Navarro-Contreras, A. Lastras-Martínez, R.C. Powell, J.E. Greene, Temperature-dependent optical band gap of the metastable zinc-blende structure β-GaN. Phys. Rev. B **50**, 8433 (1994)
82. J.R. Lakowicz, *Principles of Fluorescence Spectroscopy* (Springer, New York, 2006)
83. M.E. Levinshteĭn, S.L. Rumyantsev, M. Shur (eds.), *Properties of Advanced Semiconductor Materials: GaN, AlN, InN, BN, SiC, SiGe* (Wiley, New York, 2001)
84. D.J. As, S. Potthast, J. Fernandez, K. Lischka, H. Nagasawa, M. Abe, Mechanism of current leakage in Ni schottky diodes on cubic GaN and $Al_xGa_{1-x}N$ epilayers. MRS Online Proc. Lib. (2005). https://doi.org/10.1557/PROC-0892-FF13-04
85. P. Siddiqua, M.S. Shur, S.K. O'Leary, Electron transport within bulk cubic boron nitride: A Monte Carlo simulation analysis. J. Appl. Phys. **128**, 185704 (2020)
86. B. Ding et al., Alloy segregation at stacking faults in zincblende GaN heterostructures. J. Appl. Phys. **128**, 145703 (2020)
87. C. Camas, J.E. Conde, M.A. Vidal, H. Vilchis, Molecular dynamics simulation of cubic InxGa (1-x)N layers growth by molecular beam epitaxy. Comput. Mater. Sci. **193**, 110387 (2021)
88. C. Humphreys, Carrier localisation, efficiency droop, cubic GaN and the green gap, in *DoE SSL R&D Workshop*, (Long Beach, 2017)
89. S. Li, J. Schörmann, D.J. As, K. Lischka, Room temperature green light emission from nonpolar cubic InGaN/GaN multi-quantum-wells. Appl. Phys. Lett. (2007). https://doi.org/10.1063/1.2475564
90. S. Marcinkevičius et al., High internal quantum efficiency of long wavelength InGaN quantum wells. Appl. Phys. Lett. **119**, 071102 (2021)
91. M.-S. Oh et al., Improvement of green LED by growing p-GaN on In0.25GaN/GaN MQWs at low temperature. J. Cryst. Growth **289**, 107 (2006)
92. Z. Xu et al., Fabrication of normally off AlGaN/GaN MOSFET using a self-terminating gate recess etching technique. IEEE Electron Device Lett. **34**, 855 (2013)
93. G. Greco, F. Iucolano, F. Roccaforte, Review of technology for normally-off HEMTs with p-GaN gate. Mater. Sci. Semicond. Process. **78**, 96 (2018)

94. G. Bauer, F. Kuchar, H. Heinrich, *Two Dimensional Systems, Physics and New Devices: Proceedings of the International Winter School, Mauterndorf, Austria, February 24–28, 1986* (Springer-Verlag, Berlin, 1986)
95. Y.-C. Tsai, C. Bayram, Band alignments of ternary wurtzite and zincblende III-nitrides investigated by hybrid density functional theory. ACS Omega **5**, 3917 (2020)
96. B.E. Foutz, L.F. Eastman, U.V. Bhapkar, M.S. Shur, Comparison of high field electron transport in GaN and GaAs. Appl. Phys. Lett. **70**, 2849 (1997)
97. N. Chowdhury et al., P-channel GaN transistor based on p-GaN/AlGaN/GaN on Si. IEEE Electron Device Lett. **40**, 1036 (2019)
98. K. Zhang, M. Sumiya, M. Liao, Y. Koide, L. Sang, P-channel InGaN/GaN heterostructure metal-oxide-semiconductor field effect transistor based on polarization-induced two-dimensional hole gas. Sci. Rep. **6** (2016)
99. A. Krishna, A. Raj, N. Hatui, S. Keller, U.K. Mishra, Investigation of nitrogen polar p-type doped GaN/AlxGa(1-x)N superlattices for applications in wide-bandgap p-type field effect transistors. Appl. Phys. Lett. **115** (2019). https://doi.org/10.48550/arXiv.1910.06421
100. J. Lee, C. Bayram, *Removal of Electron Blocking Layer from InGaN-Based Visible Light Emitting Diodes.*, U. S. Patent Application 18/480,714
101. C. Bayram, J. Lee, *Removal of Electron Blocking Layer from InGaN-Based Visible Light Emitting Diodes.*, U. S. Patent Application 63/413,731
102. Y.-C. Tsai, C. Bayram, J.-P. Leburton, Interplay between Auger recombination, carrier leakage, and polarization in InGaAlN multiple-quantum-well light-emitting diodes. J. Appl. Phys. **131**, 193102 (2022)
103. J. Lee, J.-P. Leburton, C. Bayram, Design tradeoffs between traditional hexagonal and emerging cubic $In_XGa_{(1-X)}N$/GaN-based green light-emitting diodes. J. Opt. Soc. Am. B **40**, 1017 (2023)
104. C.J.M. Stark, T. Detchprohm, S.C. Lee, Y.-B. Jiang, S.R.J. Brueck, C. Wetzel, Green cubic GaInN/GaN light-emitting diode on microstructured silicon (100). Appl. Phys. Lett. **103**, 232107 (2013)
105. D.J. Binks, P. Dawson, R.A. Oliver, D.J. Wallis, Cubic GaN and InGaN/GaN quantum wells. Appl. Phys. Rev. **9**, 041309 (2022)
106. M.T. Durniak et al., Green emitting cubic GaInN/GaN quantum well stripes on micropatterned Si(001) and their strain analysis. Adv. Electron. Mater. **2** (2016). https://doi.org/10.1002/aelm.201500327
107. C. Bayram et al., *Polarization Free Gallium Nitride-Based Photonic Devices on Nanopatterned Silicon*, US9608160B1, (2017)
108. M. Pattison et al., *Solid-State Lighting R&D Opportunities.*, U. S. Department of Energy, DOE/EE-2542, 1862626, 8851, (2022)
109. J. Simon et al., Direct comparison of recombination dynamics in cubic and hexagonal GaN/AlN quantum dots. Phys. Rev. B **68**, 035312 (2003)
110. H. Machhadani et al., Intersubband absorption of cubic GaN/Al(Ga)N quantum wells in the near-infrared to terahertz spectral range. Phys. Rev. B **83**, 075313 (2011)
111. J. Lee, Y.-C. Tsai, C. Bayram, Green light emitting diodes for the ultimate solid-state lighting, in *Compound Semiconductor Week (CSW)*, (IEEE, Ann Arbor, 2022), p. 2022
112. S.A. Church et al., Photoluminescence efficiency of zincblende InGaN/GaN quantum wells. J. Appl. Phys. **129**, 175702 (2021)
113. S.F. Li et al., Strong room temperature 510 nm emission from cubic InGaN/GaN multiple quantum wells. MRS Online Proc. Lib. **831**, 613–618 (2004)
114. G. Chen et al., Performance of high-power III-nitride light emitting diodes. Phys. Status Solidi (A) **205**, 1086 (2008)
115. M. Peter et al., New developments in green LEDs: New developments in green LEDs. Phys. Status Solidi (A) **206**, 1125 (2009)

116. H. Amano, Progress and Prospect of the growth of wide-band-gap group III nitrides: Development of the growth method for single-crystal bulk GaN. Jpn. J. Appl. Phys. **52**, 050001 (2013)
117. C. Lynsky et al., Barriers to carrier transport in multiple quantum well nitride-based c -plane green light emitting diodes. Phys. Rev. Mater. **4**, 054604 (2020)
118. H.I. Solís-Cisneros et al., Theoretical and computational analysis of a Wurtzite-AlGaN DUV-LED to mitigate quantum-confined Stark effect with a Zincblende comparison considering Mg- and Be-doping. Nanomaterials **12**, 4347 (2022)
119. S. Kako, M. Holmes, S. Sergent, M. Bürger, D.J. As, Y. Arakawa, Single-photon emission from cubic GaN quantum dots. Appl. Phys. Lett. **104** (2014)
120. A.M. Berhane et al., Bright room-temperature single-photon emission from defects in gallium nitride. Adv. Mater. **29** (2017)
121. V.A. Chitta et al., Room temperature ferromagnetism in cubic GaN epilayers implanted with Mn+ ions. Appl. Phys. Lett. **85**, 3777 (2004)
122. S.J. Pearton et al., Wide bandgap GaN-based semiconductors for spintronics. J. Phys. Condens. Matter **16**, R209 (2004)
123. J.H. Buß, A. Schaefer, T. Schupp, D.J. As, D. Hägele, J. Rudolph, High temperature electron spin dynamics in bulk cubic GaN: Nanosecond spin lifetimes far above room-temperature. Appl. Phys. Lett. **105**, 182404 (2014)
124. J.H. Buß, J. Rudolph, T. Schupp, D.J. As, K. Lischka, D. Hägele, Long room-temperature electron spin lifetimes in highly doped cubic GaN. Appl. Phys. Lett. **97**, 062101 (2010)
125. C. Bayram, J.L. Pau, R. McClintock, M. Razeghi, Comprehensive study of blue and green multi-quantum-well light-emitting diodes grown on conventional and lateral epitaxial overgrowth GaN. Appl. Phys. B Lasers Opt. **95**, 307 (2009)

Chapter 27
Development of Wafer-Scale *h*-BN Quasi-Bulk Crystals

Hongxing Jiang and Jingyu Lin

1 Introduction

GaN and related III-nitride wide bandgap (WBG) semiconductors have fueled a revolution in lighting and consumer electronics industries and are poised to make significant contributions to the field of power electronics [1–11]. These WBG materials are also expected to make an impact in the areas of full-spectrum solar energy conversion, sterilization, and UV curing device technologies [12–14]. Among members of the III-nitride family, the crystal growth and use of h-BN as an ultrawide bandgap (UWBG) semiconductor are much less developed. Moreover, h-BN is distinctly different in terms of its crystalline structure and hence physical properties. As illustrated in Fig. 27.1, h-BN has a layered crystalline structure and a close lattice-match to graphite and hence to graphene [15, 16]. In its two-dimensional (2D) form, h-BN has attracted tremendous amount of interest from different fields and has been studied as an ideal template, insulating layer, and barrier material for the construction of novel 2D structures such as grapheme-based heterostructures [15] as well as a host for optically stable single-photon emitters [17].

In its three-dimensional form, h-BN possesses exceptional physical properties including an ultrawide bandgap (UWBG) of ~6 eV [18–22], high breakdown field of ~12 MV/cm [23, 24], high in-plane thermal conductivity of ~550 W/m·K [25], and large interaction cross-section between thermal neutrons and its atomic constituent ^{10}B of ~3840 barns ($= 3.84 \times 10^{-21}$ cm^2) [26, 27].

Figure 27.2 shows the calculated band structure near the high symmetry K- and M-points in the Brillion zone (BZ) of h-BN [20]. The results of the band structure calculations [20, 28] and phonon-assisted photoluminescence (PL) measurements

H. Jiang (✉) · J. Lin
Texas Tech University, Lubbock, TX, USA
e-mail: hx.jiang@ttu.edu; jingyu.lin@ttu.edu

I. C. Kizilyalli et al. (eds.), *Gallium Nitride and Related Materials*, The Materials Research Society Series, https://doi.org/10.1007/978-3-031-83056-3_27

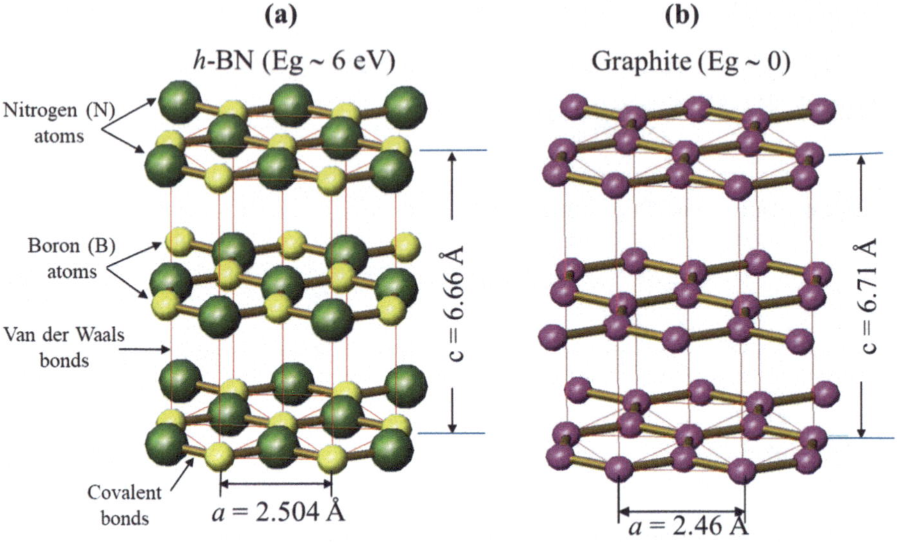

Fig. 27.1 Schematic illustration of crystal structures of (**a**) h-BN and (**b**) graphite. (Reproduced from Ref. [16])

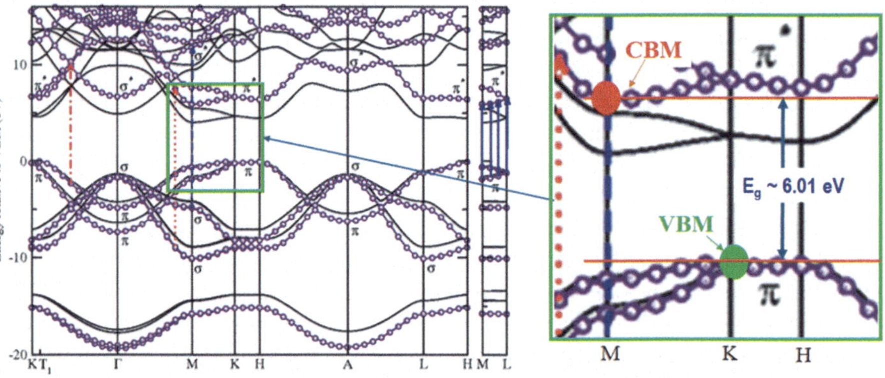

Fig. 27.2 Electronic band structure along high-symmetry lines for bulk h-BN. The thin solid lines display the LDA results, while the solid lines with open circles show the GW approximation results. The inset to the right is a zoom-in view of the band structure near high symmetry K- and M-points in the Brillion zone (BZ). (Figures reproduced from Ref. [20], Copyright 2006 American Physical Society)

[22, 29–31] have provided us with the following insights concerning the band structure near the band edges of h-BN:

(a) h-BN is an indirect gap semiconductor with a bandgap of around 6.0 eV.
(b) The conduction band minimum (CBM) is at the M-point, and the valence band maximum (VBM) is at the K-point.

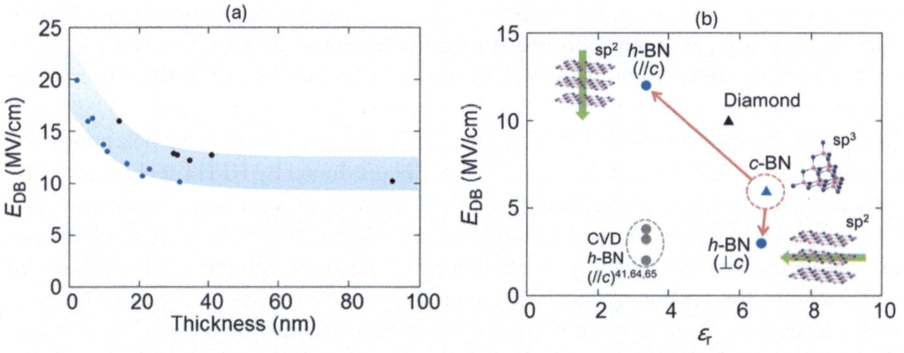

Fig. 27.3 (**a**) Breakdown field along the c-axis, $E_{BD||c}$, as a function of the *h*-BN thickness. (**b**) E_{BD} as a function of dielectric constant ε_r for *h*-BN, *c*-BN, and diamond. (Figure reproduced from Ref. [23], Copyright 2015 American Chemical Society)

(c) The slope of the dispersion of the M-K section of the conduction band seems slightly larger than that of the valence band [20], implying that the minimum direct energy bandgap of h-BN is most likely at the M-point. The M-K section is most relevant for carrier generation using the above bandgap photoexcitation and subsequent light emission.

Despite its indirect bandgap nature, h-BN exhibits higher emission efficiency than AlN in the c-direction [32, 33] as well as an exceptionally large optical absorption coefficient of 7.5×10^5 cm^{-1} [18]. These features can be attributed to several unique properties of h-BN:

(a) The optical absorption of h-BN follows that of graphene and is around 2.3% per layer (3.33 Å) for the above bandgap excitation [34].
(b) Unusually strong $p \rightarrow p$-like transitions due to the short bond length of B-N (1.55 Å) [35].
(c) 2D nature induces a sharp "step"-like density of states (DOS), g(hω), around the absorption edge [35].

Due to its layered crystalline structure, most physical properties of h-BN are also anisotropic. The breakdown field, E_{BD}, and the anisotropy in the dielectric breakdown of high-quality exfoliated single-crystal *h*-BN grown by high pressure and high temperature (HPHT) techniques have been investigated [23, 24]. Under a well-controlled relative humidity, the values of E_{BD} in the directions both normal and parallel to the c-axis ($E_{BD}\perp c$ and $E_{BD}||c$) were measured to be 3 and 12 MV/cm, respectively. The breakdown field along the c-axis seems to exceed that of diamond as shown in Fig. 27.3 [23, 24].

The thermal conductivity and the anisotropy of the thermal conductivity of high-quality single-crystal *h*-BN grown by a metal flux solution method have been measured for natural and ^{10}B and ^{11}B isotope-enriched h-BN [25]. The results revealed that natural h-BN, containing 20% ^{10}B and 80% ^{11}B, exhibits an in-plane thermal conductivity, $\kappa \perp c$, of about 550 W/m·*K*, whereas the out-of-plane thermal conductivity, $\kappa \| c$, is about 2 orders of magnitude lower [25].

The overall outstanding properties make h-BN an attractive UWBG material for applications for deep-ultraviolet (DUV) emitters and detectors [18–22, 29–37], solid-state neutron detectors [38–44], host for quantum defects, and potentially for power electronic devices. Bulk crystals of h-BN produced by HPHT techniques [45–48] and metal flux solution methods [49, 50] possess very high crystalline and optical qualities but are impractical to serve as substrates or for device implementation as their sizes are typically in millimeters. The development of crystal growth technologies for producing thick epitaxial films (or quasi-bulk crystals) in large wafer sizes with high crystalline quality is a prerequisite for utilizing h-BN as a UWBG electronic material.

Because of the large interaction cross-section between thermal neutrons and ^{10}B of ~3840 barns ($= 3.84 \times 10^{-21}$ cm^2) [26, 27], as compared to traditional III-nitrides, BN has another unique application as a solid-state direct-conversion neutron detector material, which will be discussed in more detail in Sect. 5. The neutron detector application also requires the development of quasi-bulk crystals to provide high detection efficiencies. This is because the theoretical efficiency follows the neutron absorption in the detection material, $\eta_i = 1 - e^{-d/\lambda}$, where d denotes the detector thickness and λ the neutron absorption length. For thermal neutron detection, λ is 47μm (237μm) for ^{10}B isotope-enriched (natural) h-BN [33, 40–43]. For fast neutron detection, λ in h-BN is on the order of centimeters [43].

Here, we provide a brief overview of the recent progress toward synthesizing h-BN quasi-bulk crystals and the realization of high-efficiency thermal neutron detectors based on ^{10}B isotope-enriched h-BN (h-^{10}BN) achieved by our group.

2 Wafer-Scale Crystal Growth

2.1 HVPE Growth

HVPE is a proven epitaxial growth technique capable of producing high-quality GaN and AlN quasi-bulk wafers at high growth rates, up to 100 s of microns per hour [51–53]. More recently, it was also demonstrated that the HVPE growth technique is capable of producing GaN with significantly reduced background impurity concentrations [54] and improved p-type conductivity [55] over those produced by MOCVD because precursors for HVPE growth are free from carbon impurities.

To grow h-BN wafers, natural *boron trichloride* (BCl_3) and NH_3 were used as precursors [56–58]. The growth was conducted on c-plane sapphire of 2 inches or 4 inches in diameter at a temperature above 1400 °C. As illustrated in Fig. 27.4a, due to its layered crystalline structure, h-BN self-separates from sapphire to form a freestanding wafer during cooling down after growth [40–43, 56, 57]. The thickest HVPE h-BN quasi-bulk wafers produced by our group, as of this writing, have a thickness of ~500μm, as probed by a thickness profilometer. An example of thickness calibration is shown in Fig. 27.4a.

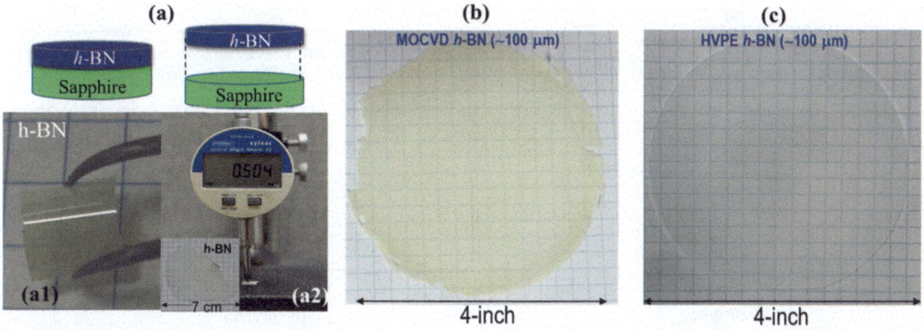

Fig. 27.4 Freestanding h-BN wafers: (**a**) schematic illustration for realizing freestanding h-BN wafer by self-separation during cooling down after growth. Insets are photos of a flexible 50-μm thick wafer produced by MOCVD (a1) and a 504-μm thick wafer produced by HVPE (a2). Figure reproduced from Ref. [40], Copyright 2017 AIP Publishing LLC. Figure (a2) reproduced from Ref. [58]. (**b**) B-10-enriched h-BN (h-^{10}BN) wafer of 4 inches in diameter grown by MOCVD. Figure reproduced from Ref. [41], Copyright 2019 AIP Publishing LLC. (**c**) h-BN wafer of 4 inches in diameter grown by HVPE

Figures 27.4b, c compare optical images among representative wafers: (b) grown by MOCVD using trimethylboron (TMB) metal-organic (MO) source as a precursor for B [41] and (c) grown by HVPE using BCl_3 as a precursor for B [56, 57]. As can be seen from this side-by-side comparison, the h-BN wafer grown by HVPE using BCl_3 gas as a precursor exhibits much better transparency and less yellowish color than those produced by MOCVD using a TMB source. The improved transparency is related to the fact that the BCl_3 precursor contains no carbon impurities. Theoretical work indicates that carbon impurities in h-BN can occupy both B and N sites as well on an interstitial site [59]. When occupying the B site, C_B is a deep donor with an energy level of 2.2 eV below the conduction band edge, and when occupying an interstitial site, C_i is a deep acceptor sitting at an energy level of 2.4 eV above the valence band edge [59], and the presence of both of which will render a yellowish colored wafer. When occupied on the N site, C_N is a deep-level acceptor sitting at a position of 3.2 eV above the valence band maximum [59].

2.2 Crystalline Quality Characterization

Although bulk h-BN single crystals grown by HP/HT and metal flux solution methods are small in crystallite size [19, 21–25, 45–50], they possess higher crystalline and optical qualities than materials produced by epitaxial growth techniques. On the other hand, MOCVD is an established technique for producing high-quality III-nitride thin epitaxial layers. Benchmarking the structural and optical qualities of h-BN quasi-bulk crystals synthesized by HVPE against those of the state-of-the-art mm-sized bulk crystals and MOCVD thin epilayers provides valuable insights regarding key areas for further improvement [58].

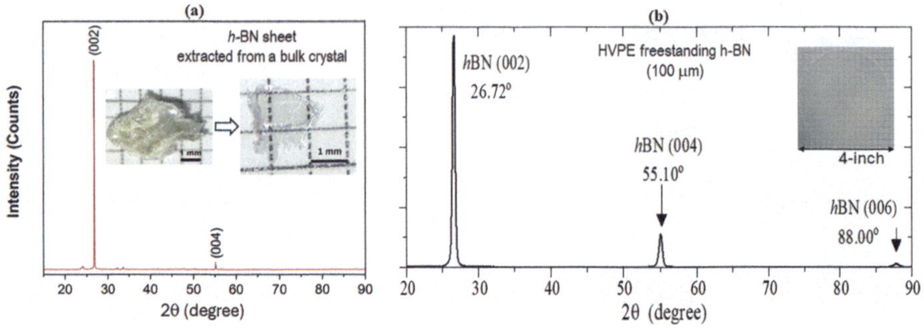

Fig. 27.5 Comparison of XRD patterns in 2θ-ω scans in a large angle range: (**a**) h-BN sheet extracted from a bulk crystal synthesized by the polymer route combined with a sintering process at an HPHT condition. Figure reproduced from Ref. [48], Copyright 2020 American Chemical Society. (**b**) 100-μm thick freestanding and flexible h-BN quasi-bulk wafer synthesized by HVPE. Figure reproduced from Ref. [58]

Shown in Fig. 27.5, XRD 2θ-ω scans in a large angle range are compared between (a) a h-BN sheet extracted from a bulk crystal synthesized by a polymer route combined with a sintering process in a HPHT condition (1800 °C and 180 MPa) and (b) a 100 μm thick freestanding h-BN quasi-bulk wafer produced by HVPE [58]. Diffraction peaks from h-BN (002) and (004) planes (stacked planes in the c-direction) were clearly observable in both samples. Moreover, the h-BN (006) peak is also observable in the freestanding h-BN quasi-bulk wafer. From the h-BN (002) peak position, an interplanar spacing of 3.328 Å was deduced for the h-BN sheet extracted from a bulk crystal [46], whereas the h-BN (002) peak in the HVPE h-BN quasi-bulk wafer is centered at $2\theta = 26.72°$, providing an interplanar spacing of 3.333 Å [58]. Since the c-lattice constant of a perfect h-BN bulk crystal is 0.666 nm [60], the XRD results imply both types of materials have a c-lattice constant that is very close to the ideal value and have a very good long-range order in the c-axis. On the other hand, the h-BN sheet extracted from a bulk crystal flake exhibits a narrower full width at half maximum (FWHM), which implies that the bulk crystals have a superior structural ordering in the c-plane over those of the quasi-bulk crystals synthesized by HVPE.

Figures 27.6a, b compare the Raman spectra obtained for a h-BN sheet extracted from a bulk crystal [48] and a freestanding h-BN quasi-bulk wafer [61]. The observed mode in both types of materials is related to the E_{2g} vibration mode (in-plane stretch of B and N atoms, as illustrated in the inset of Fig. 27.6b). The observed mode at $\Delta\sigma = 1366\ cm^{-1}$ in both types of samples corresponds well with the expected Raman peak of bulk h-BN [62], suggesting that these samples are strain-free. However, the observed spectral linewidth of 25 cm^{-1} in the quasi-bulk wafer is much broader than the value of 8 cm^{-1} observed in the h-BN sheet extracted from a bulk crystal. Since this Raman mode results from the in-plane stretch of B and N atoms, the linewidth is directly correlated with the structural ordering in the c-plane. The comparison results corroborate the XRD results shown in Fig. 27.5

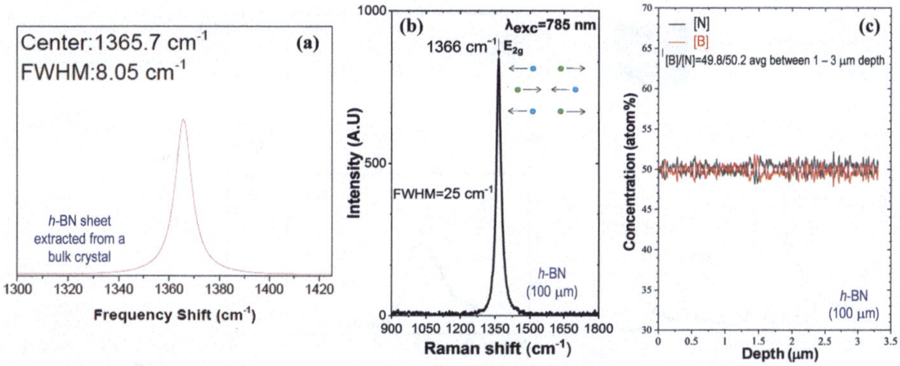

Fig. 27.6 Comparison of room temperature Raman spectra: (**a**) h-BN sheet extracted from a bulk crystal. Figure reproduced from Ref. [48], Copyright 2020 American Chemical Society. (**b**) 100-μm thick freestanding and flexible h-BN quasi-bulk wafer synthesized by HVPE. Figure reproduced from Ref. [61], Copyright 2024 AIP Publishing LLC. (**c**) B and N contents in an h-BN quasi-bulk wafer produced by HVPE probed by secondary ion mass spectroscopy (SIMS) measurements. Averaging data points in the wafer's interior revealed the ratio of boron to nitrogen, [B]/[N], to be 49.8/50.2. Figure reproduced from Ref. [57]

and again imply that the bulk crystals possess a better structural ordering in the c-plane over that of the quasi-bulk crystals. There is still a lot of room for improvement in the crystalline quality of the h-BN quasi-bulk crystals synthesized by HVPE.

The stoichiometry of our quasi-bulk wafers was assessed by the secondary ion mass spectrometry (SIMS) measurement technique, which is capable of probing into the sample's interior. Figure 27.6c shows B and N concentrations, [B] and [N], in a representative wafer. At first sight, SIMS results in Fig. 27.6c hardly reveal any difference between B and N concentrations. We therefore calculated the ratio of [B]/[N] by averaging data points in the wafer's interior, which revealed that the boron concentration is slightly less than the nitrogen concentration with a ratio of [B]/[N] of 49.8/50.2 [57]. Knowing that the atomic density of h-BN is $1.1 \times 10^{23}/\text{cm}^3$, this small deficiency in B content (0.2%) could potentially render a concentration of boron vacancies (V_B) and its complex on the order of $\sim 10^{20}$ cm^{-3}. The presence of V_B is likely related to the fast growth rate of the HVPE growth technique.

3 Photoluminescence Emission and Transport Properties

3.1 Photoluminescence (PL) Emission Properties

Photoluminescence (PL) emission spectroscopy is an effective technique for identifying the overall quality of new materials under development. For instance, by minimizing the ratio of the impurity/defect-related emission to the band-edge

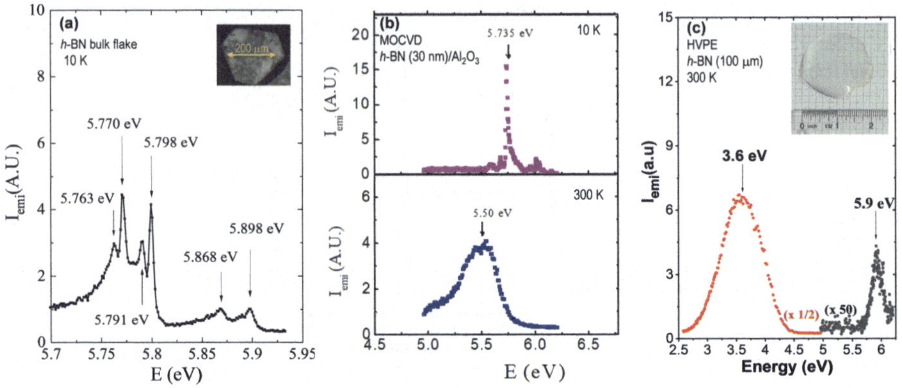

Fig. 27.7 Comparison of PL emission spectra among h-BN samples produced by different growth techniques: (**a**) Bulk crystal flake synthesized by the metal flux solution method. Figure reproduced from Ref. [50], Copyright 2016 AIP Publishing LLC. (**b**) h-BN (30 nm)/sapphire epilayer synthesized by MOCVD using a pulsed precursor flow scheme. Figure reproduced from Ref. [63], Copyright 2016 AIP Publishing LLC. (**c**) 100-μm thick freestanding flexible h-BN quasi-bulk wafer synthesized by HVPE. Figure reproduced from Ref. [61], Copyright 2024 AIP Publishing LLC

emission intensity by varying the key growth parameters (growth temperature and rate, pressure, V/III ratio, etc.), materials with enhanced quality can be obtained.

Figure 27.7 compares PL emission properties among a bulk crystal flake produced by the metal flux solution method [50], a 30-nm thick epilayer grown by MOCVD [63], and a 100-μm thick quasi-bulk wafer grown by HVPE [61]. In the PL spectrum of the bulk crystal [50], six sharp emission lines were resolved between 5.75 and 5.90 eV, which have been attributed to the intrinsic or free-exciton (FX) transitions and labeled as S-series lines [22, 64–67]. A hall mark of high-quality materials is the observation of FX recombination lines. The results shown in Fig. 27.7a are therefore an indication that the bulk crystal has a relatively low density of defects.

The top panel of Fig. 27.7b shows the low temperature (10 K) PL spectrum of a 30-nm thick h-BN epilayer produced by MOCVD, which shows that the free exciton emission line near 5.735 eV is the dominant transition line, whereas the bottom panel of Fig. 27.7b is the PL emission spectrum obtained at 300 K showing a dominant emission line near 5.5 eV [63]. The 5.5-eV emission line exhibited by MOCVD epilayers has been attributed previously to a combination of quasi donor-acceptor pair (q-DAP) and impurity-bound exciton transitions in h-BN with the most probable candidate of the impurity involved being carbon which is introduced from the metal-organic (MO) precursor for B [68].

In comparison, the room temperature PL emission spectrum of the h-BN quasi-bulk sample produced by HVPE presented in Fig. 27.7c exhibits a dominant emission line near 3.6 eV and a weak band-edge emission peak near 5.9 eV. The broadness of the PL emission peak near 3.6 eV is a typical characteristic of a donor-acceptor-pair (DAP) transition. The origin of this DAP emission line was thought to be the transition between O donors, O occupying nitrogen site (O_N), and deep-level acceptors associated with boron vacancy-hydrogen complexes (V_B-H) [57]. The assignment of V_B-related emission is also supported by the SIMS results shown in Fig. 27.6c.

The observation of the weak band-edge emission near 5.9 eV suggests that these quasi-bulk crystals have a decent crystalline quality, in agreement with the XRD results shown in Fig. 27.5. However, the DAP recombination line being dominant in h-BN quasi-bulk crystals [57] also indicates that the overall material quality is still some distance away from those of the state-of-the-art h-BN bulk crystals and thin epitaxial layers.

3.2 Mobility-Lifetime Product Characterization

UWBG materials generally possess extremely high electrical resistivities. As a result, it is difficult to characterize their electrical properties using the traditional Hall-effect measurement technique. To characterize the electrical properties, lateral devices were fabricated to take advantage of h-BN's superior transport properties in the c-plane over its transport properties along the direction of the c-axis [56, 69]. The fabrication processes include the following steps: (1) dicing freestanding h-BN wafer into detector strips, (2) mount detector strips on sapphire using a highly resistive adhesive material, and (3) a mask was fabricated and used to deposit metal contacts consisting of a bilayer of Ni (100 nm)/Au (40 nm) on the clipped edges of the h-BN strips using e-beam evaporation, leaving around ~100μm of metal covering on the two edges [41–43, 56, 57, 69]. The schematic illustration of these lateral detectors is depicted in Fig. 27.8a. The inset of Fig. 27.8b shows a micrograph of a fabricated detector strip (2 mm in width) mounted on sapphire via a layer of highly resistive and adhesive polyimide. The dark I-V characteristic shown in

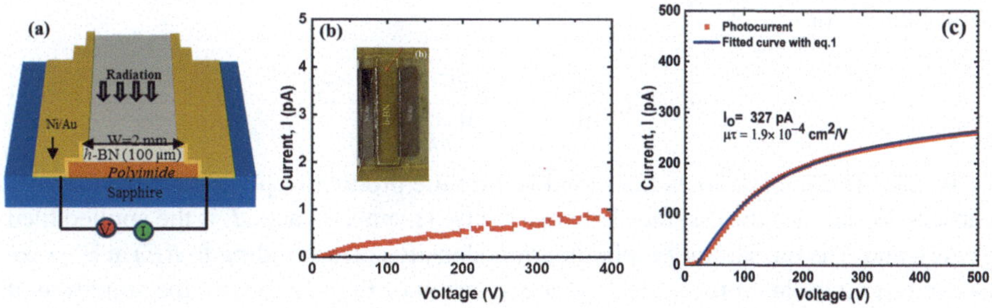

Fig. 27.8 (**a**) Schematic of a h-BN device in a lateral geometry. (**b**) Room temperature I-V characteristics of a fabricated h-BN lateral device shown in the inset. (**c**) I-V characteristics of the detector shown in (**b**) under the illumination by a broad-spectrum UV (185–400 nm) light source, from which a mobility-lifetime product ($\mu\tau$) value of 2×10^{-4} cm^2/V is deduced. (Figure reproduced from Ref. [56], Copyright 2023 AIP Publishing LLC)

Fig. 27.8b yields an electrical resistivity of ~1 × 10^{14} Ω·cm at room temperature [56].

One of the most important transport parameters for determining the device performance is its carrier mobility and lifetime product ($\mu\tau$) [33, 40–43, 69]. For instance, to use h-BN as a detector material, most of the radiation-generated charge carriers inside a detector can be collected when the carrier recombination time (τ) is greater than the transit time (τ_t), $\tau > \tau_t$, or equivalently, the charge carrier drift length ($= \mu\tau E$) is greater than the carrier transit distance (or the width of the detector strip, W). This means that the condition to ensure charge carrier collection follows:

$$\frac{W}{E\mu\tau} = \frac{W^2}{V\mu\tau} < 1, \text{or} \tag{27.1a}$$

$$W < \sqrt{V\mu\tau} \tag{27.1b}$$

where W is the width of the detector strip and E (V) is the applied electric field (bias voltage). The quantity of $\mu\tau$ is strongly influenced by the overall material quality. For MOCVD grown h-^{10}BN, the $\mu\tau$ values measured under photoexcitation have been improved by several orders of magnitude from 10^{-8} cm^2/V [70, 71] to >10^{-4} cm^2/V [42], leading to the realization of h-^{10}BN thermal neutron detectors with a record high detection efficiency of ~60% [42, 58].

Since the growth of HVPE of h-BN is at an initial stage, it is desirable to benchmark the μτ parameter against those of MOCVD-grown materials. The photocurrent-voltage (I-V) characteristics under photoexcitation were utilized to extract μτ value using the classical Many's equation for insulating semiconductors [72], which was derived by assuming that the contacts are perfectly blocking type for simplicity, i.e., Schottky contacts with an infinitely large barrier height [72, 73]. Since the contact area is small in a lateral detector, taking into consideration only the bulk trapping effect by neglecting the effect of surface recombination, we utilized a modified Many's equation to describe the I-V characteristics under uniform illumination [72],

$$I(E) = I_o \frac{\mu\tau E}{W}\left[1 - \frac{\mu\tau E}{W}\left(1 - e^{-\frac{W}{\mu\tau E}}\right)\right] \tag{27.2}$$

In Eq. (27.2), $\mu\tau$ denotes the mobility-lifetime product of photogenerated charge carriers. W denotes the distance between the two metal contacts, E is the applied filed (V/W), and I_0 is the saturation photocurrent density corresponding to $I(E)$ at E → ∞, as can be extrapolated from the Taylor expansion of Eq. (27.2) with the condition of the drift length being much greater than the transit distance of the photogenerated carriers, μτE ≫ W [73].

The measured I-V characteristics under illumination by a broad-spectrum UV (185–400 nm) light source of a representative quasi-bulk h-BN sample are shown in Fig. 27.8c. The fitting results provided a μτ value of 2 × 10^{-4} cm^2/V [56], which is in the same order as the values of MOCVD-grown quasi-bulk wafers [40–42, 74, 75]. This $\mu\tau$ value is sufficiently large to satisfy the charge collection condition of Eqs. (27.1a) and (27.1b), $\mu\tau E > W$, if we operate the detector having a width of W = 3 mm at a bias voltage of ~500 V.

3.3 Probing the Nature of Dominant Defects and Impurities

To investigate the nature of impurities/defects that control the dark resistivity of HVPE-grown quasi-bulk wafers, we conducted electrical resistivity (ρ) measurements at varying temperatures. Figure 27.9a plots the dark I-V characteristics measured at several representative temperatures, and Fig. 27.9b is an Arrhenius plot of the measured temperature-dependent electrical resistivity [57]. The energy level of the impurity/defect which determines the dark resistivity can be estimated from Fig. 27.9b and is ~1.6 eV. We were unable to conduct Hall-effect measurements due to the extremely high resistivity of ~1×10^{14} Ω·cm of the material. As such, the energy level estimated here does not include the effects caused by temperature dependence on the carrier mobility. Moreover, the dark I-V characteristics at higher temperatures of 650 K and 700 K are not perfectly linear. This is because the contacts fabricated on UWBG in the dark state are not ohmic. Realizing high-quality ohmic contacts on highly resistive UWBG semiconductors is another challenging research topic, and much more work is needed. However, we believe that the results shown in Fig. 27.9b provide a reasonable estimate of the energy level of the impurity/defect involved [57].

The properties of native defects and impurities in h-BN have been studied using density functional theory [59]. The results indicated that V_B-hydrogen complexes form deep-level acceptors with a (0/−) charge state and have an energy level ranging from 1.54 eV for V_B-2H to 1.65 eV for V_B-H [59]. Overall, the energy level obtained

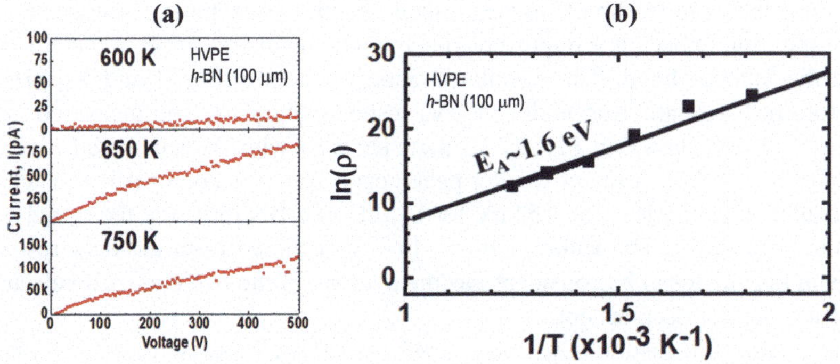

Fig. 27.9 (**a**) Plots of dark I-V characteristics measured at several representative temperatures. (**b**) Arrhenius plot of the measured electrical resistivity (ρ) of the h-BN device. The energy level of the impurity/defect which determines the dark resistivity was estimated to be around 1.6 eV. (Figure reproduced from Ref. [57])

from the dark resistivity measurements shown in Fig. 27.9 is very close to the calculated energy levels of V_B-2H and V_B-H complexes, suggesting that most likely it is the V_B-related deep-level acceptors that control the dark resistivity of quasi-bulk crystals synthesized by HVPE. This contrasts with the results of MOCVD-grown thick wafers, in which the dark current (or the background carrier concentration) is generally controlled by oxygen impurities, which are donors with an activation energy of about 0.6 eV to be discussed in more detail in Sect. 4.4 [69, 74].

The growth rate for HVPE is much faster than that of MOCVD, making the HVPE quasi-bulk materials more prone to the formation of V_B-related defects. On the other hand, the longer growth time required for MOCVD makes thick films grown by MOCVD more likely to contain oxygen impurities which are diffused from the sapphire substrate during growth at high temperatures [69, 74].

To probe the presence of V_B-related acceptors as well as hole conduction in these highly resistive materials, a photocurrent excitation spectroscopy measurement technique was employed, which is a powerful technique for measuring the impurity/defect energy levels, conduction carrier type, and optical energy bandgap in a semiconductor. Photocurrent generation involves a two-step process: (1) direct excitation of carriers from the impurity/defect levels to a band and (2) photoexcited charge carriers drift in the applied field and are collected by the electrodes. Therefore, the observed spectral peak positions in a photocurrent excitation spectrum correspond directly to the impurity/defect energy levels in a semiconductor [57].

The photocurrent excitation spectroscopy measurement setup is schematically shown in Fig. 27.10a [57]. In this experimental setup, a 300-μm wide metal slit was used to allow light to enter and illuminate only near the metal contact on the left side of the detector strip. Under a bias voltage application with a polarity shown in Fig. 27.10a, only photoexcited free holes will be able to drift across the detector and be collected by the electrode on the right. Figure 27.10b plots the measured photocurrent excitation spectrum in the excitation photon energy range of 1.55 and 1.75 eV.

Shown in Fig. 27.10b is a dominant spectral peak at 1.66 eV. Because we are measuring the hole transport, the dominant spectral peak must correspond to the photoexcitation of acceptor impurities/defects with an energy level at $E_A = 1.66$ eV above the valence band. The calculated energy levels of V_B-H and V_B-2H deep acceptors in h-BN are 1.65 and 1.54 eV, respectively [58]. The measured energy level of 1.66 eV shown in Fig. 27.10 matches well with the calculated ionization energy of V_B-H [57]. Another weaker peak near 1.62 eV is not a perfect match with the calculated energy level of 1.54 eV for V_B-2H in h-BN [59], and the signal is also near the noise level. The value of $E_A = 1.6$ eV obtained from the dark resistivity results in Fig. 27.9 can be considered as the average of the two energy levels among V_B-H and V_B-2H deep acceptors.

The predicted/measured O_N donor level is at 0.6 eV below the CBM [59, 69, 74]. These together provide a band diagram for the dominant impurities and defects in h-BN, as shown in Fig. 27.10c. Figure 27.10c also suggests that a DAP recombination involving O_N donors and V_B -H/V_B -2H acceptors is expected in quasi-bulk crystals produced by HVPE. The observed impurity emission peak near 3.6 eV in

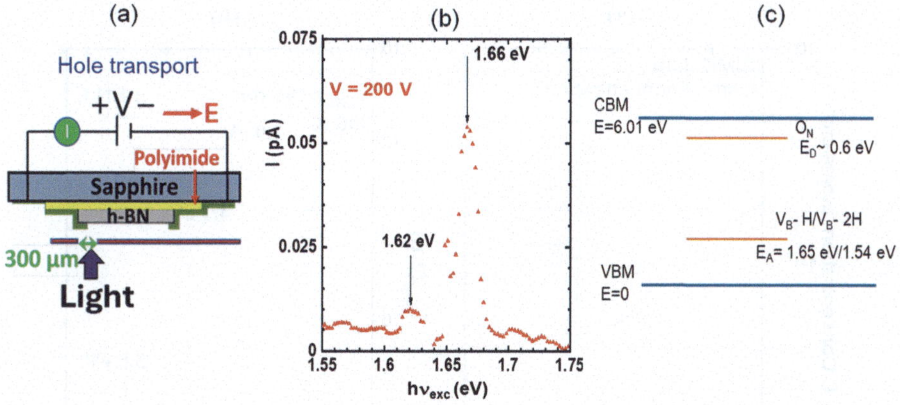

Fig. 27.10 (**a**) Schematic of the photocurrent excitation spectroscopy experimental setup. (**b**) Photocurrent excitation spectrum of hole transport in a h-BN lateral device measured in the excitation photon energy range of 1.55 and 1.75 eV. Under a bias voltage applcation with a polarity shown in (**a**), only photoexcited free holes will be able to drift across the detector and be collected by the electrode on the far right so that the spectral peaks must correspond to the photoexcitation of acceptor impurities/defects. A laser driving light source covering a wavelength range between 190 and 2400 nm was used as an excitation light source. (**c**) Energy level diagram showing the V_B-H and V_B-2H deep level acceptors in h-BN based on calculation Ref. [59] and O_N donor level based on experimental measurement [72]. (Figure reproduced from Ref. [57])

Fig. 27.7c supports this interpretation. It was shown that the peak position of this DAP emission line varies from 3.8 eV to 3.5 eV depending on the concentration of DAP present in the samples [76]. These results together reinforce that the V_B-H complexes are the dominant acceptor-type deep-level defects in h-BN quasi-bulk crystals produced by HVPE.

4 Prospects of Doping

4.1 Carbon Doping

Carbon (C) is a common impurity in group-III nitride semiconductors, which can be introduced unintentionally from reaction precursors for thin film growth techniques such as MOCVD or bulk growth techniques, including HPHT or metal flux method, and could significantly impact their electrical and optical properties. Based on the development of III-nitride semiconductors, gaining a better understanding of the properties of intentionally doped carbon impurities will be useful to provide insights for their utilization or elimination so that the full potential of h-BN as a UWBG semiconductor can be exploited for emerging practical applications [61].

Compared to MOCVD, precursors for HVPE growth are carbon-free, making intentional and controlled doping with carbon impurities possible. Figure 27.11a

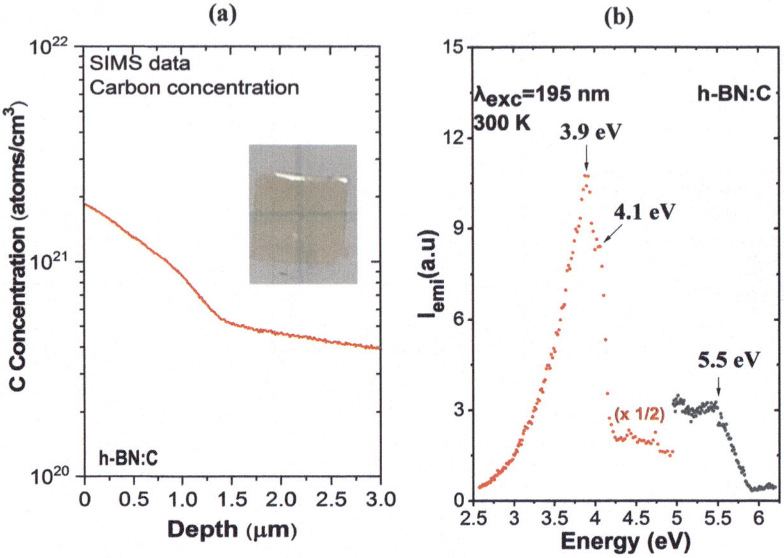

Fig. 27.11 (**a**) Carbon concentration profile in a carbon doped h-BN freestanding sample (h-BN:C) probed by SIMS. The inset shows a micrograph of a freestanding h-BN:C sample of 100 μm in thickness used in this study. (**b**) Room temperature PL emission spectra obtained under an above bandgap excitation ($\lambda_{exc} = 195$ nm) for an h-BN:C sample measured in uv-vis spectral region. (Figures reproduced from Ref. [61], Copyright 2024 AIP Publishing LLC)

plots the carbon concentration as a function of the sample depth, probed by *SIMS* of an intentionally carbon-doped quasi-bulk h-BN (h-BN:C) sample produced by HVPE [61]. Figure 27.11b shows the PL emission spectrum measured for a h-BN: C sample, revealing two emission lines near 4.1 eV and 3.9 eV and another much weaker emission line near 5.5 eV. The 4.1-eV line has been identified to be associated with the presence of C_N [78–80]. However, recent theoretical studies suggested that C_N may not be the origin of the 4.1-eV line and that its origin is more likely related to the carbon dimer defect C_2 [81, 82]. A hall mark of optical emissions in h-BN is the involvement of a very strong carrier-phonon interaction to conserve the momentum and to enhance the transition probabilities because of its indirect energy bandgap nature [20, 22, 28–31, 83]. A lower energy peak appearing around 3.9 eV is most likely a phonon replica of the zero-phonon line (ZPL) at 4.1 eV. The phonon involved is close to the known value of the 200-meV longitudinal optical (LO) phonon mode in h-BN [30, 83]. The weak and broad emission line near 5.5 eV seen in the C-doped sample was observed in MOCVD grown epilayer shown in Fig. 27.7b.

While PL emission properties of h-BN in the UV region have been extensively investigated, emission characteristics in the infrared (IR) spectral region need to be measured to probe possible transitions between defect levels. Room temperature PL emission spectra of a h-BN:C sample measured in the IR spectral region using above ($\lambda_{exc} = 195$ nm) and below ($\lambda_{exc} = 405$ nm) bandgap excitation are shown in

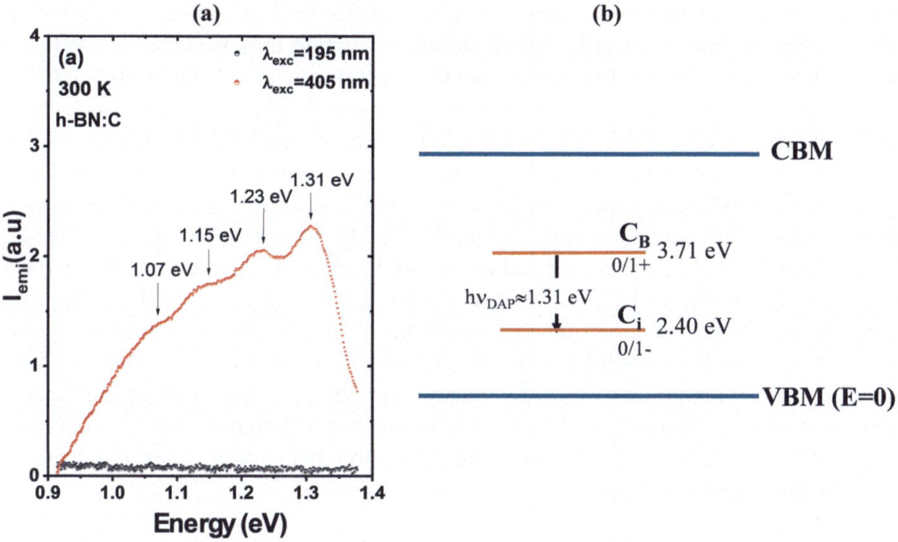

Fig. 27.12 (**a**) Room temperature PL emission spectra of a h-BN:C sample measured in the IR spectral region using above ($\lambda_{exc} = 195$ nm) and below ($\lambda_{exc} = 405$ nm) bandgap excitation. Energy diagram of carbon impurities relevant to the observed IR emission in h-BN. (**b**) Energy levels are measured with respect the valence band maximum (VBM), reinforcing the perception that C-doing tends to introduce deep level centers in h-BN. (Figures reproduced from Ref. [61], Copyright 2024 AIP Publishing LLC)

Fig. 27.12a [61]. The spectrum contains no visible features in the IR region between 0.9 and 1.4 eV under an above bandgap excitation ($\lambda_{exc} = 195$ nm) for both C-doped and undoped h-BN samples. In contrast, the emission spectrum for the C-doped sample exhibits a broad emission band with the main peak appearing at 1.31 eV under a below bandgap excitation ($\lambda_{exc} = 405$ nm), while no emission lines were evident in undoped h-BN in the same IR region under the same experimental measurement conditions [61]. More careful inspection of the IR spectrum shown in Fig. 27.12a reveals that the observed emission band comprises multiple emission lines with an equal energy separation between two adjacent peaks; a multiple Gaussian peak fitting provides a separation energy of ~78 meV. It has been shown that deep-level impurity emission lines in h-BN are strongly assisted by the intervalley phonon scattering processes [31, 83]. We thus attribute this series of emission lines to the same C impurity-related emission, which has its primary peak at 1.31 eV along with its higher-order phonon replicas displaying a successive reduction in the emission intensity with increasing the numbers of phonons involved [61]. The detailed phonon dispersions for the longitudinal acoustic (LA), transverse acoustic (TA), and out-of-plane optical (ZO) phonon branches in h-BN reveal a substantial overlap among LA, TA, and ZO phonon modes at the middle of the BZ, near the T-point [83].

The broad spectral line shape of the observed 1.31-eV emission line suggests that its origin is most likely related to DAP recombination. Based on the first principle calculation of impurities and native defects in h-BN [59], we have constructed in Fig. 27.12b, an energy diagram by including the carbon impurity energy levels for C_B (carbon occupying the boron site) and C_i (carbon interstitial), showing that C_B is located at 3.71 eV above the valence band maximum (VBM), whereas C_i is located at 2.4 eV above VBM [59]. The emission line observed near 1.31 eV in Fig. 27.12a corresponds well with the calculated energy level separation between C_B and C_i of 1.31 eV in h-BN and hence can be conclusively assigned to a DAP transition between C_B donors and C_i acceptors, as indicated by the arrow in Fig. 27.12b [61].

The results thus verify the calculated energy levels of C_B and C_i and that carbon impurities form C_B and C_i, which are deep-level impurities in h-BN [59]. The absence of the 1.3-eV emission line under the above bandgap excitation can be accounted for by the possibility that other optical recombination channels are more efficient and dominant than this DAP transition under the above bandgap excitation. The excellent agreement between the measured and calculated energy separation between C_B and C_i in h-BN reinforces the perception that C-doing tends to introduce deep-level centers [59, 61].

4.2 Acceptor Doping

The measured acceptor energy levels range from 0.3 eV to 0.15 eV for Be and Zn doped BN films of mixed phases prepared by CVD, sputtering, or ion implantation [84, 85]. More recently, p-type h-BN films have been produced by HVPE using bis (cyclopentadienyl)magnesium as a doping gas [86]. The Mg-doped h-BN thin films produced by HVPE exhibited a free hole concentration of $\sim 10^{15}$ cm^{-3} and a resistivity of about 1000 Ω·cm.

The potential of magnesium (Mg) as a p-type dopant has also been investigated in thin epitaxial layers synthesized by MOCVD [36, 37]. A Mg doping level as high as 1×10^{19} cm^{-3} was attained, as verified by SIMS measurement. It was found that like Mg-doped AlGaN alloys, post-growth thermal annealing was helpful in activating the Mg acceptors in h-BN. The "optimal" p-type material annealing conditions included an annealing temperature of about 1150 °C and an annealing time of about 45 minutes, yielding a room temperature *p*-resistivity of about 12 Ω·cm. P-type conduction of Mg-doped epilayers was confirmed by Hall mobility and Seebeck effect (hot probe) measurements [36], leading to the demonstration of a p-BN/n-AlGaN heterostructure [32, 37]. AlGaN nanowire-based deep UV LED structures utilizing h-BN as a p-type current injection layer have also been demonstrated [87]. However, it is interesting to note that the nanowire h-BN/n-AlN DUV LED structures are Mg-dopant free, in which the p-type conduction was attributed to the formation of native defects such as boron vacancies which act as acceptors [87].

4.3 Silicon Doping

The suitability of Si as an n-type dopant in h-BN has been investigated in h-BN thin epilayers grown by MOCVD [77]. Although the n-type resistivity was too high to allow Hall effect measurements at room temperature, the resistivity probed by I-V characteristics can be varied by varying the silane flow rate [77]. To probe the nature of Si in h-BN, Hall effect measurements were carried out at 850 K for an "optimized" sample and yielded an average resistivity (ρ) of ~12 Ω·cm with a free electron concentration (n) of ~1×10^{16} cm^{-3} and mobility (μ) of ~48 (± 24) cm^2/V·s. The mobility data shown in Fig. 27.13a is accompanied by a large fluctuation; however, it unambiguously confirms the n-type conductivity of h-BN:Si and that Si acts as a donor dopant in h-BN. The temperature-dependent resistivity shown in Fig. 27.13b yielded an energy level of Si in h-BN of E_D ~1.2 eV, which is quite deep. The measured Si energy level of 1.2 eV agrees well with that of a calculated value of 1.1 eV in h-BN [88]. The results shown in Fig. 27.13 seem to suggest that Si is not a suitable n-type dopant for h-BN for room-temperature applications.

However, n-type conductivity in h-BN polycrystalline thin films synthesized by magnetron sputtering on quartz and Si substrates has been obtained via Si ion implantation [89]. A density functional theory (DFT) calculation based on a h-BN monolayer model containing 1000 atoms by the same authors suggested that the substitutional Si induces a state at ~177 meV [89]. *N*-type conduction was also observed in the h-BN monolayer by in-situ Ge-O doping using low-pressure CVD [90].

Following the footsteps of III-nitrides, in order to obtain highly conductive *n*-$Al_xGa_{1-x}N$ with high Al-contents, it was critical to minimize the density of native

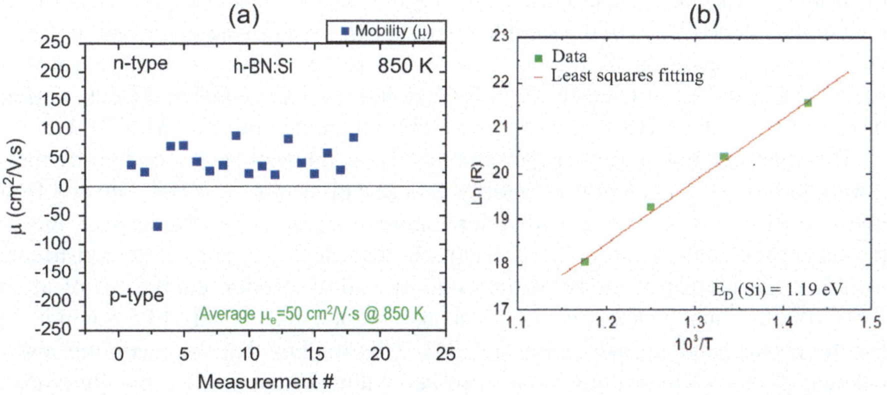

Fig. 27.13 (**a**) Mobility data obtained by Hall effect measurements carried out at 850 K for an h-BN:Si sample, thermally annealed at a temperature of $T_a = 1100$ °C after growth. In the experimental setup, $\mu > 0$ indicates electron conduction, and $\mu < 0$ indicates hole conduction. (**b**) N-type resistivity as a function of temperature of the same h-BN:Si sample measured at T > 700 K, yielding an activation energy of about 1.19 eV. (Reproduced from Ref. [77])

defects (mainly group III vacancies) and carbon impurities occupying the nitrogen sites (C_N), which act as free electron compensating centers (or electron traps) [91–99]. Therefore, it may still be worthwhile to further investigate Si or Ge doping for n-type conductivity control in h-BN under different doping and growth conditions.

4.4 Probing Oxygen Impurities

Oxygen impurity diffusion in *h*-BN has been studied in a quasi-bulk freestanding wafer grown by MOCVD at 1450 °C, by probing the emission intensity of the DAP transition involving O_N and V_B -H/V_B -2H as a function of the etching depth from the sapphire/h-BN interface [76]. The depth was controlled by successively removing the surface that was in contact with the sapphire substrate by an SF_6 chemistry-based inductively coupled plasma (ICP) dry etching technique [100]. The PL results together with the O concentration profile probed by SIMS provided a diffusion coefficient (D) of O impurities of ~2×10^{-13} cm^2/s at 1450 °C.

The measured oxygen impurity diffusion constant in *h*-BN is in the same order as those of substitutional impurities in Si (on the order of 10^{-15} – 10^{-13} cm^2/s at 1300 °C) but is about 8 orders of magnitude lower than the diffusion constants of interstitial impurities in Si (on the order of 10^{-6} to 10^{-5} cm^2/s) [101]. Moreover, the measured diffusion coefficient of O in *h*-BN is also very close to (about 4 times larger than) those of Si and O substitutional impurities in AlN of about 5×10^{-14} cm^2/s at 1450 °C [102]. These results support that O in h-BN tends to form a substitutional donor, namely O_N [59].

To obtain the energy level of O_N, dark I-V characteristics of a 50-μm thick h-BN sample grown by MOCVD at 1500 °C, such as that shown in Fig. 27.14a, were measured at varying temperatures [74]. Figure 27.14b and inset are plots of the temperature dependence of the sample resistance (R) or equivalently resistivity (ρ) in the linear and semi-log (Arrhenius plot) scale, respectively, providing a value of $E_D = 0.56\ (\pm\ 0.03)$ eV. Clearly, it is this O_N donor level that controls the dark current or resistivity of the *h*-BN epilayers grown at high temperatures by MOCVD.

This interpretation is further corroborated by an interesting observation by monitoring the μτ products that the majority charge carrier type in *h*-BN converts from holes to electrons as the growth temperature increases. To characterize the μτ products for electrons and holes, detectors in a vertical geometry were constructed to utilize the unique property of the large optical absorption coefficient of about 7.5×10^5 cm^{-1} and hence a small optical absorption length of only 14.5 nm in *h*-BN for the above bandgap excitation [18, 34]. This implies that the incoming above bandgap photons will be completely absorbed within 70 nm (~5λ) of the illuminated surface of *h*-BN, enabling the characterization of μτ products for holes and electrons separately in thick *h*-BN epilayers, as illustrated in the insets of Fig. 27.15.

Depending upon the polarity of the illuminated surface, a specific charge carrier (hole or electron) can be selected for transport in the vertical direction (along the *c*-axis of *h*-BN). A negatively (positively) biased illuminated surface collects holes

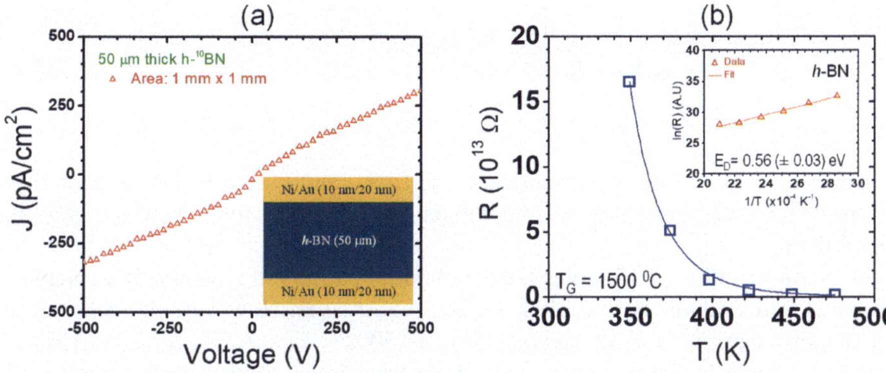

Fig. 27.14 (**a**) Room temperature current density vs. voltage (J-V) characteristic of a freestanding *h*-BN sample grown by MOCVD at ~1500 °C. The inset is the schematic of the sample used. (**b**) Temperature dependence of the dark resistance of the same sample measured in the vertical direction (along the *c*-axis), and the inset is the Arrhenius plot of the measured resistance providing an activation energy of $E_D \approx 0.56$ eV for O_N donors. (Figures reproduced from Ref. [74], Copyright 2018 AIP publishing LLC)

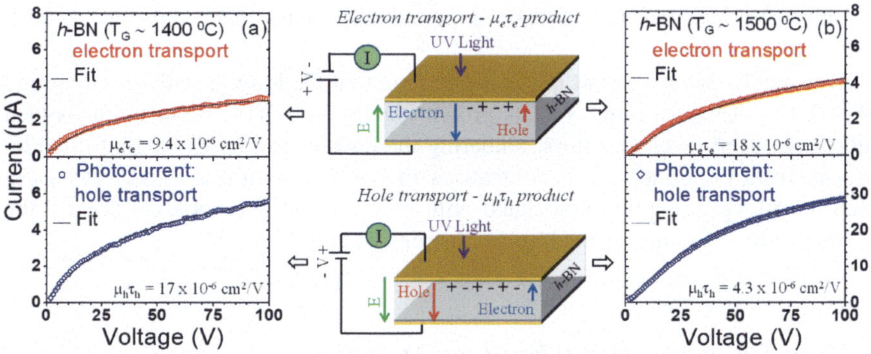

Fig. 27.15 Photocurrent-voltage characteristics of freestanding *h*-BN samples grown at (**a**) ~1400 °C and (**b**) ~1500 °C and the solid curves are fittings using Eq. (27.3), which reveal that $\mu_h\tau_h > \mu_e\tau_e$ for the sample grown at ~1400 °C and $\mu_e\tau_e > \mu_h\tau_h$ for the sample grown at ~1500 °C. The inset shows the experimental geometries which enable the characterization of the electron and hole transport properties separately. (Figures reproduced from Ref. [74], Copyright 2018 AIP publishing LLC)

(electrons) immediately upon photoexcitation and allows only electrons (holes) to transport between the electrodes. Figures 27.15a, b compare the photocurrent-voltage characteristics of two *h*-BN samples produced by MOCVD at 1400 °C and 1500 °C, respectively. Fitting the photocurrents with a modified Many's equation by including both effects of bulk and surface trapping [72] below provides the μ-τ-products for holes and electrons for both samples [74],

$$I_i(V) = I_{0,i}\left[\frac{V\mu_i\tau_i\left(1 - e^{-\frac{L^2}{V\mu_i\tau_i}}\right)}{L^2\left(1 + \frac{s_iL}{\mu_iV}\right)}\right].(= e, h) \tag{27.3}$$

Here, I_0 is the saturation current, and $\mu_e\tau_e$ ($\mu_h\tau_h$) and s_e (s_h) denote the mobility-lifetime product and surface recombination velocity for electrons (holes), respectively.

As shown in Fig. 27.15, holes are the majority carriers in typical h-BN epilayers grown at temperatures below 1400 °C, i.e., $\mu_h\tau_h > \mu_e\tau_e$. However, electrons act as the majority carriers in h-BN layers grown at 1500 °C, i.e., $\mu_e\tau_e > \mu_h\tau_h$. We believe that this change in the majority carrier type from holes to electrons is due to the increased concentration of oxygen donors in h-BN samples grown on sapphire at high temperatures for a long growth time. It is reasonable to assume that the process of oxygen diffusion always takes place during growth at temperatures $T < 1400$ °C. However, for samples deposited at $T < 1400$ °C, the total concentration of acceptor-like defects is still higher than that of donors, whereas, for samples grown on sapphire at high temperatures (T > 1500 °C), the total concentration of O_N is higher than that of acceptor-like defects. This is expected since the diffusion rate of oxygen impurities is expected to increase exponentially with increasing the growth temperature.

These results again support our interpretation that the long growth time required makes MOCVD quasi-bulk crystals grown above 1400 °C contain more oxygen donors than deep-level acceptors, rendering the dark resistivity to be controlled by electrons from O_N donors. This contrasts with HVPE-grown quasi-bulk crystals in which deep-level acceptors associated with boron vacancy complexes control the dark resistivity, as demonstrated by results presented in Fig. 27.9.

5 High Efficiency h-BN Neutron Detectors

5.1 Theoretical Efficiency of BN Thermal Neutron Detectors

Because neutrons are neutral, they interact with matter only weakly. The isotope ^{10}B possesses an unusually large thermal neutron cross-section of $\sigma \sim 3840$ barns ($= 3.84 \times 10^{-21}$ cm^2) [26, 27]. This property makes BN an outstanding material for the construction of direct conversion thermal neutron detectors. Direct conversion means that neutron absorption, energy conversion, free charge carrier generation, and free charge carrier collection all occur in the same BN material. The operating principle of BN thermal neutron detectors is based on the absorption of a thermal neutron by a ^{10}B atom. After absorbing the thermal neutron, the isotope ^{10}B would be transformed into an α particle (^{4}Helium) and a recoiled lithium nucleus (^{7}Li) [27] with a Q-value (the amount of energy released during the nuclear reaction) of 2.792 MeV in the ground state (6% probability) and 2.310 MeV in the excited state (94% probability), respectively.

The intrinsic detection efficiency (η_i) of neutron detectors as a function of the detector's layer thickness, *d*, can be described by the standard absorption equation,

$$\eta_i = 1 - e^{-d/\lambda}. \tag{27.4}$$

where *d* denotes the detector thickness and λ the thermal neutron absorption length. As a semiconductor, the density of B atoms (N) which can interact with thermal neutrons in 100% ^{10}B-enriched BN is $5.5 \times 10^{22}/cm^3$. This provides a thermal neutron macroscopic absorption coefficient (α) of 211.2 per cm and a thermal neutron absorption length (λ) of 47.3μm in h-^{10}BN. Elemental B exists in two main isotopes: ^{10}B and ^{11}B in a natural abundance of approximately 20% and 80%, respectively. It is only the isotope ^{10}B that can interact with thermal neutrons. Therefore, λ is 5x larger or 237μm in natural h-BN. Here, α and λ are obtained from the relations of $\alpha = N\sigma = 5.5 \times 10^{22} \times 3.84 \times 10^{-21} = 211.2\ cm^{-1}$ and $\lambda = \alpha^{-1}$ with $\lambda = 47.3$μm (237μm) in h-^{10}BN (in natural h-BN). Eq. (27.4) implies that:

(a) The intrinsic efficiency can be increased by increasing the wafer thickness.
(b) h-^{10}BN (h-BN) wafers with a layer thickness $d > \lambda$ of 47.3μm (237μm) are needed to achieve thermal neutron detectors with an intrinsic detection efficiency (or neutron absorption) exceeding 63% (=1-e^{-1}).

5.2 h-BN Thermal Neutron Detectors with High Efficiencies

The detection of thermal neutrons by a h-BN detector is accomplished by a three-step process: (a) the first is the neutron absorption in which the nuclear reaction creates Li and α daughter particles with large kinetic energies; (2) the second process is the subsequent charge carrier [electrons (e^-) and holes (h^+)] generation by Li and α particles; and (3) the final process is the charge carrier (electrons and holes) separation and collection by the electrodes with the aid of an applied bias voltage or electric field. Therefore, achieving a high charge collection efficiency is also necessary to ensure an overall high detection efficiency.

As shown schematically in Fig. 27.1, h-BN has a layered crystalline structure. As such, its physical properties are highly anisotropic with its carrier mobility in the c-plane (μ_a) being ~2 orders of magnitude larger than its mobility in the c-direction (μ_c) [69, 75]. In addition to the more traditional vertical detectors shown in Fig. 27.15, we have also developed a lateral detector architecture by taking advantage of h-BN's superior lateral transport properties [33, 41–43, 69, 75]. For instance, for detectors with the same dimension of 0.2 cm (width, W) × 2.0 cm (length, L) × 100 μm (thickness, *d*), by fabricating the detector in a lateral geometry, the carrier mobility can be enhanced by nearly 2 orders of magnitude, the device capacitance can be reduced by a factor of d^2/W^2 (≈400), while the surface

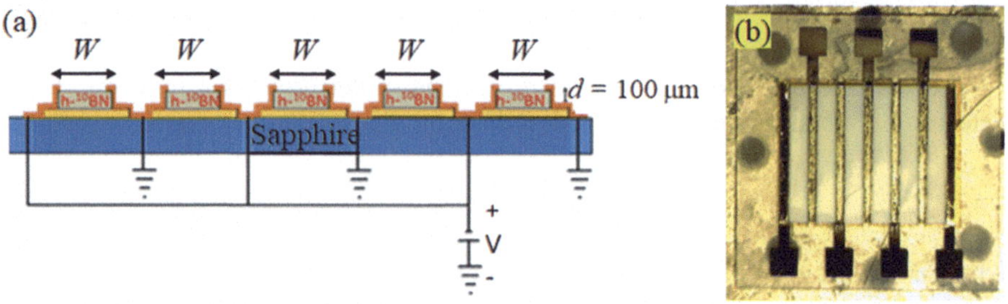

Fig. 27.16 (**a**) Schematic of combining multiple lateral strips to form a large area detector. (**b**) Optical image of a finished 1 cm^2 thermal neutron detector fabricated from an h-^{10}BN quasi-bulk wafer (100μm thick) by combining six lateral detector strips in parallel. (Figures reproduced from Ref. [58])

recombination field can be reduced by a factor of $(d/W)(\mu_c/\mu_a)$ (≈2000) and can be neglected [58].

The most important parameter for determining the charge collection efficiency of a lateral neutron detector is its carrier mobility and lifetime product, $\mu\tau$. In the first order, the condition for charge collection is described by Eqs. (27.1a) and (27.1b), $W < \sqrt{V\mu\tau}$, where V is the applied bias voltage and W is the spacing between the two electrodes or the detector's width. Based on the results shown in Fig. 27.8, the typical $\mu\tau$ values of these quasi-bulk materials are on the order of 2×10^{-4} cm^2/V. This means that the detector width should be designed to be smaller than 3 mm, $W < 3$ mm, for an operating voltage of 450 V.

The schematic of combining multi-strip detectors in parallel to form a large area detector is shown in Fig. 27.16a [41–43, 58]. It is important to note that these multi-strip lateral detectors differ from the traditional "interdigital finger" detectors fabricated from thin films. The "interdigital finger" detectors have contacts typically fabricated on the sample's top surface [34, 38, 70, 71], whereas each detector strip within the large detector shown in Fig. 27.16a has metal contacts that cover the entire vertical edges of the detector of 100 μm deep to ensure that the electric field is uniformly applied in the c-plane through the bulk of the detector material. An optical image of a finished 1 cm^2 thermal neutron detector formed by combining six lateral detector strips fabricated from a ^{10}B-enriched h-BN (h-^{10}BN) quasi-bulk wafer (100μm thick) is shown in Fig. 27.16b.

A Californium-252 (^{252}Cf) source from Frontier Technology was used as a neutron source. The number of neutrons as a function of the energy of the ^{252}Cf fission source follows a mathematical function, the Watt function of $P(E) = 0.4865 \left(\sinh\sqrt{2E}\right) \; e^{-E}$ MeV^{-1}, with an average neutron energy of 2.13 MeV and the most probable neutron energy of 0.7 MeV [103, 104]. A cube made of high-density polyethylene (HDPE) is used to moderate the fast neutrons via elastic scattering between neutrons and hydrogen atoms inside HDPE. Incoming fast neutrons from the ^{252}Cf source lose their energies after elastic scattering with hydrogen atoms. The neutron moderator setup used in this experiment was adopted from previous works [105–108]. When the ^{252}Cf source is placed at the center location of the HDPE moderator, the probability of neutron emission reaching outside of the HDPE block is nearly zero. When the ^{252}Cf source is placed at 2.5 cm from the front surface of the HDPE block, a maximum flux of thermal neutrons is obtained [105–108]. Detection electronics consist of a charge-sensitive preamplifier, pulse shaping amplifier, and multichannel analyzer (Amtek MCA 8000D) [107, 108].

The thermal neutron detection efficiencies were measured by placing the detectors at 30 cm from the HDPE block front face and side-by-side with a ^{6}LiF filled micro-structured semiconductor neutron detector (MSND) with a known detection area (4 cm^2) and detection efficiency (30%) from Radiation Detection Technologies. By measuring the counts against that of the MSND detector, the detection efficiencies of h-^{10}BN detectors can be obtained. The sensitivity of h-^{10}BN to gamma photons was tested by measuring pulsed height spectra (PHS) of a representative h-^{10}BN detector in the presence of a 662-keV ^{137}Cs source. Shown in Fig. 27.17a are PHS measured at 500 V, in which the green curve is the response to gamma photons emitted from a 662-keV ^{137}Cs source measured at 500 V, and the blue curve is the background (or dark) counts measured at the same bias voltage. It can be seen that the sensitivity of h-^{10}BN detectors to gamma photons from 662-keV ^{137}Cs source is at the background level. This is because BN is composed of low atomic number elements B and N.

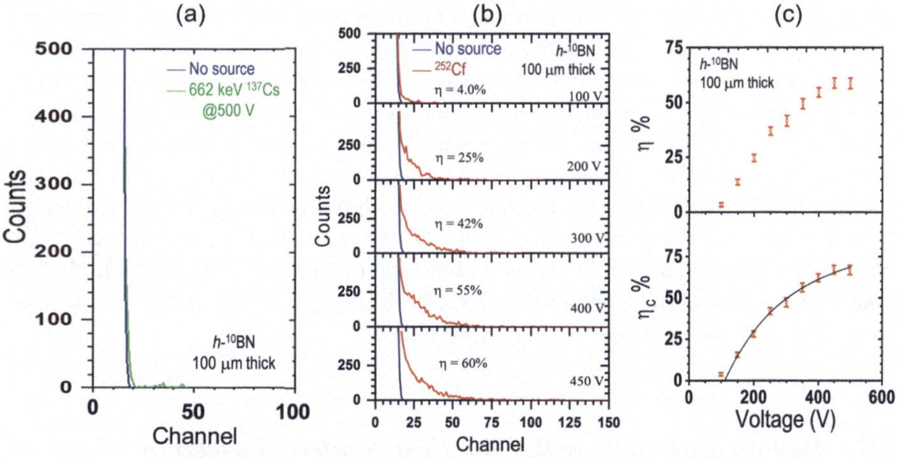

Fig. 27.17 Pulse height spectra (PHS) of a 1-cm^2 lateral detector fabricated from a 100-μm thick ^{10}B-enriched h-BN (h-^{10}BN) quasi-bulk wafer by connecting six lateral strips in parallel. (**a**) The green curve is the response to gamma photons produced by a 662-keV ^{137}Cs source, and the blue curve is the background (or dark) counts measured at 500 V. (**b**) The red curves are the responses to thermal neutrons, and the blue curves are the background (or dark) counts, measured at varying voltages. (**c**) Plot of the measured thermal neutron detection efficiency (η) and effective charge collection efficiency (η_c) as functions of the applied voltage. (Figures reproduced from Ref. [58])

Figure 27.17b shows the PHS of a 1-cm^2 lateral detector fabricated from a 100-μm thick h-^{10}BN wafer, measured at several representative bias voltages. By setting the lower-level discrimination (LLD) in the detection electronics corresponding to the response of the detector to 662-keV ^{137}Cs gamma photons as shown in Fig. 27.17a, any response to gamma photons was successfully filtered out. The measured detector efficiency (η) as a function of the applied voltage (and electric field) is plotted in the top panel of Fig. 27.17c, which shows that this 1-cm^2 detector delivers an overall thermal neutron detection efficiency of 60% at a bias voltage of 450 V. The measured detection efficiency presents the highest among all solid-state neutron detectors [59].

By defining the effective charge collection efficiency (η_c) as the ratio of the measured efficiency to the theoretical efficiency, we have,

$$\eta_c = \eta / \eta_{i,} \tag{27.5}$$

where η are measured detection efficiencies shown in Fig. 27.17b, and $\eta_{i,}$ is defined in Eq. (27.4). The bottom panel of Fig. 27.17c plots η_c as a function of the applied voltage (and electric field), yielding a maximum charge collection efficiency of 68%. Based on the XRD, PL, and photocurrent excitation spectroscopy results, the presence of native defects plays a significant role in limiting the charge collection efficiency, as they can trap free carriers.

We also noted that the charge collection efficiency is nearly zero when the applied voltage is below 100 V. We believe this is caused by two factors. First, forming ohmic contacts to the as-grown h-BN with a typical electrical resistivity of $\rho \sim 10^{13}$ Ω·cm is very challenging. The fabricated contacts are mostly blocking type which induces a surface barrier that needs to be overcome by applying a minimum electric field before a sizable charge collection by the electrodes can occur. Forming high-quality metal contacts, either ohmic or Schottky type, on UWBG semiconductors requires much more work. Second, it is well-known that the exciton binding energy in h-BN is exceptionally large, with reported values on the order of 100 s of meV [20, 22, 32, 35, 49, 50, 63–68]. The large exciton binding energy is expected to have a large impact on the performance of h-BN neutron detectors, which rely on the processes of charge carrier collection with the aid of an applied electric field. In fact, both scenarios result in an enhanced charge collection efficiency at elevated temperatures [58, 109] and deserve further exploration.

5.3 Demonstration of h-BN for Fast Neutron Detection

The physics principle of h-BN fast neutron detectors is based on charge generation by recoiling B and N ions upon elastic scattering by incoming fast neutrons and collection of these charge carriers in h-BN. The dominant elastic scattering cross-sections of fast neutrons with ^{10}B, ^{11}B, and ^{14}N in h-BN in the energy range between

0.5 and 20 MeV are all very small (~ 1.3 Barns on average). Using an average cross-section value of 1.3 Barns and the B and N atomic densities in BN [N (^{10}B + ^{11}B) = N(^{14}N) = 5.5 × 10^{22}/cm^3], the macroscopic cross-section for fast neutrons can be estimated as $\alpha = N\sigma = 5.5 \times 10^{22} \times 2 \times 1.3 \times 10^{-24} = 0.143\ \text{cm}^{-1}$, which yields a mean free path of $\lambda = \alpha^{-1} = 7$ cm in h-BN.

To verify this estimated mean free path of fast neutrons in BN, the transmission of fast neutrons (from a bare Cf-252 source) in pyrolytic BN (pBN) films has been measured [43]. pBN films have a similar structural property as h-BN semiconductors, except that they don't possess the necessary electronic properties to support the collection of neutron-generated free charge carriers as the h-BN semiconductors do. Thus, the results of fast neutron transmission measured in pBN represent those in h-BN semiconductors. The relative fast neutron fluxes (T) passing through pyrolytic BN films of different thicknesses (d), shown in Fig. 27.18a, follow the relation of T ~ $e^{-d/\lambda}$, yielding a value of $\lambda \approx 7.6$ cm, which agrees well with the expected value of 7 cm calculated from the cross-section of fast neutrons and atomic density of B and N atoms in h-BN semiconductors. In estimating this λ value, we assumed that the sum of probabilities of elastic scattering (S) and fast neutron transmission (T) is 100%, i.e., $S + T = 1$.

Figure 27.18b plots the layer thickness (d) dependence of the intrinsic efficiency of the BN detector for fast neutrons in the energy range between 1 and 10 MeV according to Eq. (27.4), using the measured value of $\lambda = 7.6$ cm. The result shown in Fig. 27.18b states that a thickness of several centimeters is required to obtain highly efficient BN detectors for detecting fast neutrons [43].

To demonstrate the feasibility of BN detectors for fast neutrons detection, a BN detector with a total detection area of 2.1 cm^2 was fabricated from a 90-μm thick h-BN wafer to directly detect fast neutrons emitted from a bare Cf-252 source with the use of a thick pBN as a thermal neutron filter [43]. Figure 27.18c shows the

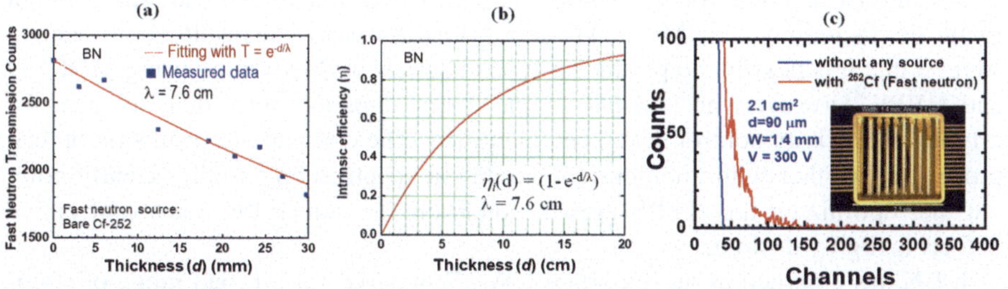

Fig. 27.18 (**a**) Relative fast neutron transmission (T) as a function of h-BN layer thickness (d) from a bare Californium-252 (Cf-252) source. Dots are experimental data, and solid curve is a fit with $T \sim e^{-d/\lambda}$, yielding $\lambda \approx 7.6$ cm. (**b**) The layer thickness dependence of the intrinsic efficiency of h-BN detector for fast neutrons if we assume that the sum of probabilities of elastic scattering (S) and fast neutron transmission (T) is 100%, i.e., $S + T = 1$. (**c**) Pulse height spectra of a 2.1-cm^2 h-BN detector in response to fast neutrons from a bare ^{252}Cf source without the use of an HDPE moderator, covering the energy range from 1 to 9 MeV (red curve) and in the presence without any source (blue curve), all measured at 300 V. The inset is a photo of the h-BN detector used. (Figures reproduced from Ref. [43], Copyright 2022 AIP Publishing LLC)

pulsed height spectra obtained with and without the Cf-252 neutron source under a bias voltage of 300 V. The average counts acquired by the detector was 1937 over a 15-minute counting time with the BN detector placed at 3 cm from the bare Cf-252 neutron source. Based on the known neutron flux at 3 cm from the bare Cf-252 neutron source and measured count rate, a detection efficiency for fast neutrons from the ^{252}Cf source is estimated to be nearly 0.1% for a 90-μm thick BN detector.

Together with the capability of BN for thermal neutron detection, the results shown in Fig. 27.18 indicate that by incorporating BN with a large thickness, BN neutron detectors are expected to possess the unique capability of directly detecting thermal to fast neutrons as well as outstanding features resulting from its UWBG. Identification of a single material that is sensitive to both thermal and fast neutrons is valuable for the development of novel neutron detection technologies [43].

6 Concluding Remarks

We provided an overview of the status of h-BN quasi-bulk crystal development. The crystalline and optical qualities of h-BN quasi-bulk crystals produced by HVPE have been compared to those of the state-of-the-art millimeter-sized bulk crystals synthesized by the HPHT and metal flux solution methods. The results revealed that h-BN quasi-bulk crystals with a thickness of around 100 μm have a good stacking sequence and long-range order in the c-axis.

In terms of transport properties, the as-produced quasi-bulk wafers exhibit a carrier mobility and lifetime product of $\mu\tau = 2 \times 10^{-4}$ cm^2/V, which is sufficiently large for the construction of detectors with high charge collection efficiencies. Thermal neutron detectors of 1 cm^2 in the detection area fabricated from a 100-μm thick ^{10}B-enriched h-BN wafer were shown to deliver record high detection efficiency of 60% at a bias voltage of 450 V. Properties of intentional and unintentional impurities including C, Si, Mg, and O have been discussed. The results so far seem to suggest that it is possible to produce p-type materials by Mg doping using MOCVD and HVPE. However, no suitable simple donor dopants, other than oxygen, for n-type conductivity control have been identified. The comparison results on optical properties together with photocurrent excitation spectroscopy studies clearly point out that optimizing the HVPE growth conditions to reduce the density of native defects is highly desired.

h-BN has emerged as an important UWBG material for a broad range of applications, from deep UV photonics, high-power electronics, high-efficiency neutron detectors, substrates for 2D devices, and solid-state hosts for single photon emitters and quantum defect qubits. These applications would benefit greatly from the availability of high-quality quasi-bulk crystals in large wafer sizes. For electronic device applications, most figures of merit scale with the bandgap. For instance, assuming a comparable carrier mobility for different materials for simplicity, Baliga's figure of merit (BFOM) scales with E_C^3, or equivalently the ideal specific on-resistance decreases with E_C^3. More recently, based on an extensive examination of numerous references, it was concluded that $E_C \propto E_g^{1.83}$ for both direct and indirect semiconductors [110]. This means that BFOM scales with $E_g^{5.5}$, where E_C and E_g denote the critical field and bandgap, respectively. A 6-eV bandgap for h-BN translates to a BFOM of ~11,000 (with respect to Si), clearly pointing to the great potential of h-BN as well as AlN and cubic BN. The ability to produce high-quality and large-size wafers of h-BN bulk crystals to serve either as the active medium or substrate becomes increasingly important. Recent developments of the quartz-free HVPE technique for GaN growth [54, 55, 111, 112] can be adopted to reduce the dominant impurities and defects in h-BN. The progress achieved thus far serves as a stepping stone for further advancing the crystal growth technology to produce electronic grade h-BN quasi-bulk crystals.

Acknowledgments The efforts of h-BN quasi-crystal growth by HVPE and high-efficiency neutron detector developments are supported by ARPA-E (DE-AR0001552, DE-AR0001257, and DE-AR000964). The efforts of h-BN epilayer growth by MOCVD were supported by DARPA, DHS, and NSF.

References

1. The Nobel Prize in Physics (2014), Available online: https://www.nobelprize.org/prizes/physics/2014/press-release/. Accessed 18 Aug 2023
2. S. Nakamura, S.J. Pearton, G. Fasol, *The Blue Laser Diode* (Springer, Berlin/Heidelberg, 2000)
3. H. Amano, N. Sawaki, I. Akasaki, Y. Toyoda, Metalorganic vapor phase epitaxial growth of a high quality GaN film using an AlN buffer layer. Appl. Phys. Lett. **48**, 353 (1986)
4. S. Nakamura, T. Mukai, M. Senoh, Candela-class high-brightness InGaN/AlGaN double-heterostructure blue-light-emitting diodes. Appl. Phys. Lett. **64**, 1687 (1994)
5. T. D. Moustakas, *Highly Insulating Monocrystalline Gallium Nitride Thin Films*, U.S. Patent No. 5,686,738
6. H.X. Jiang, J.Y. Lin, How we made the microLED. Nat. Electron. **6**, 257 (2023)
7. P.J. Parbrook, B. Corbett, J. Han, T.Y. Seong, H. Amano, Micro-Light Emitting Diode: From Chips to Applications. Laser Photonics Rev. **15**, 2000133 (2021)
8. H. Amano, Y. Baines, E. Beam, et al., The 2018 GaN power electronics roadmap. J. Phys. D. Appl. Phys. **51**, 163001 (2018)
9. I.C. Kizilyalli, O.B. Spahn, E.P. Carlson, (Invited) Recent progress in wide-bandgap semiconductor devices for a more electric future. ECS Trans. **109**, 3 (2022)
10. Ultrawide Bandgap Semiconductors, Vol. 107 in Semiconductors and Semimetals, edited by Yuji Zhao and Zetian Mi; Academic, an imprint of Elsevier (2021)
11. J.Y. Tsao, S. Chowdhury, M.A. Hollis, D. Jena, et al., Ultrawide-bandgap semiconductors: Research opportunities and challenges. Adv. Electron. Mater. **4**, 1600501 (2018)
12. J. Wu, W. Walukiewicz, K. Yu, J.W. Ager III, E.E. Haller, H. Lu, W.J. Schaff, Y. Saito, Y. Nanishi, Small band gap bowing in $In_{1-x}Ga_xN$ alloys. Appl. Phys. Lett. **80**, 4741 (2002)
13. S. Vanka, B. Zhou, R.A. Awni, Z. Song, F.A. Chowdhury, X. Liu, H. Hajibabaei, W. Shi, Y. Xiao, I.A. Navid, A. Pandey, R. Chen, G.A. Botton, T.W. Hamann, D. Wang, Y. Yan, Z. Mi, InGaN/Si double-junction photocathode for unassisted solar water splitting. ACS Energy Lett. **5**, 3741 (2020)

14. B.C. Letson, S. Barke, S.P. Kenyon, T. Olatunde, G. Mueller, P. Wass, F. Ren, S.J. Pearton, J.W. Conklin, High volume UV LED performance testing. Rev. Sci. Instrum. **93**, 114503 (2022)
15. A.K. Geim, I.V. Grigorieva, Van der Waals heterostructures. Nature **499**, 419 (2013)
16. M.R. Uddin, T.C. Doan, J. Li, K.S. Ziemer, J.Y. Lin, H.X. Jiang, Electrical transport properties of (BN)-rich hexagonal (BN)C semiconductor alloys. AIP Adv. **4**, 087141 (2014)
17. R. Bourrellier, S. Meuret, A. Tararan, O. Stephan, M. Kociak, L.H.G. Tizei, A. Zobelli, Bright UV single photon emission at point defects in h-BN. Nano Lett. **16**, 4317 (2016)
18. T. Sugino, K. Tanioka, S. Kawasaki, J. Shirafuji, Characterization and field emission of sulfur-doped boron nitride synthesized by plasma-assisted chemical vapor deposition. Jpn. J. Appl. Phys. **36**(Part 2), L463 (1997)
19. K. Watanabe, T. Taniguchi, H. Kanda, Direct-bandgap properties and evidence for ultraviolet lasing of hexagonal boron nitride single crystal. Nat. Mater. **3**, 404 (2004)
20. B. Arnaud, S. Lebe'gue, P. Rabiller, M. Alouani, Huge excitonic effects in layered hexagonal boron nitride. Phys. Rev. Lett. **96**, 026402 (2006)
21. Y. Kubota, K. Watanabe, O. Tsuda, T. Taniguchi, Deep ultraviolet light-emitting hexagonal boron nitride synthesized at atmospheric pressure. Science **317**, 932 (2007)
22. G. Cassabois, P. Valvin, B. Gil, Hexagonal boron nitride is an indirect bandgap semiconductor. Nat. Photonics **10**, 262 (2016)
23. Y. Hattori, T. Taniguchi, K. Watanabe, K. Nagashio, Anisotropic dielectric breakdown strength of single crystal hexagonal boron nitride. ACS Appl. Mater. Interfaces **8**, 27877 (2016)
24. Y. Hattori, T. Taniguchi, K. Watanabe, K. Nagashio, Comparison of device structures for the dielectric breakdown measurement of hexagonal boron nitride. Appl. Phys. Lett. **109**, 253111 (2016)
25. C. Yuan, J. Li, L. Lindsay, D. Cherns, J.W. Pomeroy, S. Liu, J.H. Edgar, M. Kuball, Modulating the thermal conductivity in hexagonal boron nitride via controlled boron isotope concentration. Commun. Phys. **2**, 43 (2019)
26. O. Osberghaus, Die Isotopenhäufigkeit des Bors. Massenspektrometrische Untersuchung der Elektronenstoßprodukte von BF_3 und BCl_3. Zeitschrift fuer Physik **128**, 366 (1950)
27. G.F. Knoll, *Radiation Detection and Measurement*, 4th edn. (Wiley, 2010)
28. L. Artús, M. Feneberg, C. Attaccalite, J.H. Edgar, J. Li, R. Goldhahn, R. Cuscó, Ellipsometry study of hexagonal boron nitride using synchrotron radiation: transparency window in the far-UVC. Adv. Photonics Res. **2**, 2000101 (2021)
29. G. Cassabois, P. Valvin, B. Gil, Intervalley scattering in hexagonal boron nitride. Phys. Rev. B **93**, 035207 (2016)
30. T.Q.P. Vuong, G. Cassabois, P. Valvin, V. Jacques, A. Van Der Lee, A. Zobelli, K. Watanabe, T. Taniguchi, B. Gil, Phonon symmetries in hexagonal boron nitride probed by incoherent light emission. 2D Mater. **4**, 011004 (2017)
31. V. Alexander, I. Weinstein, D. Zamyatin, Electron-phonon interactions in subband excited photoluminescence of hexagonal boron nitride. J. Lumin. **208**, 363 (2019)
32. H.X. Jiang, J.Y. Lin, Review—hexagonal boron nitride epilayers: Growth, optical properties and device applications. ECS J. Solid State Sci. Technol. **6**, Q3012 (2017)
33. A. Maity, S.J. Grenadier, J. Li, J.Y. Lin, H.X. Jiang, Hexagonal boron nitride: Epitaxial growth and device applications. Prog. Quantum Electron. **76**, 100302 (2021)
34. J. Li, S. Majety, R. Dahal, W.P. Zhao, J.Y. Lin, H.X. Jiang, Dielectric strength, optical absorption, and deep ultraviolet detectors of hexagonal boron nitride epilayers. Appl. Phys. Lett. **101**, 171112 (2012)
35. B. Huang, X.K. Cao, H.X. Jiang, J.Y. Lin, S.H. Wei, Origin of the significantly enhanced optical transitions in layered boron nitride. Phys. Rev. B **86**, 155202 (2012)
36. R. Dahal, J. Li, S. Majety, B.N. Pantha, X.K. Cao, J.Y. Lin, H.X. Jiang, Epitaxially grown semiconducting hexagonal boron nitride as a deep ultraviolet photonic material. Appl. Phys. Lett. **98**, 211110 (2011)

37. S. Majety, J. Li, X.K. Cao, R. Dahal, B.N. Pantha, J.Y. Lin, H.X. Jiang, Epitaxial growth and demonstration of hexagonal BN/AlGaN p-n junctions for deep ultraviolet photonics. Appl. Phys. Lett. **100**, 061121 (2012)
38. J. Li, R. Dahal, S. Majety, J.Y. Lin, H.X. Jiang, Hexagonal boron nitride epitaxial layers as neutron detector materials. Nucl. Instrum. Methods Phys. Res. Sect. A **654**, 417 (2011)
39. K. Ahmed, R. Dahal, A. Weltz, J.-Q. Lu, Y. Danon, I.B. Bhat, Growth of hexagonal boron nitride on (111) Si for deep UV photonics and thermal neutron detection. Appl. Phys. Lett. **109**, 113501 (2016)
40. A. Maity, S.J. Grenadier, J. Li, J.Y. Lin, H.X. Jiang, Toward achieving flexible and high sensitivity hexagonal boron nitride neutron detectors. Appl. Phys. Lett. **111**, 033507 (2017)
41. A. Maity, S.J. Grenadier, J. Li, J.Y. Lin, H.X. Jiang, High sensitivity hexagonal boron nitride lateral neutron detectors. Appl. Phys. Lett. **114**, 222102 (2019)
42. A. Maity, S.J. Grenadier, J. Li, J.Y. Lin, H.X. Jiang, High efficiency hexagonal boron nitride neutron detectors with 1 cm^2 detection areas. Appl. Phys. Lett. **116**, 142102 (2020)
43. A. Tingsuwatit, A. Maity, S.J. Grenadier, J. Li, J.Y. Lin, H.X. Jiang, Boron nitride neutron detector with the ability for detecting both thermal and fast neutron. Appl. Phys. Lett. **120**, 232103 (2022)
44. A. Mballo, A. Ahaitouf, S. Sundaram, A. Srivastava, V. Ottapilakkal, R. Gujrati, P. Vuong, S. Karrakchou, M. Kumar, X. Li, Y. Halfaya, S. Gautier, P.L. Voss, J.P. Salvestrini, A. Outgazed, Natural boron and ^{10}B-enriched hexagonal boron nitride for high-sensitivity self-biased metal–semiconductor–metal neutron detectors. ACS Omega **7**, 804 (2022)
45. T. Ishii, T. Sato, Growth of single crystals of hexagonal boron nitride. J. Cryst. Growth **61**, 689 (1983)
46. T. Taniguchi, K. Watanabe, Synthesis of high-purity boron nitride single crystals under high pressure by using Ba BN solvent. J. Cryst. Growth **303**, 525 (2007)
47. N.D. Zhigadlo, Crystal growth of hexagonal boron nitride (hBN) from Mg-B-N solvent system under high pressure. J. Cryst. Growth **402**, 308 (2014)
48. Y. Li, V. Garnier, P. Steyer, C. Journet, B. Toury, Millimeter-scale hexagonal boron nitride single crystals for nanosheet generation. ACS Appl. Nano Mater. **3**, 1508 (2020)
49. J. Li, J. Wang, X. Zhang, C. Elias, G. Ye, D. Evans, G. Eda, J.M. Redwing, G. Cassabois, B. Gil, P. Valvin, R. He, B. Liu, J.H. Edgar, Hexagonal boron nitride single crystal growth from solution with a temperature gradient. Chem. Mater. **32**, 5066 (2020)
50. J. Li, X.K. Cao, T.B. Hoffman, J.H. Edgar, J.Y. Lin, H.X. Jiang, Nature of exciton transitions in hexagonal boron nitride. Appl. Phys. Lett. **108**, 122101 (2016)
51. E.A. Preble, J.H. Leach, R. Metzger, E. Shishkin, K.A. Udwary, Hydride-vapor-phase epitaxial growth of highly pure GaN layers with smooth as-grown surfaces on freestanding GaN substrates. Phys. Status Solidi C **11**, 604 (2014)
52. K. Yamane, Y. Hashimoto, H. Furuya, T. Inagaki, N. Okada, K. Tadatomo, Fabrication of freestanding {20-21} GaN substrates by HVPE using SiO_2 masked GaN templates. Phys. Status Solidi C **11**, 401 (2014)
53. Y. Kumagai, K. Goto, T. Nagashima, R. Yamamoto, M. Bockowski, J. Kotani, Influence of growth rate on homoepitaxial growth of AlN at 1450 °C by hydride vapor phase epitaxy. Appl. Phys. Express **15**, 115501 (2022)
54. S. Kaneki, T. Konno, T. Kimura, K. Kanegae, J. Suda, H. Fujikura, Record high electron mobilities in high-purity GaN by eliminating C-induced mobility collapse. Appl. Phys. Lett. **124**, 012105 (2024)
55. K. Ohnishi, N. Fujimoto, S. Nitta, H. Watanabe, S. Lu, M. Deki, Y. Honda, H. Amano, Tuning the p-type doping of GaN over three orders of magnitude via efficient Mg doping during halide vapor phase epitaxy. J. Appl. Phys. **132**, 145703 (2022)
56. Z. Alemoush, N.K. Hossain, A. Tingsuwatit, M. Almohammad, J. Li, J.Y. Lin, H.X. Jiang, Toward achieving cost-effective hexagonal BN semi-bulk crystals and BN neutron detectors via halide vapor phase epitaxy. Appl. Phys. Lett. **122**, 012105 (2023)

57. Z. Alemoush, A. Tingsuwatit, J. Li, J.Y. Lin, H.X. Jiang, Probing boron vacancy complexes in h-BN semi-bulk crystals synthesized by hydride vapor phase epitaxy. Crystals **13**, 1319 (2023)
58. Z. Alemoush, A. Tingsuwatit, A. Maity, J. Li, J.Y. Lin, H.X. Jiang, Status of h-BN quasi-bulk crystals and high efficiency neutron detectors. J. Appl. Phys. **135**, 175704 (2024)
59. L. Weston, D. Wickramaratne, M. Mackoit, A. Alkauskas, C.G. Van de Walle, Native point defects and impurities in hexagonal boron nitride. Phys. Rev. B **97**, 214104 (2018)
60. S. Pease, An X-ray study of boron nitride. Acta Cryst **5**, 356 (1952)
61. M. Almohammad, Z. Alemoush, J. Li, J.Y. Lin, H.X. Jiang, Carbon-related donor–acceptor pair transition in the infrared in h-BN. Appl. Phys. Lett. **124**, 102106 (2024)
62. Q. Cai, D. Scullion, A. Falin, K. Watanabe, T. Taniguchi, Y. Chen, E.J.G. Santos, L.H. Li, Raman signature and phonon dispersion of atomically thin boron nitride. Nanoscale **9**, 3059 (2017)
63. T.C. Doan, J. Li, J.Y. Lin, H.X. Jiang, Bandgap and exciton binding energies of hexagonal boron nitride probed by photocurrent excitation spectroscopy. Appl. Phys. Lett. **109**, 122101 (2016)
64. K. Watanabe, T. Taniguchi, Jahn-Teller effect on exciton states in hexagonal boron nitride single crystal. Phys. Rev. B **79**, 193104 (2009)
65. L. Wirtz, A. Marini, A. Rubio, Excitons in boron nitride nanotubes: dimensionality effects. Phys. Rev. Lett. **96**, 126104 (2006)
66. K. Watanabe, T. Taniguchi, Hexagonal boron nitride as a new ultraviolet luminescent material and its application. Int. J. Appl. Ceram. Technol. **8**, 977 (2011)
67. L. Museur, G. Brasse, A. Pierret, S. Maine, B. Attal-Tretout, F. Ducastelle, A. Loiseau, J. Barjon, K. Watanabe, T. Taniguchi, A. Kanaev, Exciton optical transitions in a hexagonal boron nitride single crystal. Phys. Status Solidi (RRL) **5**, 214 (2011)
68. X.Z. Du, J. Li, J.Y. Lin, H.X. Jiang, The origins of near band-edge transitions in hexagonal boron nitride epilayers. Appl. Phys. Lett. **108**, 052106 (2016)
69. S.J. Grenadier, A. Maity, J. Li, J.Y. Lin, H.X. Jiang, Chapter 12: Electrical transport properties of hexagonal boron nitride epilayers, in *Ultrawide Bandgap Semiconductors*, Volume 107 in Semiconductors And Semimetals, ed. by Y. Zhao, (2021). Hardcover ISBN: 9780128228708.
70. T.C. Doan, S. Majety, S. Grenadier, J. Li, J.Y. Lin, H.X. Jiang, Fabrication and characterization of solid-state thermal neutron detectors based on hexagonal boron nitride epilayers. Nucl. Instrum. Methods Phys. Res. Sect. A **748**, 84 (2014)
71. T.C. Doan, S. Majety, S. Grenadier, J. Li, J.Y. Lin, H.X. Jiang, Hexagonal boron nitride thin film thermal neutron detectors with high energy resolution of the reaction products. Nucl. Instrum. Methods Phys. Res. Sect. A **783**, 121 (2015)
72. A. Many, High-field effects in photoconducting cadmium sulphide. J. Phys. Chem. Solids **26**, 575 (1965)
73. A. Tingsuwatit, N.K. Hossain, Z. Alemoush, M. Almohammad, J. Li, J.Y. Lin, H.X. Jiang, Properties of photocurrent and metal contacts of highly resistive ultrawide bandgap semiconductors. Appl. Phys. Lett. **124**, 162105 (2024)
74. S.J. Grenadier, A. Maity, J. Li, J.Y. Lin, H.X. Jiang, Origin and roles of oxygen impurities in hexagonal boron nitride epilayers. Appl. Phys. Lett. **112**, 162103 (2018)
75. S.J. Grenadier, A. Maity, J. Li, J.Y. Lin, H.X. Jiang, Lateral charge carrier transport properties of B-10 enriched hexagonal BN thick epilayers. Appl. Phys. Lett. **115**, 072108 (2019)
76. M. Almohammad, A. Tingsuwatit, Z. Alemoush, J. Li, J.Y. Lin, H.X. Jiang, Probing and controlling oxygen impurity diffusion in h-BN semi-bulk crystals. Appl. Phys. Lett. **123**, 252106 (2023)
77. S. Majety, T.C. Doan, J. Li, J.Y. Lin, H.X. Jiang, Electrical transport properties of Si-doped hexagonal boron nitride epilayers. AIP Adv. **3**, 122116 (2013)
78. M.R. Uddin, J. Li, J.Y. Lin, H.X. Jiang, Probing carbon impurities in hexagonal boron nitride epilayers. Appl. Phys. Lett. **110**, 182107 (2017)
79. K. Era, F. Minami, T. Kuzuba, Fast luminescence from carbon-related defects of hexagonal boron nitride. J. Lumin. **24**, 71 (1981)

80. T. Pelini, C. Elias, R. Page, L. Xue, S. Liu, J. Li, J.H. Edgar, A. Dréau, V. Jacques, P. Valvin, B. Gil, G. Cassabois, Shallow and deep levels in carbon-doped hexagonal boron nitride crystals. Phys. Rev. Mater. **3**, 094001 (2019)
81. M. Maciaszek, L. Razinkovas, A. Alkauskas, Thermodynamics of carbon point defects in hexagonal boron nitride. Phys. Rev. Mater. **6**, 014005 (2022)
82. M. Mackoit Sinkevičienė, M. Maciaszek, C.G. Van de Walle, A. Alkauskas, Carbon dimer defect as a source of the 4.1 eV luminescence in hexagonal boron nitride. Appl. Phys. Lett. **115**, 212101 (2019)
83. J. Serrano, A. Bosak, R. Arenal, M. Krisch, K. Watanabe, T. Taniguchi, H. Kanda, A. Rubio, L. Wirtz, Vibrational properties of hexagonal boron nitride: Inelastic X-ray scattering and Ab initio calculations. Phys. Rev. Lett. **98**, 095503 (2007)
84. K. Nose, H. Oba, T. Yoshida, Electric conductivity of boron nitride thin films enhanced by in situ doping of zinc. Appl. Phys. Lett. **89**, 112124 (2006)
85. B. He, W.J. Zhang, Z.Q. Yao, Y.M. Chong, Y. Yang, Q. Ye, X.J. Pan, J.A. Zapien, I. Bello, S.T. Lee, I. Gerhards, H. Zutz, H. Hofsass, P-type conduction in beryllium-implanted hexagonal boron nitride films. Appl. Phys. Lett. **95**, 252106 (2009)
86. X. Liu, S. Fan, X. Chen, J. Liu, J. Zhao, X. Liu, L. Hou, Y. Gao, Z. Chen, p-type hexagonal boron nitride films with bis(cyclopentadienyl) magnesium as a doping gas in halide vapor phase epitaxy. Appl. Phys. Lett. **123**, 202102 (2023)
87. D.A. Laleyan, S. Zhao, S.Y. Woo, H.N. Tran, H.B. Le, T. Szkopek, H. Guo, G.A. Botton, Z. Mi, AlN/h-BN heterostructures for Mg dopant-free deep ultraviolet photonics. Nano Lett. **17**, 3738 (2017)
88. F. Oba, A. Togo, I. Tanaka, K. Watanabe, T. Taniguchi, Doping of hexagonal boron nitride via intercalation: A theoretical prediction. Phys. Rev. B **81**, 075125 (2010)
89. B. He, M. Qiu, M.F. Yuen, W.J. Zhang, Electrical properties and electronic structure of Si-implanted hexagonal boron nitride films. Appl. Phys. Lett. **105**, 012104 (2014)
90. S. Lu, P. Shen, H. Zhang, G. Liu, B. Guo, Y. Cai, H. Chen, F. Xu, T. Zheng, F. Xu, X. Chen, D. Cai, J. Kang, Towards n-type conductivity in hexagonal boron nitride. Nat. Commun. **13**, 3109 (2022)
91. C. Stampfl, C.G. Van de Walle, Doping of $Al_xGa_{1-x}N$. Appl. Phys. Lett. **72**, 459 (1998)
92. T. Mattila, R.M. Nieminen, Point-defect complexes and broadband luminescence in GaN and AlN. Phys. Rev. B **55**, 9571 (1997)
93. I. Gorczyca, N.E. Christensen, A. Svane, Influence of hydrostatic pressure on cation vacancies in GaN, AlN, and GaAs. Phys. Rev. B **66**, 075210 (2002)
94. J.L. Lyons, A. Janotti, C.G. Van de Walle, Effects of carbon on the electrical and optical properties of InN, GaN, and AlN. Phys. Rev. B **89**, 035204 (2014)
95. K.B. Nam, J. Li, M.L. Nakarmi, J.Y. Lin, H.X. Jiang, Achieving highly conductive AlGaN alloys with high Al contents. Appl. Phys. Lett. **81**, 1038 (2002)
96. K.B. Nam, M.L. Nakarmi, J.Y. Lin, H.X. Jiang, Deep impurity transitions involving cation vacancies and complexes in AlGaN alloys. Appl. Phys. Lett. **86**, 222108 (2005)
97. N. Nepal, M.L. Nakarmi, J.Y. Lin, H.X. Jiang, PL studies of impurity transitions in AlGaN alloys. Appl. Phys. Lett. **89**, 092107 (2006)
98. M.L. Nakarmi, K.H. Kim, K. Zhu, J.Y. Lin, H.X. Jiang, Transport properties of highly conductive n-type Al-rich $Al_xGa_{1-x}N$ ($x \geq 0.7$). Appl. Phys. Lett. **85**, 3769 (2004)
99. K. Zhu, M.L. Nakarmi, K.H. Kim, J.Y. Lin, H.X. Jiang, Silicon doping dependence of highly conductive n-type $Al_{0.7}Ga_{0.3}N$. Appl. Phys. Lett. **85**, 4669 (2004)
100. S. Grenadier, J. Li, J.Y. Lin, H.X. Jiang, Dry etching techniques for active devices based on hexagonal boron nitride epilayers. J. Vac. Sci. Technol. A **31**, 061517 (2013)
101. S.W. Jones, *Diffusion in Silicon* (IC Knowledge, LLC)
102. H. Okumura, Y. Watanabe, T. Shibata, K. Yoshizawa, A. Uedono, H. Tokunaga, S. Koseki, T. Arimura, S. Suihkonen, T. Palacios, Jpn. J. Appl. Phys. **61**, 026501 (2022)

103. F.D. Becchetti, M. Febbraro, R. Torres-Isea, M. Ojaruega, Cf-252 fission-neutron spectrum using a simplified time-of-flight setup: An advanced teaching laboratory experiment. Am. J. Phys. **81**, 112 (2013)
104. W.A. Noonan, Neutrons: It is all in the timing—The physics of nuclear fission chains and their detection. J. Hopkins APL Tech. Dig. **32**, 762 (2014)
105. J. Clinton, *Optimization and Characterization of a Novel Self Powered Solid State Neutron Detector*, PhD thesis, Rensselaer Polytechnic Institute (2011)
106. T.C. Doan, *Thermal Neutron Detector Based on Hexagonal Boron Nitride*, PhD thesis, Texas Tech University (2016)
107. A. Maity, *Highly Sensitive Hexagonal Boron Nitride Thermal Neutron Detectors*, PhD thesis, Texas Tech University (2020)
108. S.J. Grenadier, *Advancements in Hexagonal Boron Nitride for Thermal Neutron Detection Applications*, PhD thesis, Texas Tech University (2020)
109. J. Li, A. Maity, S.J. Grenadier, J.Y. Lin, H.X. Jiang, Charge collection in h-BN neutron detectors at elevated temperatures. Appl. Phys. Lett. **118**, 092102 (2021)
110. O. Slobodyan, J. Flicker, J. Dickerson, J. Shoemaker, A. Binder, T. Smith, S. Goodnick, R. Kaplar, M. Hollis, Analysis of the dependence of critical electric field on semiconductor bandgap. J. Mater. Res. **37**, 849 (2022)
111. H. Fujikura, T. Yoshida, M. Shibata, Y. Otoki, Recent progress of high-quality GaN substrates by HVPE method, in *Proceedings of the SPIE 10104, Gallium Nitride Materials and Devices XII, 1010403*, (2017)
112. H. Fujikura, T. Konno, T. Kimura, Y. Narita, F. Horikiri, Homo-epitaxial growth of n-GaN layers free from carbon-induced mobility collapse and off-angle-dependent doping variation by quartz-free hydride vapor phase epitaxy. Appl. Phys. Lett. **117**, 012103 (2020)

Part VIII
Future Outlook

Chapter 28
Outlook of Wide-Bandgap Semiconductors and Power Electronics in the Energy Transition Toward a More Electric Future

Isik C. Kizilyalli, Jung Han, James S. Speck, and Eric P. Carlson

1 Future Outlook

The Intergovernmental Panel on Climate Change (IPCC), prior to the adoption of the 2015 Paris Agreement, estimated that without any additional efforts to reduce greenhouse gas emissions, global temperatures would increase between 2.5 and 7.8 °C (with 90% confidence interval) by 2100. Progress in technology (decarbonizing power generation and transportation) and policy over the last 10 years have significantly improved the global temperature outlook. One recent study from the Rhodium Group projects a very likely temperature increase of 2–4 °C by the end of the century with a mean of 2.8 °C [1]. While this avoids catastrophic projections, it is still a dire climate future and falls short of the Paris Agreement goal of limiting warming to well below 2 °C and certainly significantly above the aggressive target of <1.5 °C. The culprit for the worse outcomes comes about because there are industries that are difficult to decarbonize or are growing at a pace much faster than originally imagined [2].

I. C. Kizilyalli (✉)
Stanford University, Palo Alto, CA, USA
e-mail: kizilyalli@ieee.org

J. Han
Yale University, New Haven, CT, USA
e-mail: jung.han@yale.edu

J. S. Speck
University of California, Santa Barbara, CA, USA
e-mail: speck@mrl.ucsb.edu

E. P. Carlson
Booz Allen Hamilton, Washington, DC, USA
e-mail: carlson_eric2@bah.com

I. C. Kizilyalli et al. (eds.), *Gallium Nitride and Related Materials*, The Materials Research Society Series, https://doi.org/10.1007/978-3-031-83056-3_28

Electricity continues to be the fastest-growing form of end-use energy. Advances in power electronics have been one of the primary enablers of the unprecedented growth of renewable energy sources on the electrical power grid over the past several decades along with the current evolution toward ever-higher levels of electrification, the process of replacing fossil fuels with electricity as the source of energy. Power electronics are responsible for controlling and converting electrical power to provide optimal conditions for transmission, distribution, and load-side consumption. The fraction of energy that flows through power electronics (including generation and consumption) is expected to increase dramatically by 2050 as the growth of renewable energy sources continues to increase.

New opportunities for higher efficiency power electronics have emerged with the development of wide-bandgap (WBG) power semiconductor devices, driven by the fundamental differences in material properties between silicon (Si) and WBG semiconductors such as silicon carbide (SiC) and gallium nitride (GaN) [3]. Achieving high-efficiency power conversion in a reduced form factor requires low-loss power semiconductor switches. Today's incumbent power switches, typically metal oxide semiconductor field effect transistors (MOSFET), insulated gate bipolar transistors (IGBT), and thyristors, are Si-based and are quickly approaching their limits due to the fundamental material properties of Si [4]. WBG semiconductors, with their superior electrical properties, are strong candidates to replace silicon.

Widespread utilization of high-performance power converters offers substantial energy-saving opportunities both directly via inherently more efficient designs and indirectly by facilitating higher levels of adoption for fundamentally higher-performing WBG materials. High impact opportunities exist across a wide variety of applications as discussed below.

Industrial Applications Industry represents 30% of US primary energy-related carbon dioxide (CO_2) emissions or 1360 million metric tons of CO_2 (2020). Five of the highest CO_2-emitting industries where industrial decarbonization technologies can have the greatest impact across the US are petroleum refining, chemicals, iron and steel, cement, and food and beverage. These industries represent approximately 51% of energy-related CO_2 emissions in the US industrial sector and 15% of US economy-wide total CO_2 emissions [2]. There is a rapidly growing interest in decarbonizing these areas by moving away from direct heating or reaction using fossil fuels to the electrification of the processes, where carbon-free power comes from renewable sources. For instance, US steel production was over 85 Mt. in 2021 and accounted for greater than 5% of industry emissions. Iron and steel manufacturing is widely recognized as among the most difficult industrial sectors to decarbonize, owing both to the sheer tonnage of products and the dependence on carbon at every step of the process. Current blast furnace technologies—responsible for ~70% of global iron and steel greenhouse gas emissions—require carbon to reduce iron oxide ores to pig iron, a materials additive, as well as a source of heat. Electrically powered, alternative technologies that can reduce primary iron with lower heat than the incumbent processes will attack both emission sources. Recent advances in electrolysis and plasma technology provide a potential pathway to convert iron ore to iron and steel without the need to use reducing coke [5]. No matter what industry is electrified, all will require efficient, high-power (multi-megawatt) converters and controllers. As a result, power electronic circuits will play an important role in industrial decarbonization and future sustainability in this area.

Data Centers and Artificial (General) Intelligence Energy consumption in data centers accounts for approximately 4% of electricity use in the US in 2022 and is expected to reach 6% by 2026 [6]. The power delivery architecture of most modern data centers consists of a line frequency transformer, a low-voltage power distribution network, a centralized backup unit, and inefficient voltage regulators. Strategies to improve energy efficiency range from integration of lower loss power converters to complete redesign of the power delivery network. The latter approach often involves down-converting higher voltages at the rack level to reduce transmission losses and the number of conversion stages, but major technical challenges must be overcome due to congested rack space and concentrated heat dissipation. High power density converters based on WBG semiconductor devices can be key enablers for these more efficient systems, and operation at higher temperatures can reduce cooling loads and further boost data center grid-to-chip efficiency. Furthermore, there is an order of magnitude increase every 2 years in training compute of deep learning models going back to 2010. By some accounts (and based on trendlines in AI progress and development), training clusters requiring 1–100GW of electrical power buildout at a cost of 10–100 s of billions of US dollars will be deployed by 2030 [7] with power appearing to be the major constraint on the supply side. Demand for power contracts, new power sources (nuclear and geothermal), power transformers, and switchgear will grow, and total electricity consumption for all AI may require a doubling of the total electricity generation in the USA by 2030 [7]. Power electronics will be required at all voltage levels (grid scale down to the rack and IC), and GaN devices (lateral or vertical architectures) along with other power semiconductors (Si and SiC) will have a key role to play.

Electrified Aviation Longer, thinner, and lighter wings can reduce fuel consumption and emission by 50% relative to modern designs in commercial aircraft, which would save approximately 1 quadrillion Btu of energy per year across the US fleet at current demand. Achieving this transformative wing design requires electromechanical actuators that are small and lightweight with robust operation over a wide temperature range. Moreover, electrification of environmental controls, fuel pumps, brakes, and de-icing systems can further reduce weight and increase efficiency through the elimination of engine bleed and pneumatic/hydraulic systems. WBG-based converters, with high gravimetric and volumetric power density and high-temperature operation, offer a pathway to achieving significant energy savings in air transport by reducing weight in more electric aircraft and enabling new paradigms in body design. Ultimately, the electrification of aviation propulsion systems will have the largest impact on CO_2 emissions. Here, power requirements at the multi-10 MW scale will require power converters that can achieve high efficiency with low thermal losses to drive the motors of narrow-body (e.g., Boeing 737) and wide-body (e.g., Boeing 787 and Airbus A380) passenger airplanes. These

WBG converters along with electrical motors, batteries/fuel cells, and the power distribution network (cables, connectors, and circuit breakers) must also comply with safety requirements for low electrical noise and challenges of operating at high altitude without electrical breakdown/discharge [8, 9]. Development of hybrid or all-electric aircraft will require the building of the support infrastructure on the ground, namely the electrification of airports in the way of providing clean energy generation and storage for fast charging. Compact and efficient power converters deploying WBG solutions will be utilized along the power flow chain to enable this energy transition.

Distributed Energy Resources In grid applications, such as solar PV and wind, as well as the emerging fields of high voltage direct current and flexible alternating current transmission systems, power conditioners are required to process and control the flow of electricity by supplying voltages and currents in a form that is optimally suited to the load. Power converters are responsible for a loss of approximately 4% of all the electricity generated in these applications and are the dominant point of failure for the installed system. For instance, a typical maximum conversion efficiency for a silicon-based PV inverter is approximately 96% (AC output/DC input), which drops significantly at operating temperatures above 50 °C. Novel WBG electronic circuits present a route to lower system-level costs by operating at higher switching frequencies that reduce the size of passive components and lower the overall system footprint. In addition, WBG circuits will increase system-level efficiency by allowing PV arrays to operate at higher voltages (e.g., medium voltage levels), enabling DC systems with fewer voltage conversions/transformers, replacing traditional combiner boxes with DC/DC converters, eliminating the need for on-site AC transmission lines, and ultimately allowing easier integration of energy storage solutions in the central substation [10]. Together with a higher semiconductor operating temperature, the advantages of WBG electronics offer a pathway to more robust power converters with mean time before failure (MTBF) commensurate with the generation system lifetime (PV, wind, etc.). This will lower the equipment replacement cost and total plant operation and maintenance (O&M) and have a significant impact on the levelized cost of energy in distributed resource applications providing DER system-level advantages in efficiency and MTBF. In short, advances in power electronics have been one of the primary enablers of the immense growth of renewable energy sources on the electrical power grid over the past several decades along with the current evolution in electrification.

Motor Drives Across all sectors, electric motors account for approximately 40% of total US electricity demand. It is estimated that 40–60% of currently installed electric motors could benefit from variable frequency drives (VFDs), which enable efficient adaptation to speed and torque demands [11]. Depending on the application, incorporation of VFDs can reduce energy consumption by 10–30%. Conventional VFDs for high-power applications are bulky and occupy significant space. Power density and efficiency can be improved, and the overall system cost is reduced, by using WBG-based VFDs.

Automotive Power electronics have a major impact on the efficiency of electric vehicles (EVs) in two ways: directly through their internal losses attributable to switching and ohmic dissipation and indirectly by adding volume and weight. WBG inverters can reduce both direct and indirect losses by operating at higher switching frequencies, efficiencies, and temperatures [12]. As a result, a 15% improvement in energy efficiency has been predicted for representative hybrid EVs employing SiC inverters, with even larger energy savings possible given greater degrees of drive-train electrification. Assuming aggressive market adoption of EVs in the USA, the use of WBG vehicle power electronics could save as much as 1 quadrillion Btu per year by 2050 relative to conventional Si-based systems. Additionally, efficient and lightweight DC fast charging infrastructure (≥120 kW) enabled by WBG converters can further accelerate the adoption of EVs, which, in conjunction with a cleaner electricity generation portfolio, will significantly reduce US and global emissions.

However, even with the considerable materials advantages, many challenges have been preventing the widespread use of power electronics using WBG semiconductors such as cost, availability, and proven reliability. To "unlock" the enormous potential of WBG power electronics and promote widespread adoption, devices need to be fabricated cost-effectively, and confidence in the reliability of WBG power devices in actual power electronics is needed. Failure mechanisms of WBG devices need to be understood, and a fundamental understanding needs to be developed that links the basic material properties of commercially available WBG semiconductors to the reliability characteristics of power devices in power electronics. Over the last decade, significant progress has been made in these areas, and some of the barriers mentioned above have been sufficiently overcome to the point that there are currently numerous suppliers of WBG components for power electronics use [13]. It should be noted that WBG devices cannot simply be swapped for Si devices in a circuit due to the dramatically different electrical properties of the devices. Circuit designers are obliged to redesign systems to account for the WBG devices. Gate-drive circuits for SiC and GaN are more complex than those required for Si devices due to the high switching speed, and common-mode noise is typically much more of an issue for WBG circuit designs. This has led to a large design inertia or reluctance to change by circuit designers. Systems with a significantly superior price to performance need to be demonstrated to overcome this inertia. That is, power electronic systems need to be designed to take full advantage of the superior performance of WBG devices to realize more efficient, lighter, smaller, and higher temperature-capable solutions in a cost-efficient manner. The Creating Innovative and Reliable Circuits Using Inventive Topologies and Semiconductors (CIRCUITS) focused-area program was launched by the US Department of Energy's Advanced Research Project Agency-Energy ARPA-E in 2017 to pursue that objective [14]. Topic areas covered by the CIRCUITS program included the development of innovative general-purpose converters utilizing novel circuit topologies, advanced control and drive electronics, electromagnetic interference mitigation, innovative packaging, and thermal approaches, all working toward a common set of performance metrics. Technological breakthroughs in these areas were specifically encouraged to catalyze the adoption of higher-performance power converters in many of the critical applications discussed above.

Among the WBG semiconductors available in the market, GaN has the widest energy gap, the largest critical electric field, and the highest saturation velocity. The higher critical electric field (>3MV/cm) of GaN enables thinner, more highly doped voltage-blocking layers, which can reduce on-resistance by an order of magnitude in the majority of carrier architectures relative to equivalent Si devices. The high breakdown electric field and low conduction losses mean that GaN devices can achieve the same blocking voltage and on-resistance with a smaller die size. This reduces the gate capacitance and allows for higher frequency operation compared with a Si device. The significantly larger bandgap in GaN relative to Si results in a low intrinsic carrier concentration in GaN and enables reduced leakage currents and robust high-temperature performance. These properties permit GaN devices to operate at much higher temperatures, voltages, and frequencies providing a pathway to a more efficient, lighter, smaller, and higher temperature-capable power electronics than those made from conventional semiconductor materials.

However, the majority of GaN power device development to date has been directed toward lateral architectures, such as high-electron-mobility transistors (HEMTs), fabricated in thin layers of GaN grown on foreign substrates (including sapphire, Si or SiC). Such lateral devices suffer from well-known issues such as current collapse, dynamic on-resistance, inability to support avalanche breakdown [15], and inefficient thermal management. To broaden and unlock the true potential of GaN for power electronics, vertical device architectures are needed. To achieve vertical GaN power devices, progress is required in bulk substrate technology, thick epitaxial growth with controlled low doping levels with negligible compensation, low damage etch and regrowth technologies, ion implantation and annealing processes, and most importantly in the general area of selective area p-type doping capability that yields high-quality p-n junctions. Advances in all these areas are essential at the fundamental material level to achieve the desirable device and system-level performance with near-perfect reliability. Previous publications have focused on optical and lateral GaN devices looking at device operation, characterization, and reliability. While the information generated in these areas is useful, GaN vertical device architectures face other unique challenges.

As detailed in the previous chapters, very good progress has been made in many areas to accelerate the development of high-performance and reliable vertical GaN power electronic semiconductor devices to address markets requiring technology nodes as broad as 600 V–10 kV. This includes large-area bulk GaN substrate development, epitaxial growth of thick GaN layers, and processes for selective area doping of GaN. Novel materials characterization techniques and advanced microscopy have permitted fundamental learning of the impact of defects, impurities, and dislocations on the GaN device processing and performance. Still, more work remains needed to understand the impacts of native defects such as vacancies and interstitials.

Unfortunately, even with all the progress to date, a selective area doping technique has not been developed that can produce a junction that is electrically equivalent to a state-of-the-art as-grown p-n junction. More work is needed to achieve a reliable selective doping process to enable vertical GaN-based power devices. In-situ etching with reactive gases including chlorine and fluorine molecules remains promising to yield defect-free, impurity-free GaN surfaces for regrowth p-n junctions. In Si CMOS fabrication, low temperature (350–450 °C) post-metal anneals in a hydrogen/deuterium ambient (forming gas) have been successfully used to passivate the silicon dangling bonds and consequently to reduce the Si/SiO_2 interface trap charge density. This treatment is imperative from a fabrication standpoint since a large density of dangling bonds at the Si/SiO_2 interface are electrically active and lead to the reduction of channel conductance and deviations from the ideal capacitance-voltage characteristics (large variations in threshold voltages). Without this hydrogen defect-passivation anneal process, Si CMOS would not be the dominant technology that it is today, or not possible at all. The development of this breakthrough and the role of hydrogen in the passivation process were elucidated by characterization techniques such as electron spin resonance (ESR) measurements performed in conjunction with deep-level transient spectroscopy (DLTS) and capacitance–voltage (C–V) measurements [16]. An analogous passivation/annihilation process is needed for GaN devices to realize selective area p-type doping in GaN without electrically active defects. Defect physics in GaN is essentially unknown when the junction is nonplanar with semipolar and nonpolar interfaces. Refined techniques to probe native defects with micrometer-scale spatial and depth resolutions will help to determine native defects present at lateral junctions toward their passivation/annihilation. Beyond the selective area doping, more work is needed in vertical GaN device processing, including fabricating high-quality Schottky contacts (including understanding of interface degradation mechanisms) and backside contacts along with wafer thinning for device packaging.

GaN power devices are far less mature than their Si counterparts with limited availability and have yet to be proven reliable in the field which further limits their adoption. Failure mechanisms of GaN devices need to be understood, and a fundamental understanding of the reliability characteristics of the power devices needs to be developed. To this end, the art of failure analysis of vertical GaN structures could benefit from future funding opportunities. Power systems need to be designed and demonstrated (applications engineering) that take full advantage of the superior performance of vertical GaN devices to provide evidence that more efficient, lighter, smaller, and higher temperature-capable power electronics than those made from conventional semiconductor materials can be cost-effectively fabricated.

Despite the remarkable advancements in GaN devices achieved to date, significant work still remains to realize their full potential in power electronics on the path toward a more electric and sustainable future. Fundamental research into material properties and processing and continued development down the power electronics value chain into circuits and systems are vital steps to reap the benefits in a wide range of applications. Achieving a reliable selective doping process will allow for vertical GaN-based devices that offer the best performance to meet the increasing demands of evolving power systems, operating at higher efficiencies, power densities, and temperatures compared to existing Si power devices.

In summary, this book aims to fill the gap by addressing in-depth the materials processing challenges for vertical devices based on bulk GaN substrates as well as describing innovative materials characterization techniques that elucidate the processing challenges. We intended this book to be an essential reference for anyone working in either basic research or advanced development of vertical architecture GaN power electronics and technologies, more than likely fabricated on native bulk GaN substrates. In a broader sense, we hope that this book becomes a go-to reference to any scientist and engineer working in any nitride semiconductor material seeking an updated coverage of the state-of-the-art processing and characterization techniques that will push GaN know-how to new materials and device frontiers.

References

1. K. Larsen, H. Pitt, M. Mobir, S. Movalia, A. River, E. Rutkowski, M. Tamba, K. McCusker, T. Houser, *Rhodium Climate Outlook: Probabilistic Projections of Energy, Emissions and Global Temperature Rise* (Rhodium Group, 2023)
2. S.J. Davis et al., Net-zero emissions energy systems. Science **360**, eaas9793 (2018). https://doi.org/10.1126/science.aas9793
3. V. Veliadis, R. Kaplar, J. Zhang, M. Bakowski, S. Khalil, P. Moens, IEEE ITRW Working Group Position Paper— Materials and devices. IEEE Power Electron. Mag. **5**, 45–48 (2018)
4. I.C. Kizilyalli, E.P. Carlson, D.W. Cunningham, Barriers to the Adoption of Wide-Bandgap Semiconductors for Power Electronics, in *Proceedings of IEEE International Electron Devices Meeting*, (2018), pp. 19.6.1–19.6.4
5. ARPA-E DE-FOA-0003117 Revolutionizing Ore to Steel to Impact Emissions, 2023
6. https://iea.blob.core.windows.net/assets/6b2fd954-2017-408e-bf08-952fdd62118a/Electricity2024-Analysisandforecastto2026.pdf
7. L. Aschenbrenner, *Situational Awareness—The Decade Ahead* (2024)
8. ARPA-E DE-FOA-000195 Connecting Aviation by Lighter Electric Systems, 2020
9. P. Wheeler, The More Electric Aircraft: Why Aerospace Needs Power Electronics. Available at: http://www.lboro.ac.uk/microsites/research/iemrc/Events%20write%20up/Power%20Electronics%2014.05.09/More_Electric_Aircraft_000.pdf
10. D.W. Cunningham, E. Carlson, J.S. Manser, I.C. Kizilyalli, Impacts of wide band gap power electronics on photovoltaic system design. IEEE J. Photovoltaics **10**, 213–218 (2020)
11. P. Waide, C. Brunner, *Energy-Efficiency Policy Opportunities for Electric Motor-Driven Systems* (IEA, 2011)
12. K. Hamada, M. Nagao, M. Ajioka, F. Kawai, SiC—Emerging power device technology for next-generation electrically powered environmentally friendly vehicles. IEEE Trans. Electron Devices **62**, 278–285 (2015)
13. GaN & SiC Power Semiconductor Market to Hit $15 Bn by 2032 (Predicts Global Market Insights Inc., 2022)
14. ARPA-E CIRCUITS Focused Area Program, https://arpa-e.energy.gov/technologies/programs/circuits
15. Y. Zhong et al., A review on the GaN-on-Si power electronic devices. Fund. Res. **2**, 462–475 (2022). https://doi.org/10.1016/j.fmre.2021.11.028
16. K. Hess, I.C. Kizilyalli, J.W. Lyding, Giant isotope effect in hot electron degradation of metal oxide silicon devices. IEEE Trans. Electron Devices **45**, 406–416 (1998)

Index

I. C. Kizilyalli et al. (eds.), *Gallium Nitride and Related Materials*, The Materials Research Society Series, https://doi.org/10.1007/978-3-031-83056-3

S

T

U

V

W

X

Z

The manufacturer's authorised representative in the EU is Springer Nature Customer Service Centre GmbH, Europaplatz 3, 69115 Heidelberg, Germany. If you have any concerns regarding our products, please contact ProductSafety@springernature.com

Printed and bound by CPI Group (UK) Ltd, Croydon, CR0 4YY
30/09/2026
02225598-0017